T0406107

Sustainable Education and Development – Sustainable Industrialization and Innovation

Clinton Aigbavboa · Joseph N. Mojekwu · Wellington Didibhuku Thwala · Lawrence Atepor · Emmanuel Adinyira · Gabriel Nani · Emmanuel Bamfo-Agyei
Editors

Sustainable Education and Development – Sustainable Industrialization and Innovation

Proceedings of the Applied Research Conference in Africa (ARCA), 2022

Set 2

Editors
Clinton Aigbavboa
CIDB Centre of Excellence
University of Johannesburg
Johannesburg, South Africa

Joseph N. Mojekwu
University of Lagos
Lagos, Nigeria

Wellington Didibhuku Thwala
Department of Civil Engineering
University of South Africa (UNISA)
Gauteng, South Africa

Lawrence Atepor
Cape Coast Technical University
Cape Coast, Ghana

Emmanuel Adinyira
Department of Construction Technology
and Management
Kwame Nkrumah University of Science
and Technology
Kumasi, Ghana

Gabriel Nani
Department of Construction Technology
and Management
Kwame Nkrumah University of Science
and Technology
Kumasi, Ghana

Emmanuel Bamfo-Agyei
Cape Coast Technical University
Cape Coast, Ghana

ISBN 978-3-031-25997-5 ISBN 978-3-031-25998-2 (eBook)
https://doi.org/10.1007/978-3-031-25998-2

This Springer imprint is published by the registered company Springer Nature Switzerland AG
The registered company address is: Gewerbestrasse 11, 6330 Cham, Switzerland

Preface

Research's contribution to the continent's development faces a formidable obstacle. Africa appears to be a continent where research is unrelated to development. However, it is believed that research could contribute to the continent's socio-economic development.

This book contains ninety-three peer-reviewed papers on the United Nations Sustainable Development Goal 9, with a particular emphasis on the five targets listed below. i. Develop infrastructure that is high-quality, dependable, sustainable, and resilient, including regional and transborder infrastructure, to support economic development and human well-being, with an emphasis on affordable and equitable access for all. ii. Promote inclusive and sustainable industrialization and, by 2030, significantly increase the industry's share of employment and gross domestic product in accordance with national conditions and double its share in the least developed countries. Increase small-scale industrial and other enterprises' access to financial services, including affordable credit, and their integration into value chains and markets, particularly in developing nations. iv. By 2030, upgrade infrastructure and retrofit industries to make them sustainable, with increased resource-use efficiency and greater adoption of clean and environmentally sound technologies and industrial processes, with each nation acting in accordance with its own capabilities. v. Enhance scientific research and modernise the technological capabilities of industrial sectors in all nations, especially developing nations, by 2030, including fostering innovation and substantially boosting productivity.

Clinton Aigbavboa
Joseph N. Mojekwu
Wellington Didibhuku Thwala
Lawrence Atepor
Emmanuel Adinyira
Gabriel Nani
Emmanuel Bamfo-Agyei

Organisation

Conference Chair

Joseph N. Mojekwu	University of Lagos, Lagos, Nigeria

Local Organising Committee

Gabriel Nani	Kwame Nkrumah University of Science and Technology, Kumasi, Ghana
Emmanuel Adinyira	Kwame Nkrumah University of Science and Technology, Kumasi, Ghana
Cynthia Amaning Danquah	Kwame Nkrumah University of Science and Technology, Kumasi, Ghana
Rexford Assasie Oppong	Kwame Nkrumah University of Science and Technology, Kumasi, Ghana
Christiana Okai-Mensah	Kwame Nkrumah University of Science and Technology, Kumasi, Ghana
Abdulai Sulemana Fatoama	Kwame Nkrumah University of Science and Technology, Kumasi, Ghana
Lawrence Atepor	Cape Coast Technical University, Cape Coast, Ghana
Sophia Panarkie Pardie	Cape Coast Technical University, Cape Coast, Ghana
Emmanuel Bamfo-Agyei	Cape Coast Technical University, Cape Coast, Ghana

Editors

Clinton Aigbavboa	University of Johannesburg, Johannesburg, South Africa
Joseph N. Mojekwu	University of Lagos, Lagos, Nigeria
Wellington Thwala	University of South Africa (UNISA), South Africa
Lawrence Atepor	Cape Coast Technical University, Cape Coast, Ghana
Emmanuel Adinyira	Kwame Nkrumah University of Science and Technology, Kumasi, Ghana

Gabriel Nani	Kwame Nkrumah University of Science and Technology, Kumasi, Ghana
Emmanuel Bamfo-Agyei	Cape Coast Technical University, Cape Coast, Ghana

Review Panel

D. W. Thwala	University of South Africa (UNISA), South Africa
Clinton. O. Aigbavboa	University of Johannesburg, South Africa
Samuel Sackey	Kwame Nkrumah University of Science and Technology, Ghana
William Kodom Gyasi	University of Cape Coast, Ghana
Lazarus M. Ojigi	National Space Research and Development Agency, Nigeria
Rufus Adebayo Ajisafe	Obafemi Awolowo University, Ile–Ife, Nigeria
Gabriel Nani	Kwame Nkrumah University of Science and Technology, Ghana
J. Smallwood	Nelson Mandela Metropolitan University, South Africa
David J. Edwards	Birmingham City University, UK
R. Assasie Oppong	Kwame Nkrumah University of Science and Technology, Ghana
Emmanuel Adiniyra	Kwame Nkrumah University of Science and Technology, Ghana
Godfred Darko	Kwame Nkrumah University of Science and Technology, Ghana
Phil Hackney	Northumbria University, Newcastle upon Tyne NE1 8ST, UK
Richard Osae	Cape Coast Technical University, Ghana
Eric Danso	University of Leeds, UK
Eric Simpeh	Kwame Nkrumah University of Science and Technology, Ghana
Christopher Amoah	University of the Free State, South Africa
Frederick Simpeh	Appiah-Menka University of Skills Training and Entrepreneurial Development, Ghana
Emmanuel Bamfo-Agyei	Cape Coast Technical University, Ghana

Scientific Committee

Joseph N. Mojekwu	University of Lagos, Nigeria
Lawrence Atepor	Cape Coast Technical University Ghana

D.W. Thwala	University of South Africa (UNISA), South Africa
Clinton. O. Aigbavboa	University of Johannesburg, South Africa
Jim Ijenwa Unah	University of Lagos, Nigeria
L.O. Ogunsumi	Obafemi Awolowo University, Ibadan, Nigeria
Emanuel Amaniel Mjema	College of Business Education, Tanzania
Bashir Garba	Usmanu Danfodiyo University, Sokoto State, Nigeria
Samuel Sackey	Kwame Nkrumah University of Science and Technology, Ghana
Gabriel Nani	Kwame Nkrumah University of Science and Technology, Ghana
R. K. Nkum	Kwame Nkrumah University of Science and Technology, Ghana
A.N. Aniekwu	University of Benin, Nigeria
J. Smallwood	Nelson Mandela Metropolitan University, South Africa
Lazarus M. Ojigi	National Space Research and Development Agency, Nigeria
R. Assasie Oppong	Kwame Nkrumah University of Science and Technology, Ghana
Emmanuel Adinyira	Kwame Nkrumah University of Science and Technology, Ghana
Cynthia Amaning Danquah	Kwame Nkrumah University of Science and Technology, Ghana
Emmanuel Bamfo-Agyei	Cape Coast Technical University, Ghana
William Kodom Gyasi	University of Cape Coast, Ghana

Contents

Usage Behaviour of Electronic Information Resources Among Academicians' in Tertiary Institutions of Tanzania

L. L. Nkebukwa(✉)

Department of Business Administration, College of Business Education (CBE) , Dar es Salaam, Tanzania
l.nkebukwa@cbe.ac.tz

Abstract. Purpose: The paper aimed at assessing academicians' usage behaviour of electronic information resources in tertiary institutions in Tanzania. Specifically, the study focused on four factors, namely; awareness, experiences, competencies, as well as technological blockade factors.

Design/Methodology/Approach: The study conducted a questionnaire survey to 357 academicians from seven tertiary institutions in Tanzania, with a rate of return of 79.33%. Generally, study associating experiences, competencies and technology as factors reduces usage behaviours among academicians in tertiary institutions. For example, only eight (30.7%) out of 26 databases were reported to be frequently used.

Findings: The findings revealed the majority were aware of popular databases available, however, the results showed the underutilization of them. Nevertheless, the study identified among prominent barriers reducing usage behaviour morale among academicians in tertiary institutions were: technological difficulty, slow internet connectivity, inadequate searching skills, and inadequate information literacy training programmes.

Research Limitation/Implications: The study involved academics; however, during the process of data collection, it was found that some of them were not easily reached, since they were involved in field supervision, resulting in the researcher narrowing a scope based on respondents' conveniently available and reachable.

Practical implication: The study recommended the following; a need for the usage capacity build among academicians; the improvement of Wireless Networks services; and the establishment of well-equipped computer laboratories to promote EIR usage habits among users.

Social implication: the survey will add literature on the topic as well as bring new knowledge to the policy makers to amend policies to make it mandatory for quality service and system provision, this will ensure a supportive environment to motivate acceptance and usage of e-information resources in tertiary institutions.

Originality/Value: This is an inclusive study that assesses the usage behaviour of EIR among academic members of staff which have not been conducted in

C. Aigbavboa et al. (Eds.): ARCA 2022, *Sustainable Education and Development – Sustainable Industrialization and Innovation*, pp. 611–622, 2023.
https://doi.org/10.1007/978-3-031-25998-2_47

tertiary institutions of Tanzania. The study provided empirical demographic characteristics evidence that influences EIR usage behaviour among academicians in Technical institutions of Tanzania. Further, this study will have wider implications for academic libraries that are engaged in the EIR's provision of services.

Keyword: Academicians · Electronic databases · Information · Service quality · Tanzania

1 Introduction

1.1 Background of the Problem

The development of the information and communication technologies (ICT) have enhanced the access of electronic information resources (EIR). EIRs are now becoming primary sources of information in the higher learning institutions. The usage of EIR in the higher learning institutions have great academic impacts, as according to Ani et al. (2015), it promotes research performance (Kinengyere et al. 2012), increases levels of publications and teaching as well as consultancy activities (Amusa and Atinmo 2016). This was the open access initiatives aimed at improving access to the EIR's which previously were not easily accessible in low-income countries. The open access movements in the developing countries involved research4life initiatives and the library consortia, all these two initiatives aimed at promoting access to EIR in the country. For example, the Research4Life programme enhances the access of 75,000 peer-reviewed journals, scholarly databases and ebooks (Research4Life 2019). Amusa and Atinmo (2016) underscore that researchers and faculty who utilize these databases are more frequently publishing their research outputs compared to others who are not accessing EIR. In addition to that, Kinengyere et al. (2012) revealed that usage of EIR in the higher learning institutions aimed to sustain curriculum development. The role of EIR in enhancing research, teaching and consultancy activities have gained prominence globally (Amankwah 2014). This marvel in proven by some authors in various studies for example; Kuwait (Al-Ansari 2006:12), France (Vibert et al. 2007:26), Australia (Deng 2010:2) and Pakistan (Tahir et al. 2010:34), West Indies (Renwick 2005). These studies reported ease of use, freely available access, currency, adequate infrastructure and awareness are among factors that influence high usage of electronic -resources in developed countries. Similar results were found in studies conducted in the developing countries including the eastern and western African universities, including Ghana, Nigeria, and Tanzania. These studies revealed advanced levels of users' awareness of the available subscribed electronic databases (Lwoga and Sife, 2018; Kwadzo, 2015; Dadzie, 2015; Ani et al. 2010). However, these studies denoted that, still there is low usage of EIR in the developing countries compared to the developed countries (Adeniji 2014; Smeda et al. 2014 Nkebukwa, 2016; Kwadzo, 2015; Lwoga and Sukums, 2018; Lwoga and Sife, 2018). Furthermore, studies signified some institutional factors that reduced EIR usage behavior among users; these were the limited access and usage of EIR, limited funds for subscriptions and the poor ICT Infrastructures. Despite the long term adoption of EIR in the Technical institutions of Tanzania, still little research has been conducted to assess factors fostering usage behavior of academicians towards effective utilization of EIR in the Tertiary Institutions of Tanzania, this study therefore intends to bridge the existing gap. This study aimed at assessing academicians' usage behavior of electronic information resources in tertiary institutions of Tanzania. The specific objectives were to; identify the levels of awareness among academicians towards the effective utilisation of subscribed EIR; assess levels of competences among academicians towards the effective utilisation of subscribed EIR; determine the extent in which academicians are utilising the subscribed EIR; and find the challenges decreases usage behaviour among academicians towards utilisations of subscribed EIR.

2 Research Methodology

Settings the Sampling Procedures

This study employed a questionnaire survey to academicians in seven tertiary institutions in Tanzania, namely; The College of Business Education (CBE), Tanzania Institute of Accountancy (TIA), Dar es salaam Institute of Technology (DIT), Institute of Finance Management (IFM), National Institute of Transportation (NIT), Tanzania Public Services College (TPSC) and Institute of Social Works (ISW). The stratified random sampling procedure was used to select the sample of the study. Based on financial resources and time factor, these institutions were selected because of its convenient accessibility and proximity to the researcher.

2.1 Sampling Size

Study Population and Sampling

The Kish's (1965) formula guiding the cross-sectional studies was employed to determine the sample size. The sample calculated at the confidence interval of 95 per cent to estimate margin of error which was equal to four (4). This study was carried out in seven Tanzania technical institutions. The total population of the selected institutions were 857 as obtained from the respective authorities of such institutions. In this regard, the sample was calculated based on the Kish's (1965) formula, of which the researcher reached data from the sample of 450 faculty members. However, during the survey only 357 out of 450 questionnaires were returned and analysed, this was 79.33%. Data for a Study Population and Sampling shown in Table 1.

2.2 Data Collection

Research Instrument and Data Analysis

According to Saunders et al. (2012) a questionnaire is a formal instrument associated with tools used for data collection. As, Smeda et al. (2014) & Adeniji (2014) added that, thc rcsearcher may develop questionnaires by adapting techniques of the existing tools that are used by other scholars. This technique was adopted in this study and the questionnaire comprised the closed and open questions. Of which the print questionnaires were physically distributed in all selected institutions, where the librarians from the respective institutions assisted the researchers to distribute and collect them timely. Based on the research ethics a permission was sought from the respective authorities in all surveyed institutions and all of the information obtained was treated confidential. The questionnaire had the following themes:

i. Demographic characteristics of academicians
ii. Awareness levels of EIR among academicians in the surveyed area
iii. Levels of experiences and competences among academicians towards usage of subscribed EIR.
iv. Barriers affecting academicians towards acceptance and usage behavior of subscribed EIR.

This study employed the statistical package and service solution (SPSS) version of 22.0 for the descriptive analysis. The study assessed the frequency of usage of EIR among academicians in tertiary institutions of Tanzania. In this respect the Likert scale ranged from (1) never to (4) were used to measure the academicians' usage behavior trends on EIR. The study also assessed the IL competencies by using Likert scale ranging from (1) strongly disagree to (5) strongly agree. Thereafter, the data calculated using the average score of self-reported on the usage of EIR. The study therefore used one-way ANOVA for more than two groups of independent samples in exploring the influence of individual characteristics, awareness and the levels of competence among academicians on acceptance and usage of EIR.

2.3 Demographic Characteristics on Usage of EIR

The study investigated the effects of individual characteristics on academicians" acceptance and usage behavior of EIR. These individual characteristics included: (gender, age, education, teaching experience, discipline and number of publications). The computation was on the average score of acceptance and usage behavior of all EIR. In this manner, one way ANOVA was computed, and the results are indicated in Table 4. The results of the mean difference indicated that the academicians of faculty members with Master's degree used EIR at a higher level (3.26) as compared with the other educational categories. The results indicated that there were significant differences between differences of education levels at the p-value of 0.011. Further, the mean difference indicates that academicians with teaching experience between 11 and 15 years were highly used of Electronic Informational Resource (3.36) compared to other faculty members with less than 11 years, or above 15 years. A significant mean difference was observed among the five groups of the individual experiences on the EIR usage behavior at the p-level of 0.042. Other demographic characteristics were not significant (gender, age, discipline and number of publications).

Table 1. Demographic characteristics on usage of EIR

Individual characteristics		Count	Mean	Standard deviation	F	Sig
Gender	Female	84	3.20	.26	1.252	0.097
	Male	273	3.26	.22		
Age	35 years and below	35	3.25	.24		
	36–40	77	3.25	.20	1.177	0.172
	41–45	80	3.25	.26		
	46–50	123	3.26	.25		
	51–55	32	3.17	.19		
	56 and above	10	3.30	.21		
Discipline	ICT & Mathematics	16	3.32	.22	0.807	0.869
	Education	40	3.28	.22		
	Accountancy	30	3.25	.19		
	Procurement and Supplies	20	3.24	.23		
	Legal Industrial Metrology and Engineering	25	3.30	.23		
	Business Administration	50	3.17	.23		
	Social Sciences	60	3.18	.25		
	Law	22	3.26	.27		
	Banking and Finance	25	3.31	.21		
	Transportation	30	3.25	.18		
	Records Management	21	3.26	.24		
	Natural Science	18	3.35	.31		
Education	Bachelor	0				
	Postgraduate	45	3.22	.25	1.512	0.011
	Masters	250	3.26	.22		
	PhD	10	3.16	.44		
Experience	1–5 years	0				
	6–10 years	45	3.22	.25	1.355	0.042

(*continued*)

Table 1. (*continued*)

Individual characteristics		Count	Mean	Standard deviation	F	Sig
	11–15 years	250	3.26	.22		
	16–20 years	0				
	20 and above	52	3.24	.23		
Publication	1–3 publications	116	3.28	.21		
	4–6 publications	190	3.22	.25	0.849	0.804
	7–10 publications	23	3.25	.18		
	10 and above	10	3.27	.23		
	No any publication	18	3.30	.30		

3 Results and Discussion of the Study Findings

3.1 Demographic Information

The study examined factors fostering usage behavior among academicians towards effective utilization of EIR in tertiary institutions of Tanzania. The majority of respondents who were involved in this study were males (76.6%, n = 274). The implication of the finding advocated that male still predominate the academic cadre all over the world. The same situation was also noted by Paechter (2003) in the Technical and Vocational Education Training systems, particularly in Sweden and the other Nordic countries, where gender segregation was reported in the technical occupations. In the same vein, According to Shepherd (2017), the gender imbalances was reported even in the executive management level in UK businesses, such was also replicated in the education sector, where women are underrepresented in leadership roles in schools and universities (Morley 2013; Chard 2013).

Based on the findings, the majority of participants who were involved in this study possessed a Master's Degree (70%, n = 250). It has been noted that in tertiary colleges the academicians with masters' degrees predominate other careers as compared to the Universities. The judgement originates from the fact that most of the technical colleges still have limited number of postgraduate programmes offered compared to the universities. Based on teaching experiences, it was also noted that most of the faculty members who were involved in this survey had 11–15 years of teaching experience in various courses. The association of the finding is that, academicians having such working experience teaching various courses have a great significant contribution on using EIR to promote their publications. The findings correspond to Luambano (2013) who denoted the work experience or specialisation are among factors determining the extent in which academicians seek and use electronic information resources. In this regard therefore, it was definitely essential to determine experiences and specializations among factors that influence self-reported usage of e-resources. Ahmed (2013) signified those two factors with both positive and negative effects on individuals' usage of EIR as well influences to the institutional plans. Regarding the number of publications, the findings revealed the majority of faculty members who were involved in this study had published less than six papers; however the further the findings showed few of them published more than ten (10) publications. The results correspond with other scholars who asserted that in developing countries there is low research and publications (Amusa and Atinmo 2016; Adeniji 2014; Smeda et al. 2014). In some cases, low publications in academic institutions is associated with the extent in which scholar's access and use EIR (Amusa and Atinmo (2016; Nkebukwa 2016; Luambano 2013; Lwoga 2016). Similarly, according to Tripathi et al. (2016) there is a direct relationship with the lack of searching skills and under utilisations of EIR. Ani et al. (2015) also concluded that ineffective use of EIR promotes research and publications.

3.2 IR Awareness, Competences and Usage Among Academicians in Surveyed the Institution

3.2.1 Awareness of EIR in the Surveyed Institutions

The findings showed faculty members who involved in this study were mainly aware with various database namely; Google search engine 100%, n = 357), Google Scholar (100%, n = 357), Wikipedia (99.2%, n = 354) and Directory of Open Access journal (89%, n = 321). Based on the findings most of the academicians are obtaining information on databases from librarians, staff mails, and college social media networks as well as the departmental meetings. The implication of the finding is that academicians in the surveyed institutions are familiar with EIR subscribed by such institutions. However, it has been noted underutilisation of databases, which was intensified by limited availability of ICT facilities. It has been also noted by Anunobi (2006) that, uses of internet informational sources in academic environments particularly in the developing countries are commonly limited with the availability and stability of information infrastructures. In this regard, the stable IT infrastructures allows an adjustment of individual modern life (Ozkisi and Topaloglu 2015 & Askoy 2012). Based on the above arguments, various studies concluded that, awareness of Internet sources bring comfort zone for information seeking and sharing among academicians or other faculty members as well as promoting online teaching and learning environments (Cravener 1999; Bavakutty and Salih 1999; Ojedokun and Owolabi 2003; El-Berry 2015; Azubuike 2016; Akpojotor 2016; Chirra and Madhusudhan 2009 & Atakan et al. 2008.

3.2.2 Usage of EIR in the Surveyed Institutions

Bases on usage of EIR, the results showed the most academicians in surveyed institutions relied on Google scholar, google search engine, Wikipedia, DOAJ and AJOL as compared to scholarly other scholarly databases such as research4life, Emerald, Ebsco host, and IFL databases. Study by Lwoga and Sukums (2018) conducted in Tanzanian Universities revealed similar results that most of 85% of the faculty members relied on Google search engine and Wikipedia 77%. The prominence of Google search engines still raises a question on the reliable information sources retrieved (Hider et al., 2009; Norbert and Lwoga 2013. This study therefore, revealed most of the academicians are using only eight out 26 databases which is 30% of the subscribed databases. The implication of the findings, there is the underutilisation of EIR in the surveyed colleges. Low usage of EIR in academic institutions are reported in various studies (Adeniji 2014; Smeda et al. 2014). Lazarz and Baroug (2008) emphasised that the knowledge on usage of EIR for lecturers, students and librarians is concurrently with their level of IT skills they have. This was also noted in this study, the academicians who had limited searching skills were among those who accessed few relevant online databases ''occasionally'' compared with those having skills. Based on findings, faculty members with PhD are using more EIR compared to other categories. Undoubtedly, this was due to their demand of doing research and publications. This circumstance, according to Kerins et al. (2004), is closely linked by the study subject or working experiences of someone who wants to solve problems or uncertainties. The need for information for researchers is also noted by Wilson (2005:45), who denoted that ''researchers are driven to seek information to satisfy their publications' needs''. This is so-called information behaviour which put forwards on the way people approach and handle information (Wilson 1995), as a way of promoting and sustaining personal motivation towards academic growth (Luambano 2008 and Heinström 2005).

Nevertheless, similar findings for the underutilisation of EIR reported in Nigerian universities, according to Ani & Edem (2012) the reasons behind were lack skills and lack of funds for the College libraries subscribing to relevant educational databases. Other reasons for low usage of EIR revealed in African countries were higher cost for establishing ICT infrastructures, lack of searching skills, inadequate internet connectivity as well as lack of the support from institutions (Bashorunh et al. (2011). Furthermore, the findings revealed academicians who were frequently using EIR had exceedingly teaching experiences ranging to 11 and 15 years, this is (3.36) compared to other faculty members with less than 11 years, or above 15 years. According to Lwoga and Sukums (2018) the faculty members with a wide range of skills expertise and experience better users of e-resources, and in most cases having a reasonable number of publications. Yet, the diversity of the specializations and work experiences depicts the usage behaviour of EIR (Luambano 2013; Akpojotor 2016).

3.2.3 Competence on Usage of EIR in the Surveyed Institutions.

The results showed that academicians were competent on using eight out of 26 databases, which were Google search (98%); Wikipedia (93.4%,; African Journals Online (AJOL) (84%, DOAJ (100%), Oxford Journal (70%), Google scholar (57%) and HINARI (55.6%). Based on the finding, academicians in the surveyed institutions have competencies on few popular databases compared with the number of the available databases. Kinengyere (2007) revealed that individual competencies concurrently with computer knowledge becomes an important asset for the exploitation of EIR in teaching and research. In this regard, Otokunefor (2005) added that scholars lack computer skills enabling them to manipulate computer systems for acquiring all relevant informational databases. Lastly, Saadi (2002) advised that there is the need for improving our computer skills for taking advantage of ICT via the information literacy skills programmes (Sukums 2018).

3.3 Barriers Affecting the Effective Utilization of EIR

Based on the open-ended question, the findings show that the faculty members were facing the following issues that inhibited them to the effective use of EIR; these are; technical difficulty, slow internet connectivity, inadequate searching skills, and lack of information literacy training. These results are corroborated with a study by Nawe and Kiondo (2005) and Sukums (2018) who denoted similar significant barriers that hinders academicians for effective utilisation of EIR during teaching and research. Other empirical evidences on factors for underutilization of EIR revealed 'inadequate infrastructure' (Okoroafor 2010:8); 'inadequate skilled human and material resources' (Okebukola 2012:16), 'inadequate planning skills, institutional politics, financial constraints, lack of information literacy among librarians' (Nawe 2013:26), and 'poor information processing and stocking of library materials' (Nyerembe 2004:4). As, Mcharazo and Olden (2000) concluded that, the poor library collection and lack of search skills reduce the academicians' morale for the library visits as well as for the poor information retrieval.

4 Conclusion and Implications of the Study Findings

4.1 Conclusion

In conclusion, the findings showed academic staff in a surveyed area are mainly aware with only eight out of 26 popular scholarly databases. Based on the findings librarians, library notice boards, staff mails, college social media and the departmental meetings are sources used to create awareness on the available databases. Moreover, the findings revealed the underutilisation of scholarly databases among academicians in the surveyed area, but results showed faculty members with PhD are using more EIR compared to other categories, undoubtedly, this was due to their demand of doing research and publications. The finding further shows the direct and close relationship among the awareness, experience and competencies towards the intention of an individual usage of EIR usage. Regarding barriers affecting the effective utilization of EIR, the findings revealed; technical difficulty, slow internet connectivity, inadequate searching skills, and lack of information literacy training.

5 Implications of the Findings

The study has practical implications, as it will stimulate libraries to carry out scholarly database awareness raising campaigns for all faculty members including students. If these categories of users' are made aware, it will be possible for them to utilise fully the available resources fully. Secondly, the study raised the need for the parental organization increasing budgetary allocation for its academic libraries, adequate funds in libraries will facilitate the subscriptions and acquisition of enough/adequate sources of information relatively to meet the information needs. Thirdly, this study raised the need for the respective parental organizations establishing and improving effective information infrastructures that will enable improvement of the Local Area Network (LAN) for free access and easy connectivity to make them learn in the available electronic environments.

6 Area of Further Study

This study only focused on academicians in conventional treasury institutions in Tanzania, instead further studies should be conducted to assess facothe usage and competences of accessing EIR among faculty members, librarians and students. Further, they should use a combination of methods, such as questionnaire survey should be combined with qualitative methods.

References

Adeniran, P.: Usage of EIR by undergraduates at the Redeemers University. Nigeria. Acad. J. **5**(10), 319–324 (2013)

Akpojotor, L.O.: Awareness and Usage of E- information resources among postgraduate students of library and information science in Southern Nigeria. Library Philosophy and Practice (e-journal). Paper 1408 (2016). http://digitalcommons.unl.edu/libphilprac/1408

Amankwah, P.B.: Use of electronic resources by undergraduate students of the Ghana Institute of Management and Public Administration. MA dissertation, Dept. of Information Studies, University of Ghana, viewed 20th January 2021 (2014). http://hdl.handle.net/123456789/6989

Amusa, O.I., Atinmo, M.: Availability, Level of use and Constraints to use of EIR by Law Lecturers in Public Universities in Nigeria. Italian '' Journal of Library and information science. ISSN (online) 2038-1026 (2016)

Angello, C., Wema, E.: Availability and usage of ICTs and EIR by Livestock researchers in Tanzania: challenges and ways forward. Int. J. Educ. Dev. Inf. Commun. Technol. **8**(1), 34–47 (2010). www.editlib.org/p/42309/. Accessed 4 Feb 2019

Ani, O.E., Ngulube, P., Onyancha, B.: Perceived effect of accessibility and Atram Ku. P.N. (2017). Digital Library Services in the Digital Age. Int. J. Library Inf. Sci. (IJLIS) **6**(1), 79–82 (2015). http://www.iaeme.com/IJLIS/issues.asp.

Askoy, R.: Internet Ortamında Pazarlama. Ankara: Seckin Yayınları. Azubuike, C. O. (2016)'' Information Literacy Skills and Awareness of E- information resources as influencing factors of their Use by Postgraduate Students in Two Universities in South-West Nigeria. Library Philosophy and Practice (e-journal) (2012). http://digitalcommons.unl.edu/libphilprac/1407

Bosah, L.E., Amadasu, M.E.: Influence of student's characteristics in the use of library resources in college of Education in Edo State, Nigeria. J. Res. Educ. Soc. **5**(1) (2014)

Bashorun, M., Tunji, I.A., Adisa, M.U.: User perception of EIR in the University of Illorin, Nigeria. J. Emerging Trends Comput. Inf. Sci. **2**(11), 554–562 (2016)

Chard, R.: A study of current male educational leaders, their careers and next steps. Manage. Educ. **27**(4), 170–175 (2013). Google Scholar | SAGE Journals

Deng, H.: Emerging patterns and trends in utilising electronic resources in a higher education environment: An empirical analysis. New Libr. World **111**(3/4), 87–103 (2010). https://doi.org/10.1108/03074801011027600

Lwoga, E.T., Sukums, F.: Health sciences faculty usage behaviour of electronic resources and their information literacy practices. Global Knowl. Memory Commun. **67**(1/2), 2–18 (2018). https://doi.org/10.1108/GKMC-06-2017-0054

Fokomogbon, M., Bada, A., Omiola, M.: Assessment of school library resources in public secondary schools in Ilorin Metropolis. Interdisciplinary J. Contemporary Res. Bus. **3** (2013)

Kinengyere, A., Kiyingi, G.W., Baziraake, B.B.: Factors affecting utilisation of electronic health information resources in universities in Uganda. Ann. Libr. Inform. Stud. **59**(2), 90–96 (2012)

Kwadzo, G.: Awareness and usage of electronic databases by geography and resource development information studies graduate students in the University Of Ghana. Library Philosophy and Practice (e-journal). Paper 1210 (2015). http://digitalcommons.unl.edu/libphilprac/1210

Luambano, I.: Information Seeking Behaviour of Distance Learning Students in the Hybrid Environment", Case Study of the Open University of Tanzania; Ph.D. thesis: University of Dar es Salaam (2013)

Lwoga, E.T., Ngulube, P., Stilwell, C.: 'Managing indigenous knowledge for sustainable agricultural development in developing countries: knowledge management approaches in the social context. Int. Inf. Library Rev. **42**(3), 174–175 (2010)

Lwoga & Chigona: Characteristics and factors that differentiate Internet users and non-users as information seekers: the case of rural women in Tanzania; SAGE JOURNAL, 2016 (2016)

Lwoga, E.T., Sife, A.S.: Impacts of Quality Antecedents on Faculty Members' Acceptance of Electronic Resources. Library Hi Tech (2018)

Norbert, G., Lwoga, E.: Information seeking behaviour of the physicians in Tanzania. Inf. Dev. **29**(2), 172–182 (2013)

Mbangala, B., Samzugi, A.: The role of telecentres in Tanzania's rural development: a Case study of Sengerema district council, Mwanza region. Library Philosophy and Practice (e-Journal), 1–26., available in Google Scholar (2014)

Mehta, B.S., Shree, M.: Impact of ICT in small towns in India: a case of public access to Internet. Knowl. Horizons–Econ. **7**(4), 28–36 (2015). Available in Google Scholar

Morley, L.: Lost leaders: Women in the global academy. High. Educ. Res. Dev. **33**(1), 114–128 (2014). Google Scholar | Crossref | ISI

Mtega, W.O., Mgoepe, M.: Factors influencing access to agricultural knowledge: the case of smallholder rice farmers in the Kilombero district of Tanzania: Southern Africa. J. Inf. Manage. (2016). http://www.sajim.co.za/index.php

Mtega, W., Nyinondi, P., Msungu, A.: Access to and usage of EIR in Selected Higher Learning Institutions in Tanzania. In: Thanuskodi, S. (ed.) Challenges of Academic 15 Library Management in Developing Countries (2013)

Mtega, W.P., Dulle, F., Malekani, A.W.: The usage of EIR among Agricultural Researchers and extension staff in Tanzania. Libr. Inf. Res. **38**(119), 47–66 (2014)

Msagati, N.: Awareness and use of scholarly electronic journals by members of Faculty: a case study of Dar es Salaam University College of Education (DUCE). Library Philosophy and Practice (E-Journal) **1124**, 2014 (2014)

Nkebukwa, L.L.: Paper title: status on the usage of electronic-resources by students at the college of business education. Bus. Educ. J. (BEJ) **I**(2), 1–13 (2016). www.cbe.ac.tz/bej

Shepherd, S.: Why are there so few female leaders in higher education: a case of structure or agency? Manag. Educ. **31**(2), 82–87 (2017). https://doi.org/10.1177/0892020617696631

Okebukola P.: Education, human security and entrepreneurship. 7th Convocation Lecture of Delta State University, Abraka, University Printing Press (2012)

Okoroafor, C.: The role of Vocational and Technical Education in manpower development and job creation in Nigeria in the 21st century: The Journal of Assertiveness (2011)

Ozkisi, H., Topaloglu, M.: the university students' knowledge of Internet applications and usage habits. Procardia – Soc. Behav. Sci. **182**, 584 (2015)

Ramayaha, T., Leeb, J.: System characteristics, satisfaction and e-learning usage: a structural equation model (SEM)", The Turkish Online J. Educ. Technol. **11**(2) (2012)

Saunders, M., Lewis, P., Thornhill, A.: ' Research methods for business students', 6th edn. Pearson, London (2012)

Tahir, M., Mahmood, K., Shafique, F.: Use of electronic information resources and facilities by humanities scholars. Electron. Libr. **28**(1), 122–136 (2010). https://doi.org/10.1108/02640471011023423

Wu, M., Yeh, S.: Effects of undergraduate student computer competence on Usage of library electronic collections. J. Libr. Inf. Stud. **10**(1) (2012)

Factors for Information Seeking and Sharing During Accomplishig Collaborative Activities in Vocational Schools

L. L. Nkebukwa[1](✉) and I. Luambano[2]

[1] College of Business Education, Business, Po Box 1968, Dar es Salaam, Tanzania
l.nkebukwa@cbe.ac.tz

[2] University of Dar es Salaam (UDSM), Dar es Salaam, Tanzania

Abstract. Purpose: The study investigated factors fostering collaborative information-seeking behaviour of students in the completion of group tasks: a case of Tanzanian Vocational Training Institutions (TVET).

Design/Methodology/Approach: This study employed a purely qualitative research design, where the naturalistic observation and face-to-face interviews were treated as prominent tools for data collection without manipulating work environments. Eight (8) TVETs institutions in five (5) regions from Zanzibar and the mainland were included in the study. Therefore, thematic analysis was used to analyse collected data, in which ''patterns across data sets'' describe all phenomena associated with specific themes and questions.

Findings: Factors fostering CISB were; lack of awareness; the need to save time; the lack of experience; lack of skills or domain expertise and the need to minimise costs.

Research Limitation/Implications: The study involved student groups however, during the process of data collection, it was easy to find that all groups because a similar schedule assigned for completing their assignments, resulting in the researcher narrowing to scope to only 8 groups conveniently available.

Practical Implication: The study recommends the following; the need for librarians to create an environment for collaborative information seeking and sharing; the need for TVET management to improve information infrastructures; the need for teachers to allocate adequate time enabled students seeking and sharing information; the need for the Government ensuring all TVET centres establishes libraries as well as having qualified librarians.

Originality/Value: This is an inclusive study assessing the collaborative information-seeking behaviour of student groups in task completion, this study also has wider implications for TVEs management in the provision of better collaborative learning environments.

Keyword: Behaviour · Collaborative · Groups learning · Information sharing · Seeking

C. Aigbavboa et al. (Eds.): ARCA 2022, *Sustainable Education and Development – Sustainable Industrialization and Innovation*, pp. 623–635, 2023.
https://doi.org/10.1007/978-3-031-25998-2_48

1 Introduction

Collaboration is the action of working together and sharing knowledge for creating the mutual understanding needed in solving problems" (Schrage, 1999). Similarly, Foster (2010), emphasises the value of collaboration as a useful and essential tool for undertaking any difficult task, as human beings by nature need to collaborate in activities that cannot be carried out by a single individual. The value of teamwork is an essential tool useful for undertaking any difficult task. Collaboration therefore is among the prominent characteristics of a group's information-seeking behaviour (Golovchinsky & Pickens, 2011). Generally, students' intends to team up during accomplishing tasks which are not easily carried out individually (Nkebukwa, 2018). Working in groups, students pair their minds to evaluate the information needed in solving the existing problems. Based on the value of collaboration, researchers have recently begun to examine its essentials for the information seeking behaviour (CISB) in order to start planning better procedures and tools that will facilitate collaborative information seeking behaviour (Foster, 2010; Jansen, 2009; Hansen and Jarvelin, 2005). These studies are needed in developing countries like Tanzania, where proper information seeking among students in the vocational training institutions (TTVET) is a major challenge. The Tanzanian Vocational Training institutions (TTVET's) established to prepare specialised specific vocational occupations in a country. The vocational training aimed at promoting the industrial innovative skilled labour or self-employment (Ishumi, 1998; Babette & Ewald, 2000; Kigadye, 2004; Mutarabukwa, 2007; URT, 2010; Kafyulilo et al., 2012; Nkebukwa & Luambano, 2018). Tanzania experienced such a situation in the early 1990's after the government retrenchment which left some citizens unemployed (Rugumyamheto, 2000). During such a period the private sectors took hold for employing citizens because the government was no longer employer; (Ishumi, 1998; Kigadye, 2004; Mutarabukwa, 2007; URT, 2010). In this vein the government introduced the Competence Based Education Training (CBET) and Competence Based Assessment (CBA) systems in the vocational institutions for preparing occupations and skills for self-employment. According to Kigadye (2004) & Kafyulilo et al. (2012). Problems raised specifically during the implementation of CBET and CBA that required libraries to uphold new trends. Unfortunately, TVET's libraries were unprepared to meet the new demands of promoting collaborative information seeking processes. Specifically, study aimed at establishing factors fostering collaborative information-seeking behaviour of student groups during accomplishment of the collaborative tasks. Being aware of factors will enable librarians to come-up with better strategies suitable for providing information services in collaborative environments.

2 Factors Promoting CISB Among Students

Collaborative is a situation cut across gender, age, cultural and other differences, which are triggered by various factors. Various studies investigated aspects of collaboration in organisation settings; for example a study on multi-disciplinary by Reddy and Spence (2008) in a context of Medical Care of Queensland Centre recognised factors such as; the lack of immediately accessible information, complexity of information needs, lack of domain experts and the fragmented of information resources. Hansen and Jarvelin

(2005) in a study on collaborative information retrieval activities involves document-based related seeking behaviour identified factors such as; necessities of sharing limited resources based on facts of scarce resources available. Bruce et al., (2008) in a study on collaborative information seeking behaviour between two design teams in the medical field on patient-related information needs identified factors of complexity of the task ahead and limitations in time to access information. A study by Bruce et al., (2008) and Fidel et al., (2004) exploring collaborative information retrieval of design teams, a study revealed communication patterns and team work activities influences collaboration during information search and sharing. Fidel et al., (2013) underscore a number of factors triggering people to seek information collaboratively are lack of expertise in a specialised area or the need for tacit knowledge. Reddy and Jansen (2008) concludes managers seek information together for making wise organisational decisions for strategic issues.

3 Methodology

3.1 Research Design

This study employed qualitative research design, which according to Collis and Hussey (2003), is a set of non-statistical techniques used to gather data in a social phenomenon. According to Creswell (2003) qualitative research design enabled researchers to provide comprehensive and exhaustive descriptions of what have been observed during its natural occurrences. Kim (2013) suggested using observation and interview techniques for qualitative data. In this regard, all-inclusive and complete protocols on using observation and interviews had been observed. This study therefore employed purely qualitative research design, where the observation and interview were treated as prominent tools for collecting data, as also recommended by (Reedy & Jansen, 2008 and Kim, 2013). So, the naturalistic observation used to collect data from groups during task accomplishments, without manipulating work environments as proposed by (Sonnenwald, 2000; Poltrock et. al., 2003; Bruce et al., 2003; Hansen & Jarvelin, 2005 & Reedy and Jansen, 2008). Observing tasks in its natural settings increases obtaining site-specific data (Shah, 2008). Moreover, face to face interviews conducted to group leaders and subject teachers for getting clarification on issues that have been observed. Further, FGDs with second year students' studded similar courses involved 12 students in each trades. Therefore, the thematic analysis used to analyse collected data, in which ''patterns across data sets" describe all phenomena associated with specific themes and questions.

This study was conducted in eight (8) TVET institutions found in five (5) regions of Tanzania, namely Dar es Salaam, Morogoro, Dodoma and Zanzibar (Urban West Region and Southern Pemba). Dar es Salaam is the headquarters of TVET's schools in Tanzania, so its four institutions selected purposely to make comparisons with other institutions selected from their proximity or remoteness from headquarters. Other criteria used to select TVET's institutions grounded into adopting and implementing the Competence-Based Education and Training (CBET) system, and later Competence Based Assessment (CBA). The CBET and CBA insisted students to be involved in hands-on activities in collaborative environments. However, the inclusion of the institution from Zanzibar based on the facts that it is an integral part of the United Republic of Tanzania, with different vocational training systems of the Mainland. The study also involved both government (public) and private, the inclusion of public and private TVET's institutions aimed at making a comparison between commercial and non-commercial TVETs institutions in Tanzania Table 1 .

Table 1. Names of Institutions and Courses involved in the study

Name of Institution	Course involved	Regions
Lugalo Military Vocational Training Centre (LMVTC)	Electrical installation	Dar es Salaam
Msimbazi Vocational Training Centre (MVTC)	Tailoring & Cookery	Dar es Salaam
Chango'mbe Vocational Training Centre	Mechanics	Dar es Salaam
Dodoma Vocational Training Centre	Masonry	Dodoma
Kipawa TVETA –ICT	Electronics	Dar es Salaam
Morogoro VTTC	Carpentry	Morogoro
Amali Pemba	Tailoring	Southern Pemba
Amali Unguja	Electronics	Urban West Region

Sources: Field data, 2018.

3.2 Findings and Discussion of the Study

Factors fostering collaborative information seeking among TVET student groups

The objective of this study sought to identify factors fostering collaborative information seeking of student groups in the vocational training institutions. The findings as obtained through Focus Group Discussions, interviews, as well as observation revealed the following factors such as; lack of awareness or complexity of information needs among group members, lack of skills or domain expertise among individual group members, lack of experience, the need to save time, the need to minimise costs, as well as the need to develop stronger communication skills.

Name of Institution	Course involved	Factors revealed
Lugalo Military VTC (LMVTC)	Electrical installation	Lack of awareness among individuals or complexity of information needs
Msimbazi VTC (MVTC)	Tailoring	The need to save time and the lack of experience
Chango'mbe VTC	Mechanics	Lack of skills or domain experience among Individual group members
Dodoma VTC	Masonry	Lack of domain expertise and the need to save time

(continued)

(*continued*)

Name of Institution	Course involved	Factors revealed
Kipawa TVETA–ICT	Electronics	Lack of skills or domain expertise among individuals
Morogoro VTTC	Carpentry	The need to need to save time
Amali Pemba	Tailoring	The need to need to save time
Amali Zanzibar	Electronics	The need to minimise costs

Sources: Field data, 2018/19.

Lack of awareness and the complexity of information needs and

Findings of the electrical students group revealed lack of awareness among individuals or complexity of information needed were among of factors fostering or promoting them to collaborate in seeking information during the accomplishment group tasks, these task were not routine by its nature, so they were needed to employ multiple sources of information that necessitate them to collaborate. Awareness, which means "knowledge or perception of a situation or fact" (www.merriam-webster.com/awareness), being aware of sources and tools helps to meet required information needs.

> *....they were unaware of a task assigned and the assignment was complex which required them to use multiple source information, this involved approaching various books, as well as asking themselves on what was a reliable source to be employed but finally decided to use the internet sources......*

The researcher observed student group members who were assigned to accomplish the electrical installation task at Lugalo Military VTC (LMVTC).

As it was confirmed by a researcher during observing one group who was accomplishing an electrical wiring installation task. Students told a researcher that;

> *...this task is very complex that needs to seek information through various sources which requires us to team up.....*

The findings correspond with a study by Skyrius & Bujauskas (2010) who revealed that, lack of awareness or complexity of information needs necessitates users to search from multiple sources for producing the better results. The finding was similar to what a researcher revealed when observing electronics course students who were repairing a CD player, they also relied on multiples to bring them awareness. It has been noted that in collaborative learning students' awareness creation firstly relied on student group members themselves and extends to the internet, teachers, and lecture notes. When an individual is not aware or lacks information on a proper source or tools required, s/he would not be able to work properly, so according to Skyrius & Bujauskas (2010) individuals should team up to simplify a task and find solutions. In this respect, teamwork helps in creating an individual awareness through sharing knowledge and expertise. Studies by Bruce et al., 2008 & Fidel et al., 2004 have also shown lack of awareness among other factors for students formulating group discussions to collaborate. Nkebukwa & Luambano (2018) concluded that, students who engaged into collaborative information seeking processes easily minimise or overcome knowledge gaps.

Lack of Skills or Domain Experience Among Individual Group Members

Findings revealed lack of skills or domain expertise among individuals is among factors fostering collaborative information seeking. The finding obtained when a tailoring student group members accomplished a task of preparing a bridal wedding dress. The lack of skills or domain expertise resulting in students' failure accomplishing activities as well. The evidence was made during observing other group members from the cookery course who had been assigned a task on preparing commercial food, specifically making a birthday cake, they were deficient in understanding of the right ingredients required, hence required to collaborate asking each other. During their discussions safety precautions were also observed, because cooking involves manipulating the temperature. As group members were doing the task collaboratively, it became an opportunity for them to share expertise. Similar incident was noted while observing mechanics students who had a collaborative task of servicing a car engine that had the problem of overheating. At the beginning group members lacked the necessary skills on the reasons for the engine overheating as well as ways of fixing the problem. Findings shows that they have to exchange ideas among themselves as to what might be the problem and the proper procedures to follow in undertaking the task. They began their diagnostic procedures. However, they desperately failed to identify the cause of the problem, as echoed by one of the group members who explained the problem to colleague who was passing:

> "….*The engine has overheating problems, and we are unable to detect the cause despite spending a lot of time and efforts…can you assist our team …*"

It was only after the teacher came to advice on how they should start diagnosing a problem, but after the teacher brought certain working skills, they came to realise that there was a leakage at the lower part of the radiator, which meant there was no circulating water to cool the engine system. After completion of the servicing, the engine problem was solved, and the customer drove away after paying an agreed sum of money. Thereafter, the group leader explained to the researcher what had happened:

> "….*the cause of the problem was a leaking radiator. We fixed it by and added the coolant, then the engine was okay. If we had managed to get needed information timely we would have saved our customer's time as well as our own...*"

The findings collaborate Majidan and Ai (2002) in a study on the use of information resources by computer engineering students in Singapore, who found that in order to accomplish the task in a similar project the information needed was best met through subject teachers, as well as textbooks and the internet sources. However, their study advocated print sources of information are the best for technical students, despite their inadequacy and lack of library support.

Lack of Experience

Findings revealed lack of experience among factors promotes collaborative information seeking. It has been observed that student groups had been assigned to do various tasks in a bid to apply what they learned in class. However, as the group tasks are done after theoretically learning in class, students lack experience in terms of doing those activities. Seeking information collaboratively rather than individually, therefore, helps to bridge or minimise the gap on lack of experience among group members. Experience, defined as "the process of doing and seeing things and of having things happen to you" or "skill or knowledge obtained by doing something" (www.merriam-webster.com/experience) this is crucial because it involves ability to do things well, that is, those who are more experienced are likely to do things in a better way. For example, in this study, tailoring course students who were making a bridal wedding gown were doing it for the first time, which obviously is a lack of experience. Doing the task collaboratively was useful in ensuring that any weaknesses in terms of skills or sources and tools to consult to meet required information needed are accordingly addressed. This finding was substantiated during interviews with group leaders while group activities were being carried out and observed by the researcher.

The researcher observed student group members who were assigned to accomplish the car engine service at Chango'mbe VTC, among other things the researchers noted the lack of skills or domain experience among individual group members.

Need to Save Time

Adequate time, just like other resources such as requisite skills, and proper tools, is indispensable in ensuring successful accomplishment of tasks. Findings revealed shortage of time for seeking and accomplishing tasks is among factors fostering collaborative information seeking. When a particular activity is done by a group the time used in completing the task will be shorter than when the task is done by a single individual. That has been notable throughout the findings of this study. In addition to that, when an activity is done by a group (collaboratively), time used to seek information will also be minimised, as group members will enable able share and exchange information on sources and tools to be used in meeting information needs as well as collaboratively use the information in ensuring successful accomplishment of the group tasks. Moreover, given the nature of collaborative tasks, it can also become possible for group members to seek information from various sources simultaneously, each member seeking from a different source, while in the case of an individual; he/she would have to seek from one source at a time, in a linear way. As the nature of information is complex, a group environment will make it possible to get information from diverse sources and hence be able to complete the group task sooner than when the task was accomplished by a single individual. For example, in this study, tailoring course students who had an assigned task of preparing a bridal wedding could not complete the task on the first day due to lack of information, and so decided to consult various sources in the evening before resuming the task on the following day. In the process, they saved time.

The same information was obtained during the FGD session at Morogoro Vocational Teachers Training Centre (MVTTC), when one participant informed the researcher that,

> In addition to theoretical learning in classes, we appreciate working in groups in order to save time. Every individual in the group has the responsibility to seek specific types of information needed on the project in hand…".

However, some arguments against collaboration in relation to saving time were made during Focus Group Discussions and interviews. For example, during interviews with student group leaders (at Amali Pemba who servicing television and Msimbazi who were preparing a bridal wedding dress, the following was said:

> "*….working in groups might take less time, save resources, than working individually……..this challenge affects members who are working individually especially when a member does not put in his or her best efforts in a collaborative task …*".

The picture shows the tailoring student group members who were assigned to accomplish the bridal wedding dress. It was noted by the researcher that they could not complete the task on the first day due to lack of information until they approached other sources of information.

A Need to Minimize Costs

Effective accomplishment of tasks requires access to information, and yet such access has cost implications. Findings revealed that minimising costs is a factor that fosters collaborative information seeking. Due to inadequacy of learning resources, as noted in this study, individuals working collaboratively help to bridge the gap, through making it possible to share the resources for the benefit of their groups. As individuals working collaboratively in a team, any source of information obtained would be used by the whole group accomplishing that task. For example, electronic course student groups who were repairing a CD player had only one textbook which was shared by the whole group, and it was easier because they were doing the same task together. It was also noted that in this study, if one group member had access to the Internet, such as through a smartphone, the results or information obtained through searching were for the benefit of the whole group, including those who did not have smartphones. Similarly, when a teacher provided information to a group member, the information was for the mutual benefit of the group rather than one group member, because it was a group task. It would have been difficult if the group tasks were individual based. Adeogun (2001) similarly revealed that, lack of relevant and up to date reading materials to support vocational training is a major challenge hampering teaching and learning in TVET institutions, negatively affecting academic performance. So, when individuals are working together any source of information obtained has to be used by all members. During FGD with students they gave an analogy, indicating that there was also inadequacy of working tools that made it necessary for students to work together in groups for the purpose of sharing. An example given was four carpentry students, each having to construct a table, while there is only one hammer and one saw. Rather than working individually which would mean students waiting for each other to complete the task, which is a waste of time, the students could work in a team to complete each activity or task in turn. In that way the tasks would be completed effectively, in good time.

A need to develop stronger communication skills

Communication, which involves sending and receiving information between two or more people, is important because without it, people would not effectively work together towards a common goal (http://study.com/academy/lesson/what-is-communication-def inition importance.html). The information conveyed can include facts, ideas, concepts, opinions, beliefs, attitudes, instructions and even emotions. Methods of communication vary, ranging from verbal communication, written communication as well as body language which are a form of non-verbal communication. Communication is important because without it, people would not effectively work together towards a common goal. In this study, findings have shown that developing stronger communication skills is one of the factors that foster collaborative information seeking. As group members work and seek information together, they engage in communication and so develop stronger skills in communicating, which make it possible for them to effectively accomplish their group task. In this study, it was noted that communication took place between group members themselves, group members and colleagues outside the team, as well as group members and subject teachers, done through face to face conversations, telephone calls and text messages, among others. Their engagement in group tasks which involved seeking information collaboratively made it possible for them to develop stronger communication skills. Without communication there would have been chaos which would consequently result in failure in accomplishment of assigned group tasks.

The picture shows the electronic student group members who were assigned to accomplish the electronic task, it was noted by the researcher that though collaboration members were able to serve cost and time of accomplishing a task.

4 Conclusion

The government introduced the Competence Based Education Training (CBET) and Competence Based Assessment (CBA) systems in the vocational institutions for preparing occupations and skills for self-employment. Problems raised specifically during the implementation of CBET and CBA that required the library to uphold new trends. Unfortunately, TVET's libraries were unprepared to meet the new demands for promoting collaborative information seeking. This study employed qualitative research design, which is a set of non-statistical techniques used to gather data in a social phenomenon. This approach enables a comprehensive and exhaustive description of what has been observed during its natural occurrences. This study therefore employed purely qualitative research design, where the observation, interview and FGD were treated as prominent tools for collecting data. Therefore, the thematic analysis used to analyse collected data, in which ''patterns across data sets" describe all phenomena associated with specific themes and questions. This study was conducted in five (5) regions of Tanzania, namely Dar es Salaam, Morogoro, Dodoma and Zanzibar (Urban West Region and Southern Pemba). The findings as obtained through Focus Group Discussions, interviews, as well as observation revealed the following factors such as; complexity of information needs, lack of awareness among individual group members, lack of skills or domain expertise among individual group members, lack of experience, the need to save time, the need to minimise costs, as well as the need to develop stronger communication skills. Various recommendation were made as follows;

Recommendations.

Librarians in TVET's schools

Librarians need to create a conducive environment that will enable TVET's students tackle the complexity of their information needs, as well as ensuring effective provisions of the information literacy training programs that will assist students in being aware of the available information sources. As noted in this study, TVET's libraries have scarce print resources, of which awareness on usage of various databases would help the on usage of multiple kinds of information sources during accomplishment collaborative tasks.

Improvement of the information infrastructure

There is a need for the improvement of the information infrastructure that will enable students in TVET institutions to obtain adequate information sources in libraries to accommodate both print and e-information sources.

TVET's Management

Management in TVET's schools needs to establish suitable information infrastructures to enable students to minimise costs of access to information. As noted, sometimes students were required to use their personal smartphones to access the internet. Establishing free Wi-Fi would assist students to have free access while seeking or sharing information using their gadgets.

TVET's Teachers

Teachers in the vocation institutions needed to allocate adequate time for students accessing information needed to accomplish tasks. As not in this study some task was scheduled to be done out of the classes or school environment. Of which students were needed to travel from working sites to visit libraries which were distant located for where tasks have been accomplished.

Government or Policy Makers

The Government should ensure all established TVETs centres in the country have libraries hand in hand with the library automation plans. Some recommendations were made that policy makers amend policies to make it mandatory for all TVETs schools needing registration to be mandatory for having a library or a computer laboratory as well as qualified librarians.

References

Babette, P., Ewald, G.: Concepts and approaches to vocational training in the informal sector the case of Tanzanian. Report resulted from a reviewed of various reports and other documents and Discussions with staff and experts at TVETA; GTZ/TVETA, Dar es Salaam: Tanzania (2000)

Bruce, H., Fidel, R., Pejtersen, A., Dumais, S., Poltrock, S.: A comparison of the collaborative information retrieval behaviour of two designs Teams. Review of Information Behaviour Research 4, no. 1:13, p. 153, (Make a difference), San Francisco: Jossey-Bass (2008)

Creswell, J.W.: Educational Research: planning, conducting, an evaluation of quantitative and qualitative research, 2nde edn. Prentice-Hall, Columbus, OH (2004)

Fidel et al.: The many faces of accessibility: engineers' perception of information sources. Inf. Process. Manag. 40(3), 563~581 (2004). http://onlinelibrary.wiley.com/doi/10.1002/meet.2009.1450460215/pdf (Retrieved 5 Dec 2021)

Fidel, R., Mark Pejtersen, A., Cleal, B., Bruce, H.: A multidimensional approaches of a study for the human information interactions. A case study of collaborative information retrieval, J. Am. Soc. Inf. Sci. Technol. (2014)

Foster, J.: Collaborative information seeking and retrieval. Ann. Rev. Inf. Sci. Technol. **40**, 329–356 (2010)

Golafshani, N.: Understanding Reliability and Validity in Qualitative Research. Qualitative Report 8(4), 597–607 (2003). http://www.nova.edu/ssss/QR/QR8-4/golafsheni.pdf (Accessed 2 Apr 2021)

Golovchinsky, G., Pickens, J.: Designing for collaboration in information seeking. A paper presented in the 74th Annual Meeting of the Association for Information Science and Technology-[Bridging the Gulf: Communication and Information in Society, Technology, and Work] held in New Orleans, Louisiana, October 9–13, 201 (2011)

Hansen, P., Jarvelin, K.: Collaborative information retrieval in information intensive Domain. Inf. Process. Manage. **41**(5), 1101–1119 (2005)

Ishumi, A.: Vocational Training as an educational and Development strategy. Int. J. Educ. Dev. 4(1), 9 (1998)

Kafyulilo, A.C., Rugambuka, I.B., Ikupa, M.: The implementation of competency based teaching approaches in Tanzania. In the case of pre-service teachers at Morogoro Teacher Training College, Universal Journal of Education and General Studies 1 (II), 339 –347 (2013)

Kigadye, M.: The role of the library in the provision of Technical Information in TVETA. A Case study of three TVET centres in Dar es Salaam region: MA Dissertation, UDSM (2004)

Kihindi, M.: Vocational education and employment in Tanzania: a study of the role and contribution of vocational training centers towards solving the youth employment problems: unpublished M.A (DS) Dissertation. University of Dar es Salaam (1996)ss

Kim, J.: Collaborative information seeking: a theoretical and methodological critique. In: CIS 2013: 3rd Workshop on Collaborative Information Seeking At the ACM CSCW 2013 Conference, San Antonio, TX, USA (2013)

Manda, P.A.: Electronic resources usage in academic and research institutions in Tanzania. Inf. Dev. 21 (4), 269–282 (2005)

Msagati, N.: Awareness and use of scholarly electronic journals by members of academic staff: a case study of Dar es Salaam. University College of Education. Library Philosophy and Practice (2014)

Mutarabukwa, G.H.: Practical implications of the management of the competence based education and training curriculum in the vocational training institutions. The J. Iss. Pract. Educ. Dev. 27(4) (2007)

Msuya, J.: Information seeking behavior of library users in changing library environments: the case of faculty of law staff members, university of Dar es Salaam. University of Dar es Salaam Library Journal, 4(1): 58–74 (2003)

Nkebukwa: Assessment of library weeding as a tool for the sustainable information services provisions in Tanzania. Volume III, Issue II, page 11 (2020)

Schrage, M.: No more teams: mastering the dynamics of creative collaboration. New York: Currency and Doubleday (1999)

Shah, C.: Toward Collaborative Information Seeking (CIS). In: 1st International collaborative Search Workshop, Pennsylvania, Pittsburgh, USA (2008)

United Republic of Tanzania: The economic survey 2009 (2010). http://www.tanzania.go.tz/economicsurvey1/2009/the economic survey 2009.pdf

Expenditure on Education, Capital Formation and Economic Growth in Tanzania

K. M. Bwana(✉)

Accountancy Department, College of Business Education, P. O Box 2077, Dodoma, Tanzania
kembo211@gmail.com

Abstract. Purpose: Capital formation and Government spending on education has been examined to establish their relationships and how they impact the economic growth of Tanzania. Gross domestic products were used to measure economic growth while government spending on education and capital formation were used as independent (explanatory) variables.

Design/Methodology/Approach: Time series research design was adopted to analyse data extracted from world development indicator-database, data ranging from the year 1990 to 2020 were used. Before testing, econometric estimates data were subjected to normality, multicollinearity and stationarity test. The study used Johansen co integration to determine existence of co-integrating equations.

Findings: Findings of this study revealed no long-run causal relationship among the variables since Johansen co-integration test shows no co integrating equations among the variables. Findings from VAR revealed that Tanzania's economic performance in one previous year have a major positive effect on current performance. While the past year's economic performance and capital accumulated in previous years do not have any significant impact on government spending on education in the short run. Capital accumulated one year ago has a significant influence on the status of current economic performance.

Research Limitation/Implications: The study could give more meaningful results if the human development index (HDI) was used rather than using the gross domestic product (GDP) as the dependent variable. HDI captures key aspects that measure social services such as health and education. The study could also employ independent variables such as the number of enrollment in a primary as well as secondary schools as well as the number of graduates at tertiary and higher learning institutions to see the output of the spending on education in terms of the skilled labour force.

Practical implications: Results would be useful when it comes to the implementation of education policy (ministry of education) while keeping the view that economic performance during one previous year has a significant positive influence on current performance. While the past year's economic performance and capital accumulated in previous years do not have any significant impact on government spending on education in the short run.

Originality: Previous similar studies have not used capital formation as one of the independent variables in such studies in the Tanzania context. Wider understanding of the existing relationship between spending on education and growth is derived from the fact that the previous year's economic performance and capital formation do not have any impact on current government spending on education.

C. Aigbavboa et al. (Eds.): ARCA 2022, *Sustainable Education and Development – Sustainable Industrialization and Innovation*, pp. 636–649, 2023.
https://doi.org/10.1007/978-3-031-25998-2_49

Keywords: Capital formation · Economic growth · Education · Expenditure

1 Introduction

Education has been regarded as proxy measure of human develoment index (HDI) making different governments (particualry in growing economies) to keep on increasing their annual budget on education in order to keep pace with ambition of rapid economic development. In view of the same argument the relationship between government spending (investiment) on education, economic development and capital formation has drawn interest of different researchers from different countries.

As conteded in the study by Afzal, Rehman, Farooq, and Sarwar. (2011) economic growth and development largely relies on the education opportunities made available to the population. Therefore investment in education is believed to have several benefits which are not limited to preparing youth to cope with dynamics and challenges of economic growth, ensure the quality of human life which provides basis for general growth in the country. Literature revealed that rapid economic growth experienced in countries like China, South Korea and India in recent past decades have been largely caused by heavy investiment in mass education and several reforms in production of key sector (i.e. industrial and agricultural production).

Mosha (2018) contended that Major philosophy or national policies have strong influence on the country's education system. For example Tanzania has had three different national policies in different time. *Capitalism was pursued shortly after independent (1961–1967) but did not last longer; Self relience (1967–1985) and liberalization (vision 2025)*, on the other hand the country has had three different edcuation policies in different times such as*; integration education policy (1961–1967); education for self relienace policy (1967–1985); ability education and training policy (1995).*

Considering the fact that education contributes significantly to peoples well being, Tanzania has tried to complied with important international requirements in the efforts to improve its education system, example of such requirements is world declaration on education for all and millenium declaration or universal primary education. The government of tanzania in 2015/2016 removes fees in primary as well as secodary schools, particulary in public schools. According to data from world bank development indicator (WDI) databases, population of the country has doubled from 25.2million in the year 1990 to 59.7million in the year 2020. However, there is little increase in the goverment spending on education during the same period. For example in the year 2000 the government spent 2.13% of its GDP on education compared to 3.10% of GDP in the year 2020. As far as budget allocation to the education is concerned Tanzania allocated 17.9%; 19.8%;17.2% and 14.8% in the year 2013/14; 2014/15;2015/16 and 2016/17 respectively (Action aid report, 2021) which is less than the recommended (20% of the budget).

In recent years the country has witnessed increase of 38 percent enrollment in primary as well as 45 percent increase in first year of secondary education enrollment following introduction of free ordinary secondary education and primary education (URT,2019). Despite the deliberate move to increase in the enrollment in primary as well as lower level secondary there have been some challenges experienced in the education system as the result of the said increase of enrollment, for example the problem of recommended class pupils ration (PCR) of 40:1 has been difficult to achieve. It has lso been reported that avarage pupil teacher ratio (APTR) been increasing from 47:1 in 2017 to 54:1 in 2018/19. Therefore it is estimated in order to realise the recommended PCR and APTR around 47,229 new teachers and 44,982 classrooms are needed (URT, 2019).

Action plan report (2021) indicate that sustainable develoment goals number four (SGD 4) advocates that both girls and boys equal access to free and quality primary as well as secondary education. However, the report of implementation of SDG 4 in tanzania revealed that the country almost achieved universal to primary education in 2007. However, it is estimated that 3.5 million children and young aged between 7–17 years were not in school. Report further revealed that only 80% of all enrolled complete primary school while 70% of those enrolled do not finish ordinary level secondary education and only 8% Action Aid Report (2021). According to action aid report (2021) United Nations recommends that in order to realise SGD 4 a country should allocate about 20% of the budget and 6% of GDP as investiment on education. Report from world development indicator database shows that over the period of ten years spending on education as percentage of GDP was almost around 4.5% in the year 2010 only *(see Fig. 1)*.

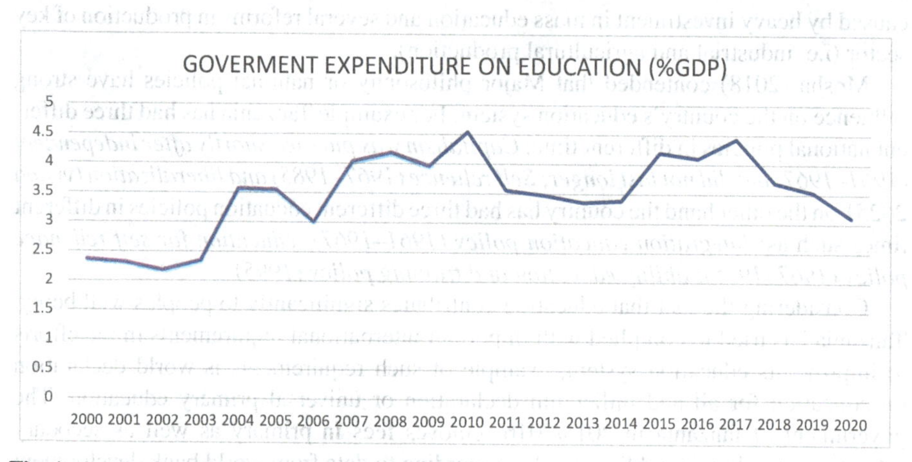

Fig. 1. Public Spending on Education (from 2000–2020). *Source: World Bank Development Indicator Database*

Vandenbussche, Aghion and Meghir (2006) contended that skilled human capital engaged in innovation usually have great impact on country's economic performance as it adjust towards technological frontier. Therefore causal relatioship manifesting the impact of spending on education on economic progress is built upon the endorgeneous growth theory during 1980s (Afzal et al. (2011); Lucas (1988); Romer (1990), Endogenous growth is manifest long-run economic progress largely caused by internal forces to the economic system, specifically the said forces should be able to govern the opportunities as well as motivations to generate technological understanding.

Rate of advancement in technology has influence on the development of total factor productivity (TFP) an outcome of long run economic performance which reflect growth rate of economic production per person countrywide. Lucas (1993) examined East Asia miracle while focusing on part of human and capital formation through learning while doing. His findings record that the engine of economic performance is effective formation and accumulation of proper human capital measured by knowledge. Findings further revealed that variability in standard of living among nations is largely based on the difference of quality of human capital.

Education is viewed as very important ingradients in preparation of human capital and cosequently very important driver of the sustainable economic development (Goode, 1959; Schultz 1961). According to Annabi, Harvey and Lan (2011) expenditure on education and preparation of human capital have emerged as the important deriver of the long term growth. Literature show that findings from recent studies argued that countries that are close to technological frontier should spend heavily on higher education in order to improve productivity as well as economic performance. Barro and Lee(2000) contended educated human resources are likely to help on the proper use of advance technology. Because of the importance of education in supporting technology Deniz and Dogruel (2008) asserted that new technology and support to proper human capital formation have been extensively discussed on studies of economic development. Annabi (2017) stressed that governments in different countries usually provide funds and allocate the funds in education and research with aim of moulding the human capital properly, it is obvious that by having skilled human capital will evenutally derive economic performane. Aghion, Boustan, Hoxby, and Vandenbussche, (2005) also argued that difference in education composition (primary, secondary and tertiary) to human capital in two different countries may result in different economic growth rates.

Gross capital formation (GCF) in a country over accounting period is termed as gross capital formation. The term capital formation include capital goods (equipment, electricity, tools and infrastructure. Countries require capital goods in order to make ensure sustainability of production and hence economic growth. Other factors being constant the higher the capital accumulated the quicker the country can increase its economic growth and aggregate income. In order to increase production of goods and services growth in capital formation is needed which eventually will promote increase in national income levels. The country need to accumulate more additional capital through household saving and new investment or through government policy. Usually countries which manifest high rate of saving at household level have high chance to accumulate more funds to manufacture capital goods faster, and government with surplus have high chance to spend (devote) the surplus in manufacturing of capital goods. According to the World Bank gross capital formation is extra amount added to long term assets (fixed assets), it also include net change in current assets. Fixed assets are include plant, machinery, equipment, as well as buildings, all these are useful in production of goods and services. Inventories involves raw materials and goods available for sale. It is argued that if saving rate increases at household level, then household may opt to invest additional savings to invest in stocks and bonds. If this done at large scale (more households are saving) in a country, then that particular country may experience a cash surplus, which is a good indication of capital formation. Country's rate of capital formation determines growth of country's GDP.

The rationale of this study built on the ongoing serious challenge of budget deficit that face many developing countries, Tanzania being one of them. Economic literature added that a country that runs surpluses (which is one of key element of gross capital formation) will have high chance to invest in capital goods and hence production of goods and services. Literature further contended that countries that have invested heavily on education have high chance to experience rapid economic growth. Therefore, major objective is to examine causal relationship between governments spending on education, gross capital formation on economic growth. Specifically, the study aims to:

i. Analyze short run causal relationship between past economic performance and current economic performance
ii. Analyze short run causal relationship between previous gross capital formation and current economic performance
iii. Analyze short run causal relationship between previous government spending on education and current economic performance

The remainig parts of this paper have been section as follows, section two covers methodology (analytical techniques as well as model) used in the study. Third section presents the findings as well as discussion while section four presents conclusion and policy recommendations to the authorities responsible for education and economic policy.

2 Methodology

Secodary data covering the year 1990 to 2020 were obtained from World Bank Development database. Variables employed were Gross Domestic Product (GDP) (dependent variable), proxy measure of economic development, while Government expenditure on education (GEED) and gross capital formation (GCF) were used as independent variables. Functional relatioship between the variables can be explained by the mathematical model below:

$$GDP = f(GCF, GEED)$$

In this equation Gross Domestic Product is adopted as proxy gauging tool for economic performance, and it is expressed as a function of gross capital formation (denoted by GCF) and Government expenditure on education (denoted by GEE) Before spefifying the econometric relationship among the variables we subjected the data into diagnostic tests in order to come up with the most appropriate model specification for the study. We started by testing normality of the variables using *kernel density graph* and found that all variables were not normaly distributed. We therefore norlamized the varibales by applying natural logarithms.

2.1 Unit Root and Correlation Test

Before checking whether our variables are stationary or not, we declare that our data are the time series data. Literature records that there are several methods that can be used to stationarity test of the variables. However, Augumented *Dicky fuller* and *Philip Perron* are the most widely used techniques (Dickey and Fuller 1979; Dickey and Fuller 1981). Therefore this study also applied the Augumented Dickey Fuller to establish stationarity of the variables and confirmed the results using *Phillips Perron.* If the absolute value of the test statistics is less than the 5 percent critical value it implies that the variables are not stationarity and vice-versa is true. The tests results reporteed that variables were not stationary at level and we therefore needed go for first difference to make them stationary. We further noticed that variables only become stationary after the first difference. This prompted us to opt for co-intergration test where we employed the Johansen's co-intergration test (Johansen and Joselius, 2000).Since the study involves multivariate time series where there are three variables, we tested if there is existence of multicollinearity problem. Multicollinearity implies the independent variables are correlated and if the degree of correlation is above 80% (Gujarati, 2003) then if may cause problem when the model is fitted or during interpretation of result. Therefore we run correlation matrix in order to test degree of multicorlianerality between independent variables. The results from the correlation matrix showed that there was no serious multicollinearity since no set of variables had a correlation exceeding 80%.

2.2 Johansen Cointegration

Before testing co-integration of variables we use Akaika creterion (AIC) Schwarz Bayesian Information Creterion (SBIC) and Hannan Quinin Information Criterion (HQIC) lag selection creteria to determine optimal lag length. We found that the optimum lag length was 2. After determining the optimal lag length, we determined how many cointergrating vectors, the *trace statistics and eigenvalue* were used for this purpose. Performing a cointergartion test was necessary so as to establish a long run relationship. In the long run the relationship may be presented despite the fact that series either drifting apart, or trending either upwards or downwards. Cointergartion test is done on their level form instead of the first difference, nevertheless variables may also be used in their log form. The findings from co-intergration test showed zero (0) rank of co-intergration. This being the case, we then opted for the Vector Autoregressive Model (VAR) to determine short run relationship between variables.

2.3 Vector Autoregressive Model (VAR)

Literature argue that VAR model provide a creditble technique to data descriptions, forecasting and structural inferences as well as policy analysis. The model assumes that variables evident stationarity after first difference. Autoregressive implies availability of the lagged value of the explained variable (independent variable), we used VAR model in this study because there was no cointergrating equation. VAR has also been used in this study because this study focuses on policy analysis which tries causal relationship between Gross Domestic Product (GDP) and government expendiure on education (GEED) and capital formation as independent variables (GCF).

This study employed Vector Autor Regressive (VAR) model in multivariate time series, VAR model implies that n- variables, n- equations model precisely define each variables as direct funtion of its former values, the former values of remaining variables considered and a serially uncorrelated shocks terms.

$$lnGDP_t = \beta 1 + \sum_{i=1}^{n} \varphi_i lnGDP_{t-i} + \sum_{j=1}^{n} \phi_j lnGCF_{t-j} + \sum_{m=1}^{n} \psi_m lnGEED_{t-m} + \mu_{1t} \quad (1)$$

$$lnGCF_t = \beta 2 + \sum_{i=1}^{n} \varphi_i lnGDP_{t-i} + \sum_{j=1}^{n} \phi_j lnGCF_{t-j} + \sum_{m=1}^{n} \psi_m lnGEED_{t-m} + \mu_{2t} \quad (2)$$

$$lnGEED_t = \beta 3 + \sum_{i=1}^{n} \varphi_i lnGDP_{t-i} + \sum_{j=1}^{n} \phi_j lnGCF_{t-j} + \sum_{m=1}^{n} \psi_m lnGEED_{t-m} + \mu_{3t} \quad (3)$$

where $\beta 1$, $\beta 2$ and $\beta 3$ are constant while $\mu_{1t,} \mu_{2t} \mu_{3t}$ are called innovations or shocks terms. φ, ϕ **and** ψ are coefficients. VAR model requires all variables in the model to have equal lags and that variables must be specified in their level forms or log forms and not at their difference. *Equation 1* simply means that gross domestic product (GDP) can be explained by constant $\beta 1$, its past value, the past values of GCF and GEED as well as serially uncorrelated error term μ_{1t}. *Equation* 2, implies that Gross capital formation (GCF) can be explained by constant $\beta 2$, its past value, past values GCF and GEED as well as serially uncorrelated error term μ_{2t}. *Equation 3* implies that Government expenditure on education (GEED) can be explained by constant $\beta 3$, its past value, past values GCF and GDP as well as serially uncorrelated error term μ_{3t}.

2.4 Stability Conditions and Diagnostic test

In orger to confirm robustness of the result from VAR model the model need to meet stability condtions. Therefore when all roots are within the unit circle (less than one) it implies the model stabiltiy conditions has been met. To confirm existence of autocorrelation in residual we need to generate residual values and establish the mean of residual whether is close to zero or not, the requirement is that the residuals should be random and have very minimum mean. Autcorrelation in the model at selected lags is tested using Lagrange multiplier test, and if the at two lags (according to Hannan Quinin creterior) the p-values is below 5 percent then we reject the null hyptohesis.

3 Findings and Discussion

Findings revealed that all variables Gross domestic products (GDP), Government expenditure on education (GEED) and gross capital formation (GCF) were not normally distributed,therefore we normalize the variables by applying natural logarith to the variables. Correlation matrix was used to test multicornealrity among variables, result show that there was no serious correlation as there was less than 0.80 correlation between variables (as general rule of thumb require correlation not to exceed 80%).

Table 1. Summary statistics

Variable	Obs	Mean	Std. Dev.	Min	Max
lngdp	31	23.61416	.9069792	22.172	24.85699
lngcf	31	3.300284	.2978761	2.701344	3.714017
lngeed	31	1.106531	.2518728	.7599036	1.513301

3.1 Unit root and correlation test

The study used Dickey Fuller test to test for stationarity of the variables and then confirm the result using Philips Perron.

Table 2. Dickey Fuller Unit Root Test

Variables	At Level		First Difference	
	Test statistic	5% critical value	Test statistic	5% critical value
LnGDP	−0.864	−2.986*	−4.198	−2.989
LnGCF	−0.572	−2.986*	−4.617	−2.989
LnGEED	−1.857	−2.986*	−6.056	−2.986

* implies that the null hypothesis which states that the variables are not stationary at 5 percent significance level has been rejected in favor of alternative hypothesis. We were compelled to confirm the results using Phillips Peron test, the results were shown on Table 4.

Table 3. Philips PeronUnit Root Test

Variables	At Level		First Difference	
	Test statistic	5% critical value	Test statistic	5% critical value
GDP	−0.844	−2.986*	−4.222	−2.989
LnGCF	−0.749	−2.986*	−4.611	−2.989
LnGEED	−1.814	−2.989*	−6.107	−2.989

* implies that the null (variables are not stationary at 5 percent level of significance) has been rejected in favor of alternative hypothesis.

Findings from the Phillips Peron test (Table 4), the variables convert to stationery after first difference (stationary at order one). Since all the variables were stationary at order one, we find it necessary to test for long run relationship using Johansen co integration test to test. The co-integrating rank is confirmed using Trace statistics as well as the maximum Eigen value as proposed by Johansen and Juselius,(Johansen & Juselius, 1990).

AIC - Akaika creterion; SBC - Schwarz Bayersian Creterion; HQIC - Hannan Quinin.

Table 4. Optimal Lag selection

```
. varsoc lngdp lngeed lngcf, maxlag(4)

   Selection-order criteria
   Sample:  1994 - 2020                         Number of obs      =        27
```

lag	LL	LR	df	p	FPE	AIC	HQIC	SBIC
0	-5.68716				.000382	.643494	.686307	.787476
1	75.5629	162.5	9	0.000	1.8e-06	-4.70837	-4.53711	-4.13244*
2	87.7425	24.359	9	0.004	1.5e-06*	-4.94389	-4.6442*	-3.93602
3	94.2713	13.058	9	0.160	1.9e-06	-4.76084	-4.33271	-3.32102
4	105.879	23.216*	9	0.006	1.8e-06	-4.95401*	-4.39743	-3.08224

```
   Endogenous:  lngdp lngeed lngcf
    Exogenous:  _cons
```

Findings of ther optimal lag selection suggest the use of two lags in this study, according to HQIC Criteria.

Table 5. Johansen co-intergration result

```
. vecrank lngdp lngeed lngcf, trend (constant)

                     Johansen tests for cointegration
Trend: constant                                        Number of obs =      29
Sample:  1992 - 2020                                            Lags =       2
```

maximum rank	parms	LL	eigenvalue	trace statistic	5% critical value
0	12	72.669252	.	23.8315*	29.68
1	17	80.476713	0.41635	8.2165	15.41
2	20	83.791518	0.20436	1.5869	3.76
3	21	84.584984	0.05325		

* implies proposed the number of cointegrating equations which corresponds to the row of in the table table (Johansen. 1995). In this case we tested the hypothesis that: Ho: no cointrgrating equation against the alternative hyptohesisi that there is contegration. *R*esults for the Johansen co integration are shown in Table 6 and the result revealed at rank 0, the trace statistics value is less than 5 percent critical value meaning there existence of zero co integrating equation and there is no long run relationship. This implies that we failed to reject the null hypothesis.

3.2 Vector autoregressive model (VAR)

Findings of VAR indicate that both first lag of government expenditure on education (GEED) does not have any significant impact on Gross Domestic Product (GDP), while second lag of government spending on education have positive significant influence on economic development at significance level of ten percent. First lag of Gross capital formation (GCF) have negative influence at significance level of one percent (1%) on average *Ceteris paribus.* Second lag of GCF do not have any significant impact on GDP. First lag of GDP have positive influence on GDP at one percent significance on average ceteris paribus while second lag of GDP do not have any signigicant influence on GDP (Appendix 1). This finding conforms with result in Afzal et al. (2011) which reported presence of feedback causation between education (all levels) and economic development. Labour force was found to be the main factor in explaining the realtioship between education and economic development as compared to gross capital formation. Specifically their findings revealed that primary education influence economic development while economic development only influence high education apart from primary education.

3.3 VAR Stability Conditions and Residual Diagnostic

To test relevancy of the model used in the study, there is a need to test stability of the model. Stability of the VAR model in this study implies stationarity and it is normally tested to establish if all inverse roots of the characteristics AR polynomial have modulus less than one and lies inside the unit circle, then the estimate VAR is stable and if the VAR is not stable diverse test conducted on the VAR model may be invalid and the impulse responses standard error are invalid. In this study we tested the stability ccondition of the VAR model (refer Table 9) and find that model is stable (sastisfy stability conditions) meaning all roots are lying within inside the unit circle (less than one).

Table 6. Eigenvalue stability condtion result

Eigenvalue	Modulus
.9580293	.958029
.7300468 + .2562432*i*	.773711
.7300468 - .2562432*i*	.773711
.1190406 + .3174189*i*	.339007
.1190406 - .3174189*i*	.339007
.0530212	.053021

All the eigenvalues lie inside the unit circle.
VAR satisfies stability condition.

We further tested the autocorrelation in the residual of the estimated model and find that the mean is very close to zero which can also be reflected in the Fig. 1 below. Result shows that error terms of residual are random and moving around the.

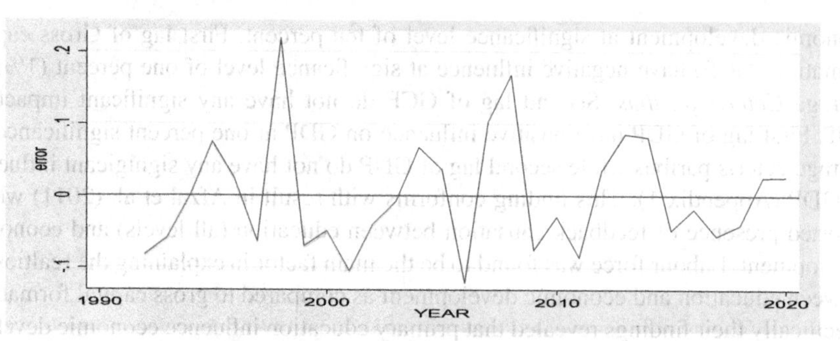

Fig. 2. Graph showing behavior of residual

Further diagnostic test involved checking for autocorrelation, where we tested the null hypothesis that there is no autocorrelation against the alternative hypothesis which states there autocorrelation. Findings shows that at lag two there is no autocorrelation since the *P value (0.11198)* is greater than 0.05 therefore we failed to reject the null hypothesis.

Table 7. Lagrange Multiplier test

lag	chi2	df	Prob > chi2
1	12.3359	9	0.19503
2	14.3019	9	0.11198

H0: no autocorrelation at lag order

4 Conclusion and policy implications

The study examines the relatiohsip between investiment in education or gverment expenditurre on education, capital formation and economic performance in tanzania. The study concludes that tanzania economic performance in one previous year have significant positive impact on current performance. However, the past years economic performance and capital accumulated in previous years do not have any significant impact on goverement spending on education in the short run. Amount of capital accumulated in one previous year affect negatively the current economic performance. It was also revealed that capital accumulated one year ago have significant influence on the status of current economic performance, this implies capital accumulation which involves saving elements (at household level) and surpluses (at national level) of which if not immediately invested may affect future economic performance and production of capital goods as well as goods and services in the short run.

The study recommends that policy makers should come up with economic policy that will provide steady economic growth through production of capital goods which will eventually promote investiment in manufacturing of goods as well as provision of services due to the fact that the economic performance of one previous year have significant impact on the current economic performance in the short run. Policy makers and responsible ministry should have long term view when crafting or implementing tanzania economic and education policy since there is no significant impact of one previous year investiment on education and current economic performance in the short run. Households, corporation and govement agency are reminded that savings and surpluses set aside in previous one year should be plough back to investiment(production) immediately to reduce negative impact on current year economic performance in the short run.

Appendix 1: VAR Result

```
Vector autoregression

Sample:  1992 - 2020                          Number of obs    =        29
Log likelihood =  84.58498                    AIC              = -4.385171
FPE            =  2.58e-06                    HQIC             = -4.075081
Det(Sigma_ml)  =  5.88e-07                    SBIC             = -3.395061

Equation           Parms      RMSE     R-sq      chi2     P>chi2
----------------------------------------------------------------
lngdp                7      .091802   0.9910   3205.151   0.0000
lngeed               7      .136342   0.7523   88.09693   0.0000
lngcf                7      .109076   0.9016   265.6535   0.0000
----------------------------------------------------------------
```

	Coef.	Std. Err.	z	P>|z|	[95% Conf.	Interval]
lngdp						
lngdp						
L1.	1.194777	.1731951	6.90	0.000	.8553209	1.534233
L2.	-.2301581	.1683765	-1.37	0.172	-.56017	.0998537
lngeed						
L1.	.1803263	.1304905	1.38	0.167	-.0754303	.436083
L2.	.2399894	.1349318	1.78	0.075	-.0244721	.5044508
lngcf						
L1.	-.4490118	.1525879	-2.94	0.003	-.7480787	-.1499449
L2.	.149314	.1619482	0.92	0.357	-.1680988	.4667267
_cons	1.423719	.6984837	2.04	0.042	.0547157	2.792721
lngeed						
lngdp						
L1.	-.0099699	.257224	-0.04	0.969	-.5141197	.4941799
L2.	.060823	.2500675	0.24	0.808	-.4293003	.5509463
lngeed						
L1.	.5302225	.1938004	2.74	0.006	.1503807	.9100643
L2.	-.1172387	.2003965	-0.59	0.559	-.5100086	.2755313
lngcf						
L1.	.3111713	.2266189	1.37	0.170	-.1329936	.7553361
L2.	-.060739	.2405205	-0.25	0.801	-.5321505	.4106725
_cons	-1.353116	1.037366	-1.30	0.192	-3.386316	.6800846
lngcf						
lngdp						
L1.	-.2988586	.2057829	-1.45	0.146	-.7021858	.1044685
L2.	.3472455	.2000576	1.74	0.083	-.0448602	.7393513
lngeed						
L1.	.2326432	.1550431	1.50	0.133	-.0712357	.5365221
L2.	.0233189	.1603201	0.15	0.884	-.2909028	.3375405
lngcf						
L1.	.9842259	.1812984	5.43	0.000	.6288876	1.339564
L2.	-.3550132	.1924199	-1.84	0.065	-.7321493	.0221228
_cons	-.1676379	.829908	-0.20	0.840	-1.794228	1.458952

References

1. Action Aid Report. Financing the Future: Delivering SDG4 in Tanzania (2021). Accessed on 26 Apr 2022. https://www.google.com/search
2. Afzal, M., Rehman, H.U., Farooq, M.S., Sarwar, K.: Education and economic growth in Pakistan: a cointegration and causality analysis. Int. J. Educ. Res. **50**(5–6), 321–335 (2011)

3. Aghion, P., Boustan, L., Hoxby, C., Vandenbussche, J.: Exploiting states' mistakes to identify the causal impact of higher education on growth. Mimeo, Department of Economics, Harvard University (2005)
4. Annabi, N.: Investments in education: what are the productivity gains. J. Policy Model. **39**(3), 499–518 (2017)
5. Annabi, N., Harvey, S., Lan, Y.: Public expenditures on education, human capital and growth in Canada: an OLG model analysis. J. Policy Model. **33**(6), 852–865 (2011)
6. Barrow, R.T., Lee, J.W.: International data on education attainment: updates and implications' NBER Working Paper No W7911 (2000)
7. Deniz, Z., Dogruel, A.S.: Disagregated education data and growth: some facts from Turkey and MENA Countries. Topics in Middle Eastern and North African Economies, 10 (2008)
8. Dickey, D.A., Fuller, W.A.: Distribution of estimator for autoregressive time series with a unit root, Journal of America statistical association, 74(366a), 427–431 (1979)
9. Dickey, D.A., Fuller, W.A.: the likelihood ratio statistics for autoregressive time series with a unit root. Econometrica **49**(4), 1057–1072 (1981)
10. Goode, R.B.: Adding to the stock of physical and human capital. Am. Econ. Rev. **49**(2), 147–155 (1959)
11. Johansen, S.: Modelling of co-integration in the vector autoregressive model. Econ. Model. **17**(3), 359–373 (2000)
12. Johansen, S.: Likelihood-Based Inference in Co-integrated Vector Autoregressive Models. Oxford University Press, Oxford (1995)
13. Johansen, S., Juselius, K.: Maximum likelihood estimation and inference on co integration - with applications to the demand for Money. Oxford Bulletin of Economics and Statistics, 52(2) (1990)
14. Lucas, R.E., Jr.: On the mechanics of economic development. J. Monet. Econ. **22**(1), 3–42 (1988)
15. Lucas, R.E.: Making a miracle. Econometrica: J. Econometric Society, 251–272 (1993)
16. Mosha, H.: The state and quality of education in Tanzania: a reflection. Papers in education and development (31) (2018)
17. Romer, P.M.: Endogenous technological change. J. Political Econ. 98(5, Part 2), S71-S102 (1990)
18. Schultz, T.W.: Investment in human capital. Am. Econ. Rev. **51**(1), 1–17 (1961)
19. URT: Education sector performance report 2018/19 (2019)
20. Vandenbussche, J., Aghion, P., Meghir, C.: Growth, distance to frontier and composition of human capital. J. Econ. Growth **11**(2), 97–127 (2006)

Mitigating the Challenges of Academic Information Systems Implementation in Higher Education Institutions in Tanzania for Their Continuous Development

A. M. Kayanda(✉)

Department of Mathematics and ICT, College of Business Education, Dar Es Salaam, Tanzania
a.kayanda@cbe.ac.tz

Abstract. Purpose: Finding out in depth the problems being faced by HEIs in the Tanzanian context on Academic Information Systems (AISs) adoption in their daily activities.

Design/Methodology/Approach: Qualitative methods were used to explore the information from the users whereby interviews and observations were used to get the information. The analysis of the collected data was done thematically whereby five themes were obtained.

Findings: Based on the obtained themes, the findings show that poor system functionalities, poor user support, lack of enough training, lack of experts and poor systems security are the main challenges faced in using the implemented AISs in Tanzanian HEIs context. The causes of the problems are also highlighted, and some improvement measures have been proposed.

Research Limitation/ Implications: This study involved only one Tanzanian HEI and involved only one AIS therefore more studies are invited to improve the findings for generalization.

Practical Implications: The findings of this study are important to the Tanzanian HEIs management to think about revising their ICT policies and their budget for information systems projects to overcome the highlighted challenges for better outcomes for their continuous development.

Originality/Value: Due to the fast-growing technology, frequent research is inevitable and assessment of technology challenges in developing countries is crucial. This study, therefore, contributes to the body of knowledge of some common challenges encountered by HEIs in the Tanzanian context on the implementation of AISs. Based on the highlighted challenges, the study also found that trust affects the success of AISs implemented in the Tanzanian HEIs context.

Keyword: Academic · Challenges · Information systems · Higher education · Tanzania

C. Aigbavboa et al. (Eds.): ARCA 2022, *Sustainable Education and Development – Sustainable Industrialization and Innovation*, pp. 650–658, 2023.
https://doi.org/10.1007/978-3-031-25998-2_50

1 Introduction

Based on the current world of technology, economic development of any country can not run away from the use of Information and Communication Technology (ICT) in its different organization (Bilan, Mishchuk, Samoliuk & Grishnova, 2019). Higher Education Institutions (HEIs) in Tanzania like any other organizations around the world needs to use ICT for their sustainable development.

On ensuring their continuous development, HEIs in Tanzania have been working so hard on adoption of ICT in doing their daily activities (Mtebe & Raphael, 2018; Mwandosya & Mbise, 2019). The adoption has led to the computerization of their different academic processes starting from students' application process to their graduation. In most of these HEIs the common academic processes include students' application process, registering of the admitted students after application, examination control and results processing and students' payments management. Most of these HEIs are now using different Academic Information Systems (AISs) in performing these academic processes.

According to Ko, Kim and Kim, (2021) the application of ICT improves the working efficiency to the organization. With the improved working efficiency it is expected that user satisfaction is attained due to the quality of service provided by reducing isolation in operations (Papanthymou & Darra, 2018). With the use of information systems (IS) these HEIs have experienced quick delivery of services to both students and staff than before; and more data is available for decision support. Better decision making and high customers satisfaction in these HEIs ensures their continuous development.

On theoretical views, Petter, DeLone and McLean (2013) highlighted six dimensions of information systems success. The dimensions included quality of the system, quality of service, information quality, users satisfaction, users intentions to use of the system and net benefits from using the system. Relying on this IS success theory, it is therefore necessary to monitor the progress of the implemented AISs and solve the experienced challenges so as to meet user needs. This study therefore intended on finding out in depth the problems being faced by these HEIs in Tanzanian context on the use of the AISs so as to ensure a successful adoption. Specifically, the study aims on encountered in the implementation process and also finding out the causes of the problems so as the situation can be improved. On fulfilling the highlighted objective, this study answered the following research questions:

1. What are the main problems encountered in the Tanzanian HEIs context in the implementation of AISs?
2. What are the causes of the problems encountered and how to improve the situation?

1.1 Literature Review

Studies have proved that the use of ICT helps in work simplification in HEIs and hence performance improvement (Đorđević et. al., 2021). According to Woyo, Rukanda & Nyamapanda (2020), proper implementation of the ICT services is in HEIs is necessary so as to get better outcomes. Some Tanzanian education institutions have taken some initiatives to collaborate with international agencies so as to improve their ICT services for better outcomes (Chirwa, 2018). Mfaume (2019) noted some benefits and challenges on the implementation of ICT in Tanzania education system which might be the same in Tanzanian HEIs. Studies shows that, the implementation of information systems in HEIs are complex due to the frequently changing requirements (Pereira, de Carvalho & Rocha, 2021) hence it is necessary to frequently research so as to get feedback for improvement and therefore this study is important.

Having better information systems in HEIs helps in providing quality services and therefore high customer satisfaction (Okenyi, Mulili & Kiflemariam, 2019). Due to the current competitive environment among Tanzanian HEIs, customers satisfaction is very important for their survival. As any other business organizations, the frequent changes on the working protocols in these HEIs is inevitable for development and hence care should be taken to ensure quality of these systems (Chaudhary, Hyde, & Rodger, 2017). A need to maintain the development flexibility for users satisfaction is always challenging (Maheshwaran, Kumar, Rajeswari & Mungara, 2017) but proper management of frequent user requirements should be considered to encourage innovation (Fitzgerald & Stol, 2014) for better AISs. More effort is needed to overcome the different challenges for successful implementation of the AISs in Tanzanian HEIs so as to meet the user expectations and therefore frequent research is inevitable.

2 Methodology

Qualitative methods were used to explore the information from the users whereby a semi-structured interview and observation were involved.

In this study, one of the Tanzanian HEI was involved. The selection of the college was done purposely based on researcher's experience with the college considering its effective use of an AIS known as Student Academic Register Information System (SARIS) for more than five years (Creswell, 2013).

The interview included five (5) lectures, two(2) ICT staff and four (4) administrative staff who use the system in different ways in managing students.

The observation involved different user activities using the AIS at different times for one month, where by the observation guide was used.

The participants were selected purposely based on their ability to provide relevant information for the study (Creswell, 2013). Both local language (Kiswahili) and English were involved to explore more information from the participants. The interview time ranged between 30 to 45 min.

To ensure trustworthiness of the data collected different techniques were used. Sometimes the same question was asked several times in different phrases to get more understanding of what the participant was saying (Shenton, 2004). At some point the respondents were required to explain by practice through using the system so as to capture the information clearly (Shenton, 2004).

Participation in the study was voluntarily. No participant was forced to participate and every information about the participants were anonymous in the data collected.

The analysis of the collected data was done thematically (Creswell, 2013). The recorded interviews were first transcribed. Thereafter all the collected data from interview and observation were translated to English and then coded. All the codes yielded five themes.

3 Findings and Discussion

The analysis of the collected data yielded five themes which have been used to answer the study questions. The themes includes:

i. Poor system functionalities
ii. Poor user support
iii. Lack of enough training
iv. Lack of experts on AIS
v. Poor systems security

The highlighted themes generally answers the first research question on what are the main problems encountered by the Tanzanian HEIs in the implementation of AISs. The themes are thereafter discussed more to answer the second research question of this study. The explanations show specifically the causes of the noted problems and how the situation can be improved based on the findings.

3.1 THEME 1: Poor System Functionalities

The study found that poor functionalities of the AIS is challenging the use of these systems in different higher learning institutions in Tanzania. The respondents think that the systems should accommodate more functionalities based on their daily activities and not only the functionalities to support the common known operations in higher learning institutions. It is clearly that system quality is very necessary to ensure user satisfaction with the system by implementing what the users request (Chaudhary et al., 2017). Based on this finding the users' explanations can be divided into three sub-themes; *missing functionalities, poor existing functionalities and systems integration.*

3.1.1 Missing Functionalities

The results highlighted that the system misses some important functionalities which users think that they could be useful in their daily activities. The results shows that users found the system useful but still they have some special needs based on their daily operations on managing students. The respondents said several statements based on missing functionalities:

> *"Although we can see students' results are now easily processed but we still need more functionalities to be added to the system so we can use it in doing more tasks in our daily activities."*
>
> *"The system should be improved to accommodate other things which cannot be done by the system to make things easier and we will like to use the system, still the system cannot do a lot of things."*
>
> *"What i can say is that, these days our tasks have become significantly simple, though there is still a need for more improvements and addition of many other things to the system."*

3.1.2 Poor Existing Functionalities

It has been found that although the system is helpful still most of the existing functionalities does not work properly. They have reported that they get a lot of problems with the system due to some system errors and they think the system should be improved to support the better. The users also highlighted that, apart from using the system they still have to do a lot of manual work which could be done with the system. The respondents said:

> *"The system has been helping us a lot but still most of the reports we need to compile them manually because the system cannot give us a report with all the required information."*
>
> *"Until now the system has been very helpful in making our work simple but it still need many improvements..."*
>
> *"I really like it when the results get uploaded smoothly. it is really upsetting when the system tells you 'did not register' or 'does not exist'..."*

3.1.3 Systems Integration

Systems integration has also been shown to be one of the requirements which users think should be considered. It was shown that there have been different systems like accounting system, timetabling system, library systems and others which they are used separately hence unnecessary manual work is required most of the times. All this systems are used in managing students in different activities. Users shown that if they are integrated they would simplify their work very much since the functionalities would be improved for better daily operations support. Different user statement showed this:

> *"It is boring when you have to download the list from the system to upload in the accounting package while the integration is possible, the systems should be integrated. Because of having different separate systems, still we cannot rely on the system since some of our tasks cannot be done using the system as we may fail. Some reports need data from another systems and hence we must do some manual work to make sure we get the correct reports."*
>
> *"We have many systems but each works independently; this makes it difficult to easily get reports. If these systems are integrated, it will be very easy to get reports, manual works which usually have many errors which many times remain unnoticed and lead to wrong decisions will be eliminated."*
>
> *"You may for example, be asked how many students have finished paying fees and you cannot know it unless you start compiling manually or you ask for a report from accounts office where the compilation is also done manually but we have SARIS here which can help us to easily get all those reports if that functionality is added to the system... If our systems would be integrated things would become very simple."*

Looking on the use claims on the poor system functionalities, the HEIs should make sure that the AISs have all the necessary functionalities based on their daily activities. It was also observed that, based on the problem of poor functionalities, the HEIs operates on unnecessary costs and waste a lot of time to solve customers issues and hence low customers satisfaction.

Based on the frequent changing environments in these HEIs, the developers should consider adopting a proper methodology on handling the user requirements so as to make systems fit their daily activities. The system development process should allow frequent requirements updating and implementation to accommodate the new requirements provided by the users daily which are inevitable for successful implementation of AISs in Tanzanian HEIs.

3.2 THEME 2: Poor User Support

The results from this study shows that there is poor support on the use of the AIS. Users claimed that they do not get help on time when they get stack. The respondents said:

"For sure, support on the use of the system is a problem here every day... and we really don't like this situation..."

"You start looking for help and if you are lucky to get the it support staff you will get an instant help, but most of the times the it support is problematic."

"...the system itself is somehow good but it support is very poor."

User support is very necessary to ensure successful implementation of any information system. Poor user support may lead to other issues like making users hate to use the system especially when they stack and fail to get support. Having a good system without an effective user support will demotivate users on using the system and hence they will eventually prefer their manual ways of working as they are comfortable with them. Good user support increases user satisfaction and hence they will like to use the system.

3.3 THEME 3: Lack of Enough Training

The findings also shows that there have been no time to time training to the users while there have been a lot of development activities taking place every day on improving the system. User find it difficult to use the new implemented features without been training. Sometimes they don't know if the new feature is added and hence they do their work manually while the system can do the task in a very simple way. The following statements shows how the respondents said:

"It is very important that we have frequent training because it is hard to know by yourself when and what changes and updates have been made in the system. Frequent training are very important to make us more familiar to the system especially when there are new features added."

"May be frequent training may help to make us comfortable with the system."

"Sometimes we do not even know that a certain report can be generated by the system, all because there is no frequent training on using the system."

Change management is one of the necessary issues to consider in any IT project. It is necessary to know what change was done, when and on what issue. It is also necessary to know who is going to be affected by the changes. All these are necessary to ensure awareness of the changes to both users and the developers for successful implementation of the change.

3.4 THEME 4: Lack of Experts on AIS

The study found that there is no enough experts on the AIS deployed and hence managing the system become a challenge which directly affect the users. It has been observed that lack of enough experts led to poor user support and even lack of time to time training. The respondents said:

"There is no enough experts on aiss to support us, something must be done on this."

"The management should put more effort to bring more experts not only to improve the system but also to assist and support us when we face challenges on using it."

"…there is no enough experts on saris … there should be more systems experts."

It was observed that the available experts are overwhelmed with work and hence it is difficult to support users on time and therefore low users satisfaction.

Based on the frequent changes in the Tanzanian HEIs operations, it is very important for the institutions to have their enough local information systems experts. Calling the external experts frequently can be very expensive which is normally not affordable by the institutions and hence poor information systems are experienced. Having local experts allows them to clearly see the reality on the daily activities and hence easy implementation of the user requirements on its correctness.

3.5 THEME 5: Poor Systems Security

It was also found that the security of these AISs implemented in Tanzanian HEIs is not very well considered and therefore some malicious activities may happen and make users found them not reliable for their daily activities. Some respondents said:

"It is very boring when you upload students results and found that they have been changed."

"…. I think we better go back to excel than using the systems which will make us repeat doing the work several times."

"I am worried that these systems may put us into troubles as the student results are changed and we may be accused for these changes…."

It was also observed that some users have lost trust on the systems and they have to use excel in some of their activities together with the system and this is really time consuming. HEIs should now understand that these AISs are currently very important and they should be made reliable by ensuring their security. The internal experts should get frequent training so as to make them aware of different security threats and how to implement them.

Surprisingly, it is noted that all the highlighted main challenges from this study leads to lost of trust on the system and therefore more research can be done to find out the effect of trust on information systems success in Tanzanian context.

The findings from this study can be useful to the Tanzanian HEIs in ensuring a successful implementation of AISs for better operations in their daily activities. In general, the highlighted challenges should be considered by any institution on implementation of AIS so as to take proper measures to avoid them as early as possible during the implementation process. To concur with Forbrig and Herczeg (2015) this study showed that new user requirements should be implemented as soon as possible every time they are provided to ensure user satisfaction. Though it is shown by Mamba and Isabirye (2015) and Govindaraju, Bramagara, Gondodiwiryo and Simatupang (2015) that to meet user expectations by implementating their frequently changing requirements is challenging, time consuming and costful; users satisfaction is crucial for AISs implementation success.

4 Conclusion

Generally the study noted that, a successful implementation of AISs in HEIs especially in Tanzania needs more investment and proper management. The institutions need to think broadly on improving the information systems computerization process by using the available advanced technologies. The findings call for the Tanzanian HEIs management to think on revising their ICT policies and their budget on information systems projects for better outcomes for their continuous development. The IT managers should put emphasize on having enough internal experts on information systems for better developments and user support.

Based on the findings of this study, it is noted that all the encountered challenges leads to users loose trust on the AISs implemented in their HEIs. In theoretical perspectives, this study found that trust has effect on the success of AISs implemented in Tanzanian HEIs and hence trust may be added as a researchable dimension on the dimensions highlighted by Petter, DeLone and McLean (2013). This study was conducted on only one Tanzanian HEI and involved only one AIS and therefore more studies are invited to improve the findings for generalization to other places.

References

Bilan, Y., Mishchuk, H., Samoliuk, N., Grishnova, O.: ICT and economic growth: links and possibilities of engaging. Intell. Econ. **13**(1), 93–104 (2019)

Chaudhary, P., Hyde, M., Rodger, J.A.: Exploring the benefits of an agile information system. Intell. Inf. Manag. **9**(05), 133 (2017)

Chirwa, M.: Access and use of internet in teaching and learning at two selected teachers' colleges in Tanzania. Int. J. Educ. Dev. Using ICT 14(2), (2018)

Cresswell, J.W.: Research Design: Qualitative, Quantitative, and Mixed Methods Approaches, 4th edn. Sage Publications Inc., Thousand Oaks, CA (2013)

Đorđević, A., Klochkov, Y., Arsovski, S., Stefanović, N., Shamina, L., Pavlović, A.: The impact of ict support and the efqm criteria on sustainable business excellence in higher education institutions. Sustainability **13**(14), 7523 (2021)

Fitzgerald, B., Stol, K.J.: Continuous software engineering and beyond: trends and challenges. In: Proceedings of the 1st International Workshop on Rapid Continuous Software Engineering (pp. 1–9) (2014)

Forbrig, P., Herczeg, M.: Managing the Agile process of human-centred design and software development. In: INTERACT (pp. 223–232) (2015)

Govindaraju, R., Bramagara, A., Gondodiwiryo, L., Simatupang, T.: Requirement volatility, standardization and knowledge integration in software projects: an empirical analysis on outsourced IS development projects. J. ICT Res. Appl. 9(1), 68–87 (2015)

Ko, E.J., Kim, A.H., Kim, S.S.: Toward the understanding of the appropriation of ICT-based Smart-work and its impact on performance in organizations. Technol. Forecast. Soc. Chang. **171**, 120994 (2021)

Maheshwaran, P., Kumar, R., Rajeswari, S., Mungara, D.: A Review on Requirement Engineering in Rapid Application Development. Int. J. Sci. Res. Comput. Sci. Eng. Inf. Technol. 2(3), 742–746 (2017)

Mfaume, H.: Awareness and use of a mobile phone as a potential pedagogical tool among secondary school teachers in Tanzania. Int. J. Educ. Dev. Using Inf. Commun. Technol. **15**(2), 154–170 (2019)

Mtebe, J.S., Raphael, C.: A critical review of elearning research trends in Tanzania. In: 2018 IST-Africa Week Conference (IST-Africa) (pp. Page-1). IEEE (2018)

Mwandosya, G.I., Mbise, E.R.: Evaluation feedback on the functionality of a mobile education tool for innovative teaching and learning in a higher education institution in Tanzania. Int. J. Educ. Dev. Using Inf. Commun. Technol. **15**(4), 44–70 (2019)

Okenyi, T.C., Mulili, B., Kiflemariam, A.: Academic management system finance module and customer satisfaction: a study of CUEA and USIU. Journal of Education **2**(3), 36–65 (2019)

Papanthymou, A., Darra, M.: Assessment of the quality of electronic administrative services in a greek higher education institution: α case study. International Journal of Higher Education **7**(2), 15–27 (2018)

Pereira, R.H., de Carvalho, J.V., Rocha, Á.: Architecture of a maturity model for information systems in higher education institutions: multiple case study for dimensions identification. Comput. Math. Organ. Theory , 1–16 (2021). https://doi.org/10.1007/s10588-021-09342-z

Woyo, E., Rukanda, G.D., Nyamapanda, Z.: ICT policy implementation in higher education institutions in Namibia: a survey of students' perceptions. Educ. Inf. Technol. **25**(5), 3705–3722 (2020). https://doi.org/10.1007/s10639-020-10118-2

Petter, S., DeLone, W., McLean, E.R.: Information systems success: the quest for the independent variables. J. Manag. Inf. Syst. **29**(4), 7–62 (2013)

Examining the Challenges of Price Quotation as a Procurement Method in Tertiary Institutions in Ghana

G. Nani[1], S. F. Abdulai[1](✉), and J. A. Ottou[2]

[1] Department of Construction and Technology Management, Kwame Nkrumah University Science and Technology, PMB, Kumasi, Ghana
sfatogma98@gmail.com

[2] Department of Management Science, Ghana Institute of Management and Public Administration, P. O. Box 50, Achimota, Accra, Ghana

Abstract. Purpose: The study identifies challenges encountered in using price quotations (PQ) as a procurement method in government tertiary institutions. This is expected to create awareness of the challenge associated with the use of PQ in government tertiary institutions and enhance the possibility of achieving value for money.

Design/Methodology/Approach: Quantitative approach was adopted for data collection and analysis. The questionnaire was administered to purposively sampled respondents, analysed and conclusions were drawn.

Findings: The findings revealed 22 challenges in using the price quotations method in government tertiary institutions. The top three challenges were: (i) Poor records keeping for procurement and expenditure documents; (ii) Lack of planning for the required goods; and (iii) Embezzlement of funds under the pretext of low-value items.

Research Limitation/Implications: The study was limited to six regions with accredited government tertiary institutions in Ghana. A study of this nature would have been more informative if it had covered all government institutions, including local institutions (civil society organisations), across the sixteen (16) regions of Ghana.

Practical implication: The study results inform procurement practitioners of the challenges associated with the use of PQ as a method of procurement. This can serve as a basis for recommending effective ways of addressing challenges and thus, ensure maximum value for money and fairness in the process.

Originality/Value: The study contributed to knowledge by filling in the lacuna in research by identifying the challenges of PQ as a procurement method in government tertiary institutions. This is expected to set the pace for further research on how these challenges could be addressed.

Keywords: Embezzlement · Goods · Price quotations · Procurement · Services

C. Aigbavboa et al. (Eds.): ARCA 2022, *Sustainable Education and Development – Sustainable Industrialization and Innovation*, pp. 659–672, 2023.
https://doi.org/10.1007/978-3-031-25998-2_51

1 Introduction

Procurement remains critical to supply chain strategy organization with need to buy goods, works and services. Thus, there are many project procurement systems that have been introduced for better performance and outcome of the procurement system. It has been observed that major stakeholders need the desire to execute their projects timely ensuring better performance in term of cost, time, value for money, minimum possibility of risk, and early confirmed design and prices (Bhutto et al. 2019). However, Mardale (2016) opined that the main difference between public and private procurement is the inflexibility of public procurement system discourages corrections to the original conditions during procurement process implementation. Procurement remains to be a key sector in Ghana, which facilitates the acquisition and disposal of goods and services, there by leading to smooth running of various institutions both public and private (Adjei-Bamfo et al. 2019). Burto (2005) affirms that, public procurement is the central instrument to assist the efficient management of public resources.

The public procurement act of Ghana which was established in 2003 and amended in 2016 sets out the guidelines for public procurement and the various procurement methods used in Ghana to ensure a judicious, economic and efficient use of state resources in public procurement (Public Procurement Act 2003). PQ as one of such methods is mostly used by public institutions to procure goods, works and services given a certain threshold of the value of goods, works and services in question (Kissi et al. 2020).

Every Procurement activity within the government agencies in Ghana must conform to the guidelines set out in the Act. Tertiary institutions which are inclusive as public institutions in Ghana is no exception. Thus, Tertiary institutions have procurement unit or department who are charged with the responsibility to procure goods, works and services. The units ensure that the procurement guidelines and procedures are followed to ensure both satisfaction on the side of the various institutions and the also conforming to the ultimate objective of the Act, that is, "to ensure cost effectiveness and efficiency and promote fairness, transparency and ensure that public procurement is non-discriminatory" (Public Procurement Act 2003).

However, in its quest to maintain and ensure the ultimate aim of the procurement act, it is chalked with several challenges during the procurement process, such as; excessive delays in procurement processes, untimely acquisition of funds leading to prolonged delivery time, lack of proper data base, bribery and corruption (Mrope 2018). There has been extant literature on procurement system in Ghana and the outside world but very little focus on PQ as a method of procurement, its usage and the challenges in the public sector procurement (Ameyaw et al. 2012; Ruparathna and Hewage 2015; Ogunsanya et al. 2016; Dzuke and Naude 2015; Anane and Kwateng 2019), therefore, the purpose of this research is to fill a knowledge gap by examining the challenges of PQ at government tertiary institutions in Ghana to procure goods, works, and services. The next section of the paper reviews relevant literature on the subject matter followed by the research methodology adopted to achieve the aim of the study, presentation of data and analysis and later concludes the findings and make relevant recommendations and direction for future studies.

2 Literature Review

This section reviews literature on key research themes to contextualize the study and enhance the research gap.

2.1 Public Procurement and Procurement Process

'Procurement' is a contemporary term, which is known to many construction practitioners and researchers by different terms. These include terms such as project approach, procurement methods, procurement delivery methods or project delivery systems (Mathonsi and Thwala 2012). The Public Procurement Act 2003 (Act 663) specifies several procurement methods such as sole source procurement, price quotation, restricted tendering and competitive tendering. This study however focuses on price quotation as one of the procurement method. Procurement systems have received well deserved attention in countries such as Australia, United Kingdom, United States of America, New Zealand and Japan but this has never been the case locally as well as in many other African countries (Mathonsi and Thwala 2012). Manyenze (2013) defined procurement as, the acquisition of merchandise or services at the most favorable total cost in the right amount and quality. Avotri (2012) also believes Procurement is the process by which organizations acquire goods, works and services. The process in procurement starts from the initiation of the need by user department and it is approved by the head of the department, as a way of acknowledging the requirement (Amemba et al. 2013). The procurement cycle involves planning, beginning with needs assessment through needs preparation, inviting offers, contractor selection, awarding contracts, executing and managing contracts, as well as final accounting and auditing. Procurement can therefore be viewed as a strategic component of supply chain management relating to satisfying customer or buyer needs. Procurement practice should therefore be responsive to aspirations, expectations and needs of the target society (Amemba et al. 2013).

2.2 Concept of Price Quotation

The term PQ which is otherwise known as Request for quotations or invitation to quote or shopping are used interchangeably (Benton and McHenry 2010). According to Lynch (2013) PQ is a procurement method utilized for small value procurements of promptly accessible off-the shelf goods, small value construction works, or small value services. This procurement method utilizes limited competition in light of the fact that the invitation is generally not advertised publicly in the newspapers. The procuring entity is normally expected to welcome quotations from a restricted number of contractors, suppliers or service providers, typically at least three. This procurement method is utilized under conditions described in the procurement rules and, in like manner, quotations may be requested in writing, by email, fax or courier, yet generally not by telephone (World Health Organization 2002).

In the United States, when procuring readily available commercially standard goods that are not specially manufactured to the specifications of the procuring entity, PQ is usually used, (US Public Procurement Bill 2017). USA and the Kenyan Procurement Act allows for the award of contract to the lowest bidder, the procuring entity shall place a purchase order with the bidder providing the lowest-priced quotation meeting the delivery and other requirements of the procuring entity. A procuring entity must prepare a list of qualified persons, submit the list to the tender committee for approval, and ensure a fair and equal rotation amongst the persons on the list to give requests for quotations (Kenyan Procurement and Disposal Act 2010). A request for quotations prepared by a procuring entity of the Act shall set out a requirement that quotations be submitted in sealed envelopes; and the mode of delivery of the sealed envelopes to the procuring entity. The opening, evaluation and comparison of quotations shall be carried out jointly by the procurement unit and the user department of the procuring entity. The Nigerian procurement Bureau does not allow for negotiation of quotations submitted by the supplier or contractor to the procurement entity (Nigerian Public Procurement Act 2007).

The PQ approach of procurement involves adopting practices with the goal of maximizing certain advantages, including maximizing value for the money spent, minimizing financial waste, and maximizing equitable and fair access for suppliers and contractors (Jones 2009). Even though, there are several benefits of PQ, the method is also characterized by several challenges, which this study seeks to examine and explore.

2.3 General Challenges of Procurement

Academics have suggested several ways of overcoming the different barriers that public officers face in reaching procurement objectives (van Berkel and Schotanus 2021). Public procurement generally is faced with lots of challenges. According to Kusi and Nyarko (2014) Not just as regulators, but also as customers and significant purchasers with the ability to influence market decisions, governments throughout the globe have become active stakeholders in the market economy. Developed and developing world governments are currently among the top purchasers of products and services. Amemba et al. (2013) studied the challenges of public procurement. Choosing the right person to award a contract is a major challenge for the public sector, they believe, because very few public institutions have adequate records. Thai (2009) postulated that the challenges in Kenyan procurement include, according to, the misuse and mismanagement of contract variations, the absence of effective checks and balances with regard to authorizing the various procurement and expenditure steps, the absence of fair and transparent competition, the absence of sufficient evidence of full receipt of the goods and services paid for, the embezzlement of funds under the guise of low-value items, and the improper filing of procurement and related expenditure documents. Formerly, all public goods purchases had to go via the Ghana Supply Commission, a government institution. However, lengthy delivery timeframes resulted from a lack of trained employees, poor preparation for the necessary commodities, an inadequate database, and difficulties in the timely collection of funding. The Finance Ministry has been working on a national Procurement Code since 1999. Accordingly, several ministries across various sectors have begun implementing their own procurement systems.

According to Asian Development Bank (ADB) and Organization for Economic Cooperation Development (OECD) (2006) government procurement is vulnerable to a wide range of corruption kinds. One of the most prevalent is bribery (Sanyal 2005), After the contract has been signed, the supplier may pay bribes to the client to increase the price, deliver cheaper materials, and utilize inferior designs that may even go against the original specifications.

Corruption through supplier cooperation is another possibility. This occurs when a group of rival vendors (the collusion ring) colludes to either reduce competition for a tender or to set the price (ADB and OECD 2006). In furtherance, Unrealistic short bidding process is also identified as one of the challenging factors of government procurement process. To this end the procurement frameworks thus prescribe a sufficient|| period for the submission of bids. Many countries also set minimum periods for the preparation of bids (ADB and OECD 2006).

3 Methodology

To accomplish the study's aim of examining the challenges of PQ in tertiary institutions in Ghana, a quantitative approach was used to conduct the research. Because of its usefulness in obtaining a broad perspective on a phenomenon, this method was given a lot of credence (Creswell 2014). The cross-sectional survey type of research design was adopted for the study to obtain a representative sample by picking a cross-section of the population (Sedgwick 2014). The questionnaire was distributed to tertiary institutions in selected regions of Ghana. They include; the Greater Accra region, Ashanti region, Bono region, Northern region, Upper East, and the Upper West. These regions were selected on the basis that they represent the southern belt, middle belts, and northern belt of Ghana, and they are accredited tertiary institutions with respondents that agreed to participate in the study.

A good quite number of the target respondents exist in the institutions considered accredited selected for the study in Ghana. Readily available participants were located using purposive sampling in the target institutions and complemented by snowballing sampling techniques to reach out to all participants across the country in the institutions. Hence, purposive and snowballing became relevant because the researchers did not have a clear-cut view of the population of the target respondents for the study. A similar approach was adopted by Agyekum et al. (2021) in their study titled "Environmental performance indicators for assessing the sustainability of projects in the Ghanaian Construction Industry". 150 questionnaires were self-administered through face-to-face and online surveys (google forms) to the target respondents made up of procurement officers, purchasing officers, stores keepers, and accountants across the tertiary institutions in Ghana. Purposive sampling was used to harness the expertise and knowledge of respondents on the research objective, purpose sampling also enhanced their willingness and ability to participate in the research. The questionnaire was designed to collect demographic data of the respondents and their responses to the main aim of the study.

3.1 Survey Administration

The questionnaire was administered through face-to-face (drop and pick) and online surveys (google forms). There were two major components of the questionnaire. Respondents' personal information was collected in Section I of the instrument. This was essential for validating the participants, and the researcher was cautioned that more information would be needed to prevent interlocutor bias without it (Pandey and Pandey, 2015).

The second section of the paper focused on the obstacles encountered while attempting to implement PQ at Ghana's public tertiary institutions. Participants were asked to use a 5-point scale to rate various aspects of their experience engaging with professionals' perspectives on the difficulties of implementing PQ in Ghana's government-run postsecondary institutions. Likert scale ranging from 1 (least severe), 2 (less severe), 3 (Neutral), 4 (severe), 5 (more severe). Google Forms collected 44 replies, whereas the drop-and-pick approach yielded 53. The response rate was 65%, with 97 questionnaires returned out of a total of 150 sent. Jackson-smith et al. (2016) state that high response rates have been achieved through the use of self-administered questionnaires, the drop and choose a technique, and online surveys (google forms). A rigorous method of follow-up was responsible for the impressive 65% response rate. 60% is deemed adequate, acceptable, or moderate; 70% is preferred; 80% is outstanding; and 90% is excellent in surveys (Davidoff et al. 2002).

3.2 Test for Reliability and Validity

The Cronbach alpha coefficient was used to evaluate the scale's internal consistency and reliability. If the Cronbach alpha coefficient is 0.700 or higher, the results can be considered credible (Muijs 2010). A Cronbach's alpha of 0.78 is thus reliable for further analysis. The pilot study was used to test the validity of the questionnaire (Mathers et al. 2007). The main purpose of the pilot study was to find out the challenges of PQ used for procuring goods, work, and services gathered from the interviews with the professionals in procurement who are equally faced with the challenges of PQ in other identified institutions in Ghana. The professionals that were engaged in the piloting stage included procurement officers, purchasing officers, storekeepers, and accountants. The response from these professionals was positive, i.e., they agreed to the inclusion of all variables in the questionnaire. A generic questionnaire was then designed to assess the level of severity of the identified challenges.

3.3 Data Analysis

Descriptive statistics (means, frequencies, and standard deviation) were employed for the biographical data, and the Relative Importance Index (RII) was utilized for the examination of the enumerated factors. Statistical Package for the Social Sciences (SPSS) for Windows version 26 was used for the analysis. SPSS is the standard for statistical analysis in the academic world (Muijs 2010).

The relative importance index (RII) allows the identification of the most important criteria based on the responses of the survey participants, and it is also an appropriate tool to prioritize the indicators (identified challenges) (Rooshdi et al. 2018), in this case, the level of severity of the challenges was identified and rated on the Likert scale adopted for the study. As adopted by Owusu-Manu et al. (2019) the RII was calculated using the formula below. [W-weighing given to each statement by the respondents ranging from 1 to 5; A—higher response integer (5); N—total number of respondents.

$$RII = \frac{\sum W}{A \times N}$$

4 Analysis and Results

4.1 Descriptive Analysis of Respondents' Demographic Profile

This section of the questionnaire comprised questions seeking basic information and some related issues in order to provide detailed respondent characteristics. The importance of knowing the profile of the respondents is to help have confidence in the reliability of the data collected. Data included type of profession, the engagement in the procurement of goods, works, and services, years of experience, and the level of education.

According to Table 1, the vast majority of respondents were procurement officers (61.9%), and the rest equally had their respective percentages as indicated in Table 1. Table 1 shows how often the respondents have been engrossed with the practice of procuring goods, work, and services, ranging from rarely, occasionally, never, very frequently, and always. However, the majority of the respondents were frequently engaged in the procurement of goods, work, and services. The years of experience of the respondents were assessed by getting a response from how long they have been engaged in procurement as professionals since the period of engagement in service can be an indication of the level of experience. Most of the respondents (percentage) had practiced in the field of procurement for 16 years or more. As part of determining the capacity and credibility of the respondents to understand the survey, they were required to indicate their level of academic qualification. This is in line with the views of Hegarthy (2011), who argues that a person's degree of education may be used as an indicator of their competence in the workplace. All responders clearly had a solid grasp of procurement procedures.

Table 1. Demographic data of respondents

Type of profession	Frequency	%
Procurement officer	60	61.9
Accountant	22	22.7
Purchasing officer	8	8.2
Stores	7	7.2
How often do you engage in procurement?		
Rarely	9	9.3
Occasionally	14	14.4
Never	17	17.5
Very frequent	50	51.5
Always	7	7.2
Years of experience		
1–5 years	11	11.3
6–10 years	21	21.6

(continued)

Table 1. *(continued)*

Type of profession	Frequency	%
11–15 years	54	55.7
16 years and above	11	11.3
Level of education		
Diploma	21	21.6
Bachelor's degree	61	62.9
MPhil/MSc	11	11.3
Doctorate	4	4.1

4.2 Understanding the Challenges of Price Quotation Used in Government Tertiary Institutions in Ghana

The results of the field survey were analyzed using the mean score and relative relevance index. We analyzed the replies and found the degree of agreement by using standard error and standard deviation. Univariate skewness and kurtosis were used to examine whether or not the data were normally distributed and skewed (Kline 2015).

The table below shows the results of an analysis utilizing the RII to evaluate the qualities in order of perceived severity in light of the difficulties encountered by government tertiary institutions. When comparing two or more variables with the same RII, the opinion that the variable with the highest mean is rated higher is consistent with the findings of Owusu-Manu et al. (2019) and Kissi et al. (2020b). When two or more variables have the same mean value, the one with the lower standard deviation is given preference in ranking. This is because standard deviation shows the consistency of agreement between the respondents' interpretation, therefore a lower standard deviation is preferable (Ahadzie 2007; Owusu-Manu et al. 2019) (Table 2).

Table 2. RII of the challenges of PQ

S/N	Challenges	Mean	Standard error	Standard deviation	Skewness	Kurtosis	RII	Rank
1	Poor records keeping for procurement and expenditure documents	4.01	0.097	0.952	−0.687	−0.432	0.802	1st
2	Lack of planning for the required goods	4.00	0.102	1.000	−1.149	1.428	0.800	2nd

(continued)

Table 2. (*continued*)

S/N	Challenges	Mean	Standard error	Standard deviation	Skewness	Kurtosis	RII	Rank
3	Embezzlement of funds under the pretext of low value items	3.99	0.100	0.984	−0.849	0.376	0.798	3rd
4	Lack of proper data base	3.98	0.104	1.020	−1.159	1.263	0.796	4th
5	Corrupt practices	3.97	0.119	1.168	−1.021	0.072	0.794	5th
6	Delay on processing payment for supplier or contractor	3.95	0.097	0.951	−1.382	2.318	0.790	6th
7	Local suppliers' inability to deliver on request orders	3.93	0.106	1.043	-0.584	-0.852	0.786	7th
8	Nepotism/favoritism	3.87	0.114	1.126	−0.979	0.171	0777	8th
9	Untimely acquisition of funds leading to prolonged delivery time	3.82	0.108	1.061	−0.924	0.427	0.765	9th
10	Excessive delays in procurement process	3.76	0.115	1.134	−0.698	−0.310	0.753	10th
11	Lack of follow up skills as an organization	3.75	0.122	1.199	−0.987	0.156	0.751	11th
12	Incomplete evidence of full receipt of goods and services paid	3.71	0.104	1.020	−0.652	-0.208	0.742	12th
13	Competent suppliers not registered with Public Procurement Authority despite their ability to supply/provide goods to meet specifications and value for money	3.70	0.117	1.156	−0.752	−0.287	0.740	13th

(*continued*)

Table 2. *(continued)*

S/N	Challenges	Mean	Standard error	Standard deviation	Skewness	Kurtosis	RII	Rank
14	Interference of procurement process by powerful stakeholders in the institutions tends to affect the overall process	3.68	0.122	1.204	−0.747	−0.296	0.736	14th
15	Presence of middlemen thereby increasing or inflating prices of goods to be procured	3.64	0.115	1.129	−0.709	−0.262	0.728	15th
16	Value for money assessment usually not done	3.63	0.130	1.277	−0.738	−0.530	0.726	16th
17	Unethical practices	3.62	0.120	0.177	−0.859	−0.062	0.724	17th
18	User departments inability to undertake market survey to enquire the prices of goods to be procured	3.53	0.105	1.031	−0.448	−0.146	0.705	18th
19	Inability of the user department to write appropriate specifications	3.53	0.109	1.081	−0.472	−0.438	0.705	19th
20	Suppliers not agreeing in giving out proforma invoices to user departments	3.51	0.114	1.119	−0.423	−0.501	0.701	20th
21	Technological challenges	3.49	0.130	1.257	−0.634	−0.595	0.699	21st
22	Shorter duration of the prices on the proforma invoice eligibility and application	3.27	0.119	1.177	−0.071	−0.892	0.654	22nd

With a RII of 0.802, a high mean score of 4.01, and a standard deviation of 0.952, respondents identified inadequate documentation of purchases and expenditures as the most pressing problem facing Ghana's government universities and colleges. It was followed closely in terms of the ranking by lack of planning for the required goods with the RII of 0.800 and a bit lower mean score of 4.00 and an SD of 1.000. Embezzlement of funds under pretext of low value items [RII = 0.798; Mean = 3.99 and SD = 0.984], lack of proper data base [RII = 0.796; Mean = 3.98 and SD = 1.020], and corrupt practices [RII = 0.794; Mean = 3.97 and SD = 1.168] ranked third, fourth and fifth respectively.

The least ranked challenges were suppliers not agreeing in giving out proforma invoices to user department [RII = 0.701; Mean = 3.51 and SD = 1.119], technological challenges [RII = 0.699; Mean = 3.49 and SD = 1.257], and shorter duration of the prices on the proforma invoices eligibility and application [RII = 0.654; Mean = 3.27 and an SD = 1.177]. These difficulties were given lower RII and mean ratings, indicating that they are minor obstacles that government tertiary institutions face while engaging in procurement.

5 Discussion of Findings

Poor records keeping for procurement and expenditure documents was ranked first as the challenges of PQ in government institution. It can therefore be concluded that the most severe challenges of PQ in government tertiary institutions in the poor records keeping of procurement and expenditure documents. Information is a primary organisational asset that is needed now and into the future. Good recordkeeping underpins the provision of good business information. However, the lack of good record keeping or inefficient record management can lead to costly consequences in an organization. Several scholars, including those cited above, have pointed to sloppy record keeping for procurement and expenditure documentation as the primary difficulty of using PQ (Rodden and Bell 2002; Cornock 2019; Cornock 2020).

Lack of planning for the required goods was ranked second to the challenges of PQ in government tertiary organizations. Planning means analyzing and studying the objectives, as well as the way in which we will achieve them. It is a method of action to decide what to do and why. Most tertiary institutions in Ghana do not plan before engaging in PQ procurement, an organization should have a procurement plan developed at the beginning of the business year. Consistent with extant literature, Ayarkwa et al. (2020) shows that public procurement faces difficulties due to inadequate preparation for goods.

Embezzlement of funds of fund under the pretext of low value items was ranked third as one of the challenges of PQ in government tertiary institutions in Ghana. As acknowledge by Secretariat (2008), due to the nature of low value items, the cost involve is usually small and for that there is a perception that persons involved tend to embezzle the money allocated for the procurement of goods, works and services. In so doing they procure goods, works and services of lesser value to the allocated fund for such prescribed procurement of the goods, works and service. This therefore possess a serious challenges of PQ. Other major challenges of price quotations as used in government tertiary institutions include; Lack of proper data base, corrupt practices, delay on processing payment for supplier or contractor and nepotism/favoritism as some of the challenges of PQ in government tertiary institutions. Consistent with current literature, Douh et al. (2013) acknowledged the untimely acquisition of funds leading to prolonged delivery time, excessive delay from procurement process, lack of follow up skills as an organization and Incomplete evidence of full receipt of goods and services paid are some of the challenges of PQ as a method of procurement in government tertiary institutions.

Ebekozien (2019) also asserts that competent suppliers not registered with Public Procurement Authority despite their ability to supply/provide goods to meet specifications and value for money, Interference of procurement process by powerful stakeholders in the institutions tends to affect the overall process, Some of the challenges to the current use of PQ in the procurement of goods, works, and services in Ghana include the presence of middlemen, which increases or inflates the costs of commodities to be procured, the failure to conduct a Value for money assessment, and the employment of unethical activities. Similar problems were found in a research done on public sector procurement in South Africa (Ambe and Badenhorst-Weiss 2012).

6 Conclusions and Recommendations

The study has examined the challenges of PQ as a method of procurement in government tertiary institutions in Ghana. This led to revealing the various challenges encountered in using PQ in procuring goods, works and services in Ghana. This research adds to the existing literature and discussion on the topic of public procurement and the procurement process, before narrowing the focus on the concept of PQ. The severe challenges as identified from the study were; poor record keeping for procurement and expenditure documents, lack of planning for the required goods and embezzlement of funds under the pretext of low values items. The implications of these findings are in two folds. Theoretically, the study reveals the key challenges with the use of PQ in government tertiary institutions in Ghana, currently under reported in literature. The examining of these challenges advances knowledge within the subject area. Practically, the study results inform procurement practitioners of the challenges associated with the use of PQ as a method of procurement in government tertiary institutions. It also serves as a basis for recommending effective ways of addressing and finding solutions to the identified challenges in order to ensure maximum value for money and fairness in the procurement process.

A study of this nature would have been more informative if the study had covered all government institutions including local institutions (local government institutions) across the sixteen (16) regions of Ghana. However, the significance of the outcome is not undermined. Future studies could collect data from respondents in the remaining cities of Ghana to provide additional empirical realities on the challenges confronting professionals in using PQ in procuring goods, works, and services in government tertiary institutions. Again, the study was limited in finding only the challenges encountered, however, future studies could seek to explore the measures/strategies in mitigating the challenges of PQ. The study further recommends all government institutions to adopt electronic procurement systems and that PQ be integrated into the e-procurement platform being developed for procuring goods, work, and services in the various government tertiary institutions in Ghana.

References

Ahadzie, D.K.: A model for predicting the performance of project managers in mass house building projects in Ghana (2007)

Ambe, I.M., Badenhorst-Weiss, J.A.: Procurement challenges in the South African public sector. J. Transp. Supply Chain Manage. **6**(1), 242–261 (2012)

Amemba, C.S., Nyaboke, P.G., Osoro, A., Mburu, N.: Challenges affecting public procurement performance process in Kenya. Int. J. Res. Manage. **3**(4), 41–55 (2013)

Ameyaw, C., Mensah, S., Osei-Tutu, E.: Public procurement in Ghana: the implementation challenges to the public procurement law 2003 (Act 663). Int. J. Constr. Supply Chain Manage. **2**(2), 55–65 (2012)

Anane, A., Kwarteng, G.: Prospects and challenges of procurement performance measurement in selected technical Universities in Ghana. Asian J. Econ. Bus. Account.1–18 (2019)

Agyekum, K., Botchway, S.Y., Adinyira, E., Opoku, A.: Environmental performance indicators for assessing sustainability of projects in the Ghanaian construction industry. Smart and Sustainable Built Environment (2021)

Avotri, N.S.: Assessment of the prospects and challenges of procurement reforms in Ghana: the case of Volta River Authority (Doctoral dissertation) (2012)

Ayarkwa, J., Agyekum, K., Opoku, D.G.J., Appiagyei, A.A.: Barriers to the implementation of environmentally sustainable procurement in public universities. Int. J. Procurement Manage. **13**(1), 24–41 (2020)

Benton, W.C., McHenry, L.F.: Construction Purchasing & Supply Chain Management. McGraw-Hill, New York (2010)

Bryman, A.: Social Research Methods. Oxford University Press, Oxford (2016)

Burton, R.A.: Improving integrity in public procurement: The role of transparency and accountability. Fighting Corruption and Promoting Integrity in Public Procurement, pp. 23–28 (2005)

Cornock, M.: Record keeping and documentation: a legal perspective. Orthop. Trauma Times **35**, 34–38 (2019)

Cornock, M.: A summary of law and ethics for the new health care practitioner. Orthop. Trauma Times **36**, 30–37 (2020)

Creswell, J.W.: Research Design: International Student Edition (2014)

Davidoff, F., Gordon, N., Tarnow, E., Endriss, K.: A question of response rate. Sci. Ed. **25**(1), 25–26 (2002). https://www.councilscienceeditors.org/wp-content/uploads/v25n1p025-026.pdf. Accessed 7 Feb 2021

Douh, S., Badu, E., Adjei-Kumi, T., Adiniyira, E.: An appraisal of challenges facing competitive tendering implementation in public works procurement in chad republic. In: West Africa Built Environment Research (WABER) Conference, p. 123 (2013)

Dzuke, A., Naude, M.J.: Procurement challenges in the Zimbabwean public sector: a preliminary study. J. Transp. Supply Chain Manage. **9**(1), 1–9 (2015)

Ebekozien, A.: Unethical practices in procurement performance of Nigerian public building projects: mixed methods approach. Theor. Empirical Res. Urban Manage. **14**(3), 41–61 (2019)

Håkansson, A.: Portal of research methods and methodologies for research projects and degree projects. In: The 2013 World Congress in Computer Science, Computer Engineering, and Applied Computing WORLDCOMP 2013, Las Vegas, Nevada, USA, 22–25 July 2013, pp. 67–73. CSREA Press USA (2013)

Jackson-Smith, D., et al.: Effectiveness of the drop-off/pick-up survey methodology in different neighborhood types. J. Rural Soc. Sci. **31**(3), 3 (2016)

Kissi, E., Ahadzie, D.K., Debrah, C., Adjei-Kumi, T.: Underlying strategies for improving entrepreneurial skills development of technical and vocational students in developing countries: using Ghana as a case study. Educ. + Train. **62**(5), 599–614 (2020a)

Kissi, E., Adjei-Kumi, T., Twum-Ampofo, S., Debrah, C.: Identifying the latent shortcomings in achieving value for money within the Ghanaian construction industry. J. Public Procurement (2020)
Kline, R.B.: Principles and Practice of Structural Equation Modelling. Guilford Publications, New York (2015)
Kusi, L.Y., Aggrey, G.A., Nyarku, K.M.: Assessment of public procurement policy implementation in the educational sector (A case study of Takoradi polytechnic). Int. J. Acad. Res. Bus. Soc. Sci. **4**(10), 260 (2014)
Lani, J.: Statistics solutions: advancement through clarity", Obtenido de (2016). http://www.statisticssolutions.com/theoretical-framework. Accessed 10 Feb 2022
Manyenze, N.O.E.L.: Procurement performance in the public universities in Kenya (Doctoral dissertation, University of Nairobi) (2013)
Mathers, N., Fox, N., Hunn, A.: Surveys and questionnaires. The NIHR RDS for the east midlands/yorkshire and the humber (2007). https://www.academia.edu/11450102/The_NIHR_Research_Design_Service_for_Yorkshire_and_the_Humber_Surveys_and_Questionnaires_Authors. Accessed 10 Feb 2022
Mathonsi, M.D., Thwala, W.D.: Factors influencing the selection of procurement systems in the South African construction industry. Afr. J. Bus. Manage. **6**(10), 35–83 (2012)
Muijs, D.: Doing Quantitative Research in Education with SPSS. Sage Publications, Thousand Oaks (2010). ISBN 144624234X978144624234
Mrope, N.P.: Determinants of performance of procurement departments in public entities in Tanzania (Doctoral dissertation, JKUAT-COHRED) (2018)
Neely, A., Richards, H., Mills, J., Platts, K., Bourne, M.: Designing performance measures: a structured approach. Int. J. Oper. Prod. Manage. **17**, 1131–1152 (1997)
Ogunsanya, O.A., Aigbavboa, C.O., Thwala, W.D.: Challenges of construction procurement: a developing nation's perspective (2016)
Owusu-Manu, D.G., Edwards, D.J., Kukah, A.S., Parn, E.A., El-Gohary, H., Hosseini, M.R.: An empirical examination of moral hazards and adverse selection on PPP projects. J. Eng. Des. Technol. **16**(6), 910–924 (2018)
Pandey, P., Pandey, M.M.: Research Methodology: Tools and Techniques, Romania Bridge Center (2015). http://www.euacademic.org/BookUpload/9.pdf. Accessed 2 July 2021
Parse, R.R.: Parse's research methodology with an illustration of the lived experience of hope. Nurs. Sci. Q. **3**(1), 9–17 (1990)
Public procurement Act: Of the Federal Republic of Nigeria Official Gazzette, p. 41 (2007)
Public Procurement and Disposal Act: Kenyan Law Reports, pp. 88–91 (2010)
Public Procurement Bill: Of the United States of America, pp. 23–24 (2017)
Rodden, C., Bell, M.: Record keeping: developing good practice. Nursing Standard (Through 2013), vol. 17, no. 1, p. 40 (2002)
Ross, A., Willson, V.L.: One-sample *t*-test", Basic and Advanced Statistical Tests, Brill Sense, pp. 9–12 (2017). https://brill.com/view/book/edcoll/9789463510868/BP000003.xml
Ruparathna, R., Hewage, K.: Sustainable procurement in the Canadian construction industry: challenges and benefits. Can. J. Civ. Eng. **42**(6), 417–426 (2015)
Sanyal, R.: Determinants of bribery in international business: the cultural and economic factors. J. Bus. Ethics **59**(1), 139–145 (2005)
C.G.I.A.R Secretariat: CGIAR procurement of goods, works and services guidelines: financial guidelines Series no. 6 (2008)
Sedgwick, P.: Cross sectional studies: advantages and disadvantages. BMJ **2014**, 348 (2014)
Thai, K.V.: International Hand Book on Public Procurement. CRC Press, New York (2009)
van Berkel, J.R.J., Schotanus, F.: The impact of "procurement with impact": measuring the short-term effects of sustainable public procurement policy on the environmental friendliness of tenders. J. Public Procurement **21**, 300–317 (2021)

Theoretical Review of Migration Theory of Consumer Switching Behaviour

L. Y. Boakye(✉)

Accra Institute of Technology, Accra, Ghana
lyawboakye@gmail.com

Abstract. Purpose: The paper proposes the *Push-Pull-Drag-Deter-Mooring (PPDDM)* migration theory as a theoretical extension to the Push-Pull-Mooring (PPM) migration theory in the consumer switching behaviour (CSB).

Design/Methodology/Approach: The proposed study employs a mixed study design involving a qualitative focus group interview and a quantitative cross-sectional survey design.

Findings: Based on a thorough review of PPM theory and relevant empirical studies, the paper identifies a theoretical gap that, the existing PPM migration theory of CSB completely ignores the simultaneous role of positive origin (firm) effects, termed drag dimension, and negative destination (competitor) effects, termed deter dimension in the switching process. While this gap has already been pointed out by previous studies, future research is yet to give attention to the PPM migration theory of CSB.

Research Limitation/Implications: Theoretically, the proposed PPDDM of CSB presents a more complete and realistic picture of the competing factors in the consumer switching process, that better explains consumer migration or switching behaviour from one service provider to another, than the existing PPM does. The main limitation is that the proposed theory is yet to be tested with empirical data.

Practical Implications: This paper contributes to theory development in the CSB literature and advances the frontiers of knowledge in Marketing theory and practice.

Originality/Value: This paper incorporates these ignored or neglected dimensions (drag and deter) in addition to those existing in the current PPM to develop a new theory of *Push-Pull-Drag-Deter-Mooring (PPDDM)* migration theory of CSB, as a major extension of the PPM theory of CSB.

Keyword: Consumer behavior · Deter · Drag · Migration theory · Switching

1 Introduction

One dominant theory that has been borrowed from human geography to explain consumer switching behaviour (CSB) in the marketing literature is the Push-Pull-Mooring (PPM) theory of migration (see Fig. 1). It was first developed by Lee (1966) as Push-Pull model, and later.

C. Aigbavboa et al. (Eds.): ARCA 2022, *Sustainable Education and Development – Sustainable Industrialization and Innovation*, pp. 673–688, 2023.
https://doi.org/10.1007/978-3-031-25998-2_52

extended by Bogue (1977) and Moon (1995) to include mooring dimension. The application of the PPM migration theory to CSB has generated much empirical evidence with its modifications in the literature (Bansal et al. 2005; Lai et al. 2012; Nimako and Ntim 2015).

The PPM theory presupposes that consumers' switching behaviour and intentions, like emigration decisions, are determined by the interplay of push, pull and mooring factors. It posits that just as negative factors (push factors) prevailing at the origin (place of current residence) push the individuals to move away to another place, so also do negative factors (push factors) prevailing at the origin (service provider) push consumers to switch to another service provider.

Moreover, just as positive factors (pull factors) at an intended destination attract intended emigrants to move to a destination, so also do positive factors (pull factors) existing with a competitor service provider attract consumer to switch to the competitor. Finally, just as migrants' decision to emigrate to another place can be influenced positively or negatively by personal and social factors (mooring factors), so also do mooring factors facilitate or inhibit consumers' switching process from a current service providers to another in a given service industry context.

As already noted, the PPM theory is a further theoretical development of an earlier migration theory, the push-pull (PP) theory of migration that excluded mooring factors. This significant theoretical extension has furthered scholars and practitioners understanding of the interplay of many factors that influence migration decision (Bogue 1977; Moon 1995) and consumers' switching behavior and intentions (Bansal et al. 2005; Lai et al. 2012; Nimako and Ntim 2015) in many research contexts. Similarly, while the current conceptualization of the PPM theory captures the role of three components (dimensions), *origin push factors, destination pull factors* and other intervening *personal and social mooring factors*, it is still critically important for researchers to attempt to provide useful conceptual and theoretical extensions to the current PPM theory based on empirical evidence. This will help scholars and practitioners better understand *other overlooked critical components (dimensions)* that influence consumer switching process and its application in marketing contexts.

The critical issue is that, in reality, intuitively there are both *negative and positive origin factors* that could affect migration decision, and there are both *negative and positive destination factors* that could also influence migration decision as noted by Nimako and Ntim (2015). However, the current conceptualization of the PPM theory includes only *negative origin factors* as *push dimension* and only *positive destination factors* as *pull dimension*. It almost completely ignores the simultaneous role of *positive origin factors* as "*drag dimension*", and *negative destination factors* as "*deter dimension*" in migration decisions, making it a one-sided theory instead of a two-sided theory that better captures the realities of life. It also fails to acknowledge the moderating role of these *positive origin factors* (drag forces) and *negative destination factors* (deterring forces). Yet, there is much evidence in the literature that some consumers with high dissatisfaction and switching intentions still do not defect or switch due to, not only personal and social mooring factors (Bansal et al. 2005) relationship strength (Nimako and Ntim, 2015), but also some other *positive origin factors* with current service provider (Baumeister et al. 2011; Ye and Potter 2011). Thus, *positive origin factors* and *negative destination factors* could act as mooring factors (deterring moderators) on the effect of *negative origin factors*, which are distinct from the personal and social mooring dimension in the existing PPM model.

In this regard, Nimako and Ntim (2015) provided some theoretical extensions to the PPM model in the CSB literature, by including competitor reputation as one more pull factor, and two new mooring factors, government regulation and relationship strength to create a Push-Pull Three Moorings *(PP3M)* Model of consumer switching. They further pointed out the omission of *drag* and *deter* dimensions as a limitation in the PPM theory and called on future research to address it to advance the PPM and its application in marketing. In particular, they posited that,

> *"…so far existing research including this study has focused on the effect of negative origin factors on migration or service provider switching decision. However, in reality a consumer's decision to switch a service provider due to some negative factors can be delayed or restrained because of some positive (origin) factors perceived by the consumer with service delivery of the current service provider. Thus, our understanding of consumer switching could be enhanced by researching into the extent to which positive origin (firm) factors can moderate or reduce the effect of negative origin (firm) factors on consumer switching intentions, as well as the extent to which negative competitor factors might affect positive (pull) competitor effects in the consumer switching process."* (p. 386).

Thus, the current conceptualization of the PPM theory has a number of significant theoretical limitations. First, it ignores the extent to which *positive origin (firm) factors* can *reduce* the effect of *negative origin (firm) factors* in the emigration process and therefore consumer switching process. In this regard, if an individual/consumer is pushed by *negative origin factors* to emigrate or switch, how do *positive origin factors* drag or discourage or negatively influence the *negative origin (firm) factors*?

Second, it also ignores the extent to which *positive origin (firm) factors* can *moderate* the relationship between negative origin (firm) and emigration (or switching) intentions. Thus, if an individual/consumer is pushed by *negative origin factors* to emigrate or switch, how do *positive origin factors (drag)* moderate the relationship between the *negative origin (firm) factors* and their intention to emigrant or switch respectively?

Third, it excludes the extent to which *negative destination (competitor) factors* might *reduce* the effects of *positive destination (competitor) effects* in the emigration and consumer switching processes. In this regard, if an emigrant or consumer is pulled or attracted by *positive destination (competitor) effects* to migrate to a new destination or switch to a competitor respectively, how might *negative destination (competitor) factors* directly deter or discourage or negatively influence the effect of the attraction or pull (*positive destination effects).*

Fourth, it fails to consider the extent to which *deterring negative destination (competitor) factors* might *moderate* the relationship between *pull destination (competitor) effects* and intention to migrate or switch. Thus, if an emigrant or consumer is attracted by *positive destination (competitor) effects* to migrate to a new destination or switch to a competitor respectively, how does *deter factors* moderate the relationship between the *pull dimension* and their intention to migrate or switch respectively.

Consequently, it is critically important to include these neglected migration (or switching) dimensions and examine their simultaneous effect to better understand the migration or switching process. Therefore, the main problem of this study is to what extent do these *positive origin (firm) dimension* and *negative competitor dimension* combine with *push* and *pull* and *mooring* effects to influence the consumer migration (switching) process? This empirical evidence is expected to significantly advance scholars' conceptualization of the realities of migration or switching dimensions and advance practitioners' role to develop effective strategies to manage migration and consumer switching process.

The main purpose of this paper is to justify the theoretical gaps for a theoretical extension in the PPM theory of migration and consumer switching by including the role of *positive origin (firm) factors* and *negative destination (competitor) effects* in the migration and consumer switching process, using the mobile telecom industry as the research context.

The reminder of the paper presents theories of CSB, justification for theoretical gaps for a theoretical extension in the PPM theory, proposed methodology for testing the proposed extensions in the PPM and implications of the new PPM theory.

2 Consumer Switching Behaviour Theories

Nimako (2012) reviews eleven CSB theories in the marketing literature as summarized in Table 1.

Out of the eleven theoretical models of CSB, this study focuses on extending the push-pull-mooring theory of CSB.

The PPM has been chosen as theoretical framework for this study because of the following reasons. First, it has been empirically validated by Bansal et al. (2005). Consequently, the PPM theory *promises* to be very useful theoretical framework in predicting consumer switching intention and behaviour and other related social phenomena in business related disciplines.

Second, it has been widely applied, making it a dominant theory in Marketing and CSB literature. As shall soon be realsed in subsequent empirical reviews in this next sections, numerous authors have applied the PPM in different contexts of consumer switching. For example, Cheng et al. (2009) applied the PPM to understand factors that affect users' switching Intentions in social networking sites (Cyber Migration), Hou et al. (2011) employed the PPM to explain the switching intentions of online gamers in gaming services.

Ye and Potter (2011) adopted the PPM to explain post-adoption switching of personal communication technologies, Fu (2011) examined IT professionals' commitment in the light of the PPM, Chiu et al. (2011) applied the PPM framework to examine the antecedents of consumer switching in multichannel services, Lai et al. (2012) applied the PPM to the study of CSB towards mobile shopping, Hsieh et al. (2012) studied bloggers' post-adoption switching behaviour for online service substitutes by employing the PPM, and Zhang et al. (2012) adopted the PPM model to understand online blog service switching behaviour. Nimako and Ntim (2015) applied it in the telecom industry in Ghana to understand context unique factors that drive CSB.

Table 1. (xxx.)

S/N	Switching model/theory	Author (s)	Industry Context
1	Product Importance Model-Based switching model	Morgan and Dev (1994)	Hospitality in a developed country
2	A Model of consumers' service switching Behaviour	Keaveney (1995)	Many service contexts in USA
3	A Catalytic Switching Model/SPAT	Roos (1999b)	Telecom in Sweden
4	Service Provider Switching Model	Bansal and Taylor (1999)	Financial Institutions in Canada
5	The Switching Process model	Colgate and Hedge (2001)	Retail Bank in Australia and New Zealand
6	Three-Component Model of Consumer Commitment to Service Provider	Bansal et al. (2004)	Auto-repairs and hairstyling services in Canada
7	Push-Pull-Mooring Theory	Bansal et al. (2005)	Auto-repairs and hairstyling services in Canada
8	General Systems Theory of Consumer switching	Njite et al. (2008)	Hospitality industry in Developed country
9	Agency Theory of Consumer switching	Aish et al. (2008)	Advertising-agent relationship in Egypt
10	Prospect theory of consumer switching	Marshall et al. (2011)	Financial market in Western and Eastern world
11	Synthesized Consumer Switching Model	Nimako and Ntim (2015)	Mobile Telecom Industry in Ghana

Source: Nimako (2012).

Third, the cross-discipline roots of the PPM in human migration has a lot rich insights to give to Marketing in broadening scholars' understanding of CSB and how it can be managed effectively by practitioners and policy makers. Fourth, the PPM is comprehensive and contextually applicable to the mobile telephone industry. Thus, this study hopes to extend the dimensions of PPM theory to advance the frontiers of knowledge in marketing theory.

Fifth and finally, the PPM theory of CSB is focused on this study because its theoretical limitations provide a novel avenue for this study to contribution to the CSB literature in a major way. Therefore, this study chooses to focus on the PPM to extend the frontiers of knowledge in Marketing.

2.1 Development of Push-Pull-Mooring Theory in Human Migration

The concept of human migration is referred to as the movement of people from one place to another in the world (Carter 2019; Sinha 2005; Strauss 2018). In the migration literature in human geography, a person moving away from his or her original place of residence to another is *an emigrant* to the place of origin and *an immigrant* to the place of destination. Within the concept of migration, there are returnees, those who return from destination to their origin after a time period. There are also refugees, those who move to new places due to serious political or economic problems to protect their lives from threat of death, perceived unlawful legal judgement or seeking political asylum (Sinha 2005). Thus, to the place of destination, the person moving into it is an immigrant.

A critical review of the human migration literature reveals that, people migrate to new places for many different reasons or factors that are either related to their origin, destination or personal and social factors (Brain et al. 2017; Moon 1995; Sinha 2005). One dominant theory in human migration that comprehensively conceptualized migration factors is the Push-Pull-Mooring (PPM) Migration Theory (Moon, 1995).

The PPM theory of human migration (see Fig. 1) was first developed by Lee (1966) as Push-Pull model, and later extended by Bogue (1977) and Moon (1995) to include mooring dimension. The PPM Theory in human migration literature explains why people move from one place of origin to another for a time period (Moon 1995). According to the theory, push effects are negative factors that force people to move away from an original residence to another, such as unemployment, wars, loss of jobs, natural disasters, poverty, political, religious, ethnic/or other forms of oppression, (Lee 1966; Moon 1995).

On the other hand, pull effects are positive factors that attract potential migrants to a new destination, such as better employment opportunities, peaceful environment, higher incomes and better economic conditions, perceived superior career opportunities in another location, Personal growth opportunities such as better education, group association, Preferable environment such as climate, housing, schools/or other institutional facilities, desire to be with kin or other favourable people in another location, lure of different physical and social activities in another place (Lee 1966; Moon 1995). The basic assumption of this theory is that negative factors at the origin push people away, while positive factors at the destination pull people toward them.

These push and pull factors do not work in isolation but interdependently with the mooring factors. The mooring factors are equivalent to the intervening and moderating variables, and act to either encourage migration or to deter the potential migrants from leaving their home or origin. Though mooring factors in migration literature were identified as personal and social factors that impact migration intentions and decisions (Moon 1995), it could be extended to include any variable that has the potential of encouraging or deterring an individual's migration decisions.

2.2 Explaining CSB with Push-Pull-Mooring Theory (PPM)

In understanding marketing phenomenon, scholars have attempted to borrow theories from geography, psychology, biology and other disciplines and applied them to study marketing problems. One such attempt was made by (Bansal et al. 2005) who borrowed the Push-Pull-Mooring theory of migration in human geography to explain consumer switching process in Marketing. They developed the PPM Migration Model of Service Switching. This was appropriate and reflected the work of Clark, Knapp, and White (1996) who had already indicated the similarity between consumer-switching and human migration.

The authors attempted to present a unifying framework for understanding the complexity of the process of consumer switching. The authors identified the issue of linearity that is displayed by existing models of consumer switching and indicated the need for a more elaborate framework that would minimize the risk of developing strategies that either overemphasize or understate the significance of certain variables as pointed out by Cronin, Brady and Hult (2000).

To them the movement of people from one place to another in human geography provided a correspondence between consumers' switching from one service provider to the other (Bansal et al. 2005). They believe that just as individuals shop for goods by comparing prices and many other features, potential migrants compare the attributes of alternative locations and express those preferences by moving to the location that best satisfies them (Bansal et al. 2005; Clark et al. 1996).

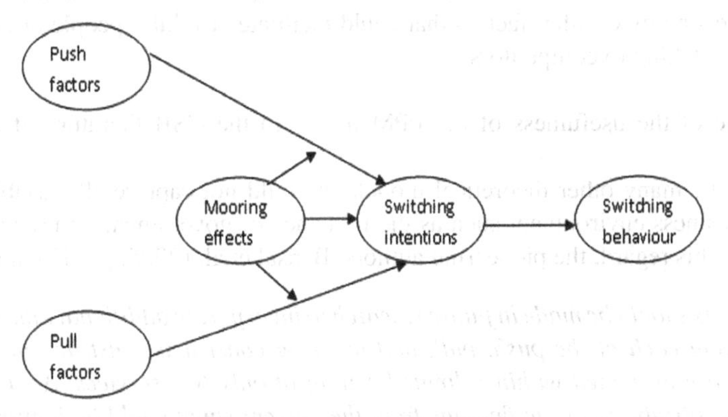

Fig. 1. Push-Pull-Mooring (PPM) Migration Model of Service Switching (Bansal et al. 2005)

According to Bansal et al. (2005), just as migrants are *pushed* by negative factors existing at their place of residence to leave their place of origin to a place of destination, so also do consumers switch from their current service providers to a competitor service provider. Then just as migrants are *pulled* by positive destination factors to move away from their place of residence to the destination, so also are consumers attracted by positive competitor factors switch from current service provider to a competitor. Finally, just as other *mooring* factors or non-firm or non-competitor factors like personal and social factors can influence individuals positively or negatively or neutrally to migrate to new places, so also can personal and social factors encourage or inhibit consumers from switching to competitors or influence their switching decision process.

The key assumptions of the PPM theory of CSB as follows:

1. It assumes that people (consumers) behave rationally in their choices and behaviour (Ajzen 1991; Bettman et al. 1998)
2. It assumes that people (consumers) make conscious effort to consider the implications of their actions before they decide to engage or not engage in certain behaviour (Ajzen 1991).
3. Other things being equal, consumer decision is influenced by three groups of factors.

a Negative factors at the origin push people away.
b Positive factors at the destination pull people toward them.
c Mooring factors either inhibit or facilitate the influence of the push and pull factors in the migration decision process.

The key components of the PPM theory of CSB as follows:

1. **Push factors:** These are factors with current service provider perceived to be basically negative by people or consumers that push or cause them to switch to competitors.
2. **Pull factors:** These are factors with competitor perceived to be basically positive by people or consumers that pull or attract them to switch to competitors.
3. **Mooring factors:** These are other non-firm or competitor factors with the consumers or the economy or other factors that could facilitate or inhibit people or consumers from switching to competitors.

In spite of the usefulness of the PPM model to the CSB literature, it has some limitations.

First, like many other theoretical models, it could not capture all variables in the external business environment such as the influence of government policy in the CSB process. In this regard, the pioneering authors, Bansal et al. (2005, p. 112), noted that,

> *"... efforts should be made in future research to incorporate additional dimensions underlying each of the push, pull, and mooring constructs. Last, the proposed models are examined within a limited setting of only two services. As a result, the generalizability of the findings from the current study could be improved by empirical examinations in other settings of service switching."*

Second the model was limited to one consequence of switching intentions, which was behaviour. It did not examine other consequences of switching intentions, such as how it affects loyalty and recommendation behaviour for current service provider (Nimako and Ntim 2015). Third, the pioneering authors also noted that the conceptualization of pull construct in their study was mainly the alternative attractiveness which they measured using only one item, and so called on future research to extend the conceptualization of the pull construct.

Fourthly, a major limitation of PPM theory of CSB stems from the fundamental limitation of its mother PPM theory of human migration which have been pointed out earlier. The PPM theory of service switching (Bansal et al. 2005) neglects the inclusion of positive origin (firm) factors and negative destination (competitor) factors that could influence consumer switching decision. The simple assumption of the PPM Theory that, negative factors at the origin push people away, while positive factors at the destination pull people toward them is problematic. This is because, as argued out in the human migration literature, migration decision and for that matter consumer switching decision could be influenced by the interplay of positive origin factors as well as negative destination factors (Sinha 2005).

In the migration literature, some authors like Sinha (2005) clearly pointed out, *the origin and destination have positive, negative and neutral characteristics, which are directly or indirectly related to the process of migration of a person* (p. 411). Sinha (2005) further agues, "*But the truth is that both the pull and push factors do exist or operate simultaneously at both the origin and destination place of a migrant. The neutral factors do not make any contribution to the decision making process of an individual, but sometimes act as a balance in the move of someone* (p. 411).

For example, the outbreak of COVID-19 pandemic in many countries of the world caused the development of government policies that discouraged migration to many destination countries in the world (Freier and Vera Espinoza 2021). Although people want to travel or migrant to new destinations in Europe and elsewhere, but due to outbreak of COVID-19 pandemic and its resultant new health, legal and immigration policies, even individuals are discouraged from travelling to certain places in the world (Freier and Vera Espinoza 2021).

For example, some people wishing to migrate to a place for better income, employment opportunities and business purposes, may not do so because unfavourable weather conditions, unfavourable social and political factors, racism, outbreak of diseases and difficulty in integrating in the social norms of the destination country (Freier and Vera Espinoza 2021; Mastrorillo et al. 2016; Saa et al. 2020). People wishing to migrant to favourite destination do not only consider the positive factors at the destination but also the negative factors. The interplay between the positive and negative destination factors determine the final migration decision to move or not.

Thus, apart from the push factors at the origin that pushes people to move away, there could also be positive origin factors that could also restrain or drag the inducement to move away. On the other hand, apart from the positive destination factors that pull or attract people to move to a destination, there could also be negative destination factors that restrain or deter emigrants' intention to migrate to the new destination. Therefore, the basic assumption of the PPM needs to be re-examined in order to better conceptualize the PPM theory. The PPM cannot always assume simply that migration decisions are based only on the negative origin factors and positive destination factors. It must now incorporate the influence of positive origin factors and negative destination factors to better conceptualize individual's migration decisions.

Similarly, in the CSB literature past researchers like Nimako and Ntim (2015) have called on future researchers to think outside the box and look at possible theoretical extensions that go beyond three-tier dimensional framework of PPM. They expounded that,

> *Finally, so far our knowledge of firm-factors that affect migration or switching decision has focused on negative (push) factors. However, logically, it is possible to have situations where positive factors at the origin say excellent functional quality, can act as moderator in reducing the strength of the negative origin factors, say low technical quality, on consumer migration or switching intentions. Our understanding of consumer switching could be enhanced by understanding the influence of positive firm factors on switching intention.* (Nimako and Ntim, 2015 p. 69).

It is argued that, just as some migrants would want to migrate to a destination due to better income but may not do so due to perceived unfavourable weather condition at the destination, so also some consumers wishing to switch to a competitor due to a particular negative factor, say high prices, with the current firm may not do so because other specific unfavourable factors with a competitor, say lack of good corporate reputation (Freier and Vera Espinoza 2021; Mastrorillo et al. 2016; Saa et al. 2020). People wishing to switch to favourite competitors may not only consider the positive factors with the competitor but also the negative factors with the competitor. The interplay between the positive and negative firm and competitor factors could determine the consumer's final decision to switch or not.

Consequently, as pointed out in the human migration literature (Sinha 2005), five groups of factors that work together to influence human migration decision are.

1. Negative origin factors (push),
2. Positive origin factors (drag),
3. Positive destination factors (pull),
4. Negative destination factors (deter),
5. Mooring factors (intervening).

But the current conceptualization of the PPM theory completely ignores the *drag* and *deter* dimensions. Similarly, in the CSB literature, as proposed by Nimako and Ntim (2015), the PPM theory in CSB also fails to capture the individual and simultaneous role of p*ositive firm factors (drag)* and n*egative competitor factors (deter)* the CSB process. Therefore, the PPM is fundamentally limited and should incorporate the influence of positive origin factors and negative destination factors to capture the complexity of factors influencing consumer switching decision to better conceptualize CSB in the Marketing literature.

As a result of its limitations, two main theoretical extensions of the PPM can be identified in the CSB literature. First, there are those studies that focus on developing models to extend the variables or constructs *within* the three-tier Push, Pull and Mooring dimensions of the PPM theory (e.g., Chiu et al. 2011; Hsieh et al. 2012; Li et al. 2018; Kim et al. 2020; Frasquet and Miquel-Romero 2021; Lin et al. (2021). Second, there are those studies that propose extensions beyond the three-tier framework of PPM (Nimako and Ntim 2015; Ye and Potter (2011).

2.3 Main Theoretical Gap and Its Significance

Thus, the limitations of the PPM offers avenues for further research to improve upon the theory to extend the frontiers of Marketing knowledge. From the discussion of the conceptual roots of PPM theory in human migration factors, we observe that, the PPM as applied to CSB simplistically incorporates *push, pull and mooring* dimensions to the total neglect of the *drag* and *deter* dimensions. As a result of this void in the PPM theory, this study hopes to contribute to filling this theoretical gap by including *drag* and *deter dimensions* in the existing PPM to develop the *Push-Pull-Drag-Deter-Mooring (PPDDM) theory*. The main problem or question from this theoretical gap then is, *to what extent do positive origin (firm) factors and negative competitor factors individually and simultaneously combine with the existing push, pull and mooring dimensions in the PPM theory to influence consumer switching intention and behaviour processes?*

This major theoretical extension is theoretically significant as it attempts to change the fundamental assumptions of the PPM. It advances scholar's knowledge as the proposed PPDDM theory by capturing both negative and positive factors at both the origin (firm) factors and destination (competitor) that combine with other mooring (intervening) factors to influence consumer switching process. In the end, the proposed PPDDM better reflects the realities and complexity of CSB than its parent PPM theory does.

Practically, it importantly brings out to practitioners that need to consider not only the push and pull factors in consumer managing CSB but also the need to consider and manage effectively the individual and simultaneous effects of *push* and *drag* factors existing at both the origin (firm), and *pull* and *deter* factors existing at the destination (competitor) that combine with other *mooring* (intervening) factors to influence CSB. Practitioners would be able to develop multi-faceted strategy that captures the complexity of CSB process in order to be effective, remain profitable and stay competitive.

2.4 Proposed Research Model: Push-Pull-Drag-Deter-Mooring Theory (PPDDM)

Based on extensive literature review the conceptual framework synthesizes a combination of theoretical constructs reviewed to help explain CSB in MTI in developing countries. The conceptual framework (Fig. 2), *Push-Pull-Drag-Deter-Mooring Theory (PPDDM)* is a proposed fundamental extension of the three-tier PPM framework. It introduces two new fundamental dimensions that have six theoretical relationships in the research model. Thus, in addition to the existing push, pull and mooring constructs in the original PPM, the PPDDM adds the *drag* and *deter* dimensions.

The proposed PPDDM theory of CSB assumes that people (consumers) behave rationally in their choices and behaviour (Ajzen 1991; Bettman et al. 1998). It assumes that people (consumers) make conscious effort to consider the implications of their actions before they decide to engage or not engage in certain behaviour (Ajzen 1991). Finally, it assumes that, other things being equal, consumer switching decision is influenced by both negative and positive factors at the origin (service provider) on one hand, both positive and negative factors at the destination (competitor), and mooring factors that either inhibit or facilitate the consumer switching process.

The key components of the proposed PPDDM theory of consumer switching process are.

a. ***Negative origin factors (push):*** These factors play one role; they encourage or push individuals to switch to competitors. They include low satisfaction, high prices, low perceived value, poor customer service inconvenience, complaints, etc.
b. ***Positive origin factors (drag):*** These factors play a dual role; first, they can drag or discourage individuals who intends to switch from their current service provider to competitors. They include positive tangibles of service quality, product warranty, good customer relationship management, and the like.

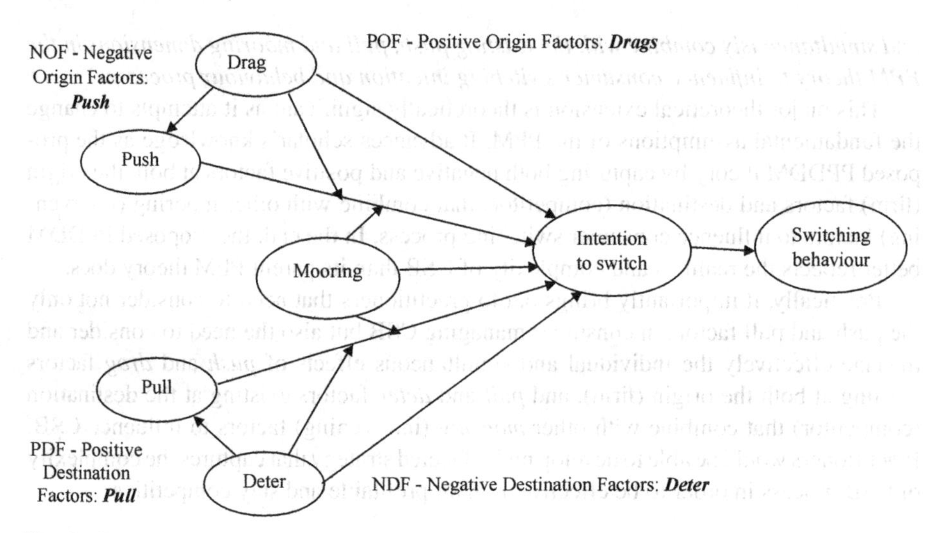

Fig. 2. Conceptual model for the proposed theoretical Model: Push-Pull-Drag-Deter-Mooring (PPDDM) Model

c. ***Positive destination factors (pull):*** These factors pull or attract individuals to switch from their firm to competitors. They include good competitor reputation and better competitor offer for services.
d. ***Negative destination factors (deter)***: These factors discourage individuals from switching to a competitor due to perceived negative factors existing with the competitor etc. They includes poor complaint management, unethical business behaviour, etc.
e. ***Mooring factors (intervening):*** These are factors outside firm and competitor factors that encourages or inhibits individuals from switching from their place of origin to a new place or destination. They include personal and social factors, relationship factors, distance, government policy etc.

3 Key Relationships/hypotheses in the Proposed PPDDM Theory of CSB

The PPDDM theory proposes the following 12 key relationships in the switching process: It proposes that,

1. Push factors could directly influence switching intentions
2. Pull factors could directly influence switching intentions
3. Mooring factors could directly influence switching intentions
4. Drag factors could directly influence switching intentions
5. Deter factors could directly influence switching intentions
6. Drag factors could directly influence push effect
7. Deter factors could directly influence pull effects

8. Mooring factors could moderate the influence of push factors on switching intentions
9. Mooring factors could moderate the influence of pull factors on switching intentions
10. Drag factors could moderate the influence of push factors on switching intentions
11. Deter factors could moderate the influence of pull factors on switching intentions
12. Switching intentions could influence switching behaviour

In sum, the proposed PPDDM theory (Fig. 2) consists of five types of antecedents that are conceptually categorized into push, pull, drag, deter, and mooring factors that influence CSB.

3.1 Objectives of the Study

Based on the theoretical gap discussed and the proposed theory and its relationship, the following are the specific objectives of the proposed study to test the proposed theoretical extensions to the existing PPM of CSB:

1. To validate the existing PPM theory of migration to consumer switching in the mobile telecom industry in Ghana.
2. To validate an extended model of CSB, the push-pull-drag-deter-mooring (PPDDM) model, which incorporates the effects of *positive origin (firm)* and *negative destination (competitor) factors* in the switching process.
3. To determine whether the PPDDM better explains consumer switching process than the existing PPM model.
4. To assess the effects of newly proposed *drag* and *deter* dimensions on the relationship between push, pull, mooring and switch intentions in the switching process.

3.2 Proposed Methodology and Analysis

In order to test the new dimensions in the extended PPM and empirically confirm the PPDDM migration theory of consumer switching, a mixed study design study will be conducted. It will involve a qualitative focus group interview and a quantitative cross-sectional survey design. Due to the multi-dimensionality of the constructs, model development approach and complexity of the inter-relationships in the proposed research model, a Partial Least Squares (PLS) structural equation modelling (SEM) approach will be used to analyse data using SmartPLS software.

4 Theoretical and Managerial Implications of the Study

Theoretically, apart from it helping scholars deepen understanding of CSB to contribute to the CSB literature in general, this paper specifically finds its relevance in proposing a new migration-based CSB theory that extends the PPM to better explain the complex realities of consumer switching process than the existing PPM theory does. In particular, the proposed PPDDM theory would incorporate in CSB process, the overlooked relevant dimensions of *drag* and *deter* effects/role to advance the frontiers of Marketing knowledge by empirically validating the reality or truth that consumers' migration/switching decision is not only influenced by negative origin or firm factors (push effect) and positive destination or competitor factors (pull effects), but also positive origin factors (drag effects) and negative destination factors (deter effects) together with the mooring effects.

This new PPM model is therefore expected to advance scholars' knowledge on the relevance of *both* positive and negative factors at *both* the origin (firm) and at the destination (competitor) places in the consumer migration/switching process. It will highlight the novel scholarly insights from the discouraging or counter role played by positive origin factors (drag effect) when consumers are pushed to switch due to negative origin factors (push effect), the counter role of negative destination factors (deter effect) when consumers are pulled to switch to competitors due to positive destination factors (pull effect), and the moderating role of drag and deter effects in the switching process. Thus, theoretically, this research will advance our knowledge in marketing theory. It finds its relevance in extending the PPM theory to PPDDM theory, and adds to new knowledge and theories in Marketing.

Managerially, this paper is relevant to practitioners in the mobile telephony industry in two main ways. On the one hand, to marketing management, it will help to understand the discouraging or counter role of positive origin factors (drag effects) in dragging switching decision when negative origin factors (push) induces consumers to switch. This means marketers of a firm will be able to know what marketing strategies to implement to highlight their positive factors in order to reduce or drag consumers' intention to switch due to their negative factors in service provision.

On the other hand, marketers will also be able to understand and know the discouraging or counter role of negative destination factors (deter effects) in deterring switching decision of customers when they are pulled or attracted by positive destination factors (pull) to switch. This means marketers of a firm will be able to know what marketing strategies to implement to discourage their customers from switching to competitors, when those customers are pulled by competitors' positive factors in a competitive service environment like the mobile telecom industry in many countries such as Ghana.

References

Alhassan, A.: Telecom Regulation, The Postcolonial State, and Big Business: The Ghanaian Experience West Africa Review (2003). http://www.westafricareview.com/vol4.1/alhassan.html

Bansal, H.S., Taylor, S.F., James, Y.: "Migrating" to new service providers: toward a unifying framework of consumers' switching behaviors. J. Acad. Mark. Sci. **33**(1), 96–115 (2005). https://doi.org/10.1177/0092070304267928

Barroso, C., Picón, A.: Multi-dimensional analysis of perceived switching costs. Ind. Mark. Manage. **41**(3), 531–543 (2012). https://doi.org/10.1016/j.indmarman.2011.06.020

Bartels, R.: The History of Marketing Thought, 2nd edn. Grid Publishing, Columbus (1976)

Batra, R., Ahuvia, A., Bagozzi, P.R.: Brand Love. J. Mark. **76**(2), 1–16 (2012). https://doi.org/10.1509/jm.09.0339

Baumeister, R.F., Masicampo, E.J., Vohs, K.D.: Do conscious thoughts cause behavior? Annu. Rev. Psychol. **62**, 331–361 (2011)

Belanche, D., Casaló, L.V., Guinalíu, M.: The role of consumer happiness in relationship marketing. J. Relat. Mark. **12**(2), 79–94 (2013). https://doi.org/10.1080/15332667.2013.794099

Bell, S.J., Auh, S., Smalley, K.: Customer relationship dynamics: Service quality and customer loyalty in the context of varying levels of customer expertise and switching costs. J. Acad. Mark. Sci. **33**, 169–183 (2005)

Bogue, D.J.: A migrants-eye view of the costs and benefits of migration to a metropolis. In: Internal Migration: A Comparative Perspective, Academic Press, New York, pp. 167-182 (1977)

Bolton, R.N.: Comment: customer engagement. J. Serv. Res. **14**(3), 272–274 (2011)

Buehler, S., Haucap, J.: Mobile numbering and number portability in Ireland. a report to the ODTR, Ovum: London. J. Ind. Competition Trade **4**(3), 223–238 (2004). https://doi.org/10.1023/B:JICT.0000047299.13443.5a

Frempong, G., Henten, A.: Telecom developments and investments in Ghana, Discussion Paper W D R 0 3 0 5, W D R Dialogue Theme 2003 (2004)

Gummerus, J.: Value creation processes and value outcomes in marketing theory: strangers or siblings? Mark. Theory **13**(1), 19–46 (2013)

International Telecommunication Union. ICT facts and figures (2013). http://www.itu.int/en/ITU-D/Statistics/Documents/facts/ICTFactsFigures2013.pdf on 6 November 2013

Keelson, S.A., Odei, A.J.: Relationship between mobile number portability and consumer choice of active multiple mobile phone numbers in Ghana. Global J. Bus. Res. **8**(4), 99–119 (2014)

Lai, J.Y., Debbarma, S., Ulhas, K.R.: An empirical study of consumer switching behaviour towards mobile shopping: a push-pull-mooring model. Int. J. Mobile Commun. **10**(4), 386–404 (2012). https://doi.org/10.1504/IJMC.2012.048137

Lee, E.S.: A Theory of Migration. Demography **3**(1), 47–57 (1966). https://doi.org/10.2307/2060063

Moon, B.: Paradigm in migration research: exploring 'moorings' as a schema. Prog. Hum. Geogr. **19**, 504–524 (1995)

National Communication Authority. Mobile Number Portability in Ghana – First Year Report – 18 July 2012 (2012). http://nca.org.gh/downloads/MNP_at_2Years_Report_3rd_September_2013.pdf on 20 August 2013

Nimako, S.G.: Consumer switching behaviour: a theoretical review and research agenda. Res. J. Soc. Sci. Manag. **2**(3), 74–85 (2012)

Nimako, S.G., Ntim, B.A.: Modelling the antecedents and consequence of consumer switching intentions in Ghana mobile telecommunication industry. Int. J. Bus. Emerg. Markets **7**(1), 37–75 (2015). https://doi.org/10.1504/IJBEM.2015.066093

Ranganathan, C., Seo, D., Babad, Y.: Switching behavior of mobile users: dousers' relational investments and demographics matter? Eur. J. Inf. Syst. **15**(3), 269–276 (2006)

Roos, I., Edvardsson, B., Gustafsson, A.: Customer switching patterns in competitive and non-competitive service industries. J. Serv. Res. **6**, 256 (2004). https://doi.org/10.1177/1094670503255850

Rust, R.T., Zahorik, A.J.: Customer satisfaction, customer retention, and market share. J. Retail. **69**, 193–215 (1993)

Sathish, M., Kumas, K.S., Jeevanantham, K.J.V.: A study on consumer switching behaviour in cellular service provider: a study with reference to Chennai far east. J. Psychol. Bus. **2**(2), 71–81 (2011)

Saunders, M., Lewis, P., Thornhill, A.: Research Methods for Business Students, 6th edn. Pearson, London (2012)

Schiffman, L.G., Kanuk, L.L.: Consumer Behavior, (Chapter 5, 10). Pearson Prentice Hall, New Jersey (2009)

Tonglet, M.: Consumer misbehavior: an exploratory study of shoplifting. J. Consum. Behav. **1**(4), 336–354 (2002)

Wangenheim, F.V., Bayón, T.: The effect of word of mouth on services switching: measurement and moderating variables. Eur. J. Mark. **38**(9/10), 1173–1185 (2004)

Tertiary Students' Accommodation Affordability in Ghana

C. Amoah[1(✉)], E. Bamfo-Agyei[2], and F. Simpeh[3]

[1] Department of Quantity Surveying and Construction Management, University of the Free State, Bloemfontein, South Africa
amoahc@ufs.ac.za

[2] Department of Building Technology, Cape Coast Technical University, Cape Coast, Ghana
emmanuel.bamfo-agyei@cctu.edu.gh

[3] Department of Construction and Wood Technology Education, Akenten Appiah-Menka University of Skills Training and Entrepreneurial Development, Kumasi, Ghana

Abstract. Purpose: The rental values of student housing are becoming a significant issue for tertiary institutions in Ghana as private developers are heavily investing in hostels due to the inability of the government to meet accommodation demand in the various public universities. This study aimed to do a comparative analysis among selected tertiary institutions in Ghana concerning accommodation affordability.

Design/Methodology/approach: A quantitative research approach was used to collect data from 626 off-campus and on-campus tertiary students randomly selected. Statistical Package for the Social Sciences (SPSS) was used to analyse the data collected.

Research Limitation/Implications: Data was collected from all the tertiary universities in Ghana. However, the findings may apply to other universities.

Findings: The findings show that both main and technical universities perceived their rental values as expensive. Again, both on-campus and off-campus students at various tertiary consider the rental values expensive. However, whilst a significant number of the on-campus students believe their rent to be not too expensive, many off-campus students consider the rent to be too expensive.

Practical Implication university institutions should provide more on-campus accommodation as students perceive that to be cheaper than off-campus accommodation.

Social Implication: There is an urgent need for the government and tertiary institution managers to institute an arrangement with private student housing developers to subsidise or regulate the rental amount to make it more affordable.

Originality/Value: The paper provides an analysis of students' perceptions concerning the affordability of their houses, thus guiding the university authorities on the accommodation provision model to meet the student's means.

Keyword: Accommodation · Affordability · Off-campus · On-campus · Tertiary students

C. Aigbavboa et al. (Eds.): ARCA 2022, *Sustainable Education and Development – Sustainable Industrialization and Innovation*, pp. 689–697, 2023.
https://doi.org/10.1007/978-3-031-25998-2_53

1 Introduction

Adequate and effective housing has been considered fundamental for modern living and essential for people's daily lives (Klis and Karsten 2008). This assertion is indicative of the importance of housing and student housing by extension. Student housing should, therefore, be perceived as a fundamental student's need. Proper student housing stimulates a conducive study environment, provides adequate security, enhances friendships among students, helps students to develop and maintain a vibrant student culture, and assists university housing administrators (Hassanain 2008; Addai 2013). Najib et al. (2011) suggest that providing suitable environments in student houses will help improve students' intellectual capabilities. Therefore, both off-campus and on-campus student accommodation should aim to accomplish student living and other social goals during their studies. Despite these benefits, many developed and developing countries are increasing tertiary student enrolment without a corresponding increase in the provision of accommodation, thus generating accommodation shortages (Ackermann and Visser 2016). According to Hubbard (2009), private developers are considered a solution to students' housing shortages in many universities across the globe. This is due to various government and public university management's inability to meet increasing student enrollment accommodation needs.

The lack of enough residence at the university premises, as experienced in other countries, is not different in Ghana. According to Ghana Education Performance Report (2010), student enrolment at tertiary institution averagely grow by 9.7% annually. The Free senior high school introduced by the New Patriotic Party government in 2018 would lead to increase in student enrolement. Despite that this education policy at the secondary level is expected to increase enrolment in the tertiary institutions in Ghana in the subsequent years, not much has been done by the government to provide additional housing at the various universities. Thus, the university authorities rely on private developers to provide the accommodation shortfalls (Nimako and Bondinuba 2012).

Pickren (2012) suggests that when demand for housing exceeds the supply, property owners usually increase the rental values and vice versa. Students might be paying significant rental values currently at their residences, which might not have been the case if there were enough on-campus houses to accommodate them. Ghana Governments, over the years, have understood the need to provide accommodation at the tertiary institutions to meet the students' demands; however, these have not materialised (Bondinuba et al. 2013). Many studies in Ghana have looked into student accommodation selection factors (Nimako and Bondinuba 2012; Bondinuba et al. 2013), student satisfaction (Danso and Hammond 2017) and facilities and services available at the student housing (Simpeh and Shakantu 2019; Simpeh and Shakantu 2020) with little attention to the rental values charged by the university authorities and private student housing developers at the various tertiary institutions in Ghana. This gap has prompted the need for this study. It is on this premise that this study determines students' views on the affordability of the rental values of their hostels with the ultimate aim of providing an overview of student housing affordability within the country. This information would help university authorities, policy makers and other stakeholders within the tertiary education sector when making decisions on matters related to student housing.

2 Students' Accommodation and Affordability

Student accommodation is housing students use as a place of abode during their studies. People ascribe many names to student accommodation. These include student housing, dormitory, campus apartment, student hostel, hall of residence, off-campus residences and student accommodation housing (Sawyerr and Yusof 2013; Khozaei et al. 2014). However, Najib et al. (2011) define student accommodation as learning hostels provided with conducive learning facilities and amenities to accommodate undergraduate and postgraduate students. According to Insch and Sun (2013), student accommodation could be off-campus or on-campus; however, the latter provides students with access to various learning facilities offered by the institutions compared to the former. Students' housing incorporates social and psychological needs to satisfy students' needs and expectations and create an environment suitable for learning. It should also provide a comfortable atmosphere, safety and convenience for favourable studies to achieve the desired educational outcome for the students (Khozaei et al. 2014). Moreover, student housing should be integral to developing students' intellectual argument, personal development and associated academic benefits (Nimako and Bondinuba 2012). Therefore, students' housing availability at tertiary institutions is necessary for effective learning outcomes. Nevertheless, the lack of adequate housing puts pressure on students in looking for a place to stay during school. This has an implication on the accommodation cost available to students as private developers have to make available housing available to students in need of housing for profit. According to Ghani and Suleiman (2016), although private housing developers are essential stakeholders in tertiary education development, their main goal is to provide housing purposefully to make a profit.

It has also been observed by Sage et al. (2013) that student-dominated capitalize experience significant rental values as property owners hike property prices because of the acute shortages resulting from the high demand for these properties. Towns and communities with higher education institutions often have higher rental values for studios, one-bedroom and two-bedroom apartments because the property owners have hiked the rental prices in student-populated areas than in the surrounding neighbourhoods (Ong Petrova and Spieler 2013). This exploitative nature of property owners and landlords in student-populated areas means the rental values will become unaffordable. However, students are compelled to occupy these properties due to the lack of alternative accommodation. Again landlords increase the rental values arbitrarily and outrageously and make students pay upfront, thus maximising their profits astronomically. Therefore, students are left at the mercy of the property owners with little or no intervention by the heads of the institutions and government heads because of the notion of the free-market economic system (Ghani and Suleiman 2016). Thus student housing sector has been described as the most "exploited housing market" (Donaldson et al. 2014). The recent global economic situation which has resulted in high cost of construction will worsen the student housing market. This indicate that some private developers will take advantage of the high construction cost and student demand to charge exorbitant rent.

Although students' accommodation choice is price sensitive, due to lack of cheaper accommodation options, students are left witout any option and are forced to pay high prices. Those who are unable to afford the high rental amount opt to share a room to spread the cost of the apartments (Munro and Livingston 2012). For instance, Garmendia et al.

(2011) suggest that most students in the UK share three and four-bedroom apartments to make them affordable. Thus affordability is paramount to both on-campus and off-campus students, plays a significant role in students' accommodation demands, and encourages comfortable living. This phenomenon is not exclusive to the UK but also in Ghana, where students share resources to stay in an apartment to make it affordable due to excessive rental values.

3 Research Methodology

The study adopted a quantitative research approach. A survey was carried out in 9 tertiary institutions in Ghana, namely the University of Cape Coast (UCC), Cape Coast Technical University (CCTU), University of Education Winneba (UEW), Accra Technical University (ATU), University of Ghana (UG), University of Professional Studies (UPSA), Kwame Nkrumah University of Science and Technology (KNUST), Kumasi Technical University (KTU) and University for Development Studies (UDS). These nine institutions were purposely selected to ensure that both the southern and northen part of the country are represented. A convenience sampling method was used to distribute structured survey questionnaires to 740 off-campus and on-campus students within these tertiary institutions. According to Blumberg et al. (2008), convenience sampling allows the research to access as many participants as possible who are within the target population. Thus any student readily available at the selected institutions staying in rented apartments either off-campus or on-campus was given a questionnaire. The questionnaires were distributed to the participants at their institutions; for those who were willing to complete, the research assistants waited for them to fill in and collected them. For the others who wanted more time, the questionnaires were left with them and collected a week later. Out of 740 questionnaires distributed, 626 were received and used for the analysis, representing an 85% response rate. The questionnaire was divided into two sections, the first was about participants' demographics, and the second was about accommodation affordability levels. The data received were analysed using Statistical Package for the Social Sciences (SPSS). Descriptive statisctics (i.e., frequencies and percentages) was used. The demographic features of the respondents are shown in Table 1.

Most (65%) respondents are male students from the demographic features, and the majority (84%) are within the school-going age of 18 to 25 years. Most (42%) are in the first year of study, 95% are full-time students, and 88% are unemployed. Again, 62% of the students are from the main universities, and 57.7% of the students are staying off-campus, buttressing the inadequacy of the on-campus accommodation to meet student enrolment.

4 Findings and Discussion

4.1 The institution type and perception of the accommodation rental value.

The respondents were asked to indicate the level of expensiveness concerning their current accommodation rent. The responses were then analysed based on the respondent's institutions to identify the perceptions of accommodation rental values among students

Table 1. Respondents' demographic features

	Nature of Respondents	Frequency	Percentage
Gender	Male	407	65.0%
	Female	219	35.0%
	Total	**626**	**100%**
Age	18–25	503	80.4%
	26–30	101	16.1%
	31–40	20	3.2%
	40 and over	2	0.3%
	Total	**626**	**100%**
Level of education	First Year	265	42.4%
	Second-Year	188	30.0%
	Third-Year	157	25.0%
	Honours	9	1.4%
	Masters	7	1.1%
	First Year	265	42.4%
	Total	**626**	**100%**
Nature of studies	Part-time basis	31	4.9%
	Full-time student	595	95.1%
	Total	**626**	**100%**
Employment status	unemployed	553	88.4%
	Working on a part-time basis	47	7.5%
	Fully employed	26	4.1%
	Total	**626**	**100%**
Educational institution	Main Universities	387	61.8%
	Technical Universities	239	38.2%
	Total	**626**	**100%**
Place of accommodation	On-Campus	265	42.3%
	Off-Campus	361	57.7%
	Total	**626**	**100%**

from these institutions. As indicated in Table 2, the findings show that students from both institutions consider the rental values expensive. Most (46%) respondents from the main universities and 44% from the technical universities stated that the rent they pay for their accommodation is expensive.

Again, whilst 28% of the students from the main universities consider their rent not too expensive, 35% of the students from the technical universities consider their

Table 2. The institution type and perception of the accommodation rental value

Variables	Main Universities			Technical Universities			Overall	
	Count	Percentage	Ranking	Count	Percentage	Ranking	Freq	Percentage
Expensive	179	46.25%	1	106	44.35%	1	285	45.53%
Not too expensive	109	28.17%	2	84	35.15%	2	193	30.83%
Too expensive	99	25.58%	3	49	20.50%	3	148	23.64%
Total	**387**	**100%**		**239**	**100%**		**626**	**100%**

accommodation rent not too expensive. At the same time, 25% of the main universities' students stated that their rents are too expensive, and 20% of the technical universities stated the same. This implies that rental values for apartments at the main universities are much more expensive than that of the technical universities. This finding is surprising as both main and technical universities are located within Ghana's cities and urban areas. Perhaps the differences in the rental values might have been occasioned due to the institutions' location and accommodation availability. However, the findings correspond with Garmendia et al. (2011) that students at the tertiary institutions in the UK consider their rental values expensive, forcing them to share rooms in apartments to make them affordable. Again, most (95%) of the respondents are full-time and unemployed (88%); hence, they are economically inactive, affecting the rental amount they can pay. Thus rental amounts beyond their means will be considered expensive. Munro and Livingston (2012) study made this observation that due to the economic incapacity of students, affording high rent is an arduous task, thus employing strategies such as room sharing to reduce accommodation costs.

4.2 Residential Type and perception of the Accommodation Rental Value

The students' perceptions regarding their accommodation rental values were then analysed based on where they stay, either on-campus or off-campus. This was to identify whether these two different accommodation types differ in rental values and the implication of the student accommodation provision. The findings of this analysis are shown in Table 3. The findings indicate that most on-campus (51%) and off-campus (42%) stated that their rent is expensive. However, whilst a significant number (40%) of on-campus students consider their accommodation not too expensive, a considerable number (35%) of the off-campus, on the other hand, believe their accommodation rental values are too expensive.

These findings imply that off-campus students are more expensive than on-campus student accommodation. This may be because off-campus accommodations are privately owned; thus, the university authorities and the government have no control over the rental charges, unlike the on-campus accommodation provided by the government. The government controls the rent paid by on-campus students; thus, the rent is usually subsidised

Table 3. Residential Type and perception of the accommodation rental value

Variables	On-Campus			Off-Campus			Overall	
	Count	Percentage	Ranking	Count	Percentage	Ranking	Freq	Percentage
Expensive	135	50.9%	1	150	41.6%	1	285	45.53%
Not too expensive	107	40.4%	2	86	23.8%	3	193	30.83%
Too expensive	23	8.7%	3	125	34.6%	2	148	23.64%
Total	**265**	**100%**		**361**	**100%**		**626**	**100%**

and often below the open market-related rental values. Private developers provide off-campus accommodations to make profits; therefore, they capitalise on accommodation shortages to increase property prices to reap more profit without any interference from the government. It has been observed by Ong et al. (2013) that private property owners hike property prices in student-populated areas, making rental values unaffordable to students. However, students have no alternatives and thus pay any amount the landlords charge. Again Ghani and Suleiman (2016) opine that due to the free-market economy practised by countries, the government has no control over property rental values; thus, landlords have free will to determine any amount payable for using their properties. Ghana practices free-market economic systems; therefore, the government has no control over the private property prices, making it difficult to control the rent payable by students at public universities. This means the government should construct more accommodations at the university premises to make accommodations more affordable.

5 Conclusion and Recommendation

Students' accommodation cost is becoming a global issue as institutions cannot provide enough to accommodate increasing student enrollment. This situation has brought about accommodation shortages across many institutions. Therefore, private developers have entered the market to make accommodation available to students in the communities on the periphery of the institutions. Therefore, the study examined tertiary students' perceptions concerning the affordability of their accommodation on-campus or off-campus. The study's findings indicate that although on-campus and off-campus students perceive their rental values as expensive, off-campus housing is more expensive than on-campus accommodation. This may imply that students are paying exorbitant prices for their accommodation, especially in accommodations outside the domain of the universities. This situation may negatively affect students who may not afford the off-campus housing and have also been denied on-campus accommodation to access university education. Therefore, it is recommended that government and university managers implement strategies to construct more affordable student accommodation. Government and universities could partner with private developers to provide reasonable accommodation to the students. The government may institute rent regulation laws for private developers to prevent exorbitant rent developers charges due to accommodation shortages.

Government can also give tax incentives or subsidies to private student accommodation developers to avoid the total transfer of development costs to student tenants to make the accommodation more affordable. Future studies should look at why the university and government cannot use innovative ways to provide enough housing for enrolled students at the various tertiary institutions.

References

Ackermann, A. Visser, G.: Studentification in Bloemfontein, South Africa. In: Szymańska, D. and Rogatka, K. editors, Bulletin of Geography. Socio-economic Series, No. 31, Toruń: Nicolaus Copernicus University, pp. 7–17 (2016)

Addai, I.: Problems of non-residential students in tertiary educational institutions in Ghana: a micro-level statistical evidence. J. Emerg. Trends Educ. Res. Policy Stud. (JETERAPS) **4**(4), 582–588 (2013)

Blumberg, B., Cooper D.R., Schindler, P.S.: Business Research Methods. Second European edn, McGraw-Hill Higher Education, UK (2008)

Bondinuba, F., Nimako, S., Karley, N.: Developing student housing quality scale in higher institutions of learning: a factor analysis approach. Urban Stud. Res. **1**(1), 1–11 (2013). https://doi.org/10.1155/2013/383109

Danso, A.K., Hammond, S.F.: Level of satisfaction with private hostels around knust campus. Int. J. Sci. Technol. **6**(3), 719–727 (2017)

Donaldson, R., Benn, J., Campbell, M., Jager, A.: Reshaping urban space through studentification in two South African urban centres. Urbani izziv, 25, supplement – 013, (special issue), S176 - S188 (2014)

Ghani, Z.A., Suleiman, N.: Theoretical underpinning for understanding student housing. J. Environ. Earth Sci. **6**(1), 1–15 (2016)

Garmendia, M., Coronado, J.M., Urena, J.M.: Students sharing flats: when studentification becomes vertical. Urban Stud. **49**(12), 2651–2668 (2011)

Ghana Education Performance Report, pp: 38. Retrieved from: http://www.idpfoundation.org/Ghana%20MoE%20Ed%20Performance%20Report%20210.pdf. Accessed 14 Jan 2021

Hubbard, P.: Geographies of studentification and purpose-built student accommodation: leading separate lives? Environ. Plan. **41**, 1903–1923 (2009). https://doi.org/10.1068/a4149

Hassanain, M.A.: On the performance evaluation of sustainable student housing facilities. J. Facil. Manag. **6**(3), 212–225 (2008)

Insch, A., Sun, B.: University students' needs and satisfaction with their host city. J. Place Manag. Dev. **6**(3), 178–191 (2013)

Klis, M.V., Karsten, L.: Commuting partners, dual residences and the meaning of home. J. Environ. Psychol. **29**(2), 235–245 (2008)

Khozaei, F., Hassan, A.S., Kodmany, K., Aara, Y.: Examination of student housing preferences, their similarities and differences. Facilities **32**(11/12), 709–722 (2014)

Munro, M., Livingston, M.: Student impacts on urban neighbourhoods: policy approaches, discourses and dilemmas. Urban Stud. **49**(8), 1679–1694 (2012)

Najib, N., Yusof, N., Abidin, N.: Student residential satisfaction in research universities. J. Facil. Manag. **9**(3), 200–212 (2011)

Nimako, S.G., Bondinuba, F.K.: Relative importance of student accommodation quality in higher education. Curr. Res. J. Soc. Sci. **5**(4), 134–142 (2012)

Ong, S.E., Petrova, M., Spieler, A.C.: Demand for University Student Housing: an empirical analysis. J. Hous. Res. **22**(2), 141–164 (2013)

Pickren, G.: Where Can I Build My Student Housing?' The politics of studentification in athens-clarke County Georgia. Southeast. Geogr. **52**(2), 113–130 (2012)

Simpeh, F., Shakantu, W.: On-campus university student housing facility services prioritisation framework. Facilities **38**(1/2), 20–38 (2019)

Simpeh, F., Shakantu, W.: An on-campus university student accommodation model. J. Facil. Manag. **18**(3), 213–229 (2020)

Sawyerr, P.T., Yusof, N.: Student satisfaction with hostel facilities in Nigerian polytechnics. J. Facil. Manag. **11**(4), 306–322 (2013)

Sage, J., Smith, D., Hubbard, P.: New-build studentification: a panacea for balanced communities? Urban Stud. **50**(13), 2623–2641 (2013)

Disease Pandemics in Africa and Food Security: An Introduction

E. Baffour-Awuah[1,2], I. N. Amanor[1], and N. Y. S. Sarpong[1,3(✉)]

[1] Department of Mechanical Engineering, Cape Coast Technical University, Cape Coast, Ghana
{ishmael.amanor,serwaah.sarpong}@cctu.edu.gh

[2] Agricultural Engineering Department, University of Cape Coast, Cape Coast, Ghana

[3] Department of Agricultural and Biosystems Engineering, Kwame Nkrumah University of Science and Technology, Kumasi, Ghana

Abstract. Purpose: The purpose was to make available the related information to researchers and industry players such as health officials and workers as well as environmentalists in their operational, tactical and strategic operations.

Design/Methodology/Approach: Content analysis was employed for this study, with particular reference to a systematic review, relying on manifest content. Twenty years were extensively considered; from 2002 to 2021 using the Microsoft Bing database. The manifest contents of 18 out of 120 research papers were then selected. The documents were manually coded, intensively reviewed and finally analyzed.

Findings: The paper found that hundreds of thousands have died in Africa as a result of these pandemics although various interventions have been employed to manage them. These interventions include hand-washing; social distancing; quarantine; face masking; disinfection; personal distancing; lockdowns; enhancement of testing; and vaccination among others. The paper also found that although these interventions, including knockdowns in Africa, were advantageous, they could contribute to negative outcomes such as food insecurity in nations within the continent.

Research Limitation/Implications: Manual coding method used has drawbacks though the computer-based analytical technique could have been more efficient because with a greater number of electronic data sets study time could be reduced and also reduce the number of human coders to achieve inter-coder reliability. The study was also limited by dwelling on the only nineteen-year period from 2002 to 2020. Relying on as many documents as possible could be more reliable for studies of this nature.

Practical Implication: The paper is intended to guide both industry players and researchers in terms of the techniques available in dealing with disease pandemics regarding the most suitable procedures and techniques in the prevention and treatment of these diseases.

Social Implication: The study implies that the information available in this paper when adopted could facilitate the achievement of the Sustainable Development Goals (SDGs) (Agenda 2030) by making easy accessibility of knowledge relating

C. Aigbavboa et al. (Eds.): ARCA 2022, *Sustainable Education and Development – Sustainable Industrialization and Innovation*, pp. 698–710, 2023.
https://doi.org/10.1007/978-3-031-25998-2_54

to the management of present and future pandemics. The study also implies that the most reliable, potent, effective and efficient management technique for pandemics is to vaccinate as many individuals as possible to attain herd immunity among the populace within individual countries in Africa.

Originality/Value: The uniqueness of this study lies in the fact that it establishes the characteristic commonalities and differences regarding coverage, mode of transmission, management techniques, and the challenges associated with the most concerned pandemic diseases of recent history.

Keywords: Covid-19 · Food security · HIV/AIDS · Lockdowns · Spanish influenza · Vaccinations

1 Introduction

The fact that the occurrence of pandemic diseases have wreaked havoc on many continents, including Africa, cannot be overemphasized. A pandemic may be defined as a quick infection of disease that cover a large area of people, animals or plant species within a relatively short period of time. For instance, if a disease spreads to fifteen out of one hundred thousand, amounting to 0.00015% of a population within fourteen consecutive days, this may be considered as a pandemic. Pandemics have the capacity to spread past a lot of countries into other countries. They may even spread across different continents. Popular pandemics that have affected many nations include the Spanish influenza, HIV/AIDS and Covid-19. Table 1 shows global pandemics with high rate of infections, the death toll, period of infection and location on the African continent (Green et al. 2002; World Population by Year 2021; World Population History 2021).

Though there are several other pandemics in world history, lack of precise data, period of occurrence and death rate precludes their inclusion in such records as recorded in Table 1. Examples of such pandemic include tuberculosis, hepatitis B and hepatitis C (World Population by Year 2021; World Population History 2021). For similar reasons, non-communicable diseases such as cancer may not be considered as pandemics (World Population History 2021). Research has shown that epidemics have negative effects on countries affected, and for that matter, the global economy (FAO 2008). Disease pandemics have therefore affected societies and health care systems. Particularly, among the adverse effects of pandemic diseases is food insecurity. In Africa where most of the population is rural and agricultural, the effect of disease pandemics on food security have always had severe hardships on the continent and its populace. (African Center for Strategic Studies 2020). The paper therefore sought to answer the following questions: What are the most outrageous human pandemics as recorded in recent human history as far as Africa is concerend? Where were they concentrated within the African continent in terms of countries and sub-regions? What were the mode of transmission? How were the pandemics managed? How long did it take to completely or partially manage the individual pandemics? Could the differences, commonalities and externalities of the pandemics be identified? To what extent did these pandemics briefly influcnce food security in Africa?

Table 1. Some global pandemics of high infection rate and their location of infection on the African continent, global death toll and period of infection

Epidemic/Pandemic	Disease	Death Toll (Global)	Date	Location
1. Black Death	Bubonic Plague	75–200M	1346–1353	North Africa
2. Spanish Influenza	Influenza A/H1N1	17–100M	1918–1920	Africa
3. Plague of Justina	Bubonic Plague	15–100M	1541–1549	West Africa
4. HIV/AIDS	HIV/AIDS	35M+	1981-Present	Africa
5. Third Plague Pandemic	Bubonic Plague	12–15M	1855–1960	Africa
6. Covid-19 Pandemic	Covid-19	4.3–9.2M	2019-Present	Africa

Source: Extracted from Green et al. 2002; World Population by year 2020

2 Methodology

This review paper applied content analysis as the study methodology, with particular reference to systematic review and manifest content. Content analysis is the study of materials which might be documents of different formats, words, pictures, audio or video. It is usually engaged to examine the arrangements in information and communication in a systematic manner Alan (2011). The method makes use of past research reading material to be reviewed to find gaps with regards to the study under contemplation. For this reason the five-step technique established by Denyer and Tranfield (2009) to systematically review literature was employed. The technique involves the formulation of questions, selection of documents, assessment of documents, analysis of documents, and finally, documentation of result.

In order to achieve the objectives of the paper, the period from 2000 to 2021 was considered, relying on Google scholar database. The period considered was based to the fact that over 90% of the documents related to this study were published within the period.

An extensive search on Google scholar was established by using the title-abstract-keyword" technique. The search keywords applied were "Covid-19"; "HIV/AIDS and "Spanish Influenza". The search provided 26,571 documents. Upon further screening using the keywords "lockdown"; "vaccination"; and "food security", 2133 documents were retrieved. Furthermore, restrictions were made using the keywords "Africa" and "African continent" with particular reference to article documents, yielding 88 documents. Fifteen research documents were finally selected and coded manually for intensive review and analysis based on the manifest content technique.

3 Findings and Discussion

3.1 The Spanish Influenza Pandemic

It is estimated that the global infection of the Spanish influenza ("Spanish flu") was as high as 500 million people; about one-third of the global population, annihilating between 20 and 50 million. At that time during which most of the African countries were under colonial regime, the toll was quite severe in Africa. Almost 2% of the population perished within a period of 180 days. It was estimated that about 2.5 million of 130 million Africans died as a result of the Spanish flu infections. In some geographical locations the disease was estimated to have infected about 90% of the populace and killing about 15%. For example, South Africa was among "the hardest hit" on the African continent (African Center for Strategic Studies 2020; Juergen 1995).

Studies have shown that about 5% of the South African population was mortalized as a result of the disease. In West Africa, Sierra Leone was the hardest hit perishing about 4% within 21 days. The pandemic also killed between 4 and 6% of Kenya's population within a period of 9 months in the East African sub region. In coastal East Africa the raw death rates and health utilization jumped by between 300% and six hundred percent and persisted at very high levels until around 1925 with concomitant serious impacts on the citizenry and economy. It must be noted that the Spanish flu had a short incubation period between one and two days with the virus spreading with asymptotic dimensions. The disease affected the younger and healthy generation as well as the old. This is as a result of the fact that the virus straightaway deeply enters the lungs of the individual. The virus then initiates the immune system to overreact and accumulates in the lungs with antibodies that generate severe respiration inability (African Center for Strategic Studies 2020; Patterson and Pyle 1991; Patterson and Pyle 1983; Sandra 1994).

The Spanish flu was first detected in March 1918. The pandemic covered large areas throughout the world as a result of troop movement from the battle fields to their countries of origin and as refuges from one country to a host nation. Documents indicate that the disease started covering North African countries from May 1918. Areas such as the coastal cities along the Mediterrean Sea were among the victims. In East Africa, Ethiopia was a victim and Portuguese East Africa as well. Sub-saharan Africa, however, was somehow free from the ravaging speed of the virus (African Center for Strategic Studies 2020; Ohadike 1983; Patterson and Pyle 1983).

The second wave of the disease, records indicate, commenced in August 1918, infecting Sub-Saharan countries such as Nigeria, Ghana, Sierra Leone, Gambia and Cameroon, all being West African countries at the time. The port cities of these countries, though, were the victims of the second wave. In East Africa the port cities of Djibouti and Mombasa were also affected with the pandemic between September and October, within the same period of time. By December 1918, the second wave was dying down. However, in January 1919 a third wave reared its head, reinforcing the already deadly infection and annihilation of human lives. Countries infected were some parts of the Sahel regions, East Africa and Madagascar (African Center for Strategic Studies 2020; Patterson and Pyle 1991; Patterson and Pyle 1983). Thus, generally, Africa experienced all the three waves of the various strains of the pandemic. The second wave variant was found to be rather more virulent.

The spread of the Spanish flu in Africa was attributed to several reasons. One, the movement of about 150,000 African servicemen and 1.4 million hands that was supporting labor and logistics during the First World War (1914–1988) in Europe to West Africa, East Africa and South Africa brought the disease to port cities. Two, the well-developed transportation system between the port cities and the urban towns compounded the transmission of the disease to the inland urban settlements. It is on record that areas along rivers, roads and railways network systems could spread the disease to a distance of about 100 km every day. Three, out of panic, due to death and illness, people along the coastal areas including demobilized soldiers, men, women and children migrated to rural areas. Thus deep inland, the pandemic got to the inland rural settlements of the continent. Peoples' fears were compounded when the disease was even infecting health workers such as doctors and nurses. Fleeing to rural areas was therefore intensified, aggravating the situation further. Thus, soldiers, migrant workers and sailors became medium of the spread in countries with coastal boundaries such as Gold Coast (Ghana); the Gambia, Tangayika (Tanzania); Nyasaland (Malawi), Cameroon; Kenya and South Africa, into intra-national rural settlements (Ohadike 1983; Patterson and Pyle 1991; 1983).

It will be naive, ignorant and irresponsible to assume that in the face of such a virulent pandemic, authorities would sit and look unconcerned for human lives to be lost at such a ravaging speed. Authorities at the time, both colonial and indigenous, made various attempts, within the continent, to mitigate the situation. Measures that were taken include social distancing and quarantines. There were orders to close down churches, schools and markets. Public gatherings were banned. As the need may be, school buildings and churches were employed to be used as hospitals. Inspections of ships and country borders were also ordered and implemented at site. Contact-tracing was also employed to reduce the spread of the disease on the continent. In the absence of vaccines, other methods employed include sharing of information using radio and telegraph, as well as gong-gong and drums in the rural settlements; in the urban areas and in ships. The gravamen of the situation at certain periods of the pandemic and at certain localities brought the involvement of chiefs and newspapers to fight the flu. This particularly occurred in Lagos and Mombassa where racial considerations initially prevented chiefs from getting involved in the fight of the pandemic (Philips 2014; Sandra 1994).

Other interventions that were employed include the activities of public health workers to open emergency relief depots, hospitals, and soup kitchens that made provision of medicine and food. Cleaning of public places with outbreaks of the disease and division of cities and towns into medical zones with assigned doctor(s) were also some of the mitigating interventions. One unpopular intervention introduced in Lagos whereby British medical practitioners were legally mandated to search of suspected infected victims created fear and panic. This brought about people vamoosing from their homes thus, rather aggravating the already threatening virulent situation. Thus the house-to-house visits could not bring about its intended purpose due to misunderstanding and mistrust, lack of lucid communication techniques and suspicion in spite of its objective to bring about some relief to infected individuals and families; and slow down, if not the stoppage of the spread of the disease (Oluwasegun 2015). Despite the various interventions adopted during the pandemic, food security continued to be threatened, even after the devastation of the pandemic. It took several years for the food situation to come to normalcy within the African continent pre-pandemic situation (Andayi et al. 2019).

3.2 The HIV/AIDS Pandemic

The HIV pandemic is one pandemic that has and continue to have devastating toll on the global populace. This is because victims do not get better but die slowly. The growth pattern of HIV/AIDS varies from region to region; and country to country with an initial growth rate being slow; followed by an accelerated rate from stage 1 to stage 2 where the virus spreads at a faster rate among the population. At the last stage the growth slows from Stage 2 to stage 3 at a slower acceleration before the final stage is reached. The final stage is the stabilized stage, stage 3, where a large percentage of the population who were at risk had been infected. Eventually, there is decrease in new infections as well as the infected fraction.

Globally, the countries with the epidemic, would be stabilized in certain countries at moderate rates; while in others a significant fraction may acquire it before reaching the maximum limit on the stylized-shaped or logistic distribution curve. For example, in West Africa, though the pandemic appeared to be stabilizing, it was later found that the current prevalence rate of HIV/AIDS infections was rising (Baffour-Awuah 2014). In certain locations over 40% of all pregnant women were found to be infected. By December 2002, over 42 million people of the global population was infected with the HIV virus with about 95% living in the developing world (UNAIDS 2020). It is estimated that over 160 million people were likely to be infected by the epidemic. About two-thirds of this population was estimated to be living in 25 most infected countries in rural Africa. With the prevalence of HIV/AIDS, the life expectancy in sub-Saharan Africa has reduced from 62 years to 47 years since 1950 (UNAIDS 2020b). In Botswana, a record low level of life expectancy has been observed within this period of time, for example.

It is estimated by the FAO (2001b) that, AIDS has mortalized about 7 million agricultural workers in the 25 most hard-hit countries on the continent since 1985, and is likely to mortalize 16 million in addition before the end of 2020. In Zimbabwe for instance, the 31–41-year group was the hardest hit (Ncube 1999). A 2000 study in Zimbabwe also showed that over 31% of pregnant women in rural settlements were infected with HIV with women farm workers registering a prevalence rate of 43.7%. The study also showed an 82.7% infection level of rural pregnant women with an infection level of rural subsistence women being as high as 25%. The infection levels and rate could be staggering considering the spread of the pandemic in Africa. Indeed a study in Ghana revealed that there could be a rebound of increase in infection rate in spite of the fact that the prevalence rate was reducing (Baffour-Awuah 2014). The prevalence of the pandemic has affected many people between the ages of 15 and 50 years. As at 2002 about over 13 million children had got one or both parents dead as a result of the pandemic. It was projected that by 2010, the number could increase to 25 million. In 2001, 70% of these orphans were found in 12 countries within Sub-Saharan Africa.

A similar study in Zambia gave a startling revelation: while 68% of rural orphans were not enrolled in school, 48% of non-orphans found themselves in a similar situation. Though the HIV/AIDS pandemic has affected all countries and regions of the globe, the effect has been disproportionate with sub-Saharan Africa being the hardest. Although Africa constitutes only 10% of the global population it is involved with 75% of AIDS-related death, according to a 2003 estimates. The case for Asia and Pacific accounted for only 19.5% or 7.4 million infections of the global infections in spite of the fact that they have the world's populous and largest countries. Jointly, Africa and Asia have 85% of current global infections of HIV and AIDS-related mortality rate. From 1997 to 2001, 57% of women within Sub-Saharan Africa were living with HIV while 75% of young men and women were infected with the virus. Within the same period the global infection of women living with HIV/AIDS increased from 41% to 50%. (UNAIDS 2020b).

The HIV/AIDS pandemic has been found to be transmitted through three fundamental mechanisms. These include one: sexual mechanism; two, exposing individuals to blood or blood products, and three, pre-natal mechanism or post-natal mechanism. Other secondary factors that influence transmission are cultural, social and environmental mechanisms. These secondary mechanisms are the main contributing factors to the differential occurrences and influences among countries and continents. It is worth noting that viral, molecular and environmental factors also exist. It is also worth noting that other host mechanisms could also characterize the possibility of the transmission of the virus (Askew and Bever 2003; Chaisson et al. 1989; Gregson et al. 2001; John et al. 2001; Read 2003; Royce et al. 1997; World Bank (2020a).

Since the beginning of the HIV/AIDS pandemic in 1983, various interventions have been adopted to mitigate its spread and morbid lethargy. The first was the introduction and maintenance of appropriate public health surveillance system for proper, efficient and effective preventive techniques to deal with the menace. The second was a reliable communication system, targeting the needed and pertinent information to high-risk areas and domains. However, in certain areas, the reverse was upheld, indicating that, for example in Latin America the rather lower infected populations and areas were the target. In other words, information, communication and education is very relevant in the fight against the pandemic. Sex education in schools was one of the important preventive and mitigation efforts to fight the pandemic (Peersman and Levy 1998).

Another preventive effort as post infection technique adopted was voluntary counselling and testing which enabled willing population to deliberately avail themselves to find out their HIV status with consequential counselling effort to assist them appreciate the results of their health status if found infected or otherwise. Peer-based techniques are interventions that employ voluntary or involuntary members of the community to pass on relevant and related information to infected population and geographic settlements with the aim of changing the behavior or attitudes of residents. Information may also be directed at infected and non-infected individuals or group of individuals within the target community (Hutton et al. 2003).

The promotion, distribution and social marketing of condom has also been found to be effective in fighting the menace. Condom has also been useful among infected and non-infected persons in fighting other sexually transmitted infections (STIs), apart from HIV. The Screening and treatment of sexually transmitted infections has also been employed to fight the HIV/AIDS pandemic. The application of condom and the screening and treatment of STI can go a long way to prevent the spread of the HIV/AIDS menace (Oroth et al. 2003).

In order to prevent mother-to-child transmission (MTCT) three techniques have been used to contain the pandemic: introduction of good contraceptives to prevent unwanted pregnancies; use of anti-retrovirals; and breastfeeding substitutes. The introduction and use of contraception methods and services whereby infected and non-infected women use these methods and services has been found to be an effective technique and strategy in the fight against the HIV menace. Rely et al. (2003) have shown that vertical transmission of the pandemic is truncated partially or wholly when antiretroviral drugs and medications are provided to infected mothers. Nduati et al. (2000) have also shown that when breast feeding is extensive, the probability of MTCT is doubled. Thus, when cost is favorable; breast feeding is substituted and sustained; potable water is available; health care is favorable; child spacing is practiced and socio-cultural factors are favorable, breastfeeding avoidance of HIV-infected mothers to babies could be a most powerful reliable technique to manage the spread of the HIV/AIDS pandemic (Coutsoudis 2002).

Prevention techniques that have been employed globally to reduce or prevent blood borne transmissions to African countries include harm reduction for injecting drug users; implementation of blood safety practices; and universal precautions (Mesquita et al. 2003). Harm reduction involves the employment of multiple techniques and strategies to deal with the menace. For example, techniques such as easy availability and accessibility to effective medication; treatment and drug substitution; availability and accessibility of condoms; availability and accessibility of counselling services; counselors as well as syringe and needle exchange programs. A classic example where this technique has proved very successful in Africa is the Southern Africa situation. With regards to the implementation of blood safety practices, a good blood safety program may include the employment of voluntary donors who have low risk of the infection; the avoidance of irrelevant and improper transfusions; a reliable national or regional blood transfusion organization in relevant and appropriate screening of blood and products donated for the HIV virus. Universal precautions as a preventive strategy in the mitigation effort is to prohibit the reuse of needles, syringes, gloves, unsterilized equipment such as scissors, gowns, goggles etc.; and reduction of injection-related infections by over 96% between 2000 and 2030 (WHO and UNAIDS 2003).

In spite of the various methods, technique and strategies that have been employed to deal with the HIV/AIDS pandemic its effect on food security continues to negatively influence Africa and its various countries.

3.3 The Covid-19 Pandemic

Getting close to the end of 2019 a new coronavirus, branded SARS-Cov-2 reared its head in the Hubei province in China, specifically in the city of Wuhan, with an associated acute respiratory illness. By the end of February 2020 the world health organization (WHO) branded the disease as Covid-19 meaning coronavirus disease 2019 (Güner et al. 2020). The viral infection is in three stages. It begins with an asymptomatic reaction. This implies that it exhibit no sign of infection with no symptoms when the virus is incubating in the victim. The second stage is when the victim show signs or symptoms of infection. This occurs after 14 days of the viral infection. The third stage is the acute respiratory distress syndrome which is associated with septic shock and multiple-organ failure. During these stages, if remedy is not sought, the third stage may lead to death (Guan et al. 2020). The WHO announced officially and publicly that Covid-19 is a public health emergency of international dimension. On March 11, 2020 the WHO finally declared covid-19 as a pandemic. This showed the gravamen of the pandemic and therefore proclaimed the need for countries to make the effort to detect and prevent the spread of the menace.

Covid-19 spreads through various routes. One, the virus may spread from person to person through coughs and sneezes. By this mode of transmission, respiratory droplets from an infected individual passes through the mouths or noses of people close by or through inhalation and deposition in the lungs. Two, through the contact of contaminated fomites and inhalation of aerosols, when aerosols are being produced. Three, during the incubation period of the virus when victims are asymptomatic (Wei et al. 2020).

Like all disease pandemics, covid-19 has no approved medication during the early years of outbreak. However, the disease can be cured using unapproved medications, which when tried are able to save the lives of infected individuals. On March 7, 2020, the WHO published on interim guidance entitled "Responding to community spread of covid-19" which dwelt on the need to prevent the transmission of the virus. The guidance indicate the need to prevent the disease from spreading by coordinating and integrating efforts via public health, commerce and business, transportation and transport, finance and banking, security and protection. Generally, the guidance emphasized on the need for entire society to prevent the spread of the virus through a holistic and integrated approach on all facets of the disease (World Bank 2020).

Prevention techniques as recommended by the WHO include screening of individuals before symptoms show early diagnosis; early isolation of infected individuals; and early treatment of victims. In sum, prevention and control measures within communities include voluntary quarantine (self-quarantine and mandatory quarantine in private residence, hospitals and public institutions, education institution, etc.). Other measures incudes crowding avoidance; isolation; use of personal protection gadgets; and hand hygiene.

Among the various strategies in the prevention of the spread of Covid -19, hand hygiene has been found to be the most important. Hand hygiene involves frequent washing of the hands diligently with potable water and hand sanitizer. It also involves avoidance of touching the inner part of the nose, ears, face and mouth with the hand. Respiratory hygiene which involves the covering of one's cough and sneeze is another preventive mechanism. The WHO also recommended crowd avoidance and close contact with individuals, particularly infected victims. Various organizations, as a means to educate and effectively communicate to communities, have designed and produced posters and brochures as well as bill boards on covid-19 protection and prevention. Various health organizations as well as the WHO have also published various visual accoutrements including posters and videos as well as audio materials to help deal with the situation.

Social distancing (physical distancing) and quarantine as a strategy in the prevention of the fight of the pandemic includes: total or partial lock down of educational institutions, business organizations and industries (hotel industry); reducing the number of guests, visitors and strangers; as well as reducing the contact between individuals residing in a confined premises such as prison, health care facilities, children's homes and orphanages, old-age care facilities, etc. Other measures include restricting, cancelling and prohibiting mass gathering and meeting, compulsory quarantine of infected persons in buildings, houses and homes. The rest are the closure of internal and external borders as well as the restriction of entire regions, districts, towns and cities to stay at home (knock down). The use of protective equipment has also been recommended by the WHO and ECDE Particularly among protective equipment is the face mask for both health workers and potentially symptomatic individuals. Cloth face covering such as home-made masks, bandanas and handkerchiefs may also be used to protect the mouth and nose when people find themselves in public places (ECED 2020).

Studies have shown that at the pre-symptomatic stage social distancing and the use of face mask are some of the critically important mechanisms to control the spread of infection (Wei et al. 2020). Studies have also shown that quarantine can reduce the rate of infection by 37% and the number of death as much as 30% (Nussbaumer-Streit et al. 2020); and that it is the most effective technique which can reduce the infection and death rates (Iwasaka and Grubaugh 2020; Pan et al. 2020). Cleaning, disinfection, and increasing testing capacity, have also been recommended by the WHO and therefore employed by various governments, government agencies and private organization as a means of controlling the epidemic.

As at August 13, 2021, the Covid-19 map, which shows the coronavirus cases, death and vaccinations by country showed that covid-19 had infected 211.8 million people; caused 4.4 million death; 452,504 new cases; and 7,561 new death within the previous 24 h. India and Brazil had experienced the highest number of confirmed cases, followed by Russia, France and UK, with 200 countries having experienced both confirmed cases and death. The 100 millionth case was recorded on the 31st January 2021. Though there are no officially present medications for controlling the pandemic, vaccines have been developed and being used by various countries. Among the 197 countries that are administering vaccines, available data shows 67 are from high-income country, 103 from middle-income countries and 26 from low-income countries with China and India having administered the highest number of jabs, at 2 billion and 581 million doses respectively. The third country is the United States with Over 362 million doses. In Latin America countries with the highest death tolls are Brazil, Mexico and Peru at 574,000, 253,000 and 100,000 death casualties respectively. Africa has seen 7.5 million cases with 190,000 deaths with South Africa experiencing 2.7 million cases and 80,000 deaths, being the highest hit on the continent, followed by Morocco with 810,000 cases and Tunisia with 640, 000 cases (John Hopkins University 2021).

Even before the pandemic, Africa was experiencing food insecurity and its consequential influence on the continent and its populace. With knockdowns as a management technique to deal with the pandemic, food security became worsened. Thus food availability; accessibility; stability; utilization and sustainability became negatively affected. The precautionary techniques and agencies available for dealing with food insecurity were similarly affected negatively.

4 Conclusion

It is an undeniable fact that the appearance of pandemic diseases have wreaked untold havoc on many continents, including Africa. Common among these pandemics that have affected many nations include the Spanish influenza ("Spanish flu"), HIV/AIDS and Covid-19. Pandemic diseases take a long time to find to get a medication, if even one can be found. For example, at the moment, no found and approved medication has been reliably found for the various variants of the Covid-19 pandemic. It is speculated that the global infection of the Spanish influenza was about one-third of the global population and killed about 20–50 million. The HIV pandemic is also a pandemic that has and continue to have a devastating impact on the global citizenry. This is because individuals infected do not get cured but perish slowly though within a relatively long period of time. Various interventions have always been adopted to manage these pandemics. Particular among the interventions include the use of face mask; hand-washing; social distancing and quarantine. Others are cleaning; disinfection; personal distancing and enhancement of testing. The rest include testing capacity, knockdown and vaccination. Among these interventions, the most potent, effective and efficient is the vaccination technique. Though these management techniques could be reliable, they could also be disadvantageous. One such technique, knockdown, could negatively affect food security as far as the African continent is concerned.

References

Africa Center for Strategic Studies. Mapping Risk factors for the spread of COVID-19 in Africa. Infographic, 3 Apr 2020

Alan, M.: Imperfect competition in the labor market. In: Handbook of Labor Economics, vol. 4, pp. 973–1041. Elsevier (2011)

Andayi, F., Sandra, S. Widdowson, M. A.: Impact of the 2018 influenza pandemic in coastal kenya. Trop. Med. Infect. Dis. **4**(2), 91 (2019)

Askew, L., Berer, M.: The contribution of sexual and reproductive health services to fight against HIV/AIDS: a review. Reprod. Health Matters **11**(22), 51–73 (2003). [PubMed]

Baffour-Awuah, E.: Attitude of the youth towards people living with HIV/AIDS (PLWHA): the case of cape coast polytechnic. J. Biol. Agric. Health Care **4**(28), 130–114 (2014)

Chaisson, R.E., Bacchetti, P., Osmond, D., Brodie, B., Saude, M.A. Moss, A.R.: Cocaine use and HIV infection in intravenous drug users in San Francisco. J. Am. Med. Assoc. **261**(4), 561–65 (1989) [PubMed]

Coutsoudis, A.: Breastfeeding and HIV transmission. In: Public Health Issues in Infant and Child Nutrition, (48). Eds. R.E., Black and K.F. Michaelson. Nestle Nutrition Workshop Series. Philadelphia: Lippincott Williams & Wilkins (2002)

Denyer, D., Tranfield, D.: Producing a systematic review (2009)

ECED. Advice on the use of masks in the context of COVID-19: Interim guidance (2020). https://apps.who.int/iris/handle/10665/331693.2020

FAO. An introduction to the basic concepts of food security. Available at: http://www.fao.org/3/a-a1936e.pdf. Accessed 7 Dec 2020

FAO. Gendered impacts of COVID-19 and equitable policy responses in agriculture, food security and nutrition. Policy brief (also available at: http://www.fao.org/policy-support/tools-and-publications/resources-details/en/c/1276740/ (2020b)

Food and Agriculture Organization: 'The Impact of HIV/AIDS on Food Security' Food and Agriculture Organization of the United Nations (FAO). Committe on Food Security. 27 th Session. Rome. CFS: 2001/3. Rome (2001b)

Green, M.S., et al.: When is an epidemic an epidemic?. Isr. Med. Assoc. J. IMAJ, **4**(1), 3–6 (2002)

Gregson, J., Foerster, S.B., Orr, R., Jones, L., Benedict, J., Clarke, B., Hersey, J., Lewis, J., Zotz, K.: System, environmental, and policy changes: using the social-ecological model as a framework for evaluating nutrition education and social marketing programs with low-income audiences. J. Nutr. Educ. **33**, pp. S4–S15 (2001)

Guan, W.J., et al.: Clinical characteristics of coronavirus disease 2019 in China. N. Engl. J. Med. **382**(18), 1708–1720 (2020)

Güner, R., Hasanôglu Firders, A.: COVID-19: prevention and control measures in communities. Turk J. Med. Sci. **50**(3), 571–577 (2020)

HLPE. Food security and nutrition: Building a global narrative towards 2030. Report 15. Rome, HLPE (also available at: http://www.fao.org/3/ca9731en/ca9731en.pdf) (2020b)

Hutton, G., Wyss, K., N'Diékhor, Y.: Prioritization of prevention activities to combat the spread of HIV/AIDS in resource constrained settings: a cost-effectiveness analysis from Chad, Central Africa. Int. J. Health Plann. Manage. **18**(2), 117–136 (2003)

Iwasaki, A., Grubaugh, N.D.: Why does Japan have so few cases of COVID-19? EMBO Mol. Med. **12**(5), e12481 (2020) [Google Scholar] [PubMed]

John Hopkins University. COVID-19 Data repository for systems Science and Engineering (CSSE) at John Hopkins University. An interactive web-based dashboard to track COVID-19 in real time. https://engineering.jhu.edu/covid-19/support-the-case-covid-19-dashboard-team/. Accessed 23 Aug 2021

John, G.C., et al.: Correlates of mother-to-child human immunodeficiency virus type 1 (HIV-1) transmission: association with maternal plasma HIV-1 RNA load, genital HIV-1 DNA shedding, and breast infections. J. Infect. Dis. **183**(2), 206–212 (2001)

Juergen, D. M.: Patterns of reaction to a demographic crises: The Spanish influenza pandemic of 1918–1919 in Sub-Saharan Africa. A research proposal and preliminary regional and comparative findings. Staff Seminar Paper No. 6. Apr 1995

Katsande, T.C., More, S.J., Bock, R.E., Mabikacheche, L., Molloy, J.B., Ncube, C.: A serological survey of bovine babesiosis in northern and eastern Zimbabwe (1999)

Mesquita, F., et al.: Brazilian response to the human immunodeficiency virus/acquired immunodeficiency syndrome epidemic among injection drug users. Clin. Infect. Dis. **37**(Supplement_5), S382-S385 (2003)

Nduati, R., et al.: Effect of breastfeeding and formula feeding on transmission of HIV-1: a randomized clinical trial. JAMA **283**(9), 1167–1174 (2000)

Nussbaumer-Streit, B., Mayr, V., Dobressu, A.I., Chapman, A. Persad, E.: Quarantine alone or incubation with other public health measures to control COVID-19: a rapid review. Cochrane Database Syst. Rev. **4**(4), CD013574 (2020) [PubMed]

Ohadike, D.C.: The influenza pandemic of 1918–19 and the spread of cassava cultivation on the Lower Niger: a study in historical linkages. J. Afr. Hist. **22**(3), 379–391 (1983)

Oluwasegun, I.M.: Managing epidemic: the british approach to 1918–1919 influenza in lagos. J. Asian Afr. Stud. **52**(4), 412–424 (2015)

Oroth, K.K., Korenromp, E.L., White, R.G., Gavyole, A., Gray, R. H., Muhaugi, L.: Higher-risk behavior and rates of sexually transmitted diseases in Mwanza compared to Uganda may help explain HIV prevention trial outcomes. AIDS, **17**(18), 26 53–60 (2003) [PubMed]

Pan, A., Liu, L., Wang, C., Guo, H., Hao, X.: Association of public health interventions with the epidemiology of the COVID-19 outbreak in Wuhan, China. J. Am. Med. Assoc. **323**(19), 1915–1923 (2020)

Patterson, K.D. Pyle, G.F.: The Diffusion of influenza in sub-Saharan Africa during the 1918–1919 pandemic. Soc. Sci. Med. **17**, 1299–1307 (1983)

Patterson, K.D. Pyle, G.F.: The Geography and Mortality of the 1918 Influenza Pandemic. Bull. His. Med. **65**(1), 4–21 (1991)

Peersman, G., Levy, J.: Focus and effectiveness of HIV-prevention efforts for young people. AIDS 1998:12 (Suppl. A): S191–96 (1998) [PubMed]

Philips, H.: Influenza pandemic (Africa), International Encyclopedia of the First World War, 8 Oct (2014)

Read, J.S.M.: Milk, Breast feeding, and transmission of Human Immunodeficiency Virus Type 1 in the United States. Am. Acad. Pediatr. HIV/AIDS. Pediatr. **112**(5), 1196–1205 (2003) [PubMed]

Rely, K., Bertozzi, S.M., Avila-Figueroa, C., Guijarro, M.T.: Cost-effectiveness of strategies to reduce mother-to-child HIV transmission in Mexico, a low-prevalence setting. Health Policy Plann. **18**(3), pp. 290–298 (2003)

Royce, R.A., Sena, A., Cates, W., Cohen, M.S.: Sexual transmission of HIV. N. Engl. J. Med. **336**(15), 1072–78 (1997) [PubMed]

Sallent, M.: External debt complicates Africa's COVID-19 recovery debt relief needed. Africa Renewal, July 2020. UN Economic Commission for Africa (2020)

Sandra, M.T.: Colonial administration in British Africa during the Influenza Epidemic of 1918–1919. Can. J. Afr. Stu. **28**(1), 60–83 (1994)

Stover, J., Glaubius, R., Kassanjee, R., Dugdale, C.M.: Updates to the Spectrum/AIM model for the UNAIDS 2020 HIV estimates. J. Int. AIDS Soc. **24**, p. e25778 (2021)

Sweat, M.D., Gregorich, S., Sangiva, G., Furlonge, C., Balmer, D., Kamengo, C. et al.: Cost-effectiveness of voluntary HIV-1 counselling and testing in reducing sexual transmission of HIV-1 in Kenya and Tanzania. Lancet **356**(9224), 113–21 (2000) [PubMed]

UNAIDS: Eight case studies of home and community care for and by people with HIV/AIDS. UNAIDS, Geneva (2001)

UNAIDS. AIDS epidemic update. Dec 2004. Geneva: UNAIDS

UNCTAD. International production beyond the pandemic. World Investment Report 2020, Geneva, United Nations Organization (UN) (2020b). (Also available at https://unctad.org/en/PublicationLibrary/wir2020/en.pdf)

Wei, W.E., Li, Z., Chiew, C.J., Yong, S.E., Toh, M.P., Lee, V.J.: Singapore Pre-symptomatic transmission of SARS-Cov-2. MMWR Morb. Mortal. Wkly Rep. **69**, 411–415 (2020)

WHO (World Health Organization) & UNAIDS: Expert group stresses that unsafe sex is primary mode of HIV transmission in Africa, p. 14. Press Release, Geneva, March, Joint United Nations Programme on HIV/AIDS (2003)

World Bank. Global Economic Prospects, June 2020a. Washington, DC. World Bank (2020a)

World Bank. Potential responses to the COVID-19 outbreak in support of migrant workers (2020b). (Also available at https://openknowledge.worldbank.org/handle/10986/33625)

World Food Programme (WFP). Global monitoring of school meals during COVID-19 closures (2020a). https://cdn.wfp.org/2020/school-feeding-map.

World Population by Year. World Population. Accessed 1 Mar 2021

World Population History. World Population. Accessed 1 Mar 2021

An Image-Based Cocoa Diseases Classification Based on an Improved Vgg19 Model

P. Y. O. Amoako[1(✉)], G. Cao[1], and J. K. Arthur[2]

[1] School of Computer Science and Engineering, Nanjing University of Science and Technology, Nanjing, China
719106020081@njust.edu.cn

[2] Department of Computer Science, Valley View University, Accra, Ghana

Abstract. Purpose: The focus of this study is to provide accurate detection of cocoa diseases based on image analysis using a deep learning model.

Design/Methodology/Approach: Transfer learning based on a convolutional neural network such as VGG19 provides significant accurate results in image classification. This paper proposes an image-based cocoa diseases classification based on an improved VGG19 model. A comparison is made with other pre-trained models such as VGG16 and ResNet50.

Findings: The results indicate that VGG19 outperforms the other pre-trained models.

Research Limitations/ Implications: The income obtained from cocoa production is one of the bedrock of the economies in some west African countries. Cocoa production has been increasing steadily globally in recent years; however, there are high disease pathogens in most areas of production. It is estimated that the black pod disease causes 30% to 90% losses in annual cocoa production. Consequently, the global losses of cocoa production due to diseases are estimated at 20% to 25%, which is about 700,000 metric tons of global production.

Researchers have proposed varying methods for the classification of cocoa diseases; however, the identification and classification of cocoa diseases still remain a challenge. Deep learning has been very promising in its application in various fields.

Practical Implications: Accurate prediction of cocoa disease will provide stakeholders, especially farmers to provide appropriate remedies and improve productivity.

Originality/Value: The model is very effective and performs better than the state-of-the-art techniques employed on the public dataset of cocoa diseases.

Keyword: Cocoa diseases · Image classification · Pre-train model · VGG 19

C. Aigbavboa et al. (Eds.): ARCA 2022, *Sustainable Education and Development – Sustainable Industrialization and Innovation*, pp. 711–722, 2023.
https://doi.org/10.1007/978-3-031-25998-2_55

1 Introduction

Cocoa is considered one of the most valuable crops, which has established a multibillion-dollar confectionary trade. The increasing demand for cocoa production by chocolate product processing industries has made it a source of weekly or oven daily income for farmers. The cocoa plant is not an annual fruit-bearing tree and can therefore flower and bear fruit throughout the year if well cultivated and managed. Globally, cocoa infections cause 40% of annual output losses globally (Ploetz, 2016). Research by Amon-Armah et al. on cocoa disease prevention and treatment found just 5% adoption (Amon-armah et al. 2021). Cocoa is grown in traditional production zones. It produces the most lucrative crops worldwide, ensuring food security. Phytopathogenic bacteria, fungus, viruses, and viroid cause various illnesses.

This research focuses on Phytophthora palmivora (black pod) and Monilia roreri (referred to as frosty pod, monilial pod rot, or water pod rot). The phytophthora causes a transparent patch on cocoa pods that turns black and rigid. After 14 days, the entire pod turns black and necrotic. On dark parts, white to yellow downy growth causes dry, shriveled internal tissues, mummifying the pod. Monilial only affects growing pods. Immature cocoa pods have brown patches that quickly spread and cover the entire pod.

Traditional naked-eye field observation detects cocoa illnesses. Spore traps and molecular technology predict epidemics. Remote sensing, GIS, and GPS are used to monitor and predict plant diseases. The existing method known for cocoa disease detection is the traditional naked-eye field observation is used to detect cocoa diseases. Spore trap and molecular technologies are used to forecast epidemic diseases. Remote sensing, GIS, and GPS are used to monitor and predict plant diseases. Other plant disease control methods include infrared thermal imaging and drone monitoring (Shuai-qun et al. 2022). Early disease identification and warning are practical and economically feasible solutions for addressing all plant-based illnesses. Deep learning's progress in agriculture is encouraging. Deep convolutional neural network models are successful in detecting and diagnosing plant illnesses (Shuai-qun et al. 2022).

Developing a deep learning model from scratch is a significant challenge coupled with the task of annotating a massive dataset for supervised training implementation. According to Rostami et al., several factors contribute to the infeasibility of developing new deep learning models, including:

- The challenge in annotating data for fine-grained multiclass classification demands training annotators to provide accurate labels and requires too much time.
- Data sharing in many domains, such as the medical domain with annotators, may be prevented by privacy issues. Such limitations make it challenging to contract skilled annotators to process data.
- Some specialized domains, such as synthetic aperture radar images, require annotators with training experience to provide proper annotation. Hence, the difficulty in obtaining qualified annotators.
- The persistent and dynamic emergence of new classes, such as variance viral infections, complicates the annotation procedure since it becomes a continual process. Consequently, incorporating the new classes into the model becomes computationally expensive.

- Highly infrequent class members exist in some applications, such as rare event classification, which cause challenges in preparing training instances for rare event annotation (Rostami et al. 2022).

Upon these challenges, this paper uses transfer learning to detect Monilia and Phytophthora infections on an existing cocoa image. Transfer learning helps classify new developing classes, reduces the need for data annotation and model training, and leverages past knowledge (Rostami et al. 2021). This paper's remaining sections are as follows: Sect. 2 presents related papers, Sect. 3 describes the technique and experimental setup, Sect. 4 discusses the results, and Sect. 5 draws conclusions.

2 Related Works

Researchers have proposed techniques for detecting root-to-leaf cocoa diseases. An android-based expert system used Forward Chaining to diagnose cocoa plant health (Ariandi et al. 2019). A machine learning-based smart cocoa health system detects, diagnoses, and predicts disease (Gyamfi et al., 2020). Machine learning identifies cocoa tree illnesses. Histogram of Oriented Gradient, Local Binary Pattern, and Support Vector Machine are machine learning techniques (Rodriguez et al. 2021). To reduce crop loss, it's important to identify diseased and healthy cocoa plants.

2.1 VGG19 and Its Implementations

Oxford University introduced the VGG19 Convolutional Neural Network design in 2014. It placed second in ImageNet Large Scale Visual Recognition Competition 2014's Image classification task (Vijayan et al. 2020). VGG19 has improved implementation results consistently.

Early disease diagnosis improves disease control. VGG19 is superior at detecting and diagnosing the globally destructive maize foliar disease northern corn leaf blight (Shuai-qun et al. 2022). Behera et al. trained VGG19 for 1.52 min to classify papaya fruit ripeness (Kumari et al. 2021). VGG19 and Convolutional Neural Networks are used to diagnose COVID-19, pneumonia, and lung cancer from chest x-ray and CT images (Ibrahim et al. 2021) (Lucca et al., 2021) (Arias-garzón et al. 2021). VGG19 studies to diagnose human and animal disorders (Vijayan et al. 2020) (Ricciardi et al. 2021) (Lang et al. 2022)[9]. VGG19 detects machine defects better (Kreutz et al. 2020).

3 Methodology and Experimental Setup

This section discusses the materials and methods employed for cocoa disease detection to clarify readability. The intent for which the chosen classifiers and their representation architectures are discussed.

3.1 Transfer Learning

According to Pan and Yang (Pan and Yang 2010), transfer learning is the technique used to train an existing machine learning model applied in a particular domain that has sufficient training data to solve a problem in another domain of interest, of which the data may be in a different feature space or with different distribution. They further state that transfer learning can be uniformly defined as follows (Pan and Yang 2010):

Given; a source domain D_S with the learning task T_s^L,
a target domain D_T with the learning task T_T^L,
the transfer learning manages to improve target learning prediction function f_T in D_T,
base on the knowledge in D_S and T_s^L,
where; $D_S \neq D_T$ or $T_s^L \neq T_T^L$.

According to Rajinikanth et al., building a new deep learning model for a specific task is challenging (Rajinikanth et al. 2020). Implementing an existing model requires an understanding of its structure, implementation complexity, initial adjustment, and validation. Most deep learning works modify high-performing existing models. Deep learning uses transfer learning extensively. It entails training an existing network on a base dataset for a particular problem and re-purposing it for a secondary problem. Transfer learning reduces training time and improves network performance prediction (Vijayan et al. 2020). This paper uses a fine-tuned VGG19 model to detect Monilia and Phytophthora on cocoa pods.

3.2 VGG19 Model Fine-Tuning

During fine-tuning, the first three fully connected CNN layers were frozen to control weight updates. The VGG19 model for cocoa disease is transformed by adding two completely connected layers. Five classes depict cocoa pod diseases to decrease the complexity of VGG19's 1000 output classes. The weights before fully-connected layers stay unchanged.

3.3 Proposed Model Architecture

Deep Convolutional Neural Network (DCNN), which evolved from the traditional Artificial Neural Network (ANN), is the most used neural network architecture. DCNNs have convolutional, nonlinearity, pooling, and fully linked layers. DCNNs perform well in image classification, image processing, speech recognition, and natural language processing. Albawi et al. say DCNNs reduce the number of ANN parameters, allowing big models to solve challenging tasks. Figure 1 shows the cocoa disease DCNN framework. For image classification and recognition, many DCNN architectures have been suggested and implemented. This paper employs DCNN VGG19 to detect cocoa illness. During fine-tuning, the first three fully connected CNN layers were frozen to control weight updates. The VGG19 model for the cocoa disease is transformed by adding two completely connected layers. Five classes depict cocoa pod diseases to decrease the complexity of VGG19's 1000 output classes. The weights before fully-connected layers stay unchanged.

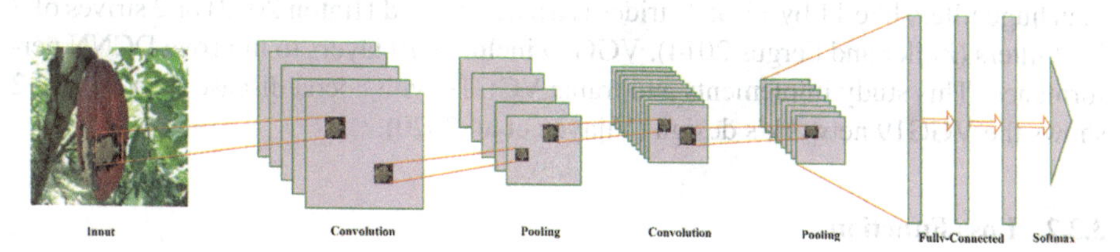

Fig. 1. General conceptual framework of DCNN diagnosis of the cocoa disease.

3.3.1 VGG19 Architecture

VGG19 is the most promising DCNN for image localization and classification. VGG19 was used to study neural network depth's effect on picture recognition accuracy (Simonyan and Zisserman 2015). Input photos were 224 × 224 RGB and preprocessed

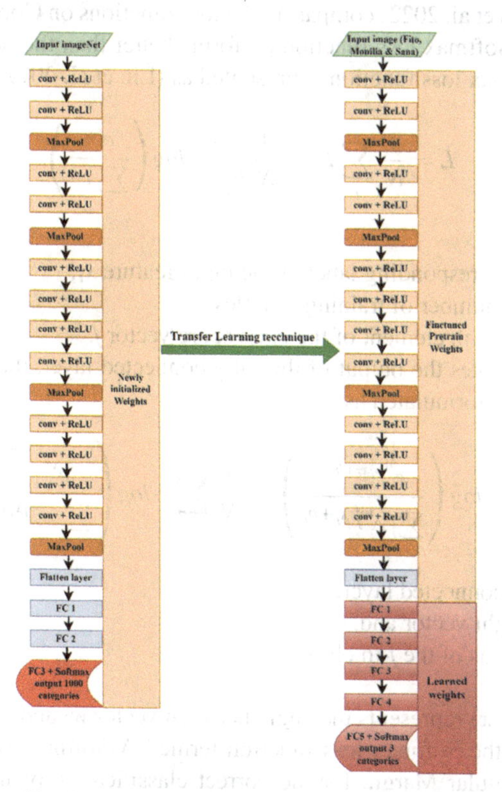

Fig. 2. The architecture of the VGG19 network

by removing the mean RGB value from each pixel. The images are then sent through a stack of 3 by 3 filtered convolutional layers and 5 max-pooling layers with stride 2. It has a Softmax output layer with ReLu activation on all hidden layers. VGG architecture ensures the consistency of the modest 3 by 3 receptive field in the full network, rather than huge filters like 11 by 11 of 4 strides (Krizhevsky and Hinton 2012) or 2 strives of 7 by 7 filters (Zeiler and Fergus 2014). VGG19 includes 19 layers to improve DCNN performance. This study implements and trains VGG19 with cocoa disease data. Figure 2 shows the VGG19 network's design (Vijayan et al. 2020).

3.3.2 Loss Function

The main aim of improving the deep learning model is to reduce the complexity of the algorithm specifically, and by this making its weight much lighter for easy implementation in the industry while not compromising its accuracy. Although more complex architecture may improve the classification and recognition capabilities. Researchers have explored varying loss functions for DCNN models to achieve better classification accuracy. Some existing loss functions include ArcFace, CosFace, A-Softmax, and Softmax loss.

The most common loss function for classification problems is the Softmax loss function realized to be performing well on separable features that are less discriminative. The work of (Shuai-qun et al. 2022) comparing the loss functions on GoogleNet architecture indicated that the Softmax loss function performs better than the others. According to Liu et al., the Softmax loss function is presented as (Liu et al. 2017):

$$L = \frac{1}{N}\sum_i L_i = \frac{1}{N}\sum_i -log\left(\frac{e^{f_{y_i}}}{\sum_j e^{f_j}}\right) \tag{1}$$

where;

y_i denotes the corresponding label of the input feature x_i,

N indicates the number of training samples.

f_i represents the j-th element of the class score vector f.

Given that f denotes the output of the fully connected layer, therefore the Softmax loss function is then formulated as:

$$L = \frac{1}{N}\sum_i -log\left(\frac{e^{w^T_{y_i}x_i+b_{y_i}}}{\sum_j e^{w_j^T x_i+b_j}}\right) = \frac{1}{N}\sum_i -log\left(\frac{e^{w_{y_i}x_i \cos\left(\theta_{y_i,i}\right)+b_{y_i}}}{\sum_j e^{w_j x_i \cos(\theta_{j,i})+b_j}}\right) \tag{2}$$

where on the fully connected layer;

w_j denotes weight vector and.

b_j denotes the bias of the j-th class.

also;

$\theta_{j,\,i}$ $(0 \le \theta_{j,\,i} \le \pi)$ represents the angle between vector w_j and x_i.

A variation of the Softmax loss function termed A-Softmax is proposed with an introduction of Angular Margin for the correct classification of the learned features optimized in DCNNs. It generalized the definition range of $\cos(\theta_{yi,\,i})$ to a monotonically decreasing angle function. This paper employed the primary Softmax loss function based on the better performance achieved in the comparison made where the dataset application domain is related to the dataset on the cocoa diseases.

3.4 Model Performance Evaluation

Estimating the model's performance in the diseases and healthy cocoa multiclass classification task employed accuracy, sensitivity, and specificity metrics. The significance of such metrics is demonstrated in the works of (Shu et al. 2017) (Kumar 2018).

- True Positive (TP) signifies correctly predicted diseased or healthy instances.
- False Positive (FP) denotes incorrectly predicted diseased or healthy instances.
- True Negative (TN) denotes correctly predicted healthy instances.
- False Negative (FN) signifies incorrectly predicted healthy instances.
- Sensitivity is defined as the ability of the model to detect the diseased cocoa pods and is determined as:

$$\text{Specitivity} = \frac{TP}{TP + FN}$$

- Specificity is defined as the ability of the model to detect the healthy cocoa pods and is determined as:

$$\text{Specitivity} = \frac{TN}{TN + FP}$$

- Accuracy is the ratio of the number of correct predictions to the total number of input samples and is determined as:

$$\text{Accuracy} = \frac{TP + TN}{TP + FN + TN + FP}$$

- Precision is defined as the percentage inaccurate prediction of the model and determined as:

$$\text{Precision} = 100 - (\mathit{Accuracy}\text{x}100)$$

3.5 Data Description and Preparation

The dataset employed in this research is obtained from Kaggle (Serrano & Gomez, n.d.) to analyze and identify cocoa diseases. The dataset was composed of three packages of images with cocoa pods: phytophthora disease labeled as "Fito", monilial disease indicated with "Monilia", and health pods identified as "Sana". This Kaggle dataset on cocoa diseases constitutes a total of 627 files distributed as Fito containing 215 files, Monilia containing 211 files, and Sana containing 201 files. This dataset provides a large set of high-resolution cocoa pod images taken under various spatial conditions. Images are labeled with a subject id in addition to the category Fito, Monilia, or Sana.

For Deep Transfer learning identification of phytophthora, monilial or healthy cocoa pod occurrence, 80% of the dataset is used for training, and 20% of samples are used for testing.

Kaggle public dataset of cocoa diseases images comprises different dimensions for each image, hence, preprocessed by resizing each input image size to 224 by 224 pixel resolution, fitting the default image size for a VGG19 model.

(a) Fito (b) Monilia (c) Sana

Fig. 3. Sample cocoa pods from the dataset. (a) shows samples of pods with phytophthora disease, (b) sample monilia infested cocoa pods, and (c) sample healthy cocoa pods.

Figure 3 indicates the pictorial view of the distribution of the original dataset.

3.5.1 Image Preprocessing and Data Augmentation

Image preprocessing is an activity applied to an image dataset to produce appropriate categories of images for analysis. Data augmentation is a technique aimed at increasing the data diversity necessary for providing required learning features for model training to improve classification (Sladojevic et al. 2016). In augmenting the cocoa disease dataset considered for this research, the images were slightly distorted in different dimensions and resolutions to reduce model overfitting during training significantly. Overfitting is realized based on the overly complicated model fitting features of the training images exceedingly specific to those images with less generalizability on the untrained images. Overfitting commonly occurs in a model that performs excellently on the training set and provides minimal accuracy on the test set. Table 1 presents the distribution of original images of the dataset and the corresponding augmented outcome.

Table 1. Distribution of image of the cocoa dataset with augmentation outcome

Class	Original images	Images after augmentation
Phytophthora	107	2996
Monilia	105	2940
Healthy pods	100	2750

Several methods are employed in image augmentation, including transformation, histogram equalization, adaptive equalization, and varying resolution. Rotation and adjusting resolution were used to augment the cocoa disease dataset. Figure 4 shows samples of augmented images.

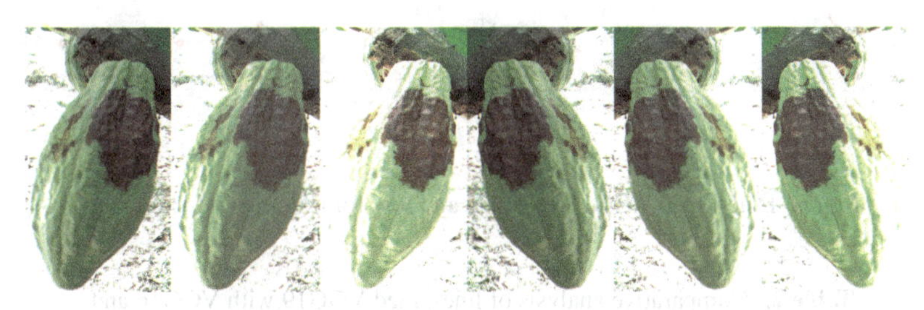

Fig. 4. Sample image augmentation of the phytophthora disease.

The first from left is the original image, followed by two resolution-adjusted images, and the next three indicate rotation with resolution-adjusted images. The model's performance evaluation was conducted later after training and further employed a stratified shuffle-split based on Scikit-learn to maintain class proportions. For evaluation, 20% of the dataset was held. The ImageDataGenerator class found in the Keras model was used where the images were left to be in their respective cocoa pod image directories.

The data augmentation was implemented using the generator after splitting the dataset, then deduced a subset of the training data for validation during the training.

4 Experimental Evaluation and Discussion

This section provides the detailed experiment and presentation of the model performance outcome benchmarked with other state-of-the-art CNN architectures. A single strategy is employed for training the models of which the initial stage of the CNN part is frozen. The weights and dense layers were trained with a high learning rate of 0.01; afterward, the whole model was fine-tuned with a minimal learning rate ranging from 0.001 to 0.0001.

A bi-level classification method is employed to predict cocoa pod diseases. The first level classified the subjects into diseased or healthy classes, with the diseased pods class comprising phytophthora or monilia. The second level provides a distinction between phytophthora and monilial diseases. Figure 5(a) presents the model's accuracy for training and validation for 10 epochs, and Fig. 5(b) shows the model's loss for training and validation.

Figure 5 describes the results of 10 iterations of the VGG19 model with training accuracy of 84.50% and validation accuracy of 84.75%. It obtained a training loss of 0.24 and a validation loss of 0.22. Table 2 compares the fine-tuned VGG19 with VGG16 and ResNet50 applied to the cocoa pod disease dataset. It is realized that the fine-tuned VGG19 outperformed the VGG16 and ResNet50 models.

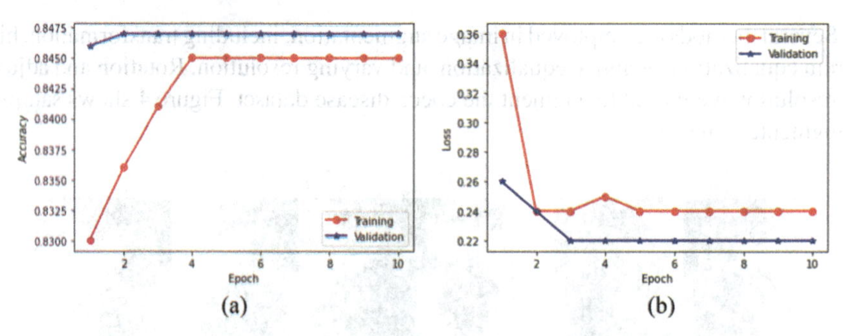

Fig. 5. Fine-tuned VGG19 accuracy and loss for training and validation

Table 2. Comparative analysis of fine-tuned VGG19 with VGG16 and

Model	Training		Validation	
	Accuracy (%)	Loss	Accuracy (%)	Loss
VGG16	76.73	0.7647	76.94	0.5647
ResNet50	78.35	0.4704	79.25	0.4114
VGG19	84.50	0.2434	84.75	0.2242

5 Conclusion

The contributions of cocoa to the world economy are very significant, and more importantly, the support for food production cannot be overemphasized. Nevertheless, the common cocoa pod diseases such as phytophthora and monilial are causing a great cocoa reduction in all the cocoa-producing countries. Determining the diseases at early stages is paramount to boosting cocoa production.

Several experiments have proven VGG19 to perform well in detecting several plant diseases. This paper successfully trained a fine-tuned VGG19 to predict the two common cocoa pod diseases on an augmented publicly available dataset. The proposed model achieves excellent performance in detecting cocoa pod diseases with an accuracy of 84.75% on such a small dataset. The success of this technique has a tremendous significance for cocoa pod disease prevention, aiding farmers and agricultural researchers in detecting and diagnosing phytophthora and monilial diseases. To increase and sustain cocoa production it is imperative to employ intelligent detection and diagnosis technology for cocoa diseases in modern sustainable agriculture.

It will be very much expedient to increase the dataset by including images of cocoa plants, diseased and healthy in Ghana to aid in accurately recognizing cocoa diseases in the Ghanaian terrain. Also, the application of this work can be expanded to cover multi-classification detection of cocoa and other plant diseases to impact agriculture and food production positively.

References

Amon-armah, F., Domfeh, O., Baah, F., Owusu-ansah, F.: Farmers ' adoption of preventive and treatment measures of cocoa swollen shoot virus disease in Ghana. J. Agric. Food Res. 100112 (2021), https://doi.org/10.1016/j.jafr.2021.100112

Ariandi, V., Kurnia, H., Marry, H.: Expert system for disease diagnosis in cocoa plant using android-based forward chaining method. Int. Conf. Comput. Sci. Eng. (2019). https://doi.org/10.1088/1742-6596/1339/1/012009

Arias-garzón, D., et al.: COVID-19 detection in X-ray images using convolutional neural networks. Mach. Learn. Appl. **6**(August), 100138 (2021). https://doi.org/10.1016/j.mlwa.2021.100138

Gyamfi, A., Iddrisu, S.A., Adegbola, O.: Machine learning-based Cocoa E-Health system. In: 13th CMI Conference on Cybersecurity and Privacy (CMI) - Digital Transformation - Potentials and Challenges (2020). https://doi.org/10.1109/CMI51275.2020.9322689

Ibrahim, D.M., Elshennawy, N.M., Sarhan, A.M.: Deep-chest: multi-classification deep learning model for diagnosing COVID-19, pneumonia, and lung cancer chest diseases. Comput. Biol. Med. **132**, 104348 (2021). https://doi.org/10.1016/j.compbiomed.2021.104348

Kreutz, M., Stief, P., Dantan, J., Etienne, A., Siadat, A.: Ice detection on rotor blades of wind turbines using RGB images and convolutional neural networks. In: 53rd CIRP Conference on Manufacturing Systems, vol. 93, pp. 1292–1297 (2020) https://doi.org/10.1016/j.procir.2020.04.107

Krizhevsky, A., Hinton, G.E.: ImageNet classification with deep convolutional neural networks. Adv. Neural. Inf. Process. Syst. **25**, 1097–1105 (2012)

Kumar, S., Kumar, B.: Diabetic retinopathy detection by extracting area and number of microaneurysm from colour fundus image. In: 2018 5th International Conference on Signal Processing and Integrated Networks (SPIN), pp. 359–364, February (2018). https://doi.org/10.1109/SPIN.2018.8474264

Kumari, S., Kumar, A., Kumar, P.: Maturity status classification of papaya fruits based on machine learning and transfer learning approach. Inf. Proc. Agric. **8**(2), 244–250 (2021). https://doi.org/10.1016/j.inpa.2020.05.003

Lang, A., et al.: Model architecture and tile size selection for convolutional neural network training for non-small cell lung cancer detection on whole slide images. Inf. Med. Unlocked **28**, 1–8 (2022). https://doi.org/10.1016/j.imu.2022.100850

Liu, W., Wen, Y., Yu, Z., Li, M., Raj, B., Le, S.: SphereFace : deep hypersphere embedding for face recognition. In: IEEE Conference on Computer Vision and Pattern Recognition, pp. 212–220(2017). https://doi.org/10.1109/CVPR.2017.713

Lucca, G., Junior, G.B., Cardoso, A., Paiva, D., Acatauassú, R., Gattass, M.: Automatic method for classifying COVID-19 patients based on chest X-ray images , using deep features and PSO-optimized XGBoost. 183(February). (2021). https://doi.org/10.1016/j.eswa.2021.115452

Pan, S.J., Yang, Q.: A survey on transfer learning. IEEE Trans. Knowl. Data Eng. 22(10), 1345–1359 (2010). https://doi.org/10.1109/TKDE.2009.191

Bailey, B.A., Meinhardt, L.W. (eds.): Cacao diseases. Springer, Cham (2016). https://doi.org/10.1007/978-3-319-24789-2

Rajinikanth, V., Noel, A., Raj, J., Thanaraj, K.P., Naik, G.R.: A customized VGG19 network with concatenation of deep and handcrafted features for brain tumor detection. Appl. Sci. MDPI **10**, 1–13 (2020)

Ricciardi, R., et al.: A deep learning classifier for digital breast tomosynthesis. Physica Med. **83**(February), 184–193 (2021). https://doi.org/10.1016/j.ejmp.2021.03.021

Rodriguez, C., Alfaro, O., Paredes, P., Esenarro, D., Hilario, F.: Machine Learning Techniques in the Detection of Cocoa (Theobroma cacao L) Diseases. Annals of R.S.C.B, **25**(3), 7732–7741 (2021)

Rostami, M., Kolouri, S., Murez, Z., Owechko, Y., Eaton, E., Kim, K.: Zero-shot image classification using coupled dictionary embedding. Mach. Learn. Appl. 100278 *8* July 2021 (2022). https://doi.org/10.1016/j.mlwa.2022.100278

Rostami, M., Spinoulas, L., Hussein, M., Mathai, J., Abd-Almageed, W.: Detection and continual learning of novel face presentation attacks. In: IEEE/CVF International Conference on Computer Vision, pp. 14851–14860 (2021)

Serrano, S., Gomez, J. F. H.: (n.d.). *Cocoa Diseases (YOLOv4)*. Kaggle. Retrieved April 30, 2022, from https://www.kaggle.com/datasets/serranosebas/enfermedades-cacao-yolov4?resource=download

Shu, D., et al.: Development and validation of a deep learning system for diabetic retinopathy and related eye diseases using retinal images from multiethnic populations with diabetes. JAMA **318**(22), 2211–2223 (2017). https://doi.org/10.1001/jama.2017.18152

Shuai-qun, P.A.N., Jing-fen, Q., Rui, W., Hui-lin, Y.U., Cheng, W., Taylor, K.: Intelligent diagnosis of northern corn leaf blight with deep learning model. J. Integr. Agric. **21**(4), 1094–1105 (2022). https://doi.org/10.1016/S2095-3119(21)63707-3

Simonyan, K., Zisserman, A.: Very deep convolutional networks for large-scale image recognition. In: Proceedings International Conference on Learning Representations, pp. 1–14 (2015) ArXiv Preprint ArXiv:1409.1556

Sladojevic, S., Arsenovic, M., Anderla, A., Culibrk, D., Stefanovic, D.: Deep neural networks based recognition of plant diseases by leaf image classification. Comput. Intell. Neurosci. 1–11(2016). https://doi.org/10.1155/2016/3289801

Vijayan, T., Sangeetha, M., Kumaravel, A., Karthik, B.: Fine tuned VGG19 convolutional neural network architecture for diabetic retinopathy diagnosis. Indian J. Comput. Sci. Eng. **11**(5), 615–622 (2020)

Zeiler, M.D., Fergus, R.: Visualizing and understanding convolutional networks. In: European Conference on Computer Vision, pp. 818–833 (2014)

Impact of Promotional Strategies on Sustainable Business Growth in a Selected Wine Processing Companies in Dodoma City, Tanzania

K. Seme and P. Maziku(✉)

College of Business Education Dodoma Campus, P. O. Box 2077, Dodoma, Tanzania
pemaziku@yahoo.co.uk

Abstract. Purpose: Evaluated the impact of advertising strategies on sustainable growth of Dodoma wine makers in Tanzania. In particular, the study achieved the following objectives to: Evaluate the role of promotional strategies on international operations for Dodoma wine processors and identify the challenges faced by the Dodoma wine processors.

Design/Methodology: Random sampling techniques applied in selection forty participants from four selected wine companies in Dodoma city and interviewed at a single point of time. Interviews, questionnaires and written review methods were used for data collection. To assess the contribution of sales strategies on the growth of wine industry, the multiple regression models was applied.

Findings: They found that, discount coupons, average customer order size, and goal attainment had a positive impact on the wine store's performance. They show that changing the homogeneity of the three metrics improved performance by 71.2%, 18.9%, and 39.4%, respectively. On the other hand, poor application of global pricing strategies, lack of response to market impulses, inadequate means of marketing products, high export costs, unfriendly policies related to exports and cultural differences, and ultimately unfavorable global pricing strategies were challenges facing countries.

Research Limitation/Implications: The survey results included only four selected wineries in the city of Dodoma, not all wineries in the country. Using a case study design limits our ability to generalize insights on advertising strategy across the industry.

Field Implication: Insights gained from the study is useful for all Tanzanian manufacturing wine companies regarding the impact of promotional strategies to improve export and market strategies within and outside the wine sector.

Social Implication: Education gained from the research is helpful to governments to improve their current foreign trade policies and regulations targeting on the increase exports, reduce imports, encourage higher wine production and boost business performance. Also, wine processing companies should design and implement proper promotion strategies that create customer awareness and loyalty on their products at the global market.

C. Aigbavboa et al. (Eds.): ARCA 2022, *Sustainable Education and Development – Sustainable Industrialization and Innovation*, pp. 723–730, 2023.
https://doi.org/10.1007/978-3-031-25998-2_56

Originality/Value: The study value stays on its ability to explain the marketing promotion stimulus that can be instituted by wine producers and exporters to capture the foreign markets through increased sales.

Keywords: Business · Performance · Promotional strategies · Sales · Tanzania

1 Introduction

Global production and market demand for wine and its products have increased dramatically in recent years. This argument is supported by research by Rakula (2016), who said that the wine market worldwide grew at a rate of 6.4%. The surge in demand was coupled with projected revenues of $417.85 billion in 2020 to $444.59 billion in 2028. However, about 77% of global wine production originated from only eleven (11) giant countries from Europe, America and Asia (Festus 2016; Lee et al. 2013). This market is dominated by these big countries and limiting other countries to venture in the market. This situation creates a hardship for a small county like Tanzania to penetrate into such international market with a throat competition. Despite of all these, some wine companies in developing countries including Tanzania are employing varieties of sales promotion strategies as means of penetrating in this competitive market (Festus 2016; Lee et al. 2013).

The production and market demand for wine and its products are increasing all over the world. According to Festus (2016), merchandising has been a key tool and competitive strategy for marketers during his 20 years. This is because they often act as a tool to create additional reasons for a target customer to choose one brand over another. To survive in a competitive business environment, it is essential to use your advertising strategy as a means of attracting and retaining customers that leads to long-term connections and increased productivity. I have emphasized one thing. Additionally, promotion, defined by Kotler (2013) and Barry and Howard (2015), is a marketing technique intended to inform, influence, and remind potential customers of the advertised product. However, finding by Kipturgo (2019) further define merchandising as the best measure of merchandising strategy performance because it is cost-effective and time-efficient. A well-designed and executed promotional strategy can therefore determine market performance and company growth.

Tanzania is her second largest among the major wine-producing countries in Africa. Currently, the Dodoma region has an annual grape production capacity of 9,361 tons (Epaphra 2016; Kipturgo 2019). Despite this production capacity, the country yet is not doing well when it comes to exporting wine. For example, only 54,118 tons (6,435 tons) of wine were exported from Tanzania in 2018 (TRA 2018; Trend Economy 2018). The decrease in exports of wine may be related to the low use of marketing strategies to international markets by domestic wine makers. This means that Dodoma City wine processors are not taking full advantage of the market opportunities available within the region and the world region. Poor application of these promotional strategies will prevent wine processing companies from generating more sales from the available markets at both the regional and global levels.

Moreover, review on the past studies on the impact of promotion tools on the business growth concentrated only answering questions in a simple manner (Kipturgo 2019; Kipkirui (2017). We've been focused, and we need to be. It does not provide sufficient literature for development. The limited literature makes it difficult to understand how promotional methods are linked to wine processor growth. Therefore, the research examined the impact of marketing measures on the firm growth for the wine companies in Dodoma.

2 Review of Literatures

2.1 Theoretical Framework Review

The theories underpinning was used to support the study about influence of promotion tools to the business success of the wine makers in the study region. His two theories on secondary research included the market mix theory and the resource his base theory. They explain how wine processing companies can use available resources to develop successful commercialization strategies and take advantage of international market opportunities.

2.1.1 Marketing Mix Theory

In the 1940s, James Culliton proposed the marketing mix theory, under which the concept of the 4P marketing mix was introduced. In this theory, the four P's (product, price, promotion, and location) are considered to be the most important tools for competitive advantage and firm performance in improving service or product sales. (Borden 1964; Grönroos 1994). The theory suggests, revenue growth, new customers, goal attainment, social media engagement and brand awareness as Key metrics for measuring performance. Under this study, sales volume was used as indicator for measuring the business performance. However, this theory does not take into account the availability of in-house resources and skills. Therefore, the need for the theory of resource-based was added.

2.1.2 Resource-Based Theory (RBT)

The theory that a firm's competitive advantage is mainly determined by its endowment in natural factor is considered as a key factor in decisive for comparative advantage in the external market of competitors (Grant 1991). Based on the argument, it explains how a firm or business can use the resources and skills they already have or can acquire to maintain financial performance (Barney 1991). It therefore emphasizes the external area of the firm as the basis of its commercial interests and the resources built to compete in that environment. Moreover, this way of thinking explains why certain companies outperform others in the same industry. This means that variation in organizational performance are linked to changes in resources and skills.

It further stipulates that, a firm's competitive advantage is determined primarily by its endowments and ability is the key factor in assessing its comparative advantages (Grant 1991). The model explains how firms and businesses can use the resources and skills they already have or can acquire to maintain financial performance (Barney 1991).

It therefore emphasizes the company's external environment as the basis of its economic interests and the resources built to compete in this environment. Moreover, this way of thinking explains why certain companies outperform others in the same industry. This means, resource endowments differences in organizational performance are linked to skills and availability of natural attributes (Grant 1991).

3 Methods

3.1 Location

Dodoma the capital of the Republic of Tanzania was the location of the research, and included four wine processing companies as case studies. This area was chosen because it is the center of Tanzania's large-scale viticulture and wine processing operations (TRA 2018). Moreover, Dodoma is the area whereby grapes are grown in the two harvest seasons, March and August/September each year. In this regard, the region was thought to good for the study under question.

3.2 Survey Design and Sampling Procedures

Information on merchandising technology and international business performance was gathered at a single point. Targeted method was employed in selecting the participants from the four wine makers in Dodoma. The selection of the five wine companies was determined by the volume of international transactions in relation to the rest companies (TRA 2018; Epaphra 2016). Moreover, these companies have reported to underperform for years (Kulwijila et al. 2018; Kipturgo 2019).

A cluster sampling method was applied in selecting participants from the unit of every industry. Tiering was based on the number of employees and those working in external business units. A proportional random sample from each stratum was used to select respondents. Meanwhile, targeted selection was used for the selection of key informants, including the Dodoma wine company's top her manager. Targeted sampling has created a chance for the study to focus on specific aspects on the research work.

3.3 Data Collection and Analysis

Collected data related to promotion measures and business growth were gathered through questionnaires. The questionnaire included both open and closed questions. The Key Whistleblower Interview Guide was used to gather key whistleblower views on the impact of marketing tools on business growthy. Collected data were analyzed speculatively and descriptively using (SPSS). Results from simple descriptive statistics were presented using descriptive measures and features.

Empirical Model Presentation

Valuation of the impact of merchandising levels on the improvement of business of a wine maker of wine in Dodoma, the multiple regression was employed. This equation of analysis was chosen because business success was a non-categorical. The following equation was used:

$$\text{Performance}(Y) = \beta 0 + \beta 1DC + \beta 2AO + \beta 3GA + \varepsilon \quad (1)$$

From where:

DC = Discounted coupons
AO = Average Order Size
GA = Goals Achievement in % and
Ɛ = error.

4 Results and Discussion

4.1 Socio-economic Features of Respondents

Findings (Table 1) indicate that, male respondents were 58% while female were 42%. This meant that most respondents who worked in wine processing were men. This is because the nature of wine industry need person with more efforts like men.

We also found that about 30% of respondents were under the age of 30, followed by participants between the ages of 31 and 35, accounting for about 20% of the respondents. Additionally, only 17% of respondents said she was 46–50 years old or over 51 years old. This means that the wine processing activities carried out in the five wine companies require a young and strong workforce. Increase.

The results in Table 1 also show that, respondent with first degree were 30%, with approximately 18% and 13% of the respondents having a certificate and a master's degree, respectively. Only 15% of respondents had different levels of education. On the other hand, the 63% said to have 1 to 4 years old, while 23% have 5 to him 7 years old. This means that the seniority of employees in the five companies is relatively good, which help much to the proper implementation of the strategies.

4.2 The Impact of Sales Techniques on the Business Growth of Wine Makers

Model suitability tests, including the use of R-squared and variance inflation factor (VIF) calculations, were performed to measure the effectiveness of multicollinearity issues before conducting an analysis of the impact of promotions on business performance. The results in Table 3 show an R-squared estimate of 93%, indicating that the model is sufficient to explain the dependent variable. We also found that, the variation in business performance (sales volume) was 93% clarified by the determining variables. On the other hand, the VIF values are 1.76 and below 10, indicating no problem of multicollinearity between the independent variables. This allowed us to proceed with our analysis using multiple regression models.

The results, shown in Table 2, indicate that all three variables included in the analysis significantly explain changes in sales volume. Retail price discount coupons were positively correlated with performance with a factor of 0.22. This means that, a unit change in discount price units can improve business performance by 22%. This result

Table 1. Socio-economic features of respondents

Characteristic	Reply	Frequency	(%)
Gender	Male	23	57.5
	Female	17	42.5
Age	Less than 30 years	12	30.0
	31 to 35 years	8	20.0
	36 to 40 years	7	17.5
	41 to 45 years	6	15.0
	46 to 50 years	4	10.0
	Above 51 years	3	7.5
Education level	Certificate	7	17.5
	Diploma	10	25.0
	Bachelor's degree	12	30.0
	Master's degree	5	12.5
	Others (PhD)	6	15.0
Working experience	Less than a year	7	17.5
	1–4 years	25	62.5
	5–7 years	9	22.5
	8–10 years	5	12.5
	Above 10 years	4	10.0
	Total	**40**	**100.00%**

Source: Field Data (2021).

is consistent with his Kipkirui (2017) who found that most promotional strategies (free samples, discounts, bonus packages, etc.) influenced positively the habit of consumer.

Similarly, it was found that, order size was correlated positively to the performance. Implying that, for every unit change, the order size for a wine processing company could improve the performance by 36%. Therefore, the proper application of promotional strategies by business organizations is expected on attracting more customers from the global market who order more wine products from that country and improve the performance of their respective companies. These findings are consistent with those of Kipturgo (2019) who found that promotional strategies influenced directly the branding performance.

However, achievement of the company's target showed a significantly positive correlation with the performance of the wine maker industry factored at 0.21. Indicating company's performance improved by an average of 21% by changing the units of measure of achievement. This also shows that achieving the goals set is highly relevant to business success in the wine industry. These results are consistent to Li (2017), who found that Ningxia wine was difficult to perfect due to the lack of promotional strategies such as

Table 2. Relationship between promotional techniques and Business performance

Variables	Unstandardized coefficients		Standardized Coefficients	t	Sig.	VIF
	B	Std. error	Beta			
(Constant)	.201	.122		1.413	.141	1.716
Discount coupons	0.221	.075	0.394	5.253	.012	
Average order size	.363	.055	0.712	12.945	.000	
Goals achievement	.213	.074	0.189	2.554	.040	
Adjusted R2	0.934					

[a]Dependent Variable: Business Growth (sales volume).

publicity, advertising and direct his marketing. Personal sales and promotions influence business success. This conclusion was similar to his Tandoh and Sarpong (2015) study in Ghana. Also, the study by Tandoh and Sarpong (2015) supported finding from this study, that there was an association between promotions and firm performance.

5 Recommendation, Conclusion and Policy Implications

Marketing strategies in relation to company growth was observed in Dodoma wine processing companies. The study recommends that the provision of subsidies by the Tanzanian government to both wine producers and companies is essential to empower them and encourage exports to international markets. In addition, wine processing companies should be providing with adequate funds so as they can develop a sound able advertising strategy.

References

African Business: Tanzania: developing a good nose for wine. London headquarters, 7 coldbathcold bath square. J. Afr. Bus (2014). Approaches; Nairobi, Act Press. Print ISSN: 1522-8916

Agbonifoh, B.A, Nnoli, D.A., Knamnebe, A.D.: Marketing in Nigeria Concepts, Principles & Decisions, 2nd edn. Afritowers Limited, Aba (2007)

Barry, T.E., Howard, D.J.: A review and critique of the hierarchy of effects in advertising. Int. J. Advert. **9**(2), 121–135 (2015)

Daniel, C.O.: Effects of marketing strategies on organizational performance. Int. J. Bus. Mark. Manag. (IJBMM) **3**(9) (2018)

Diego, B., Stefano C., Gaeta, D.: Wine and web marketing strategies: the case study of Italian speciality wineries. In: 4th International Conference of the Academy of Wine Business Research, Siena, 17–19 July 2018 (2018)

Epaphra, M.: Determinants of export performance in Tanzania. J. Econ. Libr. **100**(110) (2016)

Festus, W.: The impact of sales promotion on organizational performance: a case study of Guinness Ghana Breweries Limited. Unpublished thesis submitted to Kwame Nkrumah University of Science (2016)

Kalimang'asi, N., Majula, R., Nathaniel, N.K.: The economic analysis of the smallholders grape production and marketing in Dodoma municipal: a case study of Hombolo ward. Int. J. Sci. Res. Publ. **4**(10), 1 October 2014, Spain (2014). ISSN 2250-3153

Khaniwale, E.: Advertising exposure and advertising effects: new panel based measurement: scale development and validation. J. Consum. Res. **19**, 303–316 (2015)

Kipkirui, J.B.: The effect of innovation strategies on market share of small scale tea packers in Kenya John Bett Kipkirui: a research project submitted in partial fulfilment of the requirements of master of science in entreprenuership and innovations management, school, November 2017

Kipturgo: Strategic role of branding in building competitive advantage at Kenya Wine Agencies Limited. Int. Acad. J. Hum. Res. Bus. Adm. (IAJHRBA) (2019). ISSN 2518-2374

Kolmar, E.: 10 largest wine companies in the world (2021). https://www.zippia.com/advice/largest-wine-companies/. Accessed 20 June 2021

Kotler, P., Armstrong, G.: Marketing principles, translated by Bahman (2006)

Kotler, E.: The effect of outdoors advertisements on consumers: a case study. Stud. Bus. Econ. **5**(5), 10–16 (2013)

Kulwijila, M., Makandara, J., Laswai, H.: Grape value chain mapping in Dodoma region, Tanzania. J. Econ. Sustain. Dev. **9**(2) (2018). https//www.iiste.org/journals/index.php/JEDS/article/view/40701

Lee, C.H., Hwang, F.M., Yeh, Y.C.: The impact of publicity and subsequent intervention in recruitment advertising on job searching freshmen's attraction to an organization and job pursuit intention. J. Appl. Soc. Psychol. **43**(1), 1–13 (2013)

Mbaga, M.H.: A research proposal submitted in partial fulfillment of the requirements for the degree of master of business management (MBA-corporate management) (2016)

Mkanda, M.: The role of packaging in positioning an orange juice. Journal of food mobile phone market: the Nigerian experience. Bus. Intell. J. Model. Electr. Mark **11**, 17–25 (2009)

Mugo, S.N.: Innovations and performance of Kenya's wine industry Silas Njeru Mugo a research project submitted in partial fulfilment of the requirements for the award of the degree of master of business administration (MBA), School of Business, University of Nai (2015)

Odhiambo, V.A.: Effects of marketing strategies on the performance of retail stores in footwear sector in Nairobi City County. Doctoral dissertation, University of Nairobi (2015)

Onyejiaku, C.C., Ghasi, N.C., Okwor, H.: Does promotional strategy affect sales growth of manufacturing firms in south east nigeria? Eur. J. Manag. Mark. Stud. (2018)

Pembi, S., Fudamu, A.U., Adamu, I.: Impact of sales promotional strategies on organizational performance in Nigeria. Eur. J. Res. Reflect. Manag. Sci. **5**(4) (2017)

Rakula, E.O.: Effects of marketing practices on the performance of Phoenix of east Africa Assurance Company Ltd. Doctoral dissertation, University of Nairobi (2016)

The Role of Land Demarcation in Addressing Conflicts Management Between Farmers and Pastoralists for Sustainable Agriculture in Kiteto District, Tanzania

P. Maziku(✉) and M. Mganulwa

Department of Business Administration, College of Business Education-Dodoma Campus, Box 2077, Dodoma, Tanzania
pemaziku@yahoo.co.uk

Abstract. Purpose: The research evaluated land parcels use to resolve conflicts between farmers and nomadic cattle keepers in Kiteto district located in Manyara region of northern Tanzania, in the pursuit of sustainable agriculture. In particular, their contribution to facilitating the dispute resolution process between the two groups was investigated.

Design and Research Methodology: A single point in time survey design was used, with 80 respondents randomly selected from the Kiteto district. Questionnaires, interviews, and documentary reviews were the methods of data collection, and a binomial logistic regression model was used in examine roles of land boundaries in resolving struggles between growers and nomads within districts.

Findings: Results of the study showed that, Land gazetting, grazing land plotting, availability market and cattle security were positively and significant related with the land demarcation.

Research Limitation/Implications: Conflict between farmers and pastoralist in Tanzania exist in many regions ranging from the central zones to the southern zones, therefore dealing only with one district out of many districts in Tanzania limits its coverage and generalization.

Practical Implication: The information and skills obtained from this research is resourceful to all government, pastoralists and farmers.

Social Implication: Land demarcation strategies presented from this study will be resourceful to policy maker and government in dealing with conflict resolutions management among farmers and pastoralists which has been the common phenomenon in the developing countries including Tanzania. Moreover, establishment and construction dams and other social services in the demarcated areas will help the pastoralist from retaining their animal and reduces the chance of moving a long distance looking for pastures and water.

Originality/Value: Use of field data and nature of respondents carries the originality and value of this study.

C. Aigbavboa et al. (Eds.): ARCA 2022, *Sustainable Education and Development – Sustainable Industrialization and Innovation*, pp. 731–738, 2023.
https://doi.org/10.1007/978-3-031-25998-2_57

Keyword: Conflict management · Farmers · Land demarcation · Pastoralists · Tanzania

1 Introduction

In the recent years, the world peace has been faced with many conflicts of different kinds and dimensions both at the local and global level. Conflicts of different sorts are common phenomena in many countries ranging from Europe to America, Africa to Asia, with Africa experiencing a lot of farmers and pastoralist's conflicts (Davis 2015). Finding from the study by Asaaga and Hirons (2019) indicated that, here is an increasing evidence of land conflict among famers and pastoralist in several SSA countries which threaten the world efforts of creating peace among citizens.

Similar to other SSA, land conflicts in Tanzania are on the increasing rate in many parts since during the planned economy era lasted between 1961 to 1985 which escalating of the same trends (Mwasha 2016). Most of the land conflicts in Tanzania are reported to be resource-based conflicts whereby different groups in the society especially in the rural areas are competing over resources like land, water, grazing pastures. The main sources of all these conflicts is based on accessibility and control power on major resources among the two groups (Massawe and Urassa 2016). Resource-based conflicts are the commonly experienced conflict in Tanzania mostly involved pastoralists and farmers or between natives and investors, reserved areas authorities and community. These conflicts have been reported to affect people's livelihoods in which some of them left with homeless and some have lost their lives (John and Kabote 2017).

In Manyara region for example, many conflicts have involved villages and Lake Manyara National Park (LMNP). Damage of agricultural crops and animals were the outcomes of these conflict which ended with restriction of grazing areas and limited buffer zone (Kaswamila 2010). Similarly, to other regions in Tanzania such as Morogoro and Manyara, Dodoma region has experienced an increase in land conflicts between farmers and pastoralists in the kongwa District (Mbonde 2015). Results from these conflicts were more deaths of people and animals from farmers and pastoralists in the way of struggling on land ownership.

To reduce the effects of these conflicts, the government of Tanzania has been implementing different projects in the district. For example, the government collaborated with International Livestock Research Institute (ILRI) which helped to obtain for about 150,000 hectares of grazing land for livestock in Kiteto District in Manyara region. The implementation of village land use plans involved four villages namely Amei, Loolera, Lembapuli and Lesoit (ALLOLE). This is one of the gazette grazing areas in Kiteto shared with four villages namely Amei, Loolera, Lembapuli and Lesoit. The names of the four village constitute the name of the area-ALOLLE. To legalize the ALOLLE grazing land, the joined villages signed the agreement in September 2018 which secured the area of 98,000 hectares for village livestock and other uses (Kasyoka 2018). Community members were hopeful that this agreement could potentially bring peace to their villages.

Despite, all these efforts made by Government and other stakeholders to secure grazing lands through demarcation and other agreement between farmers and pastoralists in Kiteto District, yet conflicts still exist and the district is viewed as the leading in farmers - pastoralists' conflicts in Tanzania. For stance, over 40 deaths of people in the late 1990s were witnessed and 20 deaths of people in 2000s and properties of farmers burnt to ashes, dozens of cattle stolen in 2008 (Baha et al. 2008; URT 2005). Together with a well demarcated grazing land, reports indicated that, there is ongoing conflicts the two groups in Kiteto district (Inter Press Service 2014). This contradicting situation poses a worry regarding the well-being in the villages which could disturb the rural livelihoods. This also creates the need of a scientific enquirer to find out the extent of the contribution of land demarcation strategies in solving among cultivators and nomads in Kiteto District. Findings from the research will help the policy maker and government to implement the right strategies in resolving various conflict in the country.

2 Theoretical Review

The study was guided by the Theory of conflict as developed by Karl Marx.

Conflict Theory
This theory states that, the constant increase in conflict in any society stems from struggles over scarce resources including land (Karl Marx 1883). Security and atmosphere in society are maintained by power and order as well as regulation (Chappelow 2019). Moreover, more power and more resources give groups an edge over the poor. This theory argued that maximizing utility is the primary goal of individuals and groups within society. According to Chappelow (2019), the theory is based on his four assumptions related to revolution and competition, structural inequality, and war struggle. Four assumptions solidify the causes of conflict in society, which stem from the unequal distribution of key resources and power between individuals and groups. Conflict is likely to be at a minimal level when rebels are evenly distributed in society in terms of gender, culture and religion (Idou 2017). This theory is said to be applicable to the present study, especially as it reveals the conflict between his two groups in the Kitet district. However, in this theory, the struggle between farmers and pastoralists over scarce resources, especially land, is the main source of conflict. Low-privileged groups that prevent groups from accessing available resources, especially land. Most of the time, it turned out that shepherds in the Kitet district were feeding crops to their livestock, thus destroying the crops and breaking the pact.

3 Research Methods

3.1 Location of the Study

Kiteto district was selected as a study area located on the northern part of Tanzania in Manyara region in which two wards namely Partimbo and Name lock were selected. The rationale of selecting Kiteto district lies on the fact that, a district had already experienced a number of conflicts between the two groups whereby in 2014 and 2018 deaths of 10 people leaves others 20 injured people witnessed. Moreover, about 60 houses, 6 motorbike and 53 bicycles were bunt (Saruni 2018).

3.2 Study Design and Sampling Method

The research used single point in time design in collecting information among farmers and pastoralists from Kiteto district. To select the two wards, non-probability techniques was used and the stratified proportion method employed in the selection of farmers and pastoralist in which the proportion random sampling was used in the selection of 80 respondents.

3.3 Data Analysis

To analyze the role of land demarcation, the Binary Logit regression model was adopted. The use of the model is due to reason that the dependent variable (Conflict Management) was a dummied into 1 = effective, 0 = Not effective). The Binary Regression equation was expressed as:

$$\Pr(Y = 1/X) = \beta_0 + \beta_1 X_1 + \beta_2 X_2 + \beta_3 X_3 + \beta_4 X_4 + \beta_5 X_5 + \ldots\ldots\ldots\ldots\ldots\ldots \varepsilon_i$$

where:

Y – Dependent variable.
X_1 to X_5 – Independent (explanatory) variables
X_1 – Gazetting of grazing land
X_2 – Plotted grazing areas are divided into plots used one time at a time.
X_3 – Existence of a market for cows and their products
X_4 –Security in the demarcated area
X_5 – Availability of good climate that avails water and pasture in the demarcated area
ϵ – Residual (error)

4 Findings and Discussion

4.1 Descriptive Findings

4.1.1 Characteristics of Respondents

Results on gender indicate that majority (61.3%) of the sampled respondents were males while 38.8% of them were females. Indicating that, most of the decisions at the families related to farming and livestock were governed by males. These finding are similar to those of Mwalimu and Matimbwa (2019) who found that, the gender of nomads and famers in the study area were 55% and 46% for male and female respectively.

The results from Table 1 indicated that, 40% of the respondents had attained secondary education. About 30% of the respondents had acquired a primary education while only 10% of the total number of respondents had obtained other forms of education such as college, university and professional diploma education. These results indicate that, in pastoralist and farming community's education is yet given a higher priority as compared to farming.

These results are consistent with those of Mwambashi (2015) who found that approximately 33.75% of respondents had formal education and 28.75% had primary

Table 1. Demographics features of respondents

Variables	Frequency	Percentage
Sex		
Male	49	61.3
Female	31	38.8
Level of education		
Informal education	16	20
Primary education level	24	30
Form four and Six education	32	40
University/College	8	10
Age category (Years)		
Below 20	9	11.2
20–29	23	28.8
30–39	28	35.0
40–49	12	15.0
50 and above	8	10.0
Marital status		
Single	24	30.0
Married	50	62.5
Widowed	4	5.0
Divorced	2	2.5
Main occupation		
Farming	38	47.5
Livestock keeping	42	52.5

Sources: Field data (2020).

education. Only 16.25% of her and 21.25% of her have completed secondary and university education. As a result, most of the farmers and pastoralists in the region were found to be poorly educated. This may also be the cause of ongoing land disputes. This result supports his Adisa (2012) claim that the level of education of farmers and nomads determines the level of conflict within a district.

In respect to age, majority (88.8%) of respondent were aged above 29 while 11.2% were aged below 20 (Table 1). This implies that, farming and pastoralist activities were dominated by youth and adults as it is a labour intensive and need energetic person. It is at this age that people are very strong and do carry out serious crop growing or pastoralism but at the same time that age bract easily gets involved in fighting and conflicts. These results are in line with those of Mwasha (2016), who found that approximately 40.0% of respondents belonged to the 29–39-year-old category and 20.8% of respondents follow under the rest categories.

The results further stipulate that about 62.5% of respondents are married and 30% are single, compared to only 5% who are widowed and 2.5% who are divorced. The results show that there is a large disparity in marital status between the two group in the study area, with a largest proportion of respondents being married. The findings are similar to those of her Mwasha (2016), who found that the majority of respondents (72.5%) were married and only a minority (0.8%) were divorced. The findings reflect that in African conditions, especially in peripheral location, marriage is seen as an important which allow two people to make a family. This will be the main labor force for agricultural activities. This finding is also supported by her Mutayoba (2011), who found that stable families are more productive than unstable ones, and thus may affect agricultural production.

Regarding occupation, the results in Table 1 show that 47.5% of respondents are farmers and 52.5% are pastoralists. This means that within the field of study, the two activities were the primary occupations for the majority of respondents. This was reflected in a small gap in numbers growers and nomads. Furthermore, the findings indicated that most respondents were pastoralists, although the numerical differences between pastoralists and farmers were minimal. This findings concur to that of Mwalimu and Matimbwa (2019) that 50.9% of respondents were engaged in agriculture and 49.1% were nomads in Kambala village, Morogoro. Suggesting that two groups in the study area found to be approximately evenly distributed. This means that he has a year-round competition between the two groups for arable and grazing land. This finding is also in line to Mwasha (2016), who investigated the battle between growers and nomads in the Kilosa district of Tanzania, finding that use of land plan is directed towards two main groups.

4.2 Roles of Demarcation of Land on Resolving Conflicts the Two Groups in the Study Area

To determine the impact and depiction of conflict management among nomads and growers in the study area, binary logit regression was employed (Table 2). However, Hosmer and Lemeshow's tests were performed to test the relevant of the model. The results showed that the predictor gave x2 (1) out of 12, which is insignificant (p = 1,000) and means the data fit the binary logistic model well. Table 2 shows the results of the model showing that all variables are significant and illustrates the effect on the dependent variable.

Land gazetting is positive (0.284) and significant in terms of land boundaries in resolving farmer-nomadic disputes. This means that a 28.4% increase in surface surveys translates into an increase in effective conflict management when dealing with land disputes in the Kiteto district. Additionally, an odds ratio of 1,000 was determined. This indicates that land is one times more likely to be sighted to improve conflict management. These findings are consistent with Ibrahim and Chaminda (2017), who often complained that clear demarcations of farmland and grazing lanes throughout the region had no indication of where the actual farmland and grazing land were. It confirms that it was an eye-opener for land users. It was a route to follow.

For managed rangelands, the coefficient of managed rangelands was found to be positively correlated with land demarcation in conflict resolution (0.207) and statistically significant ($p < 0.05$). This means that increasing the use of allotted pasture by 1 unit will

Table 2. Binary logistic regression results on effects of demarcation of land in addressing conflicts (N = 80)

Variables	B	S.E.	Sig.	Exp(B)
Gauting of land	0.284	0.482	0.000	1.000
Plotted grazing land	0.207	0.120	0.001	1.000
Market presence	0.182	0.610	0.046	1.458
Animal Safety	0.191	0.953	0.015	1.614
Climate	0.136	0.775	0.058	1.000
Constant	-0.452	0.644	0.035	0.000

Omnibus Test ($\chi2$) = 21.5(3), p= 0.000

Hosmer and Lemeshow ($\chi2$) = 12(1), p=1.000

Pseudo R^2: Cox and Snell R square = 0.616, Nagelkerke R square = 0.712

-2 Log Likelihood =85.28

Dependent variable (1= If conflict management is effective, 0 = If conflict management is not effective

help the dispute resolution process by 20.7%. The odds ratio is therefore 1.000, indicating that managed rangelands are about 1 time more likely to address land disputes (Table 2). The results further indicate that cultivated pastures play an important role in resolving land disputes in the Kiteto district.

Furthermore, regarding the presence of markets, the coefficient was found to be positively correlated (0.182) with land representation in conflict management and statistically significant ($p < 0.05$). Land dispute resolution increased by 18.2%. The odds ratio value is 1.458, suggesting that there is a cattle market to resolve land disputes about 1.5 times more likely (Table 2). This result suggests that the existence of the livestock market has a positive impact on land demarcation for resolving land disputes in the Kiteto district.

The results, shown in Table 2, show that livestock safety is positively associated with conflict management with a factor (0.191). This means that a unit increase in livestock safety results in a 19.1% increase in effective conflict management. Additionally, an odds ratio of 1.614 was found, suggesting that livestock were 1.6 times more likely to be protected to reduce land disputes. These results suggest that enhancing livestock security will have positive consequences for land demarcation to address land disputes between farmers and nomads in the Kiteto district. The findings are consistent with those found in the 1999 Tanzania Land Policy: 5 Sect. 2, which states that the original of the conflict between producers and livestock keepers is the unsecured in terms of the land.

5 Conclusions and Policy Implications

The study concluded that the disclosure of land, cultivated rangelands, the existence of markets, and animal safety contributed significantly to the solving land disputes among two parts in the Kiteto district. Increase. She therefore advocates for the Ministry of Agriculture and Livestock and other stakeholders to develop plotted grazing areas in the Kiteto district to save nomads who have to travel long distances to access some of the most important social services. It recommended the establishment of social services around the area.

References

Asaaga, F.A., Hirons, M.A.: Windows of opportunity or windows of exclusion? Changing dynamics of tenurial relations in rural Ghana. Land Use Policy **2019**(87), 104042 (2019)

Chappelow, J.: Conflict Theory (2019). https://www.investopedia.com/terms/c/conflict-theory.asp. Accessed 23 Nov 2019

Davis, M.: Persistence of farmer-herder conflicts in Tanzania. Int. J. Sci. Res. Publ. **5**(2) (2015)

John, P., Kabote, S.J.: Land governance and conflict management in Tanzania: institutional capacity and policy-legal framework challenges. Am. J. Rural Dev. **5**(2), 46–54 (2017). https://doi.org/10.12691/ajrd-5-2-3

Ibrahim, B., Chaminda, A.: Effective strategies for resolution and management of farmers-herdsmen conflict in the north central region of Nigeria. Global J. Hum. Soc. Sci. Polit. Sci. **17**(2), 12–24 (2017)

Idowu, A.O.: Urban violence dimension in Nigeria: farmers and herders onslaught. AGATHOS Int. Rev., 187–206 (2017)

International Fund for Agricultural Development (IFAD): Women and pastoralism. Livestock Thematic Papers Tools for project design (2010). http://www.ifad.org/lrkm/factsheet/women_pastoralism.pdf. Accessed 11 May 2020

Kasyoka, S.: Securing rangelands and settling conflicts through village land use planning in Tanzania (2018). https://www.ilri.org/research/annual-report/2018/securing-rangelands-and-settling-conflicts-through-village-land-use. Accessed 19 2020

Massawe, G., Urassa, J.: Causes and management of land conflicts in Tanzania: a case of farmers versus pastoralists. Uongozi J. Manag. Dev. Dyn. **27**(2), 45–68 (2016)

Mwalimu, W., Matimbwa, H.: Factors leading to conflicts between farmers and pastoralists in Tanzania: evidence from kambala village in morogoro. J. Bus. Manag. Econ. Res. (JOBMER) **3**(8), 26–39 (2019)

Mwambashi, R.: Assessing the impact of land conflict between farmers and pastoralists in Tanzania: a case of ulanga district council. A dissertation submitted in partial fulfillment of the requirements for the award of Master's degree on public administration of Mzumbe University (2015)

Mwasha, D.: Farmer-pastoralist conflict in kilosadisrtict, Tanzania: a climate change orientation. A dissertation submitted in a partial fulfilment of the requirements for the degree of master of arts in rural development of Sokoine University of Agriculture. Morogoro, Tanzania (2016)

Saruni, P.: Post conflict coping strategies and well-being of farmers and pastoralists in kilosa and kiteto districts, Tanzania. A thesis submitted in fulfillment of the requirements for the degree of doctor of philosophy of Sokoine University of Agriculture. Morogoro, Tanzania. (Unbublished manuscript) (2018)

United Republic of Tanzania (URT): Statistical Abstract 2017. NBS, Ministry of Finance, Dar es Salaam, 106 (2016). istmat.info/files/uploads/statistical.pdf

Users' Satisfaction of Autorickshaw Transport Operations Towards Sustainable Intra-city Mobility, Cape Coast, Ghana

S. B. Adi[1(✉)], C. Amoako[3], and D. Quartey[1,2]

[1] Regional Transport Research and Educational Centre (TRECK), Department of Civil Engineering, Kwame Nkrumah University of Science and Technology, Kumasi, Ghana
solomon.adi@cctu.edu.gh

[2] Department of Civil Engineering, Cape Coast Technical University, Cape Coast, Ghana

[3] Department of Planning, Kwame Nkrumah University of Science and Technology, Kumasi, Ghana

Abstract. Purpose: In the global south, the gap left by a declining public transport service provision is currently being filled by the use of autorickshaw as an intermediate public transport mode. This study aimed to assess commuter satisfaction of autorickshaw in Cape Coast city.

Design/Methodology/Approach: The study employed a descriptive survey design of which a total of 384 commuters were selected purposively in the study area.

Findings: The findings show that females' commuters (61%) outnumber their male counterparts in the study area and majority of the respondents were students (50%), having low fares (43%) and ease of accessibility (91%) as principal user satisfactory reasons for its patronage relative four wheelers (taxis) for intra-city transport.

Implications/Research Limitations: Concerns of safety and discomfort were expressed by some commuters but could not be investigated; however, their mobility needs and the readily available autorickshaw to meet that need was paramount. Further studies on operatives' characteristics and operations could further add to the data build-up for review and decision making.

Practical Implications: An amendment of the policy on autorickshaws to have their operations regulated and restricted to some parts of the city with very low accessibility will provide a secure and sustainable intermediate public transport and will aid training, regulation and monitoring of their operations to meet city basic mobility needs. In addition, regulations pertaining to permits to operate and safety also need urgent attention.

Originality/Value: No known study has been conducted in Cape Coast on the subject matter and past studies have not extensively covered the user's satisfaction of autorickshaw transport operations in the Ghanaian context.

Keyword: Auto-rickshaw · Commuters · Patronage · Transport mode · User satisfaction

C. Aigbavboa et al. (Eds.): ARCA 2022, *Sustainable Education and Development – Sustainable Industrialization and Innovation*, pp. 739–751, 2023.
https://doi.org/10.1007/978-3-031-25998-2_58

1 Introduction

Cities in the global south have been flooded with emerging autorickshaw transport services at the backdrop of current systems failure to provide adequate, comfortable and safe intra-urban public transport services (Abane 2011; Agyemang 2015). Viewed as the new players in Africa's public transport sector, accommodating the influx of autorickshaws transport in cities of the global south need to be based on plan, considering road capacity and future travel demand. (Pramanik and Ashrafuzzaman Rahman (2019); Obiri-Yeboah et al. (2021)).

Modes of travel, duration of travel and respective distances travelled within a city vary widely across the globe (Heblich et al. 2020). Most people in least-developed countries continue to walk to work, followed by the use of bicycles and while in the middle-income countries, motorcycles are a commonplace (Paumgarten 2007). The next relative affordable technology adopted is the use of autorickshaw. An autorickshaw is a motorized version of the pulled rickshaw or cycle rickshaw. Most have three wheels and do not tilt. They are known by many terms in various countries including auto, auto-rickshaw taxi, baby taxi, pigeon, bajaj, chand gari, Lapa, tuk-tuk, Keke-napep, 3wheel, bao-bao or tukxi (Kumar et al. 2016). In Ghana an auto rickshaw is referred to as *Pragya* (Tuffour et al. 2014). There seem to be an increase in the use of auto rickshaw in Ghana, especially in towns with issues of timely availability, accessibility and affordability of other traditional mode of transport.

An auto-rickshaw is a common form of intra-urban transport in Cape Coast city, the regional capital of the second most densely populated region in Ghana (2910 pp per sq km), as a vehicle for commercial uses though they are licensed for private use by the Driver Vehicle and Licensing Authority (DVLA). The city, by its geographical location, is the centre of major tourist sites in Ghana and is home to a lot of distinguished educational institutions (Porter 2013). Porter (2013) again posit that Cape Coast has three principal motorized modal choices: private cars, taxis, and trotros, however, preliminary surveys conducted at the central business districts have seen autorickshaws transport pitching one-third of the total motorized transport modes on the Cape Coast town roads. Cape Coast therefore was a good testing ground for this study There are many different auto rickshaw types, designs, and variations. The most common types in Cape Coast are characterized by a sheet metal body or open-frame resting on three wheels; a canvas roof with drop-down side curtains. Despite the safety concerns due to its design, there has been a steady increase in the commuter patronage of autorickshaw. A reconnaissance survey and physical evidence on Cape Coast city roads show a high density of commercial autorickshaw plying major roads around Abura and Kotokuraba, the major business centres of the city.

There are perceived concerns associated of autorickshaw transport services. They are perceived to be unsafe, because the vehicle itself is seen as hazardous and the drivers, as poor vehicle operators having little or no formal driver training. These drivers are willing to overload their vehicles and at the same time make many trips in short periods. Autorickshaws are seen as unstable and liable to turn turtle, due to a thrust from another vehicle, a heave or pothole in the road, or overly rapid speeds by the drivers, with the passengers being thrown against the sparse, hard metal interior, whose sole soft surfaces are the rear passenger bench and the driver's seat (Kunle and Farah 2014). The lack of doors means that the occupants could be thrown onto the road, with the potential for serious injury, even at low speed. (Khayal 2019; Harding et al. 2016). Autorickshaw accidents are caused by many factors such as driver's poor level of education and substance use (Manglam et al. 2013). To reduce injury risks and increase the safety of autorickshaw, velocity must be reduced, because drivers who over speed the limits, break rules, and carry extra persons are prone to accidents (Kunle and Farah 2014). Accidents that occur in developed countries such as Europe are comparatively smaller than in developing countries because of city planning and urban infrastructure (Khaled et al. 2021). Indiscriminate parking and picking passengers, their increasing numbers, over-charging, rude behavior, continuous availability in some villages, safety concerns, poor suspension, and its higher susceptibility to rain, drivers being potential threat to women are some other concerns observed in their operations (Harding et al. 2016; Shlaes and Mani 2014).

These externalities notwithstanding, autorickshaw transport services cater for the transport needs of a large portion of populations (Shlaes and Mani 2014). The majority of people use autorickshaws to commute to work, shopping, and getting children to and from school at relatively lower cost, convenience, speed, and door-to-door service delivery (Hossain and Susilo 2011) and it's a good source of income for drivers as some earn a monthly average income of $ 150 (Rabeya and Khatun 2016). Moreover, because of its affordability, and availability, autorickshaw is highly preferred by many social groups such as middle-class people, females, and older people (Hossain and Susilo 2011).

These concerns notwithstanding, the sustainability of a product or service is the assurance of repeat business. Consumer satisfaction and possible expectation of the autorickshaw is yet to be ascertained in cities in Ghana. Understanding the needs and concerns of the commuter is essential to improve upon the nature and quality of service delivered by players in the autorickshaw business. Therefore, this study seeks to assess commuter satisfaction of autorickshaw transport operations towards sustainable intra-city mobility.

2 Research Methodology

A descriptive research design was used in this study. Of a 189,925 total population of the Cape Coast metropolis (Ghana Statistical Service 2021) is a total of 384 persons were purposively sampled for data collection (Gill et al. 2010). A structured questionnaire was administered to the respondents and personal observations were conducted from some major routes of operation of autorickshaws. Data on the demographic characteristics of commuters, autorickshaw accessibility, patronage indicators, perceived safety and comfort were collected within a six week period using similar parameters used by Karunanithy Degeras and Suliman (2021) in their study. Questions were interpreted for respondents who could not read. SPSS (Version 21) was used in the analysis of data. The road network areas for the data collection are as outlined in Fig. 1

3 Results and Discussions

3.1 Demographic and Commuter Characteristics

Cape Coast city has road transport as its sole mode of intra-city passenger and freight transport as per the 95% and 90% respectively on the national scale. Inferences from observations, and interviews confirms limited road capacity and poor transportation management and the operations are largely private sector-led, consisting of large informal operators, who account largely for the modal share with no operational standard for services (Republic of Ghana 2017). The demographic characteristics of respondents presented in Table 1 indicates, the males were in the minority constituting 39% whiles the females constituted 61% of the respondents. These results largely reflect the 2021 census data in Ghana which reports of more females than males in the country (GSS 2021). It further reflects that more of the vulnerable in society, which are women and the youth, patronize the services of autorickshaws.

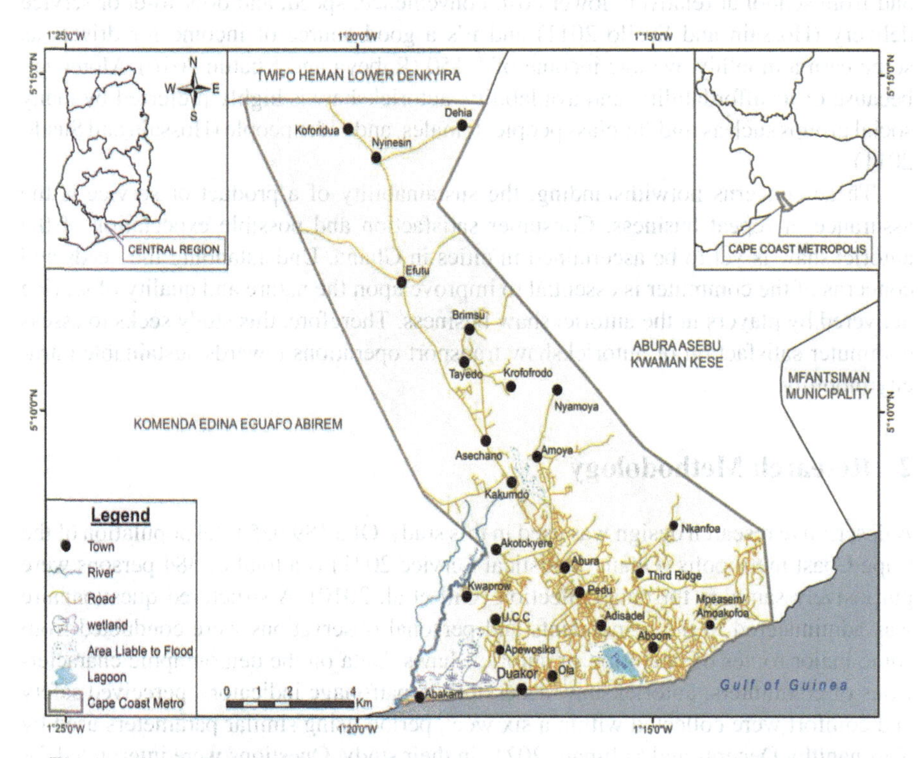

Fig. 1. Distribution of road network within Cape Coast MetropolisSource: (Andoh 2014)

Commuters were generally in the youthful age bracket of between 18 to 35 years (69%) of the respondents sampled similar to findings of Harding et al. (2016). Commuters below 18 years formed 10% of the total number of respondents and those above 50 years were the least of the respondents. Majority (50%) of these commuters were students

Table 1. Demographic & commuter characteristics

Demographic characteristics	Frequency	Percentage (%)
Gender		
Male	149	39
Female	235	61
Total	384	100
Age		
Below 18yrs	38	10
18–25yrs	172	45
26–35yrs	91	24
36–50yrs	59	15
Above 50yrs	24	6
Total	384	100
Occupation		
Student	192	50
Civil servant	86	23
Trader	32	8
Business man	39	10
Unemployed	35	9
Total	384	100
Numbers of commuter days		
1	10	3
2	4	1
3	4	1
4	14	4
5	234	61
6	67	17
7	51	13
Total	384	100
Educational level		
Basic school	23	6
Junior high school	105	27
Senior high school	100	26
Tertiary	144	38
No formal education	12	3
Total	384	100

with the unemployed being the least among the respondents (9%). A total of 41% of the respondents were employed. This implies that up to 59% of the respondents are financially dependent and may want to resort to the cheapest means of transport as much as possible. On the average, the number of commuting days for the respondents was 5.25 days. Majority (61%) of the respondents' commute to and from their works places or school at least 5 days a week while a total of only 5% of the respondents commutes up to 3 days a week. The frequency of commuting also has financial implications and transportation costs can also be driven by changes in policy according to Chingos et al. (2017). Only 3% of the respondents did not have any formal education. However, up to 97% of the respondents have at least a basic education (Table 1) which is expected to influence their modal choice according to Dingil and Esztergár-Kiss (2022) in cities. This choice was however influence largely by low fares (43%) and availability to offer trotro service to anywhere permissible (65%) and their accessibility within 5 min of waiting (91%).

An association between demographic characteristics of respondents and the most used commuting modes are presented in Table 3. There was a significant relationship between all the demographic attributes and the commuter's mode of transport ($p < 0.05$). Among the respondents, majority of the males (44.30%) commute by autorickshaw and only about 0.05% of the males walk to school and or work. Majority of the females however, mostly use taxis (44.68%) while females who mostly walk form the second largest (21.70%) among the respondents. Females who use autorickshaw constituted about 19.14% of the female respondents and unlike the findings of Shlaes and Mani (2014) did not express concerns of sexual harassment but rather of safety. It is worthy to note that walking is used mainly by female commuters than the males. This could be that most of these females who walk are traders and hawkers. Majority of the youth (18–25 years) and respondents who were educated up to the tertiary level use autorickshaw (Table 3). Usage of all the modes of transport were highest for respondents who commute at least five days in a week, giving an indication that there is more movement during the weekdays when students commute to school, workers to work, and traders to the market. Traders patronized autorickshaws as their most commuting mode relative to the other modes of transportation. This could be due to their ease of accessibility, and facilitation of the first and last mile services to and from the market (Table 2).

There was an equal patronage of the use of private vehicles and autorickshaws for business men and women which suggests their satisfaction as they may have considered it a very flexible mode of transport to aid their business endeavours. There is a sharp decline in the use of autorickshaw with increment in age and a sharp upsurge in the use of taxi within the age brackets of 26–35 years as these are perceived to be within the working class and validated by the number of civil servants (97) and many of the females within this age bracket may also have their babies on board and so their numbers fell.

3.2 Commuter Patronage Behavior

Users board autorickshaws more at undesignated stops (road side) (61%) on the Cape Coast roads, suggesting they stop anywhere at the convenience of commuters (Table 3). This is confirmed by the high number of respondents (32%) indicating they do not board autorickshaws from any designated station (Table 3, 4 and 5).

The *trotro* mode of patronage of autorickshaws (65%) tells of its accessibility for commuters to pick it as trotro. Due to the low fuel consumption of the autorickshaws to the taxis, it is able to make a lot of trips even with a single commuter on board per trip hence the findings of being accessible within 5 min of commuter waiting. The contract and calling services were rather the lowest which suggests that commuters do not go out of their way to deliberately engage them but rather find them very assessable to meet their transport needs which may not necessary mean they are satisfied with their services. The contract and calling services may refer to those engaging their first and last mile services. Mostly in the course of the trip, others board and alight making economic sense to take the risk of moving while only one person is on board. Even when no other person's board during a trip, the fuel consumption is not as the four-wheeler taxis.

The main safety concern of commuters has to do with tilting (46%) as a driver loses control and medium, high and very high noise levels (85%). These mostly results due to over speeding or distractive driving (Harding et al. 2016) and smaller and relatively shorter mufflers than 4-wheelers, however, these do not appear to deter patronage as it continues to rise comparatively to the 4-wheelers when placed on the same scale.

Commuters most of whom are literate (76%) do not know who regulates the operations of their means of mobility in the city. This, against the backdrop of the poor enforcement of road traffic regulations and the illegality of operations of autorickshaws makes it a fertile ground for anyone to start its operations without the requisite experience, training and licensing. Though its incumbent on government, metropolitan and district assemblies to evolve policies and byelaws to govern their respective jurisdictions, these mandates are implemented in an ad hoc manner (Ojo et al (2014); Andoh (2014)) and reactive manner, mostly in response to public outcry rather than a proactive

Table 2. Association between demographic characteristics and most used mode of transport of respondents

Description	Transportation mode					X^2	p-value
	Private vehicle	Taxi	Auto rickshaw	Walking			
Gender					Total		
Male	35	41	66	7	149	48.60	0.00
Female	34	105	45	51	235		
Total	69 (18%)	146 (38%)	111 (29%)	58 (15%)	384		
Age							
Below 18yrs	1	3	5	29	38	290.31	0.00
18–25yrs	45	50	74	3	172		
26-35yrs	0	81	6	4	91		
36-50yrs	13	9	17	20	59		

(*continued*)

Table 2. (*continued*)

Description	Transportation mode					X^2	p-value
	Private vehicle	Taxi	Auto rickshaw	Walking			
Above 50yrs	10	3	9	2	24		
Total	69	146	111	58	384		
Educational level							
BL	3	6	14	0	23	95.08	0.00
JHS	19	21	28	37	105		
SHS	13	52	17	18	100		
TL	31	67	44	2	144		
NFE	3	0	8	1	12		
Total	69	146	111	58	384		
Number of commuter days							
1	5	0	5	0	10	65.43	0.00
2	0	3	1	0	4		
3	3	1	0	0	4		
4	3	9	2	0	14		
5	27	106	74	27	234		
6	18	12	18	19	67		
7	13	15	11	12	51		
Total	69	146	111	58	384		
Occupation							
Student	26	23	56	13	192	138.88	0.00
Civil Servant	13	103	8	36	86		
Trader	0	7	20	5	32		
Business man	13	11	13	2	39		
Unemployed	17	2	14	2	35		
Total	69	146	111	58	384		

BL- Basic level; JHS- Junior High School; SHS- Senior High School; TL- Tertiary Level. NFE- No formal Education.

and well-thought-out plan and strategies based on scientific data and analysis (Roukouni et al. 2016) (Table 6).

The sudden rise in patronage of autorickshaws within four years (40% from 2%) of commencement of operations in the city of Cape Coast may be a confirmation of urban sprawl with a fast peripheral expansion of the city leading to an increased mobility demand, both in terms of private as well as public transportation which need to be accommodated in s sustainable manner as similarly indicated noted by Behl et al. (2018).

Table 3. Autorickshaw accessibility and safety

Description	Frequency	Percentage
Hailing location		
Road side	233	61
Loading station	151	39
Waiting time (min)		
5	351	91
10	32	9
15	1	0
Total	384	100
Commuters' nature of hailing		
Dropping	105	27
Contract service	9	2
Calling	19	5
Trotro	251	65
Total	384	100
Regulatory bodies		
CCMA	39	10
MTTD	20	5
DVLA	17	4
None	17	4
Don't know	291	76
Total	384	100

In order to sustain movement in the city in the face of urbanization, sustainability of the urban public transportation in the city which are mostly by taxis (38%) and as noted by Porter (2013) and have been complemented by the recent upsurge in autorickshaw patronage (29%) (Table 7).

Commuter accident experience (18%) with some being multiple in the past four years confirms the findings of Schmucker et al. (2011) involving motorized rickshaws in urban India and makes it a subject worth further study and examination. Commuters did appreciate risks such as tilting (46%), crash with 4-wheelers (25%) and crash with other autorickshaws (26%) but these did not deter them from continual patronage which could point to their lack of comprehensive understanding of the risk of autorickshaw transport as per similar findings of Schmucker et al. (2011) or disregard for the risks due to economic conditions (Guzman and Oviedo 2018) or ignorance.

Table 4. Patronage indicators and concerns of autorickshaw customers

Description	Frequency	Percentage
Concerns		
Tilting	178	46
Crashing with 4 wheelers	97	25
Crashing with another Auto rickshaw	99	26
Other		
Total	10	3
	384	100
Perceive noise level		
Low	56	15
Medium	119	31
High	134	35
Too High	75	20
Total	384	100
Patronage indicators		
Low price	167	43
Fast movement	108	28
Safety	24	6
Ventilation	42	11
Pleasure	39	10
Other	4	1
Total	384	100

Table 5. Commuter fares

Fare (GHS)	Frequency	Percentage
Autorickshaw		
1.00–2.00	376	98
2.10–3.00	5	1.3
3.10–4.00	3	0.7
Total	384	100
Taxi		
1.00–2.00	265	69.0
2.10–3.00	71	18.5
3.10–4.00	34	8.9
>4; <8	14	3.6
Total	384	100

Table 6. Commuters years of autorickshaw patronage

Years	Frequency	Percentage (%)
1	152	40
2	124	32
3	99	26
4	9	2
Total	384	100

Table 7. Commuters accident experience

Accident had	Frequency	Percentage (%)
0	316	82
1	33	9
2	13	3
3	12	3
4	6	2
5	4	1
Total	384	100

4 Conclusion

The paper attempts to assess the user satisfaction of auto-rickshaw transport operations towards sustainable intra-city mobility. The results indicates that there has been an increment in the patronage of auto-rickshaw in cities due to its affordability and its easy accessibility. It was derived that auto-rickshaw can further be developed into an effective and efficient sustainable transport operation system with the current decline in efficient and effective urban intra-city transport to some parts of the city as the city sprawls with its attendant externalities. That is, its availability, the potential implementation of door-to-door, first and last mile services as well as the affordability of their services makes the auto-rickshaw pivotal in the future of transportation in Ghana. The study further advances on the challenges and safety issues regarding the patronage of auto-rickshaws. It was noted that Auto-rickshaws adds to the level of congestion due to the absence of infrastructure, such as devoted auto-rickshaw parking space, which leads to reckless parking and indiscriminately picking of passengers resulting in halting traffic flow and increasing the risk of road accidents. Auto-rickshaw drivers are noted for zipping between vehicles leading to discomfort to other road users.

On this note, the government and all relevant stakeholders must consider a review of legislation governing their operations. A further study to consider their infrastructural needs; parking spaces, segregated lanes in our cities as well modalities for intensifying the enforcement of road traffic regulations for the safety of passengers and other road users in order to promote adequate, sustainable and affordable transport systems in the Cape-Coast city and Ghana at large.

References

Abane, A.M.: Travel behaviour in Ghana: empirical observations from four metropolitan areas. J. Transp. Geogr. **19**(2), 313–322 (2011). https://doi.org/10.1016/j.jtrangeo.2010.03.002

Agyemang, E.: The bus rapid transit system in the greater accra metropolitan area, Ghana: looking back to look forward. Nor. Geogr. Tidsskr. **69**(1), 28–37 (2015). https://doi.org/10.1080/00291951.2014.992808

Andoh, A.K.: Managing Traffic Congestion in the Cape Coast Metropolis, Ghana (2014)

Behl, A., Rathi, P., Ajith Kumar, V.V.: Sustainability of the Indian auto rickshaw sector: identification of enablers and their interrelationship using TISM. Int. J. Serv. Oper. Manag. **31**(2), 137–168 (2018). https://doi.org/10.1504/IJSOM.2018.094750

Chingos, M., et al.: Student transportation and educational access, pp. 1–35, February 2017. https://www.urban.org/sites/default/files/publication/88481/student_transportation_educational_access_0.pdf

Dingil, A.E., Esztergár-Kiss, D.: The influence of education level on urban travel decision-making. Periodica Polytech. Transp. Eng. **50**(1), 49–57 (2022). https://doi.org/10.3311/PPTR.16871

Ghana Statistical Service: Ghana 2021 PHC - General Report Vol 3A - Population of Regions and Districts (2021)

GSS: 2021 PHC: Provisional Results. Population and Housing Census: Provisional Results, 1–7, September 2021

Guzman, L.A., Oviedo, D.: Accessibility, affordability and equity: assessing 'pro-poor' public transport subsidies in Bogotá. Transp. Policy **68**(2017), 37–51 (2018). https://doi.org/10.1016/j.tranpol.2018.04.012

Harding, S.E., Badami, M.G., Reynolds, C.C.O., Kandlikar, M.: Auto-rickshaws in Indian cities: public perceptions and operational realities. Transp. Policy **52**, 143–152 (2016). https://doi.org/10.1016/j.tranpol.2016.07.013

Hossain, M., Susilo, Y.O.: Rickshaw use and social impacts in Dhaka, Bangladesh. Transp. Res. Rec. **2239**, 74–83 (2011). https://doi.org/10.3141/2239-09

Karunanithy Degeras, D., Suliman, K.R.: An Empirical evaluation of factors influencing the choice of mode for transportation in higher education institution using analytic hierarchy process model. IOP Conf. Ser. Earth Environ. Sci. **945**(1) (2021). https://doi.org/10.1088/1755-1315/945/1/012031

Shaaban, K., Siam, A., Badran, A.: Analysis of traffic crashes and violations in a developing country. Transp. Res. Procedia **55**, 1689–1695 (2021). https://doi.org/10.1016/j.trpro.2021.07.160

Khayal, O.M.E.S.: The History of Bajaj Rickshaw Vehicles, pp. 1–9 (2019)

Kumar, M., Singh, S., Ghate, A.T., Pal, S., Wilson, S.A.: Informal public transport modes in India: a case study of five city regions. IATSS Res. **39**(2), 102–109 (2016). https://doi.org/10.1016/j.iatssr.2016.01.001

Manglam, M.K., Sinha, V.K., Praharaj, S.K., Bhattacharjee, D., Das, A.: Personality correlates of accident-proneness in auto-rickshaw drivers in India. Int. J. Occup. Saf. Ergon. **19**(2), 159–165 (2013). https://doi.org/10.1080/10803548.2013.11076975

Kunle, M.A., Farah, M.J., Mohamed, A.H.A.: Impact of auto-rickshaw (Bajaj): a comprehensive study on positive and negative effects on burao society, somaliland. In: Paper Knowledge. Toward a Media History of Documents, vol. 7, no. 2, pp. 107–115 (2014)

Obiri-Yeboah, A. A., Ribeiro, J.F.X., Asante, L.A., Sarpong, A.A., Pappoe, B.: The new players in Africa's public transportation sector: characterization of auto-rickshaw operators in Kumasi, Ghana. Case Stud. Transp. Policy **9**(1), 324–335 (2021). https://doi.org/10.1016/j.cstp.2021.01.010

Ojo, T.K., Mireku, D.O., Dauda, S., Nutsogbodo, R.Y.: Service quality and customer satisfaction of public transport on cape coast-accra route, Ghana. Dev. Ctry. Stud. **4**(18), 142–149 (2014)

Porter, G.: Urban transport in cape coast, Ghana: a social sustainability analysis. In: Global Report on Human Settlements, pp. 3–14 (2013)

Pramanik, A., Rahman, M.S.: Operational characteristics of paratransit in medium-sized city : a case study on e-rickshaws in Rangpur City, Bangladesh. J. Bangladesh Inst. Plan. **12**, 45–62 (2019)

Basri, R., Khatun, T., Reza, M.S., Khan, M.M.H.: Changing modes of transportation : a case study of Rajshahi city corporation, January 2016. https://bea-bd.org/site/images/pdf/029.pdf

Republic of Ghana: The Coordinated Programme of Economic and Social Development Policies (2017–2024): An Agenda for Jobs: Creating Prosperity and Equal Opportunity for All, pp. 1–151 (2017). https://s3-us-west-2.amazonaws.com/new-ndpc-static1/CACHES/PUBLICATIONS/2018/04/11/Coordinate+Programme-Final+(November+11,+2017)+cover.pdf

Roukouni, A., Basbas, S., Stephanis, B., Mintsis, G.: The role of innovative transportation financing tools in achieving urban sustainability: a stakeholder's perspective. In: The Sustainable City XI, 1(Sc), pp. 585–596 (2016). https://doi.org/10.2495/sc160491

Schmucker, U., Dandona, R., Kumar, G.A., Dandona, L.: Crashes involving motorised rickshaws in urban India: characteristics and injury patterns. Injury **42**(1), 104–111 (2011). https://doi.org/10.1016/j.injury.2009.10.049

Shlaes, E., Mani, A.: Case study of autorickshaw industry in Mumbai, India. Transp. Res. Rec. **2416**(2416), 56–63 (2014). https://doi.org/10.3141/2416-07

Tuffour, Y.A., Kofi, D., Appiagyei, N.: Motorcycle taxis in public transportation services within the Accra Metropolis **2**(4), 117–122 (2014). https://doi.org/10.11648/j.ajce.20140204.12

Acute and Sub-acute Toxicity Studies of Solvent Extracts of *Crinum pedunculatum* Bulbs R.Br

P. Doe[1,2](✉), C. A. Danquah[1,2], K. A. Ohemeng[3], S. Nutakor[1], B. Z. Braimah[1], A. Amaglo[1], M. Abdul-Fatah[1], A. E. Tekpo[1], N. A. F. Boateng[1], S. N. Tetteh[1], O. K. Boateng[1], D. M. Sam[1], O. F. Batsa[1], J. T. Boateng[1], S. K. J. Gyasi[1], S. B. Dadson[1], and K. Oteng-Boahen[1]

[1] Department of Pharmaceutical Sciences, School of Pharmacy, Central University Ghana, P. O. Box 2305, Tema, Ghana
pdoe@central.edu.gh

[2] Department of Pharmacology, Faculty of Pharmacy and Pharmaceutical Sciences, College of Health Sciences, Kwame Nkrumah University of Science and Technology, Kumasi, Ghana

[3] Department of Medicinal Chemistry, School of Pharmacy, Central University Ghana, Tema, Ghana

Abstract. Purpose: *Crinum pedunculatum* R.Br. is a plant that has been used for the topical management of inflammatory diseases among herbalists in the Eastern region of Ghana. Acute and subacute toxicity studies were carried out on solvent extracts of the bulbs of *Crinum pedunculatum* to assess the safety profile for use.

Design/Methodology/Approach: Toxicity experiments were assessed according to the Organization for Economic Cooperation and Development (OECD) guidelines. Wistar albino rats were used in the acute toxicity experiments, each receiving a single dose of 2000 mg/kg of ethanol, methanol and ethyl acetate extracts of *Crinum pedunculatum* orally. 45 rats were used in the subacute toxicity research. For 28 days, the groups were given three doses of the ethanol, methanol and ethyl acetate *Crinum pedunculatum* extract (50, 100 and 200 mg/kg), whereas the control group received normal saline. Haematological, biochemical and histopathological tests were carried out at the end of the experiments.

Findings: Single dose of all solvent extracts of *Crinum pedunculatum* showed no mortality or toxicity up to 2000 mg/kg. Therefore, the median lethal dose (LD_{50}) was greater than 2000 mg/kg. In comparison to the control group (normal saline), histopathologic examination revealed no significant pathological variations in the organs of the treated group. *Crinum pedunculatum* ethyl acetate extract could produce low toxic effects with long term administration with the highest dose because a significant increase in AST levels was observed relative to the control.

Research Limitations: Chronic toxicity studies can be carried out to determine the toxic effects, if any, of the prolonged use of this plant.

Practical Implications: The results from this study show that a single dose of *Crinum pedunculatum* extract at 2000 mg/kg is relatively safe and continued oral administration of smaller doses (50, and 100 mg/kg) of the ethanol and methanol

C. Aigbavboa et al. (Eds.): ARCA 2022, *Sustainable Education and Development – Sustainable Industrialization and Innovation*, pp. 752–762, 2023.
https://doi.org/10.1007/978-3-031-25998-2_59

Crinum pedunculatum extracts showed no significant toxic manifestations. This study also enhances sustainable scientific research by maximizing observable results with justifiably reduced number of animals.

Originality/Value: To the best of our knowledge, this is the first study that evaluates the toxicity profile of the bulbs of *Crinum pedunculatum*. It also builds on several toxicity studies carried out on other *Crinum* species.

Keyword: Acute · Crinum pedunculatum · Sub-acute · Toxicity · Rats

1 Introduction

Herbal medications or medicines are currently playing a vital role in meeting the health needs of individuals around the world, particularly in developing countries, due to their accessibility and low cost. They are often regarded as safe owing to the fact that they are obtained from nature. Scientific research, on the other hand, reveals that just because a product is obtained or derived from nature does not guarantee its safety (George 2011). Crinum species are noted for forming tunicate bulbs that are fallow at certain seasons and are regarded an unexplored source of bioactive compounds (Chahal et al. 2021). Plants belonging to the Amaryllidaceae family have a long history traditionally for the treatment of diverse ailments (Fennell and Van Staden 2001). While other species from this family have been explored extensively for pharmacological activity, there is little information on the therapeutic activity of *Crinum pedunculatum*. Doe *et al.* investigated the anti-inflammatory, anti-pyretic and analgesic activity of the bulbs of this plant (Doe et al. 2021).

In 2008, the World Health Organization proposed that the scientific evaluation of the toxic effects of medicinal plants employed for the management of diseases is imperative (World Health Organization 2008). It is therefore imperative to establish the toxicity profile of these plants belonging to the Amaryllidaceae family because of their extensive traditional use. Previous studies have evaluated the toxicity profile of several *Crinum* plants; it was found that the LD_{50} of the bulbs of *Crinum ornatum* and *Crinum asiaticum* was above 3000 and 5000 mg/kg respectively (Lawal and Dangoggo 2015; Ofori et al. 2021); *Crinum jagus* bulbs at 1118 mg/kg (Azikiwe and Amazu 2015). Lower LD_{50} values were reported on the leaves of *Crinum giganteum* at 200 mg/kg (Elizabeth et al. 2016), as well as leaves and bulbs of *Crinum asiaticum* L at 243 and 507 mg/kg respectively (Riris et al. 2018). There are no studies on the toxicity profile of *Crinum pedunculatum*. Therefore, the aim of this study was to assess the oral acute and sub-acute toxicity of solvent extracts of the bulbs of *Crinum pedunculatum*.

2 Method

2.1 Plant Material Extraction

Fresh bulbs of this plant were harvested from Kwehu Asakrakra in the Eastern region of Ghana with GPS location as follows latitude: 6.62942 N 6 °37′45.9048″11; longitude:

−0.68647 W 0 ° 41′11.30253″. The identification of the bulbs was carried out by the Herbal Medicines Department, Kwame Nkrumah University of Science and Technology. A sample was stored at the herbarium of the faculty of Pharmacy and Pharmaceutical Sciences with voucher number KNUST/HMI/2021/001. To obtain the crude extracts for experiments, the harvested bulbs were cleaned, dried and pulverised into coarse powder. The coarse powder was then macerated in methanol, ethanol and ethyl acetate. The mixture was kept for 72 h with frequent agitation to facilitate extraction. After 72 h, it was filtered using Whattman filter paper with the filtrate evaporated with rotary evaporator (DW-RE-52AA Drawell Scientific China). The crude extract was then stored in the refrigerator until use.

2.2 Experimental Animals

Wistar albino rats weighing between 90–150 g were employed for both acute and sub-acute toxicity experiments. The animals were procured from the Noguchi Medical Research Institute Ghana and were fed with standard pellet diet and had unlimited access to water. They were maintained under standard laboratory conditions according to international standards (NIH Guide for Grants and Contracts 1985). All experimental methods were approved by the Institutional Review Board on Animal Experimentation Faculty of Pharmacy and Pharmaceutical Sciences with code number: FPPS/PCOL/012/2020.

2.3 Acute Oral Toxicity Studies

Acute toxicity experiments were carried out according to the OECD guidelines No. 423 Limit test (OECD 2002). Animals were fasted of food but not water for 12 h prior to the administration of crude extracts of *Crinum pedunculatum*. A total of 9 rats were kept in groups of 3 with each group receiving 2000 mg/kg of either methanol, ethanol or ethyl acetate *Crinum pedunculatum* extracts orally. Animals were observed individually for the initial 4 h following extract administration and subsequently for 14 days. On the 15th day, the liver, spleen, kidney were harvested for histopathologic examination.

2.4 Subacute Toxicity Studies

The subacute toxicity experiments were performed in accordance with the OECD guidelines No. 407 (OECD n.d.). The OECD guideline was employed for this study because it maximises observable results with a reduction in the number and distress of animals in line with enhancing sustainability of scientific research, one of the sustainable development goals. 45 Wistar albino rats were distributed into groups of 5 animals each. For 28 days, the groups were given 50, 100 and 200 mg/kg of ethyl acetate, ethanol, and methanol *Crinum pedunculatum* extract respectively, whereas the control group was given normal saline (10 ml/kg). After treatment at the end of each day, animals were periodically and individually monitored for general toxicity indicators. Animals were weighed every four days for the duration of the experiment. All rats were fasted overnight at the end of the treatment duration. Haematological and biochemical parameters were determined by taking blood samples on the 29th day. Organs (kidney, spleen, liver) were also harvested for histopathological evaluation.

2.5 Haematological and Biochemical Evaluation

All animals were fasted overnight with blood samples drawn through cardiac puncture after sacrifice. Blood collected was kept in EDTA tubes and haematological screening was carried out using an automated hematology analyser (MINDRAY BC-3000 Plus Hematology Analyzer). Red blood cell, white blood cell, haemoglobin, mean corpuscular volume and platelets were among the parameters investigated. To obtain serum, tubes containing collected blood were spun at 3000 rpm for 15 min using the centrifuge machine. Alanine aminotransferase, aspartate aminotransferase were determined.

2.6 Histopathological Evaluation

Liver, kidney and spleen of animals were harvested and weighed. These organs were subsequently preserved in 10% formalin solution for histopathological examination. 5 µm sections were cut and stained using hematoxylin and eosin before analysis microscopically. The microscopic characteristics of the organs of animals treated with solvent extracts of *Crinum pedunculatum* were compared with the control group.

2.7 Statistical Analysis

All results obtained from body weight, haematology and serum analysis are expressed as mean ± SEM. GraphPad Prism Version 8.0.2.263 was used to analyse experimental data obtained. $P < 0.05$ was considered significant.

3 Results

3.1 Acute Toxicity Studies

Animals received 2000 mg/kg of methanol, ethanol and ethyl acetate *Crinum pedunculatum* extracts and no mortality recorded during the oral acute toxicity experiments. Animals were under observation for 14 days and there were no clinical toxic manifestations in their behaviour or general attitude. Hence, the LD_{50} of the ethyl acetate, methanol and ethanol *Crinum pedunculatum* extract is considered above 2000 mg/kg.

3.2 Body Weight

Body weights of rats treated with 50, 100 and 200 mg/kg *Crinum pedunculatum* extracts is outlined in Fig. 1. There was a general dose dependent increase in body weight of animals over the 28-day treatment period with the ethanol, methanol and ethyl acetate extracts. After treatment, 200 mg/kg of the ethyl acetate extract showed highest weight gain than the methanol and ethanol *Crinum pedunculatum* treated rats.

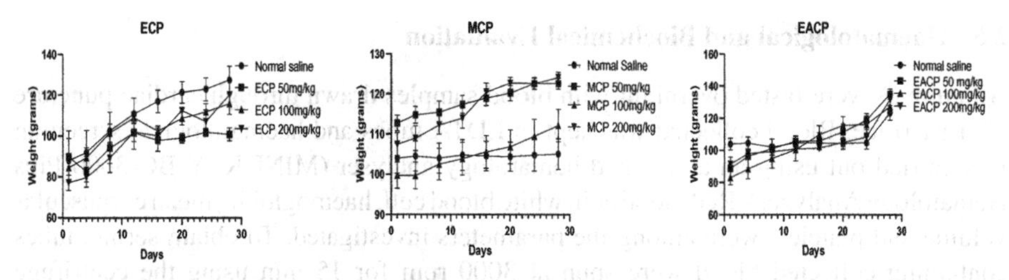

Fig. 1. Mean body weights of rats administered solvent extracts of *Crinum pedunculatum* orally for 28 days. ECP-Ethanol *Crinum pedunculatum* extract; MCP-Methanol *Crinum pedunculatum* extract; EACP-Ethyl acetate *Crinum pedunculatum* extract

3.3 Haematology and Serum Parameters

There were no significant changes seen in the haemoglobin, RBC, PLT, HCT, MCHC and MCV levels of animals treated with ethanol *Crinum pedunculatum* extract relative to normal saline. However, a significant decrease in WBC levels was seen with all solvent *Crinum pedunculatum* extracts compared to the control group (Table 1). Liver AST levels of ECP and MCP treated rats also showed no significant difference compared to control animals while a significant increase was observed with the ethyl acetate treated rats (Table 2).

Table 1. Showing the effect of solvent extracts of *Crinum pedunculatum* on haematological parameters

Treatment groups	Dose	HGB (g/dl)	RBC (x 10^{12}/L)	WBC (x 10^9/L)	PLT (x10^9/L)	HCT (%)	MCHC (g/dl)	MCV (fl)
Normal saline	10	13.5 ± 1.6	4.4 ± 1.4	22.4 ± 0.4	565 ± 221	31.6 ± 13.5	51.2 ± 14.5	69 ± 8
ECP	50	10.3 ± 2	5.4 ± 1.3	13.4 ± 0.4***	506 ± 205	32.3 ± 8.2	40 ± 2.6	58.6 ± 1.3
	100	11.2 ± 2.8	5.7 ± 1.4	2.4 ± 0.5****	419.5 ± 299.5	34.5 ± 10.5	33 ± 2	60.2 ± 3.4
	200	8.1 ± 2.6	4.4 ± 3.4	1.4 ± 0.0****	504.5 ± 257.5	20 ± 15.4	73.6 ± 43.4	54.2 ± 6.4
MCP	50	12.4 ± 0.4*	6.4 ± 0.4	13.4 ± 0.4***	711 ± 0.0****	32.3 ± 8.2	31.2 ± 0.2**	59.5 ± 0.5****
	100	16.1 ± 0.0	4.3 ± 0.4*	2.5 ± 0.5****	816 ± 0.0****	34.5 ± 10.5	52.5	65.2 ± 0.2****
	200	14.4 ± 0.4	7.2 ± 0.2	1.4 ± 0.0****	445 ± 0.0****	20 ± 15.4	52.5 ± 0.5****	57.1 ± 0****
							34.3 ± 0.3*	
EACP	50	8.4 ± 0.4	2.3 ± 0.3***	13.4 ± 0.4***	98 ± 0.0***	14.4 ± 0.4****	59.2 ± 0.2****	55.5 ± 0.5****
	100	9.4 ± 0.4	5.0 ± 0.0*	2.5 ± 0.5****	203.5 ± 3.5**	29 ± 0.1****	33.3 ± 0.3**	57.3 ± 0.3****
	200	113 ± 3****	6.1 ± 0.1	1.4 ± 0.0****	1034 ± 0.0****	38 ± 0.0*	30.2 ± 0.2***	55.5 ± 0.5

* $P < 0.05$, ** $P < 0.01$, *** $P < 0.001$, *** *$P < 0.0001$. ECP-Ethanol *Crinum pedunculatum* extract; MCP-Methanol *Crinum pedunculatum* extract; EACP-Ethyl acetate *Crinum pedunculatum* extract. HGB-haemoglobin, RBC-Red blood cell, WBC-White blood cell, PLT-Platelet, HCT-Hematocrit, MCHC-Mean corpuscular haemoglobin concentration, MCV-Mean corpuscular volume.

Table 2. The effect of solvent *Crinum pedunculatum* extracts on liver function

Treatment groups	Dose (mg/kg)	ALT (IU/L)	AST (IU/L)
Normal saline	10	42.5 ± 2.5	145 ± 5
ECP	50	16.5 ± 3.5*	144.5 ± 4.5
	100	64.5 ± 4.5*	55 ± 5
	200	76.5 ± 0.5**	102.5 ± 2.5
MCP	50	18.5 ± 5.5*	227 ± 27
	100	24 ± 1.0	308 ± 8.5*
	200	22.5 ± 2.5*	190.5 ± 0.5
EACP	50	100 ± 0.0	100 ± 0.0
	100	425 ± 25***	425 ± 25**
	200	104 ± 4	425 ± 25.5**

ALT: Alanine transaminase; AST: Aspartate transaminase. * $P < 0.05$, ** $P < 0.01$, ***$P < 0.001$. ECP-Ethanol *Crinum pedunculatum* extract; MCP-Methanol *Crinum pedunculatum* extract; EACP-Ethyl acetate *Crinum pedunculatum* extract.

3.4 Sub-acute Toxicity Studies

Effect of *Crinum pedunculatum* on Histology of Organs

Histopathological examination of the kidney in rats treated with ethanol *Crinum pedunculatum* extract showed renal tubular hyperplasia with basophilic nuclei while increased hyaline glomerulopathy was observed with MCP 100 and 200 mg/kg and intimal fibrosis with devascularisation of Bowman's capsule was observed with MCP

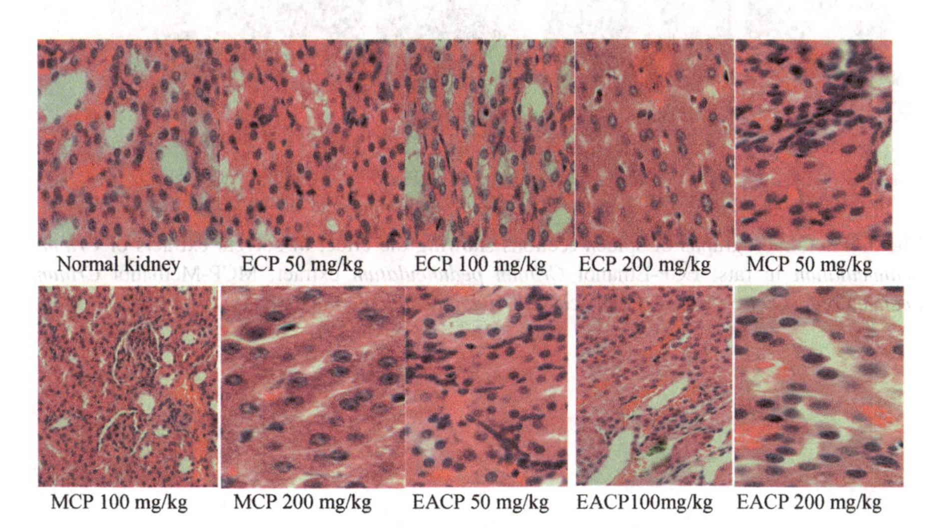

Fig. 2. Photomicrographs of kidney sections showing the effect of *Crinum pedunculatum* extracts in rats. ECP-Ethanol *Crinum pedunculatum* extract; MCP-Methanol *Crinum pedunculatum* extract; EACP-Ethyl acetate *Crinum pedunculatum* extract

200 mg/kg. The ethyl acetate *Crinum pedunculatum* extract showed mild infiltration of inflammatory cells with tubular dilatation (Fig. 2).

Histopathological structure of the spleen showed normal microstructure with normal density of splenic pulp for 50 mg/kg ethanol *Crinum pedunculatum* extract while 100 and 200 mg/kg showed splenic pulp, interstitial edema and mineralization. Treatment with the ethyl acetate and methanol extract of *Crinum pedunculatum* 50 mg/kg showed normal spleen microstructure with normal density of splenic pulp. Splenomegaly with mineral deposit was observed with methanol *Crinum pedunculatum* 100 and 200 mg/kg (Fig. 3).

Histopathological examination of the liver of rats administered with ethanol *Crinum pedunculatum* 50 mg/kg showed no evidence of necrosis and 100 and 200 mg/kg showed hepatomegaly with hepatocellular edema. Methanol *Crinum pedunculatum* extract at 50 mg/kg showed degeneration of liver parenchyma; 100 and 200 mg/kg showed kupffer cell hyperplasia with normal hepatocyte. The ethyl acetate *Crinum pedunculatum* extract showed normal liver microstructure, normal hepatocyte with normal central vein (Fig. 4).

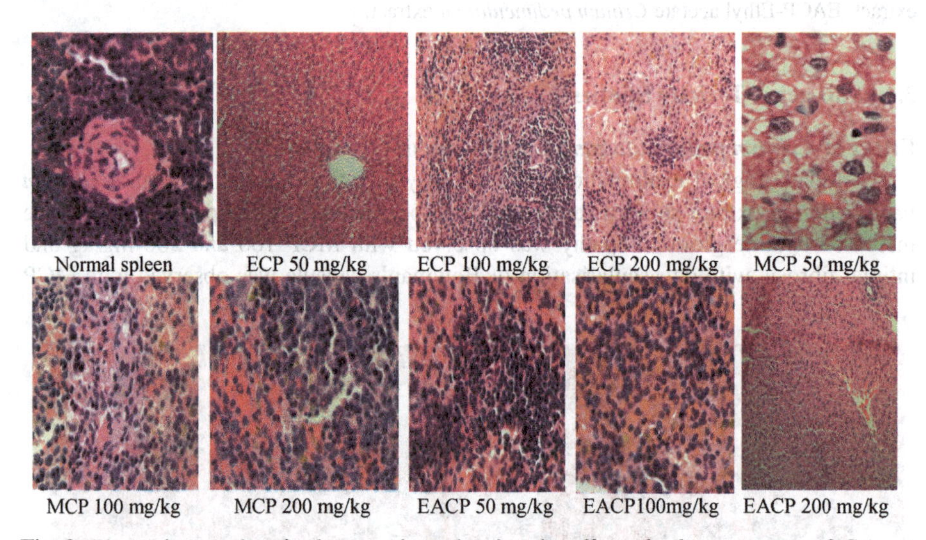

Fig. 3. Photomicrographs of spleen sections showing the effect of solvent extracts of *Crinum pedunculatum* in rats. ECP-Ethanol *Crinum pedunculatum* extract; MCP-Methanol *Crinum pedunculatum* extract; EACP-Ethyl acetate *Crinum pedunculatum* extract

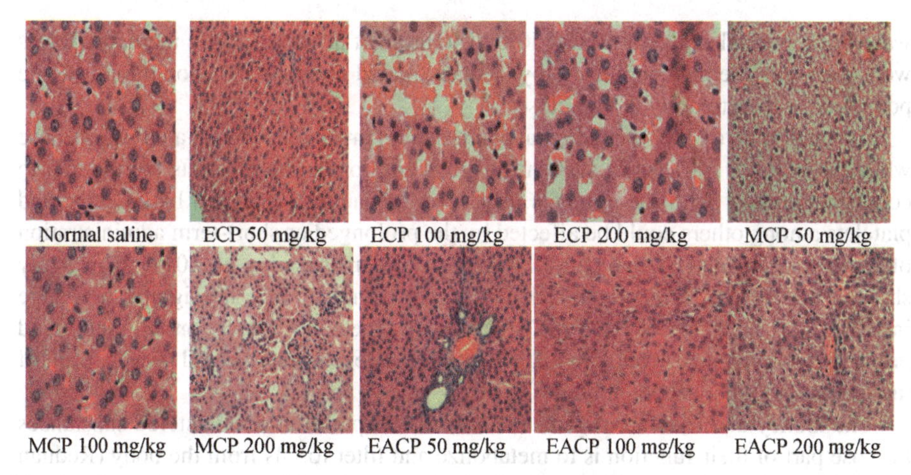

Fig. 4. Photomicrographs of liver sections showing the effect of solvent extracts of *Crinum pedunculatum* in rats. ECP-Ethanol *Crinum pedunculatum* extract; MCP-Methanol *Crinum pedunculatum* extract; EACP-Ethyl acetate *Crinum pedunculatum* extract.

4 Discussion

The general idea that medicinal plants are considered safe and can be taken without any dose or concentration monitoring is starting to loose favour. The use of some herbal preparations have led to various adverse effects, a few of which are reported (Ajose 2007). There is also a dearth of scientific literature on the toxicity profile of most medicinal plants (Saad et al. 2006) and to the best of our knowledge, there are no published studies on the acute and subacute toxicity of the bulbs of *Crinum pedunculatum. Crinum pedunculatum* has been used among people in the Eastern region of Ghana for the topical treatment of inflammation and with this use comes safety concerns.

In the acute toxicity experiment, the administration of the ethanol, methanol and ethyl acetate extract at 2000 mg/kg orally did not cause any mortality and no abnormal clinical signs of toxicity was observed compared to the control animals. Therefore, the LD_{50} was regarded to be higher than 2000 mg/kg and this result served as a basis for selecting doses for subacute toxicity determination (Nguenang et al. 2020). In subacute toxicity experiments, rats were given 50, 100 and 200 mg/kg of solvent extracts of *Crinum pedunculatum* orally for 28 days. Daily administration of *Crinum pedunculatum* showed an increase in weight at all doses and this is synonymous with other *Crinum* species (Lawal and Dangoggo 2015). No mortality was observed during the period of administration and examination. Serum levels of ALT and AST enzymes indicate the structural integrity of the liver and this can suggest the presence of any injury to the liver (Simon-Giavarotti et al. 2002). Toxicity tests after the oral administration of methanol *Crinum pedunculatum* extracts showed serum ALT levels within normal range at all doses administered, normal ALT levels were also observed for 50 mg/kg ethanol *Crinum pedunculatum* extract (Hasan et al. 2018; Sharp et al. 1998). A highly significant increase in ALT concentrations was observed by the ethyl acetate extract relative to the control (Table 2). The AST levels in the ethanol *Crinum pedunculatum* treated group was within normal range of 50 to 150 IU/L (Hasan et al. 2018) while increased levels were seen with the methanol and ethyl acetate treated groups which could indicate the potential for liver toxicity at high doses.

Evaluation of toxicity of medicinal plants should involve full blood analysis because while the blood is a transport system for drugs and toxins, its products are the first to come into contact with these agents. The level of haemoglobin, red blood cells, and platelets among others could be affected by the prolonged or short-term administration of a medicinal plant (Fazliana et al. 2008; Siti Suriani Arsad et al. 2014). This study showed that *Crinum pedunculatum* ethanol extract did not produce any significant change in RBC, PLT, HCT, MCHC and MCV levels. The methanol extract on the other hand produced a significant increase in platelet and WBC levels which could be as a result of the immunostimulatory potential of the plant.

The kidney and liver are organs that are most exposed to administered substances because part of their function is to metabolize and filter toxins from the body (Reduan et al. 2020). Therefore, the histopathological evaluation of the organs of rats orally administered with ethyl acetate, ethanol, and methanol extracts of *Crinum pedunculatum* at 50 mg/kg for 28 days showed no significant change in both treated and non-treated groups. The ethyl acetate *Crinum pedunculatum* extract at all doses produced normal microstructure in the spleen, liver and kidney as well. These results correspond with toxicity studies carried out on *Crinum asiaticum* that showed normal liver and kidney morphology after continuous administration for 28 days (Ofori et al. 2021; Riris et al. 2018).

5 Conclusion

The examination of the acute and sub-acute toxicity of the ethanol, methanol and ethyl acetate extract of *Crinum pedunculatum* was performed in rats. Results obtained show that 2000 mg/kg did not produce any mortality and no significant clinical changes were observed with repeated doses of 50 and 100 mg/kg of all extracts which suggests its safety. Lower toxic effects were observed with the highest dose with extended use and should be avoided.

Acknowledgements. The authors are grateful to Dr. Kodua of the Department of Pathology University of Ghana for his assistance with the histopathology of the organs.

Author Contribution. Conceptualization: PD and CAD; Methods on *Crinum pedunculatum* methanol extract; PD, SN, BZB, AA, MA, KOB; Methods on *Crinum pedunculatum* ethyl acetate extract; PD, AET, NAFB, SNT, OKB, DMS, KOB; Methods on *Crinum pedunculatum* ethanol extract; PD, OFB, JTB, SKJG, SBD; Data Curation, Writing, Reviewing and Editing: PD, CAD and KAO; Supervision: CAD and KAO. All authors read and approved the final manuscript.

Funding. The authors did not receive any funding for this research.

Conflict of Interest. All authors declare no competing interests that are relevant to the content of this article.

References

Ajose, F.O.A.: Some Nigerian plants of dermatologic importance. Int. J. Dermatol. **46**(Suppl. 1), 48–55 (2007). https://doi.org/10.1111/J.1365-4632.2007.03466.X

Azikiwe, C., Amazu, L.: The potential organo-toxicity safety of Morpholine and Crinum jagus in rats. Discovery **10**(25), 113–120 (2015). http://www.discoveryjournals.org/drugdiscovery/current_issue/2015/A15.pdf

Chahal, S., et al.: Unraveling the medicinal potential and conservation of Indian Crinum (Amaryllidaceae) species. S. Afr. J. Botany **136**, 7–15 (2021). https://doi.org/10.1016/j.sajb.2020.04.029

Doe, P., et al.: Analgesic, anti-inflammatory, and anti-pyretic activities of crinum pedunculatum R.Br. bulb extracts. Pharmacogn Res. **14**(1), 24–29 (2021). https://doi.org/10.5530/pres.14.1.5

Elizabeth, F.B., Obikili, E.N., Esom, A.E., Anyanwu, G.E.: Phytochemical screening and oral acute toxicity study of aqueous leaf extract of Crinum giganteum (gadalli) in Wistar rats. J. Phytol. **8**, 22–25 (2016). https://web.archive.org/web/20170922011108id_/http://scienceflora.org/journals/index.php/jp/article/viewFile/3002/2998

Fazliana, M., Muhajir, H., Hazilawati, H., Shafii, K., Mazleha, M.: Effects of Ficus deltoidea aqueous extract on hematological and biochemical parameters in rats. Med. J. Malays. **63**(Suppl. A), 103–104 (2008). https://europepmc.org/article/med/19025006

Fennell, C.W., Van Staden, J.: Crinum species in traditional and modern medicine. J. Ethnopharmacol. **78**(1), 15–26 (2001). https://doi.org/10.1016/S0378-8741(01)00305-1

George, P.: Concerns regarding the safety and toxicity of medicinal plants-an overview. J. Appl. Pharm. Sci. **01**(06), 40–44 (2011)

Hasan, K.M.M., Tamanna, N., Haque, M.A.: Biochemical and histopathological profiling of Wistar rat treated with Brassica napus as a supplementary feed. Food Sci. Hum. Wellness **7**(1), 77–82 (2018). https://doi.org/10.1016/J.FSHW.2017.12.002

Lawal, A., Dangoggo, S.M.: Phytochemical, proximate and toxicity studies of aqueous extract of *Crinum ornatum* (Toad's onion). ChemSearch J. **5**(1), 45–50 (2015). https://doi.org/10.4314/csj.v5i1

Nguenang, G.S., Ntyam, A.S.M., Kuete, V.: Acute and subacute toxicity profiles of the methanol extract of Lycopersicon esculentum L. leaves (tomato), a botanical with promising in vitro anticancer potential (2020). https://doi.org/10.1155/2020/8935897

NIH Guide for Grants and Contracts: NIH Guide for Grants and Contracts Special Edition/Laboratory Animal Welfare, vol. 14, no. (8), pp. 1–25 (1985)

OECD: OECD/OCDE 407 OECD Guidelines for the Testing of Chemicals Repeated Dose 28-Day Oral Toxicity Study in Rodents Introduction (n.d.)

Ofori, M., et al.: Acute and sub-acute toxicity studies of the chloroform extract of Crinum asiaticum bulbs in mice. S. Afr. J. Bot. **143**, 133–140 (2021). https://doi.org/10.1016/J.SAJB.2021.07.047

Organisation for Economic Co-operation and Development: Test no. 423: acute oral toxicity - acute toxic class method. In: Oecd Guideline for Testing of Chemicals, pp. 1–14, December 2002. https://doi.org/10.1787/9789264071001-en

Reduan, F.H., et al.: Acute and subacute dermal toxicity of ethanolic extract of Melastoma malabathricum leaves in Sprague-Dawley rats. Toxic. Res. **36**(3), 203–210 (2020). https://doi.org/10.1007/s43188-019-00013-5

Riris, I.D., Simorangkir, M., Silalahi, A.: Antioxidant, toxicity and antibacterial of OMPU-OMPU (Crinum asiaticum-L) ethanol extract. J. Chem. **11**(03), 1229–1235 (2018). http://rasayanjournal.co.in/admin/php/upload/476_pdf.pdf

Saad, B., Azaizeh, H., Abu-Hijleh, G., Said, O.: Safety of traditional Arab herbal medicine. Evid. Based Complementary Altern. Med. ECAM **3**(4), 433–439 (2006). https://doi.org/10.1093/ECAM/NEL058

Sharp, P., Villano, S.J., La Regina, M., Suckow, M.: The Laboratory Rat, 2nd edn. CRC Press, Boca Raton (1998). https://books.google.com.gh/books?hl=en&lr=&id=ifrxao_ZG9YC&oi=fnd&pg=PP1&ots=FfeEhmOF51&sig=yxzm2oV9hnpakG8RK32nstak1kk&redir_esc=y#v=onepage&q&f=false
Simon-Giavarotti, K.A., et al.: Enhancement of lindane-induced liver oxidative stress and hepatotoxicity by thyroid hormone is reduced by gadolinium chloride. Free Rad. Res. **36**(10), 1033–1039 (2002). https://doi.org/10.1080/1071576021000028280
Arsad, S.S., Esa, N.M., Hamzah, H.: Histopathologic changes in liver and kidney tissues from male sprague dawley rats treated with rhaphidophora decursiva (Roxb.) Schott extract. Cytol. Histol. **4**(001), 1–6 (2014). https://www.researchgate.net/profile/Norhaizan-Me/publication/262116127_Histopathologic_Changes_in_Liver_and_Kidney_Tissues_from_Male_Sprague_Dawley_Rats_Treated_with_Rhaphidophora_Decursiva_Roxb_Schott_Extract/links/0deec536b45f161dbc000000/Histopathologic-Changes-in-Liver-and-Kidney-Tissues-from-Male-Sprague-Dawley-Rats-Treated-with-Rhaphidophora-Decursiva-Roxb-Schott-Extract.pdf
World Health Organization: WHO guidelines on safety monitoring of herbal medicines in pharmacovigilance systems, pp. 105–114 (2008)

Effectiveness of E-Filing System on Improving Tax Collection in Tanzania: A Case of Ilala Tax Region

M. Jumanne and A. Mrindoko(✉)

Department of Accountancy, College of Business Education, Dar es Salaam, Tanzania
allenmrindoko@gmail.com

Abstract. Purpose: The purpose of this study was to investigate the effectiveness of the e-filing system in improving tax collection in Tanzania.

Design/Methodology/Approach: The study was conducted in Ilala tax region in Ilala district, Tanzania, which comprises four (Tanzania Revenue Authority) TRA tax centres namely, Buguruni tax centre, Samora tax centre, Shaurimoyo tax Centre and Mnazi mmoja tax centre. This study was conducted in all four tax centres. The study adopted cross-sectional descriptive design where a mixed approach was applied to collect qualitative and quantitative data using questionnaires and interview guides. The study used both primary and secondary, and the sample size was 100 taxpayers. Descriptive statistics and content analysis were used to analyse quantitative and qualitative data respectively.

Findings: The findings show that e-filling had increased tax submissions in Ilala tax region. The findings show that there is high acceptability of e-filing as a tool which has enhanced tax return submissions. Moreover, the findings of this study acknowledge e-filling to be effective in tax and duties collection at Ilala tax office. In addition, the study found that e-filling had enhanced the convenience of taxpayers. In these taxpayers were found in agreement that e-filling is reliable, easy to use, accessible, reliable, accurate and efficient in the submission of tax returns. Also, the findings indicate that e-filing has increased tax compliance as it reduces the cost of compliance.

Implications/Research Limitations: This study had some limitations which warrant rectification in future studies. Firstly, this study was conducted in a single tax correction region of Ilala, Tanzania. Therefore, future studies should consider expanding the geographical area by covering more than one tax region. Also, this study used only descriptive statistics which cannot estimate of causal-effect relationship. To gauge the effectiveness of e-filling on tax collection future study is suggested to apply inferential statistics such as regression and structural equation modelling, which are powerful models to predict causality.

Practical Implications: The findings of this paper would be useful as will awaken the Government and other institutions dealing with revenue collection to make it mandatory for tax submission to be done using an e-filling system. The findings of this study are expected to help TRA in assessing the e-filing system and make

C. Aigbavboa et al. (Eds.): ARCA 2022, *Sustainable Education and Development – Sustainable Industrialization and Innovation*, pp. 763–783, 2023.
https://doi.org/10.1007/978-3-031-25998-2_60

sure the current e-filling system is improved for an efficient and effective filing process and promotion of tax collection amongst taxpayers.

Originality/Value: Therefore, this study intended to fill this gap by analyzing the effectiveness of the tax e-filling system on improving tax collection in Ilala tax region, Tanzania by focusing on small taxpayers.

Keyword: E-filling · Filling system · Revenue · Tax collection · Taxpayers

1 Introduction

1.1 Background of the Study

Governments worldwide need resources in form of revenues to perform various functions both social and economic activities. Tax is the main source of revenue for any government. The performance of an economy is predicted on revenue collection (Malima 2013; World Bank 2021). Governments need finances to support administrative and provision of economic and social service (Komanya 2013; World Bank 2018). In less developed countries, taxation contributes close to 80% of total government revenue (Bird et al. 2008) and 15.5–18% of GDP while in developed countries tax constitute more than 20% of the GDP and higher proportional of government revenue (World Bank 2018). Virtually, performance of governments is gauged on its ability to collect tax or revenue (Akitoby et al. 2020). So any serious government is striving to increase the collection of tax so that it can afford to provide for its citizens.

Researchers have acknowledged the importance of using modern technology in tax collection and filing of tax returns (Night and Bananuka 2018; Singh 2019; Lediga 2020; Mukuwa and Phiri 2020; World Bank 2021). Rukundo (2020) contends that increased use of technology has arguably improved tax payer services, compliance and administration, which in the end increased tax collection. Among the technologies one mechanism is to have an electronic tax system (e-filing) which provides taxpayers a convenience (Nawawi and Salin 2018). In developed countries tax e-filing system has increased tax collection and reduced tax collection cost to both the tax collection authorities and tax payers (Eichfelder and Hechtner 2016; Lee 2016; Singh 2019). The advancement of tax submission has pointed at improvements in administration of fiscal systems (Rukundo 2020).

Adoption and implementation of e-services involves, among other things, transformation from manual to electronic documentation, of which acceptability varies corresponding to different users, thus, there is a need to consider all levels associated with tax administrations (Blume and Bott 2015). As such, the effectiveness of adoption and using of e-filling must be communicated to all stakeholders and especially to tax payers and practitioners who are involved in the payment of tax and management respectively. Though the problem is more prominent in developing countries, the developed countries also face problems in adoption of electronic services (Chiu and Wang 2008; Venkatesh et al. 2012; Akkaya et al. 2013; Bhuasiri et al. 2016).

The rapid growth of the digital economy in many African countries has led to concerns about whether their tax regimes are equipped to deal with this new phenomenon

(Rukundo 2020; Mukuwa and Phiri 2020). Regardless of the success in tax compliance, collection and management (Sifile et al. 2018; Night and Bananuka 2018; Efobi et al. 2019; Rukundo 2020; Mukuwa and Phiri 2021); in most African countries challenges remain, and it is unclear on whether e-filing system is appropriate mechanism governments in Africa and developing countries of the world can rely to achieve full tax collection.

In this crossroad Tanzania is no exception. Tanzania, like other nations, has considered the benefits and convenience brought about by the adoption of electronic activities. The Tanzania Revenue Authority (TRA) had been employing a paper-based tax declaration system where a number of taxpayers used to claim long queues at the TRA offices on the deadline. The Government of Tanzania through the Tanzania Revenue Authority decided to phase out manual submission of Value Added Tax (VAT) returns to e-filing since the year 2010. The delay in tax declaration and tax collection caused TRA to lose much of its tax revenue (Ernst et al. 2020). Also, TRA incurred large costs in enforcement of tax compliance by tax payers as it was cumbersome for tax payers to waste time at TRA offices for physical filling.

The Government of Tanzania through the Tanzania Revenue Authority decided to phase out manual submission of Value Added Tax (VAT) returns to e-filing since the year 2010. The TRA in a move to improve tax administration and collection, in its 3rd Corporate Plan 2008/09-2012/2013, among its strategies was the introduction of electronic operations (Sichone et al. 2017). The most important reason was to move in tandem with changes in the global technologies. In implementation of this move, among other electronic operations modified and introduced, there was the introduction of e-filing system (Sichone et al. 2017; Ernst et al. 2020; Rantasi 2021). For implementation purposes, a person who is permitted to file tax returns electronically is provided with electronic Filing Identification Number (e-FIN) (Tax Administration (General) Regulations 2016). E-filing implementation was introduced by the government to allow taxpayers to submit their tax returns on-line. The e-filing started in the administration of the Value Added Tax in 2011 with tax payers from Large Taxpayers Department (LTD) in Ilala, Kinondoni and Temeke as pilot taxpayers before rolled out to remaining tax regions.

E-filing implementation was introduced by the government to allow taxpayers to submit their tax returns on-line. This measure was seen to be cost effective for both TRA and tax payers in terms of reduced operational costs like transport and stationary costs to enhance service delivery (Liganya 2020). Moreover, the online filing of tax returns is expected to increase voluntary compliance, save time and bring in efficiency in the tax administration regime. It was viewed that online submission of VAT return would reduce compliance cost on both sides, to taxpayer and TRA by removing all challenges faced by them during physically submission of the VAT returns (Ernst et al. 2020). Despite e-filing potentials, the compliance and implementation of e-filing of tax returns as it is for other e-services is not a straight forward process (Rumanyika and Mashenene 2014; Sichone et al. 2017). The governments face challenges in implementation of e-filling. Researchers (Yonazi 2013; Matimbwa and Masue 2020; Matimbwa 2021), suggested that one of the major challenges might be the inherent obstacle to human kind of negative attitude in adopting ICT in daily activities lest online fraud due to low cyber security in developing countries.

The implementation of e-filing in any government is time consuming, multifarious and exigent (Blume and Bott 2015; Matimbwa 2021). Most of the time, in an situation that is more electronic ready, people are generally comfortable with the introduction of new information and communication technologies (ICTs) and thus e-filing inventiveness can be easily adopted and implemented (Matimbwa et al. 2020). Unfortunately, this background is lacking in Tanzania (Yonazi 2013; Rumanyika and Mashenene 2014; Sefue 2014; Malekani 2018). Since its inception in 2010, e-filing has been facing limitations, and the reasons behind such limitations are not clearly known. A broad literature on tax e-filing is available for studies conducted elsewhere but limited information is available for countries like Tanzania, and especially for small tax payers, which are peculiarly challenged by level of ICT knowledge and usage, infrastructure development to support uptake of ICT related technology, and low awareness of tax e-filling. As such, the impact of e-filing system has not been empirically determined in Tanzania. This knowledge gap is the basis of this proposed study.

Therefore, this study aims at closing the identified knowledge gap by analyzing the effectiveness of tax e-filling system on improving tax collection in Ilala tax region, Tanzania by focusing on small tax payers, which are the small and medium sized enterprises (SMEs). Therefore, this study addresses constructs of e-filing in Tanzania, and consequently fills the gap by assessing the effectiveness of e-filing system on improving tax collection in Tanzania. Hence, the study contributes to the available literature and the results will help in giving valuable information to the tax payers and practitioners as well as policy makers, to facilitate smooth transformation and implementation of e- filing. Thus, the information gathered has important implications for promoting an effective e-filing system.

1.2 Objectives of the Study

1.2.1 General Objective

The main objective of this study was to assess the effectiveness of e-filing system on improving tax collection in Tanzania.

1.2.2 Specific Objectives

Specifically, the study was guided by the following objectives:

i. To examine convenience of tax e-filing in VAT returns submission compared to paper based submission era.
ii. To assess the trend of submission for VAT returns during paper based submission and after the introduction of tax e-filing.
iii. To evaluate the trend of tax and duties collection before and after introduction of tax e-filing.

1.3 Research Questions

i. What is the trend of submission for VAT returns and revenue collection during physical submission and after the introduction of e-filing?

ii. What is the convenience of e-filing in VAT returns submission compared to physical submission era?
iii. What is the trend of tax collection before and after introduction of e-filing?

2 Literature Review

2.1 Theoretical Literature Review

2.1.1 Expected Utility Theory

The expected utility theory (EUT) model of tax evasion predicts a negative relationship between tax rates and evasion whenever fines are imposed on the evaded tax and taxpayers exhibit decreasing absolute risk aversion (Yitzhaki 1974). According to Stanford Encyclopedia of philosophy, Expected Utility Theory has been defined as a normative theory that tries to explain how people should make decision (Piolatto and Rablen 2016). The basic point emphasized by most scholars is that the rational taxpayers' main goal is to maximize their financial position (Bernasconi and Zanardi 2004; Hashimzade et al. 2013). Taxpayers are always motivated to attempt to evade tax; the theory suggests that one of the key factors in a decision of a taxpayer to comply is on the costs of compliance involved against benefits of compliance. The economic analysis thus concludes that since compliance decisions are based on assessment of cots and benefits, high probabilities of detection for noncompliance would encourage greater compliance; similarly, reduction in the general costs of compliance would directly make compliance more beneficial and hence encourage more taxpayers to comply.

2.1.2 The Theory of Reasoned Action (TRA)

Psychology theories on the other hand speculate that taxpayers are influenced to comply with their tax obligations by psychological factors (Fishbein and Ajzen 1975). They focus on the taxpayers' morals and ethics. The theories suggest that a taxpayer may comply even when the probability of detection is low. As opposed to the economic theories that emphasize increased audits and penalties as solutions to compliance issues, psychology theories lay emphasis on changing individual attitudes towards tax or revenue systems (Chaudhry 2010). According to Fishbein and Ajzen (1975) a person's behaviour is determined by their intention to perform the behaviour and that this intention is, in turn, a function of their attitude towards behaviour and subjective norms. An intention is a plan or a likelihood that someone will behave in a particular way in specific situations, whether or not they actually do so (Bagozzi et al. 1992). A tax payer who is thinking about obeying tax compliance intends or plans to obey, but may or may not actually follow through on that intent. Given the chance, a lot of tax payers will not pay taxes except there is a motivation to do so. Some accept as true that the best way is to increase incentives, while others accept as true the best way to tax compliance is to increase penalties (Chaudhry 2010).

2.1.3 The Theory of Planned Behaviour (TPB)

The Theory of Planned Behaviour (Ajzen 1985, 1987) details how the influences on an individual determine that individual's decision to follow a particular behaviour. This theory is an extension of the broadly applied Theory of Reasoned Action (TRA) by Ajzen and Fishbein (1975). The TPB proposes that the proximal determinants of behaviour are intentions to engage in that behaviour and perceived behavioural control over that behaviour. Intentions represent a person's motivation in the sense of her or his conscious plan or decision to exert effort to perform the behaviour. Perceived behavioural control refers to an individual's expectancy that performance of the behaviour is within his or her control (Ajzen 1991). Therefore, as e-filing system is a newly adopted system, the Theory of Planned Behaviour is related in understanding how tax compliance is influenced by a tax payer's behaviour which is determined by his or her intention in compliance.

2.1.4 Diffusion of Innovation Theory

Diffusion of innovation theory was developed by Rogers (1962) which explains changes in the technological acceptance over time as individuals gain experience. The theory aimed to explain how, why and at what rate new ideas and technology spread. Diffusion is considered to be an information exchange process driven by the need to reduce uncertainty. Uncertainty can be considered as the degree to which a number of alternatives are perceived in relation to the occurrence of some event, along with the relative probabilities of each of these alternatives occurring (Rogers 1962). Therefore, those who are involved in adoption of the innovation are emphasized to seek information to reduce uncertainty (Rogers 1962). Hence, as e-filing system is a newly adopted system, Diffusion of innovation theory is applicable in understanding how, why and at what rate the e-filing system will spread in compliance and ultimately promotes process of revenue collection. This study points to benefits of using e-filing in revenue collection as the new innovation which supersedes the previous system of revenue collection.

2.1.5 Technology Acceptance Model (TAM)

Technology acceptance model (TAM) is an information system theory which was developed by Davis (1989). It explains and models how a new technology and various aspects of it are received and used by the end users (Davis 1989). The actual end use is the end point where we want everybody to be able to do with the technology so as we can form the behavioural intention. The behavioural intention is the factor which led one to use the technology. This model has been one of the most influential models of technology with two primary factors influencing new technologies which are perceived ease of use and perceived usefulness (Venkatesh 2003). According to Davis (1989), perceived ease of use and perceived usefulness are the major factors that influence users' decisions about how to use the technology and when they will use it once a new technology is presented to them. Perceived ease of use (PEOU) refers to the degree to which a person believes that using a particular system would be free of effort (Davis 1989). Conversely, perceived usefulness (PU) is defined here as the degree to which a person believes that using a particular system would enhance his or her job performance (Davis 1989). Hence this theory implies that, the perceived usefulness of E-filing system is primarily predicting the belief of the government that the technology helps to collect revenue faster, more easily, and more effectively.

2.2 Empirical Literature Review

World over, taxpayers' resistance, underutilization and reluctance to use e-services remain a great concern and still plague various tax agencies which are embracing electronic tax services (Mukuwa and Phiri 2020). E-government is becoming increasingly more important in today's world due to its effectiveness and applicability in various areas (Matimbwa 2021). Tax e-filing is one of the e-government services that have been adopted by many countries today where the public has to discharge their responsibility to the government via online tax filing (Sichone et al. 2019). Tax administrations around the world have introduced e-filing of tax returns due to its potential to improve tax return filing compliance (Lediga 2020). Despite the rapid adoption of tax e-filing in many countries, researchers have argued that it is yet to establish an integrated system that is reliable, especially in developing countries due to high perceived risk by the public (Wang 2002; Azmi and Kamarulzaman 2010). This implies that it will be very difficult, if not impossible, to truly embed responsible behaviour within a community if individual perceptions of risk of the e-government service is the issue (Azmi and Kamarulzaman 2010).

Al-Debei et al. (2015) propose that consumer attitudes towards online system are positively and directly affected by trust and perceived website reputation, and this implies that if taxpayers perceive or evaluate the e-tax system to be secure, they will trust it and adopt it. Maisiba and Atambo (2016) found that taxpayers in Kenya felt uncomfortable using an electronic tax system as compared to the old manual system. Taxpayers who evaluate electronic filing system as not easy to use do not adopt it which affects tax compliance. According to Khaddafi et al. (2018), the adoption of an electronic tax system will depend on the perceived ease of use of the tax system, intensity of behaviour and user satisfaction. This implies that taxpayers must be happy and motivated to use the electronic tax system, but the tax system should as well be easy to use. The user of the electronic tax system must find it pleasant interacting with the electronic tax system. Zaidi et al. (2017) found that taxpayers with computer skills will find it easy to adopt an electronic tax system than those without.

The adoption of e-tax systems has become fundamental, as many countries adopt information systems in tax management (Ondara et al. 2016). Adoption of an electronic tax system is important not only in terms of reducing costs and taxpayer convenience but also in terms of improving tax compliance (Night and Bananuka 2020). Kimea et al. (2019) and Mukuwa and Phiri (2020) analysed factors that influence taxpayers' intention to use electronic tax filing system. The study shows that performance expectancy, effort expectancy and social influence affect behavioural intention to use e-services. A number of technology acceptance models and theories have been applied to different phenomena and varying cultural settings in many studies, producing varying results (Daka and Phiri 2019). Some of the results from these studies are consistent with the original postulations while others contradict them. It is often argued that the UTAUT model is able to explain 70% of the variance in usage intention, which is significant compared to the actual eight models used to build it. There is still lack of research about the effects of e-services on revenuc collection and tax compliance in Tanzanian context. This research plans to fill this gap.

However, importance of e-filling has been reported worldwide. Maisiba and Atambo (2016) reported that the e-tax system improves tax compliance, as it facilitates faster accessibility to tax services without a physical visit to the tax authority premises. Haryani et al. (2015) further state that a system that is easy to use, secure, and dependable, provides easy payment mode, provides a variety of services and is user-friendly boosts voluntary tax compliance. Today, electronic tax system has been adopted in many countries of the world. In African, Uganda, Nigeria, Rwanda and Kenya have embraced electronic tax system (Muturi and Kiarie 2015). Findings by Simuyu and Jagongo (2019) indicate that there is a significant relationship between the perception towards online tax filing in terms of ease and simple to file and also the system being secure, and this improves tax compliance levels. Furthermore, Ondara et al. (2016) suggest that there is a strong relationship between attitude towards electronic tax system and tax compliance. Night and Bananuka (2020) found that adoption of electronic tax system and attitude towards electronic tax system are significantly associated with tax compliance. Alcedo and Cajala (2018) reported that computerization of Bureau of Customs was effective in Philippine.

3 Research Methods and Tools

3.1 Study Area

The study was conducted in Ilala tax region due to its largest economic activities in the country and being one of the three Tax regions which were piloted and hence early adopter of e-filling system. Ilala Tax region office is in Ilala District in Dare es Salaam region which comprises four TRA tax centres namely, Buguruni tax centre, Samora tax centre, Shaurimoyo tax Centre and Mnazi mmoja tax centre. This study was conducted in all four tax centres.

3.2 Research Design and Approach

In view of the fact that the current study assess the influence of e-filling on effectiveness of tax collection in Tanzania, this study, a cross-sectional descriptive design was adopted to collect and analyse quantitative and qualitative data on the effectiveness of e-filling on the improvement of tax collection. Mixed concurrent triangulation method was used to collect both qualitative and quantitative data simultaneously. Quantitative data from staffs of domestic revenue department in Ilala region tax office and small taxpayers, while qualitative data were collected through in-depth interview. The mixed approach concurrent approach was adopted purposely to increase the validity of findings through triangulation.

3.3 Study Population, Selection of Sample, Sample Size and Key Informants

The targeted population in this paper was employees of Tanzania Revenue Authority under Domestic Revenue Department (DRD) in Dar-es-salaam plus the small taxpayers in the category. This is because DRD officers are responsible of giving out and clarifying tax clearance matters. To select a sample, the study employed a multi-stage sampling procedure as follows: Firstly, Ilala tax region was purposively selected. This selection is guided by the factor that Ilala district was among the first three tax regions to adopt tax e-filling system for small taxpayers and also has many SMEs activities. Secondly, the sampling frame was divided into four strata using stratified random sampling based on the four tax centres; Buguruni, Shaurimoyo, Mnazi Mmoja and Samora because the population from which the sample was drawn was stratified. Thirdly, calculation of the overall sample size was done by using Yamane formula which gives the degree of accuracy of the sampling technique and take into account the margin of error of 5%.

Fourthly, proportionate stratified random sampling was done using proportionate stratification formula proposed by Cochran (1977). Lastly, sub-samples in each stratum was selected using systematic sampling where sampling fraction was used to get the interval and the random start from each region, in this case simple random sampling was used to identify the random start and the remaining cases of the sample was selected systematically at fixed interval. However, in each tax centre four staffs from domestic revenue department was selected to participate in the proposed study as key informants, which is total of 16 key informants. The Yamane formula expressed as:

$$n = \frac{N}{1 + N(e^2)}$$

where;

n = Sample size
N = Total number of target population
e = Precision (5%)

Therefore, the sample size was 243 taxpayers.

3.4 Types of Data and Collection Methods

This study collected quantitative and qualitative both primary and secondary data. The primary quantitative data were collected from selected respondents using a structured interview through a questionnaire which was administered by the researcher. The questionnaire consisted of close ended questions except one question. The primary qualitative data were solicited from key informants using an interview guide. Moreover, the study used secondary data to answer the study questions and fulfil the study objectives. The secondary data were collected by reviewing and analysing different relevant documents such as tax and compliance reports released by stakeholders in Tanzania and out of Tanzania e.g. TRA, IMF and World Bank.

3.5 Data Analysis Plan

3.5.1 Data Cleaning and Processing

Data cleaning is the process that deals with detecting and removing incorrect or irrelevant parts of the data, from dataset in order to improve the quality of data. Before data entry, questionnaires and notes from unstructured interview with key informants were inspected for inconsistencies and errors. Thereafter, the, quantitative and qualitative data were analysed as described in the subsequent sections.

3.5.2 Quantitative Data Analysis

Data collected through structured questionnaires were summarized and coded. Statistical package for social sciences (SPSS) version 23.0 was employed for data analysis. The quantitative analysis was descriptive statistics such as frequency and percentage distribution for all three objectives of the study. Findings were presented in tables and graph.

3.5.3 Qualitative Data Analysis

The researcher used the content analysis approach that is considered appropriate for a qualitative study of this nature which based on narratives of data collected and review of several relevant documents. Qualitative data analysis describe patterns and trends and categorizing the data into themes in relation to the research problem, questions and conceptual framework; and finally displaying the data in an organized, compressed assortment of information that allows verified conclusions to be made (Braun and Clarke 2006). The findings of qualitative data were used to supplement explain the quantitative results for all objectives.

4 Results and Discussions

4.1 Demographic Features of Surveyed Respondents

The taxpayers' age and business experience are important when evaluating their compliance to tax. All surveyed taxpayers were divided into four age groups. Table 1 shows 37% (n = 37) were between 26–35 years, 47% (n = 47) were between 36–45 years and 11% (n = 11) were between 46–55 and 05% (n = 05) were above 56 years. The minimum age was 29 and the maximum age was 66 years. The mean age was 35 years. This implies that the majority of clients involved were engaging in labour force. These results coincide with the entrepreneurship readiness curve whereby the ideal time for one to start a business is a period between 25 years to 37 years (Helms 2006).

Moreover, the results show that most of taxpayers at Ilala tax region have experience of more than six years whereby 35.00% (n = 35) have six to 10 years experience in business and 32.00% (n = 32) and 11% (n = 11) possess 11 to 20 years and more than 20 years experience respectively. A business person (taxpayer) with more than six years experience is far better in handling business activities and is likely to work efficiently than a newly business person, and hence can manage business profitably and thus be able to pay the required tax as requirements. Generally, this observation shows that more than 70% of the participants had witnessed the previous system where VAT returns were physically submitted to TRA hence they are in position to assess the situation before and after introduction of e-filling system, and therefore data collected from them can be reliable.

Besides, Table 1 shows that out of 100 respondents 40% (n = 40) have diploma, 30% (n = 30) have degree, 22% (n = 22) had secondary education/ certificates and 8% (n = 08) had master's degree. This observation shows that about 78% of the respondents had education level of diploma and above, this is an indication that participants are well

Table 1. Distribution of demographic characteristics of respondents

Variable/parameter	Measurement	Frequency	Percentage
Sex	Male	64	72.9
	Female	36	27.1
Age	26–35	37	37.0
	36–45	47	47.0
	46–55	11	11.0
	56+	5	5.0
Business experience	0–5 Years (Low)	22	22.0
	6–10 Years (Medium)	35	35.0
	11–20 Years (High)	32	32.0
	21 + Years (Highest)	11	11.0
Education level	Secondary	22	22.0
	Diploma	40	40.0
	Bachelor	30	30.0
	Masters	8	8.0
Size of business	Small	67	67.0
	Medium	33	33.0
Have ICT skills	Yes	97	97.0
	No	3	3.0
Trained by TRA	Yes	76	76.0
	No	24	24.0

Source: Field data 2022.

educated and would be in a position to analyze critically research questions and provide a fair view. In case of type business they run, the results in Table 1 indicate that the majority 67.00%; (n = 67) were in small scale category who had capital ranging between 100,000 TZS to 2 million TZS (see Table 2). Most of these taxpayers in small scale category run businesses such as garments, cosmetics, crops, food selling and agribusiness. Also, a small proportion 33.00% (n = 33) was in group of medium scale category with capita above 2.00 million TZS to 10.00 million. For this later group the capital were for running car accessories, hardware, ICT devices, stationary, bar and restaurant and whole sale enterprises.

4.2 E-filling System on Enhancement of Convenience in Tax Collection

The results in Table 3 show that out of 100 participants who rated this statement, 46% (n = 46.) strongly agreed that e-filling is popular among taxpayers, while 40% (n = 40) agreed. On the contrary, 4% (n = 4) of participants disagreed and 10 were neutral to the

Table 2. Taxpayers businesses types

Category	Capital level	%	Types of enterprises
Small	100,000–2,000,000	67.00	Garments, cosmetics, crops and food selling
Medium	2,001,000–10,000,000	33.00	Stationary, bar and restaurant, whole sales

Source: Field data 2022.

question. This result suggests that e-filling was not a new thing at a time this research was conducted. As such, taxpayers saw it as a mean to and end of improving tax and duties collection. Thus this study's point of view is that e-filling system was very popular among its users. Moreover, during in-depth interview with key informants the study was informed that to most VAT registered taxpayers the e-filling system is not new at all, and they have mastered using it and are happy for it simplifies their tax returns reporting.

Table 3. E-filling system on enhancement of convenience in tax collection

Description	SDA n (%)	DA n (%)	NAND n (%)	AG n (%)	SAG n (%)
E-filling system is popular among tax payers	0(0.0)	4(4.0)	10(10..0)	40(40.0)	46(46.0)
E-filling system is accessible and simple to use	4(4.0)	12(12.0)	0(0.0)	48(48.0)	36(36.0)
e-filling is simple and easy to use	2(2.0)	4(4.0)	6(2.0)	48(48.0)	40(40.0)
E-filling system gives accurate tax assessments	10(10.0)	14(14.0)	4(4.0)	46(46.0)	26(26.0)
E-filling system is reliable	6(6.0)	20(20.0)	4(4.0)	38(38.0)	32(32.0)
Efficiency in filing VAT returns	0(0.0)	10(10.0)	0(0.0)	36(36.0)	54(54.0)
e-filling system reduces cost of tax return submissions	4(4.0)	4(4.0)	0(0.0)	54(54.0)	38(38.0)

Source: Researcher, 2021.

Regarding to accessibility and easy to use, the results in Table 3 show that, 36% (n = 36) of respondents strongly agreed that e-filling system is accessible 24 h while 48% (n = 48.0) agreed to the statement. Moreover, about 16% of participants disagreed to the statement (4% (n = 4) strongly disagreed and 12% (n = 12) disagreed). This generally shows that majority of e-filling system users find it is always accessible, and taxpayers can make tax submissions at their convenient time without being tied by time, and they are not required to physically appear at tax office. In the taxation world, developments in information technology have had an enormous effect in development of more flexible payment methods and more user-friendly taxation system (Lediga 2020). Relationship between accessibility and convenience in submitting generating tax income had been found to be positive and significant in many studies (Sichone et al. 2017; Kimea et al. 2019). In taxation industry when there are potential alternatives to the service, it increases accessibility and improve customers convenience. Taxpayers can access their tax count from everywhere and anytime.

With respect to simplicity in usage, the results in Table 3 show that, 40% (n = 40) of respondents strongly agreed that e-filling system was simple to use (user friendlier) while 48% (n = 48.0) agreed to the statement. Moreover, about 6% of participants disagreed to the statement (2% (n = 2) strongly disagreed and 4% (n = 4) disagreed). This generally shows that majority of e-filling system users find it is always easy to use it. Association involving ease of use of e-filling and convenience is very important for the adoption of the e-filling system. In an ICT era it is easy for the customer to choose on how and when they want to submit their tax returns (Sichone et al. 2017; Lediga 2020).

In case of accuracy of tax assessment, the results in Table 3 reveal that, 26% (n = 26) of respondents strongly agreed that e-filling improves accuracy of tax assessments, and 46% (n = 46) were happy to reported that e-filling increases accuracy of tax assessments agreed. Moreover, 4% (n = 4) of participants were undecided while 14% (n = 14) and 10% (n = 10) of participants disagreed and highly disagreed respectively. This generally shows that an e-filling system provides accurate tax assessments for submission. Thus, based on the finding it can be postulated that taxpayers have high confidence on e-filling for assessment and estimation of tax amount to be paid by taxpayer. This finding is in line with Sichone et al. (2017) and Mukuwa and Phiri (2020).

Regarding reliability of e-filling, Table 3 show that 32% (n = 32) and 38% (n = 38) of respondents strongly agreed that e-filling system is reliable. However, 4% (n = 4) of participants were undecided while 20% (n = 20) and 6% (n = 6) of participants disagreed and strongly disagreed on the statement that e-filing system is reliable respectively. The result implies that 70% of respondents have acknowledged that e-filling system is reliable. The study was informed that e-filling system is frequently used by taxpayers to file their tax returns in monthly, quarterly, biannual and annually basis. Although reliability of using e-filling is crucial, for most users of e-filling in developing world, the appeal of e-filling systems is more about accessibility, convenience and affordability than reliability. According to KIs, this mode of tax submission is an easier form, a relatively affordable, and personal and can be used anywhere and at any time at users' convenience.

With regard to efficiency in filling VAT return submission, the VAT returns submission trend is highly influenced by effectiveness in e-filling system. The results in Table 3 show that, 54% (n = 54) and 36% (n = 36) of respondents strongly agreed and agreed that e-filling system is efficient in submissions of tax returns, while only 10% (n = 10) of respondents disagreed to the statement. This generally shows that e-filling system users find it efficient thus it positively influences the trend of VAT returns electronically and stimulates revenue collection after its introduction as compared to paper based submission era. However, since efficient speaks for a lot more factors such as cost reduction, labour productivity and capital intensification, it was imperative to assess the cost reduction ability of e-filling in filling for tax returns at taxpayers' point of view.

Therefore, respondents were asked to indicate whether they agree that e-filling system has reduced taxpayers' compliance costs. The results in Table 3 showed that out of

100 respondents who rated this statement, 38% (n = 38) and 54% n = (54) of respondents strongly agreed and agreed that e-filling system has reduced compliance costs, respectively. This generally shows e-filling system has reduced compliance cost thus stimulated adherence to tax regulations by taxpayers as low compliance cost influences smooth submission to tax regulations. Therefore, e-filling system is convenient in terms of time or money as compared to physical submission eras which is inconvenient in terms of time or money ultimately resulting to tax evasion. The results suggest that saving cost by making e-filling through ICT devices is one among major determinants of tax compliance by taxpayers. The finding corroborate with Sichone et al. (2017), Kimea et al. (2019), Lediga (2020), Mukuwa and Phiri (2020), Soneka and Phiri (2019) and Matimbwa (2021).

4.3 Trend of Tax Filling for VAT Returns Before and After e-filling

The results in Table 4 indicate that 77% of surveyed taxpayers and TRA staffs were happy to report that the e-filling has effectively reduced the queue for paper based tax submissions in the four tax centres of Ilala tax region. This result implies that e-filling has been well adopted and thus tax submissions are now done online without going at the TRA offices. This suggests that though TRA has introduced e-filling still paper based tax submission continues. This can be due to low skill level of taxpayers in using the e-filling system or the rigidity to accept change. As such, this result is in line with the psychological theories of technology adoption/acceptance and behaviour. Different individuals have different propensity to technology adoption and use due to perceived uncertainties of the technology. According to diffusion of innovation theory taxpayers must seek information to reduce those uncertainties before they make decision to adopt the technology. This is supported by Venkatesh et al (2003) and Matimbwa (2021).

Further, in case of availability of e-filling system online is 24 h, Table 4 indicate 86.0% of surveyed taxpayers and TRA staffs were pleased to testimony that e-filling system online can be accessed for 24 h every day of week, month and year. Among them, 76.0% agreed and 10.0% strongly agreed that of e-filling system is available online for 24 h. That means a taxpayer can make tax submission any time he/she wishes to. It is not necessarily for tax submission to be done during working hours. In case of paper based a taxpayer must be at the TRA office physically and within working hours to be able to submit tax returns. This study establishes that e-filling is beneficial because it gives a taxpayer who also happen to be a business person a chance to do business in busy hours and do the filling activities at his/her convenient time. This finding is in line with Kimea et al. (2019), Soneka and Phiri (2019), Mukuwa and Phiri (2020) and Lediga (2020).

Besides, regarding to e-filling documentation procedures, 86.0% of surveyed taxpayers were happy to report that e-filling documentation procedures are easy to compile and send. This result suggests that some of taxpayers have ICT and basic tax compilation and submission skills. However, hiring of tax return expert is essential, and most of small and medium firms hire tax experts on job basis to compile and file their tax returns. In additional, the results in Table 4 show that 89.0% and 86.0% of TRA staffs acknowledged that e-filling saves time and financial resources. As e-filling enables a taxpayer to submit tax returns at anytime, this means that time is saved for conducting

Table 4. Distribution of tax submission (n = 100)

	Indicator/parameter	SDA N (%)	DA N (%)	NAND N (%)	AG N (%)	SAG N (%)
1	Tax filling queue has decreased	5(5.0)	10(10.0)	8(8.0)	50(50.0)	27(27.0)
2	Availability of e-filling system online is 24 h	0(0.0)	8(8.0)	4(6.0)	76(76.0)	10(10.0)
3	The e-filling documentation procedures are easy	3(3.0)	8(8.0)	3(3.0)	80(80.0)	6(6.0)
4	e-filling saves time	2(2.0)	2(2.0)	7(7.0)	81(81.0)	8(9.0)
5	e-filling is cost saving	3(3.0)	10(10.0)	1(1.0)	71(71.0)	15(15.0)
6	Tax filling submission increases after adoption of e-filling	0(0.0)	1(1.0)	0(0.0)	96(96.0)	3(3.0)

Source: Field Data 2022.

their business activities. Instead of wasting time at TRA office to submit paper based tax returns e-filling provides taxpayer an alternative which saves time but less cost as cost of transport and accommodation is eliminated. The finding corroborates with Eichfelder and Hechtner (2016), Lee (2016) and Sichone et al. (2017) and Matimbwa (2021).

Furthermore, in case of whether tax filling submissions have increased after adoption of e-filling, result in Table 4 shows 99.0% of surveyed taxpayers and TRA staffs were of the same opinion that e-filling system has increased tax submissions. Among them, 96.0% agreed and 3.0% strongly agreed that of e-filling system have increased tax submissions. The finding implies that taxpayers who were not complying with submissions of tax returns during paper based system were now in compliance with the tax requirements. Thus, e-filling system has incentivised taxpayers and changed their behaviour toward tax compliance. This might have a positive effect on revenue collection. The result confirms the theoretical foundations of expected utility theory that taxpayers compliance is subject to motivation to do so. Cost and time saved are the incentives or motivation to tax compliance. This finding is in line with Sichone (2017), Lee (2016), Kimea et al. (2019), Soneka and Phiri (2019), Mukuwa and Phiri (2020) and Lediga (2020).

In similar point of view, the findings was corroborated by key informants during the in-depth interview by informing the study that e-filling is an easy way to file for tax returns as it saves time and cost of physically appearing at TRA offices and also it serves taxpayers from inconveniences of closing business and loose income. Also, the study was informed that e-filling has been highly adopted and used by taxpayers because currently the queue in the office of tax returns is short and compared the time when e-filling system was not introduced. When asked to explain if e-filling system has improved the trend of tax submissions, key informants declared that e-filling has revolutionised the tax administration system by increasing tax submissions which reflect more revenue collection.

4.4 Trend Tax and Duties Collection Before and After e-filling

In this study improvement on tax duties collection was measured by considering trend of VAT local collection before introduction of E-filling system and after introduction of E-filling system obtained from the department of domestic revenue (TRA Revenue Reports 2011–2020). Results in Table 5 show that annual tax and duties collection before e-filling was introduced at Ilala tax office amounted to 58.2 billion in 2010/2011 financial year. After the introduction of e-filling in 2010 and its implementation in 2011, there is an improvement of tax and duties collection to 86.28 billion, which is an increase of 28.07 billion in just one year. The results in Table 5 also show that tax and duties collection were on the increase for four consecutive years from 2011/2012 to 2014/2015 financial year and drastically fell in year 2015/2016.

This could be due to deliberate effort of the fifth government under President Dr. John Pombe Magufuli to overhaul the economy, by demanding all business companies regardless of their size which had evaded tax and duties in the previous years had to repay their tax accumulations if they wanted to do business in Tanzania. This resulted to closure of many small, medium and large businesses in the country. This shows that survival of many businesses was not due to competitive advantage over competitors but unethical business practices. As such tax and duties collection fell by 88.45 billion (from 246.7 to 158.26) in 2015/2016 financial year. Afterward, in the followed year the collection of tax and duties improved from 158.26 billion to 183.14 billion and continued to rise until it reaches the highest to ever be collected in the history of tax in Tanzania in 2019/2020 financial year (606.76 billion) with annual increase of 230.3 billion, which is also a record high.

Table 5. Annual tax and duties collection before and after introduction of e-filling system

Period	Before (Million Tshs)	After	Change in Million Tshs
2010/2011	58,207.30		–
2011/2012		86,282.10	28,074.80
2012/2013		90,253.50	3,971.40
2013/2014		95,789.80	5,536.30
2014/2015		246,675.40	150,885.60
2015/2016		158,255.60	−88,449.80
2016/2017		183,138.10	24,882.50
2017/2018		312,632.40	129,494.3
2018/2019		376,423.60	63,791.20
2019/2020		606,755.80	230,332.20

Source: TRA, 2021.

Consider the Fig. 1 below presenting the trend of change in TRA annual Tax and Duties collection before and after introduction of E-filling System. The trend of tax and duties collection over the period of ten years shows a sharp increase of 28.07 billion in 2011/2012 financial year soon after introduction of e-filling. In 2012/2013 and 2013/2014 the overall tax and duties collection rose but the incremental amount fell from 28.07 to 3.97 billion and 5.54 respectively. In financial year 2014/2015 there was a sharp increase of 150.88 billion followed by a drastic fall in 2015/2106 financial year of 88.45 billion due to the reasons explained above plus it was a general election year. The economy improved in 2016/2017 and the collection rose by 24.88 billion and continued to rise for 2017/2018 by 129.5 billion. In year 2018/2019 the increase of tax and duties collection slowed down before peaking up in 2019/2020 to record the highest collection in the country.

During focus group discussion, the interviewees pegged the increment in revenue collection to among other factors including expansion of economic activities and positive results of taxes that are collected by the governments in provision of social services such as provision of free education primary and secondary, construction of health centres in every ward and district hospital in every district, roads networks, construction of SGR, Mfugale flyover, improved police and legal service in Tanzanian courts as among those mentioned during interview. Moreover, interviewees accredited their enthusiasm and patriotic charisma were also another reason the collection had increased. The most

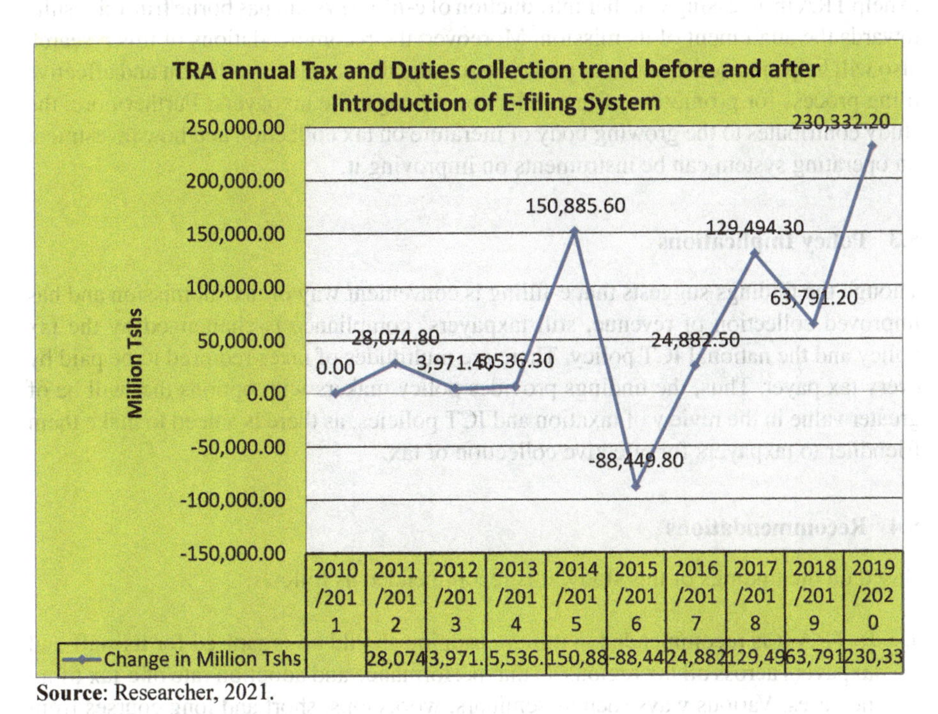

Source: Researcher, 2021.

Fig. 1. TRA annual tax and duties collection trend before and after Introduction of E-filling system. Source: Researcher, 2021.

important thing was reduction in fraud after introduction e-filling because amount of tax was automatically send to taxpayers and there is no room for negotiations and other unethical practices by the TRA staffs who are not trustworthy.

5 Conclusions and Recommendations

5.1 Conclusion

The research assessed the contribution of the e-filing system, a key modernization effort on improving tax collection. Based on the findings this study concludes that e-filing system has a positive contribution in ensuring Value Added Tax clearance is efficient and reduces compliance costs, all these are significant moves in ensuring convenience in tax compliance which generally improves tax collection.

5.2 Practical Contribution

The research assessed the contribution of the e-filing system, a key modernization effort on improving tax collection. The findings of this paper would be useful as will awaken the Government and other institutions dealing with revenue collection to make mandatory for tax submission to be done using e-filling system. The findings of this study are expected to help TRA in assessing whether introduction of e-filing system has borne fruitful results towards the attainment of its mission. Moreover, the recommendations of this research also will help enhance the current process to ensure that it delivers efficient and effective filing process for promotion of tax collection amongst the taxpayers. Furthermore, the study contributes to the growing body of literature on tax collection and how investment on operating system can be instruments on improving it.

5.3 Policy Implications

Though the findings suggests that e-filling is convenient way of tax submission and has improved collection of revenue, still taxpayers' compliance is challenged by the tax policy and the national ICT policy. There are multitudes of taxes required to be paid by every tax payer. Thus, the findings provides policy makers with options that will be of greater value in the review of taxation and ICT policies, as there is a need to make them friendlier to taxpayers for effective collection of tax.

5.4 Recommendations

Based on the findings of this study, it is recommended as follows;

i. To TRA it is recommended that more training should be organised for its staff and taxpayers across all tax regions so that performance and adoption rate of e-tax filing increases. Various ways such as seminars, workshops, short and long courses from time to time could be adopted and more education to taxpayers about benefits of e-filing system which will add more value to the performance capability and improving tax compliance.

ii. To the government it is recommended to improve the infrastructures which support e-filing system particularly internet coverage across the country for electronic filing to be of more successful on influencing tax compliance, the government should ensure efficient and robust internet connectivity in all urban and rural tax regions.

5.5 Limitation of the Study and Suggestions for Further Studies

This study had some limitations which warrant rectification in future studies. Firstly, the focus of this study was on the effectiveness of e-filing sytem on improving tax collection in a sinlge tax correction region of Ilala, Tanzania. This resulted in difficult generalization of the findings. Therefore, future study should consider expanding the geographical area by covering more than one tax region. Also, in order to generalize the results, similar studies should be conducted in other tax regions of Tanzania apart from Ilala region where this study was conducted. Moreover, this study was limited only to identification of tax payer services, examination of the influence of those tax payer services provided by the department of taxpayer services and education in Ilala tax region. As such, it is also recommended for a future study to include other factors that improving tax compliance in Tanzania as well as the probable factors that hinder the adoption of e-filling in Tanzania. Moreover, this study used only descriptive statistics which is limited in capacity especially in estimation of causal-effect relationship. To gauge the effectiveness of e-filling on the tax collection future study is suggested to apply inferential statistics such as regression and structural equation modelling, which are powerful models to predict causality.

References

Ajzen, I.: The theory of planned behavior. Organ. Behav. Hum. Decis. Process. **50**(2), 179–211 (1991)

Bird, R.M.: Tax challenges facing developing countries. Institute for International Business Working Paper (9) (2018)

Chaudhry, I.S.: Determinants of low tax revenue in Pakistan. Pak. J. Soc. Sci. **30**(2), 439–452 (2010)

Davis, F.D.: Perceived usefulness, perceived ease of use, and user acceptance of information technology. MIS Q. **13**(3), 319–340 (1989)

Fishbein, M., Ajzen, I.: Belief, Attitude, Intention and Behaviour: An Introduction to Theory and Research. Addison-Wesley, Reading (1975)

Malekani, A.A.: Access to, use and challenges of ICTs in secondary schools in Tanzania: a study of selected secondary schools in morogoro municipality. J. Inf. Knowl. Manag. **9**(2), 44–57 (2018)

Sifile, O., Kotsai, R., Mabvure, T.J., Desderio, C.: Effect of e-tax filing on tax compliance: a case of clients in Harare, Zimbabwe. Afr. J. Bus. Manag. **12**(11), 338–342 (2018)

Venkatesh, V., Morris, M., Davis, G., Davis, F.: Users' acceptance of information technology: toward a unified view. MIS Q. **27**(3), 287–294 (2003)

Sichone, J., Milamo, R.J. Kimea, A.J.: The influence of facilitating conditions, perceived benefits, and perceived risk on intention to adopt e-filing in Tanzania. Bus. Manag. Rev., 50–59 (2017). ISSN 0856-2253. eISSN 2546–213X

Azmi, A.A.C., Kamarulzaman, Y.: Adoption of tax e-filing: a conceptual paper. Afr. J. Bus. Manag. **4**(5), 599–603 (2010). http://www.academicjournals.org/AJBM

Bernasconi, M., Zanardi, A.: Tax evasion, tax rates and reference dependence. FinanzArchiv **60**(3), 422–445 (2004)

Bagozzi, R., Baumgartner, H., Youjae, Y.: State versus action orientation and the theory of reasoned action; an application to coupon usage. J. Consum. Res. **18**(3), 505–518 (1992)

Hashimzade, N., Myles, G.D., Tran-Nam, B.: Applications of behavioural economics to tax evasion. J. Econ. Surv. **27**(5), 941–977 (2013)

Piolatto, A., Rablen, M.D.: Prospect theory and tax evasion: a reconsideration of the Yitzhaki puzzle. Theor. Decis. **82**(4), 543–565 (2016). https://doi.org/10.1007/s11238-016-9581-9

Yonazi, J.: Adoption of transaction level E-government: initiatives in Tanzania. Dar es Salaam Tanzania: CLKNET The Institute of Finance Management (2013)

Sefue, O.: Electronic government. Ministry of Communication, Dar es Salaam (2014)

Rumanyika, J.D., Mashenene, R.G.: Impediments of e-commerce adoption among small and medium enterprises in Tanzania. Int. J. Inf. Technol. **32**(1), 45–55 (2014)

Lediga, C.: The impact of internet penetration on corporate income tax filing in South Africa. Ruhr Economic Papers, No. 861. RWI-Leibniz-Institut für Wirtschaftsforschung, Essen (2020). ISBN 978-3-86788-998-8. https://doi.org/10.4419/8678899

Kimea, A., Chimilila, C., Sichone, J.: Analysis of Taxpayers' intention to use tax e-filing system in Tanzania: controlling for self-selection based endogeneity. Afr. J. Econ. Rev. **VII**(2), 193–212 (2019)

World Bank: Paying taxes 2019. fourteen years of data and analysis of tax systems in 190 economies: how is technology affecting tax administration and policy? World Bank (2018)

Eichfelder, S., Hechtner, F.: Tax compliance costs: cost burden and cost reliability. Arbeitskreis Quantitative Steuerlehre (arqus), Discussion Paper, No. 212 (2016)

Mukuwa, V., Phiri, J.: The effects of e-services on revenue collection and tax compliance among SMEs in developing countries: a case study of Zambia. Open J. Soc. Sci. **8**, 98–108 (2020). https://doi.org/10.4236/jss.2020.81008

Lee, H.C.: Can electronic tax invoicing improve tax compliance? A case study of the Republic of korea's electronic tax invoicing for value-added tax. World Bank Policy Research Working Paper 7592 (2016)

Soneka, P., Phiri, J.: A model for improving e-tax systems adoption in rural Zambia based on the TAM model. Open J. Bus. Manag. **7**, 908–918 (2019). https://doi.org/10.4236/ojbm.2019.72062

Efobi, U., Beecroft, I., Belmondo, T., Katan, A.: Small businesses and the adoption of the integrated tax administration system in Nigeria. In: ICTD Research in Brief 40. IDS, Brighton (2019)

Matimbwa, H., Masue, O.S., Shillingi, V.: Technological features and effectiveness of human resources information system in Tanzania local government authorities. Am. J. Oper. Manag. Inf. Syst. **5**(3), 29–40 (2020)

Matimbwa, H., Masue, O.S.: The Influence of organizational factors on human resources information system effectiveness in Tanzanian local government authorities. ICTACT J. Manag. Stud. **6**(3), 1263–1272 (2020)

Akkaya, C., Wolf, P., Krcmar, H.: A comprehensive analysis of e-government adoption in the German household. In: 11th International Conference on Wirtschaftsinformatik, pp. 1525–1539. Technische Universität München, Leipzig, Germany (2013)

Bhuasiri, W., Zo, H., Lee, H., Ciganek, A.: User acceptance of e-government services: examining an e-tax filing and payment system in Thailand. Inf. Technol. Dev. **22**(4), 672–695 (2016)

Blume, J., Bott, M.: Information Technology in Tax Administration in Developing Countries. KfW Development Bank, Palmengartenstr, Frankfurt am Main (2015)

Chiu, C., Wang, E.: Understanding web-based learning continuance intention: the role of subjective task value. Inf. Manag. **45**(3), 194–201 (2008)

Venkatesh, V., Thong, J., Xu, X.: Consumer acceptance and use of information technology: extending the unified theory of acceptance and use of technology. MIS Q. **36**(1), 157–178 (2012)

Daka, G.C., Phiri, J.: Factors driving the adoption of e-banking services based on the UTAUT model. Int. J. Bus. Manag. **14**, 43–52 (2019). https://doi.org/10.5539/ijbm.v14n6p43

Venkatesh, V.M.: User acceptance of information technology: toward a unified view. MIS Q. **27**, 425–478 (2003). https://doi.org/10.2307/30036540

Assessing the Impact of Internal Control on the Performance of Commercial Banks in Tanzania

L. T. Bilegeya and A. Mrindoko(✉)

Department of Accountancy, College of Business Education, Dar es Salaam, Tanzania
allenmrindoko@gmail.com

Abstract. Purpose: The purpose of this study was to assess the impact of internal control on the performance of commercial banks listed on Dar es Salaam Stock Exchange (DSE) in Tanzania.

Design/Methodology/Approach: The researcher adopted a cross-sectional descriptive design where a mixed approach was applied to collect qualitative and quantitative data using questionnaires and interview guides. The study used both primary and secondary data, and the sample size was 100 commercial bank employees, and 7 branch managers as key informants. Descriptive statistics and regression models were used to analyse quantitative data, while content analysis was used to analyse quantitative and qualitative data respectively. Respondents were randomly selected. The independent variables were internal control components; control environment, information and communication and monitoring, while the dependent variable was the financial performance (ROA) of commercial banks.

Findings: The descriptive findings show that the majority of respondents to a larger extent agreed that banks implement internal control. Also, the regression results show that the control environment has a positive and highly significant (β = 1.213, t = 58.207, p = 0.000) impact on the financial performance of commercial banks. Similarly, information and communication show a positive and highly significant (β = 1.060, t = 48.989, p = 0.000) impact on the profitability of commercial banks. Furthermore, the study found that monitoring has a positive and highly significant (β = 1.049, t = 52.020, p = 0.000) impact on the financial performance of commercial banks.

Implications/Research Limitations: The study is not without limitations, the study focused on seven commercial banks listed in DSE; it is recommended that future studies consider expanding the scope to include the entire banking sector. The study further did not consider all internal control variables; there is a need for future studies to consider expanding the scope to include all control variables.

Practical Implications*:* Based on the findings this study concludes that control environment, information and communication, and monitoring are important elements which contribute to the profitability of commercial banks and thus need to be improved to ensure the system is fully safeguarded.

C. Aigbavboa et al. (Eds.): ARCA 2022, *Sustainable Education and Development – Sustainable Industrialization and Innovation*, pp. 784–806, 2023.
https://doi.org/10.1007/978-3-031-25998-2_61

Originality/Value: The study analysed the impact of control environment, communication and information and monitoring system on the performance of commercial banks. It was important because existing studies fall short of explaining these internal control components and their effect on commercial bank performance in the Tanzanian context. As such, this study has added to the body of knowledge of the importance of these internal control components to the performance of commercial banks in Tanzania.

Keyword: Control · Environment · Financial · Internal · Monitoring

1 Introduction

1.1 Background Information

The internal control system is the major part in any organization. The interests in the efficacy of internal audit and the respective elements influencing it have been a concern in the most recent years. According to Gamage, Lock and Fernando (2014), internal control is the process designed and affected by those charged with governance, management and other personnel to provide reasonable assurance about achievement of entity's objectives with regard to reliability of financial reporting, effectiveness and efficiency of operations and compliance with applicable laws and regulations. It follows that internal control is designed and implemented to address identified business risks that threaten the achievement of any of the objectives of organization (Wanjiru 2019). According to Allen (2018), an effective internal control service can, in particular, help reduce overhead, identify ways to improve efficiency and maximize exposure to possible losses from inadequately safeguarded company assets all of which can have a significant effect on the financial performance of an organization.

Banking sector is unique sector in the economy. It provides different kind of services to the customers and handles massive volume of funds daily (Gamage et al. 2014). All activities in the economy depend on the strength and stability of the banking sector (Mrindoko 2021). Due to these facts the necessity of internal control system in the banks cannot undermine. According to COSO (2004), internal control comprises of five components, namely; the control environment, the entity's risk assessment process, the information and communication system, control activities and monitoring. Only three components are covered in this study; control environment, information and communication and monitoring.

The control environment, as established by most commercial banks, sets the tone of an institution and affects the control consciousness of its human resources (Ofei et al. 2020). Control environment factors include: Integrity and ethical values; the commitment to competence; leadership philosophy and operating style; and the way management assigns authority and responsibility, and organizes and develops its people (Thornton 2004). Similarly, Ahmed and Ng'anga (2019) stated that monitoring in organizations is an essential activity that needs close attention if organizations would like to see an increase in the financial performance of their organizations. Given its importance, the management shall take full responsibility for the internal monitoring structure of the bank's financial statements and have an assessment of its efficacy (Al-waeli et al. 2020) to ensure effective financial performance.

Moreover, in managing the financial performance of commercial banks, they need to adopt effective and efficient information communication systems to ensure that timely information is generated and reported for necessary action to be taken (COSO 2004; Asiligwa and Rennox 2017). Organizations need to communicate or report to shareholders of the institution on the nature of activities and performance to create trust and confidence (COSO 2011; Hanoon et al. 2020). An effective Information and communication system ensures reliable information is identified, captured and relayed to employees to understand managements' expectations and management to understand employees concerns (Gamage et al. 2014; Asiligwa and Rennox 2017).

The banking industry in Tanzania is growing very fast following the liberalization of banking sector in 1991, and it consist of 34 commercial banks, 20 licensed financial institutions and 2 development banks. In spite of some successful reforms, the banking industry tends to play only small role to the country's economy, the fact that the industry is plagued by bad loans, the lenders have chosen to go slow in terms of credit off take (Ambrose 2017). The industry saw a combination of various internal and external events caused by poor operational risk management in banks that kept market turbulent, interest rates high and investor confidence low, resulting in shrinking investment and country's Gross Domestic Product (GDP), (Kambanga 2009; Lyambiko 2015; Kingu et al. 2018; Macha and Gwahula 2018). Besides the expansion of banking industry in Tanzania, it calls for a sound management practices and techniques for their survival, as well as, to be competitive enough in this turbulent business environment, as it is a key driver behind profitability. To do so, it is a must to identify, select and apply the appropriate measurement and management mechanisms, which are simple but effective management tools or techniques.

Over the years there has been an increase in the number of bank failures in developing and developing economies (Basel 2004; Mrindoko 2021). In Tanzania empirical evidence suggests that bank failures are due to external (macro) and internal (micro) factors (William 1995; Ng'owo 2013). Included in the internal factors are reckless lending, corruption, fraud and dishonest, embezzlement, management deficiencies, poor credit documentation, risk assessment methods, lending to the insiders and poor supervision capacity (Lyambiko 2015; Kingu et al. 2018; Lema 2019). External factors such as poor monetary and fiscal policies, bank deregulation/regulation, policies and procedures, lack of information among bank customers, homogeneity of the banking business and connections among banks do cause bank failure (Lotto and Kakozi 2016). In highly controlled economies, government and political interference in the banking policies and operations also contributed to bank failure (Murerwa 2015). For some developing countries, globalization is also contributing to the bank failure because small and inexperienced banks fail to cope with the stiff competition of the big multinational banks in terms of products and services offered Kimei (2007). This study focused on control of internal factors can improve bank's financial performance.

It is undoubted that without having proper internal control system and practicing the same, the commercial banks face numerous problems and risks. If the situation continues it will lead to disputes for bank's growth, image, employees and customers' dissatisfaction and employees' welfare, loss of investors, insolvency and finally it leads the problems to the government. The poor financial performance of many commercial banks in Tanzania is low (Kingu et al. 2018), and this is attributable to much internal fraud, inefficiencies, negligence, and risk (Kimei 2007; Ng'owo 2013; Lyambiko 2015; Mrindoko 2021). However, fraud and risk prevention, detection and control are among the functions of internal control, which aim at eliminating fraud, risk and fraudulent tendencies in the commercial banks (Lyambiko 2015). The inefficiencies of commercial banks is evident as Bank of Tanzania (BoT) have in recent years revoked license of seven banks and put others under watch (Mrindoko 2021. Therefore, internal control system is highly significant in fraud prevention and detection, risk minimization, warning the commercial bank's management of any inefficiencies and under-performance of employees. In turn contributes to the effectiveness and efficient use of organizational resources.

Despite its importance, internal control in commercial banking industry has not been adequately researched in Tanzanian context. Lyambiko (2015) conducted a study to investigate the effect of operational risk management practices on the financial performance in commercial banks in Tanzania. The author's focused on only a single variable component of internal control; risk assessment. As a result, risk assessment as component of internal control is not covered in this study. Not only Lyambiko (2015), but also Ng'owo (2013) focused on a single variable of risk assessment; fraud. Ng'owo (2013) assessed the influence of internal fraud to the performance of Tanzania commercial banks using National Bank of Commerce (NBC) as a case study. It is quite clear from the studies that none of these studies adequately discussed the impact of internal controls on the financial output of the banking sector in Tanzania. Therefore, there is relatively little evidence of the effects of internal controls on the commercial banking sector in Tanzania. Accordingly, it is imperative to conduct a study to investigate the impact of internal control on the financial performance of commercial banks in Tanzania. By addressing the question "what is the effect of internal controls on the financial performance of banks in Tanzania?" this study seeks to contribute on narrowing this dearth gap in knowledge and practice.

1.2 Statement of the Problem

According to Ng'owo (2013), Ofei et al. (2020) and Mwende and Njogu (2021), one of the main reasons of banking failures in developing country like Tanzania which results in major financial loss and even bankruptcy is due to inefficiencies and disregard to rules and regulations. In addition, failures of commercial banks are due to high risks taken by the bank management on an excessive scale and inability of controlling them (Lyambiko 2015; Wanjiru 2019). Lack of an internal control system, whose duty is to keep the risks under control or major breakdowns within an existing internal control system poses a threat against the success of the banking sector (COSO 2011; Mwende and Njogu 2021).

In the recent past there has been failure of some of the largest commercial banks in Tanzania this trend puts into question the ability of the internal control systems and the Central Bank of Tanzania prudential guidelines to steer the commercial banks stability and performance (Sadick 2020). The financial sector in Tanzania has seen some turbulence in 2018 with BOT revoking the licenses of 7 commercial banks namely Covenant

Bank, Efatha Bank, Njombe, Community Bank, Kagera Farmers' Cooperative Bank, Meru Community Bank, Mbinga Community Bank, and FBME due to undercapitalization. The well-managed control environment, communication and information system and monitoring system have positive impact on the financial performance of commercial banks, but if there is poor internal control, it will have a negative impact on the profitability of the bank (Asiligwa and Rennox 2017).

Moreover, very narrow focused research studies have been conducted in Tanzanian commercial banking sector to assess the impact of internal control on financial performance. Lyambiko (2015) and Ng'owo (2013) had focused on the effects of operational risk and internal fraud ton commercial bank performance respectively. To bridge this knowledge gap, this study has examined the effect of internal control focusing on control environment, communication and information, and monitoring. This study covered seven banks which are listed in the Dar es Salaam Stock Exchange (DSE) because they are commendable to have well developed and effective internal control.

2 Literature Review

2.1 Theoretical Literature Review

2.1.1 Agency Theory

The Agency theory advanced by Jensen and Meckling (1976) establishes that contracting is used by principals and agents to increase their resources. It suggests that there is relationship between ownership structure and banks performance. The theory was advanced by Fama and Jensen (1983) and Eisenhardt (1998). The principal (owners of business) contracts an agent (managers of business) who is mandated to run a business on their behalf through delegated decision-making authority. The key insight of Jensen and Meckling (1976) was to model the relationship between owners and managers. Therefore, at the heart of the agency theory is the separation of ownership and control. The owners contract the managers to perform the controlling tasks of a firm, and as both sought to maximize their own utility and are self-interested a conflict of interest arises, which is termed as the agency problem. As the managers have the effective control of the firm, they have the incentive and the ability to consume benefits at the expense of the owners (Coarse 1937).

In order to resolve this agency problem, internal management should be used as a safety net (Jensen and Payne 2003). Effective internal controls may address the agency problem and minimize agency costs through elimination of information asymmetry that may exist between the principal and the agent. The theory guided this study in answering the question on effect of internal control components: control environment, communication and information and monitoring on the financial performance of commercial banks listed at DSE. Moreover, this theory is adopted because, from the agency theory perspective, non-executive directors and independent directors contribute to effective corporate governance by exercising control over senior managers' decision-making. After all, they are seen as the check and balance mechanism to enhance the board's effectiveness. The theory has been generally considered the starting point of this study.

2.1.2 Institutional Theory

It originated from works done by Meyer and Rowan (1977) and DiMaggio and Powell (1983). This theory proposes that organizations develop and design structures, processes and systems not primarily on rational economic cost-benefit analysis but because they are more or less required to incorporate new practices and procedures. According to Meyer and Rowan (1977), "Organizations are driven to incorporate the practices and procedures defined by prevailing rationalized concepts of organizational work and institutionalized in society. Organizations that do so increase their legitimacy and their survival prospects, independent of the immediate efficacy of the acquired practices and procedures." Organizational structures such as internal controls are symbols of conformity and social accountability. From the literature available, firms not only compete for clients and resources but also for institutional legitimacy and social fitness (DiMaggio and Powell 1983). This theory is adopted because there is a consensus among researchers that institutional theory offers a useful lens for investigating the role and activities of internal control, particularly when used in conjunction with agency theory. This theory therefore is in tandem with this study.

2.1.3 Stewardship Theory

The stewardship theory of corporate governance holds that, because people can be trusted to act in the public good in general and in the interests of their shareholders in particular, it makes sense to create management and authority structures that, because they provide unified command and facilitate autonomous decision making which enables companies to act quickly and decisively to market opportunities (Donaldson 1991). Theory emphasize that the stewards who are board of directors and managers for the company, should work together to maximize the profits of the shareholders and should not bear any conflicting interests. Stewardship theory assumes that managers are honest, and motivated more by intrinsic rewards than extrinsic rewards, and self motivated to maximize collective interests (Nicholson and Kiel 2007; Davis et al. 2003). The Stewardship theory encourages openness in performing organizational tasks (Wanjiru 2019). This will ensure high collaboration to maximize and protecting shareholder wealth. These executives therefore protect their reputations by using all applicable internal control systems to maximize shareholder's wealth as well as improve the firm's long term performance. The stewardship theory is adopted for this study because it strongly argues that managers and boards of directors are good stewards of a firm and they should be given utmost trust.

2.2 Empirical Literature Review

2.2.1 Impact of Control Environmental on Financial Performance of Firm

Ofei et al. (2020) examined the effect of internal control environment and monitoring activities on the performance of banks in Ghana. The study found that that there was a weak significant correlation between internal control environment and financial performance of banks in Ghana. It was further observed that there was no significant effect of internal control environment on the financial performance of banks in Ghana, whereas there was no significant effect of control environment and control activities on the financial performance of banks in Ghana. Sadick (2020) assessed the effect of internal control on the financial performance of commercial banks, a case study of CRDB bank Mbeya branch. The study findings established a positive and significant relationship between components of internal control which are control environment, risk assessment, information and communication, control activities and monitoring strategies and financial performance of commercial banks.

Donati (2019) assessed the effect of internal control on the financial performance of mining industry in Tanzania. The findings of the study revealed that control environment, risk assessment, control activities and internal auditing had a positive and significant effect on financial performance of the mining company at 5% significance level. The internal control system had a positive and significant influence on financial performance. Lyambiko (2015) analysed the effect of operational risk on the financial performance of commercial banks in Tanzania. The findings from the study confirmed that the independent variables such as such as Credit risk, Insolvency risk and Operation efficiency had varying degrees of relationship with the financial performance of the commercial banks in Tanzania. The study also revealed that Operations risk management positively influenced returns of the commercial banks in Tanzania. This study also established that Operations efficiency were positively correlated with the financial performance of the commercial banks in Tanzania while the Credit risk and Insolvency risk rate negatively influenced the financial performance of commercial banks Tanzania. Based on the reviewed literature the following hypothesis was formulated;

H_1: There is no significant impact of control environmental on financial performance of commercial banks in Tanzania.

2.2.2 Impact of Communication and Information System on Financial Performance of Firm

Mbilla et al. (2020) reported that Information and communication have a weak significant effect on financial performance. There was no significant effect between Monitoring and financial performance. Mwende and Njogu (2021) investigated the effect of the internal control systems on the financial performance in the Nairobi Stock Exchange (NSE) listed banks in Kenya. The study posits that risk assessment and information and communication have a weak significant effect while the monitoring, control environment and the control activities have the weakest influence on ROA and ROE in the NSE listed banks in Kenya. Another study conducted by Asiligwa (2017) revealed that the banking sector enjoys a strong financial performance partly because of implementing and maintaining effective internal controls. The existence of effective internal control is attributed to the highly regulated and structured environment in the banking sector.

Ahmed and Ng'anga (2019) examined the effect of Internal Control Practices and Financial Performance of County Governments in the Coastal Region of Kenya. The study adopted a descriptive research design. The study found a positive and significant effect between risk assessment, monitoring, control environment, information and communication on financial performance. The study concluded that the risk identification and mitigation play the most significant role in influencing the financial performance of the County governments. William and Kwasi (2013) conducted a research on the significance of internal control in the rural banking sector, they wanted to examine whether effective internal control system is adopted by the bank in the Eastern Region of Ghana. The study concluded that the internal control systems assist management and auditors in executing their duties. The internal controls are established by management by adopting policies and procedures with each system of control having specific functions to perform. Collins (2014) examined the effect of internal control on the financial performance of microfinance institutions in Kisumu central constituency, Kenya. Findings revealed that there is a positive relationship between internal controls. Thus, this study hypothesises that;

H_2: Information and communication system does not have significant effect on financial performance of commercial banks in Tanzania.

2.2.3 Impact of Internal Monitoring on Financial Performance of Firm

Lily (2017) found that there was effect between control environment, control activities, ICT, and monitoring and the dependent variables; return on assets, net income and liquidity as was indicated in the regression analysis. The findings indicate a non-significant effect between a unit increase in control environment and an increase in financial performance of commercial banks, the banks standard operating procedures, ethics and integrity, commitment to shareholders interests and organization created a control environment aimed at increasing the banks financial performance. The findings further indicate that there was a non-significant effect between control activities and financial performance of commercial banks. The findings further indicated that improved ICT infrastructure, effective and efficient Procedures, had a significant and positive effect on increased commercial banks financial performance. Wanjiru (2019), investigated the effect of internal control on financial performance of 40 commercial banks in Kenya from 2014–2018. Descriptive statistics and correlation between the factors were done. ANOVA and F-Statistics at a significance of 5% level were utilized to decide the regression model significance. The study results revealed that how commercial banks perform is contributed by internal control.

Mbilla et al. (2020) investigated the impact of monitoring, information and communication on banks performance in Ghana. In this quantitative study, 300 representatives from twelve listed banks were engaged. Descriptive and regression analysis was performed on the field Data. The study result shows that information and communication have a weak significant effect on financial performance. There was no significant effect between Monitoring and financial performance. Hanoon et al. (2020) investigated the impact between Internal Control Components (ICC) and the Financial Performance (FP) of the Iraqi banking sector. The study results indicated that the Internal Control Components have a significant impact on Financial Performance. The positive significant relation were control activity ($\beta = 0.311$, $p < 0.05$), followed by risk assessment ($\beta = 0.203$, $p < 0.05$), monitoring ($\beta = 0.176$, $p < 0.05$), control environment ($\beta = 0.164$, $p < 0.05$) and information & communication ($\beta = 0.157$, $p < 0.05$). Based on the literature review this study hypothesized that;

H_3: There is no significant relationship between monitoring and financial performance of commercial banks in Tanzania.

2.3 Financial Performance

Financial performance refers to the ability to operate efficiently, profitability, survive growth and react to the environmental opportunities and threat (Hoskisson et al. 2011). Financial performance can be measured using return on equity, solvency, sales growth, liquidity and profitability as well as value of long-term investment, financial soundness and use of corporate assets (Huskisson et al. 2011; Sørensen 2012). Huskisson et al. (2011), mention accounting based performance using three indicators; Return on Assets (ROA), Return on Equity (ROE) and Return on Investment (ROI) as major indicators of financial performance in commercial banks. However, ROA is probably the most critical single ratio in comparing SMEs' efficiency and operating performance as it indicates the returns generated from the assets that SMEs owns (Phillips et al. 2012). The greater Return on Assets (ROA) is considered, the better the SMEs' performance because of the greater rate of return on investment. The formula for return on assets is: ROA = Net Profit After-tax/ total assets.

3 Research Methodology

3.1 Area of the Study

The study was conducted in the United Republic of Tanzania; the banks chosen were solely those which are enlisted at Dar es Salaam Stock Exchange (DSE) but the findings were inferred to all banks in Tanzania and other developing countries in Sub-Saharan Africa. The chosen banks are: Yetu Microfinance Bank, NMB Bank Plc, Mkombozi Commercial Bank Plc, Mwalimu Commercial Bank, DCB Commercial Bank Plc, CRDB Bank Plc & KCB Bank Plc.

3.2 Research design and Approach

The primary purpose of the study was to assess the impact of internal monitoring on the bank performance. Based on the nature of the research questions and specific objectives, this study used cross-section descriptive research design to determine the influence of internal control on financial performance of commercial banks listed on the Dar es Salaam Stock Exchange. This research adopted mixed concurrent approach which entails the use of both qualitative and quantitative methods simultaneously during data collection. Data were collected from different sources; quantitative data from bank staff survey and documentary review, while qualitative data were collected through in-depth interview and documentary review. However, distinct analysis was performed for quantitative and qualitative data. Quantitative results were reported first and qualitative results supplemented explanations of quantitative findings. The mixed approach concurrent approach was adopted purposely to increase the validity of findings through triangulation that provides the ground for the complementarities of one method with the other (Creswell 2014).

3.3 Population, Sample and Sampling Procedure

The population for the study was seven listed Commercial banks on the Dar es Salaam Stock Exchange. In this study the population and sampling frame coincide and so is the same, the seven commercial banks listed at DSE. In this study, a sample was drawn from seven commercial banks listed on the Dar es Salaam Stock Exchange. The study employed a multi-stage sampling procedure as follows. Firstly, commercial banks were purposively selected. Secondly, the sampling frame was divided into seven strata using stratified random sampling based on the seven banks that are listed at DSE; Yetu Microfinance Bank, NMB Bank Plc, Mkombozi Commercial Bank Plc, Mwalimu Commercial Bank, DCB Commercial Bank Plc, CRDB Bank Plc& KCB Bank Plc, because the population from which the sample was drawn did not constitute a homogeneous group (Kothari 2004). Thirdly, the calculation of the overall sample size was done by using Yamane formula which gives the degree of accuracy of the sampling technique and take into account the margin of error of 5%.

Fourthly, proportionate stratified random sampling was used where the overall sample size obtained in (2) was divided to get the sampling fraction for each listed commercial bank. Thus, drawing a sample from each stratum in proportion to the stratum share in the total population gave proper representation to each stratum and higher statistical efficiency than simple random sampling. This was done using proportionate stratification formula proposed by Cochran (1977). Fifthly, the sample in each stratum was selected using systematic sampling where sampling fraction was used to get the interval and the random start from each region, in this case simple random sampling will be used to identify the random start and the remaining cases of the sample was selected systematically at fixed interval.

To ensure that the sample size is appropriate to represent the opinions of target population, this study employed a formula developed by Yamane in 1967 with confidence level of 95%. This level of certainty represent the characteristics of the total population, therefore the marginal of error (e) was 0.05. This formula was adopted because it is simple and therefore provides accurate sample size, and it fits with the available parameter N. The formula expressed as:

$$n = \frac{\mathrm{N}}{1 + \mathrm{N}(\mathrm{e}^2)}$$

where;

n = Sample size
N = Total number of target population
e = Precision (5%)

Therefore, the sample size was 140 employees from seven commercial banks which are listed at DSE. However, the returned questionnaires were 127, which is approximately 90.3% response rate. Moreover, in every listed commercial bank one branch manager was selected to participate in this study as key informant. In total seven (7) banks branch managers were used as key informants (Table 1).

Table 1. Sample size proportionate distribution

Bank name	Proportionate sample size	Usable questionnaires
NMB Bank Plc	24	21
Yetu Microfinance	14	14
Mkombozi Commercial Bank Plc	18	17
Mwalimu Commercial Bank	22	18
DCB Commercial Bank Plc	18	16
CRDB Bank Plc	24	22
KCB Bank Plc	20	19
Total	**140**	**127**

Source: Yamane sampling formula using commercial banks employees' data, 2022

3.4 Data Analysis Plan

3.4.1 Quantitative Data Analysis

Cross-sectional data analysis largely follows OLS linear regression models (Torres-Reyna 2007) and descriptive statistics. This study adopted descriptive statistics and multiple linear regression models using SPSS. This helped to analyse the impact of internal control components on performance of listed commercial banks. Descriptive statistics mapped the extent components of internal control affect performance of commercial banks. The multiple linear regression techniques were used to estimate the impact of internal control components on performance of listed commercial banks. An F- test was used to establish the significance of the independent variables (control environment, information & communication, and monitoring) against the dependent variable (ROA). The test of significance for the linear regression was conducted using probability value (P Value) where the P – Value was $P = 0.000$ ($P \leq 0.05$ showing statistical significance). Before data analysis is done the data diagnostic tests were undertaken. Various preliminary tests were carried out to test for suitability of data for modelling and econometric analyses. There were no violations of multiple linear regression assumptions. Hence, data fitted well in the regression equation models.

3.4.2 Analytical Model

A regression analysis of the dependent and independent variables was used to predict the relationship between these variables. The linear regression model given in Eq. (1) below was adopted for this study.

$$Y = \beta 0 + \beta 1X1 + \beta 2X2 + \beta 3X3 + \beta 4X4 + \beta 5X5 + \varepsilon \quad (1)$$

where;

Y = Financial Performance (from ROA)
$\beta 0$ = Intercept (Constant)

$\beta1$–$\beta5$ = Measures of sensitivity of Variable X to Variable Y
X1 = Control Environment.
X2 = Information and Communication
X3 = Monitoring
E = Error term for the model

3.4.3 Qualitative Data Analysis

Content analysis was used to analyse the captured data from the field through semi-structured interview. Its choice was instigated by its ability to pinpoint themes contained in data (de Reuver et al. 2009). Both conceptual and relational analyses were applied. The results were presented as textual display to further explain the quantitative results. Moreover, the findings were used to explain the research hypotheses and draw implications on the performance of listed commercial banks.

3.5 Reliability

Cronbach's alpha was used to measure reliability. As a rule of thumb, the values above 0.7 represent an acceptable level of internal reliability (Hair et al. 1998). Table 2 shows that for all objectives the Cronbach's Alpha was above 0.7, hence data were reliable.

Table 2. Reliability of instrument on the basis of pilot test

Dimension	Cronbach's alpha	No. of items
Control environment	0.884	14
Information and communication	0.867	12
Monitoring	0.889	7

Source: Field Data 2022

4 Results and Discussion

4.1 Demographic Features of Surveyed Respondents

Results in Table 3 shows 44% (n = 56) were between 26–35 years, 37% (n = 47) were between 36–45 years and 14% (n = 17) were between 46–55 and 05% (n = 7) were above 56 years. The minimum age was 27 and the maximum age was 58 years. The mean age was 36 years. This implies that the majority of participants are still economically active. The results also show that, of 127 employees of commercial banks 52.7% were male and 47.3% were female. The results show that the sex of the participants was all inclusive, and there was no trace of biasness in the study.

Table 3. Distribution of demographic characteristics of respondents (n = 127)

Variable/parameter	Measurement	Frequency	Percentage
Sex	Male	67	52.7
	Female	60	47.3
Age	26–35	56	44.0
	36–45	47	37.0
	46–55	17	14.0
	56+	7	5.0
Business experience	0–5 years (low)	55	43.0
	6–10 years (medium)	51	40.0
	11–20 years (high)	17	14.0
	21+ years (highest)	4	3.0
Education level	Secondary	15	12.0
	Diploma	42	33.0
	Bachelor	62	49.0
	Masters	8	6.0

Source: Field data 2022

Experience teaches the most. The study findings in Table 3 show that most of participants 43.00% (n = 55) have experience of less than six years, and 40.00% (n = 51) have six to 10 years experience in banking sector and 14.00% (n = 17) and 3% (n = 4) possess 11 to 20 years and more than 20 years experience respectively. Therefore, data obtained from these respondents is relevant and accurate since most of them are highly experienced hence could understand the questions. Also, results show that out of 127 respondents, 49% (n = 62) have bachelor degree followed by those with diploma 33% (n = 42), then 12% (n = 15) had secondary education/ certificates and 5% (n = 6) had master's degree. This observation shows that about 88% of the respondents had education level of diploma and above, this is an indication that participants are well educated and would be in a position to analyze critically research questions and provide a fair view.

4.2 Descriptive Results

4.2.1 Control Environment

The findings in Table 4 indicate that, the statement that; organizational structure does adequately reflect chain of command scored the highest average of 4.81 (SD = 0.033). This item has very little dispersion and variability around the mean of the data set, on average. This implies that majority of the respondents agreed to a great extent that organizational structure does adequately reflect chain of command within commercial banks in Tanzania. The statement that; the bank has a code of conduct to guide behaviour, activities and decision- making recorded second highest average of 4.73 (SD = 0.087) the standard deviation value signify that data variability around the mean is very small. In addition, the statement which asks whether there is an honest and fair dealing with all stakeholders for the benefit of the organization averaged 4.69 (SD = 0.347) signifying that the item possess consistency data. These findings propose that, to a great extent commercial banks have established code of conduct to guide behaviour; activities and decision- making, and the banks are also honest and fair dealing with all stakeholders for the benefit of the organization.

Table 4. Mean standard deviation for control environment (n = 127)

Statement	N	Mean	Std. deviation
The bank has an accounting and financial management system that is adhered to	127	4.34	0.700
There is great degree of integrity in execution of the roles	127	3.67	1.120
Management of the bank create the conducive working environment for employees to act as per the companies' policies	127	4.33	0.735
Management of the bank committed to the operation of the system and closely monitor implementation of internal control Systems	127	4.32	0.700
The bank has a code of conduct to guide behavior, activities and decision-making	127	4.73	0.087
The bank has an objective, independent and active audit committee	127	3.94	0.900
The Board of Directors and its Committee are independent of Management	127	4.55	0.697
Ethical values are upheld in all the bank's decisions	127	4.18	0.785
The Board, the Management and Employees are all committed to competence and integrity	127	3.79	1.116
There is an atmosphere of mutual trust in our Bank	127	3.96	0.804
There is an honest and fair dealing with all stakeholders for the benefit of the organization	127	4.69	0.347
Management is committed to the operation of the bank	127	3.91	0.905
There are formalized policies and procedures for all major operations of the bank	127	3.85	1.102
Organizational structure does adequately reflect chain of command	127	4.81	0.033

Source: Field Data 2022

The findings show that the lowest mean score was for statement which says, the bank has integrity in execution of the roles (M = 3.67, SD = 1.120), with high variability around the population mean of data set. However, still the mean score suggests that respondents have agreed that there is great degree of integrity in execution of the roles. Therefore, from the findings of this study it can be deduce that the listed commercial banks at DSE had effective control environment. To ascertain the impact of control environment the study further tested the effect of control environment on the financial performance in listed commercial banks in DSE.

4.2.2 Information and Communication

The findings in Table 5 show that, factors measuring communication and information have positive Means of between 3.44 and 4.82 and Standard Deviations of between 0.004 and 1.201. The results also indicate that the statement which says management has identified individuals who are responsible for coordinating the various activities within the entity have highest average of 4.82, and very little dispersion and variability around the mean of the data set, on average (SD = 0.004). This implies that majority of the respondents agreed to a great extent that management has identified individuals who are responsible for coordinating the various activities within the entity.

Table 5. Mean and standard deviation for communication and information (n = 127)

Statement	N	Mean	Std. deviation
The bank has put up and acquired necessary ICT infrastructure	127	3.78	0.998
The bank provides feedback to the junior officers about the operation of the internal controls	127	3.99	0.922
All the employees understand the concept and importance of internal control including the division of responsibilities	127	4.76	0.300
Communication helps to evaluate how well the guidelines and policies of the bank are working and are implemented	127	3.89	0.900
The bank has put in place an ICT policies and procedures to guide its operations	127	4.39	0.754
The reporting system on the organizational structure spells out all the responsibilities of each department in the bank	127	4.80	0.004
The bank information systems have capability to support the overall bank strategy and strategic plans	127	4.43	0.749
All departments receive the timely information needed to achieve its objectives	127	4.75	0.304
The bank has established a continuity/disaster recovery plan for all primary information systems	127	3.62	1.001
Accurate, and useful information is identified and communicated in a timely manner to enable people perform their responsibilities	127	4.45	0.589
Communication lines are sufficient to meet bank's needs, both as senders and receivers of information in the organization	127	3.44	1.201
Management has identified individuals who are responsible for coordinating the various activities within the entity	127	4.82	0.004

Source: Field Data 2022

Besides, the statement which scored lowest mean was 'communication lines are sufficient to meet bank's needs, both as senders and receivers of information in the organization' (M = 3.44, SD = 1.201), which also suggests that the respondents have agreed. Therefore, the commercial banks in Tanzania have established both internal and external communication lines that are sufficient for bank activities. In general, the findings of this study suggest that listed commercial banks at DSE had effective information and communication systems to pass important information from top to lower level, and vice versa. To ascertain the impact of information and communication systems the study further tested the effect of control environment on the financial performance in listed commercial banks in DSE.

4.2.3 Monitoring

The findings in Table 6 point out that, factors measuring monitoring have positive Means of between 3.98 and 4.82 and Standard Deviations of between 0.006 and 0.994. The results also indicate that all statements had high mean and low variability on average. However, the item which says the bank frequently monitor financial transactions regarding bank activities scored the highest average of 4.82 (SD = .0.006), while the statement that says if There are independent process checks and evaluations of controls activities on an ongoing basis recorded lowest average of 3.98 (SD = 0.994) signifying that data variability around the mean is very small. In general, the findings show that greater number of respondents highly acknowledged that commercial banks which are listed at DSE regularly conduct monitoring to check for any discrepancies in the process of attaining bank goals and objectives.

Table 6. Mean and standard deviation for monitoring (n = 127)

Statement	N	Mean	Std. deviation
The bank has assigned responsibilities for the timely reviews of the audit reports and resolution of any non-compliance items noted in the audit reports	127	4.76	.124
Internal control system in departments is conducted periodically to ascertain its effectiveness	127	4.64	.090
There are independent processes, checks and evaluation of control activities on an ongoing basis	127	4.70	.048
Management is closely monitoring the implementation of the internal control system in our bank	127	4.54	.379
Monitoring has helped in assessing the quality of performance of the bank over time	127	4.72	.240
There are independent process checks and evaluations of controls activities on an ongoing basis	127	3.98	.994
Bank frequently monitor financial transactions regarding bank activities	127	4.82	.006

Source: Field Data 2022

4.3 Regression Results

4.3.1 Model Summary

In addition to descriptive analysis, the study conducted a cross-sectional OLS multiple regression to determine the effect of control environment on commercial banks' performance. Table 7 shows the value of R, or the *multiple correlation coefficients* of predictors is 0.979. Since this is a very high correlation, our model predicts the effect of internal control components on performance of commercial banks rather precisely. The R-Square, which is the proportion of variance in the dependent variable that can be explained by the independent variables, has a value of 0.958. This means that our predictors (control environment, information and communication, monitoring) explain 95.8% of the variability of our dependent variable, *performance commercial banks listed at DSE* whilst, 4.2% (100%-95.8%) of the variation is caused by factors other than the predictors included in this model. However, a value of adjusted R square 0.952 indicates true 95.2% of variation in the performance of commercial banks s is explained by the control environment, information and communication, and monitoring, which are to keep in the model. And since there is small discrepancy (.958-.952 = .006) between the values of R-squared and Adjusted R Square, the model fit well. The standard error of estimates shows that on average, our estimates of commercial banks' performance was wrong by 0.189 which is ignorable.

Table 7. Model summary for internal control and financial performance (n = 127)

Model	R	R square	Adjusted R square	Std. error of the estimate
1	.979[a]	.958	.952	.2050

Source: Field Data (2022)

a. Predictors: (Constant), Control Environment, Information and Communication, Monitoring

4.3.2 Anova

The results of ANOVA in Table 8 show that, the sum of squares due to regression is 611.017 while the mean sum of squares is 198.032 with 3 degrees of freedom. Also, the result shows that the sum of squares due to residual is 4.620 while the mean sum of squares due to residual is 0.39 with 124 degrees of freedom. In addition, the results in Table 8 show that; control environment, information and communication, and monitoring statistically significantly predict the financial performance of commercial banks listed at DSE, F (3, 124) = 5549.088, p (.000). The F-ratio in the ANOVA, tests whether the overall regression model is a good fit for the data. Since the p value is less than 0.05 implies that the relationship between independent variables and dependent variable is significant at 95% level of significance, the model is therefore is significant for the study and prediction. In other words, the regression model is a good fit of the data.

Table 8. ANOVA[a] for internal control and financial performance (n = 127)

Model		Sum of squares	Df	Mean square	F	Sig.
1	Regression	611.017	3	198.032	5549.088	.000[b]
	Residual	4.620	124	.039		
	Total	615.970	127			

Source: Field Data (2022)
a. Dependent Variable: Financial Performance (ROA)
b. Predictors: (Constant), Control Environment, Information and Communication, Monitoring

4.3.3 Regression Coefficients

Coefficient of independent variables (control environment, information and communication, and monitoring) and the dependent variable (financial performance) are presented in Table 9. The results show that control environment is the highest predictor contributing (0.480) to explain performance of commercial banks, followed by monitoring (0.450) and lastly information and communication (0.416). According to Dhakal (2018), standardized coefficients measure how much the outcome variable increases when the predictor variable is increased by one standard deviation assuming other variables in the model are held constant. These are useful measures to rank the predictor variables based on their contribution (irrespective of sign) in explaining the outcome variable. Moreover, the findings in the Table 9 allow us to check significance and predictive contribution of predictor variables, but also for multicollinearity. The findings show that VIF for all predictor variables was <10 and Tolerance > 0.1. According to Dhakal (2016; 2018), criteria for multicollinearity free model the VIF should be < 10 (or Tolerance > 0.1) for all variables in the regression model. Therefore, since all predictor variables in this study met the criteria, and then there was no multicollinearity in the estimated model.

Table 9. Regression results for internal control and financial performance (n = 127)

Model		Unstandardized coefficients		Standardized coefficients	t	Sig.	Collinearity statistics	
		B	Std. error	Beta			Tolerance	VIF
1	(Constant)	.141	.097		1.201	.252		
	Control environment	1.213	.020	.480	58.207	.000	.905	1.200
	Info. and Comm.	1.060	.023	.416	48.989	.000	.799	1.199
	Monitoring	1.049	.022	.450	52.020	.000	.800	1.250

Source: Field Data (2022)
Dependent Variable: Financial Performance (ROA)

Furthermore, Table 9 shows the parameter estimates, standard error and the associated p-value of the fitted regression model for the effects of internal control (control environment, information and communication, and monitoring) on the performance of commercial banks. According to the model the control environment variable was positively (B = 1.213) related to financial performance measured by ROA. Also, information and communication (B = 1.060), and monitoring (B = 1.049) were also positively related to financial performance measured by ROA. From the model, the results in Table 9 show that the constant 0.141 is the predicted value for the dependent variable if all independent variables take a value of 0. It means that an average commercial bank performance '*return on asset* would be 0.141 when all predictor variables take the value zero. The data findings analyzed also showed that taking all other independent variables at zero, a unit increase in control environment will increase the financial performance commercial banks 1.213 unit. Also, a unit increase in information and communication will increase the financial performance by 1.060 units, and also a unit increase in monitoring will lead to a 1.049 increase in financial performance. Therefore, the fitted model is:

Financial performance = 1.41 + 1.213X1 + 1.060X2 + 1.049X3 + 0.097

The results of the regression model revealed that, the effect of control environment on the performance of commercial banks was positive and highly significant (β = 1.213, t = 58.207, p = 0.000). This means that a unit increases in control environment is associated with an increase of financial performance by 1.213 units. The standard error of estimate reveals a small chance 0.020 that the estimate could be wrong. Thus, the hypotheses H_1: ***There is no significant impact of control environmental on financial performance of commercial banks in Tanzania*** was rejected. The plausible explanation is that, having integrity in execution of bank activities, audit committee, accounting and financial management system that is adhered to, formalised policies and procedures for bank operations, honest and fair dealings, committed board, management and employees, and code of conduct to guide employees' behaviour, activities and decision making process; help the commercial banks to achieve long term objectives and improve their financial performances. The findings are in line with Kinyua et al. (2015), Bayyoud (2015), Donati (2019), Sadick (2020) and Mwende and Njogu (2021). Therefore organizations need to develop their internal controls with complete emphasis on the internal control environment ensuring the potential for profitability of organisation.

In case of information and communication, the results of the regression model show positive and highly significant effect (β = 1.060, t = 48.989, p = 0.000) of information and communication on the performance of commercial banks. This means that a unit increases in information and communication is associated with an increase of financial performance by 1.060 units. The standard error of estimate reveals a small chance 0.023 that the estimate could be wrong. As a result, the hypotheses H_2: ***Information and communication system does not have significant effect on financial performance of commercial banks in Tanzania*** was rejected. The implication of this finding is that, information and communication attributes are very crucial for effective utilisation of banks' activities. When there is clear channel of communication, timely sharing of information within the bank, informing employees of every important information for their daily activities, bank has acquired the necessary ICT infrastructure for communication and information sharing, bank communicate policies and guidelines to lower level, and accurate and useful information is communicated in open communication channels; the performance of commercial banks will boosted. The findings corroborate with Lily (2017) and Nwende and Njogu (2021) who reported that ICT infrastructure is essential for effective communication and information sharing within the firm, and can improve firm's performance. This study show that the efforts done by commercial banks in order to improve the productivity through communication and information sharing is commendable. Communication and information sharing increase understanding and commitment to the attainment of firm's goals.

Furthermore, the results of the regression model show that monitoring has positive and highly significant effect ($\beta = 1.049$, t = 52.020, p = 0.000) on the performance of commercial banks. This means that a unit increases in monitoring is associated with an increase of financial performance by 1.049 units. The standard error of estimate reveals a small chance 0.022 that the estimate could be wrong. As a result, the hypotheses H_3: ***There is no significant relationship between monitoring and financial performance of commercial banks in Tanzania*** was rejected. The implication of this finding is that, there is enough evidence to categorise monitoring is among the determinants of commercial bank profitability. This finding confirm that of Magara (2013) and Muthusi (2017) who found that independent process checks and evaluations of controls activities on an ongoing basis and periodic internal control and monitoring are essential for profitability of commercial bank and they help in check and balance of bank's activities and identify inefficiencies earlier before they become harmful. Moreover, the findings corroborate with Adetiloye et al. (2016) and Lily (2017) and Mwende and Njogu (2021). Therefore, the study concludes that commercial banks listed at the Dar es Salaam Securities play crucial role in monitoring their activities, and thereby showing a clear indication of existence of sound internal control systems.

5 Conclusions and Recommendations

5.1 Conclusion

The research investigated the effect of control environment, information and communication, and monitoring on financial performance of commercial banks. Based on the findings this study concludes that control environment, information and communication, and monitoring are important elements which contributes to the profitability of commercial banks. The study further concludes that control environment information and communication need to be improved to ensure the system is fully safeguarded, the study further concludes that monitoring activities needs to be reviewed and strengthened to ensure they perform the task for which they were established. It is also concluded that there are other factors other than internal audit practices that affect the performance of commercial banks in Tanzania.

5.2 Practical Implications

The findings of this study will be useful in re-evaluation of bank's internal controls systems and strengthen their weak areas so as to improve their financial performance.

The results obtained from this study may guide building strong internal control systems, specifically control environment, control activities, information and communication technology, and monitoring. In lieu of these determinants of internal control systems, the study provides a foundation towards improving financial performance and efficiency. The findings of this study may guide commercial banks intending to strengthen their internal control systems with the view of improving their financial performance. The banks can be in a better position to know what constitutes internal control systems and their effects on financial performance. They may also be in a position to finding new ways of strengthening their weak internal control areas.

5.3 Knowledge Contribution

This study has contributed knowledge regarding commercial bank control and monitoring of activities. The findings of this study will be used as a platform for future research through an exposure of the gaps that attract further research in the area under study since its rich in literature. The results and findings of this research suggest that the management of commercial banks listed at DSE needs to incorporate monitoring in their daily operations since it is a very important factor in financial performance. This aspect presents the academicians with a task for further research.

5.4 Recommendations

Following the conducted research and the conclusions made from the study, the recommendations revolve around individual performance, better management practices, and the establishment of a better operating environment. Firstly, there is need for the internal auditors to upscale their skills by observing the changes in technology and professional regulations regarding the functioning of the internal audit department. Such an approach improves the efficiency of the team while also ascertaining better governance of the firm. Secondly, there is need for better monitoring within the department to ensure that all actions occur according to the defined procedures. Besides this, the continuous involvement lowers pressure from external factors like the top management, who could influence negative results. The manager has the capability to institute regulations that will ascertain the independence of the internal audit team. Thirdly, as part of bettering the work environment, the management should ascertain the availability of the most recent technology by conducting an audit of the ICT tools. The auditors must also understand the software in use to enable precise execution of their assigned tasks. Further, the management must be aware of the prudential regulations set by the government for improved functionality.

5.5 Limitations and Suggestion for Further Studies

The study is not without limitations, the study focused on seven commercial banks listed in DSE; it is recommended that future studies consider expanding the scope to include the entire banking sector. The study further did not consider all internal control variables; there is the need for future studies to consider expanding the scope to include all control variables.

References

Ahmed, S., Ng'anga, P.: Internal control practices and financial performance of county governments in the coastal region of Kenya. Int. J. Curr. Asp. **3**(2), 28–41 (2019). https://doi.org/10.35942/ijcab.v3iV.59

Ofei, E.F., AndohOwusu, M., Asante, C.R.: Effect of internal audit practices on financial performance of banks in Ghana. Int. J. Curr. Asp. Finance Bank. Account. **2**(2), 46–58 (2020)

Nicholson, G., Kiel, G.: Can directors impact performance? A case-based test of three theories of corporate governance. Corp. Gov. Int. Rev. **15**(4), 585–608 (2007)

Allen, B.: Internal control systems and financial performance of commercial banks in Rwanda; A case of ECOBANK. Masters dissertation, Mount Kenya University (2018)

Davis, J., Schoorman, F., Donaldson, L.: Towards a stewardship theory of management. Acad. Manag. Rev. **22**(1), 20–47 (2003)

Eisenhardt, T.: Larger boards size and decreasing firm value in small firms. J. Financ. Econ. **48**, 35–54 (1998)

Mbilla, S.A., Nyeadi, J.D., Gbegble, M.K., Ayimpoya, R.N.: Assessing the impact of monitoring, information and communication on banks performance in Ghana. Asian J. Econ. Bus. Account. 58–71 (2020)

Jensen, M.C., Meckling, W.H.: Theory of the firm: managerial behaviour, agency costs and ownership structure. J. Financ. Econ. **3**(4), 305–360 (1976)

Nicholson, G.J., Kiel, G.C.: Board composition and corporate performance: how the Australian experience informs contrasting theories of corporate governance. Corp. Gov. **11**(3), 189–205 (2003)

Lotto, J., Kakozi, E.: Determinants of financial performance of Tanzanian banks. Afr. J. Financ. Manag. **25**, 55–65 (2016)

Al-waeli, A.J., Hanoon, R.N., Ageeb, H.A., Idan, H.Z.: Impact of accounting information system on financial performance with the moderating role of internal control in Iraqi industrial companies: an analytical study. J. Adv. Res. Dyn. Control Syst. **12**(8) (2020)

Hanoon, R.N., Rapani, N.H.A., Khalid, A.A.: The correlation between internal control components and the financial performance of iraqi banks a literature review. J. Adv. Res. Dyn. Control Syst. **12**(4), 957–966 (2020)

Asiligwa, M., Rennox, G.: The Effect of internal controls on the financial performance of commercial banks in Kenya. J. Econ. Financ. **8**(3), 92–105 (2017)

Mokono, R., Njogu, D.: The effect of internal control systems on the financial performance of NSE listed banks. Int. J. Bus. Manag. Sci. **2**(4), 1–14 (2022). https://ijbms.org/index.php/ijbms/article/view/77

Hair, J.F., Anderson, R.E., Tatham, R.L., Black, W.C.: Multivariate Data Analysis, 5th edn. Prentice Hall, Englewood Cliffs (1998)

Yamane, T.: Statistics, An introductory Analysis, 2nd edn. Harper and Row, New York (1967)

Hoskissom, R.E., Covin, J., Volberda, H.W., Johnson, R.A.: Revitalizing entrepreneurship: the search for new research opportunities. J. Manag. Stud. **48**(09), 1141–1168 (2011). https://doi.org/10.1111/j.1467-6486.2010.00997.x

Ng'owo, C.: The influence of internal fraud to the performance of Tanzania commercial banks: the case of national bank of commerce. Masters dissertation, Mzumbe University (2013). http://scholar.mzumbe.ac.tz/handle/11192.1/2559

Mwende, R., Njogu, G.: The effect of internal control systems on the financial performance of NSE listed banks. Int. J. Bus. Manag. Sci. **2**(4) (2021)

Cochran, W.G.: Sampling Techniques, 3rd edn. Wiley, New York (1977)

Sørensen, H.E.: Business Development: A Market-Oriented Perspective, 1st edn. Wiley, Chichester (2012)

Murerwa, C.B.: Determinants of banks' financial performance in developing economies: evidence from Kenyan commercial banks. https://www.semanticscholar.org/paper/Determinants-Of-Banks%E2%80%99-Financial-Performance-In-Murerwa/512f1ab5b1fda3406be831280c2991b25d82cefb. Accessed 7 June 2022

Mrindoko, A.E.: Effect of credit risk and operational risk on the financial performance of banks in Tanzania. Thesis Submitted to Open University of Tanzania in Fulfilment of the Requirements for the Award of Doctor of Philosophy of The Open University of Tanzania (2021)

Wanjiru, W.: Effect of internal controls on financial performance of commercial banks in Kenya. Masters dissertation, University of Nairobi (2019). http://erepository.uonbi.ac.ke › handle

Coarse, R.H.: Nature of the firm. Economica **4**(16), 386–405 (1937)

Fama, E.F., Jensen, M.C.: Separation of ownership and control. J. Law Econ. **26**(2), 301–325 (1983)

Kingu, P.S., Macha, S., Gwahula, R.: Impact of non-performing loans on bank's profitability: empirical evidence from commercial banks in Tanzania. Int. J. Sci. Res. Manag. (IJSRM) **6**(1), 71–79 (2018)

Torres-Reyna, O.: Panel Data Analysis Fixed and Random Effects Using Stata (v. 4.2). Data & Statistical Services, Priceton University (2007)

Rogers, E.M.: Diffusion of Innovations. The Free Press, New York (1962)

de Reuver, M., Bouwman, H., MacInne, I.: Business model dynamics: a case survey. J. Theor. Appl. Electron. Commer. Res. **4**(1), 1–11 (2009)

Gamage, C.T., Lock, K.L., Fernando, A.A.J.: A proposed research framework: effectiveness of internal control system in state commercial banks in Sri Lanka. Int. J. Sci. Res. Innov. Technol. **1**(5), 25–44 (2014)

Meyer, J.W., Rowan, B.: Institutionalized organizations: formal structure as a myth and ceremony. Am. J. Sociol. **83**(2), 340–364 (1977)

COSO: Internal Control- Intergrated Framework, Committee of Sponsoring Orgarnisations of the Treadway Commission. PWC, New York (2004)

DiMaggio, P.J., Powell, W.W.: The Iron cage revisited: institutional isomorphism and collective rationality in organizational fields. Am. Sociol. Rev. **48**(2), 147–160 (1983)

COSO: Internal Control- Intergrated Framework, Committee of Sponsoring Orgarnisations of the Treadway Commission. PWC, New York (2011)

Sadick S.M.: The effect of internal control on financial performance of commercial banks. A case study of CRDB bank Mbeya branch. Masters disseertation, Mzumbe Univesrity (2020). http://scholar.mzumbe.ac.tz/handle/11192/4278

Factors Affecting Tanzanian Small and Medium Enterprises Performance in the East African Community Market: A Case of Dar es Salaam Region

S. S. Mtengela and A. E. Mrindoko(✉)

Department of Accountancy, College of Business Education, Dar es Salaam, Tanzania
allenmrindoko@gmail.com

Abstract. Purpose: The purpose of this study was to assess the extent legal and regulatory framework, cultural aspects, managerial skills and funding affect the performance of Tanzanian SMEs in EAC markets.

Design/Methodology/Approach: A descriptive case study design was adopted to collect quantitative data from 200 SME owners/managers in three districts of the Dar es Salaam region; Kinondoni, Temeke, and Ilala, using a structured questionnaire. The collected data were analysed using descriptive statistics of frequencies and percentages. Besides, the reliability test using Cronbach's Alpha Coefficient was conducted and all constructs were found to have values above 0.7.

Findings: The findings show that to a higher extent SMEs owners/managers agreed that financing does not restrict them from engaging in the EAC market. However, they dismissed the claim that they are provided with a loan by friends and relatives. Moreover, the results show that there are restricting laws and procedures to enter the EAC market. However, the current environment has been improved to hasten and reduce bureaucracy in registration and licensing. In the case of cultural values, the findings show that punishment by parents during childhood and peer group influence reduces the chances of SME owners/managers to participate in the EAC market. Lastly but not least, SMEs owners/managers were found to have moderate managerial skills.

Implications/Research Limitations: This study was concerned with the factors affecting Tanzanian SMEs' performance in the EAC market in the Dar es Salaam region. It did not cover the whole county; therefore, further research is needed covering the larger part of the country. Also, this research was limited to descriptive statistics to establish a link between factors affecting Tanzanian SMEs in the EAC market and the performance of those SMEs in the EAC market.

Practical Implications: The information that will be provided by this research will enable other researchers to develop a better understanding of SMEs' conditions and opportunities in the EAC. The findings are sources of additional knowledge regarding the accessibility and operationalization of SMEs in the EAC market. The study found that SMEs are not well protected from external competition and multinational corporations.

C. Aigbavboa et al. (Eds.): ARCA 2022, *Sustainable Education and Development – Sustainable Industrialization and Innovation*, pp. 807–826, 2023.
https://doi.org/10.1007/978-3-031-25998-2_62

Social Implications: Thus, the government of Tanzania have to formulate and enforce policies which will protect small and medium enterprises from the larger business which have financial and resources advantage over smaller businesses that are coming up.

Originality/Value: The factors that researchers have acknowledged to hinder the growth of SMEs include; legal and regulatory framework, cultural aspects, lack of management skills and lack of funding. However, past studies in the Tanzania context have not categorically assessed factors affecting Tanzanian SMEs in the EAC market.

Keywords: Cultural · Finance · Legal and regulatory · Management · Managerial

1 Introduction

1.1 Background of the Study

Small and Medium Enterprises (SMEs) have been and are still a central hub in generating income for the majority of urban dwellers with no formal paid employment (Kipilyango 2018). In sub-Saharan Africa, small and medium enterprises, or SMEs, have become a major pillar of economic growth. In fact, they make up 80% of all jobs and represent about 90% of all sub-Saharan African businesses (Igwe et al. 2018; Mashenene and Kumburu 2020). They have been the means through which accelerated economic growth and rapid industrialization have been achieved (Harris and Gibson 2016). While the contributions of small businesses to development are generally acknowledged, entrepreneurs face many obstacles that limit their long-term survival and development. Research on small business development has shown that the rate of failure in developing countries is higher than in the developed world (Mazzarol 2015; Arinaitwe 2016). Scholars have indicated that starting a business is a risky venture and warn that the chances of small-business owners making it past the five-year mark are very slim. They should therefore develop both long and short-term strategies to guard against failure (Sauser 2019).

The Africa Small and Medium Businesses Program plans to provide assistance to African financial institutions that support SME growth over a four-year period. It has received $125 million in funding from the United States (Logan 2015). Moreover, being characterised by informal practices, poor access to markets and finance, unsuitable skills, and underperformance and steady competition from large enterprises and SMEs from outside the country, the government and development partners have initiated various policies and frameworks to improve the business environment for SMEs with the intention of increasing their performance and therefore effectively participate in the mainstream of the national economy.

In the East African Community (EAC), there are many thousands of situations facing SMEs. This takes into account factors related to the type of SMEs, but also regulatory and environmental and legal aspects. A study by banks has also highlighted the skepticism of East African governments in supporting the SME sector (Pietro 2012). However, as has been seen banks have learned to adapt to their environment and to adapt through innovation and differentiation. The African Development Bank concludes that 'this practice should be encouraged through reforms to reduce the negative impact of those barriers to the continued involvement of banks with SMEs'. Promoting SMEs in East Africa is important in the workplace. Previous research has found that SMEs on average contribute to 60% of the total official jobs in the manufacturing sector in various countries (Muchai 2016; Gamba 2019; Mashenene and Kumburu 2020).

As a member of the EAC, Tanzania is committed to removing all forms of barriers, including tax-free barriers to promoting trade promotion in member states. The number of small businesses is growing rapidly in Tanzania as evidenced by the growing domestic and international activities (Namusonge 2014). All areas of operation have SMEs. These include the textile, manufacturing, finance, security, food and hotel industries, transportation; logistics sector to name a few. The business environment is full of complexities characterized by external factors and internal business features such as management, technology, operational resources, individual characteristics and international linkages (Okibo and Makanga 2014). Almost in all economies, small businesses are vital for sustained growth. Evidence available indicates that SMEs played a major role in the growth and development of all leading economies in Asia (UNIDO 2013).

In Tanzania, the SME sector is the emerging private sector and a base for private sector led growth. Though the sector possesses the highest employment potential in Tanzania, the sector is largely informal, under performing and in need of considerable assistance to overcome the entrenched disadvantages and barriers. The policy and regulatory framework for small businesses in Tanzania, however, are only in their initial stages to be implemented. Traditionally, policy makers have been overwhelmingly concerned with large businesses (CoET 2017). Tanzania's industrial growth strategy focuses on building strong SMEs that can enter export markets and establish strong links with local businesses, as well as a network of foreign firms or strategic partnerships.

Additionally, Tanzania's SME Development Policy 2003 gives its overall objective as: "to foster job creation and income generation through promoting the creation of new SMEs and improving the performance and competitiveness of the existing ones to increase their participation and contribution to the Tanzanian economy" (United Republic of Tanzania (URT), Ministry of Trade and Industry 2013). This objective will contribute towards realizing the National Development Vision 2025, which aims at transforming the Tanzanian economy from a low productivity agricultural economy to a semi-industrialized one, among other aims (URT 2018). The envisaged transformation of the economy relies on enhancing the productive activities. For the SMEs in particular, it means ensuring that as they grow, they are able to increasingly penetrate the export markets, especially in areas or activities that Tanzania has the comparative advantage. The ability to penetrate into export markets and increase that penetration over time calls for an export strategy for SMEs, which will promote both productivity and competitiveness in EAC market.

1.2 Statement of the Problem

The National Bureau of Statistics found out that three out of five businesses in Tanzania failed within the first few months of operation and even those that continued 80% failed before the fifth year (NBS 2017). According to Daily Nation Media newspaper, data from National Bureau of Statistics (NBS) released on 17, October, 2019 showed that

2.2 million micro, small and medium enterprises closed shops in the last five years. In 2019 the highest percentage of closures was recorded at 35.4% (NBS 2019). SMEs have different problems, which affect their growth and profitability and therefore reduce their ability to participate in international markets. The factors that researchers (Mazzarol 2015; Sopha and Kwasira 2016; Ndiaye et al. 2018; Mohammed and Rugami 2019) have acknowledged to hinder the growth of SMEs include; legal and regulatory framework, cultural aspects, lack of management skills and lack of funding.

SMEs continue to use inadequate technology according to Wanjohi and Mugure (2018) which is a major challenge to small businesses by reducing their performance and providing low-quality services that result in inability to enter international market which is largely driven by information and communication technologies among others. Tanzanian SMEs continue to struggle to enter international market especially EAC market (Yahya and Mutarubukwa 2015; Kapinga and Montero 2017; Kikula 2018). However, there is lack of empirical evidence to substantiate why the presence of Tanzanian SMEs in EAC market is small. Therefore, this study intended to assess the extent legal and regulatory framework, cultural aspects, managerial skills and funding affects performance of Tanzanian SMEs in EAC markets.

1.3 Objective of the Study

1.3.1 General Objective

The main objective of the study was to assess the extent legal and regulatory framework, financing process, cultural values and managerial skills affect Tanzanian SMEs performance in EAC market.

1.3.2 Specific Objectives

Specifically, this study was guided by the following objectives;

i. To examine how access to finance affect Tanzanian SMEs performance in EAC market.
ii. To assess how the legal and regulatory framework affect Tanzanian SME performance in EAC market.
iii. To evaluate how culture values of individual entrepreneurs affect performance of Tanzanian SMEs performance in EAC market.
iv. To determine how lack of managerial skills affect Tanzanian SMEs performance in EAC market.

1.4 Research Questions

i. Does access to finance affect Tanzanian SMEs performance in EAC market?
ii. To what extent legal and regulatory framework affect Tanzanian SMEs performance in EAC market?
iii. What is the role of culture on success of Tanzanian SMEs performance in EAC market?
iv. How accessible to managerial skills of Tanzanian SMEs performance in EAC market?

2 Literature Review

2.1 Theoretical Literature Review

2.1.1 Contingency Theory

According to Scott (1981) the best way to organize depends on the environment in which the organization is located. The work of other researchers including Paul Lawrence *et al.* (2017) and Thompson complements this statement. They are very interested in the impact of potential developments on the organizational structure. Their concept of organizational order was the dominant ideological notion of the organizations of the majority of the 1970s. A major art test was conducted by Penning who examined the link between environmental uncertainty, organizational structure and various functional aspects. This concept has been widely used in research to measure organizational performance and performance and states that there is no effective way to organize a company and a corporate structure (Battilana 2012).

It can be concluded that there is no other 'better way' or way of managing or doing things, different situations require a different way of handling, managing, and resolving the problem arising from acquiring three aspects of Legal and Control, management skills and business culture. Managing and ordering an 'open system', which accepts irregularities or occasional challenges, requiring an 'adaptable' and 'state' solution in order to overcome or solve a problem or problem involved (Scot 1981). The contingency vision did not go unnoticed. For example, critiques believe that this model has little or no flexibility which means it is a solid model. Fiedler is of the opinion that the natural style of leadership is consistently given and is related to the characteristics of his personality. He thought that natural leadership style would be the most effective leadership style. But he did not take into account the fact that a leader cannot always apply the natural leadership style in all situations (Mikes 2013).

Contingency Theory states that the most appropriate organizational structure is the one that best suits a given operational event, such as technology, or the environment. As each company faces its own set of internal and external issues and special environmental events affecting different levels of environmental uncertainty, there is no single good organization organization for all companies because each company has different organizational cultures and different risk perceptions (Danish 2013).

2.1.2 The Resource Based Theory

The resource-based theory was first propagated by Birger Wernerfelt in his article "A Resource Based View of the Firm" (Wernerfelt 1984). However, the resource-based view has been a common interest for management researchers and numerous writings. A resource-based view of a firm explains its ability to deliver sustainable competitive advantage when resources are managed such that their outcomes cannot be imitated by competitors, which ultimately creates a competitive barrier (Mahoney and Pandian 1992). Resource based view explains that a firm's sustainable competitive advantage is reached by virtue of unique resources being rare, valuable, and non - tradable, non-substitutable, as well as firm specific (Barney 1999). The two wrote about the fact that a firm may reach a sustainable competitive advantage through unique resources which it holds, and these resources cannot be easily bought, transferred or copied and simultaneously, add value to the firm while being rare.

A criticism that has persisted is that the resource-based theory is a duplication that has failed in its responsibility to fulfill the core criteria for a true theory argues the RBV does not contain the law like generalizations that must be expected. Rather, it assumes an analytic statement that is tautological, true by definition that cannot be tested (Kraaijenbrink 2009). However, given its simplicity to use and its direct validity, the RBV's core message is very appealing, and can easily be grasped and taught, this makes it relevant for this study. Resource based theory and contingency theory were used to enhance our knowledge and look at the importance of flexibility and organizational performance in running SMEs. Superior inter-firm flexibility is proposed to influence SME market place success which come up by concluding in one variable which are Access to Finance and Performance of SMEs,

2.2 Empirical Literature Review

SMEs account for the majority of firms in developing countries and have a large share of employment (Mashenene and Kumburu 2020), they are labor intensive, they are less efficient (Rwigema 2014) and pay low salaries and fringe benefits as compared to large firms and they lack job security (Muchai 2016) because of their high mortality rates (Alsaaty and Makhlouf 2020; Jayasekara et al. 2020). Most activities characterized as SMEs are very small, majority of them operate in rural areas, they are owned and operated by women and they tend to be concentrated in a relatively narrow range of activities such as knitting, dressmaking, retail trading (Nathaniel 2017), and they are more likely than those headed by male to operate from home (Kiggundu 2012).

In more advanced developing countries, where there is reasonable progress in the fundamental institutions, SMEs may still face challenges in accessing formal finance in the form of bank loans, guarantees, venture capital and leasing (Rwigema 2014; Storey 2017). For instance, although SMEs are by far the largest group of customers of commercial banks in any economy, loans extended to SMEs are often limited to very short periods, thereby ruling out financing of any sizable investments (Wanjohi and Mugure 2019). Moreover, due to high-perceived risks in SME loans, access to competitive interest rates may also limit (Allen 2017). Financial availability is critical to improving SME competition, as traders need to invest in new technologies, skills and strategies (Kameyama 2018). The discovery of financial problems cannot be solved by using financial strategies or plans in an idle environment (Massawe and Calcopietro 2019). There are problems in institutions that cover the range from large to small, associated with a shortage of capacity (Basil 2005).

It is also worth noting that the effort to address access to finance is not limited to government, rather SMEs need to take better action than to identify their first obstacle (Igwe et al. 2018). SMEs must apply sound business practices and continually invest in sound internal control systems in financial, planning, financial, operational and human resource management (Abuzayed 2012). Lack of access to credit/ finance affects technology choice by limiting the number of alternatives that can be considered (Wanjohi and Mugure 2019). Many SMEs may use inappropriate technology because it is the only one, they can afford. In some cases, even where credit is available, the entrepreneur may lack freedom of choice because the lending conditions may force the purchase of heavy, immovable equipment that can serve as collateral for the loan (Massawe and Calcopietro 2019). Credit constraints occur in variety of ways where undeveloped capital market forces entrepreneurs to rely on self-financing or borrowing from friends or relatives. Lack of access to long-term credit for small enterprises forces them to rely on high-cost short term finance (Allen 2013).

Apart from the various interventions aimed at improving the business environment in Tanzania, the legal and regulatory framework is structured, cost-effective and intermediate. These aspects of legal and regulatory environment adversely affect all business sizes (Ephraim 2013; Igwe et al. 2018). However, SMEs are more restricted in this area compared to large businesses due to the very complex compliance costs arising from their size (Kigguddu 2020). As a result, the majority of informal businesses have failed to operate legally and small businesses have failed to grow and graduate from small and medium enterprises (Yahya and Mutarubukwa 2015; Gamba 2019). The tax government in Tanzania is also not in favor of SMEs development. Taxes are high, instead high and collected by various authorities including the Tanzania Revenue Authority and the Regional Government (Mashenene and Kumburu 2020). In addition, entrepreneurs are not aware of the tax issue and the costs of complying with tax laws are considered high (Yahya and Mutarubukwa 2015; Kikula 2018). Although corporate tax is a necessity for national economic development, the current tax regime places a heavy burden on SMEs (Abuzayed 2018).

Moreover, to start a business in Tanzania is not easy although the government has tried to hasten the process (Yahya and Mutarubukwa 2015). There is no one-stop centre in formalisation of business since various institutions are involved (Mgulambwa 2017). The Tanzania regulatory or legal requirements include registration of the company name with the registrar of companies, acquiring a personal identification number and Value Added Tax with the Tanzania Revenue Authority, Trade License with the ministry of Trade, and finally the Local authority licenses (Kinuthia 2015). Harper (2016) observes that governments that are not concerned with the promotion of small enterprises should examine the impact of its policies and programs on the small businesses. Mann et al. (2017) makes a similar observation that government regulation about wages, taxation, licensing and others are among the important reasons why the informal sector business develops.

The management of SMEs is one of the vital ingredients for the success of the business across the world (Kolstad et al. 2016). The management skills within SMEs remain pivotal to their performance. It calls for training in how to manage your business before it becomes profitable. Also, according to Massawe and Calcopietro (2019), many entrepreneurs have a low level of formal education and lack access to adequate training in business management associated with increasing challenges in competition and technology development. In addition, research suggests that in Tanzania, legal and tax issues appear to be greater for very small firms, declining somewhat as firms grow (Levy 2019).

Similarly, better marketing services, support from technology development institutions and the introduction of tax incentives will lead to a more prosperous and competitive

SMEs sector (Mohammed and Rugami 2019). Shafeek (2016) shows that the major reasons for SME failure are lack of financial management acumen. It can be seen that in terms of skills, the weakness in strategic management, human resources management, general management and administration, are manifested in the significant weakness in financial management (Karadag 2015; Gure and Karugu 2018). Nwanko and Gbadamosi (2016) indicate that a severe weakness in the area of strategy is cause for concern.

3 Research Methodology

3.1 Description of the Study Area

The research will be conducted in Dar es Salaam regional particularly in Kinondoni, Ilala, and Temeke Districts. Dar es Salaam regional was selected because is the largest market in Tanzania where different SME's operation takes place.

3.2 Research Design and Approach

The study adopted a descriptive case study design to assess the factors affecting Tanzanian SMEs performance in EAC market. This study assessed intensively the factors affecting Tanzanian SMEs performance in EAC market. This idea was borrowed from (Mugenda and Mugenda 2003) who define case study as an in-depth investigation of an individual, group, institution or phenomenon. To carry out this study quantitative data were gathered using a questionnaire with closed end questions most of them measured in five point Likert scale.

3.3 Target Population

Since the research focused on the factors affecting Tanzanian SMEs' performance in EAC market using SMEs located in Dar es Salaam region as case studies, the population of the study comprised of all entrepreneurs in Dar es Salaam region who operate micro, small or medium sized enterprises. From the records of City Council of Dar es Salaam, there are approximately 400 registered SMEs in each tax region. Therefore, the total population is 1200 SMEs (Ilala city council report 2020).

3.4 Sample Size and Sampling Techniques

The target groups in this study are SMEs located in Dar es Salaam region in various categories of business. Purposive sampling was used to select the respondents to be surveyed. Purposive sampling was used to select the SME's owners/managers as respondents for the study. Selection of these respondents was based on the possession of special knowledge related to the study, and thus it was aimed at gathering relevant information required for this study. The sample size was estimated using Yamane formula;

$$n = N/[1 + N(e)^2]$$

where,

n = sample size,
N = target population which is 1200 SME's,
e = sample error (5%)
n = 1200/[1+1200(0.05)^2]
n = 291

Therefore, sample size is 291. However, this study used only 200 respondents considering the threshold of 30 respondents for quantitative study (Kothari 2004). Thus, a sample size of 200 respondents is more than enough for assessment of factors affecting performance of Tanzanian SMEs in EAC market.

3.5 Methods of Data Collection

In this study the researcher used both primary and secondary data sources. Primary data was collected using a questionnaire which was self-administered questionnaire, and in some occasions structured interview was applied. In the first method, a questionnaire was sent to the respondents by mail with return stamp, and a letter requesting a respondent to answer the questions and return the questionnaire to a sender address. Some respondents responded requesting to meet with the researcher for structured interview. Then a date, place and time were agreed where an interview was conducted. The use of secondary data in this study aimed at helping the researcher to get previous and present information that assist the researcher in drawing meaningfully conclusions and recommendation. Secondary data were accessed in the reports, articles and statistical data as well as from councilor's minutes, abstract accounts and academic research reports.

3.6 Reliability

Cronbach's alpha was used to measure reliability. Table 1 shows that for all constructs the Cronbach's Alpha was above 0.7; hence data collection instruments and data collected are reliable.

Table 1. Reliability of data and data collection tools

Construct	Cronbach's alpha	No. of items
Legal and regulatory framework	0.76	4
Managerial skills	0.79	3
Financing	0.73	4
Cultural values	0.77	5

Source: Researcher's Compilation, 2022.

3.7 Data Analysis Procedures

Data collected through structured questionnaires, were summarized and coded and were entered into SPSS version 21 for analysis. Data were analysed using descriptive statistics in terms of frequency and percentage. The results were presented in Tables.

3.8 Ethical Considerations of the Study

The researcher first of all ensured there a letter of authorization from the College of Business Education (CBE) Tanzania authorizing researcher to go ahead and collect research data, then additional authorization letter from government of Tanzania is required. The researcher was guided by the following ethics during the period of the study; Honesty - the research strived for honesty in all communications. Honestly report data, results, methods and procedures, and publication status. Objectivity - the researcher will be strived to avoid bias in experimental design, data analysis, data interpretation, peer review, personnel decisions, grant writing, expert testimony, and other aspects of research where objectivity is expected or required. Integrity - the researcher kept promises and agreements; act with sincerity; strive for consistency of thought and action.

4 Findings and Discussion

4.1 Demographic Characteristics of the Respondents

4.1.1 Sex of Respondents

The sex of the respondents involved in the study was gathered for the purpose of understanding respondents' participation in gender basis towards performing their daily businesses in the study area as presented in Table 2. This was important since national policies demand women and men to be given equal chances to participate in business. The findings show that most of SMEs located in Dar es Salaam, which participate in EAC market are owned/managed by females 65% against with males, which are 35%. The gender representation of the respondents indicates that, views on factors affecting small and medium enterprises in Dar es Salaam were represented by all gender, and no single opinion can be attributed to a particular gender. However, the study noted that most females were in the catrgory of small enterprises, while most males were in medium enterprises. Females were few in the medium category mainly because of their limited access to capital. Women have limited collateral and in addition they have to attend to other house chores which leave them with limited time to do business, and grow. As such, they are limited in their operations to small enterprises.

4.1.2 Age of Respondent

The ages of respondents were gathered for the purpose of understanding the SME'S age's group from youth to adult basing on their knowledge, skills and ability in performing their daily operation in the particular business. The findings in Table 2 show that age group 29 to 39 years constitute 42%, followed by 40 to 50 years which is equivalent to 34% and 18 to 29 years which constitute 24%. Therefore, considering their ages, the respondents had good understanding of EAC market, and thus the collected data are considered accurate and relevant.

Table 2. Distribution of demographic characteristics of respondents (n = 127)

Characteristic	Category	Frequency	Percentage
Sex	Male	70	35.0
	Female	130	65.0
Age	18–29	48	24.0
	29–39	84	42.0
	40–50	68	34.0
	50+	0	0.0
Business experience	<1 year (low)	34	17.0
	2–3 years (medium)	74	37.0
	4–5 years (high)	66	33.0
	5+ years (highest)	26	13.0
Education level	Primary	32	16.0
	Secondary	86	43.0
	Basic certificate	14	7.0
	Diploma	26	13.0
	Bachelor	28	14.0
	Masters	14	7.0

Source: Researchers' Compilation 2022.

4.1.3 Respondents' Level of Education

Table 2 shows the findings obtained concerning the level of education of respondents. The findings revealed that the majority (43.0%) of the respondents possessed secondary level of education followed by the 16.3% of all respondents who had the primary level education. Finally, the study findings further found that 13.3% of all respondents possessed diploma and degree only 6.7% of all respondents possessed basic technician certificate and 6.7% possessed Masters' degree level of education. The implication from what was obtained from the study area showing adequate level of education among respondents, hence implying adequate knowledge on EAC market operation.

4.1.4 Distribution of Respondents by Period of Service

The research sought to determine the length of time the SME's owner/manager have been doing business to ascertain respondents business experience. The business experiences of the SMEs owners/managers were gathered for the purpose of understanding their capacity to manage business and participate in EAC market. The findings in Table 2 indicate that 36.7% of all respondents have operated businesses for 2 to 3 years, 33.3% had 4–5 years experience, 13.7% had more than 5 years of business experience and those with least experience of less than two years were 16.7%. This implies that they have done the business for quite a while and that their experience can help them in EAC market. This shows that the respondents are competent to provide information regarding EAC market operations.

4.2 Effects of Access to Finance on Performance of Tanzanian SMEs in EAC Market

Respondents were asked several questions on access to finance as factor affecting small and medium enterprises performance in EAC Market. The findings in Table 3 show that, item that says 'on ease of access becomes considerably more important in the context of getting loans' 43% of respondents strongly agreed, 21% agreed, 5% were neutral and 13% disagreed while 9% strongly disagreed to the statement. This implies that easy access of loan is very crucial for business survival and growth. Moreover, regarding the second statement that 'restriction of finance regarding SMEs', the results show that 23% strongly agreed, 49% agreed, 6% were neutral and 13% disagreed while 9% strongly disagreed to the statement. This result suggests that, funding options for SMEs are limited.

Table 3. Access to finance among small and medium enterprises

Statement/item	SAG	AG	NAND	DA	SDA
Ease of access becomes considerably more important in the context of getting loans	106(52%)	42(21%)	10(5%)	26(13%)	18(9%)
Restriction of finance regarding SMEs	46(23%)	98(49%)	12(6%)	26(13%)	18(9%)
Provision of loans by relatives and friends for business capital	26(13%)	46(23%)	24(12%)	76(38%)	28(14%)
Provision of grants by government to SMEs	66(33%)	46(23%)	12(6%)	42(21%)	34(17%)

Source: Researcher Compilation, 2022.

Regarding the third statement which asked if they are provided with loans by relatives and friends for business capital, findings in Table 3 show that, 13% strongly agreed, 23% agreed, 12% were neutral and 38% disagreed while 14% strongly disagreed and last statement asks if SMEs are provided with grants by government, 13% strongly agreed, 23% agreed, 6% were neutral and 16% disagreed while 17% strongly disagreed. The findings imply that SMEs receive funds from relatives and friends in lower amount. It is not a common practice for relatives and friends to provide fund for SMEs. Also, to a moderate extent SMEs receive grants from government. The study findings are in agreement with other studies done across Africa, Ngobo (2017), Kibera and Kiberam (2017) and Chijoriga and Cassiman (2014) point to finance as one of the key constraints to small enterprise growth and even affects Tanzanian SMEs performance in EAC market. SMEs owners cannot easily access finance to expand business and they are usually faced with problems of collateral, feasibility studies and the unexplained bank charges.

4.2.1 Sources of Business Capital for SMEs

As shown in Table 4 respondents were asked to indicate sources of business capital, majority 48% indicated personal savings and chamaa, 28% indicated women and youth fund while 20% indicated family and friends and finally 4% indicated bank loan.

Table 4. Sources of business capital for SMEs

Category	Frequency	Percentage
Bank loan	8	4.0
Family and friends	40	20.0
Personal savings and chamaa	96	48.0
Women and youth fund	58	29.0
Total	**200**	**100.0**

Source: Researcher Compilation, 2022.

4.2.2 SMEs' Challenges of Processing and Accessing Loan

Respondents were asked the process and challenges of getting a loan from lenders. The study revealed that the SMEs were not comfortable with the interest rate and felt that this was the main financial impediment for business growth with 96% of the respondents reporting that interest rate offered by lenders is not affordable by the SMEs (Table 5). The second serious problem was that, most lenders do not provide the requested loans on timely basis. This was indicated by 94.0% of respondents. In essence, timely loan ensures fulfillment of business goals. Untimely loan affects business growth. Thus, this result implies that, loans which are provided late to SMEs affect Tanzanian SMEs performance in EAC market.

Also, findings in Table 5 shows that, parameter that affected operations of SMEs in EAC market include lack of collateral for loans (77%), which was found to be missing among SMEs; also respondents felt that the transaction costs for loans were very high (81%), and that most lenders' conditions are tough for SMEs to oblige (76%). Also, respondents were asked if government provide sources of cheaper capital to start up business, 95% said no. SMEs owners prefer to use personal savings and contributions from relatives because they find it very difficult to access financing from commercial banks due to strict requirements. The study concludes that the stringent requirements to access finances are the main reason why most SMEs cannot access finances and affects Tanzanian SMEs performance in EAC market. This is in line with a study conducted by the Financial Sector Deepening Kenya (FSD 2018), which showed that SMEs face numerous hurdles in accessing finance, denying them an important growth line at best or accessing it at a very high cost. The study alludes that access to finance is being constrained by exacting legal requirements by banks and other finance institutions, lack of a standardized and shared information registry and expensive and time-consuming enforcement mechanisms.

Table 5. SMEs' challenges of accessing loan

Category	Yes		No	
	Frequency	Percentage	Frequency	Percentage
Interest rate is affordable	8	4.0	192	96.0
Loan provision is on timely basis	12	6.0	188	94.0
SMEs have collateral	46	23.0	154	77.0
Transaction cost for loan is low	38	19.0	162	81.0
Easy conditions for accessing loan from bank	48	24.0	152	76.0
SMEs get cheap loan from the government	10	5.0	190	95.0
Total	**200**	**100.0**		

Source: Researcher Compilation, 2022.

4.3 Legal, Regulatory Framework and Tanzanian SMEs' Performance in EAC Market

Respondents were asked several questions on legal and regulatory frameworks that affect Tanzanian SMEs' performance in EAC market and the response were as follows on the ability and ease of access to license to operate SMEs 38% strongly agreed, 23% agreed, 6% were neutral while 16% disagreed and 9% strongly disagreed (Table 6). This result implies that, it is not difficult for SMEs to get license to operate in EAC market. Also, regarding to the provision of tax holidays or exemptions by the government, the findings show that 18% strongly agreed, 46% agreed, 7% were neutral and 13% disagreed while 9% strongly disagreed. This suggests that the extent the government provides tax holidays or an exemption is moderate. As such, SMEs foster their own capabilities so that they can better perform in EAC market.

In case of protection of SMEs from external competition and from multinational companies, the results in Table 6 show that, 53% strongly agreed to have been protected by the government against external competition. Moreover, 23% of respondents agreed to the statement. However, 5% were neutral, 13% disagreed and 9% strongly disagreed that government protect them from EAC market competition and from multinational companies. Also, 23% and 57% strongly agree and agree to the statement which says 'it is affordable to get license to operate SMEs in EAC' respectively. On the contrary, 6% were neutral and 13% disagreed while 9% strongly disagreed to the statement. The findings are in agreement with a study conducted by Wanjohi (2019). Unfavorable regulatory environment does not only scare away potential investors but also squeeze revenues for those in operation. These calls for County and National government to create conducive legal and regulatory business environment that will facilitate the growth and development of SMEs in order to perform in EAC market.

Table 6. Legal and regulatory framework as factor affecting SMEs

Statement/item	SAG	AG	NAND	DAG	SDAG
Ability and ease of access to license to operate of SME in EAC market	76(38%)	62(31%)	12(6%)	32(16%)	18(9%)
Provision of tax holidays or exemptions by the government to EAC market	22(11%)	46(23%)	24(12%)	76(38%)	32(16%)
Protection of SMEs from EAC market competition and from multinational companies	106(53%)	4(23%)	10(5%)	26(13%)	12(6%)
Affordability of license to operate SMEs in EAC market	46(23%)	114(57%)	12(6%)	26(13%)	18(1%)

Source: Researcher Compilation, 2022.

4.4 Effect of Cultural Values on the Performance of Tanzanian SMEs in EAC Market

The findings in Table 7 show that, 65% of respondents strongly agreed and 31% agreed that learning business skills from parents and family members contributed to SMEs performance. On the contrary, 4% were neutral and none disagreed or strongly disagreed to the statement. The result implies that SMEs owners/managers learnt to manage the business from either parents or relatives. Also, regarding the second statement which says

Table 7. Cultural factors affecting SMEs performance

Statement/item	SAG	AG	NAND	DAG	SDAG
Learning business skills from parents and family members contributed to your business performance	130(65%)	62(31%)	8(4%)	0(0%)	0(0%)
Punishment by parents during childhood for failing to be successful in some activities	38(19%)	30(15%)	6(3%)	26(13%)	100(50%)
Personal desire to be independent since childhood and its influence on your business success	142(71%)	46(23%)	12(6%)	0(0%)	0(0%)
Hard work and internal focus of control and its contribution to the success of your business	130(65%)	62(31%)	8(4%)	0(0%)	0(0%)
Peer group generates very strong influences on young people's behavior in relation to success in business	24(12%)	46(23%)	24(12%)	76(38%)	30(15%)

Source: Researcher Compilation, 2022.

'punishment by parents during childhood for failing to be successful in some activities' 19% strongly agreed, 15% agreed, 3% were neutral, 13% disagreed and 50% strongly disagreed. The findings propose that the punishment during childhood did not influence the performance of SMEs in EAC market. Moreover, third question asked if personal desire to be independent since childhood had influence on SMEs' performance in EAC market. The results show that, 71% strongly agreed, 23% agreed, 6% were neutral and none disagreed or strongly disagreed to the statement. This result implies that personal desire to be independent since childhood has impact on the performance of SMEs in EAC market.

With regard to 'hard work and internal focus of control and its contribution to the success of your business', results in Table 7 show that 65% of respondents strongly agreed, 31% agreed, 4% were neutral and none disagreed or strongly disagreed. It means that hard work and internal focus had impact on the performance of SMEs in EAC market. Besides, 13% strongly agreed, 23% agreed, 12% were neutral and 38% disagreed and 14% strongly disagreed to the statement which says 'peer group generates very strong influences on young people's behavior in relation to success in business'. This implies that peer groups generate negative behaviour in relation to business management, and hence SMEs' performance. The findings concur with Nasser (2013) and Rwigema and Venter (2014) cultures that emphasize achievement and social recognition for all forms of entrepreneurial success are more conducive to entrepreneurship. Communities with low entrepreneurial culture may discourage entrepreneurs, who fear social pressure and being ostracized and also affect Tanzanian small and medium enterprises performance in EAC market.

4.5 Effects of Managerial Skills on Performance of Tanzanian SMEs in EAC Market

Respondents were asked several questions on managerial skills that affect Tanzanian SMEs performance in EAC market. The findings in Table 8 show that, 53% of respondents strongly agreed and 20% agreed to the statement 'having studied entrepreneurship in high school, college and university and its contribution to business startup'. On the other hand, 5% were neutral and 13% disagreed while 9% strongly disagreed to the statement. This result suggests that, college and university education on entrepreneurship has positive effect on the performance of SMEs in EAC market. Moreover, 13% strongly agreed and 23% agreed to the statement which says 'the ease of access to managerial skills from government institutions'. However, 12% were neutral and 36% disagreed while 16% strongly disagreed to have easily received managerial skills from government institutions.

In case of use of the skills learnt during training in running your business contributes to its success, results in Table 8 indicate that 65% of respondents strongly agreed and 31% agreed to the statement. On the contrary, 4% were neutral and none disagreed or strongly disagreed with the statement. This means that, skills learnt during training were crucial for the performance of SMEs in EAC market. The findings are in line with findings by Thapa (2017) who indicates that there is a positive association between education and small business success. The likelihood of failure was also found to be associated with the owner/manager's work experience prior to business launch and education. Human

Table 8. Managerial skills of SMEs

Statement/item	SAG	AG	NAND	DSA	SDAG
Having studied entrepreneurship in high school, college and university and its contribution to business startup	106(53%)	46(23%)	10(5%)	20(10%)	18(9%)
Ease of access to managerial skills from government institutions	26(13%)	46(23%)	24(12%)	76(36%)	32(16%)
Use of the skills learnt during training in running your business and contributes to its success	130(65%)	62(31%)	8(4%)	0(0%)	0(0%)

Source: Researcher Compilation, 2022.

capital is the most critical agent of SME performance. Research by King and McGrath (2012) indicates that education and skills are needed to run SMEs. They further suggest that those with more education and training are more likely to be successful in the SME sector on Tanzanian small and medium enterprises performance in EAC market.

5 Conclusions and Recommendations

5.1 Conclusion

The objective of the study was to establish the factors affecting Tanzanian SMEs performance in EAC market. The study found that access to finance affect success of enterprises among Dar es Salaam markets, with most of the entrepreneurs saying their businesses are dependent on capital which they get from their saving with some respondents saying their relatives help them in raising capital. Legal regulation is another factor affecting business given that there are a lot of regulations along the way before business see the light of the day. Cultural factors play a role in one success in the enterprises with respondents saying their upbringing play important role in their success in business. The study concludes that access to finance, laws and regulation and availability of management experience are the key socio-economic factors affecting the performance of businesses among Dar es Salaam market. These three have the potential of leading to improved business performance in EAC market. The other key factor that was found to affect performance of micro and small enterprises in Dar es Salaam markets positively is cultural factors that favor business acumen.

5.2 Knowledge Contribution

The information that will be provided by this research will enable other researchers to develop a better understanding of SMEs conditions and opportunities in the EAC. The findings are sources of additional knowledge regarding the accessibility and operationalisation of SMEs in EAC market.

5.3 Policy Implication

The study found that SMEs are not well protected from external competition and from multinational corporations. Therefore, based on this result policy implication requires government of Tanzania to formulate and enforce policies which will protect small and medium enterprises from larger business which have financial and resources advantage over smaller businesses that are coming up. The laws favoring start up and growth of SMEs should be formulated.

5.4 Recommendation

Based on the findings of this study, it is recommended that; firstly, the banks and other credit giving financial institutions should come up with creative policies and procedures that make it easy for the SMEs to access financing. Secondly, to the governments of Tanzania, it is recommended that time taken to process and obtain license to export goods in EAC market should be lessen by reducing bureaucracy and illuminating some hurdles along the way. Thirdly, the government of Tanzania should also protect small and medium enterprises from big business. Fourthly, to the community, this study recommends that parents encourage their children to be independent at the younger age to avoid dependency syndrome. Any behavior that encourages laziness should be abandoned. Lastly but not least, the government should conduct seminars and trainings on basic business and financial management skills to SMEs to enable them to make informed investment decisions.

5.5 Limitations and Suggestions for Further Research Study

This study was concerned with the factors affecting Tanzanian SMEs performance in EAC market in Dar es Salaam region. It did not cover the whole county; therefore, further research is needed covering the larger part of the country. Also, this research was limited to descriptive statistics to establish a link between factors affecting Tanzanian SMEs in EAC market and the performance of those SMEs in EAC market. This study can be improved by employing inferential statistics to establish a link between those factors and performance of SMEs in EAC market. Moreover, this study did not demarcate its analysis based on status of capital structure of SMEs. Thus, in a future study, it is recommended that a comparison should be done between the financial performance of SMEs that have received microcredit and the ones that have not received the financing. This will help in shedding light on whether accessing microcredit helps the SMEs to perform better in EAC market.

References

Abuzayed, B.: Working capital management and firms' performance in emerging markets: the case of Jordan. Int. J. Manag. Finance **8**(2), 155–179 (2012)

AFDB: Meeting the Growing Demand for Retail Banking Services in Africa. Africa Development Bank (2013)

Allen, F.C.: Resolving the African Development Financial Gap A Cross Country Study of Kenya. World Bank Press, Nairobi (2013)

Allen, H.: Village Savings and Loan Associations. Intermediate Technology Publications Ltd., London (2017)

Alsaaty, F., Makhlouf, H.: The rise and fall of small business enterprises. Open J. Bus. Manag. **8**(4), 1908–1916 (2020). https://doi.org/10.4236/ojbm.2020.84116

Battilana, J.: Change agents, networks, and institutions: a contingency theory of organizational change. Acad. Manag. **60** (2012)

Ephraim, M.: Financial challenges faced by retail SMEs operating in a multi-currency environment, Zimbabwe (2013)

Gamba, F.J.: SME development policies of Tanzania and Rwanda: comparability of policy presentation on focus, significance, challenges and participation. J. Dev. Commun. Stud. **6**(1) (2019). http://www.devcomsjournalmw

Gure, A.K., Karugu, J.: Strategic management practices and performance of small and micro enterprises in Nairobi City County, Kenya. Int. Acad. J. Hum. Resource Bus. Adm. **3**(1), 1–26 (2018)

Ibrahim, M., Ibrahim, A.: The effect of SMEs' cost of capital on their financial performance in Nigeria. J. Finance Account. **3**(1), 8–11 (2015)

Igwe, P.A., Onjewu, A.E., Nwibo, S.U.: Entrepreneurship and SMEs' productivity challenges in sub-Saharan Africa. In: Dana, L.-P., Ratten, V., Honyenuga, B.Q. (eds.) African Entrepreneurship. PSEA, pp. 189–221. Springer, Cham (2018). https://doi.org/10.1007/978-3-319-73700-3_9

Jayasekara, B.E.A., Fernando, P.N.D., Ranjan, R.P.C.: A systematic literature review on business failure of small and medium enterprises. J. Manag. **15**(1), 1–13 (2020)

Kapinga, A.F., Montero, C.S.: Exploring the socio-cultural challenges of food processing women entrepreneurs in IRINGA, TANZANIA and strategies used to tackle them. J. Glob. Entrep. Res. **7**(1), 1–24 (2017). https://doi.org/10.1186/s40497-017-0076-0

Karadag, H.: Financial management challenges in small and medium-sized enterprises: a strategic management approach. EMAJ: Emerg. Mark. J. **5**(1), 26–40 (2015). https://doi.org/10.5195/emaj.2015.67

Kikula, J.S.: Challenges facing Tanzanian women entrepreneurs while managing entreprenurial ventures: a case of Mbeya city. Huria: J. Open Univ. Tanzania **25**(1), 182–208 (2018)

Kinuthia, B.: A Comparative Study Between Kenya and Malaysia on the Role of Foreign Direct Investment in Economic Development: A Survey of Literature. Africa Studies Centre, Leiden (2015)

Kothari, C.R.: Research Methodology, 3rd edn. Tata Publisher Press, New Delhi (2004)

Mashenene, R.G., Kumburu, N.P.: Performance of small businesses in Tanzania: human resources-based view. Glob. Bus. Rev. https://doi.org/10.1177/0972150920927358

Mazzarol, T.: The importance of financial management to SMEs. Centre of Entrepreneurial Management and Innovation (2015)

Mikes, A.: Towards Contingency Theory of Enterprise Risk Management. Harvard Business School (2013)

Ministry of Trade and Industry: Small and Medium Enterprises Development Policy, United Republic of Tanzania (2013)

Mohammed, R.A., Rugami, J.: Competitive strategic management practices and performance of small and medium enterprises in Kenya: a case of Mombasa county. Int. J. Curr. Asp. **3**(6), 193–215 (2019). https://doi.org/10.35942/ijcab.v3ivi.85

Muchai, K.L.: Management practices of small and medium enterprises and the challenges in socio-economic sphere: a case study of SMEs in Nairobi city. Masters dissertation, University of Nairobi (2016). http://erepository.uonbi.ac.ke/bitstream/handle/11295/99895/

Mugenda, M.O., Mugenda, G.A.: Research Methods. Acts Press, Dar es Salaam (2003)

Nathaniel, U.: The contribution of bonafide microfinance institutions in reducing income poverty among women in Chamwino District in Tanzania. Masters dissertation, St. John's University of Tanzania). Unpublished Document is available at SJUT Library (2017)

Ndiaye, N., Razak, L.A., Nagayev, R., Ng, A.: Demystifying small and medium enterprises (SMEs) performance in emerging and developing economies. Borsa Istanbul Rev. 1–17 (2018). https://doi.org/10.1016/j.bir.2018.04.003

Pietro, C.V.M.: East African Small and Medium-Sized Enterprises 'a Strategic Priority' for Banks, Finds New AfDB Study. African Developement Bank Group, Tunis (2012)

Sopha, S.I., Kwasira, J.: Influence of strategic management practices on performance of small scale enterprises in the county government of Trans Nzoia county. IOSR J. Bus. Manag. (IOSR-JBM) **18**(9), 87–103 (2016)

Yahya, M., Mutarubukwa, P.: Capacity of Tanzania micro, small and medium enterprises (MMSES) in tapping the business opportunities in the East African community. Bus. Educ. J. **1**(1), 1–20 (2015). www.cbe.ac.tz/bej

Effect of Service Quality on Customer Retention at Mount Kilimanjaro, Tanzania

R. Delphin[1] and R. G. Mashenene[2](✉)

[1] Public Service Social Security Fund, P. O. 2857, Dodoma, Tanzania
[2] Department of Marketing, College of Business Education, P. O. Box 2077, Dodoma, Tanzania
mashenenerg@gmail.com

Abstract. Purpose: The main objective of this study was to investigate the effect of service quality on customer retention at Mount Kilimanjaro National Park (KINAPA) specifically, the study intended to (i) examine the level of customer retention at Mount Kilimanjaro National Park, (ii) determine the effect of service quality dimensions on customer retention at Mount Kilimanjaro National Park.

Design/Methodology/Approach: Cross-sectional research was employed in this study. A sample of 104 respondents was secured. Data were collected using questionnaires, interviews and documentary reviews. Levels of customer retention were determined through the transformation of data. Binary logistic regression was used to determine the effect of service quality on customer retention.

Findings: The study found that all service quality dimensions such as tangibility, assurance, responsiveness, empathy and reliability were significantly affecting the customers' retention. On the other hand, the study revealed that KINAPA is faced with environmental degradation, poor tax policy, inadequate empirical studies, and financial leakages challenges in serving customers.

Implications/Research Limitations: This study was confined to the hotel services in Tanzania and ignored the specificity of Mount Kilimanjaro tourists.

Practical Implications: The study concluded that service quality significantly affects customer retention. It was recommended that the quality of service be improved to retain a reasonable number of customers.

Originality/Value: The effects of service quality on customers' retention due to its vitality as an economic source for Tanzania was established.

Keyword: Assurance · Customer · Retention · Service quality · Tangibility

1 Introduction

Business transaction worldwide depends upon customer retention. The service quality is crucial in businesses like tourism where services are intangible. Rahaman *et al.* (2020) propose that service quality increases customer retention tendencies in Bangladesh. This implies that service quality is associated with customer retention for repurchase as a result of satisfaction with the quality of service. Alketbi et al. (2020) linked the service quality with satisfaction, trust, and commitment in the United Arabs Emirates UAE hotels and

C. Aigbavboa et al. (Eds.): ARCA 2022, *Sustainable Education and Development – Sustainable Industrialization and Innovation*, pp. 827–839, 2023.
https://doi.org/10.1007/978-3-031-25998-2_63

found that service quality influences the retention and loyalty of customers. But this study focused on the hotel and hospitality industry in UAE instead of Tanzania. In supporting the findings of Alketbi et al. (2020), Fida et al. (2020) in Oman observed that empathy and responsiveness dimensions had an impact on customers' satisfaction and retention in commercial banks. The study focused on the banking industry of Oman instead of KINAPA in Tanzania. It was further recommended that banks should not neglect the importance of other variables such as reliability, assurance, and tangibles. This implies that its findings cannot be generalized to the Tanzanian context. Ali *et al.* (2021) in Iraq had a similar conclusion to that of Alketbi et al. (2020) and Fida et al. (2020). However, the study did not focus on the tourism industry, instead, it was focused on the online meeting platforms.

Likewise, Gogoi (2020) in India agrees with Parasuraman *et al.* (1998) by proposing that customers' retention is the function of service quality dimensions like service tangibility, service reliability, service assurance, empathy, service responsiveness, the competence of service providers, courtesy of services providers, the credibility of the institution, security of customers, service accessibility and communication. Although this study focused on the tourism industry, it was, however, conducted out of Tanzania with a dissimilar context to India. Similarly, Han *et al.* (2021) in China observed that the service quality of tourism in public health affects customers' satisfaction, loyalty, trust and commitment. This implies that public health services are the potential in satisfying customers, and instill trust and loyalty which are important aspects of customer retention. Despite the comprehensive findings, the study ignored other aspects of tourism like accommodation, hotels, transportation, destination attractions and communication. Moreover, this study was conducted in China, hence, it cannot exhaustively address the situation in Tanzania. Also, satisfaction was found to have a positive effect on destination loyalty. Furthermore, the findings revealed that satisfaction had a partial mediation effect on the relationship between service quality, destination image and perceived value on the one hand and destination loyalty on the other. It was concluded that the provision of high-quality services increased tourists' loyalty to the park. Abdul-Qadir et al. (2021) in Nigeria linked the service quality with customer retention of listed food and beverages companies in Kaduna State. This implies that customer retention of listed food and beverages companies in Nigeria was satisfied with service quality which derived the repurchase tendencies. Despite the clarity of findings, yet, this study had different research vicinity to the current study and had no link to tourism. Besides, did not focus on the listed food and beverages companies and other sectors.

Achieng (2021) studied a related topic in Kenya and the results revealed that there was a significant relationship between service integrity and customer retention in classified hotels in Mombasa county. The study focused on coastal tourism in Mombasa. Although, it was done in East Africa, yet, coastal tourism is different from Mountainous tourism like KINAPA in Tanzania. Kamna (2021) in Tanzania found that service quality has no impact on customer satisfaction in the retail banking sector. This study's finding is contrary to other studies and the study focused on the retail banking sector contrary to tourism which is the focus of the current study. Matolo and Salia (2021) revealed that service quality affects the customer loyalty and satisfaction of the Serengeti National Park (SENAPA). While SENAPA is a lowland tourism field based on wild animals, KINAPA is mountainous tourism. Thus, a conclusion based on SENAPA cannot suffice to address the KINAPA concerns. Similarly, Ulaya (2017) linked the service quality and customer satisfaction and retention among commercial banks with a focus on Akiba Commercial Bank (ACB). The study ignored the tourism aspect, in this regard, the findings are not sufficient to address the current study. Burhan and Kalinga (2018) noted that quality hotel services offered to tourists in Tanzania are associated with satisfaction levels. It was further learnt that tangibility, reliability, responsiveness, assurance and empathy had a positive effect on the customers' satisfaction. However, this study was confined to the hotel services in Tanzania and ignored the specificity of Mount Kilimanjaro tourists. The understanding and application of knowledge would curb state poverty.

Tanzania is one of the world's poorest countries but endowed with valuable tourist attraction destinations, whose contribution to GDP is 13%, being second to the agricultural sector (Peat 2019; World Bank. 2019). Kilimanjaro National Park (KNP) alone earns about $50 million a year which is 45% of all income generated by Tanzania's 15 national parks (TANAPA 2019). Moreover, Mount Kilimanjaro attracts over 35,000 climbers a year, plus 5000-day visitors, this is such a vital tourist attraction destination (TANAPA 2019). Previous studies link service quality and customers' satisfaction, retention and loyalty (Achieng 2021; Ali *et al.* 2021; Han *et al.* 2021; Gogoi 2020). However, these studies were not conducted in Tanzania thus, such findings cannot be generalised tothe Tanzanian context. With this in view then, it was expected that the effects of service quality on customers' retention would have been investigated and the findings emanating from such studies would be the effective implementation. However, this has not been the case, and the reasons for that have not been provided. The researcher was convinced that there was a need to study the effects of service quality on customers' retention due to its vitality as an economic source for the country.

Globally, the success of tourist attraction destinations depends on the ability to retain tourists (Gogoi 2020). Thus, customers tend to have their expectations before the visitation of tourist attraction destinations. KINAPA earns about $50 million a year which is 45% of all income generated by 15 national parks (TANAPA 2019). This is a large amount of money collected from many visitors, however, there is a great fluctuation in the number of tourists indicating the poor retention of customers (TANAPA 2019). With this in view, retaining such a great number of customers is very crucial for economic development, but there are no conclusive studies regarding the effects of service quality on customer quality. Previous studies show that quality services are linked to customers' satisfaction which in turn fosters the sales in tourism in Tanzania (Gogoi 2020; Rahaman *et al.* 2020; Mashenene 2019; Meesala and Paul 2018). Much of the previous research on the effect of service quality on customers retention in Tanzania has been done, however, not even one conclusive study has been conducted at KINAPA (Matolo and Salia 2021; Burhan and Kalinga 2018; Ulaya 2017) instead, they linked service quality and customer retention without association with the tourism industry. Although, satisfied customers are likely to be retained, however, decisive studies are missing. This elevated the urgency of investigating how the quality service links with customer retention. The question of whether tourists of Mount Kilimanjaro are being retained with quality services is not thoroughly addressed due to the insufficiency of literature. The failure to address the linkage between service quality and customers' retention with regard to Mount Kilimanjaro is likely to affect the tourism business performance and deprive the country of the tourism subsector to unleash the economic potential opportunities from the tourists' arrivals. Therefore, this study was designed to examine the effect of service quality on the customers' retention of Mount Kilimanjaro in Tanzania.

2 Theory Underpinning the Study

This study was underpinned by SERVQUAL Model which was developed by Parasuraman *et al.* (1998). The model perceives the notion of service quality into five constructs as follows: -tangibles (physical facilities, equipment, staff appearance, etc.), reliability (ability to perform service dependably and accurately); responsiveness (willingness to help and respond to customer needs) and assurance (the ability of staff to inspire confidence and trust). SERVQUAL represents service quality as the inconsistency between a customer's expectations and service quality. It is unclear if all customers will perceive and react equally to service quality. Failure to respond to such a question by the model calls for a study to be performed. The way service quality dimensions relate to the customers' retention attracted many scholars to employ in their studies (Matolo and Salia 2021; Burhan and Kalinga 2018; Ulaya 2017).

3 Research Methodology

This study used across-sectional research design in gathering information. The choice of this research design was preferred in this study due to its ability to avail vital and credible information, especially in settings with many respondents (Robson 2018; Yin 2017). Similarly, the design enables the researcher to collect data at a one-time point Kothari (2015).

The study was conducted at the Kilimanjaro National Park in Kilimanjaro region as this region has many tourists who were the major respondents of this study. Mount Kilimanjaro is markedly known for its height in Africa by being the highest mountain in Africa and the second-highest mountain in the world. This being the case now consideration of the Kilimanjaro region was worthwhile. However, a few customers were considered in this research.

The population of this study consisted of 172 potential KINAPA stakeholders. Specifically, 10 KINAPA officials, 12 TTB board members, 10 community members, 130 KINAPA tourists and 10 tourism service providers. Primarily, it is sometimes impractical to measure the entire population, thus, a sample was secured randomly to omit the possibility of getting respondents with bias.

The study employed Yamane formula (1967) to calculate the sample size of finite population.

$$n = N/\left(1 + N\,e^2\right) \quad (1)$$

N = finite (known) size of Population = 172

n = sample size
e = sampling error (5%)
$n = 172 / (1+172 \times (0.05)^2) = 120$

However, 104 out of 120 questionnaire sets were returned by respondents and used in the course of research. With this regard, the sample size of this study became 104 respondents.

The study used a proportional stratified sampling technique in selecting the sample as the study population was heterogeneous and therefore the population was stratified in form of strata of KINAPA tourists, KINAPA officials, community members, TTB members and service providers to tourists (Kothari 2015). However, due to the homogeneity of tourist lists provided by KINAPA, a simple random sampling was used to select tourists (Table 1).

Table 1. Composition of respondents by stratum

S/N	Strata	Number	Percent (%)
1	KINAPA tourists	80	76.9
2	KINAPA officials	6	5.8
3	Community members	6	5.8
4	TTB members	6	5.8
5	Service providers	6	5.8
Total		**104**	**100%**

In conducting this study, the researcher collected both primary and secondary data. Secondary data were collected from relevant documents like tourism quarterly reports and primary data were collected from tourists, community members, tour guides and government officials.

In this study, the researcher used the following data collection instruments/tools:

(i) **Questionnaires**

Primary data were collected using questionnaires and interviews (structured and non-structured). Each respondent was provided with a set of questionnaires immediately after getting service from the KINAPA. Questions were modified from the SERVQUAL model as formulated by Parasuraman *et al.* (1998).

The questionnaire set was divided into four (4) parts. Part A was general information regarding the customers' affairs. Part B included information regarding service quality with respect to customer expectation, part C comprised questions related to the effect of service quality on the customers' perception and part D was information regarding customer retention. Questionnaires are a set of questions designed to collect data from respondents. Each person was administered the same set of questions in a predetermined order (Saunders 2009). In this study, questionnaires were both open and close-ended. To simplify the data analysis the researcher used the Likert Scale with with a 5-points. This was prepared and administered to respondents selected. The questionnaire were opted due to their effectiveness in data collecting especially in the situation with large number of the respondents within shortest possible period of time and low financial cost.

A five-point Likert scale was used ranging from "1 = strongly disagree" to "5 = strongly agree" while 2, 3 and 4 are scaled as disagree, neutral and agree respectively. 4–5 indicate the positive influence of service quality dimensions on customer retention, while the range from 1–2 shows that, service quality dimension is not a factor which may influence customer retention. On the same questionnaire, respondents were required to indicate the length of stay with the KINAPA (in years).

120 questionnaires were administered to the selected KINAPA tourists, Ministry of Natural Resources and Tourism, community members, members of TTB and tourist service providers. 104 questionnaires were properly filled out and returned. This represented an overall successful response rate of 86.7% as shown in Table 2. According to Kothari (2014), a response rate of above 50% is adequate for a descriptive study. Babbie (2004) also asserted that return rates of above 50% are acceptable to analyze and publish, 60% is good and 70% is very good.

Based on these assertions from renowned scholars, an 86.7% response rate was very good for the study. Thus, the response rate of 86.7% in this study was very good for the study.

Table 2. Rate of return response of questionnaires

Rate of response	Number	Frequency
Returned	104	86.7%
Unreturned	16	13.3%
Total	**120**	**100%**

(ii) **Interview**

The interview is a systematic way of talking and listening to people to collect data. It could be structured or non-structured interviews. This method provides room for clarification to both the researcher and respondents. It guarantees a good return rate and provides more information in detail. It also helps the researcher reduce time in his data collection process. The interview method of collecting data involves oral stimuli and replying verbally. Kothari (2015) perceives that interview exists in two forms which are structured interviews and/or unstructured interviews. Although, both forms of interviews involves questions in the earlier questions are preset and deterministic meanwhile, in the latter form, involves the verbal questions which are not formalised. An open-ended instrument consisting of nine (9) questions interview guide was prepared. The questions with prompts on the effect of service quality(SQ) on the retention of customer enlisted in the interview guide (see Appendix B). Interview guide was chosen in this study based on its flexibility and ability to gather an in-depth information which are of vital role to the study. Respondents were compelled furnish the researcher with information they have and understood as enabler of KINAPA to improve performance using the service quality. Questions included stating the reason for consuming services from KINAPA, the experiences of the respondents being customers, respondents' expectations, and perceptions. Other aspects included identifying the KINAPA features that challenged them as customers.

(iii) **Documentary Review**

A checklist of documents was employed to collect secondary data vital to this research. A checklist of documents was important as not all documents related to the topic at stake were relevant and vital the current study, but this acted a tool for screening useful information for supplementing them to the primary data. Documentary review entails collection of information by rereading both internal documents on the topic of study and kept information regarding the subject matter about the performance of the organization. In this case records of the number of hotels, customers visited KINAPA and kind of tourists' products) or may be external (specifications of tourists' preferences, world annual tourism, world ecosystem and ecology). Documents take various forms of things such hardcopies or soft copies and may include reports of the year, the performance hints of the organisation, internal and external auditors reports of the year, tourists' feedback, meeting minutes, newsletters, and marketing materials (Kumar 2014). On the other hand, Kothari (2015) views documentary analysis (document analysis) as a type of qualitative research in which documents are reviewed by the analyst to assess an appraisal theme. Dissecting documents involves coding content into subjects like how focus groups or interview transcripts are investigated. The study involved a review of various documents, published and unpublished materials (Kothari 2015). The researcher managed to find information on the contribution of service quality to customer retention at KINAPA in Tanzania by responding to research questions in this study. The study used questionnaires, documentary review and interview to collect primary and secondary data for the study to bring about research triangulation using accessible and available data. The choice of the documentary review was based on the fact that it is less expensive in the sense that various documents can be accessed with certainty in various forms as per the researcher's choice (Creswell 2014).

Data analysis means translation of ran data that involves interconnecting data so that the built relationship among variables is displayed to the researcher. It encampuses assembling, feigning, and reconvening data using a chosen method under precise investigative program (Dawson 2020; Yin 2014). Based on this study, data from various sources of choice were collected summarised, coded and were analyzed.

The respondents' names in this study were concealed congruent to the ethical consideration so as to ensure that confidentiality and honour were preserved by adhering to the standards of ethics as per the university research policy. Johnson (2015) is of the opinion that SPSS software had a capacity to assign identity, analyze, store, identify the information regarding specific information for the study.

Data collected were analyzed objective wisely. To examine the level of customer retention, data from Likert Scale questions after being coded were subjected to transformation and later segregated into high or low retention rates. To determine the effect of service quality dimensions on customer retention, a binary logistic regression model was employed because the dependent variable was in binary responses (1=High, 0=low). A binary logistic regression model was applied to determine the chances of being a circumstance based on the levels of the independent variables (predictors). The chances (odds) imply the likelihood that a specific outcome is a case divided by the chance that does not occur. In this case, the responses were considered as either high or low retention levels.

$$\Pr(Y = I) = \beta_0 + \beta_1 X_1 + \beta_2 X_2 + \beta_3 X_3 + \beta_4 X_4 + \beta_5 X_5 + \varepsilon \quad (2)$$

where, $\beta_0, \beta_1, \beta_2, \beta_3, \beta_4$ and β_5 are coefficients or constants, X_1 = service tangibility, X_2 = service reliability, X_3 = service responsiveness, X_4 = Service empathy, X_5 = service assurance and $\mathcal{E}$ = error.

4 Findings and Discussion

4.1 Level of Customer Retention at Mount Kilimanjaro

The overall mean of the customer retention level is 4.436 as indicated in Table 3 implies that customer retention high.

Table 3. Level of customer retention at Mount Kilimanjaro (n = 104)

Variables	N	Mean	Std. deviation
I have made repeated purchases due to reliable hotel services	104	4.58	1.112
I have visited KINAPA for 7 years duet good treatment	104	4.21	1.334
I severally visit KINAPA because of the attractive wildlife	104	4.60	1.084
I usually purchase at KINAPA due to the responsiveness of employees to customers	104	4.21	1.334
I repeatedly purchase at KINAPA due to dependable meals and accommodation services	104	4.58	1.112
Overall mean		**4.436**	**1.1952**

Reliability of Hotel Services

Table 3 shows that the mean of respondents' responses on whether or not to make repeated purchases due to reliable hotel services was 4.58 whereas the corresponding standard deviation was 1.112. This implies that the majority of respondents believed that the repeated purchase of KINAPA was attributed to reliable hotel services. This is evidenced by a mean of 4.58 which is very close to 5andindicatesstrongly agreed with this sentiment. This has also been evidenced by the magnitude of standard deviation showed a little dispersion (1.112) from the mean. This observation is in harmony with the research findings of Gong and Yi (2018) who affirm that customer satisfaction, retention, and happiness in five Asian countries are affected. It was revealed that overall service quality has a positive influence on customers' satisfaction, which in turn leads to customer retention and customer happiness in Asian countries. The study linked the quality of service with customer retention in non-mountainous tourism in Asia. In supporting this observation, the participant stated that:

> "*.... Services that are provided by KINAPA has convinced me to keep on visiting KINAPA now and then……. this is because their service is tangible the manner that satisfies the customers compared to nearby countries of East African community* ..." (Interview held with tourist from European Union on 24th June 2021).

Good Treatment of Customers

Table 3 shows the mean and standard deviation of respondents' responses with respect to the customers' retention level as a result of good treatment is 4.21 and 1.334 respectively. This implies that the majority of respondents were impressed with the good treatment at KINAPA. The research findings are congruent to the findings of Alketbi et al. (2020) who affirmed that service quality had an impact on customer retention. Additionally, the participant had this concern concerning the impact of the service reliability:

> "*.... I have been repurchasing the service at KINAPA because their services are well offered .... there is customer service in offering their services a thing that has compelled me to keep on purchasing from KINAPA....*" (Interview held with tourist from Australia on 24th June 2021).

Likewise, John and Adebayo (2021) had a similar observation to the research findings by affirming that customers' retention is linked with service quality in terms of customer care in Malaysian rural tourism. Therefore, it was revealed that customers are sensitive to the service they receive as a result of their money being exchanged. Furthermore, Rahaman *et al.* (2020) and (Gogoi 2020) support the research findings by asserting that dimensions of service quality have a positive influence on customer retention.

Attractiveness of Wildlife

Table 3 shows that the mean and standard deviation of respondents' responses with regard to the impact of service attractiveness (wildlife) on customer retention are 4.60 and 1.084 respectively. This implies that the majority of respondents were strongly attracted to tourist destinations following the goodness of wildlife. This has also been evidenced by the size of the standard deviation showed a little dispersion (1.084) from the mean. This observation is in agreement with the findings of Rahaman *et al.* (2020) who contend that quality service correlates with customer retention in the tourism subsector. To cement the research findings, the FGD participant stated that:

".... I don't need to ask myself where I will visit next vacation, this is obvious assuredly provision of services with such quality dictates the place to go. .... KINAPA services have highly impressed me and that is why I am here today..." (Interview held with Jewish tourist on 24th June 2021).

Responsiveness of Employees to Customers Retention

Table 3 shows that the mean of respondents' responses with regard to employees' responsiveness with respect to customer retention is 4.21 and 1.334 respectively. This implies that the majority of respondents were pleased with the responsiveness of employees of KINAPA in service provision. This finding is in agreement with the research findings of Rahaman *et al.* (2020) who contend that KINAPA employees' responsiveness impacts the customers' retention in the tourism subsector. To affirm this sentiment, the participant from Key informants stated that:

"......Quality services provided to customers are the cause of the repurchases tendency at KINAPA whose services are responsively delivered to customers. .... This is crucial, especially for the visitors who severally visit KINAPA....." (Interview held with the service provider in Moshi on 24th June 2021)

Dependability of Meals and Accommodation Services

Table 3 shows that the mean and standard deviation of respondents' responses regarding the dependability of meals and accommodation services are 4.58 and 1.112 respectively. This implies that tourists were strongly attracted to the dependability of meal and accommodation services provided at KINAPA.

This finding is similar to the observation by Abdul-Qadir et al. (2021) in Nigeria who linked service quality including service dependability with customer retention in listed food and beverages companies in Kaduna State. These research findings were supported by an FGD participant who stated that:

"......Service superiority and dependability strongly attract customers to repurchase at KINAPA ....This, has convinced tourists to come several times....." (Interview held with Asian tourist on 24th June 2021).

4.2 The Effect of Service Quality Dimensions on Customer Retention

Table 4 of binary logistic regression results shows that the Chi-square was 80.114 at a *p*-value = 0.000. This indicates that service quality dimensions predict the customers' retention strongly. Cox & Snell R Square and Nagelkerke R Square were 0.537 and 0.847 respectively. This implies that 53.7% and 84.7% of the total variance in customer retention are accounted for by the service quality dimensions.

Specifically, services tangibles had a coefficient (β) of 2.231 coefficients which was significant. This finding implies that a unit change of intangibles will result in a 223.1% change in customer retention, furthermore, findings for tangibles show an odds ratio of 2.201 which implies the contribution of tangibles to the customer retention was 2.2 times. This conclusion resembled the findings of Gong and Yi (2018) who obtained similar findings where it was drawn that service quality affected customer satisfaction, retention, and happiness in five Asian countries.

Table 4. Binary logistic results

Variables	*B-Coefficient*	SE	Sig	EXP (B)
Tangibles	2.231	3.350	.000	2.201
Assurance	1.715	1.935	.000	1.004
Responsiveness	2.796	2.969	.000	2.520
Empathy	3.768	3.969	.000	1.301
Reliability	1.715	1.935	.000	2.004
Constant	3.864	.749		
Chi Square	80.114, P = 0.000			
Cox & Snell R^2	.537			
Nagelkerke R Square	.847			

Results from Table 4 further show that assurance had a coefficient (β) of 1.715 which was significant. This implies that a unit change in the assurance of services to customers will result in a 17.1.5%change in customer retention. The odd ratio of 1.004 for assurance in service provision implies that the contribution of assurance to the customers' retention was 1.004 times. The research findings are congruent to the findings of Alketbi et al. (2020) who affirmed that service assurance had an impact on customer retention.

With regard to service responsiveness, it shows that the responsive had a coefficient (β) of2.796 which was significant. This finding implies that a unit change in responsiveness will result in a 279.6% change in customer retention. Furthermore, the odds ratio of 2.52 for responsiveness shows that the contribution of responsiveness to the customers' retention was 2.520 times. The findings are congruent to the findings of Ali *et al.* (2021) who found that service responsiveness had an impact on customer retention.

Likewise, findings show that reliability had a coefficient (β) of 1.715 which was significant. This finding implies that a unit change in reliability will result in a 171.5 % change in customer retention.

Furthermore, findings for reliability show an odds ratio of 2.004which implies that the contribution of reliability to customer retention was 2,004 times. The research results are congruent with observations of Han *et al.* (2021) of China who noted that service reliability had an impact on customer retention.

Similarly, findings show that empathy had a coefficient (β) of 3.768 which was significant. This finding implies that a unit change in empathy will result in a 37.68 %change in customer retention. Furthermore, findings for a responsive show an odds ratio of 1.301 which implies that the contribution of empathy to customer retention was 1.3 times. The research findings are similar to that of Achieng (2021) in Kenya who noted that service empathy had an impact on the customer's satisfaction, royalty and retention.

Furthermore, findings for a responsive show an odds ratio of 1.301 which implies that the contribution of reliability to the customer retention was 1.3 times. The research findings are similar to that of John who noted that service reliability had an impact on the customer's satisfaction, royalty and retention.

References

Abdul-Qadir, A.B., Abubakar, H.S., Utomi, Q.A.R.: Impact of service quality on customer retention of listed food and beverages companies in Kaduna state. Gusau Int. J. Manag. Soc. Sci. (GIJ MSS) **4**(3), 213–228 (2021)

Achieng, S.A., Pepela, A.W.: Influence of service integrity on customer retention: perspectives of rated hotels along with the Kenyan Coastal Tourism Hub. J. Tour. Manag. (JTSM) **4**(2), 423–430 (2021)

Ali, B.J. et al.: Impact of service quality on the customer satisfaction: a case study at online meeting platforms. Int. J. Eng. Bus. Manag. **5**(2), 65–77 (2021). https://doi.org/10.22161/ijebm.5.2.6

Alketbi, S., Alshurideh, M., Kurdi, B.Al.: Retention and loyalty in the UAE hotel sector with respect to the impact of customer' satisfaction, trust, and commitment. J. Archaeol. **17**(4), 541–561 (2020). https://www.archives.palarch.nl/index.php/jae/article/download/390/376

Burhan, A.M., Kalinga, M.M.: The service quality analysis and satisfaction of tourists in Tanzania hotel industry. Int. J. Acad. Res. Bus. Soc. Sci. **8**(11), 423–441 (2018). https://doi.org/10.6007/ijarbss/v8-i11/4939

Dawson, F.: Research Methods for Social Sciences. New Age International Publishers, New Delhi (2020)

Fida, B.A., Ahmed, U., AlBalushi, Y., Singh, D.: Impact of service quality on customer loyalty and customer satisfaction in Islamic banks in the sultanate of Oman. SAGE Open **10**(2), 34–41 (2020). https://doi.org/10.1177/2158244020919517

Gautam, P.: The effects and challenges of COVID-19 in the hospitality and tourism sector. Sustainability **5**(3), 43–63 (2021)

Gogoi, B.J.: Service quality measures: how it impacts customer satisfaction and loyalty. Int. J. Manag. **9**(2), 64–73 (2020). https://doi.org/10.34218/IJM.11.3.2020.038

Han, J., Zuo, Y., Law, R., Chen, S., Zhang, M.: Service quality in tourism public health: trust, satisfaction, and loyalty. Front. Psychol. **5**(2), 54–63 (2021). https://doi.org/10.3389/fpsyg.2021.731279

Haradhan, M.: Qualitative Research Methodology in Social Sciences and Related Subjects, 4th edn. Munich Personal RePEc Archive, Turkey (2018)

House, M.: Tourism and COVID-19: mapping a way forward for the small states. Int. J. Acad. Res. Bus. Soc. Sci. **7**(12), 323–341 (2021)

Islam, S., Samsudin, S.: Characteristics, importance and objectives of research: an overview of the indispensable of ethical research. Int. J. Sci. Res. Publ. **10**(5), 2250–3153 (2020). https://doi.org/10.29322/IJSRP.10.05.2020.p10138

John, M.A., Adebayo, E.S.: Service quality dimensions as correlates of customer satisfaction in selected hotels in Ibadan. Int. J. Bus. Manag. **2**(1), 21–32 (2021)

Kamna, D.F.: Impact of customer service techniques on customer satisfaction in retail bank. J. Account. Mark. **10**(3), 1–7 (2021)

Kothari, C.R.: Research Methods: Methodology and Techniques. New Age International Publishers, New Delhi (2015)

Mashenene, R.G.: Effect of service quality on students' satisfaction in tanzania higher education. Bus. Educ. J. (BEJ) **2**(1), 1–8 (2019)

Matolo, R.J., Salia, P.J.: Determinants of international tourists' destination loyalty: empirical evidence from Serengeti National Park in Tanzania. Sustainability **10**(3), 821–838 (2021)
Minh, N., Quang, L.M.: Environmental protection for the sustainable development of tourism in Vietnam. J. Environ. Manag. **11**(2), 54–67 (2020)
Mishra, R.D., Rasundram, J.: Triangulation is an essential tool to enhance the validity of a case study. J. Res. Interdisc. Stud. **4**(2), 69–74 (2017)
Monteiro, A., et al.: Tourism and air quality during COVID-19 pandemic: lessons for the future. Sustainability **12**, 69–83 (2021)
Ngoa, N.M., Quang, L.M.: Environmental protection for the sustainable development of tourism in Vietnam. J. Environ. Manag. Divers. Sustain. **8**(4), 6–19 (2020)
Palil, M.: Tax knowledge and tax compliance determinants in self assessment system in Malaysia; A Thesis submitted for the fulfilment of the Award of Doctor of Philosophy at The University of Birmingham (2010)
Parasuraman, A., Zeithaml, V., Berry, L.L.: Communication and control processes and the delivery of service quality. J. Mark. **52**(2), 35–49 (1998). https://doi.org/10.2307/1251263
Park, R., Jeong, L.: Performance-only measures vs. performance-expectation measures of service quality. Serv. Ind. J. **36**, 741–756 (2019)
Peat, Y.D.: Kilimanjaro tourism and what it means for local porters and the local environment. J. Econ. Soc. Sci. **7**(6), 32–46 (2019)
Rahaman, M.A. Ali, M.J. Kejing, Z., Taru, R.D., Mamoon, Z.R.: Investigating the effect of service quality on bank customers' satisfaction in Bangladesh. J. Asian Finance Econ. Bus. **7**(10), 823–829 (2020). https://doi.org/10.13106/jafeb.2020.vol7.n10.823
Saunders, M., Lewis, P., Thornhill, A.: Research Methods for Business Students, 6 edn. Pearson, London (2009)
Tanzania National Park: Tourism Performance in Tanzania. Annual Tourist Report (2019)
Tuan, V.K., Rajagopal, P.: Analyzing factors affecting tourism sustainable development towards Vietnam in the new era. Eur. J. Bus. Innov. Res. **71**, 30–42 (2019)
Ulaya, A.Y.: Effect of service quality on customer. J. Mark. Manag. **4**(2), 211–213 (2017)
Weirmair, K., Fuchs, M.: Measuring tourist judgment on service quality. Ann. Tour. Res. **26**(4), 1004–1021 (2019)
Yamane, T.: Statistics in Introductory Analysis, 2nd edn. Harper and Row, New York (1976)
Yi, Y., Gong, T.: The electronic service quality model: the moderating effect of customer self-efficacy. Psychol. Mark. **25**, 587–601 (2018)
Yin, R.K.: Case Study Research and Applications: Design and Methods, 2nd edn. Sage, Thousand Oaks (2014)

Techno-Economic Feasibility of Hydropower Generation from Water Supply Networks in Ghana

W. O. Sarkodie[1,2](✉) and E. A. Ofosu[1]

[1] Department of Energy and Petroleum Engineering, University of Energy and Natural Resources, Sunyani, Ghana

[2] Department of Renewable Energy Technology, Cape Coast Technical University, Cape Coast, Ghana

wilson.sarkodie@cctu.edu.gh

Abstract. Purpose: This study looked at the potential energy recovery and economic feasibility of producing electricity from water distribution networks (WDN) and water transmission networks (WTN) using a Pump as a turbine (PaT) in the Western Region of Ghana.

Design: To determine energy recovery viability, seven WDNs and one WTN were evaluated in MS excel using primary and secondary data. This enabled us to estimate the power potential as well as the annual energy generation from the networks. The economic feasibility of the energy generated was determined by finding the net present value (NPV) and Payback period (PP). Discount rate variation influence on NPV was also determined.

Findings: According to the findings, Inchaban Headworks WTN has the highest annual energy generation potential, with a capacity of 71 MWh and could power 96 households. This system has an NPV of €8,130.75 and €3,231.19 at 5% and 10% discount rates, respectively. The total annual generation from the remaining systems could generate 70 MWh of electricity for about 94 households annually.

Implications/Research Limitations: The study was limited to water supply networks of the Ghana Water Company Limited (GWCL) in the Western Region of Ghana.

Practical Implication: The outcome of the study can be advanced to implement PaTs installation on water supply networks in Ghana and Africa at large for water infrastructure sustainability.

Originality/Value: Previous research has not thoroughly examined the energy recovery potential of WTN and WDN in Ghana and Africa context. As a result, the study aims to promote innovation in water supply and management for the long-term sustainability of water infrastructure in Ghana and Africa at large.

Keyword: Distribution networks · Energy recovery · Hydropower · Renewable energy · Pump-as-turbine

C. Aigbavboa et al. (Eds.): ARCA 2022, *Sustainable Education and Development – Sustainable Industrialization and Innovation*, pp. 840–853, 2023.
https://doi.org/10.1007/978-3-031-25998-2_64

1 Introduction

The sustainability of water infrastructure is largely dependent on the type of energy use. One of the goals of the 2030 Agenda is to ensure the availability and sustainable administration of water for the world (United Nations 2015). Consequently, water supply networks are faced with the difficulty of delivering safe water to consumers. Additionally, the anticipated rise in the population in the coming years and climatc change concerns exert a greater tension on water and energy resources, prompting the transition of water distribution networks to more efficient systems (García et al. 2019). The International Energy Agency (IEA) estimates that, the water sector consumed 120 Mt of oil equivalent in worldwide energy in 2014. Although water collection, pumping, and purification use only 2.6% of EU electricity consumption, it accounts for 30–40% of municipal energy expenses (European Commission 2019).

This problem has been investigated by several studies, which have suggested methods to decrease energy consumed in water supply networks by optimizing pump operations (De Marchis et al. 2014; Du et al. 2017; Fecarotta and McNabola 2017; García et al. 2019; Stefanizzi et al. 2020) and recover energy by switching pressure reducing valves (PRVs) with hydraulic turbines (McNabola et al. 2014; Samora et al. 2016). In water distribution networks, switching from a PRV to a regular hydroelectric turbine has not always been possible (Ramos and Ramos 2009). Because PRVs with low power output require a small turbine, this is the case. The installation of traditional turbines usually necessitates a significant amount of expenditure (McNabola et al. 2014). Pump-as-turbines (PATs), which run in reverse mode, have been proposed by many studies as an efficient low-cost turbine for recovering energy from WDN (De Marchis et al. 2014; García et al. 2019; McNabola et al. 2014; Morabito and Hendrick 2019; Pérez-Sánchez et al. 2017; Wallace 1996; Williams et al. 1998). PaT has been claimed to be 5–15 times less expensive than traditional turbines (De Marchis et al. 2014).

According to the Ghana Water Company Limited (2018), energy is by far the most expensive component of cost accumulation, with a rise from about 21% in 2010 to about 32% in 2018. Despite this challenge, power outages and voltage fluctuations impede operations. Because the company relies on power from the national grid, such outages disrupt operations and, as a result, have a negative impact on customers at the end of the distribution line.

To solve the aforementioned difficulties, various initiatives from industry and academia have been proposed (De Marchis et al. 2014; Du et al. 2017; Fecarotta and McNabola 2017; García et al. 2019; McNabola et al. 2014; Okedu et al. 2020; Pérez-Sánchez et al. 2017; Williams et al. 1998).

In Africa, however, there is insufficient proof of the hydropower potential of water distribution networks (WDN). As a result, the study aims to look at the hydropower potential in Ghana's WDN and Water Transmission Network (WTN). The study's findings will be useful to industry participants, policymakers, and decision makers in order to gain a better understanding of water infrastructure design in order to accelerate Sustainable Development Goals 6 and 7.

2 Energy Recovery from WDN and WTN

2.1 Hydro-turbine Selection on WDN

In hydropower plants, the traditional hydro-turbines are chosen depending on the head (H) and flow rate (Q). Impulse turbines such as Pelton and Turgo of various capacities are suitable for high-head hydroelectric projects, whereas Francis, mixed-flow, Kaplan, and bulb turbines can perform across a wide range of head and flow rate. (Morabito and Hendrick 2019; Stefanizzi et al. 2020).

PaT was discovered in micro-hydropower facilities (less than 100 kW) recenctly using miniaturized impulse turbines, reaction turbines, and Crossflow turbines. Because a PaT is a traditional pump that operates in reverse, it lacks variable guiding vanes to accommodate varying flow conditions. As a result, while the rotational speed of PaT remains constant, control can be accomplished through the use of either a valve or a bypass, depending on whether we want to reduce or increase the flow rate in the network. In order to maintain maximum PaT efficiency while also expanding the system's rangeability a speed control can be used (Stefanizzi et al. 2020). Figure 1 shows the operating conditions of a hydraulic turbine.

Giugni et al. (2009) found that when standard pumps are used in reverse mode, they are better than traditional turbines for small-scale hydropower plants in several ways. Pumps are mass-produced, whereas turbines are specially designed for each location and an integral pump and motor can be acquired and utilized as a turbine-generator unit. The unit comes in a broad spectrum of heads and flows, comes in a variety of sizes, is inexpensive, has a fast conveyance time, spare parts availability, and easy to install. Nonetheless, the fundamental limitation of PaTs is the unit's limited operating flow rate range, which is significantly narrower than that of a conventional turbine.

2.2 Energy Generation Determination

Pump characteristics curves depict the correlation between the head and flow of the pump. As the flow increases, the delivery head lowers. Understanding the most efficient point at which the pump operates is critical. The maximum efficiency varies depending on the type and size of pump, but it is normally between 40% and 80%, and the best efficiency point (BEP) of the pump is identified (Giugni et al. 2009).

The flow increases as the head increases, in a turbine mode. Some research (Morabito and Hendrick 2019; Williams et al. 1998) shows that the performance curves of PaTs can be estimated, which shows the relationship between the pump and the PaTs' BEP.

2.3 Pressure Reducing Valves and Break Pressure Tanks

The PRVs reduce the pressure in the flow in the network to the required level. In water distribution network, the pressure and volumetric flow through the network are controlled using the PRV. However, the water networks under study do not have PRVs. This is a major concern to Water Supply Company (WSC) because about 126 billion cubic meters of water is lost annually from water supply networks globally (Ahopelto and Vahala 2020). The locations of the PRVs have been considered by several researchers

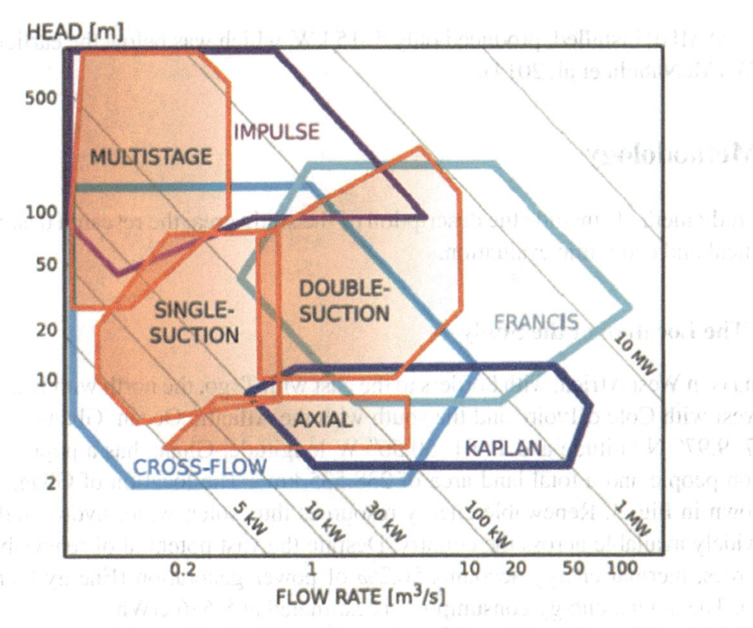

Fig. 1. Operating conditions of hydraulic turbines adopted from (Morabito and Hendrick 2019)

as potential site for energy recovery in a WDN (McNabola et al. 2014; Mcnabola et al. 2014). Flow meters at the exit of the reservoirs were considered in this study as potential site for installation of PAT systems. A typical PRV is shown in Fig. 2.

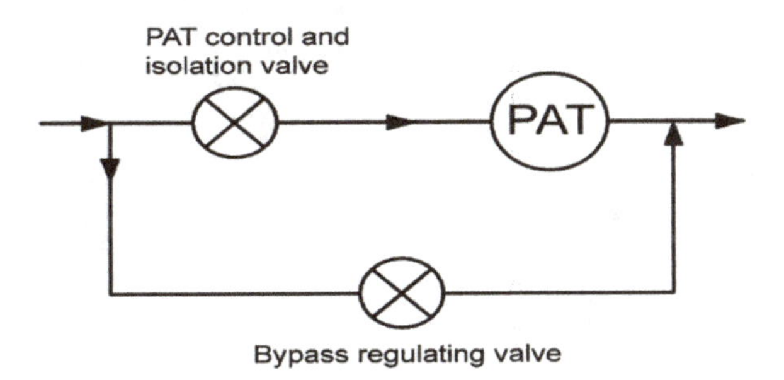

Fig. 2. Schematics of pressure reducing valve

The control valve controls the flow of water to the turbine as well as to stop the flow of water for turbine maintenance. When the turbine's required flow and pressure are exceeded, the bypass valve opens. In addition, it also opens to prevent water from entering the turbine during maintenance.

The Break Pressure Tank (BPT) on the other hand perform similar functions as the PRV. The BPT is bigger in size and also relatively expensive than PRV. Due to the extent of contamination caused by BPT in water, it is no longer a popular design choice. A market assessment of six US PRV sites found 500 kW of energy recovery potential.

The final MHP installed, produced only 5–15 kW, which was below the earlier target of 35 kW (McNabola et al. 2014).

3 Methodology

The study methods include the description of the study area, the research design and the technical and economic evaluation.

3.1 The Location of the Study

Ghana is in West Africa, with borders to the east with Togo, the north with Burkina Faso, the west with Cote d'Ivoire and the south with the Atlantic Ocean. Ghana is located at 7° 57′ 9.97″ N latitude and 1° 01′ 50.56″ W longitude. Ghana has a population of 30 million people and a total land area of 238,535 km^2. The location of Ghana in Africa is shown in Fig. 3. Renewable energy resources thus solar, wind, hydro, and biomass are widely available across the country. Despite the vast potential of renewable energy resources, thermal energy accounts 56.2% of power generation (Energy Commission 2016). The annual energy consumption is estimated at 8,646 GWh.

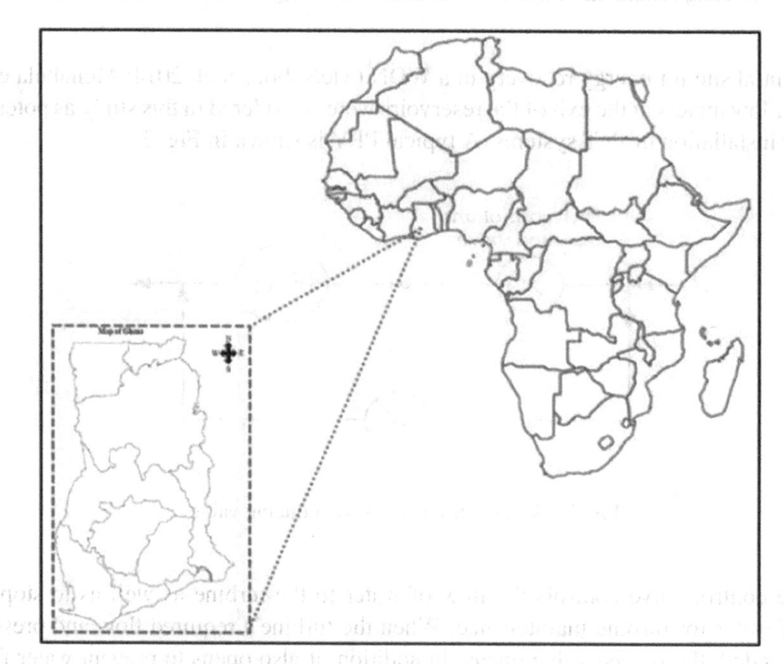

Fig. 3. Location of Ghana in the context of Africa

3.1.1 Ghana Water Company Limited

The "Ghana Water Company Limited" (GWCL) began operations on July 1, 1999, when the "Ghana Water and Sewerage Corporation" (GWSC) was transformed into a limited liability company, owned by the state under LI 1648. The company supplies urbanites with water. GWSC delivered rural and urban water until the early 1990s. The GWSC created a rural water supply section due to a paucity of water.

The GWSC developed the Community Water and Sanitation Division (CWSD) to supervise rural water and sanitation. After four years, the CWSD needed full autonomy to boost its activities. Act 564 renamed the Division the "Community Water and Sanitation Agency" (CWSA) in December 1998 to provide rural communities and small towns in Ghana with potable water and sanitation services. The Ministry of Water and Sanitation owns the water infrastructure. 77% of cities have access to water and produce 871496 m^3 daily. 80% of GCWL consumers are unmetered (GWCL 2018). The company operates in Ghana's 16 regions.

3.1.2 Western Region Water Supply Network

The study was conducted in Western Region of Ghana. The identified WDN and WTN are; Inchaban Headworks; Sekondi, Essikador, Effia, Beach Road and Harbour line. The Inchaban Headworks treats raw water and transmit it to Essikador reservoir for onward distribution to Essikador, Sekondi and Babakrom townships. Transmitted Water from Daboase Headworks are also boosted by pumps at Inchaban Headworks before transmission to Essikador reservoir. The Inchaban headworks reservoir and 32 m diameter raw water transmission pipe from the reservoir to the chemical mixing chamber is shown in Fig. 4.

Fig. 4. Inchaban headworks raw water reservoir and 813 mm diameter pipe

The WDN and WTN has no BPT and PRV to dissipate excess pressure in the network to avoid damaging the pipes. The potential site for energy recovery is illustrated in Fig. 5, a schematic flow for both the Inchaban headworks WTN and all the WDN mains across Western region.

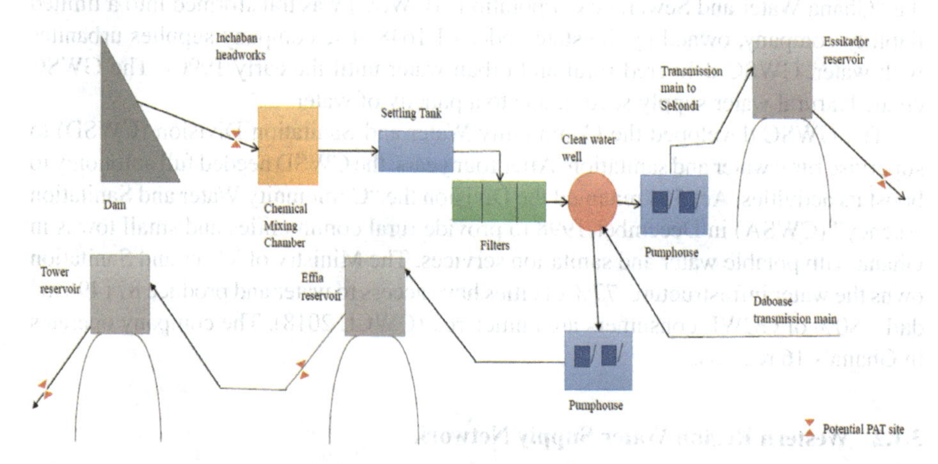

Fig. 5. Schematic view of the networks under study

The research evaluated the energy potential of networks that transmit or distribute water by gravity. Water is gravity-fed from the reservoir at Inchaban to the chemical mixing chamber through an 813 mm-diameter pipe. Inchaban Headworks' raw water transmission main offers the potential for energy recovery. Pumps transmit treated water from Inchaban Headworks to the Essikador overhead reservoir. The water is subsequently gravity-fed to Essikador, Sekondi, and Babakrom townships, which are also potential energy recovery locations. Inchaban Headworks also functions as a booster

Fig. 6. Areal view of Inchaban headworks. Extracted from google earth, 2021

station for pumping water from Daboase Headworks to the Effia overhead reservoir. Effia has two main distributions (Effia 1 and Effia 2) that provide water to the Effia population by gravity. In addition, the Effia reservoir transmits water to the Tower reservoir, which distributes water to secondi-takoradi via pumps and gravity. Water is distributed by gravity through the Harbor Line and Beach Road distribution networks at Tower reservoir. Figure 6 is the aerial view of the Inchaban Headworks showing buildings, roads, reservoir and Dam.

3.2 Study Design

The study design comprises of comprises of the evaluation of the technical viability of energy recovery from the identified networks using the primary and secondary data collected. This enabled us to estimate the power potential as well as the annual energy generation from the networks. In addition, the economic feasibility of the energy generated was estimated by finding the net present value (NPV) and Payback period (PP). The discounted PP and NPV, which are essential determinants of the viability of hydropower potential from WDN and WTN, are significantly affected by Feed-in-Tarrif (FiT) fluctuation (McNabola et al. 2014). The FiT is a policy that promotes the utilization of renewable energy sources to generate electricity. It permits power providers to sell renewable energy generated electricity to an off-taker for a set period of time at a specified pricing. The FiT varies depending on the technology, its capacity, and the location. In Ghana the FiT guarantee a sale price of 0.077 €/kWh for hydropower plant $P <= 10$ MW. In addition, discount rate variation was examined as it also influences the NPV. Figure 7 shows the proposed flow chart for the study.

The flow and hydraulic head were collected from overhead reservoirs that flows by gravity to serve the communities as shown in Table 1. The reservoir head is the elevated overhead reservoir relative to PaT. Reservoirs that are pumped for water distribution were not considered for data collection as such is not economically feasible as investigated by researchers (García et al. 2019; McNabola et al. 2014). The average flow rate and reservoir head were used to compute power potential in the WDN and WTN. Head-losses and friction losses were assumed to be zero. Microsoft excel was employed in data analysis.

Figure 8 depicts the seasonal flow variation of the Inchaban headworks WTN. The flow is seen to peak in August and then begin to drop. The minimal flow occurs in January, possibly as a result of the harmattan season (December–February) in Ghana, which lowers the reservoir's water level.

3.3 Technical and Economic Evaluation

The power output of the network was determined using Eq. (1)

$$P = \frac{\rho g Q h \eta_B}{1000} \tag{1}$$

where, P is the output power in kW; ρ is water density in (kgm^{-3}); g is acceleration due to gravity (m/s^2); Q represent the hydraulic flow in the turbine (m^3/s); h is the turbine's available head(m) and η_B is the overall power plant efficiency

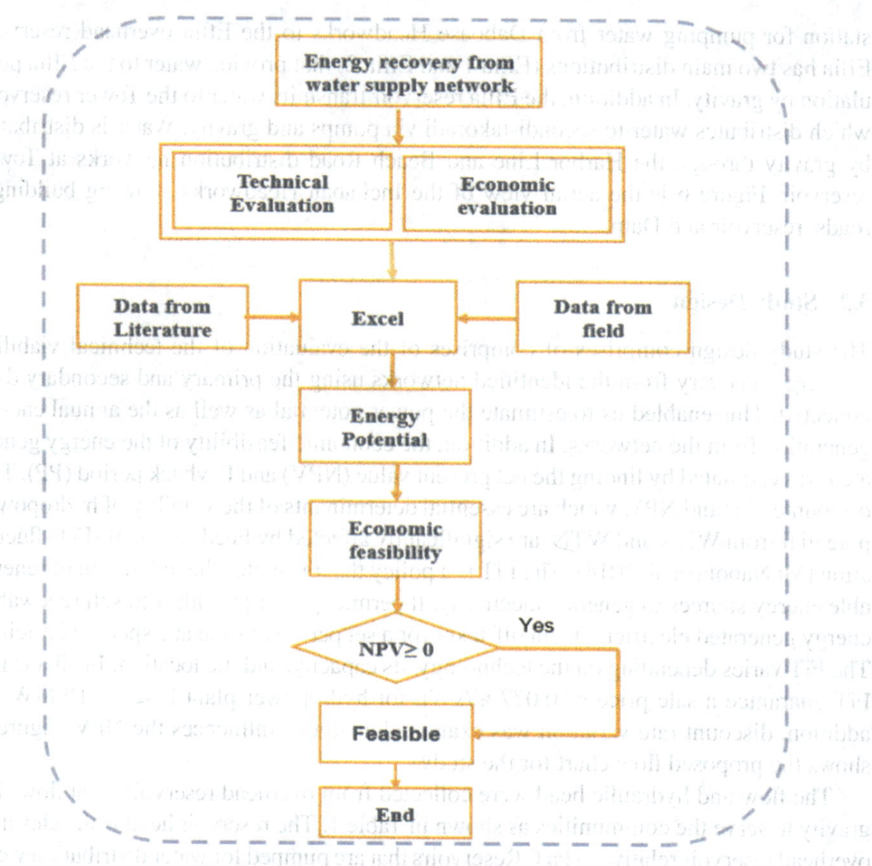

Fig. 7. Energy recovery flow chart of the study

Table 1. Main data collected from WDNs and WTN

Site/WDN	Average flow (m^3/s)	Reservoir head (m)
Inchaban	0.174	7.32
Essikador	0.044	4.92
Sekondi	0.042	4.92
Babakrom	0.04	4.92
Effia1	0.053	6.04
Effia2	0.051	6.04
Harbor line	0.014	4.88
Beach road	0.013	4.57

The assumed turbine efficiency η_B is 65% (García et al. 2019; Mcnabola et al. 2014).

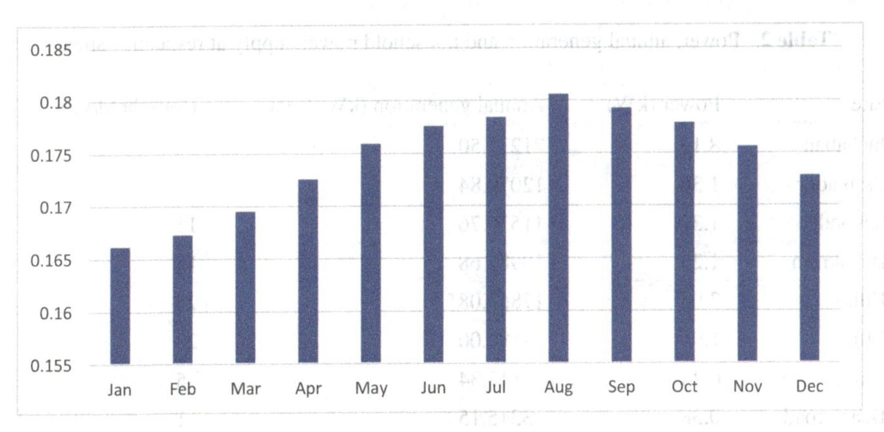

Fig. 8. Seasonal flow variation at Inchaban headwork

The economic feasibility of the hydropower investment was assessed using the Net Present Value method (García et al. 2019; McNabola et al. 2014).

The NPV was estimated using Eq. (2)

$$NPV = \sum\nolimits_{i=0}^{T} \frac{B_t - C_t}{(1+r)^t} \tag{2}$$

where, B_t is undiscounted revenue year t; C_t is undiscounted capital cost in year t; T is the project's duration and r is the discount rate

The payback period was calculated using Eq. (3)

$$PP = \frac{T_{total}}{(Hr.ie)^t} \tag{3}$$

PP is discounted payback period (years); T_{total} is total cost of PAT; Hr is hydropower recovered (W) and; ie is savings from displaced electricity costs.

4 Results and Discussion

4.1 Hydropower Potential

The hydropower potential for the Inchaban Headworks was the greatest at 8.13 kW. The average electricity consumption in Ghana per electrified household is 744 kWh per year (IFC 2020), while the Inchaban raw water distribution main has a hydropower capacity of 8.13 kW, which could provide over 71 MWh per year to about 96 households as shown in Table 2. The Effia 1 and Effia 2 sites are the second and third highest hydropower potential sites, with capacities of 2 kW and 1.9 kW with annual energy production of 18 MWh and 17 MWh respectively. Annually, Effia 1 and Effia 2 could power 24 and 23 households, respectively. The Essikador, Sekondi and Babakrom total annual electric potential of 35 MWh could power about 46 households per year. The summary of energy production and potential households' electrification from the WDN and WTN is shown in Table 2 and Fig. 9.

Table 2. Power, annual generation and household power supply at respective sites

Site	Power (kW)	Annual generation (kWh/year)	Household/year
Inchaban	8.13	71235.50	96
Essikador	1.38	12079.84	16
Sekondi	1.32	11530.76	15
Babakrom	1.25	10981.68	15
Effia1	2.04	17863.08	24
Effia2	1.96	17189.00	23
Harbor line	0.44	3812.34	5
Beach road	0.38	3315.15	1

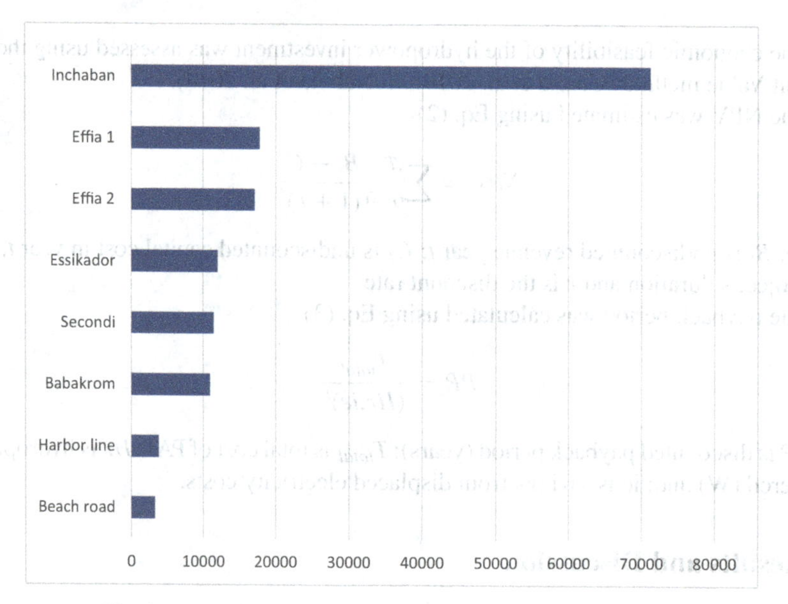

Fig. 9. Annual energy generation from water supply networks

4.2 Economic Feasibility

Calculating the NPV and capital payback time is critical from an economic standpoint. A proper analysis should take into account the following expenses: hydraulic and electric device costs; construction costs; plant O&M and administration costs; and gross sales received over the lifespan over the plant (García et al. 2019; Aonghus McNabola et al. 2014 Giugni et al. 2009). The cost of NC 150–200 pump was estimated at €1,500 per kWh, construction cost estimated at 30% of the pump cost and O&M cost is 15% of the total cost of installation (García et al. 2019; Giugni et al. 2009). Table 3 is the cost components for WDN and WTN considered in this study.

Table 3. Summary of cost components of PAT installation

Project station	Power (kW)	Pump Cost (€)	Construction cost (€)	Total installation cost (€)	O & M cost (€)/year
Inchaban	8.13	12198	3659	15857	2379
Essikador	1.38	2068	621	2689	403
Sekondi	1.32	1974	592	2567	385
Babakrom	1.25	1880	564	2445	367
Effia1	2.04	3059	918	3976	596
Effia2	1.96	2943	883	3826	574
Harbor line	0.44	653	196	849	127
Beach road	0.38	568	170	738	111

The economic viability of this project depends on the NPV. The NPVs provided were estimated over a 10-year financing period using 5%, 10%, and 15% discount rates (García et al. 2019; McNabola et al. 2014). The NPV for Inchaban headworks at 5% and 10% discount rates recorded the highest values €8,130.75 and €3,231.19 respectively. The remaining project sites also recorded positive NPV at 5% and 10% discount rates. However, at 15% discount rate, the NPV for all the project sites were negative which

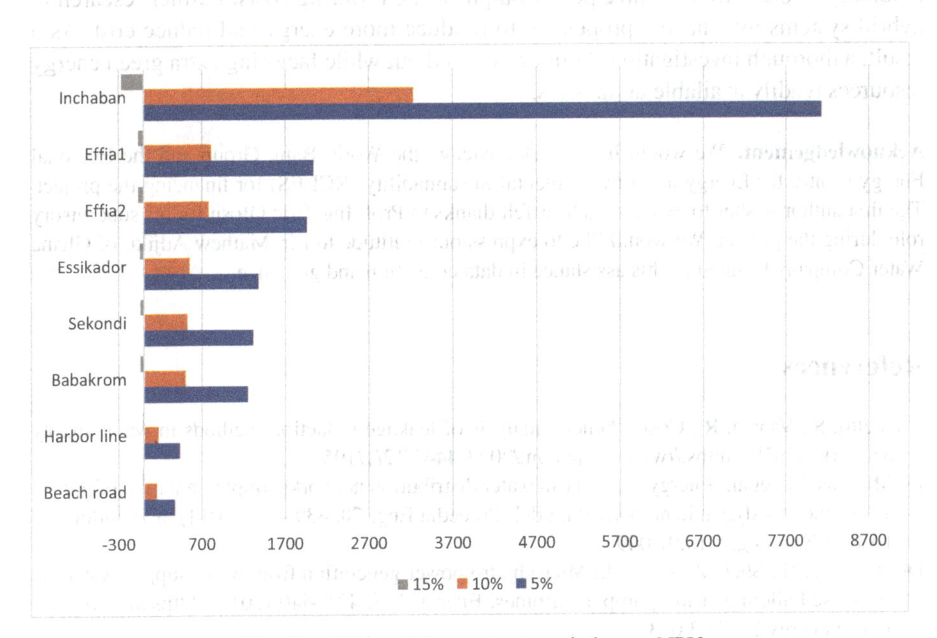

Fig. 10. Effect of discount rate variation on NPV

is unattractive for investment. Therefore, it is established that increase in discount rate decreases NPV. The effect of discount rate variation on NPV is shown in Fig. 10.

5 Conclusion

The preliminary investigation of energy recovery from Ghana's water network has revealed a great potential in the western region water supply network. Pump as Turbines (PaTs) potentially recover energy from the water supply network. The Inchaban WTN recorded the highest energy potential and could produce over 71 MWh per year to about 96 households whilst Effia 1 and Effia 2 WDNs could power 24 and 23 households with annual generation capacity of 18 MWh and 17 MWh respectively. All of the networks investigated in this study had a positive NPV when the discount rate was set at 5% or 10%. Inchaban Headworks WTN had the greatest NPV value € 8,130.75 and €3,231.19 at 5% and 10% discount rates, respectively. The outcome of the study shows substantial evidence of energy recovery potential from water supply networks in Ghana.

Implementing this in the water network has the capacity to reduce energy costs of the company while also ensuring the long-term viability of the water infrastructure. Additionally, this will create jobs for those who will install and maintain the PaT in the water supply network. The Inchaban Headworks WTN offers significant energy recovery potential, which could supplement the company's power source. Again, including hydropower in the infrastructure has the potential to offset the company's CO2 emissions.

More research is needed to improve the system by adjusting the flow and turbine efficiency in order to maximize power output while lowering costs. Further research in hybrid systems may have a propensity to produce more energy and reduce cost. As a result, a thorough investigation should be carried out while factoring extra green energy resources readily available at the sites.

Acknowledgement. We would like to acknowledge the World Bank Group and the Regional Energy Centre for Energy and Environmental Sustainability (RCEES) for financing the project. The first author wishes to express his heartfelt thanks to Prof. Ing. Eric Ofosu for his supervisory role during the project. We would like to express our gratitude to Mr. Mathew Adjiku of Ghana Water Company Limited for his assistance in data collection and guidance.

References

Ahopelto, S., Vahala, R.: Cost – benefit analysis of leakage reduction methods in water supply networks (2020). https://www.mdpi.com/2073-4441/12/1/195

De Marchis, M., et al.: Energy recovery in water distribution networks. Implementation of pumps as turbine in a dynamic numerical model. Procedia Eng. **70**, 439–448 (2014). https://doi.org/10.1016/j.proeng.2014.02.049

Du, J., Yang, H., Shen, Z., Chen, J.: Micro hydro power generation from water supply system in high rise buildings using pump as turbines. Energy **137**, 431–440 (2017). https://doi.org/10.1016/j.energy.2017.03.023

Energy Commission: National Energy Statistics (2016)

European Commission: JRC Science for Policy Report Water Energy Nexus in Europe (Issue October) (2019). https://doi.org/10.2760/968197

Fecarotta, O., McNabola, A.: Optimal location of pump as turbines (PATs) in water distribution networks to recover energy and reduce leakage. Water Resour. Manag. **31**(15), 5043–5059 (2017). https://doi.org/10.1007/s11269-017-1795-2

García, I.F., Ferras, D., Nabola, A.: Potential of energy recovery and water saving using micro-hydropower in rural water distribution networks. **145**(2016), 1–11 (2019). https://doi.org/10.1061/(ASCE)WR.1943-5452.0001045

Giugni, M., Fontana, N., Portolano, D.: Energy saving policy in water distribution networks. Renew. Energy Power Qual. J. **1**(7) (2009). https://doi.org/10.24084/repqj07.487

Ghana Water Company Limited: Proposals for Review of Aggregate Revenue Requirement and Tariff. Public Utilities Regulatory Commission - GWCL, December 2018. https://www.gwcl.com.gh/tarrif_paper.pdf

IFC: IFC in the Power Sector (2020). https://www.ifc.org/wps/wcm/connect/d8d2d63a-c877-4c6a-8aa7-7d68cb5ce389/PowerInfographic.pdf?MOD=AJPERES&CVID=lnvW0IB

McNabola, A., et al.: Energy recovery in the water industry using micro-hydropower: an opportunity to improve sustainability. Water Policy **16**(1), 168–183 (2014). https://doi.org/10.2166/wp.2013.164

Mcnabola, A., Coughlan, P., Williams, A.P.: Energy recovery in the water industry: an assessment of the potential of micro-hydropower, **2011**, 294–304 (2014). https://doi.org/10.1111/wej.12046

McNabola, A., Coughlan, P., Williams, A.P.: Energy recovery in the water industry: an assessment of the potential of micro-hydropower. Water Environ. J. **28**(2), 294–304 (2014). https://doi.org/10.1111/wej.12046

Morabito, A., Hendrick, P.: Pump as turbine applied to micro energy storage and smart water grids: a case study. Appl. Energy **241**(August 2018), 567–579 (2019). https://doi.org/10.1016/j.apenergy.2019.03.018

Okedu, K.E., Uhunmwangho, R., Odje, M.: Harnessing the potential of small hydro power in Cross River state of Southern Nigeria. Sustain. Energy Technol. Assess. **37**(October 2019), 100617 (2020). https://doi.org/10.1016/j.seta.2019.100617

Pérez-Sánchez, M., Sánchez-Romero, F.J., Ramos, H.M., López-Jiménez, P.A.: Energy recovery in existing water networks: towards greater sustainability. Water (Switzerland) **9**(2), 1–20 (2017). https://doi.org/10.3390/w9020097

Ramos, J.S., Ramos, H.M.: Sustainable application of renewable sources in water pumping systems: optimized energy system configuration. Energy Policy **37**(2), 633–643 (2009). https://doi.org/10.1016/j.enpol.2008.10.006

Samora, I., Franca, M.J., Schleiss, A.J., Ramos, H.M.: Simulated annealing in optimization of energy production in a water supply network. Water Resour. Manag. **30**(4), 1533–1547 (2016). https://doi.org/10.1007/s11269-016-1238-5

Stefanizzi, M., Capurso, T., Balacco, G., Binetti, M., Camporeale, S.M., Torresi, M.: Selection, control and techno-economic feasibility of pumps as turbines in water distribution networks. Renew. Energy **162**, 1292–1306 (2020). https://doi.org/10.1016/j.renene.2020.08.108

United Nations: Transforming Our World: The 2030 Agenda for Sustainable Development, vol. 16301, no. October. United Nations, New York (2015)

Wallace, A.R.: Embedded mini-hydro generation in the water supply industry. IEE Conf. Publ. **419**, 168–171 (1996). https://doi.org/10.1049/cp:19960142

Williams, A.A., Smith, N.P.A., Bird, C., Howard, M.: Pumps as turbines and induction motors as generators for energy recovery in water supply systems. Water Environ. J. **12**(3), 175–178 (1998). https://doi.org/10.1111/j.1747-6593.1998.tb00169.x

The Influence of Cash Management on Financial Performance of Private Schools in Tanzania

F. Johnson[1] and D. Pastory[2(✉)]

[1] Department of Business Administration, College of Business Education, Dar es Salaam, Tanzania
[2] Department of Accounting, College of Business Education, Dar es Salaam, Tanzania
passtory1@yahoo.co.uk

Abstract. Purpose: This study aims at determining the influence of cash management on the financial performance of private schools in Tanzania.

Design/Methodology/Approach: A sample of 132 respondents was selected by the researcher from five private schools which are Loyola School, Barbro Johansson Model Girls Sec. School, Dar es Salaam International Academy, International Schools of Tanganyika, and Haven of Peace Academy are both located in Dar es Salaam. The study used purposive and simple random sampling techniques while data were collected by using interviews and distribution of questionnaires and analyzed quantitatively by using descriptive statistics through Statistical Packages of Social Science (SPSS). The content analyses were run by using MAXQDA10 qualitative data to conclude.

Findings: The findings of this study revealed that cash management influences the financial performance of private schools in Tanzania. However, there is inefficiency in implementing cash management, particularly (cash planning, cash budgeting, cash collection as well as cash control) in private schools. This is influenced by various reasons which include the absence of an Accounting Staff, limited sources of financial resources, lack of Transparency and Accountability, absence of bank accounts, poor computer literacy, and poor record keeping.

Implications/Research Limitations: Five private schools from one city were used for this study hence the findings might not be a reflection of the entire country.

Practical Implications: To eradicate the above constraints, the researcher recommends increasing accounting staff, expanding sources of funds, record keeping, and provision of financial education.

Originality/Value: The study addresses how cash management influences the financial performance of private schools based in the Tanzanian context.

Keyword: Budgeting · Cash management · Performance · Planning · Private schools

C. Aigbavboa et al. (Eds.): ARCA 2022, *Sustainable Education and Development – Sustainable Industrialization and Innovation*, pp. 854–865, 2023.
https://doi.org/10.1007/978-3-031-25998-2_65

1 Introduction

At the globe level, the history of cash management in private school can be traced from the first appearance of private schools about 1660's in New Netherland (Wise 1956). Later, private schools developed in most of the American cities including private schools opened by Catholic missionaries in Florida and Louisiana in 16th century. During this era the operation of schools and cash management remained under the missionaries (Lawrence 1970). According to Lawrence (1970) school were given variable financial aid by missionaries which were usually inadequate.

In New England there were also private schools taught by individual schoolteachers who targeted to earn income and make a living out of it (Hunt 2000). The system of education was not regulated by government and the cash management system were also not considered as well. Education was provided to adults and older children's mostly at their convenient time especially in the evening (Hunt 2000).

In 19th century, private schools developed as the result of industrialization and urbanization. The increase in number of private schools welcomed massive investments in education (Carnoy 2000). Many schools were built and shortages of classrooms were gradually resolved. Many private schools started to employ cash management system for purpose of meeting their financial needs, reducing cash deficits and cash balances as well as increasing profitability (Carnoy 2000). The private schools created management system so as to plan, organize, direct, monitor and control the financial activities for their progress (Denis 2018). This was for ensuring of availability of financial resources which is vital for private school's productive process and quality of education provided (Yizengaw et al. 2021). However, unlike the public schools which has government financial regulations, auditing and financial accountability, in the private school's cash management remained under the option of the private owners of the schools (Tooley and Longfield 2015).

In African context, the history of private schools can be traced from colonial regime where many private schools operated by religious missionaries who introduced European style education (Nishimura and Yamano 2013). Between 1820 and 1881 Gold Coast (Ghana) witnessed first missionaries including Moravian, the Methodists, Bremen Mission as well as Roman Catholic who opened seminary schools (Vanqa 1995). This extended to other parts of Africa including Nigeria from 1860 to 1899 when protestant missions opened missionary schools at Badagry et al. (2004). Kenya, Uganda also followed with the establishment of the Church Missionary Society, the Universities Mission to Central Africa, the White Fathers, and the London Missionary Society opened the first mission schools between 1840 and 1900 (Chelanga 2016). The private school's financial control was under the missionaries who involved in the whole process of cash management and most of private schools lacked professional accountant and cash management system as well (Chelanga 2016).

After the World War II private education witnessed participation of Africans who gained more autonomy (Parker and Rathbone 2007). This went together with the increase of more higher learning institutions and establishment of university colleges including Accra and Ibadan in 1948, Makerere University in 1949 and Khartoum in 1951. The financial management in private schools and higher learning institutions remained in the option of private owner with had no specific policies or regulation from the governing authorities. The engagement of many private Africans operating schools, increased operation of many schools without systematic cash management as a result's some schools failed due to financial difficulties (Sunal 1998).

In Tanzanian context, the development of private schools underwent various phases which includes the nationalization era in late 1960s and 1970s where all the private schools were fully owned by the government as an implementation of free and universal public education which discouraged and restricted registration and operation of private schools (Ministry of Education and Culture 1987). The enactment of the Education Act 1978 restricted private schools except private technical education.

Later in 1980s the government allowed registration and operation of private schools as a result of increasing student flow especially in primary schools. The financial management and operation of private schools depended on the owners of private schools without monopoly of the government (Sabarwal, et al. 2020). Later in 1982 the then president of Tanzania appointed a Commission to review the education system targeting at extending provision of education, developing the curriculum, administrative structure, equipment and materials. The commission reviewed the past 19 years and made recommendation for the proceeding 20 years (Presidential Commission Report 1982). Although, the initiatives of the government were to create a systematic formal education, but there was no uniformity or regulation on the cash management in private schools.

In 1994, Tanzania experienced liberalization of the economy which influenced the operation of private schools at all levels of education and resulted to rapid increase number of private schools. The government encouraged investment in private education for expansion of economy where various individuals, Non-Governmental Organizations and private enterprises were allowed to operate their private schools (URT 1982).

Moreover, the background above reveals that private schools operate basing on its own generated revenues with no subsidies from the government (Njiri et al. 2020). Its financial sustainability depends much on its ability to control and manage financial cash flow. Although there is increasing number of private schools in Tanzania, but many schools have experienced financial challenges associated by poor cash management which includes lack of funds to ensure existence of facilities like technology, infrastructure like libraries, books and inadequate instructional materials, increase of unprecedented high fees charged on students, inadequate teachers and teaching materials, inefficient financial resources (URT 2019).

Various literatures have revealed importance of having effective cash management. Cash management involves having effective mechanisms of planning, organizing, directing, monitoring and controlling their finances (Njiri et al. 2020). According Mohamed and Omar (2016) effective management of finance is important in private schools because they depend of internal revenue especially school fees. Cash management is also a precondition for maintaining levels of quality education by ensuring availability of teaching materials, teachers and other learning resources (Wayong'o 2018).

The problem of poor cash management system causes inefficient of finance in schools as well as closure of schools (Abdullah 2008). According to Omar (2016) in Somalia private schools faces financial difficulties because of depending solely on their internal private revenue without government subsides. Ketch and Somerset (2017) state that many private schools have weak financial accountability and transparency. Omar (2016) opines that some private schools in Somalia do not have bank accounts, they have poor computer literacy among the management; and minimum involvement of accounts staffs which indicates existence of poor cash management. In Mogadishu increasing failure of students is associated by negative cash management and poor record keeping (Abdullah 2008).

Although the literature above indicates the importance cash management, but they do not address on how cash management influences financial performance of private schools basing in Tanzanian context. In Tanzania, cash management in private schools is a challenge which need to be addressed. Private schools have no accounting staffs; inadequate fund to ensure availability of teaching and learning facilities like technology, infrastructure like libraries, books and inadequate instructional materials, increase of unprecedented high fees charged on students, inadequate teachers and teaching materials, inefficient financial resources which is associated by lack of effective cash management system (URT 2019).

Now therefore, this study will address on the influence of cash management system in order to realize effective performance of private schools in Tanzania. It will particularly determine the influence of cash plan, cash budgeting, cash collection as well as cash control which are independent variables towards the performance of private schools in Tanzania which involves school profitability, existence of sufficient cash flow and better students' performance which are dependent variables.

Objectives

i. To examine the efficiency of cash management mechanisms employed by private schools in Tanzania.
ii. To point out the constraints facing private schools in their cash management.
iii. To determine strategies of ensuring effective cash management in private schools.

2 Methodology

2.1 Research Design

The study employed a mixed methods research approach in which suggests that both qualitative and quantitative data were collected concurrently, analysed leading to meaningful interpretations and findings. Mixed methods approach is useful in that qualitative do complement on the weaknesses of each specific method (Creswell et al. 2011). Accordingly, explanatory design was more relevant than other designs owing to its relevancy in explaining the phenomena, generating knowledge as well as developing new understanding on the subject matter of the study, (Adam and Kamuzora 2008). The study was carried out in Dar es Salaam, a city particularly at selected five private schools which are Loyola School, Barbro Johansson Model Girls Sec. School, Dar es Salaam International Academy, International Schools of Tanganyika, and Haven of Peace Academy. Dar es Salaam was deemed as appropriate study area considering that it is a major commercial centre in Tanzania and popular business destination and a centre for all major economic activities, hence easy to access information from targeted population (Possi and Milinga 2018).

The study drew from a sample size of 132 respondents. Data collection for this study was undertaken using questionnaire, interview and documentary review. In the first-place questionnaires were administered management of private schools. This questionnaire included five points Likert scale questions that intended to solicited quantitative data on the influence of cash management on financial performance of private schools in Tanzania. Interviews were carried to selected key informants from different schools and they were intended to explore and examine and explain the ways in which particular factors constrained management of financial performance of private schools in Tanzania. Secondary data were obtained from local newspapers, recently published research papers and magazines. Each set of data was treated separately. To both datasets, cleaning, reducing and coding were done. Data with quantitative nature were subjected to SPSS and descriptive statistics were used to make meaning out of the analysed data. Qualitative data, on the other hand was handled using MAXQDA 10 software to generate meaningful expressions. In this regard, descriptive statistics obtained were explained and cemented by quotes obtained from interviews and documentary reviews. Throughout the study, all ethical standard was adhered.

2.2 Literature Review

Njiri et al. (2020) researched on financial control and growth of private primary schools in Kenya, by employing quantitative and qualitative study design, which targeted 7,418 private primary schools in Kenya. The findings revealed that cash management have positive effect on the performance of private primary schools in Kenya. That majority of private school's success in performance were influenced by their clear budget which is basing on specific goals which are clear measurable and realistic within certain period of time. That it is important for private schools to outline clear expenditures of the school, expected income with expected expenditure which evidence expenditure control, forecast and oversight. Moreover, the authors recommend for more strong financial control which will ensure effective monitoring utilization of finances within departments and having strong communication between departments in respect to financial management and expenditures. This literature focuses on private primary schools in Kenya, it is the aim of this study to delve on assessing the effect of cost management basing in selected private primary and secondary schools in Tanzania.

Wayong'o (2018) wrote in respect to physical education safety precaution practices in private primary schools in nairobi city county. He contends that majority of private primary schools in Kenya lacks a proper financial control system which affect the performance and growth of the schools. Inefficient financial stability in many private primary schools leads to inadequacy of facilities like infrastructure, lack of teaching materials, teachers and learning resources. Majority of private schools have weak financial transparency, accountability where many incidences of fraud and misappropriation of school funds is reported more often. The authors contended further that another financial challenge facing private primary schools is absence of effective debt management and lack of proper succession planning, thus proper management of cash had positive contribution on the performance and growth of private primary schools. This literature is relevant because it addresses on the importance of cash control however, it does not elaborate categorically on the mechanisms used in cash management and how they affect performance of private schools basing on Tanzanian context as it will be covered under this study.

Yizengaw and Agegnehu (2021) carried out a study on the practices and challenges of school financial resource management implementation in Bahir Dar City administration of Ethiopia by making a comparative study between government and private secondary schools' cogent education. The findings revealed that contend that there is positive relationship between the existing financial management system and school performance. The authors contended further that the availability of financial resources is vital for private school's productive process and quality of education provided because the school ensures the availability of good infrastructures, teaching and learning materials, teachers, technology as well as cash flows this reflects the finding by Sabarwal et al. (2020) which addresses on the low-cost private schools in Tanzania and its financial challenges.

Mohamed and Omar (2016) address on the effects of cash management on financial performance of private secondary schools. They revealed that cash management strategies contribute significantly on the financial performance of private schools in Somalia. This was revealed through positive correlation between cash management as an independent variable with the financial performance which is dependent variable. The findings reflect the study of Cooke (2003) who contended that availability of cash is a lifeblood of the firm.

Tooley et al. (2018) carried out a study on school choice and academic performance focusing on developing countries. The authors contended that in many developing countries public schools are most chosen than private schools because of their affordability in terms of school fees. Although many private schools are considered to have better training facilities and performance but they are most costly because their financial sources depend on its revenues rather than government subsides. The author recommends for proper cash management system in private schools for ensuring financial sustainability and performance as well. In the study by Dong and Tay Su (2010) they findings revealed also that cash requires proper management which entails to involve cash monitoring, cash protection, effective control as well as budgeting.

Goulart and Bedi (2017) conducted a study in Portugal to examine the effectiveness of public versus private schools by taking the labor market earnings as a measure of effectiveness and controlling the personal characteristics and school choice. The findings showed that private school had an advantage of better performance as compared to public school in Portugal.

In Nigeria, Abioro (2013) opined that cash management is influenced by various factors which includes firstly, the amount of cash that a firm hold as cash balance. Secondly, short-term sources of finance which the organization use when the need arises. Thirdly, the approach and mechanisms used by a firm in collecting and disbursing funds. Lastly, the strategy used by an organization in making its cash projections and investment decisions in respect to inactive cash. The author concluded that existence of effective cash management mechanisms and policies correlates positively with business performance of an organization. Besides, according to Khoshdel (2006), who studied the relationship between free cash flows and operating earning with stock returns and growth of net market values of operating assets in Tehran Stock Exchange, found that there is a positive meaningful relationship between operating earning with return on equity, return on assets, and growing of net market values in operating assets.

World Bank (2019) addresses on the education financing in Tanzania, where contends that there is knowledge gap among the school management about financial planning, collection, organizing, budgeting and cash control which results to poor implementation of cash plans available as well as poor implementation of cash management procedures among organizations and causing the firms misuses of cash available and failure of many schools to grow and get good academic performance.

2.3 Research Gap

Many previous literatures examined the importance of cash management in general (Dodds 2009; Mohamed and Omar 2016) while very few addresses on the practice and challenges facing private school because of inefficiency finance. For instance, Wayong'o (2018) wrote in respect to physical education safety precaution practices in private primary schools in nairobi city county, where he pointed out cash management in private primary schools as one of challenges without going into details to determine the elements of cash managements and their influences on financial performance in private primary schools.

Besides, Tooley et al. (2018) carried out a study on school choice and academic performance focusing on developing countries while Nishimura and Yamano (2013) also examined the factor for school's choice and emerging of private education in Africa. However, they do not address particularly on the extent to which cash management effects positively the performance of private schools. Now therefore, it is the aim of this study to determine the extent which cash management system which involves cash planning, budgeting, collection and cash controlling facilitates financial performance of private schools in Tanzania which includes increasing students' performances, profitability and existence of sufficient cash flow.

3 Research Findings and Discussion

The Efficiency of Cash Management Mechanisms Employed by Private Schools in Tanzania

The first objective of this study was to examine the efficiency of cash management mechanisms employed by private schools in Tanzania and its efficiency of their financial performance. The findings revealed that there is a average mean score (2.78) on the efficiency of cash planning system in private schools; mean score (2.37) on the efficiency of cash budgeting system of private schools in Tanzania; mean score (3.41) on the efficiency of cash collection system of the private schools; mean score (2.24) of the efficiency of cash control system of private schools. Basing on the findings the cash management system in private schools is not efficient.

	Mean	Std. deviation
The private schools have efficient cash planning system	2.78	.868
The private schools have efficient cash budgeting system	2.37	.766
The private schools have efficient cash collection system	3.41	1.336
The private schools have efficient cash control system	2.24	1.133
Valid N (listwise)		

Although cash management have an influence in the financial performance of private schools, the findings revealed that there is still inefficiency in cash management. This reflects the findings by Mohamed and Omar (2016) who revealed that cash management strategies contribute significantly on the financial performance of private schools in Somalia. Yizengaw and Agegnehu (2021) also contend that there is positive relationship between the existing financial management system and school performance in Ethiopia. Njiri et al. (2020) which revealed that cash management have positive effect on the performance of private primary schools and majority of private school's success in performance were influenced by their clear budget which is basing on specific goals which are clear measurable and realistic within certain period of time.

According to Njiri et al. (2020) many private schools have no proper outlined clear expenditures of the school, expected income with expected expenditure which evidence expenditure control, forecast and oversight. Wayong'o (2018) also contends that majority of private primary schools in Kenya lacks a proper financial control system which affect the performance and growth of the schools. This leads to inadequacy of facilities like infrastructure, lack of teaching materials, teachers and learning resources. Majority of private schools have weak financial transparency, accountability where many incidences of fraud and misappropriation of school funds is reported more often. World Bank (2019) while addressing on the education financing in Tanzania, contends that there is knowledge gap among the school management about financial planning, collection, organizing, budgeting and cash control which results to poor implementation of cash plans available as well as poor implementation of cash management procedures among organizations and causing the firms misuses of cash available and failure of many schools to grow and get good academic performance.

3.1 The Constraints Facing Private Schools in Their Cash Management

The second objective of this study pointed out the constraints facing private schools in cash management. The findings various challenges which includes absence of Accounting Staff (mean score 4.09); limited source of financial resources (mean score (3.46); lack of Transparency and Accountability (mean score 2.70); absence of bank account (mean score 3.85); poor computer literacy(mean score 4.41); poor record keeping (mean score 4.34).

	Mean	Std. deviation
Absence of accounting staff	4.09	1.194
Limited source of financial resources	3.46	1.655
Lack of transparency and accountability	2.70	1.306
Absence of bank account	3.85	1.263
Poor computer literacy	4.41	.761
Poor record keeping	4.34	4.546
Valid N (listwise)		

Generally, there are various challenges encountering cash management in private schools. These challenges are also reflecting the findings in some literatures which includes findings by URT (2019) who opined that private schools experiences financial challenges associated by poor cash management which includes lack of funds to ensure existence of facilities like technology, infrastructure like libraries, books and inadequate instructional materials, increase of unprecedented high fees charged on students, inadequate teachers and teaching materials, inefficient financial resources.

According to Chelanga (2016) many private schools lacks professional accountant and cash management system. Ketch and Somerset (2017) state that many private schools have weak financial accountability and transparency. Omar (2016) opines that some private schools in Somalia do not have bank accounts, they have poor computer literacy among the management; and minimum involvement of accounts staffs which indicates existence of poor cash management. In Mogadishu increasing failure of students is associated by negative cash management and poor record keeping (Abdullah 2008).

According to Njiri et al. (2020) private schools faces financial difficulties because of depending solely on their internal private revenue without government subsides. This was also reflected by Tooley et al. (2018) who stated that although many private schools are considered to have better training facilities and performance but they are most costly because their financial sources depend on its revenues rather than government subsides.

Wayong'o (2018) contends that majority of private primary schools in Kenya lacks a proper financial control system which affect the performance and growth of the schools. Inefficient financial stability in many private primary schools leads to inadequacy of facilities like infrastructure, lack of teaching materials, teachers and learning resources. Wayong'o (2018) contends further that majority of private schools have weak financial transparency, accountability where many incidences of fraud and misappropriation of school funds is reported more often. The authors contended further that another financial challenge facing private primary schools is absence of effective debt management and lack of proper succession planning, thus proper management of cash had positive contribution on the performance and growth of private primary schools.

3.2 Strategies of Ensuring Effective Cash Management in Private Schools

Generally, the third objective was to determine the strategies which can be employed in ensuring effective cash management in private schools. Table 3.14 reveals the responses in respect to various strategies which includes: increase of accounting staff (mean score

4.48); expanding sources of fund (mean score 3.75); record keeping (mean score 4.23); and provision of financial education (means score 3.96).

	Mean	Std. deviation
Increase of Accounting Staff	4.48	.878
Expanding Sources of Fund	3.75	1.135
Record Keeping	4.23	.870
Provision of Financial Education	3.96	1.327
Valid N (listwise)		

Generally, the findings reflect the writing by Njiiri et al. (2020) who recommend for more strong financial control which will ensure effective monitoring utilization of finances within departments and having strong communication between departments in respect to financial management and expenditures. Dong and Tay Su (2010) also recommends that cash requires proper management which entails to involve cash monitoring, cash protection, effective control as well as budgeting for the better financial performance of an organization. Khoshdel (2006) found that there is a positive meaningful relationship between operating earning with return on equity, return on assets, and growing of net market values in operating assets and recommends for proper management of operating earnings.

According to IMF (2021) suggest for proper cash plan because cash planning ensure that expenditures are financed smoothly during the year and minimizing borrowing costs, to enable initial budget policy targets to be met and to enable smooth implementation of fiscal and monetary policy in an organization. Brigham and Houston (2014) recommend for good cash budgeting because it is useful to the organization management for purpose of determining the future financial need so as to capture estimate revenue, expenditure, assets, liability, cash-flow as well as to plan for that need and exercise control over such cash for maintaining their liquidity (Kakuru 2003).

4 Conclusion and Recommendations

Generally, the study has revealed that cash management systems in private schools are inefficient. This is after average response on the efficiency of cash planning system, cash budgeting, cash collection strategies and cash control mechanisms. Various challenges have also been pointed out which includes absence of Accounting Staff, limited source of financial resources, lack of Transparency and Accountability, absence of bank account, poor computer literacy, poor record keeping.

5 Recommendations

In order to ensure effective influence of cash management to the financial performance of private school the study recommends for the following.

1. Private schools are recommended to use cash planning system. This involves making amendment of the school policies and rules to ensure that there is a systematic estimation of flow of cash in and out of business by predicting financial requirements of the schools in future.
2. It is also recommended that cash budgeting should be ensured in private schools. The school policies and regulations should be put to ensure that the school is setting its projects time period in respect to cash coming into the business as well as the cash that leaves the business. This will help cash management because it will capture estimated revenue, expenditures, assets, liability and cash flow.
3. Proper mechanisms of collecting cash are also recommended. Private schools are recommended to employ effective mechanisms of debt collections. This is mostly to the students who do not their financial dues timely. There should be proper mechanisms of making sure all the students are paying their fees timely. This may involve writing letters to the parents, prohibiting students from attending classes if they have not paid school fees.
4. It is also recommended that proper system of cash control should be put. This involves control of the cash which is banked, cash cheque, cash balances and cash brought down. This involves also ensuring that the private schools have the bank account which is properly monitored.
5. Professional accounting staffs are recommended for ensuring that they are effectively managing the case of private schools. This involves trained professionals who are aware of the cash management. This involves also provision of financial awareness to all school staffs especially the management. This will ensure that all the financial information spent and the revenues are properly managed by trained personnel.
6. It is recommended that private school should expand the sources of their income. This involves avoiding basing solely on students' fees as a source of income. There should be other school projects for generating school funds which will increase the school revenues. For instance, establishing school agricultural projects.
7. Proper record keeping is also recommended in private schools where there should be a systematic mechanism of keeping financial records. This may involve creating a computerized program which keeps the student's financial information as well as the school financial information particularly in relation to revenues and expenses.
8. It is also recommended that there should be transparency and accountability in the private schools. This can be implemented by making an amendment of the private school's financial policies to ensure that the cash management system is transparent and all the responsible management are accountable in case of any financial misappropriation.

References

Baum, D.R., Cooper, R., Lusk-Stover, O.: Regulating market entry of lowcost private schools in sub-Saharan Africa: towards a theory of private education regulation. Int. J. Educ. Dev. **60**, 100–112 (2018). https://doi.org/10.1016/j.ijedudev.2017.10.020

Carnoy, M.: School choice: or is it privatization? Educ. Res. **29**(7), 15–30 (2000)

Chelanga, J.K.: Administrative Strategies Adopted By Private Primary Schools To Improve Academic Performance In Marakwet East And West Sub Counties, Kenya. Doctoral dissertation, Kisii University (2016)

Hair, J., Black, W., Baln, B., Anderson, R.: Multivariate Data Analysis. MaxwellMacmillan, London (2010)

Heyneman, S., Stern, J.: Low cost private schools for the poor: what public policy is appropriate? Int. J. Educ. Dev. **35**, 3–15 (2016)

Hunt, T.: The history of catholic schools in the United States: an overview. In: Hunt, T.C., Oldenski, T.A., Wallace, T.J. (eds.) Catholic School Leadership: An Invitation to Lead. Falmer Press, New York and London (2000)

Nishimura, M., Yamano, T.: Emerging Private education in Africa: determinants of school choice in rural Kenya. World Dev. **43**, 266–275 (2013)

Njiri, J.M., Mwenja, D., Kiambati, K., Mbugua, L.: Financial control and growth of private primary schools in Kenya. Int. J. Res. Bus. Soc. Sci. (2147–4478) **9**(7), 267–273 (2020)

Ocitti, J.: African Indigenous Education: As practiced by the Acholi of Uganda. East African Literature Bureau, Nairobi (1994)

Owusu, M.: Culture, colonialism, and African democracy: problems and prospects. In: Michael Jr., W.C., Plotnicov, L. (eds.) Africa in World History: Old, New, Then and Now, pp. 141–160. University of Pittsburgh, Pittsburgh (1995)

Parker, J., Rathbone, R.: African History. Oxford University Press, New York (2007)

Sabarwal, et al.: Low-cost private schools in tanzania: a descriptive analysis. policy research working paper. no. 9360. World Bank, Washington, DC. © World Bank (2020)

Sunal, S.: Schooling in Sub-Saharan Africa: Contemporary Issues and Future Concerns (ed). Overview of schooling in Sub-Saharan Africa. Garland Publishing, Inc., London (1998)

TCU: State of University Education in Tanzania 2019. Ministry of Education, Science and Technology Building (2019)

The United Republic of Tanzania (1982): "Tanzania Education System 1981–2000": A Report of the Recommendations of the Presidential Commission, Dar es Salaam, Tanzania. Jackson M. Makweta was by then the Hon. Minister for Education and the chair of the 13 people team (2019)

The United Republic of Tanzania: Education Sector Performance Report: Tanzania Mainland (2018/2019) (2019)

Tooley, J., Longfield, D.: The role and impact of private schools in developing countries: a response to the DFID-commissioned. Rigorous Literature Review. Open Ideas (2015)

Tooley, J., Bao, Y., Dixon, P., Merrifield, J.: School choice and academic performance: some evidence from developing countries. J. Sch. Choice **5**(1), 1–39 (2018)

Vanqa, T.P.: History of education. In: Abosi, C.O., Kandjii-Murangi, I. (eds.) Education in Botswana, pp. 1–41. Macmillan Botswana Publishing (Pty) Ltd., Gaborone (1995)

Wayong'o, S.N.: Physical education safety precaution practices in private primary schools in Nairobi city county, Kenya. Doctoral dissertation, Kenyatta University (2018)

Wise, C.W.: History of Education in British West Africa Toronto. Longman Green and Company (1956)

The Influence of Citizen Awareness and Willingness on Revenue Collection in Local Government Authorities: Evidence from Temeke Municipal Council, Tanzania

B. Mwakyembe and D. Pastory(✉)

Department of Accounting, College of Business Education, Dar es Salaam, Tanzania
passtory1@yahoo.co.uk

Abstract. Purpose: The study investigated the effect of citizen awareness and willingness on tax compliance and revenue collection among local Government Authorities specifically in Temeke Municipal council.

Design/Methodology/Approach: The study was quantitative in nature employing a survey design. The study contacted a sample of 100 respondents who were obtained conveniently and purposively. These included employees, normal citizens and business persons. Data were collected through questionnaires and interviews and were analyzed by using descriptive and regression analysis.

Findings: Findings of the study indicate that both citizen awareness and willingness have a statistically significant effect on tax compliance and revenue collection in Temeke municipal council (R2 29.1 and $P = 0.001 < 0.05$ and R2 30.1 and $P = 0.003 < 0.05$ for awareness and willingness respectively.

Implications/Research Limitations: The study was conducted in Dar es Salaam and specifically in Temeke Municipal council. Dar es Salaam was selected because it is a commercial metropolitan with many businesses.

Practical Implications: Municipal officers to be honest and fair in assessing taxes and administering taxes. This will raise the level of willingness among citizens.

Social Implications: The study recommends that Temeke municipal council invest in awareness campaigns and create a mutual relationship between its workers and taxpayers.

Originality/Value: Citizen awareness and willingness to tax compliance framework will lead to maximum tax compliance which will result in increased revenue collection, positive and strong relationship between payers and collectors, observance of deadlines and voluntary payments.

Keyword: Awareness · Citizen · Revenue collection · Tax compliance · Willingness

C. Aigbavboa et al. (Eds.): ARCA 2022, *Sustainable Education and Development – Sustainable Industrialization and Innovation*, pp. 866–880, 2023.
https://doi.org/10.1007/978-3-031-25998-2_66

1 Background of the Study

Different countries around the world have always practiced decentralization. It is the transfer of authority and responsibility from the central government to subordinates or quasi-independent government institutions, as well as the private sector. Although there are various types of decentralization, they all require empowering elected officials to make decisions about providing public services at various levels of government. Both developed and developing countries have practiced decentralization. Countries like Germany, Belgium, Spain among others use at least one form of decentralization (Eaton et al. 2019; Marks 2019) and have been associated to policy innovation, minimize the rate of corruption and citizen participation in government decision (Stoyan and Niedzwiecki 2018). In Africa, due to serious sensitization by donors, it is active in Uganda, South Africa, Ethiopia and Tanzania in general (Singhal and Gadenne 2014). Decentralized localities have been required to participate in providing social services to the entire citizenry. In this regard local authorities are given authority of making or taking political and financial responsibilities of collecting finance through taxes and other sources and use the income to make deliverables to the people (Elahi 2009; OECD 2019).

In Tanzania, the Local Government Acts 7 and 8 were enacted in 1982 for the same purpose. These require LGAs to facilitate easy provision of services, and are mandated by the law among other thing to raise revenue through different sources to make above aim to come true. Revenue is the income received or obtained from ones legitimate activities which involves the selling of goods and services. However with reference to the local government, revenue entails the entire receipts a local government obtains from its services provided and goods sold (Nyanumba 2010). These receipts may be from taxes, levies, user fees, fines, licenses, and permits, rent and investment along with other miscellaneous activities.

Despite the fact that all LGAs including municipal councils are mandated to collect revenue in form of taxes, levies user fees among others based on the internal revenue policies (Nuhiva 2015), collections are very low in comparison to the projected collections (Balunywa et al., 2014). For instance according to CAG reports for the five years consecutively, Temeke Municipal council, one of the LGAs in Dar-es-salaam, with a big tax or revenue base, has had deteriorating revenue collections as compared to the projected. This signifies some of the projected activities are not performed as planned (Fjeldstad 2014).

Various scholars have tried to itemize the causes of such phenomenon among local Government Authorities. A study by Twaib (2020) revealed that low capacity of municipal workers in revenue collection; in adequate tax education cripple the entire procedure. Mugambi and Wanjohi (2018) revealed that inadequate information technology, untrained and unmotivated tax collectors, corruption and tax evasion practices and lack of integrity among workers lead to the current inadequacies in revenue collection. An earlier study by Fjeldstad, (2014) revealed that the entire procedure is shuttered by uncoordinated efforts between the central and local government as well as unharmonized taxes imposed on different commodities. Cementing on this, Muturi and Adenya (2017) point that, poor law enforcement, ineffective internal control system and inadequate technology are among the key hindrances. Despite the fact that studies itemized a number of causes, most of them were based on LGAs and none of them focused on factors that might have emanated from taxpayers themselves. The existing gap has prompted the researcher it undertaking this study to unveil tax payers' related factors that hinder revenue collection among LGAs in Tanzania specifically Temeke Municipal council (TMK MC). In this study, awareness and willingness of tax payers will be the focus of the study.

In this regard therefor the study envisaged to;

i. Determine the effect of citizen awareness on tax regulation compliance and revenue collection at TMK MC;
ii. Assess the effect of citizen willingness on tax regulation compliance and revenue collection at TMK MC;
iii. Determine the relative contribution of citizen's awareness and citizen's willingness to tax compliance and TMK MC Revenue collection.

The findings of this study will benefit Temeke and other LGAs in the country and policy makers. To LGAs, findings obtained will show them the extent to which the level of awareness and willingness of citizens to comply with tax regulations affect revenue collection. Hence the study will awaken them to conduct various awakening campaigns. On the same note, Policy makers will be informed on the best way to formulate policies that do not only focus on collecting tax but providing citizen with education, showing the importance of such collections in terms of tangibles. This will awaken citizens to participate in voluntary contributing to LGAs revenues.

2 Literature Review

2.1 Theoretical Framework

2.1.1 The Regulated Enforcement Theory (RET)

The theory is based on the goal of achieving maximal legal and quality compliance. According to many authors (Mikkelsen et al. 2017), two typical approaches and styles are utilized in the process of achieving total compliance. The coercive and catalytic style. The catalytic style assumes that an individual's conformity to a circumstance is determined by the motivation he or she receives. However, because people's capacities are limited, their compliance levels are poor. To raise this level, financial assistance, education, and further incentives are required (Weske et al., 2018). The coercive style, on the other hand, expects that people would refuse to comply, therefore ideal conditions for implementing sanctions and punishment are essential (Weske et al 2018). As a result, the threat of punishment and other related consequences motivates people to follow rules. Compliance tax collection was focused on the catalytic characteristics of tax payers in this study. As a result, tax payer knowledge, education, professionalism, and willingness to pay tax (compliance) have a favorable or negative impact on tax collection in Tanzania's several LGAs. Thus by focusing on catalytic style, the theory provides three important construct which will be used to determine their effect on revenue collection in Temeke Municipality. These are tax payer's awareness, tax payer's degree of education and professionalism and willingness to pay tax.

2.2 Empirical Literature Review

2.2.1 Factors Hindering Revenue Collection in LGAs

Several researchers have tried to reveal hindrances to revenue collections amoung LGAs in Africa. Agyapong (2012) reveals that excessive reliance on District Assemblies Common Fund (DACF) and corruption are the pressing obstacles. Dada et al. (2017) on the other commented that corruption and mismtch between collections and the level of social development in Nigeria are the source of the problem. An idea the was supported by Uhunmwuangho (2013) and adds political interests and poor record keeping of tax payers to the problem, the idea that echoed Mohamed's (2017) in Mogadishu.

In another study by Fjeldstad and Heggstad (2012) while presenting the entire African picture, the researcher claimed that the reason as to why there are meagre revenue collection is that most African LGAs have outsourced the revenue collection exercise to private companies. Most of these work for their personal benefits and not otherwise. This was also supported by Mgonja (2018) who claimed that outsourcing revenue collection exercise always benefits private companies at the expense of LGAs. Mdagachule (2014) and Mwenisongole (2013) added that incompetence of staff in the revenue collection exercise, corruption, poor coordination and cooperation among actors, negative relationship existing between revenue collectors and payares, laxity among collectors and poor training of revenue collectors and payers on tax and other related payments to be made to LGAs affect the amount of revenue collections. On the other side of the coin, poor use of ICT (Kessy 2020; Olomi 2013) and unsatisfying remuneration have escalated the problem (Karimi et al. 2017).

The reality from these studies is that, in all studies, resaerchers concentrated on admninstrative issues, technology and corruptions. Little was studied with regard to citizen awreness, willingness and level of education and professionalism on revenue collectionand tax compliance. This opens a room for this study to dig deep in to the truth of these aspects on revenue collection in LGAs.

2.2.2 Citizens' Awareness on Compliance to Tax Regulations and Revenue Collection in Local Government Authority

There has grown a resinous concern among scholars that the degree of awareness among tax payers has a direct influence to revenue collection in LGAs. A study by Hasan et al. (2017) has indicated the importance of citizen awareness on tax issues and tax compliance and revealed that there is a direct and positive relationship between awareness and tax compliance as well as revenue collected. Setyorini (2016) on the other side had earlier established that tax compliance is linked to citizen's degree of awareness on tax related issues though training or mass campaigns. Prior studies of Hardiningsih and Yulianawati (2011) and Suryadi (2006) had the same findings. In line with these, Lillemets (2009) attest that, awareness of how the tax revenues are used, of the tax system of the country, of the tax policy, of the administrative policy, of the individual and the cultural characteristics of the person improves the rate of tax complance among citizens.

Paril et al. (2013) in their study revealed that awarenes on tax education and procedures is linked to tax compliance. Similar findings that were obtained by Tirada (2013) who also report that, awarenss and consciousness on tax related matters has positive effect on tax compliance and amount of revenue collection. In line with these, Narita et al. (2012) and Masruroh (2013) are all in line with other scholars that awareness of tax payers on tax related matters increases thier level of compliance. In general terms, awareness on tax matters has been revealed to be linked with tax compliance and revenue collection. However, all these studies present findings from countries out of Africa where settings and tax systems are totally different from those of Tanzania. Similarly, these studies looked in to the general tax systems of countries and not in LGAs as this study will do. From this, it is hypothesised that

Ha1: Citizens' awareness have positive and significant influence on tax regulation compliance and revenue collection at TMK MC.

2.2.3 Citizens' Willingness to Comply with the Tax Requirement on Local Government Authority Revenue Collection

Revenue payment is considered to be a function of one's willingness to pay it. As a result, a person's desire to pay taxes and other regulatory fees has an impact on revenue collection and tax compliance. Evidence from Setyorini (2016) reveals that one's willingness to pay tax leads to tax compliance. Supporting this, a study by Widayati and Nurlis (2010) report that there a direct linkage between one's willingness to pay tax and tax compiance. In the Tanzania experience a study by Kaize (2014) reveal that willingnes to pay tax improves the level of compliance. And the degree of unwillingness grows among people due to poor level of transparency in tax administration, political interfearance, corruption and unjust and unfair means of assessing and collecting tax, levies and other payments to authorities. In line with this, Kimario (2014) report that willingness to pay tax is linked to tax compliance and increased revenue collection if and only if there is adequate coordination, positive relationship between tax collectors and payers and transparency in tax procedures and expenditure. Similar findings were reported by Ally (2014) who among other things reported that poor tax collection is a result of peoples' refusal and unwillingwss to pay taxes. From this review, it is hypothesized that;

Ha2: Citizens' willingness has positive and significant effect on tax regulation compliance and revenue collection at TMK MC.

The third hypothesis of this study is a combination of citizen awareness and degree of willingness on tax regulation compliance and revenue collection. The hypothesis draws from the premise that when citizens are aware of the tax systems, taxable, amount of levies and how all these collections are used by LGAs, they will be willing to pay taxes and other forms of regulatory payments to LGAs. This will increase the amount of revenues collected and will lead to effective provision of social services to the general citizenry. Thus it is hypothesized that

Ha3: Citizen's awareness and citizen's willingness have positive contribution to tax compliance and TMK MC Revenue collection.

2.2.4 Conceptual Framework

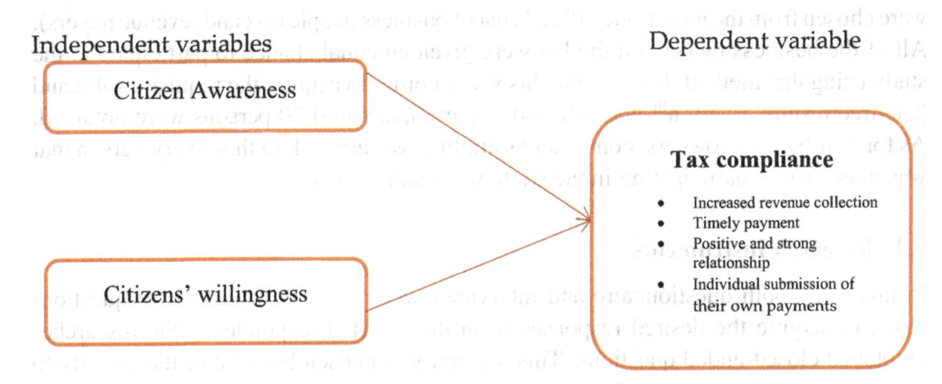

Fig. 1. Conceptual framework of citizen awareness and willingness on tax compliance

In this framework, it is expected that once the level of citizen awareness on tax related issues is high, combined with citizen willingness to pay tax, will lead to maximum tax compliance which will result to increased revenue collection, positive and strong relationship between payers and collectors, observance of deadlines and voluntary payments (Fig. 1).

3 Methodology

3.1 Research Approach and Area

A quantitative approach was used in this study with a survey design. The study was conducted in Dar es Salaam and specifically in Temeke Municipal council. Dar es Salaam was selected due to the fact that it is a commercial metropolitan with many businesses and hence it constitutes is the biggest revenue collection point in the country. Thus, anything that leads to its lower revenue collection has to be keenly observed. Specifically, Temeke Municipal council was chosen based on the reality that it is one of the biggest municipals in Dar es Salaam whose revenue collection has not been satisfying. The targeted revenue collections have not been realised for several years. Thus, choosing Dar es Salaam and Temeke specifically as the research setting is regarded as one part of successful set of the study as it was easy for the researcher to obtain data from the intended respondents so easily.

3.2 Sample Size and Sampling Technique

The sample for this population consisted of 100 respondents. For a population of less than 10,000 people, a sample of 100 people is considered an appropriate representation of the group (Saunders et al. 2007; Nirathron 2006). In this study, municipal employees working in the Municipal treasurer's office and trade offices (10 respondents); citizens owing business (70 respondents) and non-business owners (20 respondents) were used as respondents. The sampled individuals were obtained by the use of purposive technique (municipal employees) and simple random sampling (business owners) Business owners were chosen from the local trade officer's list of business people (tax and revenue payers). All of the business owners on the list were given an equal chance to participate in the study using this method. It was done this way in order to acquire the requisite valid and bias-free results. Every n^{th} was selected to participate until 70 persons were obtained. As for non-business owners, convenient sampling was utilised. In this, every person that was interested in participating in the study was taken aboard.

3.3 Research Instruments

In this study both questionnaire and interviews were used. With regards to questionnaire, to acquire the desired responses from the targeted respondents, the researcher employed closed-ended questions. This method was chosen because of its capacity to offer precise and consistent data; it also distinguishes the researchers' prejudice and minimizes their partiality, and it saves time. The researcher created the instrument based on an examination of several literatures. In this regard, seven point Likert scale with 1 = strongly disagree and 7 = strongly agree as used by previous researchers, Setyorini (2016) were developed for all the two important aspects of citizens' awareness and citizens' willingness to comply with tax regulations and payment. Each variable had four items from which measurements were made. These were administered to all business and non-business owners that participated in this study.

3.4 Data Analysis Procedures

Because the study gathered both qualitative and quantitative data, each dataset was analysed separately and then integrated during the interpretation stage. The descriptive analysis with tables and graphs was used to analyse demographic data while inferential statistics specifically multiple linear regression analysis was used to show the relationship existing among variables (citizens' awareness, willingness to pay on local government tax compliance). This was done to see if the independent factors (citizen awareness and willingness,) predicted any change in the dependent variable (LGA tax compliance). Analysis was done after successfully checking that the data satisfy the assumptions of linearity, multicollinearity, normality, and heteroscedasticity.

Thus, the regression model to be used is given as

$$Y = \beta o + \beta_1 X_1 + \beta_2 X_2 + \varepsilon$$

where Y is LGA Tax Compliance (the amount of tax paid annually), X_1-X_3 is citizen awareness and Citizen Willingness, while βo-β_3 represent the interaction term and ε stands for the error term.

4 Findings and Data Presentation

4.1 Demographic Characteristics of Respondents

The researcher wanted to learn more about the demographics of the people who took part in the survey. It was important for the researcher to reveal this information since it would help to paint a picture of the kind of people that took part in the study. Gender, age, position, and work experience are examples of such traits. This information can be used to determine the respondents' level of understanding of the research issue.

Demographic characteristics of respondents were in five components including gender, age, level of education, experience and number of years complying with tax. In terms of gender, despite the fact that there were more male respondents than female respondents, the number of women in the business community has increased recently compared to the situation in the past. This could be linked to numerous groups' local and global emphasis on female equality and empowerment programs. As a result, more women are able to start their own businesses. Abdallah (2020), who saw a growth in the number of educated women and those involved in business and other economic activities across the country, described a similar situation. Findings further indicate that among employees a reasonable number of employees were above 30 years of age with only 16% being in between 20–30 years. With regards to business persons and other citizens, the study provides the reality that most of people in business and legible to pay tax are of the old generation. As confirmed by Kirumirah and Munishi (2020) who report that most of the young persons are informal business persons working along streets.

It is further reported that, most employees have different qualifications but most of them hold a diploma onwards. This cements the reality that for one to be employed as an accountant needs a CPA. Citizens and businesspeople, on the other hand, had between no education and secondary education. This may be related to the fact that the majority of Tanzanians and businesspeople only have secondary education. A tiny percentage of respondents (34.4%), however had qualifications equivalent to college certificate and a postgraduate degree. Such findings are consistent with Greene (2015), who stated that, while education is vital in business, a significant number of businesspeople are undereducated, with some being uneducated. With regards to experience, 60% of employees were more than 5 years at the job signifying job experience. Quinones (2004) on this reported that, the more one is experienced the better his/her performance. For citizens doing business, findings reveal that there are very many persons with less than 5 years in business and a few of them 35% whose experience has gone beyond ten years. The findings could explain that due to entrepreneurship initiatives in the last five years, an increase in the number of new start-ups among young men and women have been observed. As a result, there are many more inexperienced businesspeople than experienced ones. On tax compliance, findings reveal that even those experienced in the business have observed tax regulations for only a few years. This clearly indicates that people irrespective of their experience, are ready to default paying taxes, levies rents and other forms of municipal revenues when they have a chance of doing so (Mahenge 2018). Thus, with such a situation, the level of revenue collection by Temeke Municipality is at stake, unless coercive methods are used as proposed by the regulation enforcement theory.

4.2 The Effect of Citizen Awareness and Willingness on Local Government Authority Revenue Collection

4.2.1 Descriptive Statistics

In this part, the researcher presented descriptive findings of the study with relation on the first and second objectives. The descriptive findings presented include the mean and standard deviation as presented in Table 1

Table 1. Citizen awareness on local government authority revenue collection

	N	Min	Max	Mean	Std. dev.
Awareness on TIN registration and acquisition	90	1	7	4.74	2.16
Familiarity with procedures of obtaining a business license	90	1	7	5.90	1.12
Awareness of municipal tax by-laws through training	90	1	7	4.83	2.32
Awareness on rights and obligations of a business person to the Municipal Authority	90	1	7	5.62	1.46
Awareness on tax penalties when tax laws and by-laws are violated	90	1	7	6.33	0.57
Awareness on tax laws and practices through mass dissemination campaigns	90	1	7	6.46	0.62

The objective sought to find if citizen awareness on tax matters have an effect on LGAs revenue collection. Findings indicate that respondents agreed (Mean. 4.74 and Std. Dev 2.16) that awareness on TIN registration and acquisition affects revenue collection. Further, they agreed that familiarity with procedures of obtaining a business license affects revenue collection (Mean. 5.90 and Std. Dev 1.12). With regard to awareness of municipal tax laws through training, respondents agreed that it affects revenue collection (Mean. 4.83 and Std. Dev 2.32). On other aspects, respondents revealed that awareness on rights and obligations of a business person to the municipal authority, it was recorded that it to have an effect on revenue collection (Mean. 5.62 and Std. Dev 1.46). Respondents agreed that awareness on penalties raised revenue collections agreed (Mean. 6.33 and Std. Dev 0.57) and lastly respondents assured that awareness on laws through dissemination campaigns affects revenue collection (Mean. 6.46 and Std. Dev 0.62). Such findings imply that generally, citizen awareness on tax laws, by laws; obligation and penalty threats through awareness campaigns and training have a positive effect on revenue collections by LGAs.

Similar to the awareness, in this aspect, descriptive statistics mainly focused on the mean and standard deviation of the identified aspect as presented in Table 2

With regards to willingness, findings reveal that respondents agreed that willingness to make consultation and bargaining on tax and levy payments affect revenue collection (Mean. 5.32 and Std. Dev 1.34). In the same way, willingness to search and fill tax/levy related payment documents affect revenue collection (Mean. 6.04 and Std. Dev 0.91). While willingness to find information on what, when, where and how to pay taxes has

Table 2. Citizen willingness on local government authority revenue collection

	N	Min	Max	Mean	Std. dev.
Willingness to make consultation and bargaining on tax payment	90	1	7	5.32	1.34
Willingness to search and fill tax/levy payment documents	90	1	7	6.04	0.91
Willingness to find information on what, when, how and where to pay	90	1	7	5.43	1.63
Willingness to comply with deadline for tax, rent and levy payments	90	1	7	6.57	0.72

been revealed to affect revenue collections (Mean. 5.43 and Std. Dev 1.63), Willingness to comply with deadline for tax, rent and levy payments have been revealed to affect revenue collections in LGAs (Mean. 6.57 and Std. Dev 0.72). In general findings imply that in a situation where citizens are willing to consult and bargain their tax and levy payments, search for information related to tax and levies and are willing to comply with deadlines, it is likely that revenues collected will be too much. Thus revenue collection is affected by citizens' willingness to comply with tax regulations and requirements.

4.2.2 Inferential Statistics

In this section, the researcher was interested to find out the relationship (effect) of independent variables (Citizen Awareness and Willingness) on the dependent variable (LGAs Revenue Collection). The regression analysis was done to determine the effect of the independent on the dependent variable.

4.2.3 Regression Analysis

In this, the Coefficient of determination R^2 was used to determine the effect of a change in revenue collection can be explained by changes in citizen awareness and willingness to comply with tax regulations. In the model, R^2 shows values of 0.291 and 0.306 for awareness and willingness respectively. This implies that 29.1% increase in the revenue collection in Temeke can be described by citizen awareness and 30.6% can be explained by citizens' willingness to comply with tax and levy regulations (Table 3).

Table 3. Model summary

Model	R	R Square	Adjusted R square	Std. error of the estimate
1	.497[a]	.291	.282	1.04300
2	.623[b]	.306	.301	1.00681

[a] Predictor (constant) Awareness.
[b] Predictor (Constant) Willingness.

4.2.4 Analysis of Variance

In the bid to establish the relationship existing between citizen awareness and willingness to comply with tax regulations and revenue collection in Temeke Municipal, it was important to us Anova. The analysis is presented in Table 4 in which the F statistic value of 13.878 and P value of 0.001 for citizen awareness were lower than the conventional p value set of 0.05. With regard to citizen willingness, F statistic value is depicted at 24.730 and a P value of 0.003 lower than the set 0.05. in both instances, the findings imply that there is a positive and statistically significant relationship between citizen awareness and willingness and revenue collection in Temeke Municipal.

Table 4. Analysis of Variance (ANOVA[a])

Model	Sum of squares	df	Mean square	F	Sig.
Regression	83.400	6	13.900	13.878	.001[b]
Residual	342.810	298	1.150		
Total	426.210	229			
Regression	144.240	4	36.060	24.730	.003[c]
Residual	304.021	296	1.027		
Total	448.261	298			

[a] Dependent variable: Revenue Collection.
[b]Predictor: (Constant) Citizen Awareness.
[c]Predictor: (Constant) Citizen Willingness.

Table 5. Regression analysis (coefficients[a])

Model	Unstandardized coefficients	Standardized coefficients	T	Sig.	
	B	Std. Error	Beta		
(Constant)	3.671	.286		12.320	.000
Revenue collection	.214	.266	.223	5.375	.000
(Constant)	3.414	0.26		12.137	.000
Citizen awareness	3.72	.251	.204		.001
Citizen willingness	2.48	.045	0.322		.003

a. Dependent variable: Revenue Collection.

Based on the above findings, it can be generalized that, there is a positive relationship between citizen awareness and revenue collection at Temeke Municipal council. The statistical values indicate that Awareness had R^2 .291 and $p = 0.001 < 0.05$. This implies that for any change in citizen awareness, there is an increase in 29.1% of revenue collection at Temeke Municipal council. Such findings are in line with earlier findings of Setyorini

(2016) and Hassan et al. (2017) who reported that citizen awareness is directly linked to tax compliance and revenue collection. Moreover, it is reported by Siti Musyarofah and Purnomo, (2008) that, the higher the citizen awareness the higher the level of voluntary tax payment and revenue collections. On the other side of the coin, findings contradict with those by Siti Nurlaela (2014) who report that, citizen awareness is not linked to raising the rate of tax compliance. It is only the attitude of people that raises the level of compliance. Although the findings have considerable value, however, the reality is that attitude is only built after complete awareness of an individual. With such findings therefore, the researcher concluded that citizen awareness is linked to revenue collection (Table 5).

With regard to willingness, findings reveal that there is also a positive statistical relationship between willingness and revenue collection ant Temeke Municipal. The statistical value indicate that R^2 0.306 and P value of $0.003<0.05$. From such findings, it is implied that, for any unit increase in citizen willingness to pay tax, it leads to around 30.6% increase in revenue collections. These findings confirm those in the descriptive analysis that showed a similar trend. In general findings concur with those by a number of scholars (Widayati and Nurlis 2010; Kaize 2014; Kimario 2014) who reveal that sufficient revenue collection is a function of willingness of citizens to comply with tax regulation. In an extended narration, it is reported that citizen willingness to pay tax increases the amount of revenue collection while minimizing government resources used in search of tax payers (Hardiningsih and Yulianawati 2011).

4.2.5 Hypothesis Testing

Because the researcher developed three hypotheses earlier, testing them for confirmation and rejection was of paramount importance. With regard to the obtained findings, all pre-developed hypotheses were supported and thus the researcher accepted them all. In LGAs, and specifically Temeke Municipal Council, there is a positive and considerable relationship between citizen awareness and willingness on tax compliance and the amount of revenue collections. Table 6 summarizes the hypotheses and their current state in this investigation.

Table 6. Summary of hypotheses

H_a	Hypotheses	Remarks
1	*Citizens' awareness have positive and significant influence on tax regulation compliance and revenue collection at TMK MC*	Accept
2	*Citizens' willingness has positive and significant effect on tax regulation compliance and revenue collection at TMK MC*	Accept
3	*Citizen's awareness and citizen's willingness have positive contribution to tax compliance and TMK MC Revenue collection*	Accept

5 Conclusion and Recommendations

The study concludes basing on the findings that citizen awareness on tax, levies and other related revenues payable to LGA and the level of their willingness to comply and pay such taxes have greater influence to the amount of revenue collections in Temeke Municipal. Thus the Municipal is recommended to ensure that, municipal officers should try their level best to visit different places in the municipal jurisdiction and train people about municipal taxes. Moreover this should be done in a friendly way. Such actions will apart from raising the awareness, will stimulate willingness among citizens. The study further recommends municipal officers to be honest and fair in assessing taxes and administering taxes. This will raise the level of willingness among citizens.

References

Adenya, P.K., Muturi, W.: Factors affecting revenue collection efficiency by county governments in Kenya: a case of Kiambu County. Int. J. Soc. Sci. Inf. Technol. **3**(8), 2371–2384 (2017)

Agyapong, F.M.: An evaluation of effectiveness of revenue mobilization strategies of metropolitan. Municipal and District Assemblies (MMDAs) In Ghana, A Case Study of Kumasi Metropolitan Assembly (KMA) (2012)

Ally, K.: Assessment of revenue collection efficiency in local government authorities: the case of Tanga local government authorities. Doctoral dissertation, Mzumbe University (2014)

Balunywa, W., Nangoli, S., Mugerwa, G.W., Teko, J., Mayoka Kituyi, G.: An analysis of fiscal decentralization as a strategy for improving revenue performance in Ugandan Local governments (2014)

Creswell, J.W.: Steps in conducting a scholarly mixed methods study (2013)

Creswell, J.W., Plano Clark, V.L.: Designing and Conducting Mixed Methods Research, 3rd edn. SAGE Publications Inc., Thousand Oaks (2018)

Dada, R.A., Adebayo, I.A., Adeduro, O.A.: An assessment of revenue mobilization in Nigeria local government: problems and prospects. Arch. Bus. Res. **5**(9) (2017)

Eaton, K., et al.: Measuring and theorizing regional governance. Territory Polit. Gov. **7**(2), 265–283 (2019)

Elahi, K.Q.I.: UNDP on good governance. Int. J. Soc. Econ. (2009)

Fjeldstad, O.H.: Tax and development: donor support to strengthen tax systems in developing countries. Publ. Adm. Dev. **34**(3), 182–193 (2014)

Fjeldstad, O.H., Heggstad, K.: Local government revenue mobilisation in Anglophone Africa (2012)

Gadenne, L., Singhal, M.: Decentralization in developing economies. Annu. Rev. Econ. **6**(1), 581–604 (2014)

Greene, P.: Entrepreneurs: Impact of Education on Business Growth. Forbes (2015)

Hardiningsih, P.: Faktor-faktor yang mempengaruhi kemauan membayar pajak/Factors that affect the willingness to pay taxes. Dinamika Keuangan dan Perbankan/Dyn. Finance Bank. Nopember, **3**(1) Hal, 126–142 (2011)

Hasan, J., Gusnardi, G., Muda, I.: Analysis of taxpayers and understanding awareness increase in compliance with taxpayers individual taxpayers. Int. J. Econ. Res. **14**(12) (2017)

Karimi, H., Maina, K.E., Kinyua, J.M.: Effect of technology and information systems on revenue collection by the county government of Embu, Kenya (2017)

Kessy, S.S.: Electronic payment and revenue collection in local government authorities in Tanzania: evidence from Kinondoni municipality. Tanzania Econ. Rev. **9**(2) (2020)

Kimario, P.: Challenges faced by local government authorities (LGAS) in implementing strategies to enhance revenues: a case of Dar es Salaam Municipal Councils. Master's thesis. Open University, Tanzania (2014)

Kirumirah, M.H., Munishi, E.J.: Characterizing street vendors in the urban settings of Tanzania: towards sustainable solutions to vendors' challenges. In: Mojekwu, J.N., Thwala, W., Aigbavboa, C., Atepor, L., Sackey, S. (eds.) ARCA 2020, pp. 245–261. Springer, Cham (2021). https://doi.org/10.1007/978-3-030-68836-3_22

Kothari, C.: Research Methodology: Methods and Techniques, 401p. New Age International (P) Limited., New Delhi (2004)

Lillemets, K.: Tax morale, influencing factors, evaluation opportunities and problems: the case of Estonia. Econ. Psychol. Tax Behav. Tallinn Univ. Technol. 233–569 (2009)

Mahenge, J.: Assessment on factors influencing tax compliance in Tanzania; A case of Tanzania Revenue Authority located in Mbeya Region. Bachelor Degree dissertation, Ruaha Catholic University Rucu (2018)

Mdagachule, E.J.: An assessment of factors affecting revenue collection in local government authorities: a case study of Mpwapwa district council (MDC). Doctoral dissertation, Mzumbe University (2014)

Mgonja, M.G., Poncian, J.: Managing revenue collection outsourcing in Tanzania's local government authorities: a case study of Iringa Municipal Council. Local Gov. Stud. **45**(1), 101–123 (2019). https://doi.org/10.1080/03003930.2018.1518219

Mikkelsen, M.F., Jacobsen, C.B., Andersen, L.B.: Managing employee motivation: exploring the connections between managers' enforcement actions, employee perceptions, and employee intrinsic motivation. Int. Publ. Manag. J. **20**(2), 183–205 (2017)

Mohamed, A.A.: Challenges of mogadishu local government revenue. In: Proceedings of the 2017 2nd International Conference on Humanities and Social Science (HSS 2017), vol. 83, pp. 529–533, February 2017

Mugambi, F.K., Wanjohi, J.M.: Factors affecting implementation of revenue collection systems in county governments in Kenya case of Meru county. Int. J. Res. Bus. Manag. Account. **4**(11), 525 (2018)

Mwenisongole, P.: Outsourcing revenue collection by local Government authority in Kinondoni municipal Council, Tanzania. Masters dissertation, Mzumbe University (2013)

Narita, P., et al.: Effect of tax policy and understanding of taxpayers against taxpayer compliance SMEs formal shoes and slippers in Mojokerto. GEMA J. **1**(1), 67–82 (2012)

Nirathron, N.: Fighting poverty from the street. A Survey of Street Food Vendors in Bangkok. International Labour Organization (2006)

Nyanumba, P.M.: Factors affecting revenue collection in the City Council of Nairobi. Doctoral dissertation, University of Nairobi, Kenya (2010)

Olomi, C.A.: Effectiveness of Revenue Collection in Local Government in Tanzania: The Case of Temeke Municipal Council. Masters dissertation, Mzumbe University (2013)

Quinones, M.A.: Work experience: a review and research agenda. Int. Rev. Ind. Organ. Psychol. **19**, 119–138 (2004)

Saunders, M., Lewis, P., Thornhill, A.: Research methods for business students, 652p. Financial Times Prentice Hall, Edinburgh Gate (2007)

Setyorini, C.T.: The influence of tax knowledge, managerial benefit and tax socialization toward taxpayer's willingness to pay SME's tax. Acta Universitatis Danubius. Œconomica **12**(5), 96–107 (2016)

Nurlaela, S.: Influence of knowledge and understanding, awareness, perception on willingness to pay taxes personal tax payer who perform work free. J. Paradigm. **11**(02), 72–89 (2013)

Siti Zulaikha, Masruroh: Effect of benefits TIN, understanding taxpayer, quality of service, tax sanctions against taxpayer compliance (empirical study on WP OP in Tegal). Diponegoro J. Account. **2**(4), 2337–3806 (2013)

Stoyan, A.T., Niedzwiecki, S.: Decentralization and democratic participation: the effect of subnational self-rule on voting in Latin America and the Caribbean. Elect. Stud. **52**, 26–35 (2018)

Suryadi: Model Hubungan Kausal Kesadaran, Pelayanan, Kepatuhan Wajib Pajak Dan Pengaruh Nya Terhadap Kinerja Penerimaan Pajak: Suatu Survei di Wilayah Jawa Timur/Causal relationship model awareness, service, taxpayer compliance and his influence on performance of revenue: a survey in east Java. Jurnal Keuangan Publik/J. Publ. Finance (online) **4**(1) (2006)

Twaibu, A.: Factors contributing to inefficient revenue collection in Temeke and Kigamboni municipal councils. Doctoral dissertation. Mzumbe University (2020)

Weske, U., Boselie, P., Van Rensen, E.L., Schneider, M.M.: Using regulatory enforcement theory to explain compliance with quality and patient safety regulations: the case of internal audits. BMC Health Serv. Res. **18**(1), 1–6 (2018)

Widayati, dan Nurlis: Faktor-Faktor yang Mempengaruhi Untuk Membayar Pajak Wajib Pajak Orang Pribadi Yang Melakukan Pekerjaan Bebas Studi Kasus Pada KPP Pratama Gambir Tiga/Factors that affect pay taxes for the individual taxpayer non performing works. Case study on STO Gambir Tiga. In: Proceeding Simposium Nasional Akuntansi XII/Proceeding on National Accounting Symposium XII. Purwokerto (2010)

Exploring Sustainable Agriculture Through the Use of the Internet of Things

F. O. Bamigboye[1(✉)] and E. O. Ademola[2]

[1] Department of Agricultural Sciences, Afe Babalola University, P.M.B. 5454, Ado-Ekiti, Nigeria
bamigboyefo@abuad.edu.ng

[2] Ademola Ojo Emmanuel Foundation, No 6, Ori-Ookun Street, Igbajo-Ijesa, Erinmo, Osun, Nigeria

Abstract. Purpose: The Internet of Things (IoT) in agriculture is popular in the developed world and it is gradually but slowly spreading to developing countries. However, profitable production of crops and animals in large quantities is hinged on improved practices, the use of modern equipment and in recent times, (IoT).

Design/Methodology/Approach: Cellular-IoT is designed for livestock tracking and animal health monitoring. The smart weather monitoring system is an IoT-based weather-forecast-enabled device. The drone technology is an IoT being deployed to monitor crop health, spray chemicals, and conduct soil and field analysis. RFID-based Staff Attendance and Access Control System (RFID-SAACS) is a vital tool for staff management, administration, and monitoring that impacts staff's attitude to work, as time theft by staff and other misbehaviour are eliminated.

Findings: Improved livestock production, monitoring, tracking and general management are achieved through IoT. Optimum harvest through weather forecasts and monitoring of climate changes is achievable by using IoT. Also, the best soil condition, through management, evaluation and replenishing of minerals based on crop requirements, can be achieved using IoT tech. Real-time crop and animal disease detection, diagnosis and management are enhanced through IoT. Monitoring farm activities, accurate staff assessment and evaluation, can be done without human interference by using appropriate IoT devices. Stealing at all levels of cash, time and produce on the farm is seriously minimised through the aid of IoT.

Implications/Research Limitations: Erratic power supply and inadequate availability as well as the cost of internet supply in Nigeria will have to be looked into for effective use of IoT in agriculture.

Practical Implications: The paper will be impactful to agricultural farmers, agricultural policy makers, Computer Scientists, Agricultural Engineers and unemployed interested youths. This will help to modernise and entrench agricultural production to meet the present population explosion.

Originality/Value: The use of the internet of things in agriculture is novel in Nigeria; however, IoT can be a recent technological tool that can be deployed to drive sustainable development in agriculture.

C. Aigbavboa et al. (Eds.): ARCA 2022, *Sustainable Education and Development – Sustainable Industrialization and Innovation*, pp. 881–887, 2023.
https://doi.org/10.1007/978-3-031-25998-2_67

Keyword: Artificial intelligence · Agriculture · Disease · Monitor · Tracking device

1 Introduction

Agricultural production transformation is inevitable to avert hunger and food insecurity in Nigeria. About 821 million people still suffer from hunger (FAO 2018). However, deployment of technology can be a tool in transforming agricultural production. Internet of Things in agriculture is a novel tech that can be employed in agriculture. Its use over time in agricultural production especially in Nigeria will lead to better output, zero hunger and food security. Also, in Nigeria, animal agriculture is still backward in the appropriate use of modern tech to drive production. Therefore, the abundant availability of animal products to meet the protein requirements and demand of the citizens is hindered. The use of Internet of Things in animal agriculture is yet to be popular in Nigeria. This may be due to lack of or poor awareness and information on the tech, start-up capital, inadequate internet facilities, shortage in power supply, and insufficient experts in the field.

According to Miller (2018), IoT is a "network of physical objects (such as wearable devices, home appliances, security systems, personal and commercial vehicles, nanotechnology, manufacturing equipment, and more) embedded with smart components (such as microprocessors, data storage, software, sensors, actuators, and more) and connected to other devices and systems over the internet." IoT allows objects to be sensed and controlled remotely across existing network infrastructure, creating opportunities for more direct integration between the physical world and computer-based systems, and resulting in improved efficiency, accuracy and economic benefit. IoT is any object which is capable of identifying, connecting and communicating with other objects (Santucci 2011; LOPEZ Research Series 2013; Reddy 2014).

"Things," in the IoT sense, can refer to a wide variety of devices such as heart monitoring implants, biochip transponders on farm animals, electric clams in coastal waters, automobiles with built-in sensors, DNA analysis devices for environmental/food/pathogen monitoring or field operation devices that assist fire-fighters in search and rescue operations. These devices collect useful data with the help of various existing technologies and then autonomously flow the data between other devices (Dharani *et al.* 2019). Farm management, animal monitoring, irrigation control, greenhouse environmental control, autonomous agricultural machinery, and drones are examples of IoT applications in agriculture, all of which contribute to agrarian automation. It also demands contributing to agricultural food production's long-term viability. Land appraisal, crop protection, and crop yield projection, according to these needs, are essential to world food production (Safdar et al., 2019). Farmers, for example, can manage field environments in real-time and more effortlessly regulate fields using wireless sensors and mobile networks. Farmers may also utilize IoT technology to capture essential data, subsequently creating yield maps that enable precision agriculture to produce low-cost high-quality crops (Sinha and Dhanalakshmi 2022).

Internet of Things in agriculture is the driving tool for smart farm to achieve improved agricultural productivity. This is becoming inevitably essential in developing countries like Nigeria due to increased population, insecurity suffered by farmers, destruction of crop farms by herders, climate change and looming famine. The rate at which farm workers and outsiders cart away with farm produce drastically affects the profitability of this enterprise. However, these issues can be mitigated by exploring the use of Internet of Things in agriculture. Hence, areas of relevance of IoT in agriculture that can be explored in Nigeria were discussed.

2 Internet of Things in Agriculture: Diagnosis and Monitoring of Crop and Farm Animal Diseases

The state of health of crops grown for human consumption determines yield and quality of crop produced and subsequently the farmer's profit. Proper monitoring of plant health is required at different stages of plant growth in order to prevent disease affecting plants. Existence of pests and disease affect the estimation of crop cultivation and minimizes crop yield substantially. Present day system depends on naked eye observation which is a time consuming and cumbersome process. Automatic detection of plant disease can be adopted to detect plant disease at early stages. Various disease management strategies have been used by farmers at regular intervals in order to prevent plant diseases (Rajesh *et al.* 2018). However, the use of IoT devices gives real-time essential information for early detection and proper monitoring of plant diseases.

Seelye et al. 2011 have presented low cost colour sensors for monitoring plant growth in a laboratory. An automated system for measuring plant leaf colour is developed to check plant health status. Huddar *et al.* 2013 have presented novel algorithm for segmentation and automatic identification of pests on plants using image processing.

Reddy et al. (2019) created an IoT-based system for disease and insect pest management in agriculture and the prediction of plant climatic factors. The integrated sensors help in the measurement of soil and atmospheric moisture and humidity. These features help determine the environmental conditions in which the plant flourishes and the plants' illnesses. It detects disease on the field and sprays prescribed insecticides. Web cameras take images that are then preprocessed to include RGB (Red Green Blue) to grayscale conversion, defect detection, image scaling, image enhancement, and edge detection. SVM is utilized to categorize characteristics generated from Citrus Canker diseases, such as energy, kurtosis, skewness, and entropy (damaged Lemon crop). The Arm7 microcontroller is used for hardware, power, sensors, and motor driver control. Once the illness is identified, the program will propose fertilizers and transmit the results to an LCD and the recommended fertilizers. By pump, the fertilizers will be sprayed on the diseased leaves. A solution was presented for forecasting and detecting grape disease using the CNN approach and real-time gathered data on environmental factors. First, the CNN technique is utilized to analyze the leaf images, then; different layers of the CNN method were used to create the image. Finally, it was scaled to a specific resolution before data was sent into the CNN layers for training and testing. The suggested algorithm was evaluated on four diseases known to have a higher effect on grape production. The diseases include escablack measles, anthracnose, leaf blight, and black rot. This gadget not only detects but also forecasts illnesses based on historical weather data. On the other side, the readings from the humidity, temperature, and soil moisture sensors are transferred through Rasp berry Pi to Microsoft's Azure Cloud. Following this, the sensor readings are used to anticipate the illness using a trained linear regression model. Based on the findings of the preceding detection and prediction stages, suggestions for appropriate fertilizers in the right quantities was provided to minimize fertilizer misuse and cost savings (Chavan et al. 2019).

To detect pests in rice during field production and avoid rice loss, the Internet of Things supported a model-based UAV with the Imagga cloud offered. The Internet of Things-based UAV was developed on AI mechanisms and the Python programming prototype to transmit rice disease images to the Imagga cloud and supply insect data. The Approach identifies the disease and insects by integrating the confidence ratings of the labels. The label identifies the objects in the images. To determine the pest, the tag with the greatest confidence results and more than or equal to the threshold is chosen equal to the target label. If pests are discovered in the rice, statistics will be transferred to the field owner directly to take preventative actions. The suggested method is capable of detecting all pests that influence rice production. On the other hand, this approach helped to minimize rice waste during production by conducting insect monitoring at regular intervals (Bhoi et al. 2021).

Farm animals are meant to be in a state of disease-free at all times. This is due to the fact that animal welfare will influence humans' state of health that consumes such animals, productivity of such animal and economic gains of farmers.

Moreover, about 60% of the human diseases are zoonotic in nature i.e. infection is spread from animals to human being. Zoonotic diseases include Tuberculosis, Glanders, Anthrax, Rabies, Avian and Influenza. These diseases are quite harmful so there is need to control the spread of zoonotic diseases. Highly contagious livestock diseases such as 'foot and mouth disease (FMD)', hemorrhagic septicemia (HS)', 'mastitis, peste des petits ruminant (PPR)' and "surra in cloven footed domestic animals" cause irreparable economic losses to the farming community (Singh *et al.* 2014).

These reasons necessitated a system for real time monitoring of animal health, controlling and preventing the outbursts of diseases on a large scale. Technology is already a part of modern farming and is playing an important role with the advancement in available systems and tools. Livestock farming has been one of the biggest areas of development in electronic in recent years. A lot of scholars are focused on the development of animal health monitoring system (Kumari and Yadav 2018).

Jacky *et al.* (2007) developed a "Mobile Monitoring System based on RFID" to handle the cattle efficiently with the help of "dynamic information retrieval, location identification and behavior analysis over a wireless network". Ji-De *et al.* (2013) presented the technique with an embedded system utilising "IoT sensors". The system consists of smart infrastructure which measures different parameters and communicate among them. Huircan et al. (2010) presented cattle monitoring in cropping fields based on a zigbee and utilised the scheme of localization in WSN. Lovett et al. (2009) presented a measurement technique using "infrared thermography" for detecting "foot-and-mouth disease" of livestock. Their study was focused at estimating infrared thermography as a screening technique for FMDV-infected animals and its attainable usage in the identification of suspected cattle for sampling and confirmatory diagnostic testing during FMD outbreaks.

Another smart animal health monitoring system based on IoT for real time monitoring of the physiological parameters such as body temperature, heart rate and rumination with surrounding temperature and humidity has been developed. Various sensors mounted on the body of animals were responsible for the information related to their health status and users have easy access to those data using the internet. Information can be accessed from anywhere using internet and an android app (Kumari and Yadav 2018).

These are opportunities that can be harnessed in Nigerian agriculture. However, majority of agricultural activities are carried out in the rural areas with concomitant inadequate infrastructures needed for deployment and smooth running of IoT device. Hence, the widespread use of IoT in agriculture is still a mirage in Nigeria. Most farmers are poor and can hardly afford the cost implications of setting up these facilities. However, government intervention can be a way forward. Subsidized accessible internet facilities, solar or wind powered electricity other infrastructures can be deployed to the rural areas.

3 Livestock Tracking and Weather Forecasting Powered by IoT

To prevent stock theft, animals are fitted with radio frequency identifiers (RFIDs) that enable tracking of the animal. The position of the animal can be visualised on a map in a control centre through data remitted wirelessly. In rural areas where there is communal grazing, animals tend to get lost. Livestock can be fitted with radio-frequency identifiers (RFID) chips and RFID readers are placed at various monitoring spots to transmit information to (Dlodlo and Kalezhi, 2015) the farmer.

Previous iterations of the technology have often been large and heavy, needed to be hung around the necks of cattle or other animals and costly. However, the Cellular-IoT based animal tracker is low-cost, small cellular IoT technology, a tracking device that can be attached to an animal's ear. The benefits are many, from reduced physical impact on the animal to lower cost and reduced resource usage. Other solutions may require the setup of a custom base station or gateway to communicate with the tags in the field. Apart from serving as tracker, it also monitors animal health status (Skøien 2019).

Weather is a paramount factor to be considered in crop and animal agriculture. Hence, weather forecasting and monitoring must be accurately predicted so as to carry out production activities to time for eventual improved productivity. Animal agriculture is also affected negatively by harsh weather; in terms of reproduction, performance and production. Thus, forecasting and monitoring of weather situations fortify the farmers with information that can guide them in planting, harvesting and other production activities to be carried out. Also, policy makers can be fortified with information that will guide their decisions aright (Bamigboye and Ademola 2016).

Weather forecasting can be done through analysis of weather data over long periods to reduce agricultural risk. This is referred to as big data analysis. In weather forecasts for pest management, humidity, precipitation, crop type, soil fertility, leaf wetness, temperature, winds and soil moisture are collected at local level through sensors. The life cycle of pests is monitored along with the climate data, allowing researchers to predict pest outbreaks more accurately because pest maturation depends on environmental conditions (Dlodlo and Kalezhi 2015).

Temperature, humidity, light intensity, and soil moisture can be monitored through various sensors. These can then be linked to systems to trigger alerts or automate processes such as water and air control. They can also be set up to look for early signs of pests or disease (Huang 2014).

4 Staff Monitoring Technology

The Nigerian Satellite Company Limited, has successfully designed, implemented, tested and deployed an RFID-based Staff Attendance and Access Control System (RFID-SAACS). RFID-SAACS is a vital tool for staff management, administration, and monitoring that impacts staffs' attitude to work, as time theft by staff is completely eliminated. The logged data can also serve as a means of staff monthly appraisal, while an additional utilisation of the RFID-SAACS system includes integration into the payroll system to facilitate precise salary computation and payment based upon vetting of employees' overall performance NCSL 2015). This is used in some automated farms in Nigeria (Bamigboye and Ademola 2021).

5 Conclusion

Agriculture is a profit-driven oriented business; hence, factors that influence the profitability of a farm are of great paramount and interest to the farmer. IoT can be made relevant if it can address the general needs of a locality, be made available and affordable, easy to use and packaged in the local/indigenous languages. This is in the light that farming activities are still mainly in the hands of local, illiterate-traditional farmers. The government should be involved in the transformation of Nigerian agriculture from crude/slightly mechanized to Internet of Things based.

References

Bamigboye, F.O., Ademola, O.: Internet of Things (Iot): it's application for sustainable agricultural productivity in Nigeria. In: 6th Proceedings of the iSTEAMS Multidisciplinary Cross-Border Conference, held at University of Professional Studies, Accra Ghana, 2016, pp. 309–312 (2016)

Bamigboye, F.O., Ademola, O.: Internet of things-enabled agribusiness opportunities in developing countries (Chapter 14). In: Che, F.N., Strang, K.D., Vajjhala, N.R. (eds.) Opportunities and Strategic Use of Agribusiness Information Systems, pp. 263–276. Pennsylvania, USA (2021). https://doi.org/10.4018/978-1-7998-4849-3

Bhoi, S.K., et al.: An internet of things assisted unmanned aerial vehicle based artificial intelligence model for rice pest detection. Microprocess. Microsyst. **80**, 103607 (2021)

Chavan, R., Deoghare, A., Dugar, R., Karad, P.: IoT Based solution for grape disease prediction using convolutional neural network and farm monitoring. Int. J. Sci. Res. Eng. Dev. **2**, 494–500 (2019)

Dlodlo, N., Kalezhi, J.: The internet of things in agriculture for sustainable rural development. In: International Conference on Emergence Trends Networks and Computer Communications at Windhoek, Namibia (2015)

Huang, R.: Internet of Things: 5 Applications in Agriculture (2014). http://blog.hwtrek.com/?p=626htm

Huircan, J.I., et al.: Zigbee based wireless sensor network localization for cattle monitoring in grazing fields. Comput. Electron. Agric. **74**, 258–264 (2010)

Jacky, S., Tings, L., Kwok, K., Lee, W.B., Tsang, Cheung, B.C.F.: A dynamic RFID-based mobile monitoring system in animal care management over a wireless network. In: International Conference on Wireless Communications, Networking and Mobile Computing (2007)

Huang, J.-D., Hsieh, H.-C.: Design of gateway for monitoring system in IoT networks. In: IEEE International Conference on and IEEE Cyber, Physical and Social Computing (2013)

Kumari, S., Yadav, S.K.: Development of IoT based smart animal health monitoring system using raspberry Pi. In: Special Issue Based on Proceedings of 4th International Conference on Cyber Security (ICCS) (2018)

Lovett, K.R., Pacheco, J.M., Packer, C., Rodriguez, L.L.: Detection of foot and mouth disease virus infected cattle using infrared thermography. The Veterinary J. **180**, 317–324 (2009)

Nigeria Communications Satellite Ltd. (NCSL): Staff attendance and access control system (2015). http://www.nigcomsat.gov.ng/products.php

Reddy, H.S., Hedge, G., Chinnayan, D.R.: IOT based leaf disease detection and fertilizer recommendation. Int. J. Innov. Technol. Explor. Eng. **9**, 132–136 (2019)

Rehman, A., Saba, T., Kashif, M., Fati, S.M., Bahaj, S.A., Chaudhry, H.A.: Revisit of internet of things technologies for monitoring and control strategies in smart agriculture. Agronomy **12**, 127 (2022).https://doi.org/10.3390/agronomy12010127

Huddar, S.R., Rupanagudi, S.R., Ravi, R., Yadav, S., Jain, S.: Novel architecture for inverse mix columns for AES using ancient Vedic Mathematics on FPGA. In: 2013 International Conference on Advances in Computing Communications and Informatics (ICACCI), pp. 1924–1929 (2013)

Safdar, A., et al.: Intelligent microscopic approach for identification and recognition of citrus deformities. Microsc. Res. Tech. **82**, 1542–1556 (2019)

Seelye, M., Gupta, G.S., Bailey, D., Seelye, J.: Low cost colour sensors for monitoring plant growth in a laboratory. In: IEEE International Instrumentation and Measurement Technology Conference (2011). https://doi.org/10.1109/IMTC.2011.5944221, Corpus ID: 32599104

Singh, D., Kumar, S., Singh, B., Bardhan, D.: Economic losses due to important diseases of bovines in central India. VeterinaryWorld **7**, 579–585 (2014)

Sinha, B.B., Dhanalakshmi, R.: Recent advancements and challenges of Internet of Things in smart agriculture: a survey. Futur. Gener. Comput. Syst. **2022**(126), 169–184 (2022)

Skøien, K.R.: Animal Tracking: Cellular IoT in the Field (2019). https://blog.nordicsemi.com/getconnected/animal-tracking-cellular-iot-in-the-fieldhtm

Animal Ethics and Welfare as Practised by Small Ruminant Farmers in Ado-Ekiti, Ekiti State Nigeria

F. O. Bamigboye[1(✉)], A. J. Amuda[2], J. O. Oluwasusi[1], and E. O. Ademola[3]

[1] Department of Agricultural Sciences, Afe Babalola University, P.M.B. 5454, Ado-Ekiti, Nigeria
bamigboyefo@abuad.edu.ng
[2] Department of Animal Production, Federal University Wukari, Wukari, Nigeria
[3] Ademola Ojo Emmanuel Foundation, No 6, Ori-Ookun Street, Igbajo-Ijesa, Oriade LGA, Osun, Nigeria

Abstract. Purpose: In Nigeria, animal ethics and welfare is not common issue among Small Ruminant (SR) farmers despite their importance in improving productivity. Hence, the level of compliance with ethical production techniques by SR producers in Ado-Ekiti was evaluated.

Design/Methodology/Approach: Structured questionnaire was employed to elicit information on socioeconomic characteristics and SR management from the respondents. A multi-stage sampling technique was used to select 87 respondents. Also, supplements were collected from the respondents and analyzed for their chemical composition using standard procedures (AOAC). Data were analysed using descriptive statistics and ANOVA.

Findings: Small ruminant producers in the area adhered to some ethical practices in their daily production activities; most fed their animals with supplements twice daily, treated sick animals and transported their animals for sale in the morning. All the respondents offered water to their animals but not *ad-libitum* and from any source. Some of the respondents did not provide housing facilities for their animals. The proximate composition of supplement; crude protein values ranged from 2.94 to 11.91% in cassava peel and beans chaff respectively, crude fibre: 4.06 to 24.94% and ether extract: 0.64 to 9.22% in cassava sievate and bean chaff respectively, ash content: 3.00 to 11.36% in cassava peel and banana peel, respectively. The mineral composition for sodium ranged from 37.200 to 591.267 ppm, calcium: 24.867 to 5060.33 ppm, phosphorus: 28.036 to 125.670 ppm and magnesium: 7.115 to 57.024 ppm. Iron ranged from 5.241 to 36.106 ppm, copper from 0.103 to 1.827 ppm and zinc from 0.019 to 21.186 ppm.

Implications/Research Limitations: Most supplements can complement each other incredibly when mixed. However, mixing ratio and correlation scaling can be areas for further study.

Practical Implications: The findings of the present study would be useful for small ruminant farmers in terms of requirements for animal ethics and welfare. Also, researchers/animal nutritionists will have access to the proximate and mineral compositions of the common supplements evaluated as the baseline.

C. Aigbavboa et al. (Eds.): ARCA 2022, *Sustainable Education and Development – Sustainable Industrialization and Innovation*, pp. 888–901, 2023.
https://doi.org/10.1007/978-3-031-25998-2_68

Originality/Value: Consideration of welfare and ethics among small ruminant farmers in Nigeria is not well researched. Hence, the present study enlightened farmers on the ethical management practices in small ruminant farming.

Keyword: Animal handling · Animal need · Farmers · Freedom · Supplement

1 Introduction

Handling and welfare status of farm animals is below the standards in developing countries where animals are poorly handled due to misconception and resource scarcity. In Nigeria, animal ethics and welfare is not a common issue among small ruminant farmers however, it is important in improving productivity. A study by Grunert (2000), indicated that animal welfare competes with a long range of other factors; possibly more important, quality traits, such as taste, tenderness, cut, and safety in guiding consumers' choice.

A very large amount of research has been carried out about animal welfare problems involving very specific fields of interest, such as the development of welfare assessment methods in different environments. Among the main issues involved in the concept of welfare are the concepts of 'suffering' and 'need,' as well as the 'five freedoms' which are more related to animal husbandry and management by man (Millman et al. 2004). According to Fraser and Broom (1997) "the general term 'need' is used to refer to a deficiency in an animal which can be remedied by obtaining a particular resource or responding to a particular environmental or bodily stimulus." Considering animal welfare in practice, the animal may be interacting with a variety of factors that may represent the fulfilling of the 'needs,' i.e. requirements for obtaining physical and mental health (Odendaal 1998).

Inability to satisfy the needs leads to welfare problems. In this respect, the concept of 'freedom' in animal husbandry has been introduced and plays a key role. In fact the knowledge about the needs of animals is related to the proposal of giving animals some 'freedoms' (Brambell Report 1965), revised by FAWC (1993) as follows:

Freedom from thirst, hunger and malnutrition – by ready access to fresh water and diet to maintain full health and vigour.
Freedom from discomfort – by providing a suitable environment including shelter and a comfortable resting area.
Freedom from pain, injury and disease – by prevention or rapid diagnosis and treatment.
Freedom to express normal behavior – by providing sufficient space, proper facilities and company of the animal's own kind.
Freedom from fear and distress – by ensuring conditions which avoid mental suffering.

According to Webster (1994), "absolute attainment of all five freedoms is unrealistic," but these freedoms are an "attempt to make the best of a complex and difficult situation." These have to be deeply considered in husbandry systems for farm animals, because they have to be given the possibility to adapt well to them, in order to avoid undue distress and consequently produce well in optimal conditions. In any case, animals' welfare has to be considered in a realistic way, avoiding anthropomorphism into its evaluation (Webster 1994), as well as pure mechanistic consideration.

Small ruminant are widely distributed and are of great importance as a major source of livelihood of the smallholder farmers and the landless in rural communities in tropical Africa. Sheep and goat population is higher than that of cattle in Nigeria connoting great potentials for productivity (Bamigboye 2013); sheep was reported to be 33,000,000, goats: 52,000,000 and cattle: 16,000,000 (FMA 2008). In Nigeria, small ruminants are mostly reared in small numbers around the homes as a secondary source of income. They are therefore, not accorded their expected rights and their welfare is jeopardized. In fact, some people believed that animals do not have any right. To this end, this study examined the animal ethics and welfare as practiced by small ruminant farmers in Ado-Ekiti, Ekiti Nigeria.

2 Materials and Methods

2.1 Description of Experimental Site

The study was carried out at Ado Ekiti, Ekiti state Nigeria, from the five boundaries to Ado Ekiti state which include Ado-Ikere road, Ado-Iyin road, Ado-Afe Babalola way, Ado-EKSU, Ado-Ilawe. Ekiti state lies between longitude 5°13′ 17 East of Greenwich meridian and latitude 70 37′ 16 North of the Equator, temperature is 26 ºC, humidity 74%, rainfall 300–1100 mm. The population of the area is dominated by Yorubas.

2.2 Data Collection

The instrument for data collection was formal survey with the farmers. Personal contact and oral interview were the tools employed to elicit information from the respondents. Primary data were collected by administration of both open and close ended structured pre-tested questionnaire. The questionnaire administered to the farmers centred: socio-economic characteristics, production technique, health management, feed and water management and marketing and sales.

2.3 Supplement Collection

Commonly used supplements were collected from small ruminant farmer in the study area. Crop residues, kitchen waste and agro-industrial by-products used by respondents were collected and bulked.

2.4 Proximate Analysis

Crude protein, crude fibre, ether extract, nitrogen free exact and ash contents of samples were determined by AOAC (2000).

2.5 Mineral Analysis

The samples were analyzed for mineral after wet digestion of sample with a mixture of par chloric acid and concentrated nitric acid 1:4. Potassium (K), Calcium (Ca), Magnesium (Mg), Sodium (Na) were determined using atomic absorption spectrophotometer (AAS) Buck Scientific model 210 VGP and flame photometer FP 902 PG using the calibration plot method (Greenberg et al. 1985).

2.6 Statistical Analysis

Data generated from the questionnaire were analysed using descriptive statistics such as frequencies, percentages, means and standard deviation. The data from proximate and mineral analysis were subjected to analysis of variance (ANOVA). Significant differences among means were separated using the Duncan's multiple Range Test (DMRT).

3 Results and Discussion

3.1 Socio-economic Characteristics of Small Ruminant Farmers in Ado-Ekiti

The socio-economic and personal characteristics of the small ruminant farmers randomly selected in the five boundaries of Ado-Ekiti, are presented in Table 1. Age range was from 20 and above 60 years; indicating that small ruminant production in the study areas involved both young and old and it is lucrative. A larger proportion of the respondents (40.23%) were within the age range of 36 to 50 years with mean age of 43 years. It implies that a high proportion of middle age respondents were involved in goats and sheep production in the area. Thus, small ruminant production is an adult business in the area.

Majority of the respondent were females (86.2%) while 13.8% males. This implied that small ruminant production is a female affair in the study area. It had been reported that it is a trend for women to become involved in income generating activities such as farming, trading, processing etc. to support their families (Familade *et al.* 2011). Majority of the respondents (74.71%) were married while 10.34% single and 9.2% separated. This may be due to the fact that married people were always saddled with various family responsibilities which most time calls for major attention (Familade *et al.* 2011). The high percentage of married people also may indicate that the production of small ruminants was a good source of income and food security for the families. Most (40.23%) of the respondents had 6 to 8 persons in their house hold. Respondents with large family size would have more hands to work with in rearing their animals which could aid increase in their output (Anosike *et al.* 2015). It was observed that (68.96%) of the respondents had the small ruminant production as a secondary occupation, (25.3%) primary occupation and (5.74%) had it as an alternative occupation. This may be due to the fact that the respondents do not produce in large or commercial quantity. Hence, other sources of income generation were inevitable. This is in line with observations of Kuponiyi and Sodeinde (2005) that rural dwellers engaged in livestock production as a part-time income generating activity. Most respondents (31.03%) had at least primary school education, and the least was (12.64%): tertiary education. It could be inferred that the respondents were literate and the level of basic education standard in this study area was substantially higher. For the purpose of adopting new technologies, education is an important factor which if lacking can have adverse impact on future of small ruminant production improvement (Olafadehan *et al.* 2010). The high level of literacy can provide scope for an information interface between farmers, extensionists, researchers and development agents to improve animal ethics and welfare in the area. Adekoya and Ajayi (2000) observed that the level of education affected the reception of innovation of rural dwellers.

Christianity was the dominant religion practiced by the respondents (74.71%), Islam was observed to be (20.69%) and traditional religion was (4.6%). Keeping of small ruminants had no religious undertone or bias since people of different religions reared small ruminants in the areas.

Table 1. Socio-economic and personal characteristics of small ruminant farmers in Ado-Ekiti

Variable	Frequency	Percentage (%)
Gender		
Male	12	13.8
Female	75	86.2
Age (years)		
20–35	12	13.8
36–50	35	40.23
51–60	27	31.03
60-Above	13	14.94
Marital status		
Single	9	10.34
Married	65	74.71
Divorced	5	5.75
Separated	8	9.2
Small ruminant rearing as an occupation		
Primary (major)	22	25.3
Secondary (minor)	65	74.71
Educational status		
None	23	26.44
Primary	27	31.03
Secondary	26	29.69
Tertiary	11	12.64
Household size		
1–2	8	9.2

(continued)

Table 1. (*continued*)

Variable	Frequency	Percentage (%)
3–5	35	40.23
6–8	35	40.23
Above 8	9	10.34
Location		
Ado-Ikere road	17	19.54
Ado-Iyin road	18	20.69
Ado-Afe Babalola way	17	19.54
Ado-Eksu	18	20.69
Ado-Ilawe	17	19.54
Religion		
Christianity	65	74.71
Islam	18	20.69
Traditional	4	4.6

3.2 Freedom from Thirst, Hunger and Malnutrition as Managed by Small Ruminant Farmers in Ado-Ekiti

Presented in Table 2 is the water and feed management of small ruminants in Ado-Ekiti. It was revealed that majority (93.10%) of the respondents allowed their animals to roam around and scavenge for feed, while (6.9%) of the respondents kept their animals intensively. A large number of respondents (98.85%) find it really necessary and ethical to offered feed (supplement) to their livestock. It implies that majority of the small ruminant farmers in the study area had the welfare of the animals at heart in terms of supplementing grazing. Most (88.5%) of the respondents fed their animals twice daily, while (6.9%) of the respondent fed them only once. It is unethical for small ruminant farmers to allow their animals to grazing alone. Lack of adequate feed and water supplies have been reported to be a major cause of mortality of small ruminants (Bolajoko *et al.* 2011).

All the farmers (100%) offered water to their animal; this is quite commendable. Animals have the right to be free from hunger and thirst. The livestock farmers in the study area reported that water from rain (2.29%), tap (1.15%), well (62.07%), any available water (34.49%) as sources of water offered to their animals. Well water was the most common water source utilized for small ruminants in Ado-Ekiti. This implies that well water was more accessible source of water to the respondents. The water source is paramount when purity of the water is considered. Tap water source accounted for (2.29%) of the respondents that supplied treated water to their animals. These other sources used by the framers could be contaminated by pollutants which may be from land or water bodies. Animals should be accorded the right to hygienic water and freedom from thirst (Duncan 2002). Roaming and grazing of small ruminants may not fully support improved production. Lack of adequate feed and water supplies have been reported to be a major cause of mortality of small ruminants (Bolajoko *et al.* 2011).

Table 2. Feed and water management of small ruminants in Ado-Ekiti

Variable	Frequency	Percentage (%)
Roam and graze		
Yes	81	93.10
No	6	6.9
Offered feed		
Yes	86	98.85
No	1	1.15
Number of time fed daily		
Once	5	5.75
Twice	77	88.5
When available	0	0
Others	4	4.6
Offered water		
Yes	87	100
No	0	0
Number of time offered water daily		
Always	25	28.7
Once	6	6.9
Twice	56	64.4
Source of water		
Rain	2	2.29
Tap	1	1.15
Well	54	62.07
Any available water	30	34.49

3.3 Freedom from Discomfort (Provision of Shelter and a Comfortable Resting Area)

Presented in Table 3 is the type of housing facilities offered small ruminants in the study area. Most (91.95%) of the respondents made available housing facilities to their animals while (8.05%) did not offer any form of housing facilities. This is at variance with situation of small ruminants in Iwo local government where over 90% did not provide housing facility for the animals. It is quite commendable that majority of the respondents in the present study protected their animals from inclement weather and gave them freedom from discomfort. However, about 8% subjected their animals to sleeping on the roads and verandas which are grossly inadequate for comfort; exposing them to hazards and dangers. Mainly, the housing type used was the cage (57.47%) system, followed by the pen: (22.99%) and shed: (19.54%). Farm animals should be given access to good housing facility in order to avoid undue distress and consequently reducing their optimum productivity. Inadequate housing might predispose the animals to diseases, theft, accident, discomfort, death and even predators thus hampering their right to freedom from discomfort, pain, injury and diseases. In order to maintain good animal welfare, the animal should be able to cope with its environment and satisfy biological needs (Fraser and Broom 1990). Ajayi et al. (2009) recommended that the housing type provided for small ruminants should keep them from harsh weather and theft.

Table 3. Type of housing facilities provided by small ruminant farmers to their animals in Ado-Ekiti

Variable	Frequency	Percentage
Provision of housing facility		
Yes	80	91.95
No	7	8.05
Housing type		
Shed	12	13.8
Cage	48	57.47
Pen	20	22.99

3.4 Freedom from Pain, Injury and Disease – By Prevention or Rapid Diagnosis and Treatment

Prevalent diseases of sheep and goats in the study area, is as shown in Fig. 1. They are cold, *Peste-des-Petitis Ruminant* (PPR), Foot and mouth disease and others. However, Peste-des-Petitis Ruminant (PPR) (45.97%) was the most prevalent disease. Cold (26.44%) was also reported as one of the quite prominent, while foot and mouth (19.54%) and others (8.05%). The present findings is in consonance with the report that PPR is very well known to be responsible for significant losses of small ruminants countrywide (El-Yuguda *et al.* 2013; Woma et al. 2016). FAO (1988) and SPORE Magazine (2010) cited PPR, pneumonia, foot and mouth disease, bovine virus diarrhoea (BVD) and blue tongue as the most common diseases of small ruminants.

Table 4 shows that majority (91.95%) uses veterinarian's services to diagnose and treat their animals and (8.05%) treated their animals on their own. Majority of the farmers

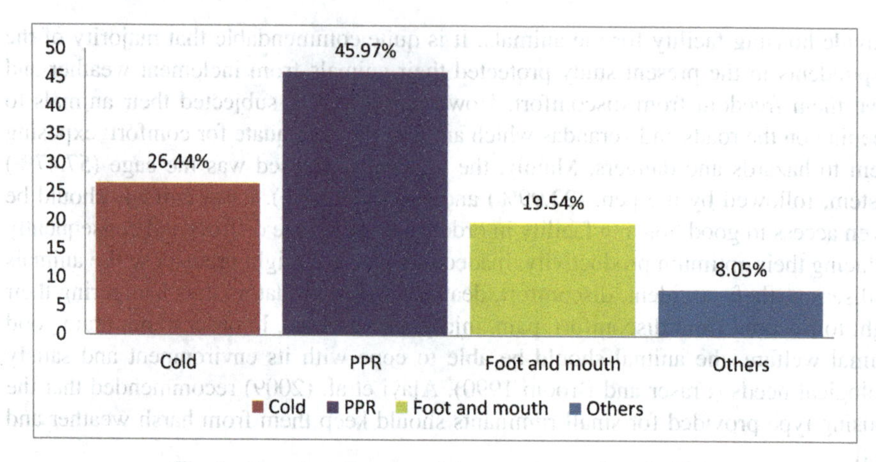

Fig. 1. Prevalent diseases of sheep and goats in Ado-Ekiti

(90.80%) use conventional drugs to treat their animals while herbs (5.75%) and other means (3.45%) used to treat small ruminants.

Table 4. Health management of small ruminants in Ado-Ekiti

Variable	Frequency	Percentage (%)
Treatment by		
Self	7	8.05
Veterinary service	80	91.95
Treatment method		
Use of herbs	5	5.75
Conventional drugs	79	90.80
Others	3	3.45

3.5 Freedom from Fear and Distress – By Ensuring Conditions Which Avoid Mental Suffering

Presented in Fig. 2 are methods of controlling and handling small ruminants in the study area. Figure 2 shows that majority (66.67%) of the small ruminant farmers talked to their animals as a method of controlling them. While (20.69) preferred beating; this is against ethics and rights of animals and it may inflict mental distress, fear and injury on the animal.

Majority of the small ruminant farmers (91.95%) used public transportation to convey their livestock to the market, while (8.05%) used their private cars to move their livestock. Most (87.4%) of the respondents transported their animals to the market in the morning

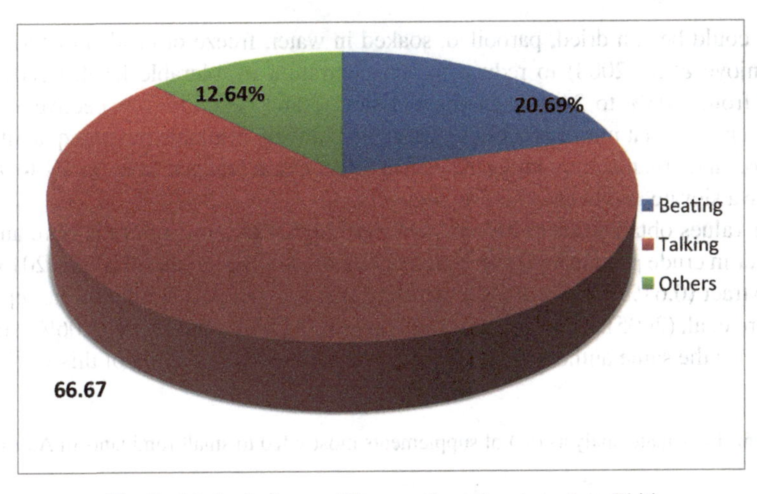

Fig. 2. Method of controlling small ruminants in Ado-Ekiti

to further reduce the stress level of the animals while (12.6%) moved their animals to the market in the evening. Animals should be subjected to less pain and stress and in terms of transportation this can be achieved by moving animals in the cool time of the day (early morning hours and evening) (Table 5).

Table 5. Transportation mode and period of small ruminants in Ado- Ekiti

Variable	Frequency	Percentage (%)
Transportation		
Private car	7	8.05
Public transportation	69	79.31
Truck	0	0
Herding	0	0
Time of transporting		
Morning	66	75.86
Afternoon	0	0
Evening	10	11.5

Presented in Table 6 is the proximate composition of supplements offered to small ruminant animals in Ado-Ekiti. The crude protein ranged from 2.94% to 11.91% in cassava peel and beans chaff respectively. The CP for beans chaff (11.91%) was the highest which is above the 7% CP requirement which will provide ammonia required by rumen microorganism to support optimum microbial activity.

Cassava based supplements had the least proximate composition in the present study. Cassava based supplement is rarely fed fresh due to the high level of HCN inherent in it so, it could be sun dried, parboiled, soaked in water, freeze or ensiled or fermented (Oladunjoye *et al.* 2008) to reduce the concentration to tolerable level. Crude fibre ranged from 4.06% to 24.94% in cassava sievate and bean chaff respectively. Okah (2004) reported that maize processing waste is relatively available in large quantities in both rural and urban communities in Nigeria. Ether extract ranged from 0.64% to 9.22% in cassava sievate and beans chaff respectively.

The values obtained from banana peel were higher in terms of crude fibre and ash but lower in crude protein and ether extract compared to Bakshi and Wadhwa 2013. The ether extract (0.64%) of cassava sievate was quite low but was similar to the report of Nwokoro et al. (2005). However, the crude protein (3.78%), crude fibre (4.06%) and ash (4.99%) by the same authors were quite comparable with the finding of this work.

Table 6. Proximate analysis (%) of supplements mostly fed to small ruminants in Ado-Ekiti

Supplement	CP	CF	Ash	EE
Cassava peel	2.94^{e}	13.43^{d}	3.00^{d}	1.71^{d}
Banana peel	6.27^{b}	15.19^{c}	11.36^{a}	4.82^{c}
Beans chaff	11.91^{a}	24.94^{a}	8.68^{b}	9.22^{a}
Cassava sievate	3.78^{d}	4.06^{e}	4.99^{c}	0.64^{e}
Pap residue sievate	4.20^{c}	18.50^{b}	9.38^{a}	5.16^{b}
SEM	0.0767	0.299	0.1007	0.409

^{a,b,c,d,e} = Means on the same column with similar superscript letters are not significantly different ($P < 0.05$).

Table 7 showed the mineral composition of the supplements mostly fed to small ruminants in the study area. It was noted that there were significant differences existed in all the measured parameters among the supplements. Sodium ranged from 11.167 ppm (cassava peel) to 591.267 ppm (banana peel); calcium: 21.567–5060.33 ppm (cassava peel and banana peel respectively). Iron was from 5.241–36.106 ppm in pap residue sievate and banana peel respectively while zinc was 0.319 -21.186 ppm (pap residue sievate and cassava sievate respectively). Banana peel exhibited the highest level of mineral assay in all the minerals examined except in Zn. Banana peel was also said to contain (mg/g) 2.9 calcium, 1.8 phosphorus, 3.0 magnesium, 0.1 sodium and 11.1 potassium macro minerals. While the micro-minerals were: (ppm) 2947.0 Fe, 386.0 Cu, 1138.0 Zn, 522.0 Mn, 3.3 Mo and 6.0 Co. However, iron, copper and zinc were beyond the maximum tolerant limits for ruminant production (Bakshi and Wadhwa 2013). Banana peels are rich in trace elements, but Fe, Cu and Zn contents are much higher than the maximum tolerance limit for ruminants, suggesting that these should not be fed *ad libitum*, but should be supplemented in the ration of ruminants as source of organic minerals (Bakshi and Wadhwa 2013).

Potassium (K) and sodium (Na) have roles in maintaining acid: base balance and the control of body fluids. Approximately 70% of a ruminant's magnesium is stored in the skeleton. Magnesium is an activator of over 300 enzyme systems and is involved in the metabolism of carbohydrates, lipids and protein. It has roles in nerve conduction and muscle contraction. Phosphorus is important for cell membranes, energy production, muscle contraction, appetite and bone formation. Ca is essential for teeth and intracellular fluids, nerve function, muscle contraction, blood clotting, activation of a number of enzymes and bone formation (Nutrimin 2016).

Table 7. Mineral composition (ppm) of supplement mostly fed to small ruminants in Ado-Ekiti

Feed	Na	Ca	K	Mg	P	Fe	Cu	Zn
A	11.167^{e}	21.567^{e}	105.033^{c}	12.491^{d}	42.769^{d}	13.369^{b}	0.174^{d}	0.734^{c}
B	591.267^{a}	5060.33^{a}	642.233^{a}	57.024^{a}	125.670^{a}	36.106^{a}	1.827^{a}	4.112^{b}
C	92.567^{c}	111.067^{b}	73.967^{d}	16.703^{c}	61.915^{c}	10.284^{c}	0.103^{c}	0.505^{d}
D	37.200^{d}	68.100^{c}	311.700^{b}	26.480^{b}	28.036^{e}	7.789^{d}	0.218^{c}	21.186^{a}
E	121.233^{b}	24.867^{d}	55.467^{e}	7.115^{e}	72.506^{b}	5.241^{e}	0.247^{b}	0.319^{e}
SEM	0.354	6.983	0.378	0.240	0.372	0.156	0.008	0.019

^{a,b,c,d,e} = Means on the same column with similar superscript letters are not significantly different ($P < 0.05$).
A- Cassava peel B- Banana peel C- Beans chaff D- Cassava sievate E- Pap residue sievate.

4 Conclusion

Small ruminant producers in the study area adhered to some ethical practices in their daily production practices. Most of them fed their animals with supplements twice daily. They treated sick animals through veterinarian services. They transported their animals for sale mostly in the morning. All the respondents offered water to their animals.

However, most of the respondents did not offer water to their animals *ad-libitum*. Also, many offered water from any available source. Some of the respondent did not provide housing facilities for their animals; exposing them to all forms of hazards.

Proximate composition of commonly fed supplements in the study area revealed that all the supplements had low fibre level; an important characteristic of a good supplement. Some were adequate in crude protein (bean chaff) and constraint in other nutrients (minerals). In the case of mineral composition, banana peel exhibited the highest mineral component except zinc.

5 Recommendation

The analysed supplements should be fed to the animals to ascertain intake and performance of the animals on these supplements. They can be individually fed or mixed on a graded level to ascertain the best option. The supplements can be recommended for small ruminant production.

References

Adekoya, A.E., Ajayi, M.A.: "An assessment of farmers" awareness and practice of land management techniques in Iddo LGA of Oyo State. J. Agric. Extension **1**(1), 98–104 (2000)

Ajayi, F.T., Salami, W.A., Lawal, B.O., Taiwo, A.A.: Rearing sheep and goats and indigenous knowledge practices. Ibadan. Tunmid Printeronic, p. 52 (2009)

Anosike, F.U., Naanpose, C.D., Rekwot, G.Z., Sani, A., Owoshagba, O.B., Madziga, I.I.: Challenges of small-holder poultry farmers in Chikun local government area of Kaduna State, Nigeria. In: Proceedings of the 20th Annual Conference of the Animal Science Association of Nigeria, 6–10 September 2015, pp. 302–306. University Ibadan, Oyo State (2015)

AOAC: Official Methods of Analysis, 21st edn. Association of Official Analytical Chemist, Arlington (2000)

Bakshi, M.P.S., Wadhwa, M.: Nutritional evaluation of cannery and fruit wastes as livestock feed. Indian J. Anim. Sci. **83**, 84–89 (2013)

Bamigboye, F.O.: On-farm use of feed-block for small ruminant production in Iwo Local Government Area, Osun state, Nigeria. Ph.D. thesis Submitted to the Department of Animal Science, University of Ibadan (2013)

Bolajoko, M.B., Moses, G.D., Gambari-Bolajoko, K.O., Ifende, V.I., Emenna, P., Bala, A.: Participatory rural appraisal of livestock diseases among the Fulani community of the Barkin Ladi Local Government Area, Plateau State, Nigeria. J. Vet. Med. Health **3**(1), 11–13 (2011)

Brambell Report: Report of the Technical Committee to enquire into the welfare of animals kept under intensive livestock husbandry systems (1965)

Duncan, I.J.H.: Poultry welfare: science or subjectivity? Brit. Poultry Sci. **43**, 643–652 (2002)

El-Yuguda Abdul-Dahiru, B.S.S., Ambali, A.G., Egwu, G.O.: Seroprevalence of Peste des Petits Ruminants among domestic small and large ruminants in semi-arid region of North-Eastern Nigeria. Veterinary World **6**(10), 807–811 (2013)

FAO (Food and Agriculture Organization of the United Nations): The development of village – based sheep production in West Africa: a success story involving women's groups, p. 90 (1988)

Familade, F.O., Babayemi, O.J., Adekoya, A.E.: Characterisation of small ruminant farmers in Iwo Local Government, Osun state, Nigeria. Afr. J. Livestock Extension **9**, 64–67 (2011)

FAWC (Farm Animal Welfare Council): Report on priorities for animal welfare research and development. FAWC, Tolworth Tower, Surbiton, Surrey KT67DX, pp. 1–24 (1993)

Federal Ministry of Agriculture (FMA): National programme on food security, FMA and WR, Abuja (2008)

Fraser, A.F., Broom, D.M.: Farm Animal Behaviour and Welfare, 3rd edn. Bailliere Tindall, London (1990)

Fraser, A.F., Broom, D.B.: Farm Animal Behavior and Welfare. CAB International, London (1997)

Greenberg, M.E., Greene, L.A., Ziff, E.B.: Nerve growth factor and epidermal growth factor induce rapid transient changes in proto-oncogene transcription in PC12 cells. J. Biol. Chem. **260**(26), 14101–14110 (1985)

Kuponiyi, F.A., Sodeinde, F.G.: Agricultural information needs of small-scale livestock farmers in Saki Agricultural zone of Oyo state, Nigeria. In: Proceedings of the 10th Annual Conference of the Animal Science, pp. 328–330. Association of Nigeria held at University of Ado-Ekiti, Ekiti state (2005)

Millman, S.T., Duncan, I.J.H., Stauffacher, M., Stookey, J.M.: The impact of applied ethologists and the international society for applied ethology in improving animal welfare applied to animal behaviour. Sci. **86**, 299–311 (2004)

Nutrimin: Mineral Requirements for Ruminants (2016). http://nutrimin.com.au › mineral-requirements-ruminants

Nwokoro, S.O., Vaikosen, S.E., Bamgbose, A.M.: Nutrient composition of cassava offals and cassava sievates collected from locations in Edo state, Nigeria. Pakistan J. Nutr. **4**(4), 262–264 (2005)

Odendaal, J.S.J.: Animal welfare in practice. Sci. **5**, 93–99 (1998)

Okah, U.: Effect of dietary replacement of maize with maize processing waste on the performance of starter broiler chicks. In: Proceedings of 9th Annual Conference on Animal Science. Association of Nigeria (ASAN), 13–16 September 2004, pp. 2–5. Ebonyi State University, Abakaliki, Nigeria (2004)

Oladunjoye, I.O., Ojebiyi, O.O., Odunsi, A.A.: Performance characteristics and egg quality of laying chicken fed Lye treated cassava peel meal. In: Proc. of the Annual Conference of ASAN, 12–19 September 2008, pp. 355–357. ABU, Zaria (2008)

Olafadehan, O.A., Adewumi, M.K.: Livestock management and production system of agropastoralists in the derived savanna of South-west Nigeria, tropical and subtropical agroecosystems. Asian J. Agric. Food Sci. **12**, 685–691 (2010)

SPORE Magazine: Small ruminants – herds and hides. Wagenigen, The Netherlands, CTA. No. 147, p. 20 (2010)

Webster, J.: Animal Welfare - A Cool Eye Towards Eden. Blackwell Science, Oxford (1994)

Woma, T.Y., et al.: Co-circulation of peste-des-petits-ruminants virus Asian lineage IV with lineage II in Nigeria. Transbound. Emerg. Dis. **63**(3), 235–242 (2016)

Challenges Facing People with Disabilities in Acquiring Equitable Employment in Small and Medium Enterprises in Tanzania

G. J. Mushi(✉), A. P. Athuman(✉), and E. J. Munishi(✉)

Department of Business Administration, College of Business Education, Dar es Salaam, Tanzania
gjeremiah16@gmail.com, amanielathuman2022@gmail.com, e.munishi78@gmail.com

Abstract. Purpose: The purpose of this study is to examine the challenges facing people with disabilities in acquiring equitable employment in small and medium enterprises in the urban context.

Design/Methodology/Approach: The study employed mixed methods research design and data were collected using a questionnaire, interview and focus group discussion. Quantitative and qualitative data were analysed using SPSS, and MAXQDA software respectively.

Findings: Findings show that challenges facing people with disabilities in acquiring equitable employment in urban-based small and medium enterprises include a lack of awareness about job opportunities and labour laws, negative perception by the community, stigma by employers and employees, financial constraints that limit them to apply for jobs, poor job application skills among others.

Implications/Research Limitation: This study focuses on urban settings notably Dar as Salaam owing to the heterogeneous nature of SMEs and employment opportunities. However, generalization of findings to another urban setting may not be possible.

Practical Implication: The study recommends improvement of policy-related labour policies and more especially equitable employment for disabled people. It is also a good source of policy advocacy material on equal employment in Tanzania.

Originality/Value: This study addresses the problem of stigma to equal employment of people with disability. The study recommends that relevant stakeholders should promote equal employment opportunities for both abled and disabled people.

Keywords: Disabilities · Employment · Equitable · SME's · Tanzania

C. Aigbavboa et al. (Eds.): ARCA 2022, *Sustainable Education and Development – Sustainable Industrialization and Innovation*, pp. 902–912, 2023.
https://doi.org/10.1007/978-3-031-25998-2_69

1 Introduction

Tanzania and other countries worldwide are responsible for ensuring employment opportunities to their citizens in equitable manner regardless of a person's condition. However, unemployment is increasingly becoming a global challenge and more especially among the people with disability owing to many countries' inability to provide enough employment opportunities to their citizens.

A recent report on the world inflation and unemployment on the year 2019–2020 indicates that the first five months of the year 2020 recorded an increase of unemployment by the average of 7.5% monthly worldwide and that this situation was further fuelled by the Covid 19 pandemic. On the other hand, the world economy in the year 2019 experienced an average unemployment rate of 8% (World Bank 2020).

The situation is worse to marginalized groups of the society and especially to people with disabilities. Studies indicate that, people with disabilities are significantly under-represented in the labour market when compared to employed social groups (Colella and Bruvere 2011; Vornholt et al. 2018). Disability refers to the loss or limitation of opportunities to take part in the normal life of the community on an equal level with others due to temporary or permanent physical, mental, or social barriers (Ntamanwa 2015).World Health Organization (2011), confirms that around 15 percent of the world population is living with different kinds of disabilities. Disabled people represent the biggest definable disadvantages group of people in the world while (ILO 2007) ascertains that over 600 million people worldwide to be having physical, sensory, mental or intellectual impairment of one form or another. People with disabilities can be found in every country with over two thirds of them living in developing world (ILO 2007).

In 2020, research in the United States of America (USA) indicated that, unemployment rate for persons with disability stood at 12.6%, having increased by 5.3% from the previous year. Their jobless rate continued to be much higher than the rate for those without a disability. The unemployment rate for persons without a disability increased by 4.4% points to 7.9% in 2020 (US Bureau of Labour Statistics).

Recognizing the value of work in people's lives various initiatives have been undertaken worldwide towards ensuring equitable job opportunities to all and more especially equal representation of the people with disabilities in the labour force. Vocational rehabilitation professionals have consistently advocated for employment as a fundamental human right. Yet, the employment rate of people with disabilities remains remarkably low compared to the general population (UN 2009).

Tanzania enacted the Disabled person, employment Act, No. 2 in 1982. This Act established a quota system requiring that two per cent of the workforce in companies with more than 50 employees must be people with disabilities. It also established the national advisory council, the role of which was to advise the minister responsible for the social welfare of people with disabilities (ILO 2009). Similarly, disabled persons employment Act, 2010, impose similar requirement with respect to the employment of people with disabilities. Specifically, Article 31 require employers to hire and maintain the employment to people with disabilities and establish workforce quota under which every employer with the workforce of 20 or more individuals must employ person with disability at the rate at least 3% of the employee's total workforce.

The country is facing growing rate of unemployed people due to the low level of economic development and rapid increase of population which does not match with few available job opportunities in public sector (Kinyondo and Ricardo 2018).

The history of Tanzania indicates that before 1985 private sector were prohibited following ideological stand of the country. The situation made the government to be the only source of employment. Due to economic hardship the country was forced by the prevailing situation to request financial assistance from Breetonwood institutions where by one of the conditions given was to liberalize the economy and promote the development of private sector of which SME's is among them (Campbell and Stein 2019).

In many countries the private sector has been the main driver of the economy as well as employment. Private sectors including Small and Medium Enterprises' (SMEs) has provided employment with a proportional of 96.5% of the total employment in 2014 with a relative same share of employment in 2018 (95.7%) (NBS 2019).

Recent employment data from 2016 show that 64.9% of Tanzania's employees were in the private sector in that year, whereas the public sector accounted for 35.1% of total employment (NBS 2018a). These data suggest that, as in many countries, the private sector in Tanzania is the main employer of the country's workforce. Interestingly, while there was an increase of workforce in the private sector, the rate of employment in the public sector dropped from 36.7% in 2015 to 35% in 2016 (NBS 2018b). This suggests that the ability of the government to create employment in the public sector is declining slightly, and employment growth in the private sector remains minimal. This further suggests that there is a need for the government to assess its employment policies and determine how the two sectors can create more equal opportunities to employment that touches the marginalized group (here means women and people with disabilities) (Mgaiwa 2021).

All in all, small and medium sized enterprise (SME) accounts for over 95% of firms, 60–70% of employment and generate large share of new jobs around the globe, majority of SME are small business (<50 employees). Most SME jobs are in the service sector, which account for two thirds of economic activity and employment in OECD countries (OECD 2000). However, many employers in SME's consider it not financially viable to retain employees who are disabled.

Tanzania and other countries worldwide are responsible for ensuring employment opportunities to their citizens in equitable manner regardless of a person condition. However, unemployment is increasingly becoming a global challenge and more especially among the people with disability owing to many countries inability to provide enough employment opportunities to their citizens. Private sector and more specifically the Small and Medium Enterprises significantly contribute to employment opportunities to all groups of people in the community and especially the marginalized (women and people with disability). Recognizing the value of work in people's lives various initiatives have been undertaken worldwide towards ensuring equitable job opportunities to all and more especially equal representation of the people with disabilities in the labour force. However, unemployment of the people with disabilities more especially in the private sector still continues at an alarming level. This suggests that a study required to inquire on the role of private sector and more especially the SMEs', in ensuring equitable employment among the people with disability. The overall objective is to explore challenges facing people with disabilities in acquiring equitable employment in small and medium enterprises. Specifically, the study addresses the following objectives: to assess challenges facing people with disabilities to acquire equitable employment in SMEs'; and to examine the ability of people with disabilities to cope with unemployment challenges in SMEs'.

2 Empirical Literature

A number of existing studies have attempted to establish constraints to employment among the disabled (UN report 2021). In South Africa the real obstacle lies with employers. Many are still hesitant to take on employees with disabilities because they believe they may create problems in the workplace. There is also an assumption that this type of appointment will incur cost as the workplace is changed to become disability friendly (UN report 2021).

In East Africa Region, nations like Tanzania are a useful case for examining the concept of disability and the difficulties in translating policy to practice because of long-standing commitment to disability. The country has improved accountability mechanism in the persons with disabilities Act of 2010 and increasing consultation with development planning officers (Alderley 2010).

The United Convention on the Rights of Persons with Disabilities (UNCRPD) which was adopted by UN General Assembly in 2006 is the first human rights convention of the 21st century. The agreement provides a comprehensive framework for protecting disability rights and emphasizes the need for states to create an enabling environment to promote full inclusion and participation of people with disabilities (PWDS) (Njue and Mburugu 2018).

In Asia for instance, India is a signatory of the convention with a large population of disabled people living in poverty, lack access to basic services and face huge barriers to participate in different activities in the society, employment being one of them (Njue and Mburugu 2018).

However, in Australia, Small and medium businesses represent a key source of employment for the communities, where 90% of the populations are employed in small to medium sized businesses. There has been limited research conducted with small and medium business owners in Australia in relation to positive employment outcomes for people with disabilities, despite being one of the largest employer groups (ILO 2007).

Moreover, Africa is represented by South Africa, where according to the Commission for Employment and Equity Annual Report 2007/08 indicated that out of 2,030,837 working population with disabilities, only 10,700 are employed by large employers which 0.52% of the total number. The report indicated further that the disadvantaged position of persons with disabilities would be diverse socio-economic and social cultural factors particularly low levels of education, discrimination in the labour market and negative attitudes of those they live amongst (UN 2010). However, employers find a challenge in hiring people with disability as per Lengnick (2008) showed that most employers are not very proactive in hiring people with disabilities and that most employers hold stereotypical beliefs not supported by research evidence. Gewrtuz et al. (2016)

from the literature and consultations found different barriers towards hiring people with disabilities. Among them included regulation versus practice, stigma, disclosure, accommodations, relationship building and use of disability organizations, information and support to employers and hiring practices that invite people with disabilities.

According to Amon (2014) factors which contribute to the exclusion of people living with disabilities from employment in mainland Tanzania include; stigma against the disabled as well as community's negative attitudes towards them. Another factor related to the entire education system that excludes the people with disabilities. This is further confirmed by Oliver (1998) who state states that the disabled people are excluded in ordinary education, public buildings, public transport and other things that may be taken for granted present obstacles to equitable employment.

Moreover, it has been established that, employments of people with disability was highly affected by the Covid 19 pandemic as affirmed by the ESRF report Covid 19 outbreak affected key sectors that employ the majority of the poor, the youth and people living with disability (ESRF 2020).

Gould et al. (2021) on their study, building sustaining and growing: disability inclusion in the society, found that there is limited information on what the organizations are doing to support disability inclusion in the workplace. Erickson et al. (2013) on employment environment, policies and practices regarding the employment of persons with disability, realized that policies are in place that favours' people with disabilities but in practice it is not well implemented.

A study by Tambwe et al. (2022) revealed that most of the respondents (67%) were not aware of the international and national laws addressing economic inequality among PWD. Based on the Human Development approach, the main causes of the persistence of economic inequalities despite the laws included inadequate empowerment, social relations and community well-being. Vornholt et al. (2013) noted that the attitude of co-workers, the employer and organizations has an impact on people with disabilities at work.

Meanwhile Sally et al. (2018) on employers' perspective of including young people with disabilities in the workforce, disability disclosure and providing accommodations realized that most employers encouraged youth with disabilities to disclose their condition and emphasized the importance of building trust and rapport. Moreover, employers created an inclusive workplace culture, diversity training, addressing stigma and discrimination, open communication, mentoring and advocacy. Employers in the Information Technology sector supported the employment of people with disabilities and not only indicated a willingness to hire qualified applicants with disabilities, but respondents also believed individuals with disabilities were able to perform as well as people without disabilities (Greenan et al. 2003).

Lindsey et al. (2018) found the benefit of hiring people with disabilities. Among the benefits includes, improvements in profitability (e.g., profits and cost-effectiveness, turnover and retention, reliability and punctuality, employee loyalty, company image), competitive advantage (e.g., diverse customers, customer loyalty and satisfaction, innovation, productivity, work ethics and safety), inclusive work culture, and ability awareness. Secondary benefits for people with disabilities included improved quality of life and income, enhanced self-confidence, expanded social network, and a sense of community.

Huang and Chen (2015) on their study in Taiwan identified four main reasons that led to hiring people with disability, including personal experience relating to people with disabilities, economic concerns, charitable perspectives, and policy implications. Although the employers were highly willing to collaborate with vocational rehabilitation systems, their needs for services rendered differed in the distinct employment processes. Employers expressed greater concern about the employability of applicants with disabilities during the recruitment and selection process than during the placement and accommodation stages.

3 Methodology

The study was conducted at Dar es Salaam Regions in Tanzania mainland and it included two organization, Jubilee Insurance Company, Tujijenge microfinance and FINCA microfinance and a list of entrepreneurs obtained from the municipal. The reason for the choice of the organization were the nature of services offered, that if it provides cover for disability benefit it is assumed that they are also employing disabled people.

A parallel convergent mixed research design was applied. Both qualitative and quantitative methods were used to achieve a better understanding concerning challenges facing people with disabilities in acquiring equitable employment in small and medium enterprises in Tanzania. We choose to use both qualitative and quantitative methods because they complement each other's weaknesses and strengths (Creswell 2014).

Two types of sampling techniques were applied; purposive sampling for the top management representatives' respondents that we know they can provide required information and simple random sampling were applied in selection of other employees in their respective organizations and entrepreneurs. The study included Finca 33, Jubilee 50, Tujijenge 10 and 50 from SMEs in the municipal. All these included a total of 143 participants, where 10 top management were included plus other employees from each organization. The data was collected by using structured questionnaire, interview and documentary review so as to provide room for more explanations to the raised questions. Data cleaning process took place, coding and finally the analysis using SPSS version 23 were used. The descriptive data were analysed by using thematic analysis.

4 Findings and Discussion

Demographic characteristics of respondents.

There were 53.3% of men who responded to the questionnaire and interviews and 46.6 percent women. Among them 27% had primary education, 27% had secondary education and 46 percent had college education. Furthermore, 31.1 percent of the respondent are aged between 20–29, 40% are aged between 30–39, 24.4% are between 40–49 and 4.4% are above 50 years of age. Moreover 66.7% of the information about job existence were obtained through relatives, while 33.3% got the information through advertising. Above all 57.8% of the respondents had physical disability percent vision disability and 20% had albinism.

A: Challenges facing people with disabilities in acquiring equitable employment
In addressing the research objective about the challenges facing people with disability to acquire equitable employment in SMEs' the study organized the collection of data under major three questions with seven categories each namely; the challenges at work place, challenges at the community level and support obtained from co-workers.

i. *Workplace related challenges*

It was established the first category of challenges to equitable employment among people with disabilities relate to the Workplace related challenges. Such challenges included little payment to people with disability compared to people without disability. The findings reveal that 28.1% of the respondents agreed with the statement while 71.9% of the population disagree that there is little payment to disabled people at work place. Lack of promotion findings also reveals that 22.2% support the statement that there is lack of promotion to people with disability while 77.8% of the respondents did not support the statement. Supportive infrastructure especially toilets and in tall buildings was also a factor impeding people with disability at work place. Also lack of participation in income related issues in the community made them feel discriminated among the group. Lack of education and lack of recognition of the participation of people with disability (Table 1).

Table 1. Challenges at work place

Item	Yes (%)	No (%)
Little payment to people with disability	28.1	71.9
Lack of promotion	22.2	77.8
Supportive infrastructure especially toilets and in tall buildings	28.9	71.1
Lack of participation in income related issues	11.1	88.9
Lack of education	37.8	62.2
Lack of recognition of the participation	35.6	64.4

ii. *Challenges at community level*

Another category of challenges to equitable employment among people with disabilities were that challenges at community level. This includes stigmatization and discrimination where by 78% of the respondents revealed that most of the disabled people face this type of challenges in the community while 22% have not come across with such challenge. Little involvement of people with disability in community activities, from the study 86.7% of the respondents faced this challenge at the community level. Lack of security and low opportunity to participate in economic activities was also a challenge that according to the findings 40% of the respondents were affected at the community level. Negative perception at the community level about people with disability had 42.2%

that it affects people with disability. This result support the study by Amon (2014) Gewrtuz et al. (2016). Negative perception among people with disability and unawareness to disabled people's law had lower percentages of 17.8 and 4.4% respectively (Table 2).

Table 2. Challenges at community level

Item	Yes (%)	No (%)
Stigmatization and discrimination	78	22
Little involvement of people with disability in community activities	86.7	13.3
Lack of security and low opportunity to participate in economic activities	40	60
Negative perception at the community level	42.2	57.8
Negative perception among people with disability	17.8	82.2
Unawareness to disabled people's law	4.4	95.6

iii. *Support obtained from co-workers*

Moreover, the findings revealed that equitable employment among the people with disability was jeopardised by lack of support from co-workers. Accordingly, findings revealed that little support is provided at parking areas and supportive infrastructure at the toilet by indicating 48.9%, reduced workload scored 35.6 percent moral support was a challenge by 40%. Meanwhile most disabled people obtained support on hard work implementation; this was indicated by results of the respondents that 57.8% got support from others. Given more time to perform duties and provided with machines to perform duties were rated low by respondents at 26.7 and 8.9% respectively (Table 3).

Table 3. Support from co-workers

Item	Yes (%)	No (%)
Parking areas and supportive infrastructure at the toilet	48.9	51.1
Reduced workload	35.6	64.4
Moral support	40	60
Support on hard work implementation	57.8	42.2
Given more time to perform duties	26.7	73.3
Provided with machines to perform duties disability	8.9	91.1

B: Strategies employed by the people with disabilities to cope with unemployment challenges

The results revealed that self-confidence has made people with disability to cope with inequality challenges by 51.1% which support the study by Tambwe et al. (2020). Another factor for coping was the ability to participate in decision making. This enabled people with disability to cope up by 48.9%. Other factors for coping were, efforts to accomplish the assigned job, self-motivation, being responsible, capacity building on job related skills, 35.6, 25.6 and 35.6% respectively (Table 4).

Table 4. Coping strategies at workplace

Item	Yes (%)	No (%)
Self-confidence	51.1	48.9
Ability to participate in decision making	48.9	51.1
Efforts on assigned job,	35.6	64.4
Self-motivated and being responsible	25.6	74.4
Capacity building on job performed	35.6	64.4

However, awareness campaign and improvement of infrastructure to people with disability who cannot cope with inequality challenges were among the strategies that were suggested by respondents with 88.9% each. Life skills and entrepreneurship education is another factor that was supported by majority of the respondents, by 42.2%, however provision of soft loans, community campaign and special areas for business and workings obtained lower rate of 13.3, 8.9 and 6.7% respectively (Table 5).

Table 5. CFactors to promote equitable employment

Item	Yes (%)	No (%)
Awareness campaign	88.9	11.1
Improvement of infrastructure	88.9	11.1
Life skills and entrepreneurship education	42.2	57.8
Special areas for business and workings	6.7	93.3
Provision of soft loans	8.9	91.1
Community campaign	13.3	86.7

5 Conclusion

As a conclusion it can said that, challenges facing people with disabilities in acquiring equitable employment are three-fold. The first category is workplace related challenges, that included little payment to people with disability, lack of promotion, Supportive infrastructure especially toilets and in tall buildings, lack of participation in income related issues, Lack of education and lack of recognition of the participation. The second category are the challenges at community level that involve Stigmatization and discrimination, little involvement of people with disability in community activities, Lack of security and low opportunity to participate in economic activities, Negative perception at the community level, Negative perception among people with disability and lack of awareness on disabled people's law. The third category, was lack of support obtained from co-workers. This was characterised by lack of or inadequate parking areas and supportive infrastructure at the toilet, reduced workload, inadequate moral support, lack of support on hard work implementation as well as inadequate time to perform duties.

The disabled coped with the workplace related challenge through self-confidence, ability to participate in decision making, efforts on assigned job, self-motivation and being responsible and capacity building on job performed. Accordingly in order to promote equitable employment among the people with disability there should be awareness campaign, Improvement of infrastructure, Life skills and entrepreneurship education, Special areas for business and workings, Provision of soft loans as well as mass community campaign on the importance of equitable employment among the people with disability.

References

Amon J.B.: The exclusion of people living with disabilities from employment in Mainland Tanzania. Master's thesis, The Open University of Tanzania (2014)

Burke, J., Bezyak, J., Fraser, R., Pete, J., Ditchman, N., Chan, F.: Employers' attitudes towards hiring and retaining people with disabilities: a review of the literature. Aust. J. Rehabil. Counsel. **19**(1), 21–38 (2013). https://doi.org/10.1017/jrc.2013.2

Creswell, J.W.: A Concise Introduction to Mixed Methods Research. SAGE Publications, New York (2014)

Erickson, W.A., Von Schrader, S., Bruyère, M.S.: The employment environment: employer perspectives, policies, and practices regarding the employment of persons with disabilities. Rehabil. Counsel. Bull. **57**, 195–208 (2013). https://doi.org/10.1177/0034355213509841

ESRF report (2020)

Gewurtz, R.E., Samantha, L., Danielle, S.: Hiring people with disabilities: a scoping review. Work **54**, 135–148 (2016)

Gould, R., Mullin, C., Parker Harris, S., Jones, R.: Building, sustaining and growing: disability inclusion in business. Equality Diversity Inclusion **41**, 418–434 (2021). https://doi.org/10.1108/EDI-06-2020-0156

Greenan, J.P., Wu, M., Black, L.E.: Perspectives on employing individuals with special needs. J. Technol. Stud. **28** (2002)

Gudex, C., Lafortune, G.: An inventory of Health and Disability-Related Survey in OECD Countries: OECD Labor Markets and social policy Occasional Papers No.44 OECD Publishing Paris. OECD iLibrary (2000)

Huang, I.-C., Chen, K.R.: Employing people with disabilities in the Taiwanese workplace: employers' perceptions and considerations. Rehabil. Counsel. Bull. **59**(1), 43–54 (2015)

ILO Report (2004, 2009)

ILO. Facts on Disability in the world of work-Geneva 22 Switzerland (2007)

Kevin, M., Jenny, C., Jessica, Z., Greig, W.: Employer engagement in disability employment; A missing link for small to medium organizations–a review of literature IOS (2018)

Lengnick-Hall, M.L.: People with Disabilities are an Untapped Human Resource, Wiley Online Library, New York (2008)

Lindsay, S., Cagliostro, E., Albarico, M., Mortaji, N., Karon, L.: A systematic review of the benefits of hiring people with disabilities. J. Occup. Rehabil. **28**(4), 634–655 (2018). https://doi.org/10.1007/s10926-018-9756-z

Murfitt, K., et al.: Employer engagement in disability employment: a missing link for small to medium organizations–a review of the literature, pp. 417–431 (2018)
Mgaiwa, J.S.: Fostering Graduate Employability: Rethinking Tanzania's University Practices, pp. 1–14. SAGE Open, New York (2021). https://doi.org/10.1177/21582440211006709journals.sagepub.com/home/sgo
Ntamanwa, F.: Factors leading to low employment rate of people with physical disability in Tanzania: a case of temeke municipal council. Dar es Salaam –Tanzania. Dissertation submitted to Open University of Tanzania (2015)
Oliver, M.: Theories of Disability in Health Practice and Research. University of Greenwich, Eltham London (BMJ, 1998 NOV 21:317 7071:1446-1449) (1998)
Lindsay, S., Cagliostro, E., Leck, J., Shen, W., Stinson, J.: Employers' pespectives of including young people with disability in the workforce, disability disclosure and providing accommodations. J. Voc. Rehabil. **50**(2), 141-156 (2019). https://doi.org/10.323-jvr-180996
Tanzania Employment Act. The Persons with Disabilities Act, 2010 (Act No.9 of 2010)-United Republic of Tanzania (2010). https://www.ilo.org/dyn/natlex/natlex4.detail?p_isn=86525&p_lang=en
United Nation, Report: United Nations (2004)
UN. The United Nations Convention on the right and dignities for persons with disability: A Panacea for ending disability discrimination? **3**(3), 266–285. Elsevier (2009)
United Republic of Tanzania. Tanzania in Figures- National Bureau of Statistics (2019)
United Republic of Tanzania. Eliminate stigma and Discrimination-National Bureau of Statistics (2018a)
United Republic of Tanzania. Women and Girls with Disability-National Bureau of Statistics (2018b)
Vornholt, K., Uitdewilligen, S., Nijhuis, F.J.N.: Factors affecting the acceptance of people with disabilities at work: a literature review. J. Occup. Rehabil. **23**(4), 463–475 (2013). https://doi.org/10.1007/s10926-013-9426-0
Vornholt, K., et al.: Disability and employment-overview and highlights. Eur. J. Work Organ. Psychol. **27**(1), 40–45 (2018). https://doi.org/10.1080/1359432X.2017.1387536
World Bank. Poverty and Equity Data Portal Tanzania (2020)
WHO. World Report on Disability-WHO Library Cataloging-in-Publication data (2011)

Competence of Traditional Automobile Practitioners in Maintenance of Automatic Transmission Drives and Implications for Transportation Planning in Ghana

G. Boafo[1], R. S. Wireko-Gyebi[2], S. K. Nkrumah[3], and F. Davis[4](✉)

[1] Department of Mechanical Engineering, Cape Coast Technical University, Cape Coast, Ghana

[2] Department of Planning and Sustainability, University of Energy and Natural Resources, Sunyani, Ghana

rejoice.wireko-gyebi@uenr.edu.gh

[3] Department of Supply Chain and Information Systems, Kwame Nkrumah University of Science and Technology, Kumasi, Ghana

sknkrumah.ksb@knust.edu.gh

[4] Department of Mechanical Engineering, Kwame Nkrumah University of Science and Technology, Kumasi, Ghana

fkdav@yahoo.com

Abstract. Purpose: This study investigates the extent of patronage of Automatic Transmission Drives (ATDs) and assesses the competence of automobile maintenance practitioners in the diagnosis and repair of automatic transmission vehicles in Ghana.

Design/Methodology/Approach: A combination of research methods comprising interviews, questionnaire and face-to-face dialogue were employed in collecting data from automatic transmission repair and servicing garages. A total of 1000 questionnaires were administered to four automobile stakeholders, with 983 responses received from vehicle operators (536), wayside mechanics (202), used vehicle importers (232) and new car dealers (13) in five cities in Ghana specifically, Takoradi, Kumasi, Tema, Tamale and Accra. The quota sampling technique was used in sampling respondents from the general population in these cities. Analysis of variance, chi-square and basic charts were used to analyse and interpret trends in the data.

Findings: The study revealed that the degree of patronage of ATDs showed a linear relation of $y = 52.239x - 105743$ over the period 2007 to 2019, an indication that the number of ATDs sold has been increasing at an average rate of 52 ATDs per year for each garage over the period. Vehicle operators classified 1.9% of wayside mechanics and 59.5% of dealers, respectively, as competent in repair of ATDs. Similarly, 66% of vehicle operators patronize wayside mechanics with the claim that their service charges are affordable, while 3% indicated that they can afford the charges of the dealers.

Implications/Research Limitations: The study focus on the challenges associated with the maintenance of automatic transmission drives in Ghana.

Practical Implications: The configurations of the electronics and mechanical components of the automatic transmission vehicles of different manufacturers are

C. Aigbavboa et al. (Eds.): ARCA 2022, *Sustainable Education and Development – Sustainable Industrialization and Innovation*, pp. 913–926, 2023.
https://doi.org/10.1007/978-3-031-25998-2_70

not standardized; however, there are basic principles in their operations which can be taught to the wayside mechanics.

Social Implications: Since the wayside mechanics are unable to detect and deal with complex problems associated with automatic transmission vehicles, the government should institute regular problem-based training and refresher courses for the wayside mechanics through engagement with the appropriate educational institutions.

Originality/Value: Although there have been some previous studies covering the automatic transmission drives in Ghana, they have not covered the competence of maintenance practitioners which is captured in this study.

Keyword: Automatic · Garages · Transmission drives · Vehicle importers · Vehicle operators

1 Introduction

An automobile may be defined as a driven vehicle predominantly powered by an internal combustion engine that is used in carrying persons and items from one destination to another (Akayeti 2014). Automobiles include motorcycles, tricycles, trucks, buses, vans, coaches, among others. When a mechanical device is said to be maintained, the notion is to keep it in a good and efficient state. Vehicle maintenance consists of a practice where an automobile is serviced on a consistent basis to avert a major breakdown. It implies that an automobile vehicle will last longer and function very well if the user follows the vehicle maintenance schedule recommended by the manufacturer (Chand et al. 2021; Karlaftis 2011; Akinola 2005; Jbili et al. 2018). The indigenous automobile repair and maintenance garages in Ghana have played a major role in the socio-economic development of the country, and can become an important contributor to the development of the country by adding value to the economy (Mitropoulos and Prevedouros 2016; Lee 2011). Modern vehicles are equipped with complex computer-controlled electrical systems while older vehicles function with simple wiring and little or no electric components (de Almeida Correia and Menendez 2017; Hirsch et al. 2020; Santini and Van Gelder 2017; Merriman et al. 2021). The improved complexity and expanded variety of makes and models have created a need for timely access to relevant, complete, and accurate information in order to carry out maintenance, diagnosis and repair activities (Mitropoulos and Prevedouros 2016).

Maintenance management is gradually moving away from the traditional skill-based management discipline which depended on experience, guts and luck, to a more modernized form and the use of sophisticated and complex machines and equipment (Mahesh and Ram 2010). Technological developments in the automobile industry have led to the modification of most vehicle systems, thus making their use very easy; such vehicle system is the transmission system which drives most vehicles today (Amjad et al. 2010; Chan 2017; Overtoom et al. 2020; Walker and Marchau 2017).

The automobile industry in Ghana in recent years is progressively experiencing an apparent increase in the number of automatic transmission vehicles (Abu-Eisheh and Mannering 2002). In spite of the number of automatic cars imported into the country, only few standard auto service workshops such as Toyota Ghana Company Limited, Japan Motors Company Limited, Rana Motors and Mechanical Lloyd are capable of offering quality maintenance and repairs for these cars. The increase in automatic transmission vehicles calls for an increase in the number of service garages to provide maintenance services. Several authors including Andoh et al. (2013); Davis et al. (2018a) and Davis et al. (2018b) have researched into some of the challenges faced by the automobile industries in Ghana, indicating that vehicular repairs have become increasingly sophisticated; new equipment, software and techniques are needed to determine the root causes of failure for many components in modern vehicles. Although automatic transmission vehicles come with user convenience, potential issues regarding the availability of skilled mechanics in maintenance and repair need to be addressed. Automobile maintenance practitioners in developing countries like Ghana are not well-equipped with the skill and knowledge in repairing faults associated with automatic transmission drives and have limited means of enhancing their servicing capabilities (Davis et al. 2018a; Sekyere et al. 2018). It is prudent to investigate the challenges associated with the maintenance of automatic transmission drives in Ghana to help mechanics as well as the users of these vehicles in servicing them.The objectives of the study are to ascertain the extent of patronage of automatic transmission drives and the challenges associated with them and to investigate the level of expertise of traditional automobile maintenance practitioners in servicing automatic transmission drives.

2 Research Methodology

2.1 Study Area

The study seeks to assess and report on the traditional practices used by mechanics in servicing automatic transmission drives in Ghana. The study area comprised five cities in Ghana namely Accra, Tema, Kumasi, Tamale and Takoradi. These cities are the main business centers of the nation where there are a lot of transportation-related activities. Tema and Takoradi are Port cities which serve as entry points for all new vehicles that come into the country. Accra and Kumasi are also the two most populated cities considered as the most vibrant economic zones in the country. Tamale is the most populated and industrious city in the Northern Part of Ghana. Due to large number of vehicles in the selected cities, they host many transport business organizations and mechanics.

The total registered vehicles in Ghana as of 2010 was approximately one million with concentration in the four biggest cities, namely Accra, Kumasi, Tema, and Takoradi (Hesse and Ofosu 2014). The four cities mentioned as well as Tamale are urban areas where this study sampled views. These cities are a rational representation of the urban geographical locations in Ghana and are among the top ten cities with high population, being major centers of economic activities in Ghana (Ghana Statistical Service 2015). Hence, the views of the respondents from these cities would suitably represent the whole population of Ghana.

2.2 Sampling Technique

The target population for the study comprised wayside mechanics, used vehicle importers, vehicle operators and new car dealers above 18 years of age and found within the study area at the time of the study. These categories of people were used for the study because they possess the information necessary for achieving the study objectives. The wayside mechanics have been major players in the automobile transportation sector of the Ghanaian economy. They have been offering services in the repair and maintenance of almost every component of automobiles including the power transmission train of the non-automatic drives. It is well known that some of the master mechanics have specialized in the repair and maintenance of gearboxes. It is therefore, prudent to engage these wayside mechanics to ascertain their competencies or otherwise with the advent of automatic transmission drives.

Vehicle operators falling into two categories, commercial vehicle drivers and private vehicle users, were targeted for the reason that, as the users of the services of wayside mechanics, they are in position to assess the quality of the service they receive from wayside mechanics in the country. A very objective and scientific assessment of the performance of a transmission drive after repair or maintenance would have required test-running the components to failure. However, this would have occupied a colossal amount of time and several design and manufacturing parameters would have to be investigated. This research did not follow such an approach and thus using the assessment of both commercial and private vehicle drivers to assess the performance of a transmission drive after repair or maintenance by wayside mechanics in the country, was seen as a credible and valid alternative assessment of the performance of wayside mechanics in their repair or maintenance of automatic transmission drives in the country. Also, new car dealers as part of the target population was justified as they are known to possess well-organized workshops for servicing vehicles. Further, the importers of used vehicles could supply valuable information in relation to the initial state of the drives before their use in Ghana.

The researchers adopted the non-probability sampling method of quota sampling in the collection of data from the respondents. The use of quota sampling ensures that there was the selection of respondents to represent the various subgroups of the target population, namely, vehicle operators, car dealers, vehicle importers, and wayside mechanics. The selection of the respondents from the various subgroups of the target population was based on which respondents were readily available and were willing to participate in the study, and had the relevant information in relation to the study (See Kumar 2014; Leedy and Ormrod 2015).

In relation to the sample size, the researchers targeted 1000 respondents f with the view that such a sample would allow the researchers to get rich and varied data to compensate for the use of a nonprobability sampling method in selecting samples for the study. The researchers also hoped that a a target of a sample size of 1000 respondents would allow the researchers to approach a saturation point in data collection where, as researchers, we are able to get adequate information from the respondents in relation to the study (See Kumar 2014; Leedy and Ormrod 2015; Schindler 2022). At the end of the survey, a total of 536 vehicle operators, 13 new car dealers, 232 used vehicle importers and 202 wayside mechanics were involved in the study, totalling 983 respondents.

2.3 Data Collection

Questionnaires were used for the collection of data. Four (4) sets of questionnaires were designed for each group of respondents (wayside mechanics, vehicle operators, used vehicle importers and new car dealers). The questionnaires were designed based on Delmar's automotive test which is a sub-section of the Automotive Service Excellence (ASE) standard (Delmar 2001). After the data collection, questionnaires were cleaned and coded for data entry. The Statistical Package for Social Sciences (SPSS) version 22 was used to analyse the data. Descriptive and inferential statistics were used to analyse the data collected, while tables and Figures were used to present the data. Additionally, the management of the transport department of the Kwame Nkrumah University of Science and Technology (KNUST) was purposively selected to get their views on the subject matter. KNUST is known to be the leading engineering and technology university in Ghana and the transport department of KNUST has a wealth of experience and knowledge in automobile engineering.

3 Results and Discussion

3.1 Profile of Respondents

The new car dealers (dealers) who participated in the study had technicians who are usually manufacturer-trained and specialize in performing repairs on special vehicle makes in their shops. They are known for providing high quality service. They import new vehicles and vehicle components that are required for repairs of vehicles and are more likely to be on hand at a dealership. They adhere strictly to the manual instructions of their parent companies. The dealers who participated in the study are Toyota Ghana Company Limited, Japan Motors, CFAO, Stallion Motors Ghana Limited (Hyundai, Honda, Audi), Mantrac Ghana Limited, Mechanical Lloyd, Joe Auto Company, Rana Motors, Delta Equipment Limited (Africa Motors Division), Silver Star and Alliance Motors Limited. Approximately 50% of the dealers had a master's degree; 33.3% had a diploma or a bachelor's degree; and 8% each had secondary/technical and basic education.

The wayside mechanics who participated in the study were observed to have low technical educational levels, lack of appropriate maintenance practices and, lack of diagnostic machines. Majority (53%), of the wayside mechanics, had Junior High School/Middle School level education; 27% of them had basic education;, and 20% had Secondary/Technical education. It is evident from the study that majority of mechanics at the various wayside garages in Ghana have lower levels of education compared to the dealers. Additionally, they lacked adequate training and re-training and were most likely unable to use computers and internet facilities for vehicle repairs. This conforms to a finding by Akpakpavi (2015) who indicated that lack of training creates an inability of mechanics to identify parts of modern vehicle engines by their correct technical names and functions. Baidoo and Odum-Awuakye (2015) in their research on the influence of service quality delivery in the small and medium enterprises (SMEs) of Motor Vehicle Repair Service industry in Ghana, concluded that practices of the SMEs could be improved by giving their personnel the needed professional training or retraining.

Out of the 232 used vehicle importers surveyed, a total of 140 (60.3%) had Secondary/Technical education and 91(39.2%) had tertiary level education. Approximately 85% of the vehicle operators who participated in the study were drivers of private cars whilst 15% were commercial drivers. Most of the vehicle operators who took part in the survey (443, representing 82.6%) had tertiary education. This gives an indication that the majority of the respondents are educated enough to have a fair opinion about the subject matter under discussion. Fifty-seven (57, 10.6%) of them had either a basic or secondary level education. Nineteen vehicle (3.5%) operators who took part in the survey had no formal education. Majority of respondents (81.7%) were males, with the remaining being females (18.3%). The distribution for the vehicle operators also showed that 39.9% of the respondents were within the ages of 36 to 40 years; 22.2% within the ages of 46 and 50 years. Additionally, 16.8% of respondents were reported to be between the ages of 31 to 35 years. 11.3% were above 51 years, 6.7% between the ages of 41–45 and 3.0% within the ages of 25–30 years.

3.2 Frequency of ATDs Sold by Dealers and Used Vehicle Importers

A time plot of the average number of ATDs sold from both dealers and used vehicle importers from 2007 to 2019 is shown in Fig. 1. This is an indication of the extent of patronage of ATDs over this period. From the figure, an upsurge in popularity of ATDs can be observed which is in agreement with Akple et al. (2013). However, the plot shows a drop in sales of ATDs between 2009 and 2010 before picking up again after 2010. This drop, according to the respondents is likely as a result of the change in government in 2009 leading to a halt in government procurement of goods and services for the public sector. The extent of patronage of ATDs revealed a linear relation of $y = 52.239x - 105743$, where y is mean of ATDs sold and x is the deviation from the base year, the slope can be taken as the rate of patronage of ATDs showing a uniform sale of ATDs at an annual increasing rate of 52 ATD cars per garage.

3.3 Expertise in Servicing of ATDs

According to the the dealers and wayside mechanics, automatic transmission drive problems mostly reported to their garages include; burnt transmission fluid, leaking transmission fluid, noisy transmission and slips between shifts as shown in Table 1. Meanwhile, the transport department of KNUST revealed that the major problems faced by ATD users in Ghana are: (1) Transmission won't engage or stay in gear; (2) Shifts are delayed or missing gears; (3) Transmission slipping or engine is revving high; (4) Transmission fluid is either low, leaking and or heating; (5) Grinding or shaking sensation in gear; (6) Makes Noises: whining, buzzing, humming or clunking, noisy in neutral; (7) There is burning smell; (8) Car has no power; (9) Check Engine Light or Over Drive Light is on. The variation in what is mentioned by the garages indicates a possibility that mechanics in the garages may have inadequate knowledge about the other problems mentioned by the Transport Department of KNUST.

This section sets out to do two major things: the first is to verify the faults mechanics of the various garages usually repair with ease; second is to check the list of tools mostly used in fixing these faults. The various faults outlined by the mechanics to have been

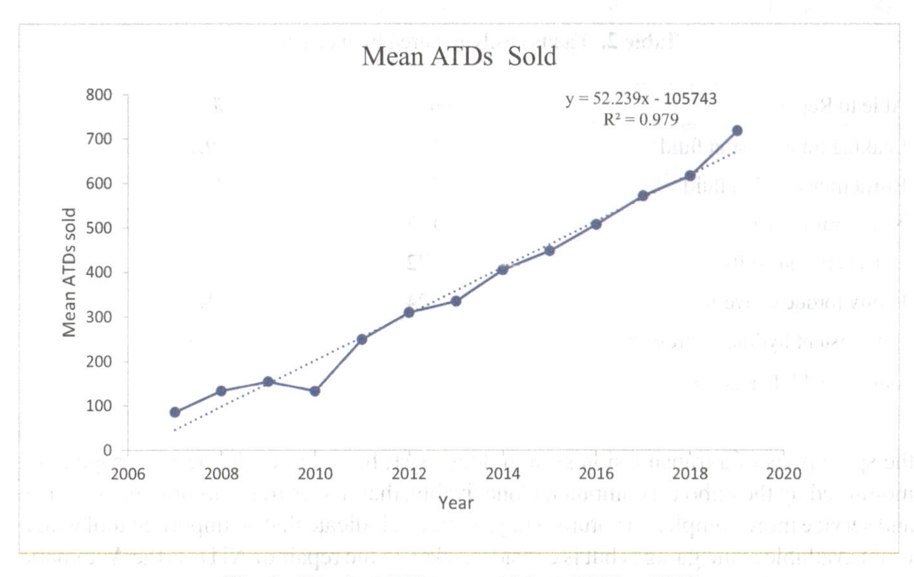

Fig. 1. Trend of ATD cars sold from 2007 to 2019

Table 1. Automatic drive problems

Automatic drive problems	n	%
Burnt transmission fluid	193	95.1
Leaking transmission fluid	182	89.7
Noisy transmission	175	86.2
Slips between shifts	163	80.3
Faulty torque converter	140	69.0
Inconsistent hydraulic pressure	111	54.7

Source: Field data 2019.

solved without difficulty include: leaking transmission fluid, burnt transmission fluid, noisy transmission, as indicated in Table 2.

The most available tools used by the mechanics in fixing ATD problems are: sets of spanners, sockets, screw drivers, pliers and cutters (72.9%); followed by the Transmission Jack and Jack Stands (30.0%). Other tools available are Torque Converter and Gasket scraper (29.6%); Automatic Transmission Fluid (ATF) Changer (29.6%); Anti-Lock Brake system diagnostic tool (18.2%); Wrenches and Flywheel Wrench (15.3%); Spline Alignment Tool (11.8%); and Circuit tester 12V test light (7.9%). The availability of these tools support easy diagnostics of problems associated with ATDs (Naunheimer et al., 2010; Tim, 2012). From the study, it is revealed that diagnostic tools for repair of ATDs at garages are not adequate for service delivery. This confirms studies by Baidoo and Odum-Awuakye (2015), and Akpakpavi (2015) that there was lack of modern equipment and logistics among mechanics. The Transport Department of KNUST asserts that

Table 2. Faults easily repaired by mechanics

Able to Repair	n	%
Leaking transmission fluid	182	89.7
Burnt transmission fluid	180	88.7
Noisy transmission	177	87.2
Slips between shifts	72	35.5
Faulty torque converter	24	11.8
Inconsistent hydraulic pressure	21	10.3

Source: Field data 2019.

the spare parts of a transmission system does exist; however, each one is managed and monitored by the onboard computer. Conceivably, that makes transmission maintenance and service more complex in nature. They, however, indicate that an important tool which is not available at the garages but is considered key to the repair of ATDs is the Automatic Transmission Diagnostics Scanner System and Pressure Gauge. This tool according to them is used with the computer running the transmission system, sending and receiving information about the transmission's performance and operation. Mechanics are able to decipher the information to determine what is the cause of any potential transmission problem.

3.4 Knowledge and Competence of Wayside Mechanics

According to the wayside mechanics, an average of 17 (SD = 11.74) vehicles are serviced in a month of which 15 (SD = 9.17) are automatic transmission drives. The number of vehicles serviced in a month ranges from 13 to 29. The average number of mechanics for a wayside garage is 5 (SD = 6.87). Vehicle operators were asked to rate the knowledge and competence of wayside mechanics using an ordinal scale ranging from 1–9 where 1 represents low knowledge or competence and 9 represents high knowledge or competence. The mean score for knowledge of ATD problems is 4.43 (SD = 1.08) representing a percentage score of 48.77%. The mean score for Competence in ATD was 3.03 (SD = 0.56) representing a percentage score of 31.27%. Table 3 shows ANOVA test results for the knowledge and competence of mechanics based on the type of transmission drives they service. Knowledge levels of mechanics varied significantly for transmission types normally serviced by the mechanics ($p = 0.00$). Mechanics who serviced manual and automatic drives had higher knowledge than those who service either manual or automatic transmission drives. For specialized transmission drives, there were significant differences between the knowledge levels of the Mechanics ($p = 0.00$). Mechanics who serviced specialized transmission drives had higher knowledge scores than the mechanics who serviced both manual and automatic. In relation to competence, no significant differences were recorded for mechanics who service automatic, manual, or both transmission drives ($p = 0.25$). Meanwhile, there were significant differences in competence in mechanics who serviced specialized transmission drives and those who serviced both manual and automatic transmission drives ($p = 0.00$). Competence levels were lower than knowledge in dealing with the specific ATD problems.

Table 3. Knowledge and competence by specialization

Transmission types	Competence			Knowledge		
	Mean	S.D	*p*-value	Mean	S.D	*p*-value
Automatic transmission	3.09	1.09	0.25	4.33	2.61	0.00
Manual transmission	1.93	0.51		3.58	0.51	
Both (automatic and manual)	3.32	1.05		4.61	1.81	
Specialized transmission	3.51	1.60	0.00	4.21	1.77	0.00
Automatic transmission	0.41	0.51		3.41	0.51	
Both (automatic and manual)	3.09	1.48		6.42	3.34	

Source: Field data 2019.

The fact that competence levels were lower than knowledge in dealing with the specific ATD problems indicates that though most of the mechanics have some knowledge about ATD problems it does not necessarily mean they can solve the problems. The unavailability of most of the ATD tools in the garages is corroborated by observations of Baidoo and Odum-Awuakye (2015) and Tim (2012) that SMEs of automobile service garages lacked several modern equipment needed to provide quality services. Similarly, Akpakpavi (2015) concluded in his study on modern automobile vehicle repair practices that the inability to acquire modern vehicle diagnostic equipment and tools by garages affects their operations. Wayside mechanics do a lot of trial and error. Though they have knowledge they are unable to deal with some problems, their training should be problem-based, practical-based and project-based.

3.5 Choice and Ratings of Garage for Servicing

Results of the study on the analysis of the choice of garage for servicing of ATD's showed that 78% of the vehicle operators preferred wayside mechanics, while 22% would normally choose dealers. From Fig. 2, factors influencing the choice of garage are cost, quality of work, and time of delivery. From the study, 69% of vehicle operators would choose a garage based on cost. Thirteen percent (13%) of operators based their choice of repair of servicing centre on quality of services rendered. Three percent (3%) of vehicle operators prefer using the services of dealers even though they spend more on them than what the wayside mechanics charge for the same services.. Out of these, 79% evaluated the dealers as providing superior services compared to the wayside mechanics. Five percent (5%) of vehicle operators based their choice of service centre on delivery time and out of these 56% were satisfied with the promptness with which they were attended to by the wayside mechanics (3%). This is consistent with findings by Luís and Brito (2007) that independent garages (wayside garages) are highly patronised compared to branded dealers (dealers). Chi-square test performed showed an association between the choice of garage for servicing vehicles and the reasons given (χ^2 (6, n = 536) = 221.27, $p = 0.000$). This indicates that respondents attributed their choice of wayside garage to the cost of servicing, which is in agreement with Meier (2010). According to him, the main consideration of consumers is the cost of service. Similarly, other vehicle operators chose dealerships for the quality of service they offer, which according to Baidoo and Odum-Awuakye (2015) is due to the high educational level of mechanics at dealerships.

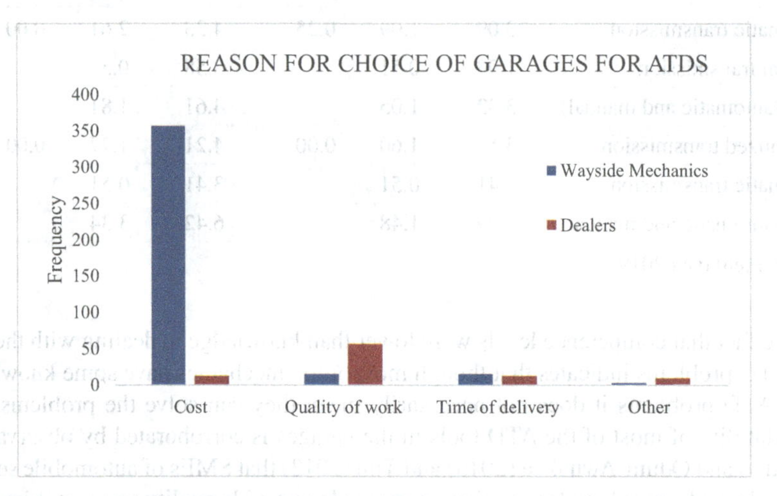

Fig. 2. Reasons for choice of garage for ATD service. Source: Field data, 2019

Vehicle operators were requested to rate the performance of garages in servicing automatic transmission drives using an ordinal scale (1 – Poor to 5 – Excellent). The results showed that for the performance of garages, dealers had a mean score of 3.51 ($sd = 0.71$) and Wayside garages a score of 1.55 ($sd = 1.00$) as indicated in Table 4. This implies that dealers scored higher ratings, an indication of better service delivery compared to the wayside garages.

Table 4. Rating of garages

Garages	Poor	Average	Good	Very good	Excellent	Mean (s.d)
	n (%)	n (%)	n (%)	n (%)	n (%)	
Dealers	5 (0.9)	10 (1.9)	8 (1.5)	194 (36.2)	319 (59.5)	3.51 (0.710)
Wayside garage	95 (17.7)	148 (27.6)	206 (38.4	77 (14.4)	10 (1.9)	1.55 (1.002)

Source: Field data 2019.

The results are a justification that knowledge and competence leads to a higher performance which confirms the earlier assertion that mechanics who operate with the dealers have higher knowledge and competence in solving ATD problems compared to mechanics at the wayside garages.

3.6 Reoccurrence of Transmission Problems

Vehicle operators were also asked to indicate whether there is a reoccurrence of the transmission problem after it has been fixed. Majority (93.7%) of the respondents indicated that there is a reoccurrence of transmission problem while 6.3% of respondents indicated that the transmission problem does not re-occur after it has been fixed. Among respondents who had a reoccurrence of transmission problem, 71.5% indicated that it takes between a month to six months for the transmission problem to reoccur. Also, 6% of the vehicle operators indicated that the transmission problem reoccurs between a week to a month, while 15.3% indicated that it takes between six months to a year before a reoccurrence of the transmission problem. About 7.2% of respondents, however, indicated that the transmission problem does not reoccur until after a year and beyond.

The reoccurrence of ATD problems could be attributed to the fact that majority of vehicle operators patronise wayside garages who are less competent in solving ATD problems. This is primarily due to the lower cost of services offered. The participating vehicle operators maintained that it is expensive to obtain service from the dealers, and that in making a choice for a garage they would prefer the affordable wayside garages.

4 Implications for Transportation Planning

Transportation planning comprises planning the entire operations, provision and management of facilities and infrastructure of the different modes of transport in a country to ensure safer, convenient and economical movement of people. The road transport system is a complex system in which road users, vehicle operators and infrastructure interact with each other. Automatic transmission drives, which could be classified as transport facilities are said to contribute to the reduction of road accidents by reducing human errors which are contributing factors in many road accidents. Unfortunately, automatic transmission drives could rather contribute and increase the rate of accidents if there is limited knowledge on their operations and repairs. For instance, the inability of mechanics to identify and fix issues related to the complexities in ATDs are likely to put human life at risk of accidents.

Indeed, the key stakeholders in the automobile industry include: regulatorory authorities (Ghana Standards Authority (GSA), and Driver and Vehicle Licencing Authority (DVLA)); industry (vehicle manufacturers and dealers); and policy makers which is largely government. The following subsections discusses the role of the key stakeholders in improving transportation planning in Ghana.

4.1 Regulatory Authorities and Industry

Rapid urbanisation has strained the capacity of local governments to provide adequate transport infrastructure to reduce road accidents, therefore, it is important for local governments, specifically, the district assemblies and the vehicle regulatory body (ie. DVLA and GSA) to ensure that vehicle operators and mechanics put vehicles in the right condition before they are driven. From the trend in the introduction of automatic transmission drive vehicles in Ghana, it is clear that sooner or later most of the vehicles with manual transmission may become obsolete. Coupled with the development of driverless and electric cars which certainly incorporate automatic transmissions, it is imperative that repair and servicing techniques will have to keep up with the changes. It is worth noting that the configurations of the electronics and mechanical components of the automatic transmission vehicles of different manufacturers are not standardized; however, there are basic principles in their operations which can be taught to the wayside mechanics. Therefore, there is the need to initiate a program for trainers to accelerate the dissemination of the knowledge and competencies. Vehicle manufacturers and dealers should therefore be willing to open up their facilities to train wayside mechanics and bring them up to speed with the modern trends in automobile diagnostics. The economic situation in Ghana makes it difficult for majority of vehicle operators to patronise the services of dealers anytime wayside mechanics fail to deal with automobile repairs. In view of this, dealers are encouraged to provide such services at affordable prices. The regulatory authorities must ensure that wayside mechanics are certified after receiving the requisite training from appropriate bodies before they operate.

4.2 Policy Makers

The government of Ghana should incorporate issues relating to the findings of the study on knowledge and competencies of vehicle operators and mechanics in its future policies and investments. Government is encouraged to adapt the principles of the transactive planning theory in the implementation of the policies. Since the wayside mechanics are unable to detect and deal with complex problems associated with automatic transmission vehicles, the government should institute regular problem-based training and refresher courses for the wayside mechanics through engagement with the appropriate educational institutions. Government should ensure that every mechanic registers with the Ghana Association of Garages to ensure that they can easily access problem-based training opportunities when organised. The association of garages should ensure standardisation of service charges for the wayside mechanics. This will guarantee value for money among vehicle operators who patronise the services of wayside mechanics. Government ought to subsidise the cost of tools and equipment used in diagnosing, analysing, and repairing ATDs.

5 Conclusions

The purpose of this work was to investigate and establish the level of expertise of traditional automobile maintenance practitioners in servicing automatic transmission drives. The extent of patronage of ATDs revealed a linear relation of $y = 52.239x - 105743$, an indication that the percentage of ATDs has been increasing at an annual rate of 52 ATD cars per garage over the period under review. It was found that Burnt Transmission fluid, Leaking Transmission Fluid, Noisy Transmission, Slips between Shifts, Inconsistent Hydraulic Pressure and Faulty Torque converters are the ATD problems brought to the garages. Out of these problems, only Burnt Transmission Fluid, Leaking Transmission Fluid and Noisy Transmission are faults the mechanics repair easily. Most of the wayside mechanics did not have modern tools and equipment for proper diagnosis and repair. Competence of wayside mechanics in solving ATD problems was poor due to electronic nature of modern ATDs. They therefore resort to trial and errors to solve these problems. Vehicle operators prefer wayside mechanics to service their ATDs due to the affordability of cost even though they have low knowledge and competence. Acquisition of requisite tools and instruments by the wayside mechanics would be a prerequisite to improving the wayside mechanics' performance.

References

Abu-Eisheh, S.A., Mannering, F.L.: Forecasting automobile demand for economies in transition: a dynamic simultaneous-equation system approach. Transp. Plan. Technol. **25**(4), 311–331 (2002)

Akayeti, A., Sackey, S.M., Dzebre, D.E.K.: Development of indigenous automobile design and manufacturing in Ghana. African J. Appl. Res. (AJAR) **1**(1), 179–192 (2015)

Akinola, B.T.O.: Basic Automobile Technology. Olajuyin Printers, Akure (2005)

Akpakpavi, M.: Modern automobile vehicle repair practices in micro, small and medium scale garages in Ghana. Int. J. Sci. Technol. Soc. **2**(6), 216–222 (2015)

Akple, M.S., Turkson, R.F., Biscoff, R., Borlu, B.K., Apreko, A.A.: Driver preference for automatic or manual transmission systems for vehicles: a case study in Ghana. J. Inf. Eng. Appl. **3**(9), 22–27 (2013)

Amjad, S., Neelakrishnan, S., Rudramoorthy, R.: Review of design considerations and technological challenges for successful development and deployment of plug-in hybrid electric vehicles. Renew. Sustain. Energy Rev. **14**(3), 1104–1110 (2010)

Andoh, P.Y., Davis, F., Fiagbe, Y.A.K., Alhassan, T.: Tyre pressure model for predicting fuel consumption of vehicles on Ghana roads. Int. J. Sci. Technol. Res. **2**(8), 120–124 (2013)

Baidoo, F., Odum-Awuakye, G. A. (2015). Influence of service quality delivery in the SMEs of the motor vehicle repair service industry in Ghana. Afr. J. Appl. Res. (AJAR) J. **1**(1), 429–439 (2015)

Chan, C.Y.: Advancements, prospects, and impacts of automated driving systems. Int. J. Transp. Sci. Technol. **6**(3), 208–216 (2017)

Chand, S., Li, Z., Dixit, V.V., Waller, S.T.: Examining the macro-level factors affecting vehicle breakdown duration. Int. J. Transp. Sci. Technol. **11**(1), 118–131 (2022)

Davis, F., Sackey, M.N., Fanyin-Martin, A.: Traditional practices of automobile maintenance in Ghana: an analysis in the removal of automobile engine thermostat. Afr. J. Appl. Res. (AJAR) **4**, 62–70 (2018)

Davis, F., Sackey, M.N., Fanyin-Martin, A., Amissah, P.K.: Road transportation safety in Ghana: an assessment of the variability of articulated trailer design parameters. Afr. J. Appl. Res. (AJAR) **4**, 14–26 (2018)

de Almeida Correia, G.H., Menendez, M.: Automated and connected vehicles: effects on traffic, mobility and urban design. Int. J. Transp. Sci. Technol. **6**(1), iii–iv (2017)

Delmar, T.: Automatic Transmission and Transaxles Test A2, 2nd edn. Delmar, New York (2001)

Fouracre, P.R., Sohail, M., Cavill, S.: A Participatory approach to urban transport planning in developing countries. Transp. Plan. Technol. **29**(4), 313–330 (2006)

Ghana Statistical Service. Ghana Statistical Yearbook 2011–2013 (2015). http://www.statsghana.gov.gh/docfiles/publications/2013%20statistical%20yearbook_website.pdf. Accessed 13 June 2017

Hesse, C.A., Ofosu, J.B.: Comparative analysis of regional distribution of the rate of road traffic fatalities in Ghana. Eur. Sci. J. **10**(4), 159–172 (2014)

Hirsch, M., Diederichs, F., Widlroither, H., Graf, R., Bischoff, S.: Sleep and take-over in automated driving. Int. J. Transp. Sci. Technol. **9**(1), 42–51 (2020)

Jbili, S., Chelbi, A., Radhoui, M., Kessentini, M.: Integrated strategy of vehicle routing and maintenance. Reliab. Eng. Syst. Saf. **170**, 202–214 (2018)

Kumar, R.: Research Methodology: A Step–by-Step Guide for Beginners, 4th edn. Sage, London (2014)

Lee, C.: The rise of Korean automobile industry: analysis and suggestions. Int. J. Multidisciplinary Res. **6**, 428–480 (2011)

Leedy, P.D., Ormrod, J.E.: Practical Research: Planning and Design, 11th edn. Pearson, Harlow (2015)

Luís, E.P.Z.B.R., Brito, B.A.L.A.L.: Customer choice of a car maintenance service provide. Int. J. Oper. Prod. Manag. **27**, 464–481 (2007)

Mahesh, P., Ram, V.K.: Plant maintenance management practices in automoblie industry: a retrospective and literature review. J. Industr. Eng. Manage. **3**(3), 512–541 (2010)

Merriman, S.E., Plant, K.L., Revell, K.M., Stanton, N.A.: What can we learn from Automated Vehicle collisions? A deductive thematic analysis of five Automated Vehicle collisions. Saf. Sci. **141**, 105320 (2021)

Mitropoulos, L.K., Prevedouros, P.D.: Incorporating sustainability assessment in transportation planning: an urban transportation vehicle-based approach. Transp. Plan. Technol. **39**(5), 439–463 (2016)

Naunheimer, H., Bertsche, B., Ryborz, J., Novak, W.: Automotive Transmissions: Fundamentals, Selection, Design and Application, 2nd edn. Springer, New York (2010). https://doi.org/10.1007/978-3-642-16214-5

Nguyen, D., Imamura, F., Iuchi, K.: Disaster management in coastal tourism destinations: the case for transactive planning and social learning. Int. Rev. Spatial Plan. Sustain. Dev. **4**(2), 3–17 (2016)

Overtoom, I., Correia, G., Huang, Y., Verbraeck, A.: Assessing the impacts of shared autonomous vehicles on congestion and curb use: a traffic simulation study in The Hague, Netherlands. Int. J. Transp. Sci. Technol. **9**(3), 195–206 (2020)

Schindler, P.: Business Research Methods, 14th edn. McGraw-Hill Education, New York (2022)

Tim, G.: Automotive Service: Inspection, Maintenance, Repair. Delmar, New York (2012)

Walker, W.E., Marchau, V.A.: Dynamic adaptive policymaking for the sustainable city: the case of automated taxis. Int. J. Transp. Sci. Technol. **6**(1), 1–12 (2017)

Zuraida, R., Iridiastadi, H., Sutalaksana, I.Z.: Indonesian driver's characteristics associated with road accidents. Int. J. Technol. **8**(2), 311–319 (2017)

Enhancing Customer Satisfaction Through Listening in Tanzanian Higher Education

A. K. Majenga[1(✉)] and R. G. Mashenene[2]

[1] Department of Development Finance and Management Studies, Institute of Rural Development Planning, P. O. Box 138, Dodoma, Tanzania
amajenga@irdp.ac.tz

[2] Department of Marketing, College of Business Education, P. O. Box 2077, Dodoma, Tanzania
g.mashenene@cbe.ac.tz

Abstract.
Purpose: This paper analyzed how to create customer satisfaction through listening in Tanzanian higher education (HE).

Design/Methodology/Approach: The study was carried out at the Institute of Rural Development Planning (IRDP) and it adopted a cross-sectional survey for a sample of 350 students in bachelor three who were stratified proportionately in departments. Key Informants (KIs) including academic staff, non-academic staff, heads of departments and top management were selected purposefully for the interview to collect qualitative data. Qualitative data were analyzed with the use of the content analysis method. Quantitative data were analyzed using descriptive and inferential statistics. Frequency, percentage, mean scores and standard deviations were computed descriptively whereas a binary logistic regression model was performed to the determination of the relationship between listening attributes and customer satisfaction.

Findings: The findings show a low level (62.00%) of customer satisfaction with listening. The findings further indicate a significant ($p < 0.01$) and positive relationship between all listening attributes (sensing, evaluating and responding) and influenced customer satisfaction.

Implications/Research Limitation: The study was conducted at one higher education institution posing bias. The inclusion of more HE is recommended for improved generalization.

Practical Implication: HE management should train staff on listening skills and emphasize effective listening as a marketing strategy to optimize the satisfaction of customers.

Social Implication: Customers will feel social highly respected and recognized.

Originality/Value: The manuscript will be subjected to Turnitin; a plagiarism

C. Aigbavboa et al. (Eds.): ARCA 2022, *Sustainable Education and Development – Sustainable Industrialization and Innovation*, pp. 927–934, 2023.
https://doi.org/10.1007/978-3-031-25998-2_71

checker aiming at maximizing originality/value and consequently contributing significantly to the existing knowledge on issues concerning listening and customer satisfaction.

Keyword: Customer · Higher education · Listening · Satisfaction · Tanzania

1 Introduction

The term listening has become popular in several disciples such as marketing, psychology, sociology, linguistics, medical and others and it been defined by various scholars. Refers to listening well and demonstrating an interest in what is being said (Yrle and Galle 1993) whereas Castleberry and Shepherd (1993) argued that listening is the cognitive process of actively sensing, interpreting, evaluating and responding to the verbal and non-verbal messages of actual or potential customers. In marketing and for the focus of this study, listening meant an active process that involves sensing, evaluating and responding to customers' messages.

The concern of listening to customers elsewhere globally has become vital for customer satisfaction and profitable relationship management. For instance, Ramsey and Sohi (1997) investigated the impact of sales representatives' listening behaviour on trust, satisfaction and anticipation of future interaction, the findings indicated that listening is the outcome of sensing, evaluation and responding. These findings imply that for customers to build up trust, be satisfied and anticipate future interaction with the business organization, service providers should demonstrate sound listening behaviour in terms of sensing, evaluating and responding.

In Tanzania, most of the previous studies in higher education have focused on the analysis of service quality and customer satisfaction while the issue of listening in relation to customer satisfaction has been ignored. For instance, Mashenene (2019) investigated on the effect of service quality on students' satisfaction in Tanzania higher education. In this study, all variables excluded the issue of listening as one of variables that can affect service quality. The study of Magasi et al. (2022) re-examined service quality and students' satisfaction in Tanzania higher education using ERVQUAL model. This study also did not take into account listening as on the variables that affect students' satisfaction. The study of Mashenene et al. (2019) focused on investigating students' retention strategies and satisfaction in Tanzania higher education, this study again did not include listening as one of important attributes in students' satisfaction. In this regard, the relationship between listening and students' satisfaction has remain unattended, this necessitated for the study to be undertaken. This study addressed the following specific objectives; i) to determine the level of customer satisfaction with listening practices demonstrated by staff and ii) to determine the relationship between listening attributes and customer satisfaction.

2 Theories Underpinning the Study

Two theories namely Steil *et al.* (1983) and Schutz (1996) were used to underpin this study. The theory of Steil et al. was developed in 1983 to address three dimensions of listening; sensing, evaluating and responding. Sensing being the first dimension of listening, according to the theory it is meant as the initiation of the listening process by the company staff whereby the incoming stimuli is sensed. The stimuli are categorized either verbal or no verbal such as words spoken, voice tone, gestures like nodding of head, use of arms and others. The customer perception on active sensing is based on company staff to sense what is being said by paying eye contact throughout the conversation, make a good focus on the conversation and use of non-verbal behaviour that help gathering of incoming stimuli (Yrle and Galle 1993). Evaluating being the second component of listening according to the theory refers to the process whereby marketers or service providers assign meaning to messages and establish their importance. The third component of listening is responding; the feedback process during listening which is responsible for informing, controlling, sharing feelings and ritualizing (Mead 1986). From this theory, three variables (sensing, evaluating and responding) for this study were established and used in this study.

3 Research Methodology

This study adopted a cross-sectional research design where a sample of 350 students from the Institute of Rural Development Planning (IRDP) in Dodoma city was involved using a questionnaire survey. IRDP was selected since it the second higher learning institutions ins Dodoma city with the highest number of students and it adopts competence-learning and teaching approach from which high level of interaction between staff and students is sought. Proportionate stratified sampling technique was used to selected a sample from the population of IRDP bachelor three students. The reason for drawing a sample from the population of bachelor three students was that bachelor three students have been at IRDP for three years. In this regard they are exposed to high level of interactions with staff to allow listening practices and their level of satisfaction to be studied. The sample size for the study was 350 though it was estimated earlier using the Cochran (1977) formula as cited by Mashenene (2016) to be 384 but due to some of questionnaires were not returned it decreased to be 350 which was 91.15%. The Cochran formula used to estimate sample size was presented by the following equation;

$$n = \frac{Z^2}{e^2} * \frac{pq}{} \tag{1}$$

whereby:

n is the sample size.

Z is the selected critical value of desired confidence level which is 1.96 for a 95% confidence level.

p is the proportion in the largest population which is 50%

q is 1-p and e is the degree of accuracy or acceptable margin of error, set at 0.05.

Then, $n = \frac{1.96^2}{0.05^2} * \frac{0.5*0.5}{} = 384.$

Quantitative data were collected using a questionnaire with 7 points Likert scale questions (1 = strongly disagree and 7 = strongly agree) that were adopted from Ramsey and Sohi (1997). Interview with key informants (KIs) (academic and non-academic staff, head of departments and top management) was carried out using interview guide so as to collect qualitative data.

Quantitative data analysis involved both descriptive and inferential statistics. To determine level of customer satisfaction, descriptive statistics involving cross tabulation to establish the association between customer satisfaction with sex and students' programmes were performed. Descriptive statistics also involved computations of frequency, percent and transformation of 7-points Likert scale data into two levels; low and high. To determine the relationship between listening attributes and customer satisfaction, binary logistic regression model was used since the customer satisfaction as dependent variable was treated as a dummy variable (1 = satisfied, 0 = dissatisfied). The indices were computed for the listening attributes as independent variables. The binary logistic regression equation was presented as Eq. 2. Qualitative data were analyzed using content analysis in which data collected from the field were first transcribed followed by categorization into relevant themes and sub-themes and further they were discussed intensively and matched with the literature.

4 Findings and Discussion

4.1 Reliability Test

The results for the Cronbach's alpha (Table 1) indicated that all variables scored the value greater than 0.70 which is statistically highly acceptable (Magasi et al. 2022; Mashenene and Kumburu 2020). The implication of these results is that the data collected had high and acceptable reliability.

Table 1. Reliability test

Variables	Number of items	Cronbach's alpha
Sensing	4	0.8450
Evaluating	5	0.838
Responding	4	0.827
Customer satisfaction	3	0.821
Overall	**16**	**0.844**

4.2 Level of Customer Satisfaction

Generally, the findings (Table 2) indicate low level of customer satisfaction (62.00%) as compare to only 38.00% of high level of customer satisfaction. The implication of these findings is that in Tanzania higher education still students are not satisfied with the way listening is practiced by staff of respective institutions. These findings were supported by qualitative findings from the interview with one of the academic staff "*...sometimes it is difficult to demonstrate good listening following all attributes since some students* *are troublesome, we are also overwhelmed with a lot of activities to do which force us to listen students in a rush...".* The qualitative findings imply that there is a gap in attitude or mindset of staff on the concept of proper listening. These findings are supported by those of Min et al. (2021) which revealed that customer satisfaction is improved with active listening.

Table 2. Level of customer satisfaction

Customer satisfaction level	Frequency	Percent
Low	217	68.00
High	133	32.00
Total	**350**	**00.0**

4.3 Relationship Between Listening Attributes and Customer Satisfaction

The findings (Table 3) present that the chi – square = 307.94 was statistically significant (p = 0.000) depicted that the model fitness was acceptable, inferring that the model predicted the relationship between listening attributes (sensing, evaluating and responding) and customer satisfaction. The value of pseudo $R^2 = 0.6625$ suggest that data for listening attributes that were entered in the model described 66.25% of variance of customer satisfaction. Statistically, the value of pseudo R^2 in the logit is not a true statistic measure like R^2 values in the OLS (Mashenene and Kumburu 2020).

The logit results (Table 3) indicate that the coefficient of age was negative (−0.2391) and significant (p = 0.008) related to customer satisfaction, signifying that any unit increase in students' age will lead to decrease in customer satisfaction by 23.91%. This reciprocal relationship between students' age and customer satisfaction implies increased knowledge, skills and attitude towards satisfiers and dissatisfiers. This means that as the age of students increase, the understanding and level of thinking of students also increase, as the result, they become more demanding and their level of satisfaction increase. These findings are dissimilar with those of Magasi et al. (2022) which revealed that students' age was directly proportional to service quality.

Further, the findings in Table 3 indicate that the coefficient of sensing was positive (2.1900) and significantly (p = 0.000) related to customer satisfaction, suggesting that any unit increase in sensing during students' listening will result into 219.00% increase in customer satisfaction. This means that students are satisfied when they see staff during conversation keenly focus on them, keep eye contact all the time and use of non-verbal gestures. These findings are in congruence with those of Ramsey and Sohi (1997) which revealed the same.

Further, the findings in Table 3 show that the coefficient of evaluating was positive (1.616) and significant (p = 0.003) in relation to customer satisfaction, implying that any unit increase in evaluating during listening will lead to 161.6% increase in customer satisfaction. This means that students are highly satisfied during conversation when staff

Table 3. Logit results

Variables	Coefficients	Std. err.	P > |z|
Sex (dummy)	0.2323	0.4977	0.641
Age (years)	−0.2391	0.0903	0.008
SE_Cust (index)	2.1900	0.4712	0.000
EV_Cust (index)	1.6159	0.5355	0.003
RE_Cust (index)	3.1557	0.4798	0.000
_Cons	2.1537	2.1850	0.324
Chi-square	307.94, p = 0.000		
Pseudo R^2	0.6625		
-2Log likelihood	78.4510		

Notes: SE = Sensing, EV = Evaluating, RE = Responding, Cust = Customer Satisfaction.

ask students for more details, paraphrase students' questions, allow students to finish without interrupting them and let the conversation continue without changing the subject matter as a sign of understanding what is being listened. These findings are in harmony to those of Ramsey and Sohi (1997) which revealed similar findings.

Moreover, the findings in Table 3 indicate that the coefficient of responding was positive (3.1557) and significant ($p = 0.000$) in relation to customer satisfaction, depicting that any unit increase in responding will result into 315.57% increase in customer satisfaction as they are directly proportional to each other. This means that when staff give full responses instead of yes or no, providing relevant information, being eager in students' responses and giving responses in appropriate time are likely to enhance students' satisfaction during listening.

5 Conclusion and Recommendations

5.1 Conclusion

From the research findings presented, it can be concluded that the level of students' satisfaction with staff' listening at IRDP is low (62.00%). These findings connote that the majority of students perceive that they are not really listened since listening is not just hearing but rather the process of trying harder to sense, evaluate what is being presented and responding to what has been presented by students. Such findings pose gaps in the listening attributes that need to be filled through improved listening. Further, it is concluded that sensing, evaluating and responding statistically have positive and significant relationship with customer satisfaction, implying that listening is one of the attributes that need to be imbedded in service quality which is a precursor of customer satisfaction.

5.2 Recommendations

Based on the conclusion, the following recommendations were put forward;

i) To the Ministry of Education, Science and Technology (MoEST), listening should be integrated in the communication skills syllabus so that graduates be equipped with listening knowledge, skills and attitude since such graduates they serve people in the labour market.
ii) To the IDP management, staff listening knowledge, skills and attitude should be intensified through regular training that will unfold the understanding of how they can practice listening. The training should cover all attributes of listening (sensing, evaluating and responding) and its important in building strong customer relationships and consequently customer satisfaction. The IRDP management also should establish complaint desk dealing with receiving and handling effectively customer complaints regarding listening. This will tend to fuel good listening.
iii) To the IRDP staff, listening to students should be prioritized since they serve people. Service offering in several cases requires maximizing interactions between the service providers (the listener) and the students (the customer), the sought interactions are accomplished through active listening.

5.3 Areas for Future Research

Similar study should be conducted in higher education by involving more than one institution and from both public and private institutions so as to allow comparison. The study should also involve different levels of education programmes i.e. diploma, bachelor, masters and PhD students, this will allow researchers to make comparison and draw conclusion based on the level of listening across different programmes. It is further recommended that systematic review can be carried out by involving higher learning institutions in Africa and across the world so as to allow comparison of listening practices.

Acknowledgements. The authors are grateful to the management of the College of Business Education (CBE) for financial support for the conference fee and publication of conference proceedings as book chapter. The authors also are grateful to the anonymous reviewers of the conference manuscript that will review the manuscript for their extremely useful inputs for the purpose of improving quality of the manuscript. Usual disclaimers apply.

Declaration of Conflicting Interests. The authors declared no potential conflicts of interest with respect to the research, authorship, presentation to the conference and/or publication of this article.

Funding. The authors declare no financial support for the research and authorship of this manuscript but they received financial support from CBE for conference presentation and publication of this article as book chapter.

References

Castleberry, S.B., Shephard, C.D.: Effective interpersonal listening and personal selling. J. Pers. Sell. Sales Manage. **13**(Winter), 35–49 (1993)

Cochran, W.G.: Sampling Techniques. 3rd edn. John Wiley and Sons, New York. 96p. (1977)

Magasi, C., Mashenene, R.G., Dengenesa, D.M.: Service quality and students' satisfaction in Tanzania higher education: a re-examination of SERVQUAL model. Int. Rev. Manage. Mark. (IRMM). **12**(3), 18–25 (2022). https://doi.org/10.32479/irmm.13040

Mashenene, R.G.: Effect of service quality on students' satisfaction in Tanzania higher education. Bus. Educ. J. (BEJ) **2**(2), 1–8 (2019)

Mashenene, R.G.: Socio-cultural Determinants of Entrepreneurial Capabilities among the Chagga and Sukuma Owned Small and Medium Enterprises in Tanzania, p. 274. Thesis for Award of PhD Degree at Sokoine University of Agriculture, Morogoro, Tanzania (2016)

Mashenene, R.G., Kumburu, N.P.: Performance of small businesses in Tanzania: human resources based view. Global Bus. Rev. (GBR – SAGE Publications), 1–15 (2020). https://doi.org/10.1177/0972150920927358

Mashenene, R.G., Msendo, A., Msese, L.R.L.: The influence of customer retention strategies on customer loyalty in higher education in Tanzania. Afr. J. Appl. Res. **5**(1), 85–97 (2019). https://doi.org/10.26437/ajar.05.01.2019.07

Mead, N.A.: Listening and speaking skills assessment. In: Performance Assessment: Methods and Applications. Ed. Ronald A Berk, pp. 509–521. Johns Hopkins University Press, Baltimore, MD (1986)

Min, K.S., Jung, J.M., Ryu, K.: Listening to their heart: why does active listening enhance customer satisfaction after service failure? Int. J. Hosp. Manage. **96**, 102956 (2021). https://doi.org/10.1016/j.ijhm.2021.102956

Ramsey, R., Sohi, R.: Listening to your customers: the impact of perceived salespersons listening behaviour on relationship outcomes. J. Acad. Mark. Sci. **25**, 125–137 (1997). https://doi.org/10.1007/BF02894348

Yrle, A.U., Galle, W.P.: Using interpersonal skills to manage more effectively. Superv. Manage. **38**(4), 4 (1993)

Technology Adoption and the Financial Market Performance in Nigeria and South Africa

O. N. Oladunjoye(✉) and N. A. Tshidzumba

North-West University, Mafikeng, South Africa
onoladunjoye@oauife.edu.ng, aaron.tshidzumba@nwu.ac.za

Abstract.
Purpose: Technology adoption is increasingly important for the effective operation of the financial market in the world. It is pertinent to investigate the nexus between technology adoption and financial market performance in Nigeria and South Africa.

Design/Methodology/Approach: Country-specific and panel econometric techniques were used to study the nexus between technology adoption and financial market performance in Nigeria and South Africa. The technology diffusion theory which states that the continuous usage of a particular technology tends to lead to an equilibrium path over a given period is the foundational theory on which this study is based. Technology adoption index (TECH) and financial market performance index (FMP) were generated using the principal component analysis while variables such as "gross domestic product (GDP), domestic credit to the private sector (CPS), bank deposit rate (BDR) and bank lending rate (BLR)" were used as control variables. Data for all variables were gotten from the World Bank and the Heritage database.

Findings: The study reveals that for country-specific analysis, technology adoption has a strong and direct impact on the financial market performance in Nigeria while in South Africa, technology adoption negatively and significantly impacts financial market performance. But, the panel estimation reveals that technology adoption positively and significantly promotes the financial market performance of both Nigeria and South Africa respectively.

Implications/Research Limitation: The study focused on the nexus between technology adoption and financial market performance in Nigeria and South Africa.

Practical Implication: The policy implication emanating from this study is that an increase in technology adoption has a robust and important influence on the financial market in Nigeria only but as soon as the two countries deepened the operation of their financial market through the removal of restrictions, then, technology adoption will significantly promote the financial market performance in both Nigeria and South Africa.

Originality/Value: This study expanded the frontiers of knowledge on how the

C. Aigbavboa et al. (Eds.): ARCA 2022, *Sustainable Education and Development – Sustainable Industrialization and Innovation*, pp. 935–952, 2023.
https://doi.org/10.1007/978-3-031-25998-2_72

removal of cross-border restrictions in the financial market through technology adoption will significantly promote the financial market of the two countries. This finding also supports the arguments for a deepened economic relationship among African economies.

Keyword: Adoption · Econometric · Financial market · Performance · Technology

1 Introduction

Technology adoption is increasingly important for the operation of the financial market since they are the first to invest and adopt new technologies in the world which serve as a platform for electronic trading of stocks and securities thereby causing a paradigm shift in the operation of the financial markets (Hitt et al. 1998; Cliff et al. 2011). Technology adoption has been identified to have the capacity to accelerate financial inclusion, economic growth, and development in an economy (Kauffman and Riggins 2012; World Bank 2014; UNCTAD 2019). Also, Igue et al. (2021) hinted that the high rate of internet access as well as mobile technology usage has greatly enhanced the activities of financial markets globally, most especially an increasing rate of internet technology penetration on the continent has been observed but Africa is still confronted with the problem of harmonization of the financial sector which can greatly impede the competitiveness of her financial markets in the global market space.

Furthermore, Tinn (2006, 2008) highlighted that equity price instability, as well as the absence of liquidity in the financial market, significantly discourages technology adoption which explains the low participation by high net worth investors in underdeveloped financial markets, especially in Africa. Whilst, Okwu (2016) argued that technology such as ICT could be adopted to enhance the competitiveness, efficiency, and liquidity in the financial market and can also be used as a tool to promote regional cooperation and integration that will facilitate the listings of stocks and securities in the major financial markets in Africa.

Nigeria as well as South Africa being the two biggest economies in Africa recently signed and ratified the execution of the African Continental Free Trade Agreement (AfCFTA) in 2019 and 2020 respectively with the major goal of promoting intra-African trade. Therefore, to have viable trade relations among the countries of Africa, between Nigeria and South Africa in particular, then, the relevance of technology cannot be over-emphasized.

The African Union has also joined the technology revolution, when the union emphasized the opportunities embedded in the adoption of digital technology in all spheres of African economies, most especially for the teeming African youthful population to be competitive in the global market. Also, the union affirms that digital technology is an impetus for innovation, sustainable and growth that is inclusive, a platform for numerous job opportunities, and poverty alleviation will assist in drastically reducing income inequality, facilitate the attainment of the 2063 agenda of the African Union and also leads to the attainment of the Sustainable Development Goals in Africa of which technology and deepening of economic integration is at its core.

Currently, Nigeria plus South Africa have the biggest financial market in Africa. This is actualized through the adoption of technology at almost every transaction by the use of phones and other digital devices which have revolutionized activities and operation of the financial market thus leading to a cashless society. Giving the individual country success hence the call for a deepened relationship in the form of trans-border stock listing between Nigeria and South. The study by Irving (2005) argued that establishing regional cooperation and integration will greatly benefit cross-border listings leading to the transfer of technology and expertise. This, among others, underscores the relevance of technology adoption on the performance of the financial markets.

Furthermore, Nigeria and South Africa have the most viable financial market in sub-Saharan Africa. Specifically, Nigeria has the most buoyant financial market in the West African sub-region with over 177 listed firms and more than US$56 billion as market capitalization as of the end of 2021 while South Africa has the best viable financial market in Africa with about 264 listed firms and US$105 billion as market capitalization at the end of 2021 (World Bank 2022). Also, the Nigerian Stock Exchange Market and the Johannesburg Stock Exchange Market are believed to be similar in relation to the kind of reforms and technological innovations they brought to bear in their respective market.

The various reforms carried out in the two stock exchanges over the years have led to continuous activities of external interests in domestic listed firms and the total overhaul of the trading platform from the "open outcry trading system" to a fully digital and computerized trading system which allows for the participation of investors and potential investors across the globe. Some specific reform was carried out in the two stock exchange markets which include the adoption of the digital system, improvements in listing requirements, and the review of laws involving capital market operations along with market divisions (see Irving 2005; Ezirim et al. 2019; Obiakor and Okwu 2011).

On one hand, extensive studies have examined the relevance of technology adoption as a tool for improved efficiency, profitability, and to gain competitive advantage in some major sector of the Nigeria's economy such as the manufacturing sector, service sector, small and medium scale enterprises (see Kajogbola 2004; Agwu 2016; Hassan and Ogundipe 2017; Okundaye et al. 2019; Ramachandran, Obado-Joel, Fatai et al. 2019; Kyari and Akinwale 2020; Ohiani 2020; Abdulquadri et al. 2021) and in South Africa, studies on the importance and benefits of technology adoption and even with some appraising the readiness of the country for the forth industrial revolution giving the criticalness of technology (see James et al. 2001; Gono et al. 2013; Pillay 2016; Jenkin and Naude 2018; World Bank 2019; Ntimane 2020; Olaitan et al. 2020; National Planning Commission (NPC) 2020; Centre of Excellence in Financial Services 2022; Smidt 2021; Slazus and Bick 2020), however, all these studies were country-specific and failed to recognize the need for a deepened financial market in Africa, especially between Nigeria and South Africa that will serve as the financial hub for the promotion and actualization AfCFTA goals of viable intra-African trade through building a sound technology driven digital single market in Africa by 2030 which will promote unrestricted movement capital, goods and services within the continent and promote individuals and businesses unrestricted access to online activities.

Similarly, the financial market is the major source of capital for investors in an economy which serves as a channel through which surplus funds are moved to where it is most needed. Africa is currently being confronted with the problem of a small and underdeveloped capital market compared to the developed countries and it is somewhat dominated by the commercial and investment banks which make it to have an implication for the performance of the real sector as well as the small scale industries on the continent. The rate of credit given to the private firms in African economies on average is less than 30 percent of the GDP of all African countries which is very low compared to what is obtainable in East Asia and the Pacific which is 138 percent of their GDP (ECA 2020). Given this challenge of low level of credit creation implies that very small capital is made available to the real sector by the financial markets. However, the AfCFTA offers boundless prospects for African capital markets to develop by producing sustainable markets, economies of scale, better effectiveness through competition, and new prospects to participate in African markets which have the capacity to attract more investment for higher productivity in Africa. This can be achieved by massively leveraging the rapid and growing rate of financial technology which will facilitate connectivity for cross-border transactions throughout the continent of Africa.

Very few studies such as Etim (2011), Soutter et al. (2019), Economic Commission for Africa (2020), International Telecommunication Union (2021), and World Bank (2021) studied the association between technology adoption and financial market performance in Africa, but studies between the two leading sub-Saharan African countries of Nigeria and South Africa are very scanty except for the study by Okwu (2016). These studies are faced with the problems of inadequacies in the measurement of variables wherein several parameters were used to capture both the technology adoption and financial markets. This made the studies suffer from the problem of serial autocorrelation and estimation bias thereby making their studies prone to faulty policy inferences.

Furthermore, the Covid-19 outbreak has taken to the fore the relevance of technology adoption where governments and several corporate organizations in Africa, Nigeria, as well as South Africa inclusive have to resort to working remotely with the aid of several digital technology platforms in order to observe the necessary social distance requirements and lockdown mandate. Further divisions of this study include the literature review, methodology, analysis of results, and conclusion.

2 Review of Literature

Technology adoption has been viewed as a driving force for global competitiveness and financial performance in organizations (Dangolan 2011; Mwashiuya and Mbamba, 2020). More particular is the submission of scholars such as Enu and Gberbi (2015), Kombe and Wafula (2015), Wasilwa and Omwenga (2016), and, Kairu and Rugami (2017) that technology adoption is a channel through which operational services and performance are actualized and enhanced in financial institutions. Thus, numerous studies have explored the relevance of technology adoption on the operational and financial performance of organizations specifically small-scale industries and the financial industries in different economies.

For instance, Chairoel et al. (2015) in a study on the factors affecting technology adoption and the latter's effect on small-scale industries in Indonesia discovered that technology adoption contributes to the efficiency and effectiveness of the organization and its performance. It also discovered that both internal (organization and managerial) and external (environmental) elements influence technology adoption in an organization. This study also found that while technology adoption causes an increase in profit margins, operational performance, and productivity, it also causes a decline in costs. Additionally, Igwe et al. (2020) studied technology adoption in the sales of small and medium manufacturing industries in Port Harcourt, Nigeria, and revealed that SMEs with seldom or no technology adoption face stiff competition and the threat or occurrence of liquidation. Using the descriptive survey research and simple regression statistical methods, this study reiterated the earlier finding by Chairoel et al. (2015) by revealing further the usefulness of technology, ease of use, and organizational and environmental variables considerably impact technology adoption and in turn drives sales performance. Thus, this study noted that investment in technology is paramount for higher technology adoption, competitive advantage, and greater sales performance.

According to Rambe and Mpiti (2017), both public and private funding is very important for promoting the financial performance of small, micro, and medium firms in South Africa. Using theoretical investigation, this study also showed that apart from public and private funds, organizational and environmental factors mediate the influence of funds and the performance of SMMIs in South Africa. More importantly, this study noted that technology acquisition which is an organizational and environmental factor plays a prominent part in the successful performance of SMMIs in South Africa. Similarly, Thapelo (2020) investigated the role of government financial backing in adopting technology in SMEs in South Africa using a quantitative research method based on a positivist research paradigm. This study agreed with the position of Rambe and Mpiti (2017) that there is a direct association amongst technology adoption and the performance of sales recorded in South Africa. It also posited that the high failure rate in SMEs is due to numerous factors such as the issue of size, lack of skills, and access to capital and technology adoption. The study using the multiple regression further posited that institutional financial assistance from the government is an important determinant of the level of technology adopted and the extent of performance in SMEs in South Africa.

Agwu (2016) also revealed that the application of technology to business organizations especially SMEs have yielded unprecedented improvement in profit maximization and service delivery. Meanwhile, in determining the effect of technology adoption on SMEs in Nigeria, this study showed that despite the apparent importance, availability, and benefit of technology, its diffusion, and adoption in SMEs are relatively low. This is due to the disconnect between the required skills of various SME operators and the adoption of technology. Therefore, for SMEs in Nigeria to derive the full benefits of technology adoption, this study suggested that the government and stakeholders embark on massive awareness. Moreover, Yusuf (2017) posited that to combat the issue of the dwindling oil revenue, high unemployment, and recession in Nigeria and achieve diversification in the economy, SMEs are necessary tools. Thus, this study opined that these tools (SMEs) can only be effective when there is proper education on skill acquisition, technology adoption, and globalization. In light of this, using linear regression and cross-sectional research design, this study concluded that with proper education, SMEs can operate effectively and competitively thereby achieving productivity, and profitability and promoting the growth and recovery of the economy.

Mathu and Tlare (2017) investigated the influence of technology adoption on the supply chains of SMEs in South Africa using structural equation modeling (SEM) and confirmatory factor analysis. This study revealed that technology adoption impacts positively on the supply chains of SMEs in South Africa. This is because its adoption greatly improved the collaboration and integration of supply chains which in turn yielded other benefits such as inventory management, relationship building, and improved customer service delivery. Likewise, Selase et al. (2019) in investigating the utilization and influence of technology adoption on Ghana's SMEs using SEM with partial least squares (PLS) agreed with other studies (Agwu 2016; Mathu and Tlare, 2017; Igwe et al. 2020) that technology adoption has a positive relationship with market performance. However, this study pinpointed that though there is a positive influence, factors such as compatibility, perceived usefulness of technology, and cost-effectiveness can hamper or aid the effect of technology adoption on financial market performance.

In addition, Dabwor et al. (2017) noted that while financial and economic environments are dynamic in nature, technology is not static. In examining the consequence of technology adoption on the viability of banks' performance in Nigeria using the t-test statistical tool and descriptive research design, Dabwor et al. (2017) found that there is a direct association among technology adoption (web-based transactions, automated teller machines, and mobile payments) and operational performance in banks of which this forms a basis for capital investments in banks. However, this study noted investment in technology as a necessity to ensure and intensify speed, convenience, clear-cut service delivery, profitability, and competitiveness. Also, Olanrewaju (2016) determined the effects of technology adoption on the banking industry's organizational performance in Nigeria using chi-square. It was noted that apart from the positive relationship, the technology adoption enhanced the performance of employees, customer satisfaction, and profitability.

Furthermore, Odum et al. (2017) looked at the influence of technology acquisition on banks' financial performance in Nigeria through a quantitative research technique, Pearson correlation analysis, percentage change trend analysis, and linear regression analysis. This study showed that investment in technology infrastructure positively promotes banks' financial performance in Nigeria in that it has a significant effect on electronic business income, net interest income, and customers' deposits whereas an insignificant effect on net profit after tax. This study also suggested that for financial performance to increase in Nigerian banks, an objective evaluation of technology infrastructure and other fixed assets is necessary to differentiate income assets from those to be divested. The study thus recommends among others that there should be an objective evaluation of the computer equipment as well as other fixed assets of the focused banks in order to distinguish the revenue assets from those that should be divested.

Also, Adeku (2020) investigated the innovation of technology and the challenges, and prospects involved in the Nigerian banking system using a cross-sectional survey design, Pearson correlation, and trend analysis. This research showed that with the aid of technology adoption, automated banking has greater potential. It also noted that in as much as technology innovation significantly improves the competitiveness and service quality (physically and virtually) in banks, it also causes cybercrime. Though no significant association was observed between the high rate of fraud and technology innovation in the financial system of Nigeria, there is a need to secure e-banking platforms and equally increase investment in technology adoption.

According to Kyari and Akinwale (2020), financial technology has revealed the significance of technology in bank service deliveries. The study adopted both ordinary least square regression, descriptive, and inferential analysis and found that the adoption of financial technology in banks is at the medium stage of which the common transactions are payment and transfer of money. Also, it was discovered that the adoption of financial technology positively affects collaboration with external companies (CEC), software technology acquisition (STA), in-house R&D activities (IRD), and hardware technology acquisition (HTA). Also, the finding revealed that a direct association exists among fintech innovation adoption and a 5% financial performance in banks, thus, investment in Fintech enhances robust service delivery and a regulatory environment, promotes wealth creation, and constant economic growth. Similarly, Gambo (2020) determined the influence of technology innovation on the performance of commercial banks in Nigeria. Using quantitative analysis, this study found that technology innovation (automatic teller machines, mobile banking, and internet banking) positively and significantly promotes commercial banks' performance in Nigeria. The study also suggested that for the effectiveness and efficiency of the financial service sector, there is a need for an improvement in technology innovation, financial deepening, and reliability of the services.

Oyewole (2019) studied the association between technology adoption and the productivity of Nigerian ports using the correlation technique. The finding from the study showed that technological adoption raises the performance of Nigerian Ports in that, it positively and significantly affects corporate performance in Nigerian ports. Furthermore, Mwashiuya and Mbamba (2020) established the connection between technology adoption and the performance of microfinance institutions in Tanzania using SEM. It showed that the adoption of technology plays a major role in the relationship of the operational performance (that is, availability convenience, and affordability) of microfinance organizations in Tanzania. It also found and agreed with the positions of Chairoel et al. (2015) and Selase et al. (2019) that investment in technology enhances business operations and communication and reduces costs. Wamboye et al. (2015) also analyzed the nexus between technology adoption and labour productivity growth in Sub-Saharan Africa using quantitative techniques. The findings of this study revealed that there is a strong association among technology adoption and labour productivity growth even though this occurs through the influence of foreign direct investment inflows and economic openness through trade. Therefore, financial development is of great importance in explaining the association between technology adoption and labour productivity growth in SSA economies.

Additionally, Ejiaku (2014) determined the challenges affecting technology adoption in developing economies as infrastructure, government policies, and culture. However, this study noted that for the extent of technology adoption to increase in an emerging economy, there must be government investment. Likewise, Okwu (2015), in a panel study on technology adoption and financial markets in Nigeria and South Africa using panel least squares, opined that technological adoption had heterogeneous effects. While the adoption of technology had a positive effect on capitalization it had a negative effect on market indices. This study likened the adoption of technology to growth and development in Nigeria and South Africa's capital markets, thus the need for more investment in technology. Toader et al. (2018) using the generalized method of moments (GMM) also analyzed the influence of technology adoption on the rate of growth in the European Union. It was revealed that though the effects varied on the basis of the type of technology examined, technology adoption had a robust influence on economic growth in the EU economies. Thus, the study posited that technology adoption is an important channel for promoting economic growth among the EU member countries. Likewise, Alshubiri et al. (2019) analyzed the effect of technology adoption on financial development in Gulf Cooperation Council (GCC) countries using fixed-effect estimation and GMM. This study showed that technology adoption (fixed broadband and Internet users) has a strong influence on financial development among the GCC member countries. Therefore, Alshubiri et al. (2019) opined that technology adoption and well-improved and effective technology infrastructure are necessities for financial development and growth of economic sectors in GCC countries. The study by Edo et al. (2019) examined the influence of the adoption of internet usage on financial development in Nigeria and Kenya using the econometric techniques of vector error correction mechanism (VECM) as well as dynamic ordinary least squares (DOLS). While this study found that financial development in Nigeria and Kenya can be further enhanced through appropriate policies which encourage internet adoption, it showed that internet adoption enhances financial development in Nigeria and Kenya. Additionally, Nigerian Communication Commission (2021) examined the issues facing technology penetration in Nigeria and the role of the infrastructure deficit. Using the descriptive research method, this study revealed that despite the tremendous growth in technology adoption, the economy experienced an infrastructure deficit because technology adoption yielded weak or no economic development in Nigeria. Thus, this study opined that government has a major starring role to execute in addressing the factors affecting technological advancement.

Using the Bayesian vector autoregressive (BVAR) estimation technique to study the association between technology adoption, innovation, and financial development in Africa, Ejemeyovwi et al. (2020) found that impulse shocks from technology innovation positively affects financial development in Africa. This study also posited that technology adoption is required to initiate financial development across all the major sectors in Africa since all sectors utilize funds to improve financial performance. Likewise, Alabi and Olaoye (2022) using a panel study, pooled ordinary least square and feasible generalized least squares estimation techniques, observed the influence of technology adoption on financial inclusion in China and Nigeria. This study showed that while mobile cellular subscriptions, internet usage, and automated teller machines have no significant effect on financial inclusion in China as well as Nigeria, technology and the gross domestic product growth rate have a direct and robust influence on financial inclusion. However, Alabi and Olaoye (2022) in reiteration of the conclusions by Edo et al. (2019) and Ejemeyovwi et al. (2020) posited that more investment is required since more potentials exist in the influence of technology adoption on financial development.

Most studies have analyzed the effect of technology adoption on the performance of small industries within various economies. However, few studies have examined the association between technology adoption and financial market performance in Africa. Hence, it is pertinent to investigate the nexus between technology adoption and financial market performance in Nigeria and South Africa.

3 Methodology

The technology diffusion theory propounded by Chow (1967, 1983) which states that the continuous usage of a particular technology tends to lead to an equilibrium S-shaped path over a given period of time is the foundational theory on which this study is based. The technology diffusion theory is further expanded by Rogers (2003) by highlighting the various elements which affect the diffusion of technology including "relative advantage, compatibility, complexity, and observation". Hence, Rogers (2003) emphasizes that the adoption of technology should lead to an improvement over the current state, be consistent and conducive to the environment where it is adopted, not be too complex to use and understand, and must promote a continuous increase in productivity and gains in the industry or sector wherein it is being used. Similarly, Anyasi and Otubu (2009) describe diffusion as a channel through which technological development is been transferred via a certain and definite path in a given system over a period of time. The study adapts the model specified by Chow (1967, 1983), Okwu (2015, 2016), and, Igwilo and Sibindi (2022) where financial market performance is a direct function of the technology adopted in the financial sector.

$$lnW_{i,t} - lnW_{i,t-1} = \delta_{i,t}\left[lnW^*_{i,t} - lnW_{i,t-1}\right] \tag{1}$$

$W_{i,t}$ is the technology adopted in the financial market i at a given period t, $W_{i,t-1}$ is the technology adopted in the financial market in the i previous year t-1, the $W^*_{i,t}$ is the post-technology adoption equilibrium outcome of the financial market i while, $\delta_{i,t}$ implies the speed of adjustment to long-run equilibrium path in the financial market.

Technology adoption in the financial market is expected to lead to improved market performance. Hence, the key variables of interest are substituted into Eq. (1) in order to reveal the dependent and the explanatory variable.

$$lnFMP_{i,t} - lnFMP_{i,t-1} = \beta_0 + \beta_1 lnTECH_{i,t} + \varepsilon_t \tag{2}$$

where: $FMP_{i,t} - FMP_{i,t-1}$ is the difference between the current and previous indicators of financial market performance outcome due to the adoption of technology, $TECH_{i,t}$ comprises the various indicators of technology adoption, ln is the natural logarithm, the constant β_0 reveals the financial market outcome in the prior to the technology adoption, β_1 stands for the coefficient of technology which measures how financial market performance respond to technology adoption while ε_t is the error term model at time t.

The Eq. (2) is transformed and expanded to accommodate other relevant intervening variables relevant for this study.

$$\Delta lnFMP_{k,i,t} = \Delta\left[\beta_0 + \beta_{k,i,t}\sum\nolimits_{i=1}^{l} TECH_{k,i,t} + \beta_{k,i,t}\sum\nolimits_{i=1}^{m} X_{k,i,t} + \varepsilon_t\right] \tag{3}$$

where: Δ is the difference, $\sum$ is the summation, $X_{k,i,t}$ is the various intervening variables, and $_k$ is a descriptor. Equation (3) is a short-run model and to obtain the long-run model, the difference component from the model must be eliminated as demonstrated in Eq. 4.

$$lnFMP_{k,i,t} = \beta_0 + \beta_{k,i,t}\sum_{i=1}^{l} TECH_{k,i,t} + \beta_{k,i,t}\sum_{i=1}^{m} X_{k,i,t} + \varepsilon_t \tag{4}$$

The Eq. (4) is then linearized and expanded to accommodate all the various variables used in this study.

$$lnFMP_{t,N} = \beta_0 + \beta_1 lnTECH_{t,N} + \beta_2 lnFFI_{t,N} + \beta_3 lnGDP_{t,N} + \beta_4 lnCPS_{t,N} + \beta_5 lnBDR_{t,N} + \beta_6 BLR_{t,N} + \varepsilon_{t,N} \tag{5a}$$

$$lnFMP_{t,SA} = \beta_0 + \beta_1 lnTECH_{t,SA} + \beta_2 lnFFI_{t,SA} + \beta_3 lnGDP_{t,SA} + \beta_4 lnCPS_{t,SA} + \beta_5 lnBDR_{t,SA} + \beta_6 BLR_{t,SA} + \varepsilon_{t,SA} \tag{5b}$$

The Eq. 5(a,b) is derived to illustrate a country-specific analysis between Nigeria and South Africa where *N* stands for Nigeria and *SA* stands for South Africa before the panel model combining the two countries is demonstrated in Eq. (6).

$$lnFMP_{i,t} = \beta_0 + \beta_1 lnTECH_{i,t} + \beta_2 lnFFI_{i,t} + \beta_3 lnGDP_{i,t} + \beta_4 lnCPS_{i,t} + \beta_5 lnBDR_{i,t} + \beta_6 BLR_{i,t} + \varepsilon_{i,t} \tag{6}$$

where: "FMP stands for financial market performance, TECH stands for technology adoption, FFI stands for financial freedom index, GDP stands for the gross national product, CPS stands for domestic credit to private sector, BDR stands for bank deposit rate and BLR stands for banking lending rate".

This study comprises of a country-specific as well as panel data analysis on the nexus between technology adoption and financial market performance in Nigeria and South Africa spanning the period of 1980 to 2021. Variables such as "fixed broadband subscription, fixed broadband subscription per 100 persons, fixed telephone subscriptions, fixed telephone, number of individuals with access to internet, number of secure servers, and secure internet servers per 1 million people have been used at one time or the other as a measure of technology adoption while market capitalization of listed domestic firms as a percentage of the GDP, market capitalization of listed firms at the current US dollar, the total value of stock traded as a percentage of GDP, total value of stock traded, turnover ratio of stock traded as well as total number of listed domestic firms" have been used as a measure of the financial market (see Okwu 2016; Igwilo and Sibindi 2022). These individual measures are susceptible to the problems of autocorrelation and estimation bias arising from both the dependent and independent variables. To correct this anomaly, the study adopts the principal component analysis to estimate an index that sufficiency captures both technology adoption (TECH) and financial market performance (FMP) in Nigeria and South Africa. All variables used in the indexes as well as the control variables like "gross domestic product (GDP), domestic credit to the private sector (CPS), bank deposit rate (BDR), and bank lending rate (BLR)" were gotten from the World Bank's development indicators database while the financial freedom index (FFI) was obtained from the Heritage database.

4 Analysis and Interpretation of Result

This section adopts both a country-specific as well as a panel analysis to establish the impact of the technology adoption on the financial market performance in Nigeria and South Africa using the generalized method of moment (GMM) approach as well as ordinary least squares (OLS) and two-stages least squares (2SLS) regression for robustness check on the estimates. The GMM method is chosen because of its superiority over other techniques of analysis like ordinary least squares, two-stage least squares, and feasible generalized least squares and it addresses the problem of endogeneity that may arise in the explanatory variables (Arellano and Bond 1991).

The coefficient of the long-run estimates of the country-specific OLS regression between Nigeria and South Africa as presented in Table 1 reveals that technology adoption (TECH) has a positive and substantial effect on financial market performance (FMP) in Nigeria ($t = 5.21$; $p < 0.05$) while for South Africa, technology adoption (TECH) has an adverse and substantial impact ($t = -2.53$; $p < 0.05$) on financial market performance (FMP). This suggests that a unit rise in technology adoption has a robust and direct influence on the financial market in Nigeria but in South Africa, it will have a negative consequence. Findings on Nigeria support studies such as Rogers (2003), Olanrewaju (2016), Dabwor *et al.* (2017), Odum *et al.* (2017), Adeku (2020), Kyari and Akinwale (2020), Igue *et al.* (2021) who argue that technology adoption promotes the viability of financial market while in the case of South Africa, the finding contradicts previous studies such as Rambe and Mpiti (2017), Thapelo (2020), and Mathu and Tlare (2017) who argue that technology adoption is of great importance to the economy, most especially, in promoting the financial market performance.

Also, the financial freedom index (FFI) has a positive and somewhat significant impact ($t = 1.84$; $p < 0.10$) on financial market performance in Nigeria but in South Africa, the financial freedom index (FFI) has a negative but insignificant impact ($t = -0.23$; $p > 0.05$) on the financial market performance. This finding implies that potential investors largely enjoy more freedom in the choice of their investment decisions in Nigeria than in South Africa. The finding is in line with study by Igwilo and Sibindi (2022).

Interestingly, the study finds that the gross domestic product (GDP) ($t = 2.07$; $p < 0.05$) and domestic credit to the private sector (CPS) ($t = 2.15$; $p < 0.05$) positively and significantly promote financial market performance in South Africa while in Nigeria, the gross domestic product (GDP) and domestic credit to private sector merely reveal a positive but inconsequential impact on financial market performance. This implies that the South African economy is more buoyant and provides more credit facilities to domestic investors than what is obtainable in Nigeria. This finding supports the study by Chairoel (2015).

The estimate of the two-staged least squares presented in Table 2 reveals that technology adoption (TECH) has a positive and somewhat significant ($t = 1.92$; $p < 0.10$) on financial market performance in Nigeria but in South Africa, technology adoption (TECH) has a negative and insignificant impact ($t = -1.04$; $p > 0.05$) on the financial market performance. This validates the finding of the country-specific ordinary least squares presented in Table 1.

The estimate of the generalized method of moment (GMM) which is the main method of analysis adopted due to its superiority over both ordinary least squares and two-staged least squares also reveals that technology adoption (TECH) has a positive and important effect ($t = 2.28$; $p < 0.05$) on the financial market performance while in South Africa, technology adoption (TECH) has a negative but insignificant impact ($t = 1.47$; $p > 0.05$) on the financial market performance. Therefore, the country-specific analysis is throwing up an interesting economic scenario in the nexus between technology adoption and financial market performance among the two leading African economies which will go a long way to influence the implementation of the AfCFTA on the continent (Table 3).

Table 1. Ordinary least squares regression

Dependent variable: FMP	Ordinary least square regression							
	Nigeria				South Africa			
	Coefficient	Standard Deviation	t-statistic	Probability	Coefficient	Standard Error	t-Statistic	Probability
C	−0.54	2.18	−0.25	0.80	−7.31	4.08	−1.79	0.08
TECH	0.89	0.17	5.21	0.00*	−0.49	0.19	−2.53	0.01*
FFI	0.05	0.03	1.84	0.07*	−0.00	0.03	−0.23	0.81
GDP	3.97	2.40	1.65	0.11	7.64	3.69	2.07	0.04*
CPS	0.11	0.09	1.26	0.21	0.05	0.02	2.15	0.04*
BDR	0.06	0.13	0.46	0.64	0.23	0.17	1.35	0.18
BLR	−0.23	0.14	−1.64	0.11	−0.13	0.10	−1.28	0.21

Source: Authors Computation, 2022.

Table 2. Two-staged least squares regression

Dependent variable: FMP	Two-staged least square regression							
	Nigeria				South Africa			
	Coefficient	Standard Deviation	t-statistic	Probability	Coefficient	Standard Error	t-Statistic	Probability
C	6.17	11.45	0.53	0.59	−1.41	48.22	−0.85	0.40
TECH	1.18	0.61	1.92	0.06*	−1.81	1.73	−1.04	0.30
FFI	0.19	0.26	0.74	0.46	0.30	0.45	0.67	0.50
GDP	−4.58	1.74	−2.26	0.79	−7.53	2.43	−0.30	0.75
CPS	−0.02	0.38	−0.05	0.95	0.19	0.21	0.88	0.38
BDR	−0.49	1.52	−0.32	0.74	−0.13	0.81	−0.17	0.86
BLR	−0.36	1.00	−0.36	0.71	0.43	0.85	0.50	0.61

Source: Authors Computation, 2022.

Given the future economic engagements in Africa, most especially, because of the important role of both Nigeria and South Africa in directing the trajectory of the development path in Africa as it concerns the objective of promoting intra-African trade and enhancing viable financial market through the AfCFTA initiative, this study investigates the possible expected outcome of pulling both Nigeria and South African economic

Table 3. Generalized method of moments

Dependent variable: FMP	Generalized method of moments							
	Nigeria				South Africa			
	Coefficient	Standard Deviation	t-statistic	Probability	Coefficient	Standard Error	t-Statistic	Probability
C	6.17	12.09	0.51	0.61	−41.13	32.48	−1.26	0.21
TECH	1.18	0.52	2.28	0.03*	−1.81	1.22	1.47	0.15
FFI	0.19	0.21	0.92	0.36	0.30	0.25	1.20	0.24
GDP	−4.58	1.51	−0.30	0.76	−7.53	2.52	−0.29	0.76
CPS	−0.02	0.32	−0.06	0.94	0.19	0.22	0.86	0.39
BDR	−0.49	1.33	−0.36	0.71	−0.13	0.72	−0.19	0.84
BLR	−0.36	1.08	−0.33	0.73	0.43	0.46	0.93	0.35

Source: Authors Computation, 2022.

indices together for a better insight using a pulled panel data estimation of the panel least squares, panel two-staged least squares and panel generalized method of the moment on Table 4.

The panel least squares estimate reveals that technology adoption (TECH) has a positive and significant impact ($t = 3.18; p < 0.05$) on the financial market performance of both Nigeria and South Africa. This implies that shared technology adoption between these two countries significantly promotes the financial market performance of the two countries. This is in line with studies by Irvin (2005), Okwu (2015) Wamboye et al. (2015), AU (2020), and Ejemeyovwi et al. (2020) who argue that technology adoption has a robust influence on the economy of Africa countries and thus should be embraced for a deepened economic relationship among African economies.

Also, the gross domestic product (GDP) ($t = 3.61; p < 0.05$) and the bank deposit rate (BLR) ($t = 5.34; p < 0.05$) in both Nigeria and South Africa positively and significantly promote the financial market performance of the two leading African economies. This implies that the volume of growth in the economies of both Nigeria and South Africa will significantly promote financial market performance and the confidence in the money markets has been enhanced which will ultimately promote the financial market. The bank lending rate (BLR) ($t = -1.95; p < 0.05$) has a adverse and substantial impact on the financial market performance. This implies that measures must be put in place to drive down the banking lending rate to a level that will ensure the promotion of the financial market performance in both Nigeria and South Africa.

The panel two-staged least squares estimate shows that technology adoption (TECH) ($t = 3.33; p < 0.05$), gross domestic product (GDP) ($t = 2.95; p < 0.05$) and the bank deposit rate (BDR) ($t = 3.23; p < 0.05$) has direct and major influence on the financial market performance in both Nigeria and South Africa. While, the panel generalized method of the moment which is the main technique of analysis due to its superiority over panel least squares and panel two-staged least squares also shows that technology adoption (TECH) ($t = 3.33; p < 0.05$), gross domestic product (GDP) ($t = 2.95; p < 0.05$) and the bank deposit rate (BDR) ($t = 3.23; p < 0.05$) has a direct and substantial impact on the financial market performance in both Nigeria and South Africa. Therefore,

Table 4. Pulled panel data estimations on Nigeria and South Africa

Dependent variable: FMP	Panel least squares				Panel two-staged least squares				Panel general method of moments			
	Coefficient	Standard Deviation	t-statistic	Probability	Coefficient	Standard Deviation	t-statistic	Probability	Coefficient	Standard Deviation	t-statistic	Probability
C	−1.89	1.55	−1.21	0.22	−0.04	2.23	−0.01	0.98	−0.04	2.23	−0.01	0.98
TECH	0.46	0.14	3.18	0.00*	0.53	0.16	3.33	0.00*	0.53	0.16	3.33	0.00*
FFI	0.03	0.03	1.06	0.29	0.00	0.04	0.15	0.89	0.00	0.04	0.15	0.87
GDP	6.90	1.91	3.61	0.00*	8.16	2.76	2.95	0.00*	8.16	2.76	2.95	0.00*
CPS	−0.04	0.06	−0.59	0.55	−0.15	0.12	−1.28	0.20	−0.15	0.12	−1.28	0.20
BDR	2.23	4.17	5.34	0.00*	2.02	6.23	3.23	0.00*	2.02	6.23	3.23	0.00*
BLR	−0.03	0.01	−1.95	0.05*	−0.02	0.02	−1.11	0.27	−0.02	0.02	−1.11	0.27

Source: Authors Computation, 2022.

this study supports the call for more integrated African economies that will promote the activities of the financial market through technological innovations. This finding validates the estimate of both the panel least squares as well as the panel two-staged least squares.

4.1 Conclusion and Recommendation

The study concludes that in country-specific analysis, technology adoption positively and significantly promotes financial market performance in Nigeria while in South Africa, technology adoption does not promote financial market performance. It has a substantial negative influence on the financial market performance. However, using the panel estimates, the study concludes that technology adoption, gross domestic product, and bank deposit rates positively and considerably promote the financial market performance in Nigeria and South Africa. However, measures should be put in place to drive down the banking lending rate to a level that will enhance the financial market performance in both Nigeria and South Africa to fully actualize a robust financial market that will support and promote intra-African trade through technology adoption.

References

Abdulquadri, A., Mogaji, E., Kieu, T.A., Nguyen, N.P.: Digital transformation in financial services provision: a Nigerian perspective to the adoption of chatbot. J. Enterpris. Commun. People Places Global Econ. **15**(2), 258–281 (2021). https://doi.org/10.1108/JEC-06-2020-0126

Adeku, S.O.: Technology innovation in the Nigerian banking system: prospects and challenges. Rajagiri Manage. J. **15**(1), 2–15 (2020). https://doi.org/10.1108/RAMJ-05-2020-0018

African Union (2020). The digital transformation strategy for Africa (2020–2030). Addis Ababa, Ethiopia

Agwu, E.: ICT diffusion, adoption and strategic importance in Nigerian SMEs. Int. J. Res. Manage. Sci. Technol. **4**(3), 1–25 (2016)

Alabi, A.W., Olaoye, F.O.: The effect of technology adoption on financial inclusion: a cross-country panel analysis between China and Nigeria. Eur. J. Bus. Manage. Res. **7**(2), 1–11 (2022). https://doi.org/10.24018/ejbmr.2022.7.2.1314

Alshubiri, F., Jamil, S.A., Elheddad, M.: The impact of ICT on financial development: empirical evidence from the Gulf cooperation council countries. Int. J. Eng. Bus. Manage. **11**, 1–14 (2019)

Anyasi, F.I., Otubu, P.A.: Mobile phone technology in Banking system. Its economic effect. Res. J. Inf. Technol. **1**, 1–5 (2009)

Arellano, M., Bond, S.: Some tests of specification for panel data: Monte Carlo evidence and application to employment equations. Rev. Econ. Stud. **58**, 277–297 (1991)

Cliff, D., Brown, D., Treleaven, P.: Technology trends in the financial markets: A 2020 vision. The Future of Computer Trading in Financial Markets Driver Review – DR 3 (2011)

Centre of Excellence in Financial Services. The impact of the 4th industrial revolution on the South African financial services market. The Centre of Excellence in Financial Services, South Africa (2022)

Chairoel, L., Widyarto, S., Pujani, V.: ICT adoption in affecting organizational performance among Indonesian SMEs. Int. Technol. Manage. Rev. **5**(2), 82–93 (2015)

Chow, G.C.: Technical change and the demand for computers. Am. Econ. Rev. **57**(1), 17–30 (1967)

Chow, G.C.: Econometrics. McGraw-Hill, New York (1983)
Dabwor, T.D., Ezie, O., Anyatonwu, P.: Effect of ICT adoption on competitive performance of banks in an emerging economy: The Nigerian experience. IOSR. J. Human. Soc. Sci. (IOSR-JHSS). **22**(8), 81–89 (2017)
Dangolan, S.K.: Impact of information technology in banking system: a case study in a Bank of Keshavarzi, Irani. Proc. Soc. Behav. Sci. **30**, 13–16 (2011)
Economic Commission for Africa. Economic report on Africa 2020: Innovative finance for private sector development in Africa. Addis Ababa (2020)
Edo, S., Okodua, H., Odebiyi, J.: Internet adoption and financial development in Sub-Saharan Africa: evidence from Nigeria and Kenya. Afr. Dev. Rev. **31**(1), 144–160 (2019)
Ejemeyovwi, J.O., Osabuohien, E.S., Bowale, E.I.K.: ICT adoption, innovation and financial development in a digital world: empirical analysis from Africa. Trans. Corp. Rev. **13**, 16–31 (2020). https://doi.org/10.1080/19186444.2020.1851124
Ejiaku, S.A.: Technology adoption: issues and challenges in information technology adoption in Emerging Economies. J. Int. Technol. Inf. Manage. **23**(2), 5 (2014)
Enu, P., Gberbi, J.T.: Effect of information and communication technology (ICT) on the delivery of banking services in Ghana: a case study of Zenith Bank Ghana Limited. Global J. Manage. Stud. Res. **2**(2), 60–82 (2015)
Etim, A.S.: Mobile technology adoption for microfinance delivery in Sub-Saharan Africa School of Business and Economics. Winston Salem State University, North Carolina Department of Accounting and Management Information Systems (2011)
Ezirim, B.C., Adebajo, U.R., Elike, U., Muoghalu, I.M.: Capital market growth and information technology: empirical Evidence from Nigeria. Int. J. Bus. Econ. Perspect. **4**(1), 1–17 (2019)
Gambo, N.: Effects of technology innovation on financial performance of commercial banks in Nigeria. J. Manage. Sci. Entrepreneursh. **15**(7), ISSN 2285-3138 (2020)
Gono, S., Harindranath, G., Özcan, G.B.: Challenges of ICT adoption by South African SMEs: a study of manufacturing and logistic firms. School of Management, Royal Holloway University of London, UK (2013)
Hassan, H., Ogundipe, A.: ICT adoption by micro and small scale enterprises in Nigeria: a case study of the Federal Capital Territory, Abuja (2017).https://doi.org/10.2139/ssrn.2951901
Hitt, L.M., Frei, F.X., Harker, P.T.: How Financial Firms Decide on Technology. University of Pennsylvania, Wharton School (1998)
Heritage Database: https://heritage.org/index/explore?view=by-region-country-year&u=637913083162570057
Igue, C., Alinsato, A., Agadjihouédé, T.: E-commerce in Africa: issues and challenges. In: Chapter 5, pp. 118–139 (2021)
Igwe, S.R., Ebenuwa, A., Idenedo, O.W.: Technology adoption and sales performance of manufacturing small and medium enterprises in Port Harcourt. J. Mark. Develop. **5**(1), 44–49 (2020)
Igwilo, J.I., Sibindi, A.B.: ICT adoption and stock market development: Empirical evidence using a panel of African countries. Risks **10**(25), 1–17 (2022)
International Telecommunication Union (ITU): Digital trends in Africa: Information and communication technology trends and development in the African region 2017–2020. ITU Publications, Geneva (2021)
Irvin, J.: Regional integration of stock exchanges in Eastern and Southern Africa: Progress and prospects. In: IMF Working Paper WP/05/122, African and International Capital Markets Department, International Monetary Fund (2005)
James, T., Esselar, P., Miller, J.: Towards a better understanding of the ICT sector in South Africa: problems and opportunities for strengthening the existing knowledge base, pp. 1–34 (2001)
Jenkin, N., Naude, R.: Digitalization: developing competencies for a just transition of the South African banking sector. In: Working Paper, University of Witwatersrand, Johannesburg (2018)

Kairu, M.M., Rugami, M.: Effect of ICT deployment on the operational performance of Kenya revenue authority. J. Strat. Manage. **2**(1), 19–35 (2017)

Kajogbola, D.O.: The impact of information technology on the Nigerian economy: a study of manufacturing and service sectors in the South Western and South Eastern Zones of Nigeria. In: ATPS Working Paper Series No. 39 (2004)

Kauffman, R., Riggins, F.: Information and communication technology and the sustainability of microfinance. Electron. Commer. Res. Appl. **11**, 450–468 (2012)

Kombe, S., Wafula, M.: Effects of internet banking on the financial performance of commercial banks in Kenya. Int. J. Sci. Res. Publ. **5**(5), 1–11 (2015)

Kyari, A.K., Akinwale, Y.O.: An assessment of the level of adoption of financial technology by Nigerian Banks. Afr. J. Sci. Policy Innov. Manage. **1**, 118–130 (2020)

Mathu, K., Tlare, M.T.: The impact of IT adoption in SMEs supply chains: a case of Gauteng and Free State Provinces of South Africa. S. Afr. J. Bus. Manage. **48**(3), 63–71 (2017). https://doi.org/10.4102/sajbm.v48i3.36

Matsepe, N.T., Van der Lingen, E.: Determinants of emerging technologies adoption in South African Financial Sector. S. Afr. J. Bus. Manage. **53**(1), 1–12 (2022)

Mwashiuya, H.T., Mbamba, U.O.: Relationship of information and communication technology adoption on microfinance institutions operational performance and access to financial services in Tanzania. Int. J. Inf. Bus. Manage. **12**(1), 214–237 (2020)

Nigerian Communications Commission: Challenges of technology penetration in an infrastructure deficit economy (Nigeria Perspective) (2021)

National Planning Commission: Digital futures: South Africa's digital readiness for the fourth industrial revolution. Draft: National Planning Commission, South Africa (2020)

Ntimane, L.M.: Disruptive innovation in the South Africa banking sector: A case study of Capitec Bank. Unpublished thesis, Stellenbosch University, South Africa (2020)

Obiakor, R.T., Okwu, A.T.: Empirical analysis of the impact of capital market development on Nigeria's economic growth (1981–2008: case study of the Nigerian stock exchange. Bus. Econ. Rev. **20**(2), 79–96 (2011)

Odum, A.N., Odum, C.G., Chukwu, E.E.: Technological infrastructure investments and financial performance of banks in Nigeria. Int. Digital Organ. Sci. Res. IDOSR J. Current Issues Soc. Sci. **3**(1), 1–22 (2017)

Ohiani, A.S.: Technology innovation in the Nigerian banking system: prospect and challenges. Rajagiri Manage. J. **15**(1), 2–15 (2020)

Okundaye, K., Fan, S.F., Dwyer, R.J.: Impact of information and communication technology in Nigeria small to medium sized enterprises. J. Econ. Financ. Administr. Sci. **24**(47), 29–46 (2019)

Okwu, A.T.: ICT adoption and financial markets: a study on the leading stock exchange markets in Africa. J. Acc. Manage. **5**(2), 53–76 (2015)

Okwu, A.T.: ICT and stock market nexus in Africa: evidence from Nigeria and South Africa. ACTA Universitatis Danubius **12**(4), 35–50 (2016)

Olaitan, O., Issah, M., Wayi, N.: A framework to test South Africa's readiness for the fourth industrial revolution. S. Afr. J. Inf. Manage. **23**(1), 1–10 (2020)

Olanrewaju, B.E.: Effects of information technology on organisational performance in Nigerian banking industries. Res. J. Financ. Acc. **7**(3), 52–64 (2016)

Oyewole, F.O.: Technology adoption and performance of Nigerian ports. RSU J. Strat. Internet Bus. **4**(2), 714–746 (2019)

Pillay, P.: Barriers to information and communication technology (ICT) adoption and use amongst SMEs: a study of South African manufacturing sector. Unpublished thesis, University of Witwatersrand, Johannesburg, South Africa (2016)

Ramachandran, V., Obado-Joel, J., Fatai, R., Masood, J.S., Omakwu, B.: The new economy of Africa: Opportunities for Nigeria's emerging technology sector. Center for Global Development (2019)
Rambe, P., Mpiti, N.: The influence of private and public finance, organizational and environmental variables on the performance of beauty salons in the Free State, South Africa: a theoretical perspective. Int. Bus. Econ. Res. J. **16**(2), 1–18 (2017)
Rogers, E.M.: Diffusion of Innovation, 5th edn. The Free Press Rouse, New York (2003)
Slazus, B.J., Bick, G.: Factors that influence fintech adoption in South Africa: a case study of consumer behavior towards branchless mobile banking. Athens J. Bus. Econ. **8**(1), 43–64 (2020)
Selase, A.M., Selase, A.E., Ayishetu, A.-R., Comfort, A.D., Stanley, A., Ebenezer, G.-A.: Impact of technology adoption and its utilization on SMEs in Ghana. Int. J. Small Medium Enterp. **2**(2), 1–13 (2019). https://doi.org/10.46281/ijsmes.v2i2.382
Smidt, H.J.: Factors affecting digital technology adoption by small-scale farmers in agriculture value chains in South Africa. Inf. Technol. Dev. **28**, 558–584 (2021). https://doi.org/10.1080/02681102.2021.1975256
Soutter, L., Ferguson, K., Neubert, M.: Digital payments: impact factors and mass adoption in sub-Saharan Africa. Technol. Innov. Manag. Rev. **9**(7), 41–55 (2019)
Thapelo, P.: The Role of government financial support and innovation adoption on the performance of SMEs in South Africa. Unpublished thesis. Faculty of Commerce, Law and Management, University of the Witwatersrand, Johannesburg (2020)
Tinn, K.: The speed of technology adoption with imperfect information in equity markets. London School of Economics, Financial Market Group (2006)
Tinn, K.: Technology adoption with exit in imperfectly informed equity markets. Stockholm School of Economics (2008)
Toader, E., Firtescu, B. N., Roman, A., Anton, S.G.: Impact of information and communication technology infrastructure on economic growth: an empirical assessment for the EU Countries. Sustainability. **10**, 3750 (2018). https://doi.org/10.3390/su10103750. www.mdpi.com/journal/sustainability
United Nations Conference on Trade and Development (UNCTAD): Création et Captation de Valeur: Incidences sur les pays en Développement, Rapport sur l'Economie Numérique 2019, UNCTAD Publications (2019)
Wamboye, E., Tochkov, K., Sergi, B.S.: Technology adoption and growth in sub-Saharan African countries. Comp. Econ. Stud. **57**, 136–167 (2015)
Wasilwa, N.S., Omwenga, J.: Effects of ICT strategies on performance of commercial banks in Kenya: a case of Equity Bank. Int. J. Sci. Res. Publ. **6**(11), 382–400 (2016)
World Bank: Global Financial Development Report 2014: Financial Inclusion. World Bank Publications, Washington, DC (2014)
World Bank: South Africa: Digital economy diagnostic. The World Bank Group (2019)
World Bank: Africa's pulse: An analysis of issues shaping Africa's economic future. Covid-19 and the future of work in Africa: emerging trends in digital technology adoption, 23, 1–90 (2021)
World Bank (2022). https://databank.worldbank.org/source/world-development-indicators
Yusuf, E.: Influence of entrepreneurship education, technology and globalization on performance of SMEs in Nigeria. Afr. J. Bus. Manage. **11**(15), 367–374 (2017)

Burnt Clay Grinding Pot Waste Powder as a Partial Replacement of Ordinary Portland Cement for Concrete Production

A. Nimo-Boakye[1(✉)], E. Nana-Addy[1], and K. Adinkrah-Appiah[2]

[1] Department of Building Technology, Sunyani Technical University, Sunyani, Ghana
nimoboakye@yahoo.co.uk

[2] Department of Civil Engineering, Sunyani Technical University, Sunyani, Ghana

Abstract. Purpose: The paper explored the potential of Burnt clay grinding pot waste powder (BCGPWP) as a partial replacement for Ordinary Portland Cement for concrete production.

Design/Methodology/Approach: Given such aim, chemical analysis was conducted on the (BCGPWP)in the Ghana Water Company laboratory. One hundred and five (105) concrete cubes were prepared and tested for compressive and split tensile strength. The mix ratio for the specimen were 0%, 5%, 10%, 15%, 20%, 25% and 30% respectively. A concrete mix ratio of 1:2:4 was used. The (BCGPWP) percentages formed part of the one-part cement. The concrete specimen was cured for twenty-eight (28) days under the shed, covered with woolen material and sprinkle with water on them every day. The compressive and split tensile tests were done on the 7, 14, 21 and 28 days.

Findings: It was established that the sample contains about eight essential oxides found in Ordinary Portland Cement (OPC). The constituents were Calcium oxide (CaO), Silica oxide (Sio2), Iron oxide(fe_2O_3), Aluminium oxide (Al_2O_3), Magnesium oxide (MgO), and Sodium oxide (Na_2O), Potassium Oxide (K_2O) and Sulphur Oxide (S_O3). On the 28th day of testing, the control specimen had 17.05 N/mm^2, 5% replacement had 16.53%, 10% replacement also recorded 16.55 N/mm^2 with 15% declining to 7.46 N/mm^2. Twenty percent went up to 17.57 N/mm^2 with 25% registering a compressive strength of 15.18 N/mm^2 while the 30% recorded 15.35 N/mm^2.

Implications/Research Limitations: The study could not ascertain the type of clay used for the production of the traditional burnt clay grinding pot. Again, the study could not unearth the temperature at which the pot/pot waste were subjected to before it came out of the fire.

Practical Implications: The usage of Burnt clay grinding pot waste powder as a partial replacement for Ordinary Portland Cement for concrete production will help to reduce the cost of construction.

Social Implications: It will help in improving the environment by recycling waste into construction materials for use.

Originality/Value: The use of burnt clay grinding pot waste powder as a partial replacement for Ordinary Portland Cement for concrete production is a novelty in construction material development.

C. Aigbavboa et al. (Eds.): ARCA 2022, *Sustainable Education and Development – Sustainable Industrialization and Innovation*, pp. 953–966, 2023.
https://doi.org/10.1007/978-3-031-25998-2_73

Keyword: Burnt clay · Curing · Compressive strength · Grinding pot · Partial replacement

1 Introduction

The need for locally manufactured building materials has been emphasized in many countries including Ghana. There is imbalance between the expensive conventional building materials coupled with depletion of traditional building materials. To address this situation, attention has been focused on low-cost alternative building materials (Agbede and Manasseh 2008). The use of locally available material can help reduce the high cost of building and reduce the current housing deficit which stands at 2 million units (Graphic online 2021) This was emphasized by the statement made by the Minister of Works and Housing, Mr Francis Asenso-Boakye. The report said, currently it is estimated that 60 percent of the country's population would need some form of government assistance to help them to get access to housing, while 35 per cent would not be able to access housing even with government support in terms of subsidy. The above information shows that there the need to get an alternative building material to support the imported ones which at the end of the day reduce the cost of constructing building to accommodate the growing population.

Cement as a binding material is very useful material in construction industry, without it most buildings may not stand and perform it expected role. It is used for the preparation of concrete, mortar and similar products. Cement is not locally available materials as fine and coarse aggregates are used to be. It is an imported material either in a finish state or its raw materials. In view of that its cost keeps on rising every day. Though there are several brands of cement at the market in Ghana, but the cost of a 50 kg bag of cement is higher. As of August 2022 bags of cement were sold between Ghs 60.00 to Ghs 68.00 in Sunyani.

Cement contributes to the high cost of cement product as noted by the research conducted by Abdulahi (2006). The report stated that, cement, as a binder, is the most expensive input in the production of sandcrete blocks. This has necessitated producers of sandcrete blocks to produce blocks with low OPC content that will be affordable to people and with much gain to the blocks manufacturers (Sholanke et al 2015). Due to the high cost of the conventional cement, there is the need to look for another alternative material which can supplement the conventional cement. Agricultural product such as rice husk ash is of the materials. When burnt under controlled conditions, the rice husk ash (RHA) is highly pozzolanic and very suitable for use in lime-pozzolana mixes and for Portland cement replacement (Sholanke et al 2015). Clay naturally has some binding properties and when heated at a high temperature, it might gain more binding properties. In the course of producing the traditional clay grinding pot, a lot of waste are produced in the process. These damaged or broken pot becomes waste; therefore, this study seeks to find out whether the waste when processes into powder form after burning could contains some cementitious compounds.

A gain to find out if the powder could be used to partially replace normal cement to produce concrete with adequate strength. The aim of the study is to assess the potentials of burnt clay grinding pot waste powder as binding material to supplement ordinary pot land cement in concrete production. The objective of the study are as follows; to conduct the chemical analysis test of the burnt clay grinding pot waste powder to find out whether it contains some cementitious compounds; and to combine the burnt clay grinding pot waste powder to OPC for the production of concrete to determine its mechanical properties such as compressive strength, split tensile and water absorption through laboratory test.

2 Materials and Methods

The materials for the preparation of the of the experiment include, burnt clay grinding pot waste, pit sand, granite chippings or stones, cement and water.

2.1 Burnt Clay Grinding Pot Waste

The burnt clay grinding pot waste was obtained from Tanoso in Ahafo region where they manufacture the products.

Ordinary Portland Cement

Ordinary Portland cement (GHACEM 32.5R) will was obtained from a hardware store in the Sunyani municipality which was used as the main binder for the preparation of the specimen or the test samples.

Fine Aggregate (Sand)

The fine aggregate used for the specimen was purchased from an aggregate seller at Sunyani near the Red Cross Eye clinic. It was checked to make sure that it does not contain much clay and any other unwanted materials.

Water

Pipe borne water was used for the mixing of the samples for the casting of the test specimen for the study. The water used was pure since it was fit for drinking.

3 Materials Preparation

3.1 Burnt Clay Grinding Pot Waste Powder (BCGPWP)

The above was the only material that required rigorous preparation since all them finished products or did not require any preparation. After gathering the burnt clay grinding pot waste (BCGPW), they were sorted to do away with any unwanted materials, then sent to a stone grinding machine where the powder was produced to get the BCGPWP. The pictures below highlight the preparation and processing methods.

3.2 Batching of the Materials for the Spacemen

Weight batching was used to determine the constituent's weight for the specimen since it gives a greater accuracy. A total of 105 specimen or samples were be prepared, 15 cubes from each percentage (0%, 5%, 10%, 15%, 20%, 25%, 30%). The measured material was mixed in a mixing bowl or pan and casted into a mould. The moulded cubes measured $100 \times 100 \times 100$mm were cured for maximum of 28 days.

Tools and Apparatus

The tools and apparatus used for the mixing included head pan, trowel, shovel and measuring cylinder. The automatic compressive strength test machine at the Building Technology Laboratory of Sunyani Technical University was used for the compressive strength test as well as the split tensile strength test of the specimen. The electronic balance apparatus was used for the moisture absorption test of the specimen.

Ratio of the mix

The ratio of the mix was 1: 2: 4, that is one part of cement, two parts of sand and four parts of coarse aggregate. The burnt clay grinding pot waste powder was included in one part of cement in the various percentages.

Mixing Procedures

The required quantities of dried materials were measured according to the mix proportion stated. The fine aggregates and the cement were first poured onto the mixing pan and mixed manually in a dry state till a uniform colour is obtained without any lumps or clusters. Water was then added to the mixture while mixing until the require paste was obtained.

Curing of the Moulding Specimen or Sample

The mould specimen was kept in the mould after which they removed for curing. The samples were cured under a shed, covered with woolen material and water them every day. The curing was done for 28 days and the testing was done intermittently for seven (7), fourteen (14), twenty-one (21) and twenty-eight (28) days respectively.

Testing of the specimen

The samples or the specimen for experiment were tested at Building Technology laboratory using compressive testing machine to determine the compressive as well as split tensile strength. Before the testing, the weight of each cube was recorded and used to set the compressive test machine. The test results were record and used for the analysis.

Water assumption test was also be conducted. The dry weight of the specimen was recorded after which they immersed in water for 72 h. They were removed from the water, allowed to drain before they were re-weigh and their values recorded. They were put in the formular;

$$\frac{\text{Wet weight-Dry weight}}{\text{Dry weight.}} \times 100.$$

Analysis of Results

The data obtained from the laboratory analyses, recorded and present in tables, graph and bar chart.

4 Data Presentation and Analysis

Data obtained from the experiment is being presented, starting from the chemical analysis of the burnt clay grinding pot waste powder (BCGPWP).

4.1 Chemicals Analysis of Burnt Clay Grinding Pot Waste Powder (BCGPWP)

Chemicals analysis of burnt clay grinding pot waste powder (BCGPWP) conducted in Ghana Water Company laboratory is presented in the Table 1. It shows the breakdown of the basic constituents or the oxide composition. The composition of the cement used for the experiment was also analysed at the same Water Company's lab.

Table 1. Chemical analysis results for (BCGPWP) and OPC

Constituents	Oxide Composition mg/l	
	BCGPWP	OPC
Calcium Oxide (CaO)	200	600
Silica Oxide (Sio2)	150	200
Iron Oxide (fe_2O_3)	0.40	30
Aluminium Oxide (Al_2O_3)	0.22	50
Magnesium Oxide (MgO)	1.94	20
Sodium Oxide (Na_2O)	22	6
Potassium Oxide (K_2O)	4.20	3
Sulphur Oxide (S_O3	1.20	20

From Table 1, it is realised that BCGPWP has almost all the constituents or the chemical composition that are found in OPC. Eight of the oxides were tested for including Calcium oxide (CaO), Silica oxide (Sio2), Iron oxide (fe_2O_3), Aluminium oxide (Al_2O_3), Magnesium oxide (MgO), Sodium oxide (Na_2O), Potassium Oxide (K_2O) and Sulphur Oxide (S_O3). The OPC showed its superiority by registering higher oxide composition than the BCGPWP, but there was appreciable quantity of the same oxide found in the BCGPWP. For instance, OPC had Calcium oxide (CaO) of 600mg/l, BCGPWP recorded 200mg/l. Again, OPC recorded 200mg/l of Silica Oxide (Sio2) while BCGPWP had 150 mg/l. Iron Oxide (fe_2O_3), Aluminium Oxide (Al_2O_3), Magnesium Oxide (MgO) and Sulphur Oxide (S_O3), all recorded oxide composition higher than BCGPWP, but with oxides like Sodium Oxide (Na_2O) and Potassium Oxide (K_2O) recorded 22mg/l and 4.2 mg/l as against 6mg/l and 3mg/l of OPC.

5 Water Absorption Test Results

The amount of water the specimen could absorb within a certain period of time was also tested. This test was to determine the specimen ability to resist the movement of water within the structure. Three specimens from each group of mix ratio were weighed in their dry state, then immersed in water for three days (72 h) before taken them out of the water. The surface area of the specimen was allowed to drain off or semi-dry up and re-weighed. The amount of water absorbed was calculated for the three specimen of each mix group and their average recorded.

The control specimen had an average of 1.63%. The remaining results from the specimen which contains Burnt Clay Grinding Pot Waste Powder (BCGPWP) are as follows; 5% replacement recorded 2.03%, 10% replacement also recorded 1.87%, 15%

replacement 2.22%, 20% replacement had 2.03%, 25% replacement 1.69% and finally the 30% replacement recorded 2.49%

Common aggregates were used for the experiment, therefore any rate of water absorption or the permeability of the specimen maybe influenced by the addition of the BCG-PWP as well as the deficiencies in the specimen production process. The control had the lowest water absorption rate (percentage) and 30% replacement on the other hand recorded the highest absorption rate which was 2.49%. Twenty-Five percent (25%) replacement recorded 1.69% which was way closer to the control specimen (1.63%). Ten percent (10%) also had 1.87% but the rest of the specimen exceeded 2%.

6 Compressive Strength

Compressive strength for the individual specimen

Compressive strength for the individual specimen for the experiment was determined by testing them at the laboratory using a compressive testing machine. The results of the various mix of the specimen are shown in Fig. 2. The mix ratios were zero percent (0%) that is pure cement as a binding material. Then 5% of burnt clay grinding pot powder replacing then ordinary Portland cement (OPC) up to 30% replacement. The specimen was in seven groups as could be seen from Fig. 1. The compressive strength was determined on the 7th, 14th, 21st and 28th days of curing.

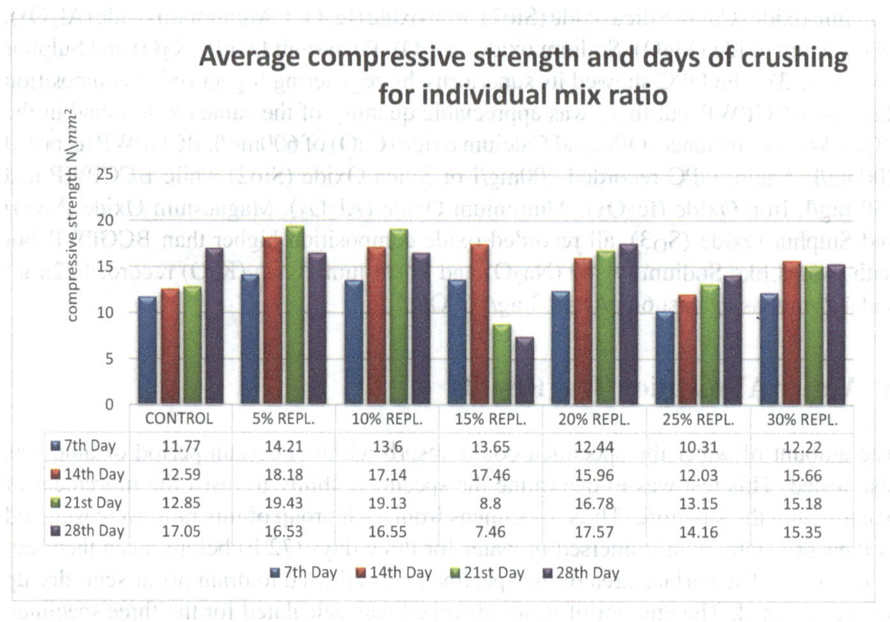

	CONTROL	5% REPL.	10% REPL.	15% REPL.	20% REPL.	25% REPL.	30% REPL.
7th Day	11.77	14.21	13.6	13.65	12.44	10.31	12.22
14th Day	12.59	18.18	17.14	17.46	15.96	12.03	15.66
21st Day	12.85	19.43	19.13	8.8	16.78	13.15	15.18
28th Day	17.05	16.53	16.55	7.46	17.57	14.16	15.35

Fig. 1. Average compressive strength of individual mix ratio specimen.

From Fig. 1, the compressive strength of control specimen (0% of burnt clay grinding pot waste powder) recorded 11.77 N/mm^2 on the seventh day. It rose to 12.59 N/mm^2 on the 14th day and then to 12.85 N/mm^2 on the 21st day of curing. The compressive strength produced on the 28th of curing was 17.05 Nmm^2. The strength difference between day 7 and day 28 was 5.28 N/mm^2, representing 30.67% increase in compressive strength.

Five percent (5%) burnt clay grinding pot waste powder exhibited similar strength development. The compressive strength recorded at week 7 was 14.21 N/mm^2, 18.18 N/mm^2 on the 14th day and went up to 19.43 N/mm^2 on the 21st day. The trend in strength development changed on the 28th day by decreasing to 16 N/mm^2.,

Ten percent (10%) replacement followed the trend of 5%. The records show a progressive strength development from the testing day 1 (7th day of curing) to testing day 3 (21st day of curing) but came down on the last day of testing (28th day of curing). The compressive strength of 14th and 21st days were 18.18 N/mm^2 and 19.43 N/mm^2 while 28th day's strength was 16.55 N/mm^2.

The compressive strength record of 15% of burnt clay grinding pot waste powder was slightly different from the previous ones. The compressive strength for the 7th day of testing was 13.65 N/mm^2 and rose to 17.46 N/mm^2. There was a progressive strength development in 1 and 2, but decreased to 8.8 N/mm^2 on the 21st day and 7.46 N/mm^2 on the 28th day. Twenty percent (20%) behaved differently from the previous ones. The compressive strength was progressive, from 12.44 N/mm^2 on the 7th day to 17.57 N/mm^2 on the 28th day. Day 14 recorded a compressive strength of 15.96 n/mm^2 and 16.78 N/mm^2 on the day 21st day respectively. Twenty-five percent (25%) replacement of the BCGPWP to the OPC had the following compressive strength, though they were reduced as compared to the 20%. Day 7 recorded 10.31 N/mm^2 while 12.03 N/mm^2 registered for day 14. 21st day of testing had an average compressive an average compressive strength of 13.15 N/mm^2 as against 14.16 N/mm^2 for the 28th day.

The last specimen to talk about is the 30% replacement of BCGPWP to the OPC. The first day of testing (7th day of curing), the compressive strength was 12.22 N/mm^2, greater than 25% replacement on the same day. Day 14 recorded 15.66 N/mm^2 while day 21 recorded 15.66 N/mm^2. They were all higher than the compressive strength recorded on 25% replacement specimen. Notwithstanding, the 14th day compressive strength was higher than the 21st day test strength. Day 28 recorded a compressive strength of 15.35 N/mm^2, also lower than week 14 and 21 respectively.

7 Split Tensile Strength of the Individual Specimen

Split tensile strength test was also conducted to determine the specimen's ability to resist the forces that tend to split the unit. The following results came out from the various mix ratio as indicated in Fig. 2.

It could be seen the Fig. 2 that, the control specimen for instance recorded 1.05 N/mm^2 for the 7th day of curing, while day 14 recorded1.6 N/mm^2. Week three (21 days of curing) recorded a split tensile strength of 1.81 N/mm^2 as against 1.89 N/mm^2 for the 28th day. There was a progressive split tensile strength development from the 7th to the 28th day.

Different split tensile strength was experienced with the 5% replacement specimen. Day 7 recorded 1.77 N/mm^2 while day 14 recorded 1.15 N/mm^2. There was a significant reduction, but rose up to 3.25 N/mm^2 on the 21st day. Meanwhile day 28 also recorded

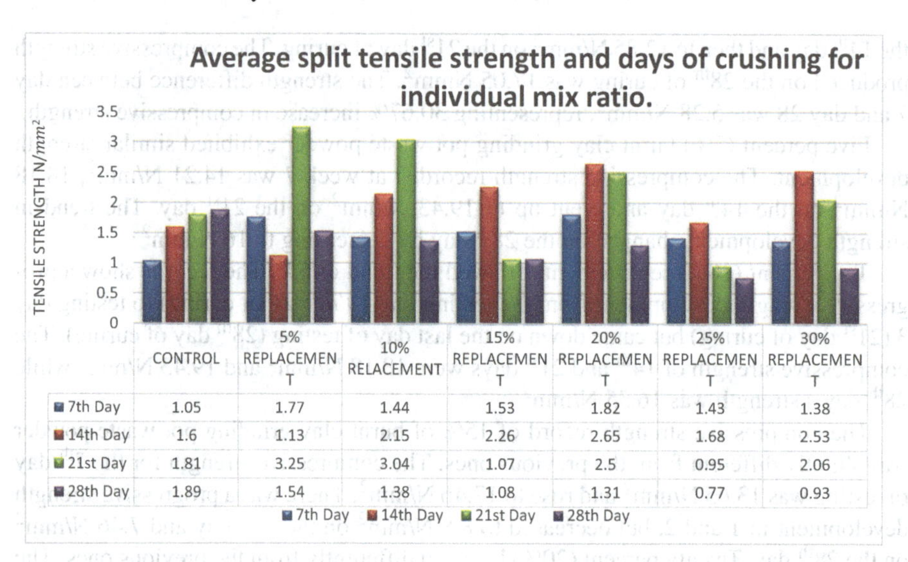

Fig. 2. Average split tensile strength for the individual mix ratio specimen

1.54 N/mm^2 which was also lower than week three (21 day of curing) but greater than week 2 (14 days of curing).

Ten percent (10%) replacement followed with the following records. Day 7 recorded 1.44 N/mm^2 while day 14 also recorded 2.15 N/mm^2. It rose to 3.04 N/mm^2 on week three and came down to 1.38 N/mm^2 on the fourth week (28days of curing). It is notice that week three for 5% and 10% recorded highest split tensile strength. Then came to 15% replacement specimen, the following records were obtained, week one (7 days of curing) recorded 1.53 N/mm^2, week 2 (14 days of curing) also recorded 2.

26 N/mm^2. The third and fourth weeks also recorded the following split tensile strength, 1.07 N/mm^2 and 1.08 N/mm^2 respectively. The last two split tensile strength for 15% replacement were lower compared to week 1 and 2 of the same specimen. Twenty percent (20%) replacement shows the following records. Week one recorded 1.80 N/mm^2 followed by week two with a split tensile strength of 2.65 N/mm^2. Week three had strength closer to week two and it was 2.50 N/mm^2. Week four had the lowest split tensile strength and it was 1.31 N/mm^2. Twenty five percent (25%) replacement exhibited different split tensile strength behaviour. Week one saw a split tensile strength of 1.43 N/mm^2 while week two recorded a strength of 1.68 N/mm^2. The third week also recorded a split tensile strength of 0.95 N/mm^2, followed with 0.77 N/mm^2 for the fourth week.

The seventh specimen (30%), produced the following split tensile strength results. Week one had 1.38 N/mm^2 while week two recorded 2.53 N/mm^2. Week three also recorded 2.06 N/mm^2 with the last week producing a split tensile strength of 0.93 N/mm^2.

8 Average Compressive Strength of Combined Mixed Ratio

The average compressive strength of the individual mix ratio has been combined here to give a clear view of the strength development among the various mix at each testing day, i.e. 7th day up to 28th day, Fig. 3 shows the details. The figures underneath the lines shows the various mix ratio and the compressive strength recorded. The lines above the figures for the various mixes shows the strength development pattern.

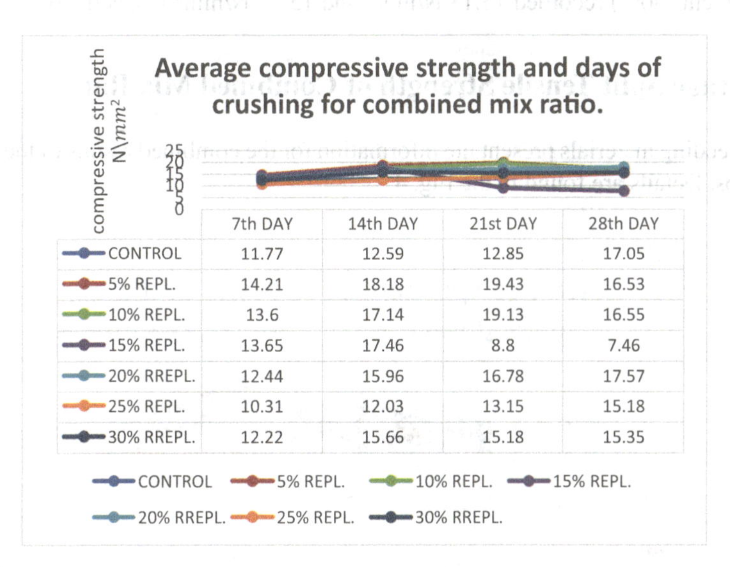

	7th DAY	14th DAY	21st DAY	28th DAY
CONTROL	11.77	12.59	12.85	17.05
5% REPL.	14.21	18.18	19.43	16.53
10% REPL.	13.6	17.14	19.13	16.55
15% REPL.	13.65	17.46	8.8	7.46
20% RREPL.	12.44	15.96	16.78	17.57
25% REPL.	10.31	12.03	13.15	15.18
30% RREPL.	12.22	15.66	15.18	15.35

Fig. 3. Average compressive strength of the combined mix ratio and testing days

From the Fig. 3, the control specimen recorded 11.77 N/mm^2, followed by 5% with a compressive strength of 14.21 N/mm^2. Ten percent (10%) recorded 13.60 N/mm^2, while 15% recorded 13.65 N/mm^2. 20% also had an average compressive strength of 12.44 N/mm^2. Twenty-five percent (25%) and thirty percent (30%) recorded 10.31 N/mm^2 and 12.22 N/mm^2 respectively, all in the first 7 days of curing.

On the fourteenth (14th) of curing, the following results came out from the compressive test. The control specimen recorded 12.59 N/mm^2 with the 5% recording 18.18 N/mm^2. Ten percent (10%) and fifteen percent (15%) recorded a very close compressive strength of 17.14 N/mm^2 and 17.46 N/mm^2. Twenty percent (20%) recorded 15.96 N/mm^2 which was lower than 15% on the same day. Twenty-five percent (25%) recorded 12.03 N/mm^2, while 30% had 15.66 N/mm^2.

The testing day three (21 days of curing) came out with the following results. The control specimen had 12.85 N/mm^2, followed by the 5% which recorded 19.43 N/mm^2 and was the highest among all the mix ratios. Ten percent (10%) replacement of BCG-PWP to the OPC had 12.85 N/mm^2 on the seventh day of curing from the test.it was followed by the 5% with an average compressive strength of 19.43 N/mm^2. Ten percent (10%) replacement had 19.13 N/mm^2 and 15% recorded 8.80 N/mm^2, which was the lowest among the group for the 21st day of testing. 20% registered a compressive strength of 16.78 N/mm^2, while 25% replacement specimen recorded 13.18 N/mm$^{2\cdot}$ The thirty percent (30%) which was the last mix ratio recorded 15.18 N/mm^2.

The week four test also yielded the following results. The control specimen recorded 17.05 N/mm^2, while the 5% had 16.53 N/mm^2. Ten percent (10%) also recorded a compressive strength of 16.55 N/mm^2. Fifteen percent (15%) specimen showed a drastic decrease in strength and the record was 7.46 N/mm^2. Then came to twenty percent (20%) with an average compressive strength of 17.52 N/mm^2. Twenty-five percent (25%) and thirty percent (30%) recorded 15.18 N/mm^2 and 15.35 N/mm^2 respectively.

9 Average Split Tensile Strength of Combined Mix Ratio

The proceeding materials present the information for the combined results of the various mix ratios. Details are found in the Fig. 4.

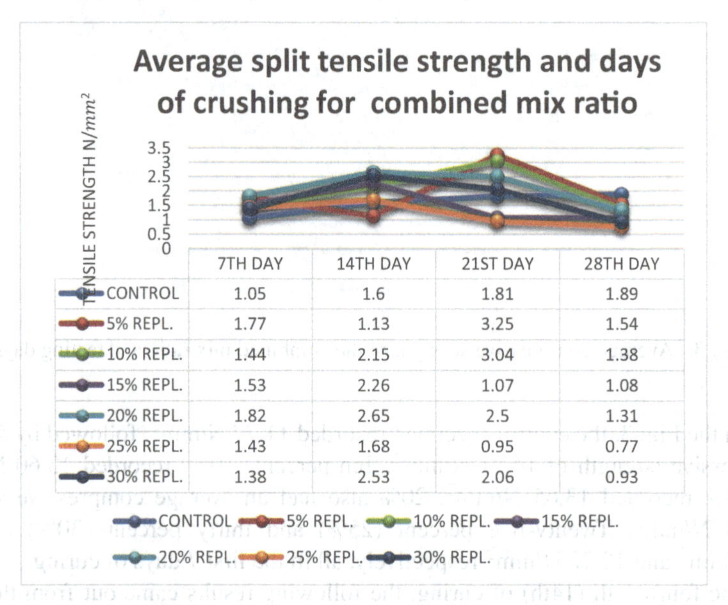

	7TH DAY	14TH DAY	21ST DAY	28TH DAY
CONTROL	1.05	1.6	1.81	1.89
5% REPL.	1.77	1.13	3.25	1.54
10% REPL.	1.44	2.15	3.04	1.38
15% REPL.	1.53	2.26	1.07	1.08
20% REPL.	1.82	2.65	2.5	1.31
25% REPL.	1.43	1.68	0.95	0.77
30% REPL.	1.38	2.53	2.06	0.93

Fig. 4. Average split tensile strength of combined mix ratio and days of testing

From the Fig. 2, all the seven-mix ratio produced different results representing the split tensile strength. Starting with the seventh day, the control specimen had 1.05 N/mm^2, while the 5% also recorded 1.77 N/mm^2, 10% partial replacement recorded 1.44 N/mm^2 as against with 15% registering 1.53 N/mm^2. Twenty percent (20%) recorded 1.82 N/mm^2 as against 1.43 n/mm^2 and 1.38 N/mm^2 for both 25% and 30% respectively.

On the second testing day (14th day) of curing, the control specimen produced a split tensile strength of 1.60 N/mm^2 with 5% recording 1.13 N/mm^2. Ten percent (10%) recorded 2.15 N/mm^2, while 15% recorded 2.26 N/mm^2. Twenty percent (20%) recorded higher split tensile strength with an average value of 2.65 N/mm^2. The twenty five percent (25%) and the thirty percent (30%) had 1.68 N/mm^2 and 2.53 N/mm^2.

From the figures and the behaviors of split tensile strength as indicated by the column bars and the line graphs, the split tensile strength for the 5% and 10% shot up with an all-time record of 3.25 N/mm^2 and 3.04 N/mm^2. It was followed by 20% with 2.50 N/mm^2.

Thirty percent (30%) also recorded 2.06 N/mm^2, while the control had 1.81 N/mm^2. Twenty five percent (25%) recorded the least value of 0.95 N/mm^2.

The fourth week of testing (28th day of curing), saw a drastic change from the previous weeks or testing days results. The control took the lead with a split tensile strength of 1.89 N/mm^2, followed by the 5% which recorded 1.54 N/mm^2. Ten percent (10%) partial replacement recorded 1.38 N/mm^2 with 20% recording 1.31 N/mm^2. Fifteen percent (15%) had 1.08 N/mm^2 while 25% and 30% recorded the least value of 0.77 N/mm^2 and 0.93 N/mm^2 respectively.

10 Discussion of Results

The succeeding materials discusses the results obtained from the laboratory test for the chemical analysis of the burnt clay grinding pot waste powder (BCGPWP) as well compressive and split tensile strength test conducted on the specimen or concrete cubes produced by combining OPC and the BCGPWP.

Chemical Analysis Test Results of BCGPWP

The motive behind this test was to ascertain the cementitious properties of the material which was intended to partially replace the Ordinary Portland Cement (OPC) as a binding material. The test came out with eight constituents including the following, Calcium Oxide (CaO), Silica Oxide (SiO2), Iron Oxide (fe_2O3), Magnesium Oxide (MgO), Aluminum Oxide (Al_2O3), Sodium Oxide (Na_2O), Potassium Oxide (K_2O) and Sulphur Oxide (So3) with the ordinary Portland cement as a basis for comparison in terms of oxide composition, the BCGPWP had Calcium Oxide (CaO) of 200mg/l as against 620mg/l indicates that the OPC contains about 67.74% of CaO higher than the BCGPWP. Silicate Oxide (SiO2) of BCGPWP had 150mg/l as against 200mg/l for the OPC. Oxide like iron, Aluminum, Magnesium, and Sulphur for the OPC was having higher values than the BCGPWP. On the other hand, BCGPWP had the following oxides higher than those found in the OPC. For instance, BCGPWP had Sodium Oxide (Na_2O) of 22mg/l greater than that of the OPC which recorded 6mg/l. Another oxide that the BCGPWP recorded higher the OPC was Potassium Oxide, OPC had 3mg/l as against 4.2mg/l for the BCGPWP. The oxide composition found in the BCGPWP clearly demonstrates that it could be used as partial replacement to OPC for concrete production and satisfy the objective which was sought to determine whether the burnt clay grinding pot waste contains some cementitious compounds similar to OPC.

Compressive Test Results Compared

The compressive strength for the control specimen that is, cement only as the binding material, 20% and 25% BCGPWP showed a progressive strength development successively. The same cannot be said about the 5%, 10%,15% and 20% replacement specimen. Their compressive strength development fluctuated as it could be seen from the bar chart in figure (a). The control recorded 11.77 N/mm^2 which served as a bench mark, rose to 12.59 N/mm^2 on the 14th day of testing and 17.05 N/mm^2 on the 28th day of testing. The five percent (5%) replacement of BCGPWP specimen also recorded 14.21 N/mm^2 and it was higher than the control's value as well as other replacement specimen for the same period. Fifteen percent (15%) was the second highest with an average compressive strength of 13.65 N/mm^2, that was also higher than the control. Twenty and thirty percent replacement specimen recorded 12.44 N/mm^2 and 12.22 N/mm^2 which also was greater than the control. Only 25% replacement recorded a compressive strength of 10.31 N/mm^2 on the same testing day and it was lower than the value of the control. Five percent (5%) replacement recorded 14.21 N/mm^2 on week one, 18.18 N/mm^2 on week two and 19.13 N/mm^2 on week three or testing day three (21 days). They were all higher than the values of the control on the same testing days. It was only the 28th day of curing that the 5% had 16.53 N/mm^2 which was lower than the control. Similar strength development behavior was exhibited with the 10% replacement with the final value on the 28th day as 16.55 N/mm^2 which was higher than both the control and 5% replacement specimen. The fifteen percent specimen's strength development was slightly different, it registered higher values for week one and two (13 N/mm^2 and 17.46 N/mm^2) and came down to 8.80 N/mm^2 and 7.46 N/mm^2 respectively. Twenty percent (20%) replacement of BCGPWP to the OPC specimen shows progressive strength development at each testing day from 12.44 N/mm^2 for week one, 15.96 N/mm^2 for week two, 16.78 N/mm^2 for week three and 17.57 N/mm^2 for the week four. They were all higher than the control. With the 25% replacement, though it recorded a lower value of 10.31 N/mm^2 which was lower than the bench mark of 11.77 N/mm^2, it showed a progressive compressive strength development for the succeeding weeks. The thirty percent (30% replacement) values were encouraging, but rose up to 15.66 N/mm^2 on the second day of testing which was the highest among its group and came down to 15.35 N/mm^2.

The comparison made from the above analysis suggest that 25% and 30% replacement are not bad, the burnt clay grinding pot waste powder can replace OPC up to 20% for concrete production.

Split Tensile Test Results Compared

Split tensile is one the basic and important test on concrete and interesting results came out from the experiment. The control specimen recorded 1.05 N/mm^2 which was lower than the mix ratio from 5%–30%. Twenty percent (20%) replacement of BCGPWP to the OPC specimen had the highest split tensile strength of 1.82 N/mm^2 followed by 5%, recording 1.77 N/mm^2. The third highest was 1.53 N/mm^2 and it was produced by 15% replacement. Ten percent (10%), 25% and 30% recorded the following split tensile strength, 1.44 N/mm^2, 1.43 N/mm^2 and 1.3 N/mm^2 respectively. The split tensile strength for testing day 2 was similar to that of day 1. All other specimen had strength greater than the control which recorded 1.60 N/mm^2 except the 5% which also recorded 1.13 N/mm^2. Testing day three (3) brought different strength development pattern, while some of the specimen's strength were going up, others were coming down. For instance, the control went up from 1.60 N/mm^2 from the previous week to 1.81 N/mm^2, 5% rose up from 1.13 N/mm^2 to 3.25 N/mm^2 while 15%, 20%, 25% and 30% came down from their previous figures.

On the 28th day of testing, the control had the highest split tensile strength than all the other mixes which BCGPWP were added to the OPC. The impression is that BCGPWP contains some compounds that improves the split tensile strength of the specimen in the early days of curing which are not sustainable as the concrete is aging, considering the period of the experiment. It cannot be concluded that such downward strength development would be experienced in long term since the strength of some of the mixes were alternating.

11 Conclusion and Recommendation

The study has met its objectives. It has confirmed that burnt clay grinding pot waste powder contains cementitious compounds or materials that can be found in ordinary Portland cement. About eight of such oxides including Aluminium oxide (Al_2O_3), Iron oxide (fe_2O_3), Magnesium oxide (MgO), Sodium oxide (Na_2O), Potassium oxide (K_2O) and Sulphur oxide (So_3) in varying proportions. Some of the oxides content in mg/l were below the OPC, others were also very closer to the OPC, for instance, BCGPWP had 150mg/l of Silica oxide as against 200mg/l of the OPC for the experiment. BCGPWP recorded oxide composition of Potassium oxide (K_2O) 4.2mg/l, higher than OPC which also recorded 3mg/l used for the study. In terms of the viability of BCGPWP as a binding material to supplement the Portland cement as the main binding materials for concrete and similar works in Ghana, the compressive and split tensile test values confirm that, it could be used, though further research may be needed to confirms the outcomes of this research where the various compounds found will be put to rigorous laboratory test. It emerged from the study that BCGPWP is a good material that can supplement the OPC for concrete works because almost all the specimen with the additions of BCGPWP to the OPC at earlier days have compressive strength higher than the control which was serving as a bench mark in terms of compressive strength. Twenty percent (20%) replacement of BCGPWP even had a successive compressive strength development higher than the control in all the testing days (7, 14, 21, and 28). Even on the 28th day where the control proved its superiority, recording 17.05 N/mm^2, the 20% replacement had 17.77 N/mm^2. That suggests and confirms the viability of the BCGPWP as a cementitious material that can supplement the OPC in concrete production.

The study recommends that the BCGPWP when well prepared, can partially be replaced the Ordinary Portland Cement (OPC) in concrete production up to about 20%. Although the final compressive strength difference between 25% and 30% replacements were closer to what the control specimen recorded, the study is limiting its suggestion to the 20% which emerged as an unquestionable percentage in terms of higher compressive and split tensile strength abilities.

This study therefore recommends that further research could be conducted to ascertain the temperature at which the pot reaches its peak where the clay particles could be fused together. It is also recommended that research could be conducted to explore the viability of the burnt clay as a supplement material for the OPC, because it can create employment avenue for the unemployed populace in the community and beyond since the working age groups are vigorously working on the pot production centers.

12 Limitation/Deficiency of the Research

The study could not ascertain the type of clay used for the production of the traditional burnt clay grinding pot. Again, the study could not unearth the temperature at which the pot/pot waste were subjected to before it came out of the fire.

References

Abdullahi, M.: Properties of some fine aggregates in minna. Niger. Environs Leonardo J. Sci. **8**, 1–6 (2006)

Agbede, I.O., Manasseh, J.: (2008) Use of cement-sand admixture in laterite bricks production for low-cost housing. Leonardo Electron. J. Prctices Technol. **12**, 163–174 (2008)

Sholanke, A.B., Fagbenl, I.O., Aderonmu, P.A., Ajagbe, M.A.: Sandcrete block and brick production in nigeria - prospects and challenges. IIARD Int. J. Geogr. Environ. Manage. **1**(8) (2015). ISSN 2505–8821

.Graphic online. Graphic online Housing deficit now 2 million units (2021). (https://www.graphic.com.gh/news/general-news/housing-deficit-now-2-million-units-minister.html).Accessed 19 Apr 2021

The Effect of Covid-19 on the Teaching and Learning Process of Entrepreneurship Education

M. C. Ntimbwa[1] and C. M. Ryakitimbo[2(✉)]

[1] Department of Business Administration, College of Business Education (CBE), Dar es Salaam, Tanzania
cntimbwa@gmail.com

[2] Department of Management Studies, Tanzania Institute of Accountancy, Dar es Salaam, Tanzania
crispinryakitimbo@yahoo.com

Abstract. Purpose: The study aims to identify the effect of COVID-19 pandemic in teaching and learning entrepreneurship in the institutions of higher learning (HLI).

Design/Methodology/Approach: The research methodology of this study was empirical review as information was collected from archives, documentaries, reports and research articles related to the subject matter. The process was formulated through designing and conducting the literature review, data review, structuring and writing the review. But also content analysis technique was applied for more knowledge.

Findings: The findings revealed three major issues related to the theory adopted; there was a gap in teaching and learning approaches, need of creativity and innovation in teaching and learning and student's mental hardship in learning process of which affected the learning process.

Research Limitations/Implications: The study adapted qualitative research methodology of which involved analysis of empirical studies through systematic and content analysis. The use of qualitative study could have added more weight in the study.

Practical Implication: The study finds indicates that the Experiential Learning Theory is applicable in HLI's even in difficulty times provided that technological network is applied. In this case technology should be observed in a larger picture.

Originality/Value: Using Experiential Learning Theory (ELT), the study has highlighted and revealed the negative effect of covid-19 being the cause-factor towards the ineffectiveness of learning by doing in HLI's. This research underpins theoretical attention whereby the applicability of the Experimental Learning Theory is relevant in the modern times but also during the time of crisis.

Keywords: COVID-19 · Education · Entrepreneurship · Learning · Teaching

C. Aigbavboa et al. (Eds.): ARCA 2022, *Sustainable Education and Development – Sustainable Industrialization and Innovation*, pp. 967–973, 2023.
https://doi.org/10.1007/978-3-031-25998-2_74

1 Introduction

China has got several provinces including Hubei its capital city is known as Wuhan with estimated population to be 11 million people. The news broke out that it has been affected by Coronavirus (COVID 19). Within a short time, the virus moved from epidemic to pandemic as it moved from Asia to Europe, America and the rest of the world (Mohammed Mustapha, et al. 2021). The pandemic caused fear and uncertainty and in certain countries at certain times and environment the health system was overwhelmed by the situation (Liguori et al. 2021). The total cases until September 2022 was 614,385,693, including 6,522,600 deaths and a total of 12,677,499,928 vaccine doses was administered (WHO 2021).

What the world is experiencing is not a new phenomenon as history identifies a number of endemics, epidemics, plague and pandemics. This includes great plague, the Justinian Plague, The Black Death, Third Plague Pandemic, The Spanish Flu, HIV/AIDS, SARS, Dengue, and Ebola. The previously mentioned diseases effects vary and are numerous including Ebola 2013–2016 had social economic impact of $53 billion across West Africa not mentioning deaths and application of total lock down. Maybe what is so different is increased science and application of technology which has made COVID-19 pandemic manageable but otherwise the world could have experienced more. In addressing COVID-19 pandemic, WHO through governments introduced various protocols popular known as restrictions of which countries implemented differently but almost the same including restrictions on body contacts, wearing facemasks and total lockdown (Pradhan et al. 2020).

The effects of COVID-19 are seen in production, supply chain, and uncertainty in financial markets (Mohammed et al 2021). The countries producing up to 65% of global manufacturing and exports that is developed world plus China, India and South Korea were affected strongly by COVID-19 (Baldwin and Evenett, 2020). The medical supply was affected to extent that some industries after reregulating the law had to be allowed to manufacture medical equipment's such as ventilators as the case of Ford and Dyson of America (Iyengar 2020). The America activated the Defense Production Act to allow motor vehicle industries to produce ventilators. Asians countries including China, Japan and Malaysia became good supplier of medical equipment's to European countries and America (Aubrecht et al. 2020). The implementations of COVID-19 protocols that is social distancing, masking and total lock down had an impact into consumer behaviour as they bought only essential products including digital accessories (Schluep 2009).

Maintaining preventive measures affected public gathering including direct physical contact in education system, closure of shops, barring entertainments activities, hotels, traveling restrictions or even stopping them. The aim is to stop spreading of coronavirus but in other way, it contributed to the destruction of economic activities i.e. markets, supply chain, and production and unemployment. In education, it contributes to issues related to teaching and learning. As learning institutions had close their physical learning location as suggested by WHO and adopted technology. In this juncture education institutions had to adjust to alternative channels (Liguori et al. 2021). Therefore, understanding coronavirus in education, especially teaching and learning entrepreneurship is paramount. This is because, maintaining strict COVID-19 protocols means no classes. This distracted education system i.e. curriculum implementation, physical classes, time rescheduling, shorter semester, and field practical. Thus, this study objectives are to identify the effect of COVID-19 in teaching and learning entrepreneurship education in higher learning institutions.

2 Kolb Experiential Learning Theory and Learning Styles

Around the world, education institutions are developing means of actively involving learner and learning environment apart from using the classes while addressing the needs of learning and society (Pokhrel and Chhetri 2021). Therefore, the guiding theory of observation for COVID-19 effect in teaching and learning process is **Experiential Learning Theory (ELT)** involves practical learning as learners has opportunity of using experience and cognitive ability in making interpretation of the environment. ELT defines learning as "the process whereby knowledge is created through the transformation of experience. Knowledge results from the combination of grasping and transforming experience" (Kolb 1984). The prevailing situation and rising demand for higher learning institutions to be more practical through integrating learning experiences and actual what is in the society. The more practical approaches which makes the learning process more participatory is demanded.

The experimental learning will be evaluated through the experiences in COVID-19 pandemic as learned turned to Blended learning of which includes televised classes, broadcast programs, internet classes and channels of that nature which does need meeting face to face (Batac et al. 2021). The number of learners affected by COVID-19 pandemic is estimated to be that 60% (UNESCO 2020).

3 Methodology

The research methodology of this study is documentary review as information was collected from archives, documentaries, reports and research articles related to the subject matter. The process is formulated through designing and conducting the literature review, data review, structuring and writing the review. This is developed first by identifying journal articles using key words. The paper adopts the critical literature review as the process allows for reviews, criticism of papers and investigative of the subject matter in relation to other literature (Grant et el. 2009). The systematic literature review adopted to control bias (Snyder 2019). Apart from that it gathers information across the studies and identify the research gap as it crosses (Davis et al. 2014). The study apart from analysis of literature in a systematic way, content analysis technique was applied to extract and organize the content as related to the study objectives (Khirfan et al. 2020). Content analysis is among the best way of drawing conclusion after analysis of facts in the text (Drisko and Maschi 2016). The content analysis makes the prediction of the future which serves the purpose of this study (Stemler 2015). This kind of methodology underscore the magnitude of the pandemic nexus entrepreneurship learning programs as variables: knowledge gained or lost, relationships, the magnitude of the problem, and the research agenda (Snyder 2019).

4 Findings

The finding through literature analysis showed that implementation WHO protocols that is lockdown, social distancing and washing hands changed how teaching and learning could be managed during the pandemic (Pradhan et al. 2020). Even the world supply chain was interrupted which affected production and financial markets (Mohammed et al. 2021) This made institutions to turn to online teaching and learning and other digital platforms (Anggadwita et al. 2017; Liguori et al. 2021; Inglês 2020; Radić et al. 2020). In running this operation, it needed creativity and innovation (Ratten 2020) in order to make learning process more effective. Batac et al. (2021) names it as blended learning. Though learning process makes the learners fill isolated from other peers (Ratten and Jones 2020) and its expensive for developing world to adapt and use. Cordiva (2021) advice for partnership with institutions from developed world.

The findings from the literature also indicated that there were challenges of application of blended learning which includes readiness, accessibility of stable and affordable internet services and health risk (Batac et al 2021; Inglês 2020). Radić et al. (2020) emphasize that online learning has some challenges to government, education and learners. The provision of higher learning education is based on law, policy and regulations. The law has created councils to oversee the operations of higher learning institutions (Radić et al 2020). To some of these institutions online certifications is not regulated. Education system was affected especial tertiary education which turned to digital platforms (Inglês 2020).

5 Discussion

There are two major findings related to experiential learning; the current entrepreneurship education is facing challenge of reducing the gap between teaching online and providing experiential learning (learning by doing) (Anggadwita et al. 2017). That is balancing experiential education with current societal needs. Online teaching poses a challenge of making application of what is learnt to reality. But also, the theory demands a prepared educator to use the system-initiated necessity of traditional reluctant educators to adapt to digital world and empress the new skills needed. On the other hand, learners are digital friendly so adapting to new means of delivering the content is accepted but are HLI ready for change? This is because, the entrepreneurship education involves learning which requires the learner to acquire the skills and competence of which makes learning effective and relevant but at the same time at the end the learner is able to perform a tangible task.

The need of creativity and innovation in teaching and learning process (Ratten 2020) is of paramount. The physical contact in teaching and learning process is still the major means of delivering learning content. But the pandemic has initiated the necessity of using alternative channels such as online learning. The question that arise focuses on the sufficient skills for instructors to make the overall process a success. The studies have indicated that this is among the area which should be addressed by scholars and HLI's, Hence HLI needs to be creative and innovative enough to address the current crisis. But also, Ratten (2020) calls it for community to mitigate the effects; government agencies, civil society, private sector, alumina, instructors, students of which makes community ought to digest together about the present and future with or without pandemic in proposing innovative initiatives.

The rise and existence of the pandemic has spiral effect in the world including education systems. The impact is extended to courses that require practical engagement of students. Literature has revealed the psychological impact of the students after the closure of HLI's. Nowadays in HLI's', entrepreneurship education embraces skills and competence for the purpose of learning and intellectual development. The learning institutions concentrates more on intellectual development. The practice of closing and opening learning institutions around the world due to COVID-19 pandemic affects the way how students learn and their ability to learn through peer networks because online learning could not fulfil the gap between teaching online and learning by doing. This has caused students to have been physically and socially isolated from their peers that has caused mental hardship (Ratten and Jones 2020). Moreover, the practice of class invitations of successful entrepreneurs, industrial visits stopped suggesting that reality practices are not entertained during pandemic.

The learning content needs to be revisited for instructors and learners to engage more on practical oriented activities such as games and other stimulus activities. For example, starting business can be turned into an online game for learners to play while developing their idea. The same as Business Plan the topic needs to be a goal oriented and not a theoretical based with expectation that after completing studies the graduates can start a business. The theories of Entrepreneurship must be learnt practical for learners to know who/how/why was it developed. How far is it realistic! These includes Schumpeter (1934), McClelland (1962), Shapero and Sokol (1972) and Ajzen (1992) aim being to develop a learner who is able to make application of the theories.

Online interaction emphasizes the digitalization of education system which is an opportunity as well as a challenge. The educator's ought to sort-out how to prepare next generation of entrepreneurs (Liguori et al. 2021). This is through engaging with foreign partnership and strengthening "virtual internationalization" which help to maintain and eventually increase accessibility. Partnership has been identified as strategy towards addressing the pandemic crisis (Cordiva et al. 2021).

6 Conclusion and Implications of the Study

In conclusion, this study has investigated the effect of teaching and learning process of entrepreneurship education in higher learning institutions and provided insights into the current teaching and learning practices during covid-19 whereby various literature was reviewed. Using Experiential Learning Theory (ELT), the study has highlighted and revealed the negative effect of covid-19 being the cause-factor towards the ineffectiveness of learning by doing in HLI's. This research underpins theoretical attention whereby the study brought together two themes by applying the teaching-learning theory to entrepreneurship education in HLI's, which had not been satisfactorily researched collectively in the past. The findings from this research add new knowledge to the existing literature on the teaching-learning theory and entrepreneurship education but more important the provision of new insights into the impact of covid-19 on teaching and learning process in higher learning institutions.

7 Suggestion

The governments and private sectors should provide financial support to education institutions to enhance professional development and continue to develop platforms for learning process to continue. Addition, the institutions mandated to supervise higher learning institutions needs to accept, regulate and support online programs. Moreover, Higher Learning Institutions ought to develop deeper approaches of integrating blended learning with needs of the learners and society.

References

Anggadwita, G., Luturlean, B.S., Ramadani, V., Ratten, V.: Socio-cultural environments and emerging economy entrepreneurship: women entrepreneurs in indonesia. J. Entrepreneurship Emerg. Econ. **9**(12), 85–96 (2017)

Ajzen, I.: The theory of planned behaviour: organisational behaviour and human decsions process. ELSEVIER **50**(2), 179–211 (1991)

Aubrecht, P., Essink, J., Kovac, M., Vandenberghe, A. -S.: Centralized and decentralized responses to COVID-19 in federal systems: A US and EU comparison. (2020). Available at SSRN 3584182

Batac, K.I.T., Baquiran, J.A., Agaton, C.B.: Qualitative content analysis of teachers perceptions and experiences in using blended learning during the COVID-19 pandemic. Int. J. Learn. Teach. Educ. Res. **20**(6), 225–243 (2021)

Cordiva, J.E.L., et al.: Policies to support business through the COVID-19 shock: a firm level perspective. World Bank Res. Observer **36**(1), 41–66 (2021)

Mohammed, T., et al.: A critical analysis of the impacts of COVID-19 on the global economy and ecosystems and opportunities for circular economy strategies. J. Resour. Conserv. Recycl. **164**, 105169 (2021)

Davis, J., Mengersen, K., Bennett, S., Mazerolle, L.: Viewing systematic reviews and meta-analysis in social research through different lenses. Springerplus **3**(1), 1–9 (2014). https://doi.org/10.1186/2193-1801-3-511

Drisko, J.W., Maschi, T.: Content Analysis. Oxford University Press (2016)

Grant, M.J., Booth, A.: A typology of reviews: an analysis of 14 review types and associated methodologies. Health Info. Libr. J. **26**, 91–108 (2009)

Inglês, A.: Trends in expert system development: a practicum content analysis in vocational education for over graw pandemic learning problems. Indonesia J. Sci. Technol. **5**(2), 246–260 (2020)

McClelland, D.C.: Achieving Soceity Business Ball. Princeton, New Jersey, New Jersey, United States of America (1961)

Iyengar, K., Mabrouk, A., Jain, V.K., Venkatesan, A., Vaishya, R.: Learning opportunities from COVID-19 and future effects on health care system. Elsevier Public Health Emerg. Collect. **14**(5), 943–946 (2020)

Pokhrel, S., Chhetri, R.: A literature review on impact of COVID-19 pandemic on teaching and learning. High. Educ. Future **8**(1), 133–141 (2021)

Pradhan, D., Biswasroy, P., Naik, P.K., Ghosh, G., Rath, G.: A review of current interventions for COVI-19 prevention. Arch. Med. Res. **51**(5), 363–374 (2020)

Liguori, E.W., Winkler, C., Zane, L.J., Muldoon, J., Winkel, D.: COVID-19 and necessity-based online entrepreneurship education at US community colleges. J. Small Bus. Enterp. Dev. **28**(6), 1462–6004 (2021)

Kolb, D.A.: Experiential Learning: Experience as the Source of Learning and Development. Englewood Cliffs, NJ: Prectice Hall, Hoboken (1984)

Khirfan, L., Peck, M.L., and Mohtat, N.: Digging for the truth: a combined method to analyze the literature on stream daylighting. Sustain. Cities Soc. **59**(4), 102225 (2020)

Schumpeter, J.A.: The Theory of Economic Development. Transaction Publisher, London, England (1934)

Schluep, M.: E-waste management in Africa – Rising up the political agenda, Recycling International, 56–61 (2009)

Shapero and Sokol.: Social Dimensions of Entrepreneurship, in Kent, C.A. Sexton, D.L and Vesper, K.H. Englewood Cliff: New Jersey (1982)

Stemler, S.E.: Content Analysis, Emerging Trends in the Social and Behavioural Sciences, Scott, R and Kosslyn (Ed), Wiley & Sons Inc., Hoboken (2015)

Snyder, H.: Literature review as a research methodology: an overview and guidelines. J. Bus. Res. **104**, 333–339 (2019)

Ratten, V.: Entrepreneurial ecosystems. Thunderbird: Int. Bus. Rev. **62**(5), 447–455 (2020)

Ratten, V. Jones. P.: COVID-19 and entrepreneurship education: implications for advancing research and practice. Int. J. Manage. Educ. **19**, 100432 (2020)

Radić, G., Ristić, B.S., Anđelić, S., Kuteto, V., Ilić, M.: "E-Learning Experiences of Higher Education Institutions in the Republication of Serbia During COVID-19, Content Analysis and Case Study ITS Belgrade", Proceeding of the 36th International Business Information Management Association (IBIMA). Granada, Spain (2020)

UNESCO. COVID-19 Education Disruption and Response (2020)

WHO. World Health Organization Dashboard (2021)

Managing Pandemic Diseases and Food Security in Africa

E. Baffour-Awuah[1](✉), N. Y. S. Sarpong[2], and I. N. Amanor[3]

[1] Mechanical Engineering Department, Cape Coast Technical University, Cape Coast, Ghana
emmanuelbaffourawuah37@yahoo.com

[2] Agricultural Engineering Department, University of Cape Coast, Cape Coast, Ghana
serwaah.sarpong@cctu.edu.gh

[3] Department of Agricultural and Biosystems Engineering, Kwame Nkrumah University of Science and Technology, Kumasi, Ghana
ishmael.amanor@cctu.edu.gh

Abstract. Purpose: This review paper considered three of the most widespread pandemics, including Covid-19, HIV/AIDS and Spanish Influenza with regards to their coverage, management methods and their generic effect on food security in Africa.

Design/Methodology/Approach: The paper adopted content analysis with specific reference to a systematic review, based on manifest content. A twenty-two-year period of documents was extensively scrutinized; from 2000 to 2021 using the Google scholar database. Manifest contents of 20 out of the 90 documents were finally selected. After manually coding the documents, they were eventually reviewed and analyzed.

Findings: The paper indicates that hundreds of thousands have perished in Africa due to these pandemics, nevertheless, several interventions have been sought to control them. The interventions include quarantine; hand-washing; face masking; social distancing; disinfection; lockdowns; personal distancing; lockdowns; lockdowns; and vaccination among others. The paper also indicates that though these interventions, including knockdowns, could be advantageous, they might contribute to unfavourable consequences including food insecurity in many countries on the continent.

Implications/Research Limitations: The study period also dwelt on only twenty-two years, spanning between 2000 and 2021. Depending on a longer period might have contributed to scrutinizing relatively more documents as might be required by studies of this sort.

Practical Implication: The paper intends to present a document that shall guide both industry players and researchers in the management of pandemic diseases regarding the most suitable procedures, methods and techniques in the prevention and treatment of such diseases.

Social Implications: The information in this paper could be relied upon to help in the achievement of the Sustainable Development Goals (SDGs) (Agenda 2030) by facilitating the accessibility of information concerning the management of

C. Aigbavboa et al. (Eds.): ARCA 2022, *Sustainable Education and Development – Sustainable Industrialization and Innovation*, pp. 974–985, 2023.
https://doi.org/10.1007/978-3-031-25998-2_75

contemporary and future pandemics. Additionally, the paper implies that the most widespread management technique of pandemics is to employ knockdowns. However, vaccinating as many individuals as possible to attain herd immunity among the people is the most potent management technique within countries on the African content.

Originality/Value: This study is unique based on the fact that it portrays the characteristic features of pandemics of recent history to the sphere of influence, mode of transmission, management techniques, commonalities, differences, externalities and challenges. The study further portrays that interventions such as knockdowns and quarantines are advantageous, as they could bring about food insecurity in Africa.

Keyword: Covid-19 · Food · Pandemic. Security · SDGs

1 Introduction

Disease pandemics including the Spanish influenza, HIV/AIDS and Covid-19 have impacted on various sectors of economies in the world, and Africa is no exception. Economic, financial, agricultural and other sectors have all been affected. Of great concern to Africa, among the rest, is food security. Even before pandemics, Africa has been plagued with the issue of food insecurity and its impact on the continent and its people. Food security involves food availability, accessibility; stability; and agency. The others are access; utilization; sustainability and precautions. Food security is also influenced by various factors. These factors are referred as the dynamics of food security. For these reasons the relationship between pandemics and food security is a very complicated one. Various interventions have been adopted to deal with the health risks and the negative effects of pandemic diseases. However, these interventions and the pandemics themselves have had untold hardships on the people of the continent. It is therefore necessary that other short-, medium-, long- and longer-term measures are employed to subsequently deal with disease pandemics and its negative consequences on the continent of Africa.

2 Methodology

This paper adopted content analysis as the method of study, with particular reference to systematic review and manifest content. According to Alan (2011), content analysis is the review and analysis of study materials such as texts of varying formats, words, pictures, video and or audio. It is employed to analyse text patterns in information and communication from a systematic point of view. The procedure utilizes former study materials for review in order to find gaps in relation to the study under consideration. The study therefore employed Denyer and Tranfield's (2009) five-step procedure for systematically reviewing literature. The procedure involves question formulation, document selection, assessment and analyses, as well as result documentation.

To achieve the objectives of study, the systematic review covered a period of 5 years, between 2017 and 2021 relying on Google scholar database. The five-year span between

2017 and 2021 was justified for the reason that a relatively larger percentage research documents in the subject discipline fell within the period.

An extensive search was carried out utilizing the "title-abstract-keyword" procedure. The search yielded 63,235 documents. Upon further search employing the keywords "Covid-19", "HIV/AIDS", "pandemics" and "Spanish Influenza". As many as 31355 documents were obtained. Upon restricting the documents to the keyword "food security", 303 documents yielded. Upon further scrutiny using the keywords "Africa" and "African continent", yielding 64 documents. Final scrutiny yielded 20 documents of manifest content which were coded manually for detail review and consequential analysis (Lee et al. 2001).

3 Results and Discussion

3.1 Meaning and Dynamics of Food Security

The Food and Agricultural Organization (2003) defines food security as the ability and capacity of individuals to have access to available physical and economic resources, such as adequate safe and nutritious food to meet their daily requirements throughout the year as well as their preferential meals and diets to satisfy active and balanced life. Various dynamics have arisen to influence food security. These include disruptions to food supply chains, loss of income and livelihoods; widening of inequality; and disruptions to social protection programmes. The rest are altered food environment; an uneven localized food prices; as well as lower food productivity and production (Arouna et al. 2020; Clap and Moseley 2020; Klassen and Murphy 2020; Laborde et al. 2020a). Pu and Zhong 2020; Van den Broeck and Maertens 2016). Figure 1 shows a schematic diagram of the relationship between COVID-19 and the various dynamics that threatens food security. Though the diagram specifically dwells on COVID 19 it could be applicable to other disease pandemics considering the nature, characteristics, rate and implications of disease pandemics in general.

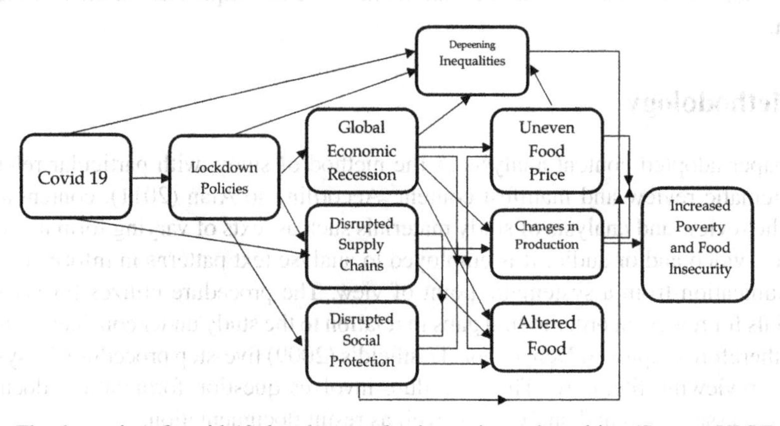

Fig. 1. The dynamics of covid-19 that threaten food security and nutrition (Source: HLPE 2020)

The main food supply chains are disrupted during pandemics, particularly when there are lockdowns. This goes a long way to affect food availability as well as food quality (Barrett 2020). Restaurants and other food service centers may be affected leading to low demand for perishable and specialty goods (Lewis 2020). Food products may also be dumped and ploughed back as a result of low demand (Yaffe-Bellany and Corkery 2020). The closure of international borders left in its wake low demand of particular food products, specifically producers who ply their trade through distant export markets. Producers of perishable food and agricultural products; and specialty crops such as fruits and vegetables and cocoa products respectively were the worst victims of supply chain disruptions (Clap and Moseley 2020). Imposition of export restrictions on essential staple foods also disrupts supply chains of these food produce and products. Food supply chains are also disrupted when food system workers become ill or restricted from migrating from one country or region to another or one location to another (Haley et al. 2020; Stewart et al 2020).

Disease pandemics also create global economic recession with related loss of income. Global economic recession affects purchasing power as a result of job losses. For example, about 451 million people have lost their jobs as a result of the COVID-19 pandemic (Torero 2020). According to Ashford et al. (2020) the spread of COVID-19 and its associated knockdowns has widened the social inequalities in most countries, particularly developing countries. Working conditions have been intense and dangerous, particularly for women as frontline health and food system workers; care work that are not paid; as well as community work done (Klassen and Murphy 2020; McLaren et al. 2020).

The fourth dynamic factor linking disease pandemic and food security is the disruptions to social protection programmes. Lockdowns result in loss of school meal programmes in both advanced and developing countries. It is estimated that, for instance over 320 million children lost opportunity to access school meals as a result of closure during the COVID-19 pandemic (WFP 2020a). Governments' capacity to implement social protection were also restrained in the provision of social protection programme due to lower income from revenues and closure of productive institutions. The best alternative of borrowing from relatively richer countries was also restrained as a result of the lending country's ability to also generate income (Salent 2020). Lockdown measures and supply chain disruptions, as a result of disease pandemics alter food environment and consequently, food security. When restaurants and food service facilities, food retail outlets are shut down leading to altered food environments, food access, availability and utilization are affected (Battersby 2020; Moseley and Battersby 2020; Young and Crush 2019).

Localized food price increase also influences food security. For example, during the initial stages of the COVID-19 pandemic, global cereal stocks were at record high and an overall low world food commodity price. However, as the crises aggravated, food such as meat and cereals (rice and wheat) saw their prices increasing; some countries as high as 50 percent; though countries such as Kenya experienced an increase of about 2.6 percent (Espita et al. 2020; FAO 2020c; Laborde et al. 2020b). Thus, localized food prices increased making food access more difficult and expensive and food insecure, especially to the people in poor countries.

Finally, disease pandemics have the potential to change food production and productivity and subsequently impact on food security. For instance, at the beginning of COVID-19, global cereal stocks were at record high while food supplies were adequate. Depending on the duration of lockdowns, different strains that may evolve, the number of waves that could erupt and the reliability with respect to the timing and extent of the virus and its evolution, production capacity and the quantity of products produced per individual unit time could change. Though production and productivity have expanded in many countries during the last 10 years or so, these factors could alter the gains made so far. This is so particularly true with regards to horticultural production, processing and export (Van den Broeck and Maertens 2016); and cereal production in industrialized countries (Schmidhuber and Qiao 2020). Disruptions of supply chains of labor, transportation, agricultural inputs such as fertilizer and seeds as a result of knockdowns could potentially change food production and productivity and consequently impact on food security (Pu and Zhong 2020; Schmidhuber and Qiao 2020).

3.2 Pandemics and Food Security in Africa

The relationship between pandemics and food security is a very complex one (refer Fig. 1). Food security has six fundamental dimensions to contend with, for that matter. Thus, food security has six dimensions which include food availability; and agency. The others are accessibility; utilization; sustainability and precautions. This implies that these dimensions have effect on food security. Thus disease pandemics do affect all these food security dimensions.

Before pandemics strike, the level of food production in Africa could be assumed to be manageable. However, food production in advanced countries which tend to be more mechanized and require little labor could not be said to be that much disposed to such vulnerabilities. In this regard, this is quite different for African countries with relatively less agricultural mechanization but rather intensive labor participation instead. It is documented that many African small-scale farmers are also women who are already weak in terms of menial agricultural work. When pandemics strike, the already weak or sick women could not involve in labor in any way. Naturally strong men when sick cannot in any way get involved in menial labor participation.

For example, when pandemics strike, in African countries where intensive labor is rife, supply chains for meat packaging, horticulture and diary are negatively influenced due to sensitivity of workers to sickness. In medium and large corporate institutions that are involved in such packaging operations, the spread of pandemics is greatly increased as a result of labor intensive farm operations. In addition to this, disease pandemics disorganize supply chains of agricultural machinery and equipment, seed production and chemicals such as weedicides, pesticides and fertilizer. This negatively affect food productivity for that matter. Consequently food availability, quality and pricing are affected to a very large extent (Anday et al. 2019; Akiwumi 2020; HLPE 2020).

Food accessibility, inarguably, is the most affected dimension of food security, and in this regard, the most influenced. In Africa, where the ability to make available enough social safety interventions with reference to short-term policies and strategies to circumvent the effect of lockdowns, accessibility of food is greatly negatively influenced by the effects of pandemics. Short-term safety interventions in Africa involve withdrawal of cash, selling of foodstuffs including fish, horticultural products, meat and animals among others. Many households in Africa live below poverty levels with strict budgets and tight opportunity to manage household spending. Little or no social safety interventions results in reduction on expenditure on food as a result of income and finance reduction. Increase in prices of food when pandemics strike results in its trail households' incapacity to purchase enough food for lively consumption.

In African countries where large portions of populations are mainly wage workers, farmers, informal traders, hawkers, and low-income workers, increase in food prices largely influence a lot of the citizenry. Moseley and Battersby (2020) have documented that, in Africa where diseases such as HIV/AIDS are prevalent, many of the citizenry live beyond the economic fringes and incomes are generally low, symptomatology and chronicity give rise to increased morbidity and mortality rates with negative implications for food, income and labor accessibility. Disruptions of livelihood and transportation to around 60% of its citizenry, who are largely farmers, who could not sell their produce; purchase farm inputs; as well as access vertinary and farm extension services, additionally influences negatively on food accessibility (Arndt et al. 2020).

Utilization of nutritious food is among one of the most significant means through which disease pandemics may be managed. For instance, the covid-19 encounter has proved that better nutrition influences the human immune system to minimize the risk of pandemic transmission. For example, Covid-19 minimized the availability and accessibility of food as a result of sicknesses and imminent lockdowns (FAO 2020a). In low- and middle-income nations of which many African countries are part, individuals' incapacity to enjoy nutritious meals was exposed, making them more amenable to health risks and pandemic infection. Availability and accessibility to organic vegetables and fruits moved to processed commodities which results in poor nutrition. Malnutrition, access to potable water and poor sanitation further aggravate individual's vulnerability to the disease, as inequalities got worse as a result of the pandemic, to poor utilization of nutritious diets (Micha et al. 2020).

Bene (2020) has reported that intense disorganization towards food supply chains influence stability to the supply of food. Espita et al. (2020) have also reported that though export restrictions on staples might be temporal due to a pandemic, the introduction of later export restrictions might be a risk to several African nations' economic and currency value uncertainties resulting from a pandemic contribution to increase in food price pressure and instability in the world food market place. Africa as part of the world market do experience price shocks due to such occurrences. An UNCTAD (2020) document has showed that a disease pandemic might prevent the citizenry and organizations to invest within the agricultural and food sectors ensuring instability in investments in several countries around the globe, specifically, low- and middle-income nations of which several are found in Africa. Low investment results in lower production and productivity with consequential effect on a cycle of low incomes and increased poverty levels in such nations.

Worker agencies such as workers unions and the ability of unions to fight and defend workers interests might be negatively influenced when pandemics strike. It has been documented that unfortunate food related employees including food producers and food system workers become greatly affected when pandemics strike. As people leading food supply and having been negatively influenced by high infection rate and supply chain disruptions food related employees are affected by job and livelihood deficiencies due to contact losses with their agents, who in divers ways, could have defended them to get their jobs back. This may result in lockdowns and social and physical distancing as well as synonymous governmental short-term policies and strategies to minimize the risk of disease infection. With the minimization of agency involvement activities, women's economic and social empowerment which demands a lot of social and contact meetings, direct and team work is greatly influenced (FAO 2020b). Estimates from the International Labor Organization show an unbelievable 200 million individuals losing their jobs globally, of which a large number locate in Africa (Blanke 2020).

Sustainability, being one of the non-conventional aspects of food security, in addition to agency, is complicatedly inter-woven with pandemic diseases in several ways. Firstly, in an effort to enhance agricultural investment, production and productivity, industrial agriculture tends to combine the involvement of both humans and animals. Through enhanced industrial agriculture, wildlife habitats are destroyed resulting in increased possibility of humans and animals coming into contact. This results in the possibility for diseases to be transmitted from animals to human beings; Covid-19 being a particular instance. The Spanish flu is also a classical instance. This implies that, in an attempt to maintain food supply through industrial agriculture pandemic disease risks and infection could be introduced (Everard et al. 2020). Secondly, knockdowns creates enhanced food waste resulting in restaurant and food service system closures as well as minimization of some other group of food products (Sharma et al. 2020). Thirdly, pandemics also creates enhanced employment of packaging and carrier bags of single-use plastics which degrades environment sustainability (Vanapalli et al. 2020). Fourthly, pandemics negatively influences longer-term sustainability of food systems by transferring financial and some other resources which could have been used in dealing with climate change and other environmental issues involving ecological challenges and biodiversity loses (Barbier and Burgess 2020).

And finally, disease pandemics bring about job and livelihood loses; social and economic challenges, diversion in production and productivity activities. This goes a long way to minimize a comparatively longer-term sustainability of food and food-related systems (HLPE 2020a). In Africa where these problems are quite severe, pandemic diseases such as covid-19, have discharged, and continue to discharge an excruciating consequence on its almost 1.2 billion populace. The classical Spanish-flu, also, was not an exception.

4 Conclusion

It has been shown that, historically, the challenge of food security has been common in Africa even before disease pandemics rear its ugly head on the populace. Examples of the worse pandemics include the Spanish flu, HIV/AIDS and Covid-19. Disease pandemic restrictions, dos and don'ts as interventions, have been the bane of food security, nutrition and health of several African nations and its citizenry. Nevertheless, African nations have sought and introduced various interventions whenever pandemics strike in order to reduce pandemic impacts on food security among others. Food security involves food availability, accessibility; agency; access; utilization; sustainability and precautions. In spite of this, there is the requirement for improved as well as more interventions. These involve short-, medium- and long-term strategic interventions. The strategic and policy interventions, in this regard, include effective social protection programmes; integrated and holistic system linkages for food chain supply and distribution systems; improvement and diversification of agricultural and productivity for essential food produce, products and associated systems. Longer term interventions include the implementation of vigorous social protection programmes; protecting the women, vulnerable, marginalized food system workers and farmers affected by the pandemic; protecting countries who largely depend on imported food products; strengthening and coordinating policy responses to pandemics; supporting heavily variegated and strong distribution systems; and enhancing support for strong food production systems The aim of these interventions are to minimize the dependence of these and substituted enterprises and commodities to minimize vulnerabilities to the advantage of African nations and its citizenry.

Recommendations.

In the short, medium and long term four major policy shifts could be adopted in the face of pandemic diseases. Africa has most of its women population marginalized. For this reason, the first policy shift is to have an integrated and holistic approach to transform food systems. This can be achieved by introducing basic transformation techniques with respect to diversification of food supply and systems; empowering marginalized and vulnerable groups, and advancing sustainability of food supply systems, from cradle to grave to support increased food production. Africa has most of its population women marginalized (HLPE 2020b).

The second shift is to avoid cross-purpose activities but to appreciate synergies among systems such as social food systems, economic ecological systems, bio-diversity systems and environmental systems among others; as well as ensure linkage among these systems. Inter system linkages are the best approach to dealing with the intricate workings of pandemics that have very intricate dynamics, bearing in mind that the link between food systems and pandemics is complex, particularly the relationship between agriculture and zoonoses. Countries in Africa with forest cover will need this policy in order to reduce the complexities between pandemics, zoonoses and risks to infections. The penultimate shift has to deal with the involvement of the comprehension of the complex interaction between obesity, micronutrient deficiencies, hunger and malnutrition in general. These challenges greatly expose poor, vulnerable and marginalized people to risks to pandemic infections. In Africa, since most people belong to these category of people, disease pandemics largely affect the populace and this policy should therefore be gladly appreciated (HLPE 2020b; Yeshanew 2018).

Finally, in an attempt to transform food system policies, African governments should take cognizance of variability in demographics such as gender; age; typology of food system workers crops, countries, highest level of education etc. Since the impact of pandemics varies in accordance with variability of these parameters. Thus transformative food system policies must be specific and contextual with reference to locations, populations and groups (HLPE 2020b).

In the longer-term, six thematic areas need to be considered by African governments. These are the implementation of vigorous social protection programmes; protecting the vulnerable, marginalized food system workers and farmers affected by the pandemic and women; as well as protecting countries who largely depend on imported food products. The rest are to strengthen and coordinate policy responses to pandemics; support heavily variegated and strong distribution systems; and support for more strong food production systems (HLPE 2020b; Yeshanew 2018).

With the inception of the African Continental Free Trade Agreement (AFCTA), African countries should provide sufficient exigent food aid from both domestic, regional and international source to supplement available food supply. African governments who are saddled with debt should negotiate for debt relief and deferment to sustain social safety nets. Social Safety nets should be as strong as possible due to undulating nature of household food expenses with reference to expenditure on health care, housing utilities and education among others. It is important that food assistance programmes must be designed to ensure sufficient nutrition; provision of options to school feeding programmes when children are at school and allowance for accessibility to health care such as physical, psychological, and mental health care among others. In order to protect the vulnerable, marginalized food system workers, farmers and women, there is the need for African governments and civil society organizations (CSO) to protect the rights of food system workers by recognizing and integrating these rights in national, regional and continental legislations and ensure their effective implementation through compliance and established corporate and national norms (Clapp 2017; HLPE 2020b).

Food system workers and farmers must be protected from hazards and risks of pandemic infection in terms of distancing systems; health, hygiene and safety measures; and personal protective equipment. Other protective mechanisms are eating and sanitary systems and allowance for sick workers and redundant workers. The rest are protection from health risks and provision for social protection. There is also the need for governments and corporate organizations in African countries to ensure migrant food system workers and farmers are protected from health risks, given access to health services and provided with social protection. Farmers and small-scale and micro-scale agricultural producers must be provided with particular insurances, protected from unwarranted transfers and redundancies, as well as ensuring adequate distribution of agricultural and farming inputs (Yeshanew 2018).

With reference to protecting countries who largely depend on imported food products there is the need to dissuade countries that impose food export restrictions in order to protect those that rely on importation of food. Countries that make the effort to enhance food production, domestically, should also be encouraged to build production capacity towards the achievement of these goals within the medium to long term frame. Production capacity, in this regard, should include human, finance, technology, infrastructure and storage (Clapp 2017; Viatte et al. 2009).

The issue of strengthening and coordinating policy responses to pandemics has to do with the impact of pandemics on food systems, food security and nutrition, both domestic and international dimensions. African nations must recognize the role of the African Union and other regional bodies such as ECOWAS to coordinate international governance reaction to pandemics. In this regard the establishment of a task force to monitor pandemic impacts on food security must be monitored. The surveillance and monitoring activities and findings should be properly reported and documented so that both domestic and other African country stakeholders can get access to them. There must be a continental campaign to communicate to the public the need for practicing good nutrition so as to prevent and manage pandemic infections by individuals and households. In terms of decision-making, there is the need to involve all stakeholders including food system workers as well as agricultural producers and farmers unions and organizations at national, international and regional levels (HLPE 2020b; McKeon 2015).

In dealing with the impact of disease pandemics on food security one remedy is to heavily support variegated chains and territorial markets. This can be achieved by African countries through the enhancement of investments in African market infrastructure at regional, national and local levels. Also at the regional, national and local levels it will be important to adopt stronger competititive policies to encourage micro-, small- and medium-scale agricultural-based food produce and products to be involved in national building. Last but not the least, policies that are inimical to informal markets in favor of retail food outlets should be reviewed by African governments. These informal markets include rural markets and street vendors who may not be regular businessmen and women but sporadic ones (HLPE, HLPE 2020b; Blay-Palmer et al. 2020).

The sixth and final longer-term measure is for African countries to support and build strong food production systems through investment in agricultural and ecological research-based projects; support the development and implementation of agro-ecologically-based curricula for institutions of agriculture at the basic, secondary and tertiary levels; support home and community gardens and the individual and community demands; adopt policies that will sustain animal production; forestry, fisheries; and aquaculture in an integrated and holistic approach to improve nutrition and livelihoods; and finally to encourage sustainable agricultural projects as well as agro-ecology interventions (Altieri and Nicholls4 2020; Love et al. 2020; Bennett et al. 2020).

References

Akiwumi, P.: COVID-19: *A threat to food security in Africa.* Available at: http://unctad.org/news/covid-19-threat-food-security-africa. Accessed 8 Dec 2020 (2020)

Altieri, M.A., Nicholls, C.I.: Agroecology and the reconstruction of a post-COVID-19 agriculture. J. Peasant Stud. **47**(5), 881–898 (2020)

Arndt, C., Davies, R., Gabriel, S., Harris, L., Makrelov, K., Robinson, R., et al.: COVID-19 lockdowns, income distribution and food security: an analysis for South Africa. Glob. Food Sec. **26**, 100410 (2020)

Arouna, A., Soullier, G., del Villar, P.M., Demont, M.: Policy options for mitigating impacts of COVID-19 on domestic rice value chains and food security in West Africa. Glob. Food Sec. **26**, 100405 (2020)

Barbier, E., Burgess, J.: Sustainability and development after COVID-19. World Dev. **136**, 105082 (2020)

Barrett, C.: Actions now can curb food systems fallout from COVID-19. Nature Food **1**, 319–320 (2020)

Battersby, J.: South Africa's lockdown regulations and the reinforcement of anti-informality bias. Agric. Hum. Values **37**(3), 543–544 (2020). https://doi.org/10.1007/s10460-020-10078-w

Béné, C.: Resilience of local food systems and links to food security – a review of some important concepts in the context of COVID-19 and other shocks. Food Secur. **12**(4), 805–822 (2020). https://doi.org/10.1007/s12571-020-01076-1

Bennett, N., Finkbeiner, E., Ban, N., Belhabil, D., Jupiter, S., Kittenger, J.: The COVID-19 pandemic, small-scale fisheries and coastal fishing communities. Coast. Manag. **48**(4), 336–347 (2020)

Blanke, J.: Economic impact of COVID-19: Protecting Africa's food systems from farm to fork (2020). http://www.brookings.edu/blog/africa-in-focus/2029/06/19/economic-impact-of-covid-19-protecting-africas-food-systems-from-farm-to-fork/. Accessed 10 Dec 2020

Blay-Palmer, A., Carey, R., Valette, E., Sanderson, M.: Post COVID-19 and food pathways to sustainable transformation. Agric. Hum. Values **37**, 513–519 (2020)

Clap, J., Moseley, W.G.: This food crisis is different: COVID-19 and the fragility of the neoliberal food security order. J. Peasant Stud. 1–25. covidwho-843508 (2020)

Clapp, J.: Food self-sufficiency: making sense of it, and when it makes sense. Food Policy **66**, 88–96 (2018)

Espita, A., Rocha, N., Ruta, M.: COVID-19 and food protectionism. Policy Research Working Paper 9253. Washington, DC. World Bank (2020)

Everard, M., Johnston, P., Santillo, D., Staddon, C.: The role of ecosystems in mitigation and management of COVID-19 and other zoonoses. Environ. Sci. Policy **111**, 7–17 (2020)

FAO: An introduction to the basic concepts of food security (2008). http://www.fao.org/3/a-a1936e.pdf. Accessed 7 Dec 2020

FAO: The state of food security and nutrition in the world, 2020: Transforming food systems for affordable healthy diets. FAO, Rome (2019)

FAO: Gendered impacts of COVID-19 and equitable policy responses in agriculture, food security and nutrition. Policy brief (2020a). http://www.fao.org/policy-support/tools-and-publications/resources-details/en/c/1276740/

FAO: Food Outlook – June 2020b (2020b). http://www.fao.org/3/ca9509en/CA9509EN/pdf

Haley, E., Caxaj, S., George, G., Hannebrey, J.L., Martell, E., McLaughlin, J.: Migrant farmworkers face heightened vulnerabilities during COVID-19. J. Agric. Food Syst. Commun. Dev. **9**(3), 1–5 (2020)

HLPE: Impacts of COVID-19 on food security and nutrition: Developing effective policy responses to address the hunger and malnutrition pandemic (2020a). www.fao.org/3/cb1000en/.pdf. Accessed 10 Dec 2020a

HLPE: Food security and nutrition: Building a global narrative towards 2030. Report 15. Rome, HLPE (2020b). http://www.fao.org/3/ca9731en/ca9731en.pdf

Klassen, S., Murphy, S.: Equity as both a means and an end: lessons for resilient food systems from COVID-19. World Dev. **136**, 105104 (2020)

Laborde, D., Martin, W., Vos, R.: Poverty and Food insecurity could grow dramatically as COVID-19 spreads. Washington DC. International Food Policy Research Institute (IFPRI) (2020a). https://ww.ifpri.org/blog/poverty-and-food-insecurity-could-grow-dramatically-covid-19-spreads. Accessed 10 Dec 2020a

Laborde, D., Martin., W., Swinnen, J., Vos, R. (2020b). COVID-19 risks to global food security. Science **369**(6503), 500–502. https://sciencemag.org/content/369/6503/500

Lewis, L.: Coronavirus serves up a surplus of Wagyu beef. Financial Times (2020). (April 3, 2020) https://www.ft.com/content/bb540839.2f63-43bc-897c-b73b2d9f6dc7

Love, D.C., et al.: Emerging COVID-19 impacts, responses, and lessons for building resilience in the seafood system. Glob. Food Sec. **28**, 100494 (2021)

McKeon, N.: Food Security Governance: Empowering Communities, Regulating Corporations. Routledge, London (2015)

McLaren, H.J., Wong, K.R., Nguyen, K.N., Mahamadachchi, K.N.D.: Covid-19 and women's triple burden: Vignettes from Sri Lanka, Malaysia, Vietnam and Australia. Soc. Sci. **9**(5), 87 (2020)

Micha, R., et al.: 2020 global nutrition report: action on equity to end malnutrition (2020)

Moseley, W.G., Battersby, J.: The vulnerability and resilience of African food systems: Food security and nutrition in the context of the COVID-19 pandemic. Afr. Stud. Rev. **63**(3), 449–461 (2020)

Pu, M., Zhong, Y.: Rising concerns over Agricultural production as COVID-19 spreads: lessons from China. Global Food Secur., **26**, 100409 (2020). 1016/i.gfs.2020.100409

Sallent, M.: External debt complicates Africa's COVID-19 recovery debt relief needed. Africa Renewal, July 2020. UN Economic Commission for Africa (2020)

Schmidhuber, J., Qiao, B.: Comparing crises: Great lockdown versus great recession. Rome, FAO. http://www.fao.org/3.ca8833en/CA8833EN.pdf

Sharma, H.B., et al.: Challenges, opportunities, and innovations for effective solid waste management during and post COVID-19 pandemic. Resour. Conserv. Recycl. **162**, 105052 (2020)

Stewart, A., Kottasová, I., Khaliq, A.: Why meat processing plants have become COVID-19 hotbeds. CNN, June 27 (2020)

Sweat, M., et al.: Cost-effectiveness of voluntary HIV-1 counselling and testing in reducing sexual transmission of HIV-1 in Kenya and Tanzania. Lancet **356**(9224), 113–121 (2000)

Torero, M.: Prepare food systems for a long-haul fight against COVID-19. Washington, DC. IFPRI (2020). https://www.ifpri.org/blog/prepare-food-systems-long-haul-fight-against-covid-19

UNCTAD: International production beyond the pandemic. World Investment Report 2020, Geneva, United Nations Organization (UN) (2020). https://unctad.org/en/PublicationLibrary/wir2020/en.pdf

Vanapalli, K.R., et al.: Challenges and strategies for effective plastic waste management during and post COVID-19 pandemic. Sci. Total Environ. **750**, 141514 (2021)

Yaffe-Bellany, D., Corkery, M.: Dumped milk, smashed eggs, plowed vegetables: the food waste of the pandemic. New York Times, April 11 (2020)

Yeshanew, S.: Regulating labor and safety standards in the agriculture, forestry and fisheries sectors. Rome, FAO (2018). http://www.fao.org/3/CA0018EN/ca0018en.pdf

Young, G., Crush, J.: Governing the informal food sector in cities of the Global South Hungary cities. Discussion Paper 30 (2019)

Dividend and Share Price Behaviour: A Panacea for Sustainable Industrialization

N. M. Moseri[1(✉)], S. I. Owualah[1], P. I. Ogbebor[1], I. R. Akintoye[2,3], and H. T. Williams[2,3]

[1] Department of Finance, Babcock University, Ilishan, Ogun State, Nigeria
ndukamosie@yahoo.co.uk

[2] Department of Accounting, Babcock University, Ilishan, Ogun State, Nigeria

[3] Department of Finance, Redeemer's University, Ede, Osun State, Nigeria
williamsh@run.edu.ng

Abstract. Purpose: We tested agency theory on capital market sustainability and we focused on the extent to which risk-averse investors have used financial resources such as dividend payment and share price to drive sustainable industrialisation and innovation by re-investing dividends received.

Design/Methodology/Approach: A diagnostic research design and probability sampling were used. 250 observations were collected on panel data from 50 quoted firms for five years (2017–2021). A quantitative research method called the Cobb-Douglas production model was used to analyse dividend payment, share price and sustainable industrialization and innovation.

Findings: The results show investors could use a potential capital market with the available financial resources such as dividends and share price to drive sustainable industrialization and innovation through the reinvestment of dividends.

Research Limitation/Implications: The limitation is pinpointed on the assumption that the share price goes up and down. A decline in share prices affects sustainable industrialization and innovation in the capital market.

Practical Implication: The financial variables used in this study would inform stakeholders and policy holders that meeting the goals for sustainable development, and financial resources are vital.

Social Implication: The macroeconomic variables used in the study implied that financial resources have a linear and long-run relationship with sustainable industrialization and innovation in the capital market.

Originality/Value: The novelty of this study lies in the financial instruments used as proxies, it informed policy makers that achieving sustainable industrialization and innovation required financial resources.

Keywords: Agency theory · Dividend · Financial resources · Share price · Innovation

1 Introduction

Achieving sustainable industrialization and innovation in any economy required a complex and long term process. However, this research focused on the capital market and the

C. Aigbavboa et al. (Eds.): ARCA 2022, *Sustainable Education and Development – Sustainable Industrialization and Innovation*, pp. 986–992, 2023.
https://doi.org/10.1007/978-3-031-25998-2_76

extent in which the financial resources provided by the capital market can contributes to an effective industrialization and innovation. Development goals does not only depends on an effective management of the natural resources available within the ecosystem but also on an effective use of financial resources. This study identify dividend and share price as two financial resources to attract investment and achieve sustainable developmental goals, industrialization and innovation in Nigeria. Building manufacturing firms and other firms that attract investment required financial resources and sustaining firms operations required financial resources.

The payment of dividends and the use of share prices for quoted firms in the capital market is one of the most complex debates in the field of finance and modern financial economics as investors are risk averse and would only maximize returns and minimize risk in any given situation. The question is would risk averse investors re-invest their returns to enhance the effective and efficient development of both industrialization and innovation for firms seeking additional capital? Whatever the answer may be revolve around a need for financial resources and potent ideas to drive industrialization and innovation. Many studies have researched dividend payment and share price behaviour but no have focused on industrialization and innovation and integrating the agency theory. Dividend payments have been used to attract investors to drive industrialization and innovation and expand the growth of firms. Much literature has argued that the payment of dividend is good as it support re-investment and re-investment aid industrialization but most time all activities of the use of financial resources are tied to an agency problem. The stated assertions are supported by the work of Williams and Ayodele (2017), Anh et al. (2021), Ilo and Olawale (2021).

Nations not effectively attaining sustainable development goals, industrialization and innovation are linked to the agency strategies and policy implementation. Aboramadan et al. (2020) stated that agency theory could be used to solve the problem of organizational leadership. Kaivo-oja, Panula-Ontto, Vehmas and Luukkanen (2013) stated that there are many agents of sustainable development, meaning that driving sustainable industrialization and innovation is dependent on humans and institutions hence humans are the major agents for sustainable development goals, sustainable industrialization and innovation. Jatmiko (2016) stated that in as much as dividend payment is important, the use of globalization for more than four decades has linked economies together with good investment return and constant payment of dividends have contributed to sustainable industrialization and innovation. Payment of dividends has been part of the system of corporate entities that transform firms to achieve industrialization and innovation (Mensah 2019).

Sustainable industrialization and innovation focused on an organized principles drawn from related fields to meet human development goals with the aim of creating a system that would shield the natural resources and ecosystem services on which the economy and society depend. Ukaga, Maser and Reichenbach (2011) stated that the term sustainable development, industrialization and innovation are universal development model catchword for international agencies on environment and economic activities. Scopelliti, Molinario, Bonaiuto, Bonnes, Cicero, Dominicis and Bonaiuto (2018), and Shepherd, Knight, Ling, Darrah, Soesbergen and Burgess (2016) supported the concept sustainable industrialization and innovation as it have attracted the world attention on development issues. Abubakar (2017) and Hylton (2019) stated that understanding the concept of sustainable industrialization and innovation would create a linkage between human action, economic and finance activities as well as the environment.

Barney (1991) stated that when a linkage exist between two independent variables that are interconnected and interdependent between each other there would exist a linear relationship that would integrated the variables. Relating this assumptions of Barney (1991) a system theory is formed on the link between sustainable development, industrialization and innovation. Mensah and Enu-Kwesi (2018) assumed that the concept of sustainable development implies improving and sustaining a system that improve the existence of human development while sustainable industrialization and innovation focused on building firms for continuous production.

Browning and Rigolon (2019) stated that sustainable development, industrialization and innovation is a platform that required the absolute attention of man to improve the basic standard of living without endangering the earth's ecosystems and firms in existence. The implication of this is that man is expected to be an agent of the environment and any actions taken by man would either have a positive or negative effect on all elements of the earth including industrialization and innovation. Therefore, the problem here is that financial resources drive economic activities but have been ignored by researchers on its effect on industrialization and innovation. Based on this premise that this study focused on attracting investors to driving sustainable development, industrialization and innovation through dividend and share price in the capital market.

2 Theories and Related Literature Underpinning the Study

System theory and the agency theory were used to lay the foundation of this study. While the system theory assumed that there is a linear relationship between man, the environment and sustaining the environment, the agency assumed man to be the primary agent of driving sustainable industrialization and innovation through the financial resources available. Jensen and Meckling (1976), stated that the agency theory is derived from the study of financial and economic activities managed by man. However, there have not been sufficient literature to link sustainable development, industrialization and innovation and the availability of financial and economic resources but other related literature exist. Abel (1983) postulated that financial resources are prone to risk and may either generate positive or negative returns. Agbebaku (2022) stated that the N12.92 billion dividend payment declared by the Nigerian Breweries (NB) Plc is a way portraying how financially sustainable the company would be in the long run. In Brazil a legislature has been enacted for a very long time to mandate firms operating in the country to pay at least 25% of their adjusted income in the form of dividend (Cristiano, Fernanda & Denis, 2015). Deakin (2018) stated that the wealth a nation is dependent on the existent the nation can go as wealth (dividend) is potential factors used by shareholders to satisfy their purchasing and consumption pattern while Williams and Ayodele (2017) stated that accomplishing any goals is dependent on the power of financial resources available and such resources can be obtained from investment return via dividend. Cox, Ross and Rubinstein (1979) stated that the inconsistency in the share behaviour may affect a stable dividend payment and thus concluded that share price move in paths, meaning it goes up and down. However, Dushko and Mico (2020) stated that the use of the trinomial models and the risk neutral probability makes the share price to either go up, go down or remain constant overtime. An inconsistency in the share price would amount to a double problem in a country that is tasked with instability and economic uncertainty and would probably slow down sustainable development goals. Ogbebor et al. (2020) assess the extent in investors' prospects and wealth have affected the trend of share prices in the Nigeria. The implication of this is that individual wealth contribute to sustainable development while Ogbebor (2019) opined that a positive and negative effect of fundamental factors on the price of shares. Mohieldin (2017) stated that the combination of economic sustainability is a modern method to development which uses all form of resources available in an effective and efficient manner to ensure that the resources continue to linger for future inhabitants.

Evers (2017) stated that the notion of sustainable industrialization and innovation is to formulate a standard for meeting the achievable human development goals and concurrently nourishing the volume of natural systems and ecosystem to afford the platform in which the economy and society depend. Goidsmiths (2018), Zhai and Chang (2019) stated the general goal of the sustainable development concept is to ensure that environmental equilibrium and economic growth are achieved.

3 Methodology

We measure the extent in which financial resources have affected sustainable industrialization and innovation in Nigeria using quantitative analysis. We studied two variable i.e. dividend and share price and relate it to the Cobb Douglas Production model. The researchers observed that replacing labour and capital with dividend and share price in the Cobb Douglas Model would yield the expected results of the study. Since financial resources are contributed by investors, we introduced the risk neutral probability to curtain risk on the premise that investors are risk averse. The data were panel data with 250 observations collected from 50 firms over five years. We assume that share price goes up and down based on the binomial model and investors would contributes to sustainable industrialization and innovation if their share prices increase and yield good dividend otherwise investors would opt out for alternatives which may negatively affect sustainable industrialization and innovation.

$$\mathrm{PRNP} - \mathrm{U} = \frac{(1+r)^t - D}{U - D}$$

$$\mathrm{PRNP} - \mathrm{D} = 1 - PRNP_u$$

where: RNP-U = Periodic Risk-Neutral Probability for Up.

RNP-D = periodic risk-neutral probability for down, D = the proportion of stock price down movement, r = risk free rate (adjusted with inflation and interest rate), t = period in which share prices changes (every 6months = 0.5) (Fig. 1).

SPBU = Share Price Behaviour going Up, SPBD = Share Price Behaviour going Down,

The Share Price Behaviour Path

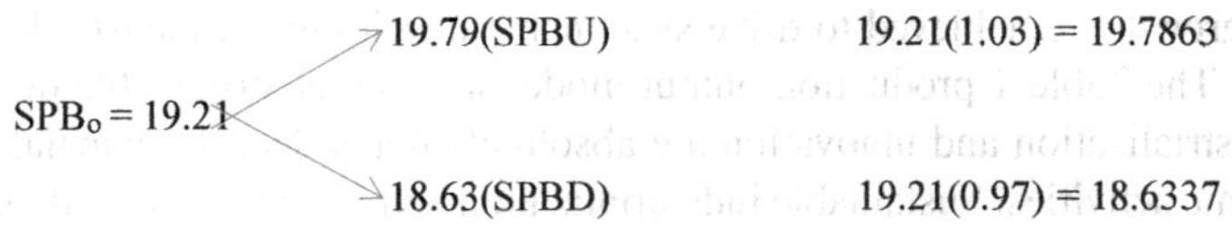

	→ 19.79(SPBU)	19.21(1.03) = 19.7863
SPB_0 = 19.21		
	→ 18.63(SPBD)	19.21(0.97) = 18.6337

4 Mathematical Model

The standard Cobb Douglas Production function model used in the study is:

$$Q = AK^{\alpha}L^{\beta} \quad (1)$$

α = human development index = 1.07% = 0.0107 (2017–2021)
β = 5years average change in capital market index (taking from 31st Dec. Each year) = 0.8%

$$SD = EIDP^{\alpha}SPB^{\beta} \quad (2)$$

$$SD = EIDP^{0.0107}SPB^{0.008} \quad (3)$$

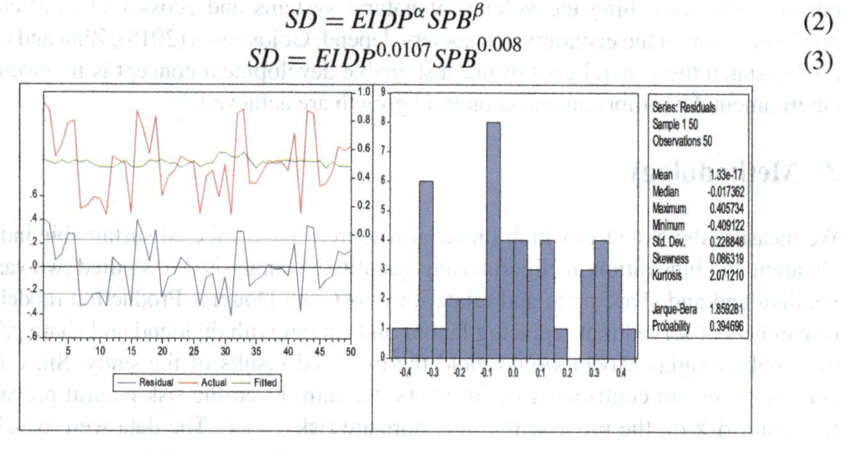

Fig. 1. Reflection of dividend payment and share price behavior of 50 listed firms

$$SDG = EIDP^{0.0107}SPB^{0.008} \quad (4)$$

5 Findings and Discussions

Achieving sustainable development and industrialization and innovation required the connection of both financial and non-financial resources within the ecosystem. Dividend and share price behaviour were identified as financial resources required to attract and sustain investors to re-invest their funds for the continuity of sustainable industrialization and innovation. However, share prices react in two directions i.e. going up and down. Cox et al. 1979 stated that share price goes up and down, which corroborates with this study. When share price behaviour react in a positive manner, it attracts investors and strengthen economic growth and sustainable industrialization and innovation (Evers, 2017, Deakin 2018). The implication of this is that share price make the financial resources available to drive sustainable industrialization and innovation. The addition of human development of 0.0107 and the capital market index of 0.008 indicates a decreasing return to scale in relations to the production model. Goidsmiths (2018), Zhai and Chang (2019) believed that a scale of 1 and above would empirically show that environmental equilibrium and economic growth are achieved to drive sustainable development, industrialization and innovation. The Table 1 production output model shows that sustainable development goals, industrialization and innovation are absolutely dependent of financial resources and economic activities. Sustainable industrialization and innovation would increase the financial activities of the capital market.

Table 1. Production function model output (SD)

Period	DP (billion)	ASPB	EI	$DP^{0.0107}$	$ASPB^{0.008}$	SD (SDG)
2021	16.5	19.21	0.0365	1.03045037	1.0239252	0.0385113
2020	15.6	17.08	−0.018	1.02983212	1.0229629	−0.0189626
2019	18.3	20.6	0.022	1.03159262	1.0244976	0.02325101
2018	17.2	20.1	0.019	1.03090858	1.0242962	0.02006316
2017	17.8	19.8	0.08	1.03128688	1.0241730	0.08449729

Source: Authors computation with excel.

6 Conclusion

Sustainable development, industrialization and innovation does not only relays to the standard of meeting human development goals and maintaining the natural resources and ecosystem but also improving the production capacity of firms to attain optimal outputs. The economy and society depend on this for a long term healthy relationship between all elements in the environment. For an effective management of the environment and the natural resources and ecosystem to attain a potent sustainable industrialization and innovation, the need for financial resources. This study concluded that financial resources such as dividend payment and a stable share price can be used to attract investors to the capital market and drive sustainable development, industrialization and innovation in Nigeria. This study quantitatively show that sustainable development goals cannot work in isolation without integrating some vital elements such as industrialization, innovation an effective capital market. Therefore, this study is essential for stakeholders to join financial resources together and invest in research and development to drive sustainable development, industrialization and innovation. Capital market activities should be reviewed and necessary policies implemented for optimal investment results that would aid sustainable development, industrialization and innovation and general meets the SDG goals in Nigeria.

References

Abel, A.: Optimal investment under uncertainty. Am. Econ. Rev. **73**, 228–233 (1983)

Abubakar, I.R.: Access to sanitation facilities among Nigerian households: determinants and sustainability implications. College of Architecture and Planning, University of Dammam, Saudi Arabia; Sustain. **9**(4), 547 (2017). https://doi.org/10.3390/su9040547

Aboramadan, M., Dahleez, K., Hamad, M.H.: Servant leadership and academics outcomes in higher education: the role of job satisfaction. Int. J. Organ. Anal. **29**(3), 562–584 (2020). https://doi.org/10.1108/IJOA-11-2019-1923

Agbebaku, U.: Capital market review. In: The Guardian Newspaper. May 9th, 2022. Cristiano, F. Fernanda, P., Denis, A. (2015). Determinant Factors of Dividend Payments in Brazil. Article presented at the 14th Brazilian Conference on Finance, Recife, Brazil, 2014 (2022)

Anh, H.N., Pham, C., Doan, N., Trang, T., Tu Van, T.: The effect of dividend payment on firm's financial performance. An empirical study of Vietnam. J. Risk Financ. Manage. **14**(353), 1–11 (2021)

Barney, J.: Firms resources and sustained competitive advantage. J. Manag. **27**, 643–650 (1991)

Browning, M., Rigolon, A.: School green space and its impact on academic performance: a systematic literature review. Int. J. Environ. Res. Public Health **16**(3), 429 (2019). https://doi.org/10.3390/ijerph16030429

Cox, J.C., Ross, S.A., Rubinstein, M.: Option pricing: a simplified approach. J. Financ. Econ. **7**(3), 229–242 (1979)

Deakin, H.: Dividend policy and share price volatility. Invest. Manage. Finan. Innov. **12**(1), 226–234 (2018)

Dushko, J., Mico, A.: A review of the binomial and trinomial models for option pricing and their convergence to the black-scholes model determined option prices. Econ. Adv. Appl. Data Anal. **24**(2), 53–85 (2020)

Evers, B.A.: Why adopt the Sustainable Development Goals? The case of multinationals in the Colombian coffee and extractive sector: Master Thesis Erasmus University Rotterdam(2018)

Gossling-Goidsmiths, J.: Sustainable development goals and uncertainty visualization. Thesis submitted to the Faculty of Geo-Information Science and Earth Observation of the University of Twente in partial fulfilment of the requirements for the degree of Master of Science in Cartography (2018)

Hull, J.C.: Options, Futures and Other Derivatives, 9th edn. Prentice Hall, Upper Saddle River (2014)

Jatmiko, D.P.: The influence of agency cost, market risk, and investment opportunities on dividend policy. Int. J. Manage. Comm. Innov. **3**(2), 68–75 (2016)

Jensen, M., Meckling, W.: Theory of the firm: managerial behavior, agency costs and ownership structure. J. Financ. Econ. **3**, 305–360 (1976)

Kaivo-oja, J., Panula-Ontto, J., Vehmas, J., Luukkanen, J.: Relationships of the dimensions of sustainability as measured by the sustainable society index framework. Int. J. Sust. Dev World (2013). https://doi.org/10.1080/13504509.2013.860056

ILo, B.M., Olawale, L.S.: The effect of dividend policy on stock price in Nigeria. J. Œcono. Acta Universitatis Danubius **14**(6) (2021). ISSN: 2065–0175

Mensah, J., Enu-Kwesi, F.: Implication of environmental sanitation management in the catchment area of Benya Lagoon, Ghana. J. Integrative Environ. Sci. (2018). https://doi.org/10.1080/1943815x.2018.1554591

Nigerian Capital Market Service Report, 2022 Q1. Equities Market Performance Review, 1 May 2022

Mohieldin, M.: The sustainable development goals and private sector opportunities. EAFIT University of Medellín (2017). http://pubdocs.worldbank.orgThe-SustainableDevelopment-Goals-and-Private-Sector-Opportunities.pdf

Ogbebor, P.I.: Fundamental factors and stock price performance in Nigeria: 2008–2017. UNILAG J. Bus. **5**(2), 69–82 (2019)

Scopelliti, M., et al.: What makes you a "hero" for nature? Socio-psychologicalprofiling of leaders committed to nature and biodiversity protection across seven; EU countries. J. Environ. Planning Manage. **61**, 970–993 (2018). https://doi.org/10.1080/09640568.2017.1421526

Shepherd, E., et al.: Status and trends in global ecosystem *Molinario et al.* 29 services and natural capital: Assessing progress toward Aichi Biodiversity Target 14. Conservation Lett. **9**, 429–437 (2016). https://doi.org/10.1111/conl.12320

Ukaga, U., Maser, C., Reichenbach, M.: Sustainable development: principles, frameworks, and case studies. Int. J. Sustain. Higher Educ. **12**(2) (2011). https://doi.org/10.1108/ijshe.2011.24912bae.005

Ogbebor, P.I., Siyanbola, T.T., Alalade, Y.S.A., Awonuga, A.R.: Individual investors' expectations and stock price behaviour: evidence from the Nigeria exchange limited. J. Solid State Technol. **63**(2), 4028–4040 (2020)

Williams, H.T., Ayodele, T.D.: An Empirical investigation of the impact of dividend Policy on performance of quoted companies in a developing economy. Singaporean J. Bus. Econ. Manage. Stud. **5**(12), 1–7 (2017)

Zhai, T.T., Chang, Y.C.: Standing of environmental public-interest litigants in China: evolution, obstacles and solutions. J. Environ. Law **30**, 369–397 (2019). https://doi.org/10.1093/jel/eqy011

Study of Social Capital and Business Performance of Micro Women Entrepreneurs in Lagos State Nigeria: Implications for Sustainable Development

J. C. Ngwama(✉) and E. E. Omolewa

Department of Business Administration, Crawford University, Igbesa, Ogun State, Nigeria
ngwamaj@yahoo.com

Abstract. Purpose: The purpose of this study is to assess the effect of social capital on the business performance of micro women entrepreneurs and contribution to Sustainable Development Goals (SDGs) using selected micro women entrepreneurs in a Local Government in Nigeria.

Design/Methodology/Approach: The study adopted a survey research design. Taro Yamane method of determining sample size was employed to select a sample of sixty (61) respondents from a population size of seventy-two (72). The study population size 72 of hair dressers were from the database of the Association of hair dressers in Ijanikin, Ojo Local Government Area of Lagos State. The study used structured questionnaire as a research instrument. Also, the reliability test using Cronbach's Alpha was conducted and the coefficient (0.872) which indicated that the data collected was reliable. The data were analysed using descriptive statistics tools analysis of variance (ANOVA). Ordinary least squares regression technique was used in order to estimate the models.

Findings: The finding of hypothesis one indicated that social capital affected business performance of micro women entrepreneurs. Also, the result of hypothesis two showed that social capital had significant effect on business sustainability of micro women entrepreneurs. The result of hypothesis three pointed out that social capital affected business trust among micro women entrepreneurs.

Iimplications/Research Limitations: Economists and geographers have always accepted that economic growth is regional, and that it is driven by and spreading from specific regions, cities, or even neighborhoods. This study was limited to a local government. Broader studies should include states in the federation to have broader understanding of the challenge of integrating resources faced by micro women entrepreneurs through social capital.

Practical implication of the study: The result derived from the study will have a broad impact on all the relevant entrepreneurs, stakeholders, and the government regarding the design of entrepreneurship programs that can enhance the development of viable business ideas at the rural level. More so, the findings from this study will enhance the state economic and development planning aimed at achieving SDGs. The results from the study will be of benefit to entrepreneurs at all levels, in regards to raising funds indirectly through social networking and trust which in turn will improve their knowledge and improve their productivity. The

C. Aigbavboa et al. (Eds.): ARCA 2022, *Sustainable Education and Development – Sustainable Industrialization and Innovation*, pp. 993–1009, 2023.
https://doi.org/10.1007/978-3-031-25998-2_77

study will broaden the knowledge of researchers and academics on the impact of social capital on business performance of micro-women entrepreneur.

Originality and value: Many researchers have carried out studies on how to raise capital for small businesses in Nigeria. However, there has been scanty literature on the knowledge and use of social capital towards enhancing the business performance of micro women entrepreneurs and its contribution towards achieving sustainable development goals (SDGs): poverty reduction, zero hunger, gender equality. This study has provided avenues for sourcing capital based on the outcome of its theoretical and empirical research on how these women can look inward, build trust, and improve their knowledge in funding their businesses.

Keyword: Micro women · Entrepreneurs · Social · Capital · Business performance · Networking

1 Introduction

One major problem associated with most women entrepreneurs is that they are capital constrained and, therefore, unable to undertake all good investment opportunities they encounter, including opportunities to expand their businesses. Micro entrepreneurs play a significant role in the development of private sector both developing and developed economies. It is observed that 99% of businesses are provided by small businesses and 70% of the employment is accounted for by these small, micro, and small businesses(QECD). This indicates that small businesses are major contributors to the objectives of Sustainable Development Goals (SDGs). Related studies such as the Enterprise Centre Fall (2019) Surveyed by the Forum for Women Entrepreneurs (FWE) in Canada, observed that most women entrepreneurs face serious limitations, ranging from access to finances, limited understanding in financial management, lack of planning and mentorship, and above all social capital. These have been problems, according to OCED (2011), impeding the growth of women entrepreneurs businesses all over the world especially in developing nations. Similarly, Idris and Agbim (2015), advocate that finance has been the major problem enhancing the development of women entrepreneurs in developing nations like Nigeria.

Social capital appears to be an alternative way in financing businesses, it is embedded in networking, trust and societal norms which have potential for people's productivity (Colemlack1988), and it is fascinating to observe how many women entrepreneurs especially in developed nations excelled in their businesses through social capital.

Social capital provides an indirect way of finance through networking which can be digitized and personalized to promote individual initiatives. It also creates interpersonal trust and norms which are acceptable within the society and creates a functioning organization not only of the society but also of the economy (Coleman, 1988).

Other than human and material, there may be need to adopt social capital in helping women entrepreneurs towards improving their economic performance (Olomola, 2002). According to OECD (2018), women appear to have significant role in the sustainable development goals, especially as the micro enterprises is fundamental in the economic development of their communities and regions, reducing poverty and hunger, ensure reduction of income gap between men and women which are critical factors of sustainable development agenda. This suggests that at the rural level of development the micro women entrepreneurs play a fundamental role particularly in generating employment. The link between social capital and economic performance of micro women entrepreneurs may not be in doubt, considering this assertion. Therefore, this study investigated the extent to which social capital influence the economic performance of micro women entrepreneurs and also its implication towards achieving Sustainable Development Goals (SDGs). The study will be using selected women from Ijanikin, Oto Awori community in Ojo LGA of Lagos State, Nigeria.

2 Statement of the Research Problem

In spite of the fundamental role women entrepreneurs plays in economic development especially in the micro enterprises at the development of their communities, most micro women entrepreneurs still face a lot of challenges in all facets of their business stemming from lack of social capital which affect their business expansion, profit, growth and performance (Neumark, Wall, & Junfu 2011).

Most women entrepreneurs especially at the micro level have attributed these shortcomings to poor funding. Apart from human capital and physical cash most micro women entrepreneurs do not understand the value of social capital as an intangible asset that creates value in growing their businesses through social networking, shared ideas, and knowledge.

Many researchers have carried out studies on how to raise capital for small businesses in Nigeria (Olomola, 2002; Oluranti, 2011 & Osunde, 2014), however, there has been little research on how to give micro women the social capital they need to improve their business performance at local government level. It is the need to fill this knowledge gap and add to literature that motivated this research.

3 Objective of the Study

The general objective of the study is to assess the effect of social capital on the business performance of micro women entrepreneurs in Nigeria.

Specifically, the study attempted to:

i examine the influence of social capital on business performance of micro women entrepreneurs.
ii determine how micro women entrepreneur can use social capital to sustain their businesses.

examine how social capital enhances trust among micro women entrepreneurs.

4 Research Questions

The study was be guided by the following research questions:

i Does social capital have any significant influence on business performance of micro women entrepreneurs?
ii What is the significant effect of social capital on business sustainability of micro women entrepreneurs?
iii To what extent does social capital enhance business trust among micro women entrepreneurs?

5 Research Hypotheses

The following Hypotheses stated were tested in this study;

Hypothesis 1

i Social capital has no significant influence on business performance of micro women entrepreneur.
ii Social capital has no significant effect on business sustainability.
iii Social capital does not enhance business trust among micro women entrepreneurs.

6 Literature Review

This section deals with the review of literature on the study. The section also presents empirical literature. The theoretical framework consists of some relevant underpinning theories.

6.1 Conceptual Framework

6.1.1 Social Capital

The concept of social capital can be described as the total resources available to group and individuals; it could be virtual or actual that built on relationship, network with friends, relations and other acquaintance (Bourdieu & Wacquant, 1992). In defining social capital, Coleman integrates the economic and societal perspective of the concept from the sociological perspective. Coleman (1988) defines social capital as a way of seeing the individual in a social and cultural environment, subjected to 'norms, rules, and obligations'. Economically individuals are seen as self-interested, independent seeking to fulfill their goals.

Putnam (1993) views social capital from societal perspective, "as features of social organizations such as networks, norms, and social trust that facilitate coordination and cooperation for mutual benefit." In addition, Donald and Laurence (2001) using organizational perspective, view social capital from the bases of mutual trust, understanding and common values underlying cooperate existence embedded in human networks among people and communities that enhances cooperative actions. For Adler and Kwon (2002), social capital could be seen from another structural perspective containing relationship of the actors that could extend to the goodwill accessible to groups and individuals.

Beni, Manggu, & Sensusiana (2018) assert social capital as the quantity of accumulated resources on an individual or group as a result of network of reciprocal relationship.

From the concepts of social capital, it can be said that social capital lies within two distinctive aspects. First, social capital is seen as a resource that is available and shared among cohesive and committed members of a society which has a contextual influence on the member. Second, social capital is seen as an area of network where individual resources are shared within the social network.

6.1.2 Micro Women Entrepreneurs

Due to the variation in conceptualizing micro entrepreneurs, Fadahusi, (1992) described micro entrepreneurs based on the characteristics they possess, these include, the size and scope of the business, number of employees, infrastructure, financial capacity, capital, unconventional mode of operations and autonomy. All these factors are used as determinants of social capital.

In an attempt to link social capital and diverse outcome in the society, Kawachi and Beckman (2000) identified eight fields of enquiry: i) democracy and governance, ii) family and youth behavior problem, iii) schooling and education, iv) community life, v) work and organization, vi) public health, vii) criminology viii) economic development.

Different bodies of research documents have shown the relationship between social capital and various variables of economic development. Burt and Burzynska (2017) assessed social capital and success of entrepreneurs in China using a comparative analysis between Chinese firms and Western firms. Burt and Burzynska (2017) from the findings of their study, provided a strong relationship between social capital and the success of the firms. Using societal norms and trust approach of social capital. Wuebker, Hampl and Wüstenhagen (2015), reported strong ties in trusted third parties playing a critical role in connecting entrepreneurs to the venture capitalists who finance them. With respect to networking. Dahl and Sorenson (2012), studied entrepreneurs in Denmark and founded that firms that survived the longest and that proved most profitable are those that have shared intellectual resource embedded in relationship.

In the same vein, studying a relationship between social capital and health, Hiroshi, Yoshinori and Ichiro (2012), found a strong relationship between social capital and individual health, which in turn is positively correlated to business performance of entrepreneurs.

6.1.3 Performance Measurement of Micro Entrepreneurs

Performance measurement is a determinant which creates an enterprise's value. Performance Measurement as clearly assessed by Amaratunga and Baldry (2002), provides the basis an organization assess on how well it is progressing towards its predetermined objectives, helps to identify areas of strength and weaknesses, and decides on future initiatives, with the goal of improving organizational performance.

6.2 Theoretical Framework

Hoselitz Socio-cultural Theory of Entrepreneurship

The tenet of socio-cultural theory of entrepreneurship was propounded by Hoselitz (1993). The tenets and principle are based on the premise that communities have in any cultural setting people endowed with creative power and consequently giving for the development of social practices and attitudes. The permission of variability of different cultural norms and in the choice of paths of life they follow is the fundamental process of socialization of the individuals although this may not be so completely standardized. Conclusively, Hoseltz (1993) submitted that "entrepreneurs develop their attitudes in the direction of productivity and creative integration".

Also, he found that when culture allows different choices and process of socialization it is flexible. This might enhance people to major and develop on the areas of interest in entrepreneurship development. The theory observed that high level of entrepreneurship development and innovations tend occur among culturally marginalized groups probably because of the precarious situation they found themselves and have to fight for survival.

6.3 Social Capital and Business Performance of Entrepreneurs

The issue of raising capital for women entrepreneurs inevitably leads to rethinking of the notions of social capital as a way of raising funds through inter- and intra-relationship of networking, trust and societal norms.

Apart from getting physical capital through some social capital channels such as the micro finance banks and cooperatives, Carl &Mayowa (2012), posited that social capital can be considered as an innovative way of increasing profit through mutual interaction among entrepreneurs and innovation. Accordingly, Schumpeter (1985), pointed out the danger of this monopoly because it could generate exploitative profit making ventures capable of destructing the process in innovation.

6.4 Social Capital and Business Sustainability

Many researchers have studied the impact between social capital and business sustainability of entrepreneurs (Van-Auken & Werbel, 2006; Chin, 2012; Dave & Sharma, 2013). The findings of the studies indicated a relationship between social capital and continuity of small and micro businesses. This also showed that social capital plays a vital role beyond financial and human factors because it galvanizes the positive relationship and goodwill that are needed to enhance business performance. The study also found out that social capital plays a fundamental role in expanding businesses from generation to generation especially in family businesses. It provides the platform for training, mentoring, and enables the father to groom the offspring who may likely take over to ensure continuity in the business Chin (2012).

However, Danes & Brewton, (1985) noted that goodwill appears to be the social capital that bonds the people together in small and micro businesses among family members and community.

Danes & Brewton opined that "Social capital can take the form of trust, mutual respect, love, selfless concern and reciprocal exchanges within family members and with their staff". This may make micro or family business gain competitive advantage and sustainability. They concluded that social capital and social network were fundamental for firms' success and sustainability. They have potential to create opportunities for businesses, create customer and workers loyalty, access to finance and translate businesses from generation to generations (Dave and Sharma 2013).

6.5 Social Capital and Sustainable Development

For sustainable development in alleviating poverty through productivity in any society to be achieved, it requires a synergistic approach of management of all resources. These approaches in most cases involve entrepreneurship ability to generate innovations which needs to be nurtured in all facets of human endeavor (Calestous & Francis 2018). Considering the importance of gender equality in the sphere of development in this 21st century, the micro women play an important role by complementing their knowledge with their local resources to provide for livelihood for their families and communities.

The quest for sustainability in the economic ability of women involvement in alleviating poverty needs to be harnessed with a structured formalized institutional networking system from both government and non-governmental organizations at different levels and locations. This requires a purposive action of mobilizing resources embedded in a social structure that are accessible through social capital (Lin, 2008).

6.6 Empirical Literature

Wenlong, Suocheng, Haiying, Zehong, Zhuang and Bing (2022), studied on the subjective well-being of herdsmen and farmers in Inner Mongolia in China; they critically looked at how their subjective well-being was influenced by social capital and their productive system. Seven hundred and thirty two (732) famers and herdsmen were examined through microscopic data points. Welong et al., employed multivariate order probity model. The finding of the study indicated that social capital exacted a dramatic influence on the herders' and farmer's subjective well-being. Also, they found that the major contributor among the individual dimension was social network. The variable of social trust played a major and significant role on the collective social capital dimension.

Similarly, Oluranti (2011) examined the role that social capital plays in determining and distributing business earnings of female entrepreneurs in selected rural communities of Ogun State, Nigeria. Oluranti studied with 275 female micro-entrepreneurs in five rural communities in Ogun State. The finding indicated that though human capital variables contribute to earnings in the usual Mincer's parlance, social capitals as well as neighborhood effect variables appear much more important determinants.

7 Methodology

This section discusses the area of study, research design, sample size and sampling techniques, research instruments and data analyses as applicable to this research work.

7.1 Research Design

The study used descriptive research design to obtain the responses of micro women entrepreneurs on the effect of social capital in enhancing their economic performance. Descriptive research design was used to describe the current practices regarding networking, trust and societal norms.

7.2 Population of the Study

The study population consists of selected (hair dressers) micro women entrepreneurs in Ijanikin Oto Awori community, Ojo LGA in Lagos state. The study adopted a purpose sampling procedure.

The study population size was taken from the database of the Association of Hair Dressers in Ijanikin, Oto Awori community, Ojo LGA in Lagos state. From the data base the total population was 72 hair dressers in Ijanikin, Oto Awori community, Ojo LGA in Lagos state.

7.3 Sample Size Determination

The sample size for this study was determined using Taro Yamane method of selecting sample selection. The formula according to Yamane is stated as follows:

$$\mathrm{N_y} = \frac{N}{1 + N*(e)^2}$$

where

$\mathrm{N_y}$ = under laying population

N = Yamane Sample size

e = determined confidence 95%

= 61

7.4 Sampling Techniques

This study employed a stratified sampling technique which was used to categorize the study population (Hair dressers). Hence all the hairdressers regardless of their age and size of business were grouped into five according to their trade. Taro Yamane method of sampling selection was used in determining the sample size. A total of 61 micro women entrepreneurs (hair dressers) were selected from the hairdressers for the purpose of the study.

7.5 Sources of Data Collection

The study used the primary source of data collection through structured questionnaire as research instrument to obtain opinion from the respondents, micro women entrepreneurs (hairdressers).

7.6 Reliability and Validity of the Research Instrument

The research instrument (questionnaire) was subjected to a pilot test. Ten (10) copies of the questionnaire was administered to micro women (hairdressers) in another local government area. Cronbach Alpha Test Reliability was used in the study to determine the reliability of the research instrument. Cronbach Alpha's overall average is 0.872 > 0.70. This implies that the measurement items are internally consistent.

The content validity was carried out through some experts and University lecturers in the field of Business Administration and Entrepreneurship study who evaluated the questionnaire. Feedback was received and after that, the questionnaire was modified and administered to the respondents.

8 Data and Measurement

In other to access social capital on micro women entrepreneur's economic performance, the social capital variables measured were: social networking, trust, and societal norms.

8.1 Business Performance

Business performance was measured using several indicators. All indicators were sourced from the association of hair dressers database in Ijanikin, Oto Awori, Ojo LGA in Lagos State and from the respondent responses in the questionnaire. Business performance was evaluated using financial measures for business success or growth with reference to sales growth, profits, and employee growth (Rahman & Ramli, 2014).

8.2 Social Capital

Measuring social capital cannot be derived without dispute and disagreement from scholars because social capital involve the integration of traditional resources (physical capital, human capital) with other resources derived from social capital (social networks, trust, norms and values) to create value and efficiency for individuals (Coleman 1988).

Since the study was to assess social capital and business performance from economist's perspectives, the beneficial impact of social capital is measured from participates expectations. Having this in mind the study used the three major indicators of social capital i) Networking, ii) trust, iii) societal norms. Data used for social capital were derived from respondent response in the questionnaire which was designed to measure change, attitudes, values and benefits of relationship.

Table 1. Social capital and business performance of micro women entrepreneurs

Model summary				
Model	R	R square	Adjusted R square	Std. error of the estimate
1	.689[a]	.475	.432	1.01116

a. Predictors: (Constant), Entrepreneurship programmes and improved performance, information flow due to networking, Entrepreneurship skill and knowledge due to network, Access to capital and networking.

Table 2. Social capital and business performance of micro women entrepreneurs

ANOVAa						
Model		Sum of Squares	df	Mean Square	F	Sig.
1	Regression	45.382	4	11.346	11.097	.000[b]
	Residual	50.099	49	1.022		
	Total	95.481	53			

a. Dependent Variable: trust and confidence due to network.
b. Predictors: (Constant), Entrepreneurship programmes and improved performance, information flow due to networking, Entrepreneurship skill and knowledge due to network, Access to capital and networking.

8.3 Data Presentation and Analysis

The section presents the analysis of inferential Data.

Hypothesis One
The result from the Model Summary table in Table 1 explained the extent of variation to which social capital influences business performance of micro women entrepreneur, which was 47.5% (R square = 0.475). This implied that 42.5% variation can be explained by other factors not accounted for in this study i.e. not only does social capital contributed to business performance of micro women entrepreneur there were also other variables that made up the remaining 42.5%. The adjusted R-Squared value was 0.432; this implied that the prediction of social capital on business performance of micro women entrepreneur accounts for 43.2% less variance. This revealed that social capital explained 43.2% of variations in business performance of micro women entrepreneur, while the rest were explained by other variables not comprised in this study.

The Anova table in Table 2 explains the statistics of the result; it showed the significance of the two variables on each other. In the Anova table, the statistical significance was less than 0.05. Therefore, since the P-Value is < 0.05, null hypothesis was rejected and accept the alternate hypothesis which stated that social capital had significant influence on business performance of micro women entrepreneurs was accepted.

The coefficients in Table 3 showed the extent to which social capital influence business performance of micro women entrepreneurs. The beta coefficient was 68.0%. This means that a unit changes in social capital would result in 68.0% change in business performance of micro women entrepreneurs. The level of significance (0.05) was 1.96 on the Ttab while, the Teal was 6.239. Therefore, when the Ttab was greater than the Teal alternate would be rejected and vis-versa. In other words, H1 was accepted since the teal (6.239) was greater than the Ttab (1.96). Since the Teal was greater than the Ttab, it implied that social capital affected business performance of micro women entrepreneurs.

Table 3. Social capital and business performance of micro women entrepreneurs

Coefficients[a]						
Model		Unstandardized Coefficients		Standardized Coefficients	T	Sig.
		B	Std. Error	Beta		
1	(Constant)	1.160	.845		1.373	.176
	Access to capital and networking	.191	.219	.113	.871	.388
	information flow due to networking	.011	.021	.059	.557	.580
	Entrepreneurship skill and knowledge due to network	−.192	.181	−.137	−1.062	.293
	Entrepreneurship programmes and improved performance	.627	.100	.680	6.239	.000

a. Dependent Variable: trust and confidence due to network.

4.3.1.2 Hypothesis Two

The result from the Model Summary table in Table 4 explained the extent of variation to which social capital affected business sustainability of micro women entrepreneur, which was 42.9% (R square = 0.475). This implied that 47.1% variation can be explained by other factors not accounted for in this study i.e., not only did social capital contributes to business sustainability of micro women entrepreneur there were also other variables that did, which made up the remaining 47.1%. The adjusted R-Squared value was 0.395; this implied that the prediction of social capital on business sustainability of micro women entrepreneur accounts for 39.5% less variance. This revealed that social capital explained 39.5% of variations in business sustainability of micro women entrepreneur, while the rest were explained by other variables not comprised in this study.

The Anova table in Table 5 explains the statistics of the result, which showed the significance of the two variables on each other. In the Anova table, the statistical significance was less than 0.05. Therefore, since the P-Value is < 0.05, null hypothesis was rejected and accept the alternate hypothesis which stated that social capital had

Table 4. Social capital and business sustainability

Model summary

R	R Square	Adjusted R Square	Std. Error of the Estimate
.655[a]	.429	.395	.72554

a. Predictors: (Constant), Social capital and product development, intra/inter relationship due to networking and sustainability, Social networking program and business sustainability.

Table 5. Social capital and business sustainability

ANOVA[a]

Model		Sum of squares	Df	Mean square	F	Sig
1	Regression	19.772	3	6.591	12.520	.000[b]
	residual	26.321	50	.526		
	total	46.093	53			

a. Dependent Variable: business expansion and competition.
b. Predictors: (Constant), Social capital and product development, intra/inter relationship due to networking and sustainability, Social networking program and business sustainability.

significant effect on business sustainability of micro women entrepreneurs. The alternate hypothesis was accepted, which stated that social capital had a significant effect on business sustainability of micro women entrepreneurs.

Table 6. Social capital and business sustainability

Coefficients[a]

Model		Unstandardized coefficients		Standardized coefficients	t	Sig.
		B	Std. error	Beta		
1	(Constant)	1.321	.534		2.474	.017
	intra/inter relationship due to networking and sustainability	.531	.103	.585	5.145	.000
	Social networking program and business sustainability	.204	.109	.223	1.865	.068
	Social capital and product development	−.058	.123	−.057	−.472	.639

a. Dependent Variable: business expansion and competition

The coefficients in Table 6 showed the extent to which social capital influence business performance of micro women entrepreneurs. The beta coefficient was 58.5%. This means that a unit changes in social capital would result in 58.5% change in business performance of micro women entrepreneurs. The level of significance (0.05) is 1.96 on the Ttab while, the Teal it was 5.145. Therefore, when the Ttab is greater than the Teal H_1 would be rejected and vis-versa. In other words, H1 was accepted since the Teal (5.145) was greater than the Ttab (1.96). Since the Teal was greater than Ttab, it established that social capital affected business sustainability of micro women entrepreneurs was accepted.

4.3.1.3 Hypothesis Three

The result from the Model Summary table in Table 7 explains the extent of variation to which Social capital enhance business trust among micro women entrepreneurs, which was 36.6% (R square = 0. 366). This implied that 66.4% variation can be explained by other factors not accounted for in this study i.e., not only does social capital contributes to business trust among micro women entrepreneurs, there were also other variables that does, which made up the remaining 66.4%. The adjusted R-Squared value was 0.609; this implied that the prediction of social capital on business trust among micro women entrepreneurs accounts for 32.8% less variance. This indicated that social capital explains 32.8% of variations in the business trust among micro women entrepreneurs, while the rest are explained by other variables not included in this study.

Table 7. Social capital and business trust

Model Summary				
Model	R	R Square	Adjusted R Square	Std. Error of the Estimate
1	.605[a]	.366	.328	.89265

a. Predictors: (Constant), Reliance on social network to solve business problem, Cooperative Society helps to grow the business, Business improvement due to social association.

Table 8. Social capital do enhance business trust

ANOVA[a]						
Model		Sum of Squares	df	Mean Square	F	Sig.
1	Regression	22.974	3	7.658	9.611	.000[b]
	Residual	39.841	50	.797		
	Total	62.815	53			

a. Dependent Variable: I don't have issue of trust with any organisation or platform.
b. Predictors: (Constant), Reliance on social network to solve business problem, Cooperative Society helps to grow the business, Business improvement due to social association.

Table 9. Social capital and business trust

Coefficients[a]						
Model		Unstandardized Coefficients		Standardized Coefficients	T	Sig.
		B	Std. Error	Beta		
1	(Constant)	.833	.521		1.598	.116
	Cooperative Society helps to grow the business	.422	.114	.486	3.695	.001
	Business improvement due to social association	.105	.158	.089	.664	.510
	Reliance on social network to solve business problem	.198	.091	.250	2.179	.034

a. Dependent Variable: I don't have issue of trust with any organisation or platform.

The Anova table in Table 8 explains the statistics of the result, which showed the significance of the two variables on each other. In the Anova table, the statistical significance is less than 0.05. Therefore, since the P-Value is < 0.05, the null hypothesis was rejected and the alternate hypothesis, which stated that social capital has significant effect on business trust among micro women entrepreneurs was accepted. This indicated that social capital enhances business trust among micro women entrepreneurs was accepted.

The coefficients in Table 9 showed the extent to which social capital enhances business trust among micro women entrepreneurs. The beta coefficient was 48.6%. This means that a unit change in social capital would result in 48.6% change in business trust among micro women entrepreneurs. The level of significance (0.05) is 1.96 on the Ttab while, the Teal is 3.695. Therefore, when the Ttab was greater than the Teal, H_1 would be rejected and vis-versa. In other words, H_1 was accepted since the Teal (3.695) was greater than the Ttab (1.96). Since the Teal was greater than the Ttab, the studies accepted that social capital affects business trust among micro women entrepreneurs.

9 Discussion of the Findings

Inferential data on hypothesis one indicated that social capital affected business performance of micro women entrepreneurs. This was consistent with (Coleman, 1988 and Olomola (2002), who pointed out that other factors than human and material, the social capital helped many women entrepreneurs to improve their economic performance. Also, Hiroshi, Yoshinori and Ichiro (2012) also, found a strong relationship between social capital and individual health, which in turn is positively correlated to business performance of entrepreneurs.

The finding of hypothesis two indicated that social capital had a significant effect on business sustainability of micro women entrepreneurs. This was consistent with Danes & Brewton (1985) who postulated that "social capital … had the potential to make micro or family business to gain competitive advantage and sustainability". They further opined that social capital was fundamental for firms' success and sustainability. This implied that micro women entrepreneurs could take advantage of social capital to build sustainable businesses.

Also, the finding of hypothesis three showed that social capital affects business trust among micro women entrepreneurs. This in concomitant with Wenlong, Suocheng, Haiying, Zehong, Zhuang and Bing (2022), who in their study, found out that "social capital significantly promote the subjective well-being of farmers and herdsmen, and social network was the leading contributor among the dimensions of individual social capital, while social trust was the leading contributor among the dimensions of collective social capital".

10 Conclusion

This study concluded that social capital had significantly helped these women to grow and sustain their businesses. The growth of micro women entrepreneurs will ameliorate the poverty ravaging women in Nigeria as it is today the government cannot provide the necessary assistance for these women. The fundamental objective of sustainable development goal (SDGs) include goal (1) poverty reduction, (2) zero hunger and (5) gender inequality. Social capital could empower women for the attainment of (SDGs) goal (No.8) which is targeted towards employment, job creation and reduction of inequality among women and youth which in turn impact on the households.

From the findings, the study concluded that social capital was an important factor in galvanizes the positive relationship and goodwill that were needed to enhance business performance. The study found out that that social capital and social network were fundamental for firms' success and sustainability. This also, showed that micro women could as well explore social capital to create opportunities, customer and workers loyalty towards improving and sustaining their businesses for generations. Improving their businesses will not only reduce poverty and hunger but will create decent job for their families and their sustenance.

Trust fosters business networks and serves as a building block for businesses and therefore, women entrepreneurs need to build trust among them to enhance and strengthen their relationship for a better network among themselves to enhance microbusinesses. Good business network will help micro women to build formidable businesses thereby helping to realize the sustainable agenda goal of empowering individual and communities for sustainable income and decent work.

11 Recommendations

The following recommendations were made based on the above findings:

i. There is the need for micro women entrepreneurs to explore social capital by net working with other business entrepreneurs, relationships, trust and informal business groups to improve their businesses performance.
ii. There is the need to create social network to facilitate micro women, it is possible to create or organize small groups, provide materials, finance and training to take advantage of social capital to nurture and grow their businesses.
iii. Women entrepreneurs need to build trust among themselves to enhance to strengthen their network.

References

Adler, P., Kwon, S.: 'Social Capital: Prospects for a new concept', Acad. Manage. Rev., **127** (1), 17–40 (2002)

Akomolafe, C.O.: Open and distance learning as a mechanism for women empowerment in Nigeria. Educational foundations and management (2006)

Beni, S., Manggu, B., Sensusiana, S.: Modal sosial sebagai suatu aspek dalam rangka pemberdayaan masyarakat. *Jurkami*, **3**(1), 18–24 (2018)

Bourdieu, P., Wacquant, L.J.: An invitation to reflexive sociology. University of Chicago press (1992).

Burzynska, K.: Chinese entrepreneurs, social networks, and guanxi. Manag. Organ. Rev. **13**(2), 221–260 (2017)

Calestous, J., Francis, M.: African Regional economic integration: The emergence, evolution, and impact of institutional innovation. Faculty Research Working Paper Series. Harvard Kennedy School. Cambridge (2018)

Chin, A.H.: The role of Family Social, Human and Financial Capital in Family Business Sustainability (2012)

Carl, O., Mayowa, G.: Microfinance and entrepreneurial development in Nigeria. JORIND **10** (3), 405–410 (2012)

Chalmeta, R., Palomero, S., Matilla, M.: Methodology to develop a performance measurement system in small and medium-sized enterprises. Int. J. Comput. Integr. Manufact. **28**(8), 716–740 (2012)

Coleman, J.S.: A rational choice perspective on economic sociology. The Handbook of Economic Sociology **2**, 166–180 (1994)

Coleman, J.S.: Social Capital in the creation of human capital. Am. J. Sociol. **94**, 95–120 (1988)

Dave, S., Sharma, A.: *Small Scale Family Business Succession and Sustainability.* Chattigarh: IL (2013)

Danes, S.M., Brewton, K.E.: Follow the Capital: Benefits of Tracking Family Capital Across Family and Business Systems. In: CARSRUD, A., Brännback, M. (eds) Understanding Family Businesses. International Studies in Entrepreneurship, vol 15. Springer, New York, (1985). https://doi.org/10.1007/978-1-4614-0911-3_14

Donald, J.C., Laurence, P.: 'in good company: how social capital makes organizations work.' Harvard Business School Press, Website (2001)

Fadahusi, O.: Entrepreneurship and small scale industry in commonwealth. British J. Market. Stud. **4**(5), 21–36 (1992)

Hofstede, G.: Constraints in cultural theories management. Acad. Manage. **7**(1), 81–94 (1993)

Lin, N.: A network theory of social capital. The Handbook of Social Capital **50**(1), 69 (2008)

Idris, A.J., Agbim, K.C.: Effect of social capital on poverty alleviation: a study of women entrepreneurs in Nasarawa State. Nigeria. JORIND **13**(1), 208–222 (2015)

Nasip, S., Fabeil, N.F., Buncha, M.R., Hui, J.N.L., Sondoh, S. L., Abd Halim, D.N.: The influence of entrepreneurial orientation and social capital on the business performance among women entrepreneurs along West Coast Sabah Malaysia. Proceedings Ice, 2017 (Ice), pp. 377–395 (2017)

Neils, B., Mirjam van, P., Roy T., Geritt de W.: 'The value of human and social capital investments for the business performance of start – ups (2002)

Neumark, D., Wall, B., Junfu, Z.: Do small businesses create more jobs? New evidence for the United States from the national establishment time series. Rev. Econ. Stat. **93**, 16–29 (2011)

Ogunrinola, I.O., Ewetan, O. & Agboola, F.A. "Informal savings and economic status of rural women in Nigeria". J. Econ. Financ. Stud., **2**(1) 10–26 (2005)

Olav, S.: Entrepreneurs and social capital in China. Manag. Organ. Rev. **13**(2), 275–280 (2017)

Olomola, A.S.:Social capital, microfinance group performance and poverty implications in Nigeria. Ibadan: Nigerian institute of social and economic research (2002)

Oluranti, O.: Social capital and earnings distribution among female micro-entrepreneurs in rural Nigeria. Afr. J. Econ. Manag. Stud. **2**(1), 94–113 (2011)

Osunde, C.: Entrepreneurs and entrepreneurship in developing countries: The Nigerian experience. Stand. Int. J. Indust., Financ. Bus. Manage. **2**, 26–32 (2014)

Waśniewski, P.: Informal performance measurement in small enterprises, institute of economics and finance. Poland J. Sci. **192**(8), 3310–3319 (2021)

Rahman, N.A., Ramli, A.: Entrepreneurship management, competitive advantage and firm performances in the craft industry: concepts and framework. Procedia Soc. Behav. Sci. **145**, 129–137 (2014)

Robert, P., Burgess, E., McKenzie, R.: The death and life of great American cities. Johns Hopkins University Press (1984)

Schumpeter, J.A.: The theory of economic development: An inquiry into profits, capital, credit, interest, and the business cycle. Harvard University Press (1969)

SMEDAN: Survey Report on Micro, Small and Medium Enterprises (MSMEs) in Nigeria. 2010 National (2012)

Wenlong, L., et al.: Influence of Rural Social Capital and Production Mode on the Subjective Well-Being of Farmers and Herdsmen: Empirical Discovery on Farmers and Herdsmen in Inner Mongolia. Int. J. Environ. Res. Public Health **2022**, 19 (2022)

Women Entrepreneurship Knowledge Hub: The state of women's entrepreneurship in Canada 2020. Diversity Institute, Ryerson University, Toronto (2020)

Van-Auken, J., Werbel, F.: Famity Dynamics and Family Business Financial Performance Spousal Committment. Family Bus. Rev. **19(1),** 49–63 (2006)

Wuebker, R., Hampl, N., Wüstenhagen, R.: The strength of strong ties in an emerging industry: Experimental evidence of the effects of status hierarchies and personal ties in venture capitalist decision making. Strateg. Entrep. J. **9**(2), 167–187 (2015)

The Impact of Access to Finance on the Micro-enterprises' Growth in Emerging Countries Towards Sustainable Industrialization

M. A. Mapunda(✉) and M. A. Tambwe

Department of Marketing, College of Business Education, P.O Box 1968, Dar Es Salaam, Tanzania
mmapunda@yahoo.com

Abstract. Purpose: The effect of access to finance on microenterprises growth was assessed using financial literacy as a moderator in emerging countries, particularly, Tanzania.

/Methodology/Approach: This research paper applied a quantitative research design and the data were collected from owners and managers of microenterprises in the city of Dar Es Salaam using a structured questionnaire. Structural Equation Modeling (SEM) using IBM AMOS version 23 and the Statistical Package for the Social Sciences (SPSS)version 26 were both used in the analysis of the 262 respondents who were randomly selected. Before starting to analyze the data, concerns about data screening, reliability, and validity were all addressed.

Findings: The results show that financial literacy has a positive and significant effect on the growth of micro-enterprises, although access to financing and terms of interaction are not statistically significant. This suggests that financial literacy directly impacts micro-business expansion rather than moderating the relationship between access to credit and micro-business growth.

Implications/Research Limitations: The study can be expanded further by combining micro, small, and medium-sized businesses and by applying a resource-based view theoretical lens to examine the moderating role of financial literacy between financial access and the growth of microenterprises. This can be equally crucial in identifying the impact of financial literacy on the relationship between financial access and business growth.

Practical Implications: The study's findings will be useful for microbusiness owners and managers, legislators, and other stakeholders in the facilitation and regulation of financial services to develop strategic training programs on financial literacy. As a result, microbusiness owners and managers may have improved their financial literacy knowledge and capabilities, giving them access to more financial resources.

Originality/Value: Past research has demonstrated that there scarce literature on the use of financial literacy to increase access to financing for micro, small, and medium enterprises in emerging countries, including Tanzania. The study is anticipated to contribute to knowledge and literature in the fields of financial literacy,

C. Aigbavboa et al. (Eds.): ARCA 2022, *Sustainable Education and Development – Sustainable Industrialization and Innovation*, pp. 1010–1025, 2023.
https://doi.org/10.1007/978-3-031-25998-2_78

access to financing, and micro-enterprise enterprises in developing nations based on past empirical and theoretical lenses. Similarly, financial literacy promotes the long-term growth of microenterprises through its impact on access to financing.

Keyword: Access. Finance. Financial literacy. Micro-enterprises. Tanzania

1 Introduction

The Small and Medium Enterprise (SME) sector has a substantial impact on the world economy. Globally, industry is crucial to the social and economic advancement of a nation, especially in terms of employment and job generation (Amoah and Amoah 2018). SMEs account for 90 percent of all business businesses in Sub-Saharan Africa, generate around 60 percent of the GDP, and produce about 70 percent of employment in emerging and developing economies (Hossain et al. 2020). (Amoah and Amoah 2018). Four million Tanzanians are employed in the sector, which accounts for 35 percent of the country's GDP (Israel and Kazungu 2019) of which most of them are engaged in micro-enterprises that employ (1–4) persons (NBS & MITI, 2016). However, the growth of SMEs faces a lot of challenges due to limitations to financial resources access (Zarrouk et al. 2020). In addition, most of their owners or managers lack financial literacy (Buchdadi et al. 2020). Therefore, according to Adomako, Danso, and Ofori Damoah (2016), these difficulties have had some impact on the growth of SMEs in developing nations.

For SMEs to grow and prosper, access to finance as a gauge of financial inclusion is crucial. This can be done by utilizing already available financial services and products from established institutions (Aduda and Kalunda 2012). The increased access to financial resources among micro, small, and medium enterprises in developing nations leads to increased employment development (Ayyagari et al. 2021) and enhances firms' growth. Similar to this, Bongomin et al. (2018) claim that having access to funding allows SMEs the chance to invest in cutting-edge technology, grow their operations, and gain market competitiveness. On the other hand, the enterprise's health is likely to be poor without proper access to capital, endangering its ability to grow (Adomako et al. 2016). This means that in order to expand and be able to explore market opportunities, including those in local and global markets, SMEs need to access funds. Therefore, external financing availability and access are crucial for SMEs' improvement and growth, especially in developing nations (Bongomin et al. 2018).

According to Bongomin et al. (2017), SME owners or managers must be financially literate in order to comprehend and carry out successful plans for accessing financing. According to Huston (2010) financial literacy is the ability and confidence of owners or managers to put their financial literacy to good use in making sound financial decisions. The relationship between business managers or owners and access to financial services is crucial. According to Susan (2020), in order to ensure their company's growth, business venture owners or managers need to be financially literate and have access to financial services. Similar to this, Mabula and Ping (2018) find that managers or owners of SMEs with financial knowledge and expertise will be more active in choosing the best source of funding by examining the costs, advantages, and dangers related to prospective funds. Therefore, financial literacy assists SMEs and their owners in making wise financial judgments and choices, minimizing loan liability and interest charges, and enhancing business operations (Bongomin et al. 2018).

Past research has shown that financial literacy plays a key role in increasing access to financing, resulting in growth for SMEs. For example, research by Susan (2020) in Indonesia found that financial literacy has stimulated access to financing, increasing the growth of SMEs. Based on the research of Bongomin et al. (2017) financial literacy improves and modifies the affiliation between the availability of finance and SME growth in Uganda. Financial literacy is a key resource that reduces information irregularities and the lack of guarantees when financial institutions assess loan applications in the United Kingdom (Hussain et al. 2018). These authors advise that increased financial literacy reduces monitoring expenses and aids in capital structure optimization for SMEs, all of which have a favorable impact on the expansion of the company. As a result, it is acknowledged that having a solid understanding of financial management is a crucial tool for SMEs, owners, and managers to use when making decisions (Hussain et al. 2018).

Besides, Malaysian research by Wasiuzzaman (2019) discovered that inter-firm association plays a significant impact in convincing SMEs to access capital in terms of resource sharing, but only for tangible resources (assets and costs) rather than intangible ones like expertise and information. Similarly, Lusimbo (2016) conducted research in Kenya, and the results show that even though the managers or owners of Micro and Small Enterprises were knowledgeable about the management of debts. Unfortunately, many of them did not know the impact that inflation and interest rates had on the loans they borrowed. This lack of knowledge ultimately affected their financial decisions on access to finance. By comparing the micro-level effects of various financial instruments and grants, Nyikos, Béres, Laposa, and Závecz (2020) carried out another study in Hungary. Their findings show that financial instruments have more direct and significant effects on SMEs' access to financing. Financing is inhibited by many barriers such as; borrowing terms and conditions, the amount and type of collateral required, the level of interest rates, the complexity of loan administration, and the lack of loans with long interest terms, to name a few.

These studies collectively provided a summary of the financial literacy impact on SMEs' growth as well as information asymmetry across developed and developing economies. Some scholars have noted various difficulties in obtaining financial resources, particularly formal credits for SMEs (Nyikos et al. 2020; Megersa 2020; Zarrouk et al. 2020; Nkwabi and Mboya 2019; Wasiuzzaman, 2019; Lusimbo 2016). For instance, Megersa (2020) confirmed that there is a financial gap of US$ 5.2 trillion for SMEs in developing countries every year. Likewise, it has been highlighted that there is limited research on financial literacy and access to finance for SMEs in developing nations (Hossain et al. 2020; Khan et al. 2022).

The research used Barney's Resource-Based View (RBV) (1991) of the enterprise to see how resources impact the growth of micro-enterprises in Tanzania and how financial literacy contributes to this effect. The fundamental tenets of the theory relating to enterprise resources are that SMEs need various skills and expertise (intangible resources) and access to finance (tangible resources) to promote their growth (Turyakira et al. 2019). Therefore, the RBV theoretical framework was used in this research to examine the influence of financial access on the growth of microenterprises in developing countries, specifically in Tanzania using financial literacy as a moderator. The study focuses on two main areas: the influence of financial access on the micro-enterprises growth, and the effects of financial literacy as a moderator.

2 Theory and Hypothesis Development

The RBV as propagated by Barney (1991) was used in the research to assess the association between the firm's access to finance and the micro-enterprises growth in Tanzania through financial literacy as a moderator. This theoretical framework considers the company as a collection of resources (Madhani 2010; Barney 1991). Conferring to the RBV model, a firm with resources that are uncommon, cherished, unique, and non-substitutable can achieve a competitive advantage by implementing tactics that produce value that competitors cannot duplicate (Barney 1991). Only businesses that have distinctive resources, value creation strategies, and the ability to seize market opportunities can improve growth and achieve sustainable performance (Khan et al. 2021). A company's resources and capabilities, such as its management abilities, organizational procedures and abilities, information, and knowledge, are what provide it a sustained competitive advantage (Ortega 2010; Runyan et al. 2006; Barney 1991). Similar to this, Adomako et al. (2016) underline the importance of internal operational skills for access to finance and the link to growth.

Resources have been divided into tangible and intangible categories based on earlier studies (Madhani 2010; Runyan et al. 2006). Such resources take on tangible and intangible forms. Tangible ones include money, access to financing, and location (Runyan et al. 2006). In order for these businesses to have a competitive advantage and experience business success, their resources must be balanced. Financial literacy is a business skill that SME owners or managers can use in conjunction with access to funding to generate better growth outcomes (Adomako et al. 2016). RBV theory is important in this study because it explains how financial literacy in Tanzania changes the relationship between access to finance and microenterprise growth. Consequently, the study used the RBV model to assess how financial literacy has affected the growth of Tanzanian micro-businesses and access to financing.

2.1 Impact of Access to Finance on the Growth of Small and Medium-Sized Enterprises

The growth of SMEs is strongly influenced by and determined by access to financing, according to empirical research (Ayyagari et al. 2021; Rahaman 2011; Hossain et al. 2020). According to Hossain et al. (2020), access to credit significantly affects how well SMEs are doing financially. Access to finance is considered a crucial component since it promotes financial inclusion, deepens the financial sector, and promotes overall economic growth (Buchdadi et al. 2020). Despite the fact that financial access is key for the development and improvement of SMEs, Megersa's study (2020) found that in developing nations, roughly half of all formal SMEs lack access to formal credit. The degree of financial constraint varies according to the type of enterprise and the level of development of the country in which the enterprise operates. In order to build physical capital, reduce shocks and switch to high-income venture capital, SMEs face fewer financial constraints (Nanziri and Wamalwa 2021). We, therefore, have the following hypothesis:

H1*. Access to finance has a significant impact on the growth of small businesses in developing countries.*

2.2 Financial Literacy as a Moderator Between Access to Finance and Micro Enterprises Growth

According to empirical research, financial literacy has a significant role in affecting the relationship between SMEs' growth and access to financing (Hossain et al. 2020; Susan 2020; Bongomin et al. 2017; Adomako et al. 2016). Financial literacy is essential, especially for owners or managers of micro, small and medium-sized enterprises. Since financial management is a connected resource that reduces information asymmetry and the lack of collateral when evaluating loan applications, SMEs, owners, and managers need to be knowledgeable and skilled in it (Hussain et al. 2018). The financial practices of SMEs, the caliber of objective reporting, and sales revenue are all improved by owners or managers who have financial literacy, as demonstrated by Drexler et al. (2014). This is corroborated by a study by Mutegi et al. (2015), who discovered that financial literacy instruction aids people in making decisions that will support their livelihood, economic progress, a stable financial system, and a decrease in poverty. This includes timely payment of invoices and proper debt management. Additionally, Susan (2020) states that regardless of the complexity of financial products, managers or owners of SMEs in developing countries who are financially literate make informed decisions.

Financial ignorance, however, could exacerbate barriers to SME access to financial services. That is backed by Adomako et al. (2016), who contend that owners or managers of financial illiterate SMEs may overlook necessary risks, be unable to translate them accurately, or fail to address them in a timely manner, which is likely to reduce the potential for positive growth associated with access to finance. Hossain et al. (2020) contend that financial literacy has to be regarded as a necessary skill for SME owners or managers in developing nations because they deal with difficult financial choices in the world of corporate finance. In order to boost business performance and growth, SME owners or managers can make strategic decisions with the help of financial knowledge and skills. As a result, we come to the following assumption:

H2*. Financial literacy moderates the relationship between access to finance and micro-enterprise growth in emerging.*

2.3 The Conceptual Framework

The conceptual model (Fig. 1) includes the dependent variable (micro business growth), an independent variable (access to financing), and a moderating variable (financial literacy). The moderator variable is linked to dependent and independent variables by an arrow that indicates the relationship between access to finance and the growth of micro-businesses.

Source: Adapted from Adomako et al. (2016).

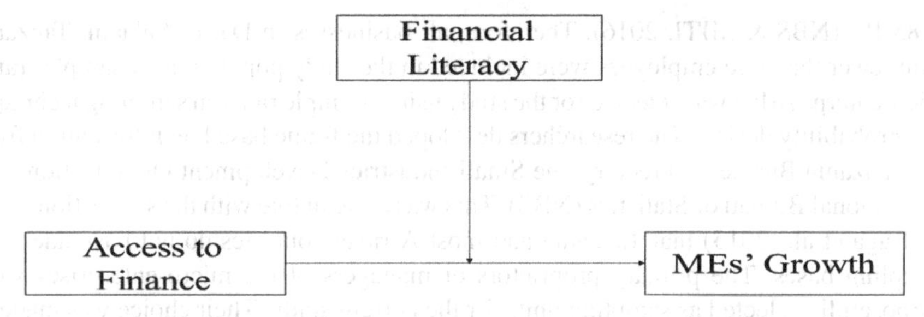

Fig. 1. The impact of financial literacy between access to finance and micro-enterprise growth

3 Methodology

3.1 Research Area

The study was carried out in the city of Dar Es Salaam in Tanzania, as there are numerous micro-enterprise establishments. Of the 41,660 microenterprises in mainland Tanzania, 5,763 are located in the region (NBS & MITI, 2016). Dar es Salaam consists of five administrative districts with a total area of 1,493 square kilometers and a projected population of 6,072,000 (WEF 2020). Tanzania's GDP increased 4.8 percent in 2020 to US$64.4 billion, up from 60.8 percent in 2019. (Tanzania Economy 2022). According to the Household Budget Survey, the economic growth rate has made progress in reducing poverty from 28.2 percent in 2011–12 to 26.4 percent in 2017–18. (NBS 2019). The Tanzanian government, along with the World Bank and donors, has taken a variety of steps to guarantee that projects and development initiatives improve the business climate and industrial growth (Hansen et al. 2018). Despite these efforts, financial constraints, followed by capital limits, inadequate technologies, and stringent laws, have the biggest impact on the growth of Tanzanian SMEs (Nkwabi and Mboya 2019). Therefore, the study was conducted in a situation where access to finance can be one of the greatest barriers to the micro-enterprises growth.

3.2 Design of the Study

The research used a quantitative research plan to analyze the impact of financial literacy on the growth of micro-enterprises in Tanzania. This research design was chosen because the researchers used planned and highly structured data collection approaches (Saunders et al. 2016).

3.3 Sampling with Sample Size

All micro-enterprises in Dar es Salaam were included in the population of the study based on the definition of the 2003 national paper on SME policy in Tanzania. But the researchers only considered the number of employees. Under this policy, enterprises with (1–4) employees are classified as microenterprises, (5–49) as small enterprises, and (50–99) as medium-sized enterprises. Additionally, the Dar Es Salaam region is in the lead with 5, 763 microbusinesses nationwide out of 41,660 establishments which is 85.1% (NBS & MITI, 2016). Therefore, all businesses in Dar es Salaam, Tanzania with fewer than five employees were included in the study population. A sample frame microenterprise list was selected for the study using a simple random sampling technique in a probability design. The researchers developed the frame based on information from the Tanzania Business Directory, the Small Industries Development Organization, and the National Bureau of Statistics (NBS). This was done in line with the suggestion made by Isaga et al. (2015) that Tanzania and most African countries do not have adequate sampling bases. The primary proprietors or managers of the micro-enterprises were purposefully selected as sampling units for the current study. Their choice was made in accordance with the suggestion made by Adomako et al. (2016) that these respondents possess an in-depth understanding of firm growth and performance. Using the table created by Krejcie and Morgan, a sample size of 262 was determined (1970).

3.4 Data Gathering

The data were gathered by means of a questionnaire. Three constructs (financial literacy, access to finance, and micro-enterprise growth) were each given a seven-point Likert scale with 1 meaning strongly disagree and 7 meaning strongly agree (Bongomin et al. 2017). The three constructs were: financial literacy, "Am aware of the costs and benefits of getting credit," (moderating variable); access to finance sample statement "The financial services provided by the bank is safe for our business"; and micro-enterprise growth. "Our company's sales have increased by a factor of two this year" (dependent variable). Three separate, qualified translators back-translated an English version of the study questionnaire into Swahili, a language widely spoken in Tanzania, to verify translation comparability. The questionnaires were distributed by research assistants with the help of business managers in each of the five districts of the Dar Es Salaam region. Clients were informed of the objectives of the study, and those who expressed interest had the opportunity to ask questions while the research assistants filled out the questionnaires. Data collection took place from Monday through Friday during the second and third weeks of May 2022. 370 questionnaires were distributed, of which 262 (71%) were complete, 86 (23%) were not returned, 9 (2%) were missing, and 13 (4%) were outliers.

3.5 Ethics

A research clearance permit was requested from the appropriate authorities after the ethical concerns were noted, as indicated by Saunders et al. (2016). Likewise, individuals were requested to give their agreement before participating in the study, and both confidentiality and voluntary participation were ensured.

3.6 Analysis of the Data

The SPSS Version 26 and the IBM AMOS version 23 were used for data analysis. The SPSS was used to filter the variable data to identify missing values, outliers, normality, multicollinearity, and homoscedasticity before the actual analysis. The screening process was done as proposed by Hair Jr, Black, Babin, and Anderson (2019) that the researcher receives important insights into the features of the data when the data screening is completed prior to the applications of any multivariate technique. Nine questionnaires with missing data were discovered in the sample and eliminated using the listwise deletion technique. According to (Mazana, Tambwe, Mapunda, & Kirumirah, 2021), removing the missing data has no effect on the study's findings when the amount of missing data is less than 2%.

Using the Mahalanobis distance value, 13 outliers were also eliminated from the analysis. As a result, just 262 responders remained. The two items (ATF1 and ATF2) were excluded from the analysis due to the multicollinearity assumption since their variance inflated factor (VIF) was larger than 10. All other assumptions, on the other hand, fell within the permissible range.

4 Results and Discussion

Demographic characteristics

The respondents' gender, age, degree of education, business experience, and ownership of businesses are all described in the demographic characteristics. The survey's majority of respondents were men, according to the findings. According to the survey, there are more males (75.2%) than females (24.8%). According to the data, the respondents' ages are concentrated most heavily in the 30–39-year-old range (37%). Only 1.5% of respondents in this study have postgraduate degrees, whereas 50% of respondents have only secondary education. Similarly, the majority (86.6%) have business training. Additionally, the majority of businesses are run by owners (82.1%), with very few being run by non-owners. As a result, Table 1 provides more information about the traits of the respondents.

4.1 Exploratory Factor Analysis

The exploratory factor analysis (EFA), one of the processes to guarantee to construct validity, was carried out after the data cleaning procedure. This was carried out in accordance with Barney's (2016) recommendation that EFA is carried out to ascertain how the observable variables are related to their underlying latent variables. Some items from the Financial Literacy (FL) and Micro-Enterprise Growth (BG)—were eliminated from the study due to cross-loading. As shown in Table 2, the factor loadings of the KMO items were satisfactory as a result.

4.2 The Measurement Model

The confirmatory factor analysis (CFA) was conducted after EFA in order to eliminate some measurements error since the EFA does not take into account. Actually, the measurement model gives empirical confirmation of the study (Kumar 2015). Some components were removed from the study's individual construct CFA during the process, including one each from financial literacy, access to financing, and microenterprise growth. The construct level goodness of fit indices are shown in Table 3, and they were within acceptable bounds.

Table 1. Demographic Characteristics of the Respondents

Category	Frequencies	Percentage (%)	Cumulative (%)
Gender	197	75.2	75.2
Male	65	24.8	100
Female	**262**	**100**	
Total			
Age	28	10.7	10.7
20 to 29	97	37.0	47.7
30 to 39	79	30.2	77.9
40 to 49	58	22.1	100
50 and above	**262**	**100**	
Total			
Education Level	4	1.5	1.5
Postgraduate	46	17.6	19.1
Undergraduate	131	50.0	69.1
Secondary	57	21.8	90.8
Primary	24	9.2	100
Informal	**262**	**100**	
Total			
Business Training	227	86.6	86.6
Yes	35	13.4	100
No	**262**	**100**	
Total			
Business Ownership	215	82.1	82.1
Owner	47	17.9	100
None-owner	**262**	**100**	
Total			

4.3 Validity and Reliability

The CFA findings facilitated the researcher in the computation of the composite reliability (CR), convergent as well as discriminant validities. In this study, Cronbach's Alpha and composite reliability both exceeded 0.7 which met the suggested rule of thumb by (Kumar 2015). Likewise, convergent validity was measured using average variance extracted (AVE) and its value should be more than or equal to 0.5. Further, discriminate validity was evaluated using shared AVE, and its value should be greater than the AVE which is also been fulfilled showing that the constructs are discriminatory to one another and do not have any linkages (Hair Jr et al. 2019). As a result, Table 4 provides an overview of the reliability and validity analyses' findings.

4.4 Structural Equation Modelling (sEM)and Hypothesis Testing

The SEM was utilized as a major statistical method in the study to examine the association between financial access and micro-business growth as well as the potential

Table 2. The Exploratory Factor Analysis (EFA) Results

Items	1	2	3
ATF4	0.860		
ATF5	0.838		
ATF3	0.826		
ATF6	0.796		
ATF7	0.738		
ATF8	0.725		
ATF9	0.640		
ATF10	0.546		
FL8		0.662	
FL1		0.638	
FL13		0.596	
FL9		0.583	
FL2		0.563	
FL11		0.519	
FL12		0.446	
FL3		0.349	
BG2			0.880
BG1			0.847
BG3			0.825
BG10			0.717
BG4			0.673
BG9			0.665
BG7			0.541
BG8			0.527
BG6			0.495
Eigen Value 6.333 3.108 2.606			
Variance Explained 25.332 12.430 10.423			
Total variance 48.186			
KMO			0.820
Bartlett's Test of Sphericity		Approx. Chi-Square	3465.450
		Df	300
		Sig	.000

Table 3. The Results of Confirmatory Factor Analysis

Constructs	Nos. Items	X^2(df)	GFI	CFI	TLI	RMSEA
Financial Literacy (FL)	7	2.856	0.966	0.972	0.951	0.082
Access to Finance (ATF)	5	4.985	0.947	0.962	0.928	0.121
Micro-enterprise Growth (BG)	4	2.374	0.989	0.992	0.974	0.071

Table 4. Reliability and Validity Results

Dimension	Alpha (α)	CR	AVE	Shared AVE
Access to Finance (ATF)	0.903	0.921	0.788	0.887
Financial Literacy (FL)	0.708	0.881	0.601	0.775
Micro-enterprise Growth (BG)	0.701	0.856	0.607	0.779

moderating role of financial literacy. This statistical method was suitable since it provides high-quality data for moderation analysis and can reduce measurement error using confirmatory factor analysis (CFA) (Afriyie, Du, & Ibn Musah, 2019). In this study, the moderation analysis was conducted using both SPSS and SEM. First, standardized variables for the three constructs—access to finance (ZATF), financial literacy (ZFin L), and micro-enterprise growth (ZBGR) were created using SPSS. The second phase involved multiplying the standardized independent and moderating factors, namely financial literacy (ZFin L) and access to finance (ZATF), to construct the interaction item. In order to determine the predicted moderating impact, the four variables that are ZATF, ZFin L, interaction, and ZBGR were included in the model. These four variables were made, and the SEM was utilized to determine the moderation effect.

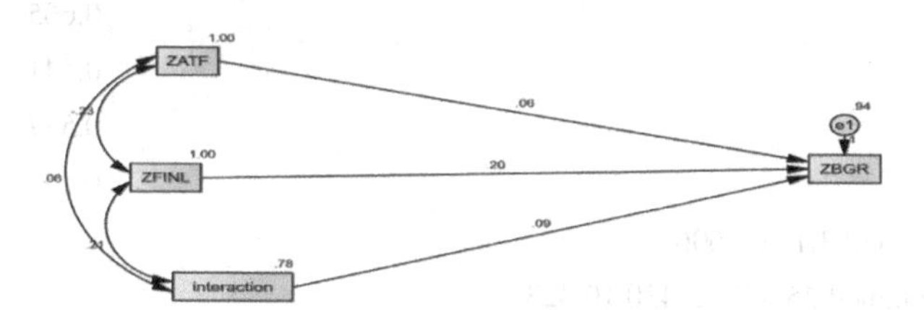

Fig. 2. The effect of Financial Literacy on Access to Finance and MEs Growth

The model in Fig. 2 was used to test the second hypothesis (H2) to find if financial literacy moderates the relationship between access to finance and micro-enterprise growth in emerging countries, Tanzania. The results showed that access to finance and interaction terms do not statistically significantly affect the growth of micro-enterprises, but financial literacy does. As a result, it is not statistically significant that financial literacy has a moderating influence on the relationship between financial access and micro-enterprise growth. The SEM results are summarized in Table 5.

Table 5. SEM Results

Path	Estimate	SE	CR	P	Remarks
ZBGR < --- ZATF	.047	.060	.778	.436	Rejected
ZBGR < --- ZFin_L	.252	.061	4.104	***	**Accepted**
ZBGR < --- Interaction	.044	.066	.671	.502	Rejected

Hypothesis Results.

H1*. Access to finance has a positive and significant impact on micro-enterprise growth in emerging countries.*

The study's first premise was that having access to financing will significantly and favorably affect the expansion of microbusinesses in Tanzania and other developing nations. As the crucial ratio value and significant level are outside of the suggested limits, the results show that access to finance does not significantly affect micro-enterprise growth. The significance level ($P = .436$) is higher than 0.05. The critical ratio ($CR = .778$) from Table 5 is smaller than 1.96. According to Hox and Bechger (1999), the relationship should be considered significant if the critical ratio (CR) is greater than 1.96, and the P-value (P) is less than 0.05. As a result, the theory cannot be proven. The findings contrast with those of (Ayyagari et al. 2021; Bongomin et al. 2017), which demonstrated a favorable and significant relationship between access to finance and the growth of businesses in emerging nations. Since our study concentrated on micro-organizations, whereas their investigations looked at the moderating effects of financial literacy on access to finance and the growth of small, and medium enterprises, this discrepancy may be caused by the size of the businesses.

H2*. Financial literacy significantly moderates the relationship between access to finance and micro-enterprise growth in emerging countries.*

The second hypothesis holds that financial literacy has a significant impact on the relationship between financial access and the growth of microenterprises in emerging countries, particularly Tanzania. According to the results, the critical ratio ($CR = .671$) and significance level ($P = 0.502$), which fall outside of the allowed range, are not significant. If the critical ratio ($CR > 1.96$) and the P-value ($P < 0.05$), the link should be regarded as significant, according to Hox and Bechger (1999). Therefore, the theory is not supported. These results differ from those of (Adomako et al. 2016; Bongomin et al. 2017), which showed that financial literacy has a favorable and significant influence on the association between access to financing and micro-enterprise growth, consequently moderating the relationship. In contrast to earlier studies, which examined the moderating impact of financial literacy on access to credit and the expansion of small and medium-sized businesses, this study's focus was on micro-enterprises, which accounts for the discrepancy.

H3*. Financial literacy has a positive and significant impact on micro-enterprise growth in emerging countries.*

For the purpose of determining the moderation effect, the moderator variable is treated as a predictor variable, and the third predictor variable (interaction term) is produced by multiplying the predictor variable by the moderating variable. This hypothesis was therefore developed in the statistical model even though it does not appear in the conceptual model. The third hypothesis made the claim that financial literacy has a considerable and positive impact on the growth of microenterprises in developing countries, particularly Tanzania. The findings indicate that the critical ratio (CR = 4.104) value and P-value (P < 0.05). The critical ratio value is within the allowed range and the value is significant, the hypothesis is accepted.

Hox and Bechger (1999) said that the association is regarded as significant if the critical ratio (CR > 1.96) and P-value (P < 0.05) are obtained. This demonstrates a positive and significant association between financial literacy and micro-enterprise growth, supporting the hypothesis. The findings demonstrate that financial literacy has a direct impact on the growth of micro-businesses, showing that it will be difficult to secure funding without this expertise and understanding. Lack of financial literacy knowledge and skills prevents entrepreneurs from accessing external financial resources, claim Ye and Kulathunga (2019).

Similar claims are made by Eniola and Entebang (2017), who assert that business training for owners or managers, such as business management skills, leadership development, networking through marketing, and management of financial literacy, may significantly affect the performance and growth of business enterprises in developing countries. The outcomes are comparable to those of (Susan 2020; Buchdadi et al. 2020), who discovered a robust and positive relationship between owners' or managers' financial literacy and their enterprises' growth in developing countries' contexts.

5 Conclusion

The study assessed how owners' or managers' financial literacy moderates the relationship between access to finance and micro-enterprise growth in emerging nations, specifically in Tanzania using the RBV theoretical framework. The study explicitly examines how financial access affects the development of micro-enterprises and how financial literacy can mitigate both of those effects. The results showed that access to finance and interaction terms do not statistically significantly affect the growth of micro-enterprises, but financial literacy does.

The report suggests that financial literacy training programs should be developed for policymakers and microbusiness owners/managers. This might improve the owners' and managers' knowledge and abilities in financial literacy, which would boost their access to financial resources. Similarly, the projected expansion of microbusinesses brought on by financial literacy will boost the nation's entrepreneur population and lower the jobless rate. Additionally, knowing how financial literacy affects microbusiness growth would help policymakers and other stakeholders provide direction for the execution of future national programs and policies.

The study is expected to add to the body of knowledge and literature on financial access, the growth of microenterprise enterprises in developing countries in general as well as financial literacy. Similar to this, financial literacy promotes microbusiness long-term growth through its influence on capital access.

Future research should focus on specific small and medium-sized companies in order to determine the moderating effect of financial literacy between access to financial resources and firms' growth by using the RBV theoretical lens.

References

Adomako, S., Danso, A., Ofori Damoah, J.: The moderating influence of financial literacy on the relationship between access to finance and firm growth in Ghana. Ventur. Cap. **18**(1), 43–61 (2016)

Aduda, J., Kalunda, E.: Financial inclusion and financial sector stability with reference to Kenya: A review of literature. J. Financ. Bank. **2**(6), 95–120 (2012)

Afriyie, S., Du, J., Ibn Musah, A.-A.: Innovation and marketing performance of SME in an emerging economy: the moderating effect of transformational leadership. J. Glob. Entrep. Res. **9**(1), 1–25 (2019). https://doi.org/10.1186/s40497-019-0165-3

Amoah, S. K., Amoah, A. K.: The role of small and medium enterprises (SMEs) to employment in Ghana. Int. J. Bus. Econ. Res.7(5), 151–157 (2018).

AU. African Union:agenda 2063, First 10 years implementation plan 2014–*2023*

Ayyagari, M., Juarros, P., Martinez Peria, M.S., Singh, S.: Access to finance and job growth: firm-level evidence across developing countries. Rev. Finan. **25**(5), 1473–1496 (2021)

Barakabitze, A.A., Kitindi, E.J., Sanga, C., Shabani, A., Philipo, J., Kibirige, G.: New technologies for disseminating and communicating agriculture knowledge and information: challenges for agricultural research institutes in Tanzania. Electron. J. Inf. Syst Developing Countries **70**(1), 1–22 (2015)

Barney, J.: Firm resources and sustained competitive advantage. J. Manag. **17**(1), 99–120 (1991)

Bongomin, G.O., Munene, J.C., Ntayi, J.M., Malinga, C.A.: Determinants of SMMEs growth in post-war communities in developing countries: Testing the interaction effect of government support. World J. Entrep. Manage. Sustain. Dev. **14**(1), 50–73 (2018)

Bongomin, G.O., Ntayi, J.M., Munene, J.C., Malinga, C.A.: The relationship between access to finance and growth of SMEs in developing economies: Financial literacy as a moderator. Rev. Int. Bus. strategy **27**(4), 520–538 (2017)

Buchdadi, A.D., Sholeha, A., Ahmad, G.N.: The influence of financial literacy on SMEs performance through access to finance and financial risk attitude as mediation variables. Acad. Acc. Financ. Stud. J. **24**(5), 1–15 (2020)

Drexler, A., Fischer, G., Schoar, A.: Keeping it simple: Financial literacy and rules of thumb. Am. Econ. J. Appl. Econ. **6**(2), 1–31 (2014)

Eniola, A.A., Entebang, H.: SME managers and financial literacy. Glob. Bus. Rev. **18**(3), 559–576 (2017)

Hair Jr, J. E., Black, W. C., Babin, B. J., Anderson, R. E.: Multivariate Data Analysis ((Eight edition). ed.). United Kingdom: Cengage Learning EMEA . (2019)

Haji, M.: Tanzania: Skills and Youth Employment A Scoping Paper. International Development Research Centre, Ottawa (2015)

Hansen, M.W., Langevang, T., Rutashobya, L., Urassa, G.: Coping with the African business environment: Enterprise strategy in response to institutional uncertainty in Tanzania. J. Afr. Bus. **19**(1), 1–26 (2018)

Hossain, M. M., Ibrahim, Y., Uddin, M. M.: Finance, financial literacy and small firm financial growth in Bangladesh: The effectiveness of government support. J. Small Bus. Entrep. 1–26. Retrieved from (2020). https://doi.org/10.1080/08276331.2020.1793097

Hox, J.J., Bechger, T.M.: An introduction to structural equation modeling. Fam. Sci. Rev. **11**, 354–373 (1999)

Hussain, J., Salia, S., Karim, A.: Is knowledge that powerful? Financial literacy and access to finance: An analysis of enterprises in the UK. J. Small Bus. Enterp. Dev. **25**(6), 985–1003 (2018)

Huston, S.J.: Measuring financial literacy. J. Consum. Aff. **44**(2), 296–316 (2010)

Isaga, N., Masurel, E., Van Montfort, K.: Owner-manager motives and the growth of SMEs in developing countries: Evidence from the furniture industry in Tanzania. J. Entrep. Emerg. Econo. **7**(3), 190–211 (2015)
Israel, B., Kazungu, I.: The Role of Public Procurement in Enhancing Growth of Small and Medium Sized-Enterprises: Experince from Mbeya Tanzania. J. Bus. Manage. Econo. Res. **3**(1), 17–27 (2019)
Khan, F., Siddiqui, M.A., Imtiaz, S.: Role of financial literacy in achieving financial inclusion: A review, synthesis and research agenda. Cogent Bus. Manage. **9**(1), 1–37 (2022)
Khan, R.U., Salamzadeh, Y., Kawamorita, H., Rethi, G.: Entrepreneurial orientation and small and medium-sized enterprises' performance; does 'access to finance'moderate the relation in emerging economies? Vision **25**(1), 1–16 (2021)
Krejcie, R.V., Morgan, D.W.: Determining sample size for research activities. Educ. Psychol. Measur. **30**(3), 607–610 (1970)
Kumar, S.: Structure Equation Modeling Basic Assumptions and Concepts: A Novices Guide. Asian J. Manage. Sci. **3**(7), 25–28 (2015)
Lusimbo, E.N.: Relationship between financial literacy and the growth of micro and small enterprises in Kenya: A case of Kakamega Central sub-county (Doctoral dissertation, cohred, JKUAT). Jomo Kenyata University of Agriculture and Technology, Nairobi (2016)
Mabula, J. B., Ping, H. D. (2018). Financial literacy of SME managers' on access to finance and performance: The mediating role of financial service utilization. ,. Int. J. Adv. Comput. Sci. Appl. 9(9), 32–41
Madhani, P., Madhani, P. M.: Resource based view (RBV) of competitive advantage: an overview. Resource based view: concepts and practices (2010).
Makina, D.: Introduction to the financial services in Africa special issue. Afr. J. Econ. Manag. Stud. **8**(1), 2–7 (2017)
Mazana, M. Y., Tambwe, M. A., Mapunda, M. A., Kirumirah, M. H. (2021). ICT and Marketing for Agricultural Products: Determinants of Mobile Phone usage to Small Scale Orange Farmers in Tanzania. J. Bus. Educ.
Mbani, M., & Mirondo, R. (2019, February 6). Tanzania Economy to Grow at 6.6 percent in 2019. Retrieved from The Citizen: https://www.thecitizen.co.tz/news/Tanzania-economy-to--grow-at-6-6pc-in-2019--/1840340-4969384-ajc6ia/index.html
Megersa, K.: Improving SMEs' Acc'4ess to Finance Through Capital Markets and Innovative Financing Instruments: Some Evidence from Developing Countries. Institute of Development Studies (2020).
Mhando, D.G., Ikeno, J.: Production and Marketing of Orange in Two Villages in Muheza District, Tanzania. African Study Monographs **55**, 85–98 (2018)
Mutegi, H.K., Njeru, P.W., Ongesa, N.T.: Financial literacy and its impact on loan repayment by small and medium entrepreneurs. Int. J. Econo. Commer. Manage. **3**(3), 1–28 (2015)
Nanziri, L.E., Wamalwa, P.S.: Finance for SMEs and its effect on growth and inequality: evidence from South Africa. Transnational Corporations Rev. **13**(4), 1–17 (2021)
NBS. United Republic of Tanzania, House Hold Budget Survey 2017–18. Dodoma: National Bur. Stat. (2019).
NBS, & MITI. (2016). *National Bureau of Statistics (NBS) and the Ministry of Industry, Trade and Investment (MITI).* The 2013 Census of Industrial Production. Analytical Report, Dar Es Salaam. Retrieved 2022
Nkwabi, J., Mboya, L.: A review of factors affecting the growth of small and medium enterprises (SMEs) in Tanzania. Eur. J. Bus. Manage. **11**(33), 1–8 (2019)
Nyikos, G., Béres, A., Laposa, T., Závecz, G.: Do financial instruments or grants have a bigger effect on SMEs' access to finance? Evidence from Hungary. J. Entrep. Emerging Econo. **12**(5), 667–685 (2020)

Ortega, M.J.: Competitive strategies and firm performance: Technological capabilities' moderating roles. J. Bus. Res. **63**(12), 1273–1281 (2010)

Prasanna, R.P., Jayasundara, J.M., Naradda Gamage, S.K., Ekanayake, E.M., Rajapakshe, P.S., Abeyrathne, G.A.: Sustainability of smes in the competition: A systemic review on technological challenges and sme performance. J. Open Innov. Technol. Markets Complexity **5**(100), 1–18 (2019)

Quartey, P., Turkson, E., Abor, J.Y., Iddrisu, A.M.: Financing the growth of SMEs in Africa: What are the contraints to SME financing within ECOWAS? Rev. Dev. Financ. **7**(1), 18–28 (2017)

Rahaman, M.M.: Access to financing and firm growth. J. Bank. Financ. **35**(2), 709–723 (2011)

Rasheed, R., Siddiqui, S.H., Mahmood, I., Khan, S.N.: Financial inclusion for SMEs: Role of digital micro-financial services. Rev. Econ. Dev. Stud. **5**(3), 571–580 (2019)

Rogath, H., Mwidege, A.M.: Analysis of Value Chain for Pigeonpea in Tanzania. Int. J. Phys. Soc. Sci. **4**(6), 419 (2014)

Runyan, R. C., Huddleston, P., & Swinney, J. (2006). Entrepreneurial orientation and social capital as small firm strategies: A study of gender differences from a resource-based view. *The International Entrepreneurship and Management Journal*(2), 455–477

Saunders, M.N., Lewis, P., Thornhill, A.: Business Research for Business Students, 7th edn. Pearson Education Ltd., Harlow (2016)

Simoes, N., Crespo, N., Moreira, S.B.: Individual determinants of self-employment entry: What do we really know? Journal of economic surveys **30**(4), 783–806 (2016)

Sitorus, D.: Improving access to finance for SMEs in Tanzania: Learning from Malaysia's experience . The world Bank (2017).

Susan, M.: Financial literacy and growth of micro, small, and medium enterprises in west java, indonesia. Int. Symp. Econ. Theor. Econometrics **27**, 39–48 (2020)

Tanzania Economy. (2022, March 2). Retrieved April 1, 2022, from Tanzania Investment: https://www.tanzaniainvest.com/economy

TanzaniaInvest. (n.d.). *SME- Tanzaniainvest*. Retrieved from TanzaniaInvest: https://www.tanzaniainvest.com/sme

Thaba, K. L., & Leshilo, T. T. (2016). An investigation into the challenges faced by micro-enterprises Turfloop Plaza, in South Africa. *Proceedings of Annual South Africa Business Research Conference 11 - 12 January 2016, Taj Hotel, Cape Town, South Africa, ISBN: 978–1–922069–95–5.* Cape Town

Torero, M.: A Framework for Linking Small Farmers to Markets. Conference on New Directions for Smallholder Agriculture. 24–25 January. Rome,: IFAD HQ (2011).

Turyakira, P., Kasimu, S., Turyatunga, P., Kimuli, S.N.: The joint effect of firm capability and access to finance on firm performance among small businesses: A developing country perspective. Afr. J. Bus. Manage. **13**(6), 198–206 (2019)

UNDP. (n.d). *Sustainable Development Goals.* Retrieved Dec 12, 2019, from https://www.undp.org/content/dam/undp/library/corporate/brochure/SDGs_Booklet_Web_En.pdf

URT. (2013). *United Republic of Tanzania Small and Medium Enterprises Development Policy, November.* Policy Document

Wasiuzzaman, S.: Resource sharing in interfirm alliances between SMEs and large firms and SME access to finance: A study of Malaysian SMEs. Manag. Res. Rev. **42**(12), 1375–1399 (2019)

WEF. (2020, 02 15). *These are fifteen fastes growing cities in the world.* Retrieved March 17, 2022, from World Economic Forum: https://www.weforum.org/agenda/2020/02/15-fastest-growing-cities-world-africa-populations-shift/

Ye, J., Kulathunga, K.: How Does Financial Literacy Promote Sustainability in SMEs? Developing Country Perspect. Sustain. **11**(2990), 1–21 (2019)

Zarrouk, H., Sherif, M., Galloway, L., El Ghak, T.: Entrepreneurial orientation, access to financial resources and SMEs' business pl4erformance: The case of the United Arab Emirates. J. Asian Financ. Econ. Bus. **7**(12), 465–474 (2020)

Human Capital Development and Economic Growth in Tanzania: Public Spending Perspectives

K. M. Bwana(✉)

Accountancy Department, College of Business Education, P. O Box 2077, Dodoma, Tanzania
kembo211@gmail.com

Abstract. Purpose: The study examines the role of investment as well as human capital development (Public spending on education and health) on the economic performance in Tanzania.

Methodology: Time series data covering 1990 to 2020 were, the data extracted from the World Bank data base. The independent (explanatory) variables used were government spending on education, government spending on health, total investment while the dependent variable was the economic growth (measured by gross domestic growth). Before subjecting the data to econometric estimation normality test was carried out and the normalized data were subjected to Augmented Dickey Fuller (ADF) and Phillips–Peron to confirm if data are stationary. The study also used Vector Error Correction (VECM) Model to estimating Long run dynamics after confirming existence co-integrated which was tested using Johansen co-integration test.

Findings: Findings from unit root test confirm that variables are stationary after first difference while test for co-intergration revealed of three co-integrated equations. Findings from VECM estimates indicates long run relationship between the economic performance (as measured by GDP) and total investment (TI), spending on health (HE) and spending on education (EE). In the long run both spending on health and education have positive effect on economic gross domestic product (GDP) per capita while Total investment have adverse impact on GDP. A percentage change in expenditure on health per capita will result on 0.1617% increase of GDP per capita. On the other hand result also revealed that a percentage change on expenditure on education (EE) per capita result in 0.186% increase in the GDP per capita while a percentage change on the total investment result in the 0.128% change in GDP per capital. Result from the long run relation which are restricted based on GDP per capital (independent variable) revealed that all three variables EE, HE and TI are relevant in predicting movement or changes on GDP per capital.

Research Limitation: Result from this study would be widely useful if the sub variables measuring human capital development would have employed issues such as attributes, competences, knowledge as well as skills embodied on individuals which eventually enables someone's creations, innovation as well as economic wellbeing.

C. Aigbavboa et al. (Eds.): ARCA 2022, *Sustainable Education and Development – Sustainable Industrialization and Innovation*, pp. 1026–1038, 2023.
https://doi.org/10.1007/978-3-031-25998-2_79

Practical Implications: Findings from this study would be useful for ministries responsible for the education and health while implementing the health and education policy of the country since the elasticity of the economic performance toward a unit change of the health and education expenditure is known. On the other hand policy makers could also set strategic health and education policies that ensure steady economic growth derived by human capital development.

Originality: Previous similar studies conducted in Tanzania did not cover government spending on health as well as education (altogether) as key measures of human capital development, past studies also ignored total investment as the key intervening variable in economic growth. Therefore, compared to previous empirical studies, findings obtained broaden the thoughtful of the human capital development and economic development based on key relevant measures of human capital.

Keywords: Human capital · Economic growth · Public spending

1 Introduction

Human capital development can be looked at different angles such as public spending on education, health as well as improvement in life expectancy (Javed et al. 2013). Improvement of education and health in the country are the key components of the human capital development which make human very productive hence contribute to the overall economic performance (growth) of the country. Economists believes that expenditure on any one of these may affect one another factor and may contribute significant influence on the prosperity of economy.

Human capital development has seriously attracted attention in the growth theories. However, measurement of human development on economic growth has not been discussed adequately in literature. Different studies have employed different measures for human capital development, *for example* Mankiew *et al.* (1992) employed enrollment of secondary education as measure of human development. Barro and Lee (1993) as well as Bosworth *et al.* (1995) adopted average years spent in school as measure of human development. Generally, economists acknowledge the role of human capital development in the form of spending on education and health. It is believed that health and education contribute significantly in the role of forming human resources, and in order to ensure improvement in productivity and economic performance, people must have good health.

It means health and education both are important ingredients of human capital development. Gupta and Varhoeven (2001) contended that amount of resources spent and how well have they been spent determine improvement in the social economic performance. This implies that the influence of education on economic will not only depend on resource (budget) increase but how efficiently are being used. In the past recent decades different researchers have shown interest to explore the relationship between human capital and economic growth. Imran et al. (2012), Bloom et al. (2001). Ngwilizi et al. (2018) examine connection between human capital development and economic development. Jung and Thorbeck (2003) conducted a study on the influence of education expenditure on human capital development and poverty status in Tanzania and Zambia. In this study per capita expenditure on education and spending on health has been used to measure human capital development while in the study by Ngwilizi et al. (2018) human capital development was measured by gross primary enrolment and public expenditure on education. Therefore this shows how spending on education is very important for human capital development.

In Tanzania trend show an increase on the spending on education as percentage of Gross domestic product (GDP) from 14% in 1996 to 21% in 2018 (Fig. 2). In recent years the country has witnessed increase of 38% enrollment in primary as well as 45% increase in first year of secondary education enrollment following the introduction of the free lower level secondary education and primary education (URT 2019). According to action aid report (2021) UN recommends that in order to realise SGD 4 a country should allocate about 20% of the budget and 6% of GDP as investiment on education. As far as budget execution is concerned, Overall budget implementation of Tanzania education sector is good. The total implementation of the education budget was 87.2% in the year 2018/19, where the recurrent expenditure was 96.4% and development expenditure financed by internal sources was 90.8% (Fig. 1).

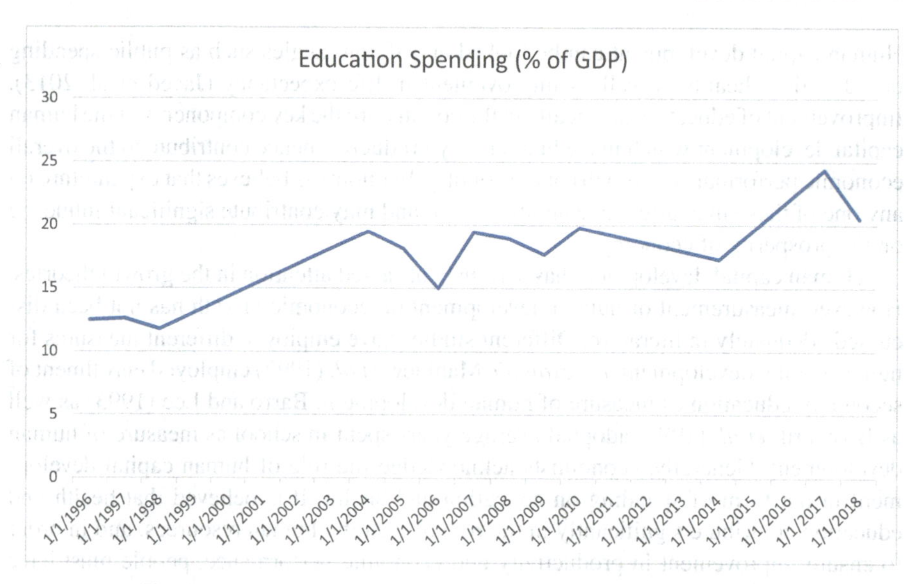

Fig. 1. Trend of public spending on education as percentage of GDP

The literature revealed that discussion on human capital-economic growth causality continues to expand as it depends on the directional of causality but as well as methodologies methodologies used and variables used as proxy measure to estimate human capital (Ngwilizi et al. 2018). There is discrepancies on the findings of the different researches on the causal relationship between the human capital and economic performance, for example Bloon et al. (2001) revealed positive impact of expenditure on health and economic performance. In a study by Jung and Thorbecke (2003) where the researchers examined the relationship between spending on education and growth in Tanzania and Zambia, their finding reveled that public spending can raise economic performance.

Ngwilizi et al. (2018) examined the influence of human capital development and growth in Tanzania by employing Granger causality (since variable were stationary at I(0) and I(1) the researchers employed ARDL (Autoregressive distributed lags), their finding report unilateral causality of human capital development toward economic performance. They further contended that human capital development yield optimistic and important result in both short and long run dynamics.

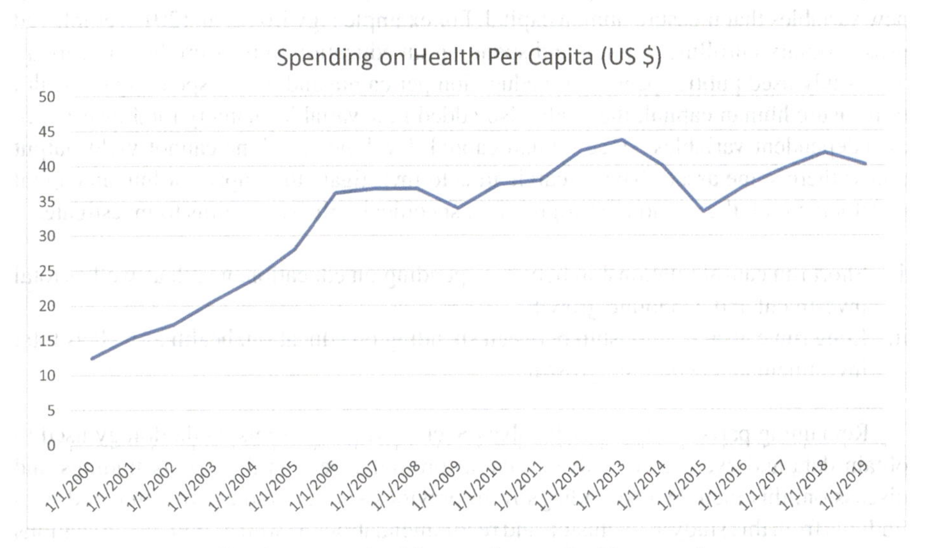

Fig. 2. Trend of public spending on Health per capita

Public spending on health is considered as one of the key element on human capital development, as it stand currently government spending on health is not enough to ensure access to quality health services to all citizens. Though the trend show that modest increase of per capita health spending from US$23.6 in 2010 to US$28.5 in 2017. However, the said modest increase does keep pace with population growth that the country experience. Therefore, it is still well below what is required to provide a complete package of adequate and quality health services to the whole Tanzania population (Moritz Piatti-Fünfkirchen Mariam Ally 2020).

According to (Moritz Piatti-Fünfkirchen Mariam Ally 2020) Tanzania has been facing difficulties to translate its rapid economic achievement and development into increased and wider access of quality health services. Over the last decade, the country has witness a decline of health expenditure as a proportion of government expenditures (decline by 3% to 6% in 2017). Since then, there has been a pressure on domestic expenditure, which implies that there are even limited chances to reallocate public spending (resources) towards health. This study employed total investment (TI) as one of the independent variables is measured using value of a manufacturer's procurements, minus removals, of fixed assets add to certain additions to the worth of non-produced assets (such as subsoil assets or quality and quantity improvement of land productivity) got by the productive activity of institutional units over the accounting period. In the year 2020, total investment (measured by gross capital formation) of Tanzania was 7.2%. The growth total investment of the United Republic of Tanzania has been fluctuating substantially in recent years, though the trend reflect increase through the year 1971–2020 period and marking 7.2% in 2020. Given the fact literature show existence of mixed result when it comes to the impact of human capital and economic performance, this study differs with the study by Ngwilizi et al. (2018) examine causal relationship between study on human capital development and economic growth in Tanzania by introducing new variables that measure human capital. For example Ngwilizi et al. (2018) employed gross primary enrollment ration and spending on education to measure human capital. This study used public spending on education per capita and public spending on health to measure human capital, the study also added new variables namely total investment as independent variables, since human capital development alone cannot yield output unless there some assets. The research aims to investigate the impact of human capital development on the economic progression, specifically the study aims to investigate:

i. Short run causal relationship between spending on education, health as well as total investment and economic growth
ii. Long run causal relationship between spending on education, health as well as total investment and economic growth

Remaining parts of the paper involves Sect. 2 which presents methodology used to obtain data and eventually analyze data accordingly. Section 3 presents findings and discussion, this include also findings from previous similar studies are compared with findings from this study. Conclusion and recommendations as well as policy implications of the findings are presented in Sect. 4.

2 Methodology

The research engaged time series data from 1990–2020 extracted from word development indicators data base. Key variables used in the study include gross domestic product per capita (GDP) as proxy measure of growth (dependent variables) while expenditure on education per capital (EE), expenditure on health per capital (HE) total investment (TI) were used as independent variables. To scrutinize the causal relationship between explained and explanatory variables the following model is specified.

$$LnGDP = \beta_0 + \beta_1 LnEE + \beta_2 LnHE + \beta_3 LnTI + \mu$$

where

LnGDP- represents natural log of gross domestic per capital (in USD)
LnEE- natural log of expenditure on education per capita (in USD)
LnHE- represents natural log of health expenditure per capital (in USD)
LnTI- represents total investments (as percentage of GDP) (in USD)

2.1 Unit Root and Co-integration Test

Literature indicate that in most cases time series data experience problem of unit root and therefore the use of Non-stationary variables in time series may cause fake results which may mislead in case of precise decision (Nabila et al. 2012).

As usual and best practice in empirical analysis it is advised to follow the required steps such as carrying out the umit root using standards test such as Argumented Dickey Fuller (ADF), Phillip-Perron (PP) and Ng-Perron. Literature revealed various studies have applied ADF and PP in testing the stationarity of ther variables (Nabila et al. 2012). Therefore this study also applied the Augumented Dickey Fuller (ADF) to establish stationarity of the variables and confirmed the results using *Phillips Perron.* If the absolute value of the test statistics is lower than the 5% critical value it implies that there is no stationarity and vice-versa is true. The study used johansen cointegration test to establish if there is cointergrating equation (Johansen and Joselius 2000). Cointegration is a statistical test explaining long-run causal relationship (Johansen 1988). We reject the *null hypothesis* of no existence *cointgrating* equation if the trace statistics is higher than the critical value at 5% and vice versa is true. Unlike the study by Ngwilizi et al. (2018) where *Bounds test approach* was used to test cointergration among the variables (gross enrollment on primary school, public spending on education and economic development) which finally led to the use of ARD Model. Javed et al (2013) also employed Johansen co intergreation to test for cointregration among the variables (public spending of education, spending on health, primary scholl enrollment, secondary school enrolment, total investiment and gross domestic products).

2.2 Vector Error Correction Model (VECM)

Both long run as well as short-run dynamics among variables can be apprehended using vector error correction model (VECM). The model is also known as the unique case of restricted Vector Autoregressive (VAR model) which is used when the variables are stationary after first difference and at the same time the variables manifest existence of long-run causal relationship. Engle and Granger (1987) evident that representation of Vector Autoregressive model in first difference when there is existence of cointergration is mis-specification therefore VECM with co-integrated systems of equation is the appropriate approach. The VECM launches the speed of modification from short-run equilibrium towards long-run equilibrium state. Therefore the following system of four equations are expressed to represents the model specifications.

$$\Delta LnGDPt = \alpha + \sum_{i=1}^{k-1} \beta_i \, \Delta LnGDP_{t-i} + \sum_{j=1}^{k-1} \sigma_j \, \Delta LnEE_{t-j} + \sum_{m=1}^{k-1} \theta_m \, \Delta LnHE_{t-m} + \sum_{n=1}^{k-1} \partial_n \, \Delta LnTI_{t-n} + \lambda_1 ECT_{t-1} + \mu_{1t} \quad (1)$$

$$\Delta LnEEt = \psi + \textstyle\sum_{i=1}^{k-1} \beta_i \, \Delta LnGDP_{t-i} + \displaystyle\sum_{j=1}^{k-1} \sigma_j \, \Delta LnEE_{t-j} + \textstyle\sum_{m=1}^{k-1} \theta_m \, \Delta LnHE_{t-m} + \sum_{n=1}^{k-1} \partial_n \, \Delta LnTI_{t-n} + \lambda_2 ECT_{t-1} + \mu_{2t} \quad (2)$$

$$\Delta LnHEt = \phi + \textstyle\sum_{i=1}^{k-1} \beta_i \, \Delta LnGDP_{t-i} + \displaystyle\sum_{j=1}^{k-1} \sigma_j \, \Delta LnEE_{t-j} + \textstyle\sum_{m=1}^{k-1} \theta_m \, \Delta LnHE_{t-m} + \sum_{n=1}^{k-1} \partial_n \, \Delta LnTI_{t-n} + \lambda_3 ECT_{t-1} + \mu_{3t} \quad (3)$$

$$\Delta LnTIt = \gamma + \sum_{i=1}^{k-1} \beta_i \, \Delta LnGDP_{t-i} + \sum_{j=1}^{k-1} \sigma_j \, \Delta LnEE_{t-j} + \sum_{m=1}^{k-1} \theta_m \, \Delta LnHE_{t-m} + \sum_{n=1}^{k-1} \partial_n \, \Delta LnTI_{t-n} + \lambda_4 ECT_{t-1} + \mu_{4t} \quad (4)$$

where Δ denote difference operators and ***k-1*** indicates optimal lag length (reduced by 1)

ECT $_{t-1}$ lagged value of residual obtained from co-integrating regression of the dependent variables on the independent. Contains long run information derived from the long run cointegrating relationships. λ is the speed of adjustment with negative sign. γ, β, ϕ, and α are constant while β, σ, ∂ and θ are short run dynamics of the model's adjustment long run equilibrium and μ stochastic of error terms (innovations shocks).

2.3 VECM causality

VECM causality presents that if two variables are co-integrated and the variables are stationary after first difference it implies that there is bidirectional causality. The study explore use of multivariate causality test to establish causal relationship. Causality in the system of equations exists only if the lagged error correction term (ECTt-1) also considered as long-run dynamics and sum of the coefficients of the lagged variables which consider short run dynamics are significant (Nabila et al. 2012). In a study by Ngwilizi et al. (2018) on the human capital development and growth, *bounds test* was used to test existence of co intergration among the variables and Granger causality was employed to test causal effect of the estimated variables. Jung and Thorbecke (2003) used general equilibrium on the impact public education expenditure on different outcomes (such as human capita, growth and improved life standards) in Tanzania and Zambia. Therefore, this study differs with abovementioned two studies in terms of methodology as well as the variables that measure human capital development.

3 Findings and Discussion

Finings revealed that all variables Gross domestic products (GDP) per capita, expenditure on education (EE), health expenditure (HE) and total investiment were not normally distributed. Therefore we normalize the variables by applying natural logarith to the variables. Correlation matrix was used to test multicornealrity among variables, result show that there was no serious correlation as there was less than 0.80 correlation between variables (as general rule of thumb require correlation not to exceed 80%) (Table 1).

3.1 Unit Root and Johansen Co Intergration Result

Findings revealed that variables were not stationary after conducting stationary test using augmented dickey fuller (ADF) test and confirm the result using Phillips Perron (PP). We make the data stationary after first difference and confirm the result using PP test, result for ADF test are shown on *Appendix 2* were we can see that at level all four variables (GDP), expenditure on education (EE), expenditure on health (HE) and total investment (TI) were not stationery. However all of them become stationery after the first difference

Table 1. Summary statistics

Variable	Obs	Mean	Std. Dev.	Min	Max
lngdpperca~a	31	6.141365	.6569465	5.028317	6.960333
lntotalinv~p	31	3.388608	.2145878	2.876442	3.680217
lngoventex~d	31	2.775577	.7180955	1.609438	3.7612
lngoventex~l	31	3.450738	.3064262	2.742906	3.779141

showing that they are I (1). The result was contrary to the result from Ngwilizi et al. (2018) in which variables were stationary at different levels.

Since all variables were stationary after first difference, we were compelled to run the Johansen co integration test to test for the long run causal relationship among the variables. In this study, the co-integrating rank was tested using the Trace statistics developed by Johansen and Juselius, (Johansen and Juselius 1990). Table 2 depicts output of the Johansen co integration test. From the Table 2 it shows that at rank 3 there is existence of three co integrating and therefore we reject the null since value trace statistics value is less than the 5% critical values.

Table 2. Johansen Co intergration test

Johansen tests for cointegration

Trend: constant — Number of obs = 29
Sample: 1992 - 2020 — Lags = 2

maximum rank	parms	LL	eigenvalue	trace statistic	5% critical value
0	20	87.92612	.	66.7019	47.21
1	27	105.54483	0.70331	31.4645	29.68
2	32	113.33793	0.41577	15.8783	15.41
3	35	120.08431	0.37203	2.3856*	3.76
4	36	121.27709	0.07897		

3.2 VECM Result

Result from VECM estimates indicates long run relationship between the economic performance (as measured by GDP) and total investment (TI), expenditure on health (HE) and expenditure on education (EE). In the long run both expenditure on health and education have positive influence on economic gross domestic product (GDP) per capita which is consistent with our prior expectation that spending on health and education have positive impact on economic performance, on the other hand Total investment have negative impact on the GDP. A percentage change in expenditure on health per capita will result on 0.1617% increase of GDP per capita. On the other hand result also revealed that a percentage change on expenditure on education (EE) per capita result in 0.186% increase in the GDP per capita while a percentage change on the total investment result in the 0.128% change in GDP per capital. Result from the long run relation are represented in Table 3, where variables are restricted based on GDP per capital (independent variable). Result have also revealed that all three variables EE, HE and TI are relevant in predicting movement or changes on GDP per capital which implies that the variables have asymmetric effect (have opposite impact) on GDP per capital in the long run on average ceteris paribus (since the p-values of the three variables is less than 1%). Findings in this study are consistent with the findings in Javed et al. (2013) conducted in Pakistan, their findings revealed that magnitude of the coefficient (of long run) was not very high as far as spending on education and health is concerned though they have positive impact and the variables are significant. Positive influence of spending on education and health implies that Tanzania can benefit economically by investing on human capital development (spending on education and health). Jung and Thorbecke (2003) also contended that spending on education will have significant impact on poverty reduction if the spending target poor household.

Table 3. Result reflecting long run relationship

Johansen normalization restriction imposed

beta	Coef.	Std. Err.	z	P>|z|	[95% Conf.	Interval]
_ce1						
lngdppercapita	1	.	.	.	.	.
lntotalinvesimentgdp	1.231806	.2197399	5.61	0.000	.8011239	1.662488
lngoventexpedpercapitalusd	-.848007	.0721379	-11.76	0.000	-.9893947	-.7066193
lngoventexphealthpercapital	-1.16742	.1717372	-6.80	0.000	-1.504019	-.8308214
_cons	-3.999615	.	.	.	.	.

The long run equation is presented in Table 3, where error correction terms are generated, findings revealed that both log of government spending on education and log of government expenditure on health have positive influence on natural log of growth per capita, on the other hand log of total investment have negative impact on the log of gross domestic product. All three variables EE, HE and TI are relevant in predicting movement or changes on GDP per capital which implies that the variables have asymmetric effect (have opposite impact) on GDP per capital in the long run on average ceteris paribus (since the p-values of the three variables is less than 1%). The coefficients all three variables are statistically significant at 1% level of significance.

Findings from VECM estimates also revealed that in the long run there is causality at 1% significance level where ECT coefficient shows that the adjustment term (−0.277) is statistically significant at !% level, implying that previous years' deviations from long run equilibrium are corrected for within current year at convergence speed of 27.7% . The extract of cointergrating equation and long run model is represented in Eq. 5.

$$\boldsymbol{ECT_{t-1}{=}[1.00000LnGDP_{t-1}{+}1.231806LnTI_{t-1}{-}0.848007LnEE - 1.16742LnHE{-}{-}3.999615]} \tag{5}$$

Result for Eq. 1 (which is generalized form of VECM) as the target variables is represented in the Eq. 6 below

$$\boldsymbol{\Delta LnGDPt = 0.0327 - 0.0561\Delta LnGDP_{t-i} - 0.0983\Delta LnEE_{t-j} + 0.2309\Delta LnHE_{t-m} - 0.1113\Delta LnTI_{t-n} - 0.2769ECT_{t-1} + \mu_{1t}} \ldots\ldots \tag{6}$$

4 Conclusion and Recommendations

The study aimed at exploring the causal relationship between human capital development (gauged by spending on education and spending on health) and economic growth (measured by GDP). The study employed VECM model after confirming existence of long run relationship among the variables. Findings revealed that public expenditure on education and health have positive influence on (growth) performance in the long run. Finding also report that spending on health and education does not have any impact on economic performance in the short run. This study appreciate investment in human capital development as one cornerstone of economic performance, and that human capital development encourage innovation and labor productivity in the long run.

The study recommends that ministries responsible for health and education should be given high priority (especially in budget allocation) in view that the country will be able to have economic growth in the long run. It is also recommended that policy makers should come up with health and education policy and guidelines which insist on the efficient use of the resources and quality of education and health services so that the expected overall economic growth could be experienced in its totality.

Appendix 1: Unit Root Test Result

```
Dickey-Fuller test for unit root                     Number of obs   =        29

                               ---------- Interpolated Dickey-Fuller ---------
                  Test         1% Critical       5% Critical      10% Critical
               Statistic           Value             Value             Value
------------------------------------------------------------------------------
 Z(t)             -4.121            -3.723            -2.989            -2.625
------------------------------------------------------------------------------
MacKinnon approximate p-value for Z(t) = 0.0009

. dfuller dlntotalinvesimentgdp

Dickey-Fuller test for unit root                     Number of obs   =        29

                               ---------- Interpolated Dickey-Fuller ---------
                  Test         1% Critical       5% Critical      10% Critical
               Statistic           Value             Value             Value
------------------------------------------------------------------------------
 Z(t)             -5.002            -3.723            -2.989            -2.625
------------------------------------------------------------------------------
MacKinnon approximate p-value for Z(t) = 0.0000

. dfuller dlngoventexpedpercapitalusd

Dickey-Fuller test for unit root                     Number of obs   =        29

                               ---------- Interpolated Dickey-Fuller ---------
                  Test         1% Critical       5% Critical      10% Critical
               Statistic           Value             Value             Value
------------------------------------------------------------------------------
 Z(t)             -6.492            -3.723            -2.989            -2.625
------------------------------------------------------------------------------
MacKinnon approximate p-value for Z(t) = 0.0000

. dfuller dlngoventexphealthpercapital

Dickey-Fuller test for unit root                     Number of obs   =        29

                               ---------- Interpolated Dickey-Fuller ---------
                  Test         1% Critical       5% Critical      10% Critical
               Statistic           Value             Value             Value
------------------------------------------------------------------------------
 Z(t)             -4.956            -3.723            -2.989            -2.625
------------------------------------------------------------------------------
MacKinnon approximate p-value for Z(t) = 0.0000
```

Appendix 2: VECM Result

Vector error-correction model

Sample: 1992 - 2020		Number of obs	= 29
		AIC	= -5.416885
Log likelihood = 105.5448		HQIC	= -5.018197
Det(Sigma_ml) = 8.11e-09		SBIC	= -4.143885

Equation	Parms	RMSE	R-sq	chi2	P>chi2
D_lngdppercapita	6	.076415	0.6769	48.17657	0.0000
D_lntotalinves~p	6	.120591	0.1640	4.510638	0.6079
D_lngoventexpe~d	6	.138652	0.2263	6.726982	0.3468
D_lngoventexph~l	6	.211295	0.1418	3.801234	0.7036

	Coef.	Std. Err.	z	P>|z|	[95% Conf.	Interval]
D_lngdppercapita						
_ce1 L1.	-.2769696	.0659259	-4.20	0.000	-.406182	-.1477573
lngdppercapita LD.	-.1113597	.2093862	-0.53	0.595	-.5217491	.2990298
lntotalinvesimentgdp LD.	.2309165	.2166076	1.07	0.286	-.1936266	.6554595
lngoventexpedpercapitalusd LD.	.0983228	.1251048	0.79	0.432	-.1468781	.3435237
lngoventexphealthpercapital LD.	-.0560722	.0792738	-0.71	0.479	-.2114459	.0993016
_cons	.032712	.0186244	1.76	0.079	-.0037912	.0692152
D_lntotalinvesimentgdp						
_ce1 L1.	-.1483744	.1040384	-1.43	0.154	-.3522859	.0555371
lngdppercapita LD.	-.4601316	.3304347	-1.39	0.164	-1.107772	.1875085
lntotalinvesimentgdp LD.	.4745666	.3418308	1.39	0.165	-.1954095	1.144543
lngoventexpedpercapitalusd LD.	.3287995	.1974293	1.67	0.096	-.0581547	.7157538
lngoventexphealthpercapital LD.	.0605587	.1251028	0.48	0.628	-.1846384	.3057558
_cons	-.0101299	.0293914	-0.34	0.730	-.067736	.0474763
D_lngoventexpedpercapitalusd						
_ce1 L1.	.0062249	.1196196	0.05	0.958	-.2282252	.2406751
lngdppercapita LD.	.0285747	.3799221	0.08	0.940	-.7160588	.7732083
lntotalinvesimentgdp LD.	-.0526996	.3930249	-0.13	0.893	-.8230142	.7176151
lngoventexpedpercapitalusd LD.	-.2738076	.2269972	-1.21	0.228	-.7187139	.1710986
lngoventexphealthpercapital LD.	-.0534483	.1438388	-0.37	0.710	-.3353672	.2284706
_cons	.0784217	.0337932	2.32	0.020	.0121882	.1446552
D_lngoventexphealthpercapital						
_ce1 L1.	.2255314	.1822918	1.24	0.216	-.131754	.5828167
lngdppercapita LD.	.0225231	.5789742	0.04	0.969	-1.112246	1.157292
lntotalinvesimentgdp LD.	-.0954989	.598942	-0.16	0.873	-1.269404	1.078406
lngoventexpedpercapitalusd LD.	.1664349	.3459275	0.48	0.630	-.5115706	.8444404
lngoventexphealthpercapital LD.	.1947788	.2192001	0.89	0.374	-.2348455	.6244031
_cons	.0313439	.0514985	0.61	0.543	-.0695912	.1322791

References

Action Aid Report: Financing the Future: Delivering SDG4 in Tanzania (2021). https://www.google.com/search. Accessed 26 April 2022

Aurangzeb, A.: Relationship between health expenditure and GDP in an augmented solow growth model for Pakistan: an application of cointegration and error-correction modeling. Lahore J. Econ. **8**(2), 1–18 (2003)

Bloom, D.E., Canning, D., Sevilla, J.: The effect of health on economic growth: theory and evidence. World Dev. **32**(1), 1–13 (2001)

Bose, N., Haque, M.E., et al.: Public expenditure and economic growth: a disaggregated analysis for developing countries. Manch. Sch. **75**(5), 533–556 (2007)

Devarajan, S., Swaroop, V., et al.: The composition of public expenditure and economic growth. J. Monet. Econ. **37**(2), 313–344 (1996)

Sangeev, G., Varhoeven, M.: The efficiency of Government expenditure. Experience from Africa, J. Policy Model. **23**(4) (2001)

Imran, M., Bano, S., Azeem, M., Mehmood, Y., Ali, A.: Relationship between human capital and economic growth: use of cointegration. Approach. J. Agric. Soc. Sci. **8**(4), 135–138 (2012)

Javed, M., Abbas, S., Fatima, A., Azeem, M.M., Zafar, S.: Impact of human capital development on economic growth of Pakistan: a public expenditure approach. World Appl. Sci. J. **24**(3), 408–413 (2013)

Jung, H.S., Thorbecke, E.: The impact of public education expenditure on human capital, growth, and poverty in Tanzania and Zambia: a general equilibrium approach. J. Policy Model. **25**(8), 701–725 (2003)

Khilji, B.A., Kakar, Z.K., Subhan, S.: Impact of vocational training and skill development on economic growth in Pakistan. World Appl. Sci. J. **17**(10), 1298–1302 (2012)

Piatti-Fünfkirchen, M., Ally, M.: Tanzania Health Sector Public Expenditure Review 2020, Open Knowledge Repository, Report No: AUS0001650 (2020)

Ngwilizi, D.N., Mwaseba, S.L., Mwang'onda, E.S: Human capital and economic growth nexus in Tanzania: an econometric analysis (2018)

Pritchett, L.: Where has all the education gone? World Bank Econ. Rev. **15**(3), 367–391 (2001)

Psacharopoulos, G.: Returns to investment in education: a globe update. World Dev. **22**(9), 1325–1345 (1994)

Samimi, A.J., Madadi, M., Heydarizadeh, N.: An estimation of human capital share in economic growth of Iran using growth accounting approach. Middle-East J. Sci. Res. **11**(1), 90 (2012)

URT: Education sector performance report 2018/19 (2019)

Coping with Crime Threat and Resilience Factors Among the Motorcycle Taxi Operators and Customers in Dar es Salaam Tanzania

E. F. Nyange[1(✉)], I. M. Issa[2], K. Mubarack[3], and E. J. Munishi[3]

[1] Department of Marketing, College of Business Education, Dar es Salaam, Tanzania
e.nyange@cbe.ac.tz

[2] Department of Procurement and Supplies Management, College of Business Education, Dar es Salaam, Tanzania

[3] Department of Business Administration, College of Business Education, Dar es Salaam, Tanzania

Abstract. Purpose: This paper aims to examine the coping strategies of customers in reducing the crime threat faced by motorcycle taxi riders.

Design/Methodology/Approach: The study used a mixed-approach research methodology based on a sample size of 287, selected through cluster and purposive sampling techniques. Relevant passengers and motorcycle taxi riders were among the participants in the survey. To supplement the data, open- and closed-ended surveys, interviews, as well as five focus group discussions, were held with motorcycle riders and consumers serving as the primary crime victims. Descriptive statistics were used to evaluate open-ended questions, SPSS version 16 was used to analyze quantitative data, and MAXQDA 10 was used to analyze the qualitative data.

Findings: Findings indicated that the crime victims managed to develop reactive and proactive strategies for coping with the threats. On the other hand, some of them experienced unsuccessful coping due to various reasons. To increase the victims' ability to cope with the threats, the study recommends equipping riders and customers with relevant knowledge and skills as well as ensuring the availability of adequate security services. Moreover, operators should be facilitated to form meaningful social networks that enable them to access security support.

Research Limitation/Implications: Motorcycle taxi business is increasingly becoming one of the key livelihood strategies in the urban settings of Sub-Saharan Africa. However, crime against operators is threatening the subsector and jeopardises the livelihood of motorcycle riders.

Practical Implication: These findings contribute to the government's initiatives towards alleviating crime related to motorcycle taxi operations.

Originality/Value: These findings shed light on practical ways of alleviating crime threats and their consequences among motorcycle taxi riders and customers in the urban setting. The paper contributes to promoting safety among young men and women who, through their creativity have devised means of promoting reliable transport systems in cities where public transport is almost paralyzed due to limited transport infrastructure and traffic congestion.

Keyword: Copping · Crime · Ridders · Strategies · Urban

C. Aigbavboa et al. (Eds.): ARCA 2022, *Sustainable Education and Development – Sustainable Industrialization and Innovation*, pp. 1039–1052, 2023.
https://doi.org/10.1007/978-3-031-25998-2_80

1 Introduction

1.1 Background Information

Motorcycle taxi operations play significant role by supporting the lives of urban migrants' subgroups who live in urban and peri-urban areas (Qian 2014). The operations further help to form alternative means of transportation specifically to the upper-class passengers who feel fed up with sitting in traffic areas in the city centre (Kumar 2011). This serves a very critical role in many developing countries, particularly in service areas where conventional public transport is not available (Nguyen-phuoc et al. 2019). Further, the business form a source of self-employment among the taxis operators (Adiambo 2020). For example, majority of the taxi riders use motorcycle taxi riding operation to afford basic needs such as education, food, shelters, clothes to mention just a few in this context (Jasna *et al.* 2015). Moreover, one of the factors catalysing motorcycle taxi riding operations in the city relates to unreliable urban public transport system characterised by delay, congestion as well as inability to support growing urban population (Mbegu & Mjema 2019; Nyachieo 2015; Opondo & Kiprop 2018). This further makes motorcycle taxi riding operations indispensable in developing countries especially Sub-Sahara African countries (AFCAP 2019). By definition, motorcycle taxi entails a form of transportation mode (bike taxi) which is composed of two tires, used to carry one passenger at any time to any convenient place (Caroline & Neil 2015a, b).

As stated earlier, motorcycle taxi business operations principally play a key role of supporting customers, transportation of goods to difficult areas where lorries or normal vehicles are unable to reach due to narrowness or poor roads passable (Gibigaye 2006; Opondo & Kiprop 2018). More specifically, motorcycle taxi riders facilitate transportation and delivery of goods, carrying passengers, picking up required goods, home groceries, carrying school pupils to and from schools to mention just a few (Caroline & Neil 2015a). However, when undertaking their role, motorcycle taxi riding operators are said to be confronted with crime threat that naturally hamper their livelihood in various ways (Tarimo 2013). The riders fall victims to motorcycle theft, kidnapping, lack of cooperation from customers as well as robbery and violence; murder; assault, abduction; mob justice; fraud and forgery; fighting; handling stolen property; indecent assault: malicious damage to property; burglary; rape to mention just a few that among other things push them to poverty as well as death (Adiambo 2020; Kariuki Nyaga & Gichuru Kariuki 2019). Thirdly, the crime is organized by customers who collude with other robbers while murder happens when the riders are kidnaped by robbers when they struggle to secure their equipment (bodaboda) (Opondo & Kiprop 2018). Consequently crime threat seriously jeopardises the riders' livelihood in such a manner that they fail to earn their daily basic needs such as food, money to pay bills, health related problem, loss of job as well as loss of confidence and trust in the motorcycle taxis means of transport (Nguyen et al. 2018; Isaac et al. 2014).

Based on the above reality, the coping capacity related to motorcycle taxi riding operations has been shown by various outlook to curb with crime. With stand to forming permanent and temporary group, motorcycle taxi operators undertake to be involved in crime and they report in their groups and group members who expect to come turn up for support or disseminating the information further (Tarimo 2013; Opondo & Kiprop 2018).

In other word, the riders cope with crime by limiting themselves on the working hours as far as carrying the customers at night time (Opondo & Kiprop 2018; Qian 2014). Further, collaborating with community members is another capacity level in curbing crime threats (Reisman et al. 2013). This is coupled by collaborating with police officers in searching as well as reporting crime (Divall *et al.* 2019; Starkey 2016). Self-regulatory mechanism is another way adopted by taxi riders in curbing crime especially at night (Ehebrecht *et al.* 2018). Literature further confirmed that riders cope with crime through mobile phone communication that is through phone calls, sms and WhatsApp communication with their friends or other groups to facilitate identification and follow up of robbers (Nyachieo 2015). To conclude, this section has explored the works related to the various strategies for coping with crime threat employed by motorcycle taxi riders in the urban settings. As çan be seen much, the works highlighted a panoramic view of the coping strategies in the in the context of Sub-Sahara Africa, it has not adequately established these strategies in the context of the motorbike riding operations in Dar es Salaam. Moreover, some works has instead provided a few examples of this subject matter from the urban settings in far east and western Europe, Asia and latin America, examples that would far differ from the real situation in the context of Dar es Salaam and coast regions in Tanzania that are the major focus for this study.

Finally, this study, among other things, intends to promote safety in the city especially among young men and women operating as Motorcycle taxi riders. This group of individuals have tried to contribute, through their innovative ideas, to easy flow of individuals and goods around the city in which limited and unfavourable infrastructure and traffic congestion are salient features. The innovation of Motorcycle taxi riders has opened up the possibility of quick transportation means in the city where the public transport means is overburdened due to depleted infrastructural network system.

Research Objectives

Following the foregoing discussion, the objective of the current study is to explore crime threat in the context of motorbike taxi riding and resilience implications for the motorcycle taxi operators and customers in Dar es Salaam-Tanzania. Specifically, the study explore the operators and customers strategies for coping with crime threat their capacity to cope and strategies for more competently cope with the threat.

2 Methodology

2.1 Description of Research Site

This work was carried out in the urban settings of Dar es Salaam and Coastal region in Tanzania. These areas were considered for they are among urban centres that have attracted huge number of the young people from across the country currently engaged in the motorbike taxi riding as their chief livelihood strategy. In Dar es Salaam city, all the five municipal councils notably Ilala, Kinondoni, Temeke, Kigamboni and Ubungo were considered while three urban settings of coastal region notably Bagamoyo, Kibaha and Kisarawe Districts were considered.

Research Design

The study adopted mixed approach research design to enable the researchers consider both qualitative data for the study depth and the quantitative data for the study breadth. This design has been utilised and proved effective in some previous studies related urban (Ngemera 2017; Nyachieo 2015; Awour 2020) on motorbike taxi riders have used the same approach. Accordingly, methods of data collections included questionnaires, interviews, focus group discussions, documentary review as well observations. Both Questionnaire and interviews were carried out with the motorbike riders and the police officers. Interviews were held face to face following prior studies approaches (Rollason 2020), with an exception of a few police officers who preferred to be interviewed on telephone due to their tight schedule from which no time would be obtained of conduction the interview face to face. Moreover, review and analysis of documents, was done in order to ensure that data obtained from respondents is validated.

The study adopted cluster, purposive and snowball sample size techniques with 287 responds from respondents. On the one hand, five administrative municipals were clustered and townships including some peri urban settings were individually termed as a cluster on its own (Olvera 2021). Researchers were forced to do so in order to ensure that respondents every place of the city and part of the coast regions were adequately and evenly represented. In this regard, we had a total of eight zones-five from the urban and three from the peri urban from which we sampled 287 respondents including riders and passengers. From each cluster, 35 respondents were obtained.

Data Analysis Technique

Qualitative data were analysed using content analysis technique that facilitated to summarize, arrange, organize and interpret the data. Based on the analysed data the findings were presented in relation to the objectives of the study as well as in line with research questions. Accordingly, codes, sub-code as well as quadrants were generated, grouped logically and leading to categories and content. In this MAXQDA 10 [VERBI Software, Marburg, Germany] tool was used to facilitate the exercise. The quantitative data, were analysed using SPSS software package version 16 for data analysis. This was done through the analysis of close ended questions by means of descriptive statistics. Moreover, comparative analysis was conducted between the data from Dar es Salaam and those from Coastal region. The comparative graphs were employed to illustrate the truth of the findings in which for these data, differences and similarities were observed thoroughly.

Validity and Reliability

Validity and reliability were maintained mainly by ensuring that all tools used in the study were examined by at least two experts to ensure that they were worded, arranged and indeed had captured all the intended aspects of data Heale & Twycross (2015). Reliability of the findings, was ensured through observing all the required procedure; notably giving clear explanation on setting of the study, the sample size used, how it was sampled and methods involved in the data collection exercise. In addition, the questionnaires were subjected to Cronbach Alpha as way of determining the correlation among variables (Price *et al.* 2015).

3 Findings and Discussion

3.1 Crime Victims' Coping Strategies and the Capacity to Cope with the Crime

This section examines coping mechanisms of the motorbike riders and the crime victims in urban environments and their ability to deal with dangers and criminality. The types of crimes experienced by motorbike rides and victims, include motorcycle theft, kidnapping, and uncooperative clients, as well as murder, general stealing, motorist assault, mob justice, fraud and forgery, handling stolen property, malicious property damage, burglary, rape, and mob justice (Nguyen et al. 2018; Opondo & Kiprop 2018; Nyagase 2017).

3.1.1 Crime Victims' Coping Strategies by the Riders

Researchers wanted to know how the riders and other victims coped with the crime that they regularly come across. In this section, riders copying strategies are presented as in Fig. 1.

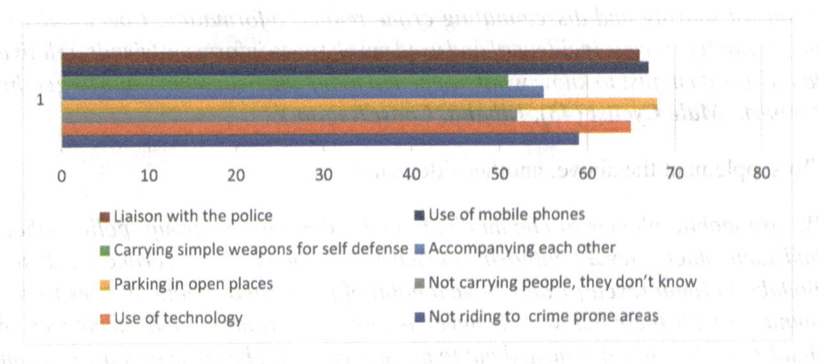

Fig. 1. Motorcycle riders' crime copying strategies

The results shown in Fig. 1 show that motorcycle riders have used a variety of techniques to cope with the crime environment. These tactics include parking in open spaces, using mobile phones and other technologies, collaborating with the authorities, avoiding crime-prone locations, avoiding attracting new clients, and carrying small arms.

Accordingly, finding demonstrated that parking in open places perceived to be more secure during nights time was strategy adopted by taxi riders as reported by around 68%. This was most popular method that motorcyclists used to emulate criminal behavior. In this instance, the riders park at bus stops, intersections, in front of businesses, in front of police stations, and in other open areas where they can be seen by oncoming traffic. Although it is a strategy for attracting customers, it is also a way for criminals to replicate. The majority of them claim that as long as you are in public view, you are safe. The emphasis is always given at night as this rider was quoted saying;

At night I normally park at well-known places sometime I will park near petrol station but during weekend I will park near night clubs because I know there I can also get customers. We use an open space with full light mainly. I park in an area where the movement of people is high. At night I normally park at a place called CCM and some of the riders' park at the police station for security purposes ***(Male Cyclist (113), Temeke, Dar es Salaam)***

Another tactic was the use of mobile phones technology as confirmed by over 67%. Riders indicated that they utilized their mobile phones to communicate crime-related information with various individuals and authorities who could help them in an emergency situation and who could also reduce crime both reactively and proactively. The respondents emphasized that they would call, send a WhatsApp message, or send an SMS to their colleagues' operators whenever riders carried questionable customers or were traveling to unsafe or suspicious locations. Second, the majority of riders had set up WhatsApp groups in which they would regularly discuss matters pertaining to crime and security. Thirdly, most riders were able to collect and save crucial mobile phone numbers. A rider was quotes saying;

Mobile phones are really of great help to us. One can imagine that without these mobile phones would have been wiped away by crime as we wouldn't have any means of sharing and disseminating crime related information. I personally got into a problem [crime incidence] and used my phone to inform my friends, relatives, family for them just to know what happened to me and where possible to get their support. ***(Male Cyclist (43), Kibaha, Coast Region)***

To supplement the above, another rider said;

We use mobile phones to communicate with other riders' group, police officers and some other relevant authorities whenever problem occurs [crime incidence]. Bodaboda [motorcycle] riders have a habit of using their mobile phones to communicate with their fellows whenever involved in crime related incidences. We thank God that, this has turned out to be very effective, because they always come to our rescue. ***(Male Cyclist (29), Ilala, Dar es Salaam).***

Also, using General Packet Radio Services (GPRS) technology was another strategy adopted by riders as confirmed over 54% of the respondents. However, this web-based wireless software can find a motorcycle's position even after it has been taken from its owner. Unless unplugged and thrown away, the device, which is mounted on the motorcycle, aids in locating the vehicle. Those who did not adopt this technology blamed their failure on a combination of technical ignorance and financial constraints. One rider confirmed this when asked:

Yes, I have personally installed it and it is working pretty well. It helps to identify your stolen property very easily even though robbers and thieves sometime remove the GPRS as soon as they steal it. The problem is that not all of us have managed to install and use the technology because we lack sensitization and also the technology is very expensive. Some of our bosses are reluctant to acquire such devices ***(Male Cyclist (63), Ubungo, Dar es Salaam)***

Correspondingly, the liaison with the police was another way of coping with crime by taxi riders. This show that riders were collaborating with the police in combating crime. The majority of them claimed that they were in frequent contact with the police in order to report crime againt boda boda riders and together come up with solutions to crime. They achieved this by giving out early warnings and information that would help in identifying and combating crime. Riders claimed that, in the context of their employment, this method was successful in deterring crime as well as dealing with it, as quoted below;

> *We share information [crime related] with the police and inform them what has happened if the information offered to them is accurate they will start looking for stolen motorcycle. We [therefore] really appreciate the support that we get from the police officers anytime we report our matter to them they assist us to look for the stolen motorcycle and the criminals* ***(Male Cyclist (88), Kisarawe, Coast Region).***

On the other hand, other riders said that although reporting the incidents as necessary and expected, they were unable to get the best and most fast assistance from the police.

> *Sometimes when we report crime scenes to the police, they don't show full support, this is the big challenge that we face from them. We can say that there is a little cooperation from them especially on making follow-up, be educated by police and sensitized on different tricks used to steal bodaboda [motorcycle taxi) among others. Other collaborative technique is just for reporting to them* ***(Male Cyclist (23), Bagamoyo, Coast Region)***

Avoid Riding to Some Specific Destinations Perceived to be Crime Prone

The riders' ability to avoid traveling to certain locations they believed to be crime-prone areas was one of their techniques for dealing with crime. This went hand in hand with not serving shady customers late at night. They also tried to exercise prudence by dropping customers who they had already accepted if they foresaw any danger or criminal activity. Near the Dar es Salaam University College of Education (DUCE), across from the national stadium, along Kinyerezi Malamba Mawili Road, in the Goba, Mbezi, and Bunju neighborhoods, as well as other peaceful regions like Masaki and Mbezi Beach, were some of the areas thought to be problematic. The forests between Dar es Salaam and Kisarawe, along Maneromango and other Kisarawe villages, along Ruangwa and slum parts of Bagamoyo Mlingotini are some other locations. One cyclist noted that they stay away from these spots;

> *During late hours, at night I don't ride to places that I believe they are not safe for me even if the customer promises to offer me some good money. At night I park at an open place like at petrol station, near night club or at police station. Absolutely at night time, we only pick the customers who are normally from this kijiwe [only people from around the vicinity] to be their own station* ***(Male Cyclist (72), Kinondoni, Dar es Salaam)***

Accompanying each other using more than one motorcycles during night or when working in some crime prone areas was another tactic. Finding show that riders also used many motorcycles to travel together at night or while working in some high-crime regions as a means of coping with crime. They claimed that this technique would deter possible robbers as well as malicious consumers from harming them. This was frequently practiced at the Mbezi Magufuli stand and Bagamoyo neighborhoods of Dar es Salaam. One motorcyclist was reported to have said;

When a customer wants you to take him or her to a very far place, or a place that we don't know more especially at night we accompany one another and charge them high in order to compensate the other rider who accompanies you. This is because it is very dangerous if we don't do that [they might end up committing crime against the riders] ***(Male Cyclist (62), Mbezi, Dar es Salaam)***

Another tactic was not carrying some people, they don't know. In this method, the riders were avoid transporting people who appeared suspicious or unfamiliar, especially late at night. This is due to the fact that riders had learned that most crimes, including robbing them and injuring or killing them, were committed during the evening hours.

We have our unwritten rule here in our parking station that we don't carry people that we don't know at night. Due to that, during night hours I personally don't carry a customer that I don't know, or going to potentially dangerous places. This practice is also accompanied by our decisions to park only in some places that are well known by ourselves and our security organs such as the police. ***(Male Cyclist (119), Kigamboni, Dar es Salaam)***

Some of them reported that as a means of protecting themselves, when taking passengers to new places they would carry them up to certain destinations, and would hand them to other riders who know such areas very well. Although this was mentioned, it scored the least 52%. Some of the cyclists mentioned that, it was very hard to implement that because if they carried those they knew, it would be difficult to get the required cash and they would end up starving. One of them said "*…we normally take risks and carry those we don't know just because we need money…*".

Carrying simple weapons for self-defense was another tactic as scored about 51%. Accordingly, findings show that One respondent in Dar es Salaam was discovered carrying a machete, but this happened frequently in Bagamoyo. Riders in Bagamoyo also mentioned carrying some basic guns for self-defense in case they became involved in any sort of criminal activity. One of them mentioned, "*…it is a normal thing here*". When asked if their passengers would not be scared, he said, "*…if I do not tell you where we keep them, you will never see them at all*". Researchers discovered a respectable number of motorcyclists using these weapons after learning the secret. The following come riders were quoted:

Self-protection is very important. I have a screw driver that I carry everyday as a weapon for self-defense. If you don't defend yourself nobody will come to your rescue ***(Male Cyclist (121), Bagamoyo, Coast Region).***

Another one supplemented;

I can witness that strategy works. One day an ill intended customer refused to pay, when I dragged my dagger, the customer immediately paid the money. **(Male Cyclist (25), Bagamoyo, Coast Region)**

Some riders and operators, primarily those in Dar es Salaam, were opposed to this practice because they believed that once clients saw such weapons, they would get frightened. In this instance, they forbade the behavior. Some of them even noted the possibility of malicious clients using such tools to damage them. It was stated about one of them as follows:

No. I am totally against the practice of carrying weapons because it will scare the passengers. They might also report you and end up opening a court case against you. We may consider opting for other precautions rather than carrying weapons. And I would say that usually, bodaboda[motorcycle] riders do not carry out weapons for protection because customers such weapons may be used against them. **(Male Cyclist (81), Kinondoni, Dar es Salaam).**

In concluding to this finding, these results are consistent with those of Opondo and Kiprop (2018), who identify the usage of phones as a key tactic that has proven beneficial. However, they also noted that in Kenya, some riders would beat up people they suspected of being thieves, raising concerns in the local community. Also, findings show that liaison with law enforcement has been among the most widely used strategies because no investigation is possible without police participation. According to several studies, enlisting the assistance of the police or taking part in community policing has helped people recover lost items or even stop some crimes from happening, despite some drawbacks. Studies by Olveira et al. (2018), Reisman et al. (2013), Divall et al. (2019), Starkey (2016), and Ehebrecht *et al.,* (2018). Thus, it implies that the current study makes a substantial contribution to our understanding of this particular study region in an urban situation. On the other hand, an interesting discussion in this cover letter was about parking in open spaces perceived to be more secure, especially at night, and avoiding traveling to some specific locations perceived to be crime-prone, whereby taxi drivers play these roles in coping with crime and threats against the robbers.

3.1.2 Crime Victims' Coping Strategies by Customers and Other Victims

Results showed that several passengers had their own experiences with how they deal with crimes. These include using of registered bodaboda, mobile phones, carrying simple weapons, avoiding the use of suspicious bodaboda, walking in groups among others. It has been observed that, a number of victims are women in comparison to male counterparts. Figure 2 shows the passengers and potential customers responses on how they cope with such crimes.

According to findings, the respondents stated that more than two in one motorbike/sharing of the motorbike is adopted by customers in curbing with crime as reported by over 81% of the passengers. They further ended up by said due of the lax regulations in the coastal zone, the system was also widely used there. They maintained that because

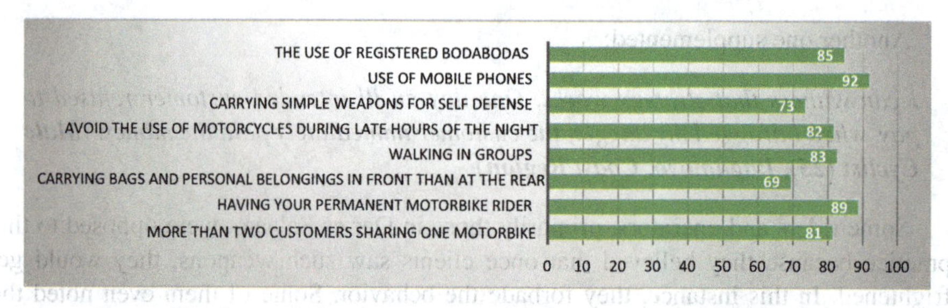

Fig. 2. Strategies used by passengers in copying with crimes

there were more of them, it was harder for them to be attacked by robbers or suffer from any other crime associated to motorbike taxis. *If we're going somewhere at night, I just ask my neighbor or relative to escort me. And we usually board one motorbike. Then we are sure that we are safe.*

Also, finding show that identifying a permanent motorbike driver was another coping capacity adopted by passengers. This further narrated that Some passengers are accustomed to recognizing a consistent motorbike driver who would be taking them to several locations. Women tended to support this more than males. They claimed that it offered them peace of mind that they would not be attacked or suffer from any motorbike-related crimes. This was also useful to them because, as one of them put it, "*Having a permanent rider and taking his phone number makes me free and protected.*" Therefore, this tactic can aid in preventing any crime that motorcycle riders are likely to commit.

Nevertheless, findings confirmed that carrying valuables in front and not at the rear was another way of counteracting with crime as dilapidated by over 69%. In this instance, carrying valuables like handbags and cell phones up front rather than in the back was one of the methods used by the consumers to combat crime. The tactic was utilized when working along the roads, streets, congested areas and purposefully for safeguarding them from robbers in downtown. For instance, the traveler reaffirms:

> *I keep it [the handbag] in front of my chest. Obviously, this is the only strategy for protecting my valuables. You see busy places like Posta and Karikoo [downtowns and shopping centers in Dar es Salaam Tanzania] are so dangerous for us. This are places where someone can be robbed by these motorbike people very easily.* ***(Female, customer (32), Kinondoni Municipality, Dar es Salaam)***

Furthermore, findings confirmed that walking in groups was another way of curbing with crime. This had emphasised that at night and in some crime-prone locations, passengers walked with groups. This was carried out in such a way that folks who lived or worked in the same location as one another or close by tended to travel home at night or when passing through some crime-prone regions together. "*It keeps us protected and no one could assault us," one of them remarked.*

In other word, avoid the use of motorcycles during late hours of the night was another strategy in curbing with crime. In this instance, findings show that women were more likely to use this tactic than males. Eighty-two percent of them agreed with the strategy. Urban dwellers used it in the wee hours of the night to stay safe, notably from criminals pretending to be motorbike taxi drivers. They insisted that they would choose a different mode of transportation whenever they could not find their unique motorbike. One of them was videotaped responding as follows when asked if she could stop from employing such methods:

Yes, yes, my husband has identified special riders for me even when I am late at night. However, whenever, I am late and I cannot find this rider, they I will always opt for another kind of transport. In this way I will be safe.

In a nut shell, carrying simple weapons for self defence was another way of coping capacity. Of which only a very small percentage of respondents indicated that they would carry guns to protect themselves from crime. Knives, metal objects, and other pointed things were some of the weapons that were frequently described. Not only that but also, the use of mobile phones technology was another tactic since the passengers enabled to contact for assistance anytime they encountered any kind of crime. They made use of their phones to call the police, snap pictures of the stolen motorbikes, and share them with social media groups in an effort to raise suspicion.

According to findings, the use of registered Boda-boda was another tactic utilized by passengers in the city. They ended up by confirmed that using licensed motorbike taxis, such as those licensed under UBER and BOLT helps them to assure safe journey *"We use it but it even though it is a bit costly. But for the sake of our security, then we have nothing to lose. To be secure is the most important thing"*.

Drawing from above findings, it can be summarized that coping strategies which relate to using a phone and locating a single rider for one's use, using motorbikes for UBER, BOLT, registered bodaboda, mobile phones, carrying simple weapons, avoiding the use of suspicious bodaboda, walking in groups among others has become standard in Dar es Salaam. In the coastal region, this has not been the case. This is due to the inaccessibility of these companies' network there. Moreover, in Dar es Salaam, where a large percentage of the population works, it is common practice to walk in groups. In the coastal region, a different approach is required. Furthermore, most women in Dar es Salaam have long been accustomed to carrying their belongings in front rather than at their sides or backs. This is due to the fact that the rate of these crimes is higher than it is in the nearby region. However, research shows that some tactics are more frequently employed in the coastal region than in Dar es Salaam. In the townships of Kibaha, Kisarawe, and Bagamoyo, sharing motorcycles with multiple passengers—a practice called as "mshikaki"—and carrying guns are both widespread. This might be because, when someone is observed riding a motorcycle with more than one passenger, police officials in those townships are less harsh than they are in Dar es Salaam. Additionally, although it is challenging for clients to carry such firearms in Dar es Salaam, it is typical for people in neighboring locations. While people are not referred to as criminals in the neighboring region as they are in Dar es Salaam.

In light of the data above, it can be concluded that registered boda-boda, mobile phones, carrying basic weapons, and other victims' coping techniques were frequently related with crime victims. Avoid riding a motorcycle after midnight; carry valuables up front rather than in the back; carry no more than two people on one motorcycle; identify a permanent motorcycle driver; Utilization of mobile phone technology and licensed

Boda-Bodas. These results usually agree with earlier research, especially that done by (Opondo & Kiprop 2018). The literature study did not, however, mention any findings regarding the use of more than two motorcycles or sharing of motorcycles, identifying a permanent motorcycle driver, or carrying valuables up front rather than in the back. These could be brand-new discoveries from these recent investigations, and they could be important for accumulating new knowledge.

4 Conclusion and Recommendations

4.1 Conclusion

It was made clear in this purpose that crimes against riders, consumers, and other persons do occur. Thus, both riders and passengers devised innovative coping strategies such as the use of tracking devices, avoiding carrying suspected customers, parking in open spots, riding together, carrying basic weapons, using mobile phones, and liaising with the authorities have been shown to be more prevalent methods among riders to curb with crime.

Although it has been claimed that riders try to prevent themselves by sharing information about hazardous areas and suspicious individuals, they have nevertheless ended up killing innocent bystanders despite the lack of any conclusive evidence that they are criminals. This raises a serious concern since it indicates that this group operates under the influence of mob psychology and, if not stopped, will lead to lawlessness.

The methods employed are region-specific with relation to travellers. For instance, people in Dar es Salaam have resorted to using registered riders under BOLT and UBER, carry their bags in front because bag snatching has been so common, work in groups to avoid attack, and sometimes have their permanent riders, whereas people in the coast region occasionally carry with them some simple weapons and share motorcycles simply because of the reluctance of traffic police in these areas.

It is harmful to the nation's economy to have a situation like this when people are constantly concerned about the circumstances. Additionally, it shows that security, despite being rated as good, has a number of flaws, particularly in urban regions where these occurrences are frequent. It informs the security agencies that there is more work to be done in order to protect citizens and their possessions.

4.2 Study Recommendations

The following recommendations should be considered in order to assist the riders in more successfully coping with the crime and the nation eradicating crimes related to motorcycle taxi riding operations in general.

While tracking devices are highly helpful, it is advised that they be made compulsory for everyone who registers to operate as a cab driver. However, the government should undertake the required measures through the ministries of trade, internal affairs, and other agencies to guarantee that the costs of these devices are kept to a minimum. Also, Riders should always be trained in handling situations involving possible criminals and abstaining from sexual harassment and defiling. This needs to go the entire nation, not just to Dar es Salaam. If such instruction is not provided, we should expect many killings, rapes, teenage girls dropping out of school, and severe sexual harassment of women—all of which are brought on by riders.

Police personnel must act swiftly to respond to crimes, conduct impartial investigations, and apprehend perpetrators. In a similar spirit, police officers should refrain from pulling over Bodaboda users for insignificant offenses. As a result, there will be more confidence between riders and police, and riders will feel free to report any questionable circumstances.

Authorities on community policing are also recommended by the report. According to recent research, it has only been used to bother riders, not assist them. As a result, the police and local government officials must restructure the community police's activities, assign them specific tasks, and instruct them on how to cope with lawlessness.

The study suggests that the TRA, the police, and the government regulate used parts. For instance, every piece of spare equipment imported must have a certain number. This number will reveal where the component was sourced. A spare part should be doubted if it lacks such a number. Dealers in spare parts will avoid purchasing parts if they are aware of their origins in this way. This will reduce motorcycle theft, which is ultimately merely marketed as used replacement parts.

Acknowledgement. We would like to recognize and appreciate invaluable financial support provided to us by Research on Poverty Alleviation (REPOA). The financial support obtained from REPOA facilitated the entire research cycle.

Declaration. This publication resulted from an ongoing research project titled ***"Urbanisation, Crime, Threat and Coping capacity in the Context of Motorcycle Taxi Riding Operations in Urban Settings of Tanzania: Evidence from Dar es Salaam and Coast Region"***. The project is supported by Research on Poverty Alleviation (REPOA) for a duration of one year from December 2021 to December 2022 through a grant **No. DFA/REPOA/1210/CST/357.**

References

Adiambo, E.: The contribution of motorcycle business to the socio-economic wellbeing of operators in Kisumu County, Kenya. Int. J. Soc. Develop. Concerns **13**(November), 55–69 (2020)

AFCAP: Facilitation Services for Consultation on Motorcycles Operations in Ghana Final Report Quality assurance and review table ReCAP Database Details: Facilitation Services for Consultation on Motorcycles Operations in Ghana ReCAP Website Procurement Method Co. ReCAP Facilitation Services for Consultation on Motorcycles Operation in Ghana, 1(July) (2019)

Caroline, B., Neil, R.: Tanzania motorcycle taxi rider training : assessment and development of appropriate training curriculum Final report Transaid May 2015. In: International Conference on Transport and Road Research (Issue May) (2015a)

Caroline, B., Neil, R.: Tanzania motorcycle taxi rider training : Assessment and development of appropriate training curriculum Final report Transaid May 2015. In: International Conference on Transport and Road Research, 1–27 May (2015b)

Gibigaye, M.: Organising informal transport workers: the informal economy and the unionisation of motorcycle taxi drivers in Benin (2006)

Isaac, K.W.O., Benjamin, O., Opondo, F.: The effect of increased investment in Bodaboda business on economic empowerment of people in Kisumu west District. Eur. J. Bus. Manage. **6**(39), 177–185 (2014)

Jasna, M., Damijan, Š.: The role and use of e-materials in vocational education and training : The case of Slovenia. The Role and Use of E-Materials in Vocational Education and Training : The Case of. Turkish Online Journal of Educational Technology (2015)

Kariuki Nyaga, J.J., Gichuru Kariuki, J.: The influence of motorcycles/boda boda on community development in Rural Kenya: a study of the challenges facing motor cycle operators in Meru South Sub-County. J. Educ. Hum. Develop. **8**(1), 2334–2978 (2019). https://doi.org/10.15640/jehd.v8n1a10

Kumar, A.: Understanding the emerging role of motorcycles in African cities: a political economy perspective. Sub-Saharan Africa Transport Policy Program (SSATP). Sub-Saharan Africa Transport Policy Program, Urban Transport Series **1**(13), 1–32 (2011). http://documents.worldbank.org/curated/en/391141468007199012/Understanding-the-emerging-role-of-motorcycles-in-African-cities-a-political-economy-perspective

Mbegu, S., Mjema, J.: Poverty cycle with motorcycle taxis (Boda-Boda) business in developing countries: evidence from Mbeya—Tanzania. Open Access Lib. J. **06**(08), 1–11 (2019). https://doi.org/10.4236/oalib.1105617

Munishi, E.: Rural-urban migration of the Maasai nomadic pastoralist youth and Resilience in Tanzania. Case studies in Ngorongoro District, Arusha Region and Dar es Salaam City. (PhD Thesis), Freiburg University, Freiburg (2013)

Munishi, E.J, Hamidu, K.M.: Urban crime and livelihood implications among the motorcycle taxi riders in Dar Es Salaam City- Tanzania Int. J. Res. Bus. Soc. Sci. **11**(4) (2022), 246–254 (2022)

Munishi, E.J.: Rural-urban migration and resilience implications on the Maasai Households' in North- Eastern Tanzania. Afric. J. Appl. Res. **5**(2), 24–44 (2019). http://www.ajaronline.com, http://doi.org/https://doi.org/10.26437/ajar.05.11.2019.0

Munishi, E.J.: Coping with urban crime and resilience factors. The case of the Maasai security guards in Dar es Salaam, Tanzania. J. Soc. Develop. **1**(1), 60–79 (2016)

Nguyen, T.D., et al.: Injury prevalence and safety habits of boda boda drivers in Moshi, Tanzania: a mixed methods study. PLoS ONE **13**(11), 1–16 (2018). https://doi.org/10.1371/journal.pone.0207570

Nguyen-phuoc, D.Q., Nguyen, A.H., Gruyter, C.D., Su, D.N., Nguyen, V.H.: Exploring the prevalence and factors associated with self-. Transp. Policy **17** (2019). https://doi.org/10.1016/j.tranpol.2019.06.006

Nyachieo, G.M.M.: Socio-cultural and economic determinants of Boda-boda motorcycle transport safety in Kisumu county, Kenya. In Hilos Tensados, vol. 1, Issue June. Kenyatta University (2015)

Opondo, V. O., & Kiprop, G. (2018). Boda Boda Motorcycle Transport

Qian, J.: No right to the street: Motorcycle taxis, discourse production and the regulation of unruly mobility. Urban Stud. J. (2014). https://doi.org/10.1177/0042098014539402

Tarimo, J.: Challenges faced by youth in conducting bodaboda business at mbezi juu ward: a case of makonde area in kinondoni district. Mzumbe University (2013)

Effects of Open Performance Review Appraisal System in Assessing and Appraising Employees' Performance at First Housing Finance Tanzania Limited

D. K. Nziku[1(✉)] and C. B. Matogwa[2]

[1] Business Administration Department, College of Business Education, Dar es Salaam, Tanzania
dismasnziku@gmail.com

[2] Accountancy Department, College of Business Education, Dar es Salaam, Tanzania
c.matogwa@cbe.ac.tz

Abstract. Purpose: This scrutiny has inspected outcomes of exercising an Open Performance Review Appraisal System at First Housing Tanzania Limited (FHTL) and provided the means to be adopted by other related organizations invested in advanced technology in their day-to-day operations.

Design/Methodology/Approach: The inquiry employed both numerical statistics and descriptive research approaches to meet the targeted goal. Data were collected by using focus group discussions, interviews and questionnaires. A purposive sampling technique was deployed to obtain the sample as the study took all 68 employees at First Housing Tanzania Limited (FHTL).

Findings: Findings revealed that organizations invested in systematic and advanced technology that integrates all the activities related to the employees' appraisal are the best means and practice of exercising fair means of assessing and appraising employees in the organization appropriately.

Research Limitation: The study focused on only employees and employers of First Housing Tanzania Limited (FHTL) found in Dar es Salaam, this scenario narrows the spread of the inquiry.

Practical Implication: The inquiry's discoveries are used by employers across the country to re-think new ways of gauging employees rather than relying on the old approach named; Open Performance Review Appraisal System (OPRAS) alone to determine the actual altitude and effectiveness of the employees' performance in the organization.

Social Implication: Findings of this study are suggested to be used by Tanzania policy makers particularly from the Ministry of Work to rectify some aspects of evaluating employees from time to time.

Originality/Value: The study on hand has challenged the old approach of evaluating employees' performance at the workplace, and has suggested employing the system-integrated method that assesses employees' performance accurately.

Keyword: Employees · Appraisal · Performance · Tanzania

C. Aigbavboa et al. (Eds.): ARCA 2022, *Sustainable Education and Development – Sustainable Industrialization and Innovation*, pp. 1053–1063, 2023.
https://doi.org/10.1007/978-3-031-25998-2_81

1 Introduction

1.1 Background of the Study

In the context of Tanzania, some years back before 2009 most of the organizations evaluated employees' performance by using old system called closed performance review appraisal system (CPRAS) which made some difficulties for the appraisee to know whether is fairly or unfairly evaluated. This situation caused shouting of employees in both private and public organization because the evaluation based on the rule of thumb and feelings of the supervisor. CPRAS used as a tool to detect the best worker to be announced and awarded incentives during the international workers day which is cerebrated every May 1st each year. It was a favoritism kind of approach since was made by the feelings of the supervisor alone, thus the appraises had unequal opportunity to become best workers.

First Housing Tanzania Limited (FHTL).

The context of employees' performance in the organization had been the major problem in both private and public organization since they lack uniformity and similar base from evaluation of the employees (Nwata et al. 2016; Hamed & Potapova 2021; Iqbal et al. 2015 and Mani et al. 2014). Most of the practicing organizations in Tanzania fail to have equal and fair approach of assessing employees' performance which is the prime base of motivating and promoting employees in the organization. This reason has pushed the researcher to carry out the study, and eventually suggesting the best way of assessing employees' at workplace and guiding the practicing organization on the best way which can be engaged to motivate and promote employees fairly and transparently for efficient organization performance at FHTL as other related organization operating in Tanzania. The Objectives of this study were to: identify methods used for employee's assessment and appraisal at First Housing Tanzania Limited (FHTL); examine employee's involvement in performance adherence on an OPRAS at First Housing Tanzania Limited (FHTL); and examine effects of exercising an OPRAS at FHTL.

This survey has helped First Housing Finance Tanzania Limited to introduced special training program to the staff evaluated negatively in terms of their performance as discovered by the system. The appraisal system installed make the exercise of assessing individual performance easy and success and failures are rapidly detected.

OPRAS comprised information linking with perfection and imperfection of the employees to be easy known to the management. The management has to bring back to the staff and everyone has to take serious remedies for the identified weakness and better performance should receive incentives and or promotion.

The automated open performance review appraisal system at the First Housing Finance Tanzania Limited has made the process identifying the candidates who requires promotion easy since Performance Anchored Rating Scale (PARS) indicates directly his/her performance and action required to him/her.

The appraisal system at the bank indicates results associated with individual performance in the system, thus has made the process of detecting the induvial performance easy particularly strengths and weaknesses and the final action required by the management to him/her immediately. From the particular organization when the individual weaknesses raise above the tolerable level the staff likely to be terminated otherwise promoted.

The process of effectiveness of employees' appraisal system helps the management of the organization to set appropriate compensation strategies as related guidelines for effective operation at the First Housing Finance Tanzania Limited.

The study focused at First Housing Finance Tanzania Limited situated in Dar es Salaam – Tanzania, which is offering the housing loans national wise.

2 Definition of Key Terms

The Open Performance Review Appraisal System (OPRAS): It is an approach of assessing employees performance in both public and private organizations operating in Tanzania which is designed to capture both core and routine activities of the organization aiming at determination of individual competences on performance and is a base for organizations' performance's improvements (Songstad et al. 2012 & Mathias 2015).

Employees' appraisal: refers to the annual managerial process of reviewing employees' job performance based on skills, achievements, growth and overall individual employee contribution to the organization (Iqbal *et al.* 2013).

Employees: Employees are the people employed by another person or organization for wages or salary and in a position below the executive level and he or she should be accountable for the results (Farooq et al. 2014).

Employees' Performance: refers to the situation at which staff fulfils properly their assigned duties by completing the required tasks with required behaviour at workplace by indicating positive effects on morale and quality of output produced (Muda et al. 2014).

3 Empirical Studies

The study done by Songstad et al. (2012); found that there was a general reluctance towards OPRAS as health workers did not see OPRAS as leading to financial gains nor did it provide feedback on performance. They realized that the aim of having OPRAS as a performance measuring tool was to seek to improve performance through setting individual goals, measuring the achievement of the goals and providing feedback to the workers in a certain organization. Dickson (2013) indicated that health workers had great expectations towards P4P but also that they were reluctant towards filing the OPRAS forms. The study pointed out that at the time of introducing OPRAS into the working systems, the primary objective was to increase openness and escaping from ill-treatment of the employees which is against the standards set by International Labour Organization (ILO) approaches on employees' evaluation. Matete (2016); pointed out that some of primary school teachers did not see practicability of OPRAS in their working environment as it had goals that did not reflect reality found in their schools. According to Brim (2012, p. 3) as quoted by Matete (2016); "*employees responses towards work depend on intrinsic or extrinsic motivation, possibilities of attaining goals, ability of the managers to support subordinates*". The working condition at workplace also had been the soft drive to the employees in attaining organization goals.

Nchimbi (2019) found out that OPRAS as used at Iramba District Council by then, it was not adequate tool to improve employee's performance. Bernard (2013) revealed that more than half of respondents viewed OPRAS as a measurement tool which has no advantage, it is a weak tool for measuring work performance, and this is because majority of respondents have heard about OPRAS, but their involvement in the whole process is minimal, less than half of respondents are involved.

Tefurukwa (2014) asserted that OPRAS is the prime tool for identifying, judging, evaluating and comparing the magnitude at which the objectives of the organization had been achieved. For more than 20 years back, there had been no OPRAS improvements in any means to make it simpler, suitable and compatible for the process of employees' evaluation in the public organizations. These weaknesses have made the process of evaluating employees in the organization to be less effective and cannot assist managers properly in the process of making decisions. However, in private organizations like FHTL there had been great changes which attracted the researchers to conduct an inquiry.

4 Conceptual Framework

OPRAS is a complicated approach that requires mutual understanding between all key players of the organization. First Housing Finance Tanzania Limited has automated the process, which easy the employees' assessment procedures. The performance indicators are adjusted based on really performance of the individuals when filling the activities done in the system. Performance Anchored Rating Scale (PARS) indicates individual performance by comparing the agreed performance criteria and the actual one where the indicators are indicated in percentages and with bar charts. The summary of flow of information indicating performance review appraisal system indicated below stage after stage, while the outcome after evaluation may be negative or positive (Fig. 1).

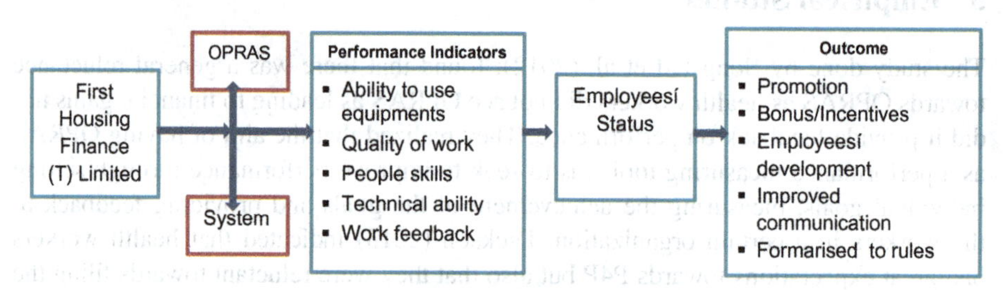

Fig. 1. Source: Research Construct (2022)

The outcome from the process may be promotion, demotion, termination, improved communication, formalized rules, and or employees' development. The electronic open performance review appraisal system does not entertain biasness; it evaluates someone based on the output of which the individual employee was assigned by the superior through system to perform in the specified period of time.

5 The Relationship Between Literature Review and Knowledge Gap

Baraka (2015), Rugeiyamu (2019), Alman & Yusuph (2020) and Macha (2014) assert that, OPRAS had been the common tool used for assessing employees' performance in the organization for decades back. It had been ineffective since has number of deficiencies which remained without improvement for quite long time despite of being in use. The existing OPRAS had been using as a tool for motivation and promotion determination, which realized to be unfair because evaluation of employees' performance sometimes is based on the feelings of the supervisor. The management at FHFTL viewed the process and came up with the modern means of assessing employees' performance (e-OPRAS) which seem to be the best, since limits interactions between supervisor and respective subordinates under appropriate span of control.

6 Methodology

6.1 Research Approach and Design

The nature of the study has influenced the researcher to engage focus group discussion, interview and questionnaires to obtain information which supported in data analysis and discussion. The inquiry aimed at determining the usefulness of the OPRAS to both management and employees at housing finance organization. The analysis, discussion and interpretation of the research information considered on what personnel say and comment. With that regards, the study is the mixture of both qualitative and quantitative study and has managed to include both descriptive techniques and statistical measures for study generality.

6.2 Case Study Area

The research has covered one organization named, the First Housing Finance Tanzania Limited situated in Dar es Salaam, Tanzania as case study. The organization is offering housing loans to all parties of Tanzania. The company is specialized to the mention activities and is the first in the country. Other financial institutions offer general loans but not specific as the First Housing Finance Tanzania Limited normally do. The company do appraise their staff by using electronic OPRAS, thus the researcher was attracted to assess the effects towards organizational performance.

6.3 Targeted Population

The FHTL has got 68 employees, including the executives. The study took all staff into account, because the performances of all staff are determined by using the OPRAS which is automated to simply the analysis and biasness during evaluation. Kothari (2017) asserts that, when the population is small, then it is possible to take all candidates into consideration by considering that all of them have information that may be useful in completion of an inquiry.

6.4 The Study Samples

The OPRAS is a tool of assessing individual employees' performance, where everyone is responsible for the process. The FHFTL has only 68 employees, therefore all the employees will be taken as sample size for study generalization since there are few employees.

6.5 Sampling Procedures

The FHFTL has only 68 employees, thus all employees were considered in data assortment for the completion of the research requirements.

6.6 Methods and Instruments for Data Collection

i. Methods for Data collection

Assessing employee's performance is a complicated process and depends on many factors such as employer-employee relations, altitude of the employer and employee, employees' knowledge and skills, level of education and specific skills, kind of activities employees assigned by superior, resources budgeted for completion of the same assignment and working environments. For the study to be conclusive and meaningful, had assumed selected approaches of data gathering of which all numerical statistics and descriptive data were gathered and analyzed. Qualitative technique involves gathering of primary through in-depth interview, structured questionnaires, site visiting, observation and analyzing data based on the views of the respondents. Quantitative research includes review of secondary data and their analysis provides descriptive results, correlation and statistical results for the study's interpretation.

ii. Instruments for Data collection

The common data collection tools in research are observations, questionnaires, interview, survey, documentary review and many others based on the specific purpose of the study (Kielhofner & Coster 2017). But for the purpose of modalities of assessment of employees' performance at FHFTL the researcher used questionnaires and documentary review to obtain data that fulfil the study.

iii. Validation of the instruments for Data collection

The instruments (questionnaires and documentary review) were tested for their validity and reliability. Nora et al. (2018), Kim (2009) and Bammann et al. (2011). Questionnaires were tested by using rating scales which are suitable and compatible to the study while documentary review were tested using performance checklist to explore achievements of the organization for the financial years 2018/2019, 2019/2020 and 2020/2021.

iv. Data processing and Analysis

Primary data were collected by using questionnaires from the respondents, assembled and interpreted with the aid of researchers' team. The primary data were analyzed by using the Statistical Package for the Social Science (SPSS) version 23.0 to get statistical distribution and the outcomes were conveyed in a form of numbers for quantitative presentations. Moreover, the qualitative data mainly obtained through review of secondary data and interviews related with employees' performance review were tabulated to make a simple presentation suitable for the report.

v. Ethical considerations

The ethical permission was granted by the FHFTL authority and releases a permission to conduct the study at their organization. Information about the application of the OPRAS and the rating mechanisms included within questionnaires as paramount tool for gathering primary data used by the researcher from the respondents. The researcher has obeyed all the ethical research principles as recommended by academicians and researchers practitioner (Resnik 2011, Resnik 2014 and Resnik & Elliott 2016).

7 Results and Discussion

Methods of employee's assessment at First Housing Tanzania Limited (FHTL).

The observation from the FHTL particularly on assessment of the employees' performance is fair, and allows very little intervention from the supervisors of all sections. The organization has developed objectives to every individual that require them to perform and feed feedback into the system; hence the system itself assesses the individual employee based on ones' performance. The paramount condition for this case, the objectives must be defined in the system as criteria for the assessment. At FHTL many different methods of employees' assessment had been employed for many different periods to various organizational levels in the order as indicated in the Table 1 from the old to the latest approach.

Table 1. Methods of employees' assessment at **FHTL**

Employees assessment methods employed at FHTL	N	Minimum	Maximum	Mean
Management By Objectives	68	1	6	4.48
360 degrees feedback System	61	1	6	4.26
Assessment Center System Method	63	1	6	4.38
Human Resources (Cost) Accounting Method	65	1	6	4.55
Psychological Appraisal System	68	1	6	4.34
Behaviorally Anchored Rating Scale System (BARSS)	68	1	6	4.34
Valid N (listwise)	57			

Source: Field data (2022).

Among the listed methods of employees' assessment in Table 1 the most popularly used approach of assessing employee's performance at FHTL is Behaviorally Anchored Rating Scale System (BARSS) and had been effective, fair and is a base tool for promotion and incentive provision in the organization as had been rated by the respondents with an average of 4.34 of all methods of employees' assessment.

Employee's involvement on the exercise of employees' assessment at FHTL.

FHTL is a private organization with its own policies, laws, procedures, direction as determined by mission and vision of the organization. It had been observed that, the issue of employee's involvement in the exercise of OPRAS in not negotiable, that is to say everyone working in that organization must adhere to the rules, laws and policies of the organization to secure his/her employment. Thus, majority had been forced to participate in the exercise vigorously.

Behaviorally Anchored Rating Scale System (BARSS) is the current approach of assessing employees' performance at FHTL, and the OPRAS are tailor-made to meet the requirement of the organization and are of different based on ones' position in the organization as indicated in the figure below (Fig. 2):

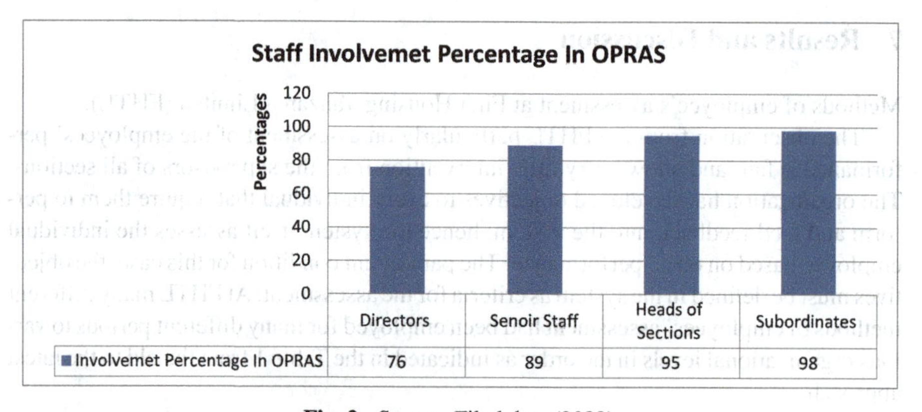

Fig. 2. Source: Filed data (2022)

The response to the involvement of electronic OPRAS (e-OPRAS) is low from the top management and is high to the subordinates. The Directors and some senior staff are exempted from filling e-OPRAS due to the specific and special tasks excluded from the assessment however they are obligated to compete their activities on time. Heads of Section and Subordinated are obligated to fill the e-OPRAS as required, because remain as a tool form promotion, incentive provision and their performance of the job bring job security, that being the reason the percentage of the involvement in OPRAS is the highest.

Outcomes of exercising OPRAS at FHTL

The engagement of sophisticated means of assessing employees' performance at FHTL had been significant approach for the development of the organization at great extent as reflected in various aspects of the organization.

i. *Productivity growth:* The organization is using electronic means of assessing employees' performance in the organization, which is assessing fairly and everyone is feeling his or her value in the organization. This brought higher employees' commitment towards work performance and eventually the profits had been increasing in the trend of 41% in the year 2018/2019; 71.6% in the year 2019/2020 and 87% in the financial year 2020/2021 with growth of an average of 15.6% annually.
ii. *Employees' motivation:* The approach of assessing employees' performance has little human supervision, and has reduced unnecessary bureaucratic and dramatics situations between the processes of employee's motivation resulted from their performance. The system had been fair all the time in its operations, the employees are motivated based on the criteria indicated form bonus, and other financial benefits including salary increase based on the frequencies of bonus an employee earnings.
iii. *Employees' promotion:* The employees' assessment approach had been the tool for employees' promotion. In the organization, the employee promotion is based on frequencies of individual employees' performance and his/her contribution in the organization. In the financial year 2018/2019, three staff were promoted to higher position; 2019/2020 four staff were promoted to higher position; and 2020/2021, two staff were promoted to higher position
iv. *Proper utilization of resources:* At FHTL the resources are allocated and utilized properly. Everyone is assigned specific objectives and supplied appropriate resources which facilitate him/her to achieve the stated objectives on time. Management by walking around techniques is the one engaged at FHTL to control the resources so that can only be used for specific purposes.

8 Conclusion and Recommendations

8.1 Conclusion

Employees' assessment and appraisal are twins which are inseparable in any formal and practicing organization. Human nature tells that, most of the bosses are selfish, group employees into classes, have both positive and negative feeling against employees at workplace. All these can be the sources of assessing employees properly or improperly as per superior's feeling if the means of assessment is a paper work used by all public organizations operating in the United Republic of Tanzania (URT), where a boss has to indorse his/her comments that can lead to someone promotion or demotion. Thus, the use of e-OPRAS as used by FHTL is the best means because is fairly done, due to the use of feedbacks fed and approved in the system and reduces unnecessary interaction between supervisors and subordinates at work place.

8.2 Recommendations

This study recommends that, all employers (private and public employers) should adopt the electronic system of assessing employee's performance at workplace for better improvements, fairness, judging employees based on their performance and motivation and promotion without formidable, with the reason that favoritism will automatically be eliminated. Through engagement of the electronic means of assessing employees' performance at workplace, it can also be recommended that, employers/organization should use the system appropriately to identify best performing workers for the particular financial year and to be publicized and awarded accordingly during the international workers day (Mei Mosi).

References

Alman, K.L., Yusuph, M.L.: Performance appraisal experiences and its implications on human resource decisions in decentralized health services in shinyanga region, Tanzania. J. Public Admin. Govern. **10**(2), 181193 (2020)

Bammann, K., et al.: The IDEFICS validation study on field methods for assessing physical activity and body composition in children: design and data collection. Int. J. Obes. **35**(1), S79–S87 (2011)

Baraka, M.: The role of performance appraisal system on individual workers performance in Public organization a case of selected public entities in Tanga Region, Doctoral dissertation, Mzumbe University (2015)

Bernard, A.: Performance appraisal and local government staff in Tanzania: a case of OPRAS in Mpwapwa district council, Doctoral dissertation, The University of Dodoma (2013)

Brim: as quoted by Matete, R. E. (2016): Implementation of management by objective through open performance review and appraisal system (OPRAS) for teachers in Tanzania. Int. J. Educ. Literacy Stud. **4**(3), 24–33 (2012)

Dickson, D.: Open performance review and appraisal system (OPRAS) in Tanzania: Case study of Monduli and Meru district hospitals, Doctoral dissertation, Mzumbe University (2013)

Farooq, M., Farooq, O., Jasimuddin, S.M.: Employees response to corporate social responsibility: exploring the role of employees' collectivist orientation. Eur. Manag. J. **32**(6), 916–927 (2014)

Hamed, S.M., Potapova, M.A.: Motivation as a mechanism for employees'performance: a case study of samberi hypermarket, vladivostok. Russia. Экономика труда **8**(5), 549–564 (2021)

Iqbal, N., Ahmad, N., Haider, Z., Batool, Y., Ul-ain, Q.: Impact of performance appraisal on employee's performance involving the moderating role of motivation. Oman Chapter Arab J. Bus. Manage. Rev. **34**(981), 1–20 (2013)

Iqbal, N., Anwar, S., Haider, N.: Effect of leadership style on employee performance. Arab. J. Bus. Manage. Rev. **5**(5), 1–6 (2015)

Kielhofner, G., Coster, W.J.: Developing and evaluating quantitative data collection instruments. Kielhofner's research in occupational therapy: methods of inquiry for enhancing practice, pp. 274–295 (2017)

Kim, Y.M.: Validation of psychometric research instruments: the case of information science. J. Am. Soc. Inform. Sci. Technol. **60**(6), 1178–1191 (2009)

Kothari, C.: Research methodology methods and techniques by CR Kothari, vol. 91. Published by New Age International (P) Ltd., Publishers (2017)

Macha, E.R.: Assessment of Open Performance Review and Appraisal System (OPRAS) in Local Government Authorities in Tanzania: A case of Kibondo district council, Doctoral dissertation, The University of Dodoma (2014)

Mani, K.P., Sritharan, R., Gayatri, R.: Effect of job stress on the employees performance. Annamalai Int. J. Bus. Stud. Res. **6**(1) (2014)

Mathias, L.S.: An assessment of the implementation of open performance review and appraisal system in local government authorities: a case of Morogoro municipal council, Doctoral dissertation, The Open University of Tanzania (2015)

Muda, I., Rafiki, A., Harahap, M.R.: Factors influencing employees' performance: a study on the Islamic Banks in Indonesia. Int. J. Bus. Soc. Sci. **5**(2) (2014)

Nchimbi, A.: Implementation of open performance review and appraisal system in Tanzania local government authorities: some observations and remarks. Int. J. African Asian Stud. **53**, 32–40 (2019)

Nora, C.R.D., Zoboli, E., Vieira, M.M.: Validation by experts: importance in translation and adaptation of instruments. Revista Gaúcha de Enfermagem **38** (2018)

Nwata, U.P., Umoh, G.I., Amah, E.: Internal organizational communication and employees' performance in selected banks in Port Harcourt. Int. J. Novel Res. Human. Soc. Sci. **3**(3), 86–95 (2016)

Resnik, D.B.: Scientific research and the public trust. Sci. Eng. Ethics **17**(3), 399–409 (2011)

Resnik, D.B.: Data fabrication and falsification and empiricist philosophy of science. Sci. Eng. Ethics **20**(2), 423–431 (2014)

Resnik, D.B., Elliott, K.C.: The ethical challenges of socially responsible science. Account. Res. **23**(1), 31–46 (2016)

Rugeiyamu, R.: The Role of OPRAS on Employees' Performance: a Case Study of Local Government Training Institute", Doctoral dissertation, The Open University of Tanzania (2019)

Songstad, N.G., Lindkvist, I., Moland, K.M., Chimhutu, V., Blystad, A.: Assessing performance enhancing tools: experiences with the open performance review and appraisal system (OPRAS) and expectations towards payment for performance (*P4P*) in the public health sector in Tanzania. Glob. Health **8**(1), 1–13 (2012)

Steinhubl, S.R., et al.: Rationale and design of a home-based trial using wearable sensors to detect asymptomatic atrial fibrillation in a targeted population: the mHealth Screening to Prevent Strokes (mSToPS) trial. Am. Heart J. **175**, 77–85 (2016)

Tefurukwa, O.W.: The paradox of the nexus between employee's performance appraisal scores and productivity in Tanzania. Int. J. Soc. Sci. Entrepren. **1**(10), 260–273 (2014)

Stakeholder's Intervention in Reducing Crime Threat Among Motorcycle Taxi Riding Operators in Dar es Salaam, Tanzania

I. M. Issa[1](✉), E. F. Nyange[2], K. Mubarack[3], and E. J. Munishi[3]

[1] Department of Procurement and Supplies Management, College of Business Education, Dar es Salaam, Tanzania
immubundo@gmail.com
[2] Department of Marketing, College of Business Education, Dar es Salaam, Tanzania
[3] Department of Business Administration, College of Business Education, Dar es Salaam, Tanzania

Abstract. Purpose: The overall objective of this study is to examine stakeholder interventions in reducing crime threats among motorcycle taxi riding operators in Dar es Salaam, Tanzania.

Design/Methodology/Approach: A mixed approach research strategy was used in the study. To obtain 287 respondents (n = 287), the authors used cluster and purposive sampling approaches. The study included relevant government employees, including police officers as well as officers from the local government and the ministry of internal affairs, citizens and *'bodaboda'* riders. Data was collected through open and close-ended questionnaires, interviews and documentary reviews. While quantitative data were analysed descriptively, content analysis was used for qualitative data.

Findings: Results showed that actors have implemented a variety of actions through their various institutions aimed at lowering the crime threat in the context of motorcycle taxi riding activities. To name a few, these include the adoption of uniforms, the registration of cyclists, the inclusion of motorcycle taxi drivers in community police, and the registration of motorcycle riders.

Research Limitation/Implications: The purpose of this study was to close the knowledge gap between what is known and what is unknown regarding the stakeholder's contribution to reducing the crime threat among motorcycle taxi operations in urban settings.

Practical Implication: These results support the stakeholder's efforts to reduce crime associated with motorcycle taxi operations.

Originality/Value: The research focuses on stakeholders' interventions in reducing crime threat, among Motorcycle Taxi riding operators in the urban setting of Tanzania. This raises a need to explore the role of stakeholders in alleviating crime. The paper contributes to promoting safety among young men and women who, through their creativity have devised means of promoting reliable transport systems in cities where public transport is almost paralyzed due to limited transport infrastructure and traffic congestion.

Keyword: Crime · Interventions · Motorcycle taxi · Ridders · Urban

C. Aigbavboa et al. (Eds.): ARCA 2022, *Sustainable Education and Development – Sustainable Industrialization and Innovation*, pp. 1064–1076, 2023.
https://doi.org/10.1007/978-3-031-25998-2_82

1 Introduction

1.1 Background Information

Motorcycle taxi riding plays significant roles. It has helped in meeting urgent needs in terms of raising income and has helped citizens to reach any place on time with minimum delay (Nyaga & Gichuru 2019). Moreover, they have helped citizens to reaching the unreached places especially where roads are not easily accessible (Isaac et al. 2014). The operations have also been used for the supply and distribution of goods and services to different destinations which enhance livelihood (Adiambo 2020; Urioh 2020; Brakel & Town 2010). With the growth of urban centres, motorcycle taxi operations have increased rapidly and the sector is considered among the economic fortunes of many (Adiambo 2020;

most common vehicles and may account for 75% of passenger and freight transport (Opondo & Kiprop 2018; Adiambo 2020). Despite the widespread use of this means of transport, the sector has of recent been characterized by crime threats and that has led affected citizens in conducting their day to day lives (REPOA, 2009). While some of these crimes are classified into major/ serious and minor offences, to this study, a crime remains to be a crime. Major/serious criminal offences are those which are serious in terms of life and public concern; and minor ones are those that annoy individuals, public life or cause minor harm to individuals (Reisman et al. 2013).

Several scholars (Opondo & Kiprop 2018; Adiambo 2020; Nyaga & Gichuru 2019) have mentioned a number of crimes associated to boda-boda operators and that harm their living as well as hinder smooth operations of their works. These include motorcycle theft, kidnapping, uncooperative customers, robbery and robbery with violence, murder, rape to mention just a few. Such crimes jeopardize victims' efforts and thus lead to negative consequences in their livelihoods (Nguyen et al. 2018). Moreover, crime incidences have been associated to crippling riders economic chances and death, permanent disabilities, loss of confidence, early pregnancies among others (Opondo & Kiprop 2018; Nyaga 2017).

Nevertheless, (Rollason 2020; AFCAP 2019; Opondo & Kiprop 2018; Nyaga 2017; Nyachieo 2015; Caroline & Neil 2015a, b) tried to mention a number of initiatives employed by different groups of riders and citizens to eradicate the increasing extent of crime threats. They mention of insurance covers, introduction of unique motorcycle plate numbers for commercial and private uses; formation of motorcycle taxi driver trade unions, community policing and nyumba kumi initiatives among the motorcycle taxi riders, self-regulation among others (Starkey 2016). However, most of these initiatives are formulated by individual riders and little or no other stakeholders' including government initiatives are mentioned. Moreover, most of such initiatives are used in other countries including Kenya and Uganda and may not apply in the Tanzania context. To the best of the researchers' knowledge, the involvement of different stakeholders in curbing down the rate of crime has not been well captured in literature in the Tanzanian context. Thus, this study intended to answer the question, who are involved in curbing down the threat of *'bodaboda'* related crimes? What initiatives have they implemented to reduce the threat of *"bodaboda"* related crimes in the urban settings? How effective are their initiatives in curbing down the threat of *'bodaboda'* related crimes?

This publication resulted from an ongoing research project titled ***"Urbanisation, Crime, Threat and Coping capacity in the Context of Motorcycle Taxi Riding Operations in Urban Settings of Tanzania: Evidence from Dar es Salaam and Coast Region".*** The project is supported by Research on Poverty Alleviation (REPOA) for a duration of one year from December 2021 to December 2022 through a grant **No. DFA/REPOA/1210/CST/357.** The study intends to expose the crime situation among the motorcycle taxi riders and other potential victims, to determine the riders and other victims coping capacity, initiatives undertaken to counteract the crime and propose strategies for alleviating the crime. In this regard, the study promotes safety in the city especially among young men and women operating as motorcycle taxi riders. This group of individuals has tried to contribute, through their innovative ideas, to easy flow of individuals and goods around the city in which limited and unfavourable infrastructure and traffic congestion are salient features. The innovation of motorcycle taxi riders has opened up the possibility of quick transportation means in the city where the public transport means is overburdened due to depleted infrastructural network system.

2 Methods

2.1 Description of Research Site

This study was conducted in Dar es Salaam city and the Coastal region. The authors carried the study of these two regions based on the fact that great percentage of youths who operate motorcycle taxi, their interaction between one municipal to another as well as crime-threats related to motorcycle taxi riders were also great significance. The city is divided in five administrative municipals which include Ilala, Kinondoni, Temeke, Kigamboni and Ubungo. On the other hand, the coast region is the gate to the city of Dar es salaam from every angle. And thus, for the purpose of this study, townships of Bagamoyo, Kibaha as well as Kisarawe were included in this study based on their proximity to Dar es salaam city.

2.2 Research Design

The study employed mixed approach research design. Based on the nature of study, the authors adopted this approach in order to avoid obtaining one group of data set. This approach was used by a number of scholars (Opondo & Kiprop 2018; Ngemera 2017; Nyachieo 2015; Olvera 2021; Awour 2020) who studied about a related phenomenon. It was confirmed that, by merging with quantitative and qualitative approach, authors were enabled and allowed to complement, corroborate and strengthen datasets and reduce the shortcomings of each approach (Creswell 2017). Moreover, in the context of qualitative approach, data were collected through interviews, open ended questionnaire and documentary reviews (Rose et al. 2015). On the one hand, this approach helped the authors to interact with respondents and capture their deep feelings, attitudes, beliefs and experiences (Astalin 2013). Quantitative data were obtained through close ended questionnaire.

Cluster and purposive sampling were used to obtain 287 respondents. Each municipal and township in the studied area was regarded as a cluster from which 35 persons were recruited. The technique was also applied by Olvera (2021). Purposive sampling was used to sample Police officer to supplement data obtained through questionnaire. Data were collected through open and close ended questionnaire and interview. Questionnaire were administered to *'bodaboda'* riders and aimed at soliciting qualitative and quantitative findings. Questionnaire were self-administered due to the fact that some boda-boda riders did not know how to read and write and some did not prefer filling the questionnaires themselves.

Interviews were administered to police officers and local government authorities' officers and were either carried face to face or on telephone as used by other scholars (Rollason 2020; Starkey 2016). Interviews lasted between 20 and 30 min and were centred on approaches and mechanisms used to topple *'bodaboda'* related crime threats. During the interview sessions, respondents were free to say what they know regarding initiatives taken by stakeholders in curbing down the crime threat to boda-boda operators. Interviews also facilitated interaction between researchers and respondents given that the main reasons of interviewing them was to know unknow fact about versatile initiatives taken to curb down crimes.

2.3 Data Analysis Technique

Qualitative-content analysis was employed to summarize, arrange, organize, interpret, analyse and present the findings in relation to the objectives of the study. The findings were provided in line with research questions which mainly wanted to identify active stakeholders and their efforts in taming down crime threats among 'bodaboda' riders in which code, sub-code as well as quadrants were generated, grouped logically and leading to categories and content. In this MAXQDA 10 [VERBI Software, Marburg, Germany] tool was used to facilitate the exercise. With regards to quantitative data, the authors employed SPSS version 16 for data analysis in which close ended questions were analysed by means of descriptive statistics. Similarly, the authors had conducted a comparative analysis between the data from Dar es Salaam and those from Coastal region. The comparative graphs were employed to illustrate the truth of the findings in which for these data, differences and similarities were observed thoroughly.

2.4 Validity and Reliability

Authors ensured that all tools applied in this study were inspected by at least two experts to check if they are well worded, arranged and really capture what was intended. Furthermore, the researcher ensured that the tools include all aspects that need to be used in measuring an item as recommended by Heale & Twycross (2015). As for reliability of the findings, the authors ensured that the entire procedure of this study is well explained. This included giving clear explanation on setting of the study, the sample size used, how it was sampled and methods involved in the data collection exercise. Further, we subjected the questionnaire to Cronbach Alpha to determine the correlation among variables (Price et al. 2015).

3 Findings and Discussion

3.1 Stakeholders and Initiatives Undertaken to Curb Down Motorcycle Taxi Riding Related Crime Threat

The purpose of this part was to identify various actors and their initiatives to reduce the number of crimes committed against motorcycle riders and potential clients. It was noticed that different actors have taken part in, and executed various initiatives to reduce motorcycle taxi riding crime in urban settings, albeit to varying degrees. The police, politicians, citizens, NGOs, motorcycle riders, and local government officials are just a few of the stakeholders mentioned here.

Beginning with police force; finding demonstrated that police are involved in ensuring that people's safety and their property. Despite the fact that they have been participating for a while, findings regarding their involvement have been conflicting. On one hand, respondents agreed that police officers typically give them a lot of assistance in dealing with the crime; they also added that the police always worked with them when they were notified of a crime. One cyclist said:

> *Police are of great hep to us. They support us whenever we present our problem to them. They will not only file the case by they follow it up and investigate it. They provide RB and Police detectives for investigation so as to find your motorcycle* ***(Male Cyclist (77), Ubungo, Dar es Salaam).***

However, some respondents felt that certain police personnel were under-supported in their efforts to combat crime. They added that unless something extremely serious had occurred, like a fatality, police officers did not typically express much worry whenever the riders were involved in criminal activity.

> *Sometimes, the police are not all that cooperative unless someone has died. Moreover, when we go to report our matter to the police, they arrest us more especially if the stolen motorcycle is not yours. In this case they will accuse you of colluding with criminals and because of such situation we are even reluctant to report such incidences to police.* ***(Male Cyclist (23), Kibaha, Coast Region).***

A respondent in Bagamoyo reported that;

> Some police officers are unethical, they collude with criminals. We have reported a criminal whom we suspected of being involved in a crime scene. The police went to search his house, they found the stolen motorcycle with him, he was detained for a few hours and released, and nothing more happened. That is just a single case but we have several cases **(Male Cyclist (139), Bagamoyo, Coast Region)**.

Another Actor were Politicians. According to findings, the respondents affirmed that politicians had made very little progress in reducing crime for this group. For example, local council members, mayors and member of parliament were advocating for motorcycle riders' access to secure parking spaces. Politicians also pushed for uniforms for the riders so that they could always be recognized when they were the victims of crime.

On the other hand, the majority of respondents believed that they did not receive much political backing for efforts to reduce crime as narrated by one of the respondents.

> *They don't support us [politicians]. They will only use us during political elections campaigns and as soon as the elections are over they forget about us completely. They don't show us any support. Only a few of them can show concern but majority of them don't support us. That is the most I can tell you* ***(Male Cyclist (92), Ubungo, Dar es Salaam)***

Nevertheless, the Citizens Were Another Actor Who Intervene the Crime. Accordingly, findings show that citizens support for cyclists who become victims of crimes. Citizens would always be the first to respond when they yelled for assistance. They further reported that numerous residents in the neighborhood would also yell for more assistance anytime they overheard bikers calling for assistance when they were involved in criminal activity. Along with showing their sympathy, people would also donate money to help the injured riders with their medical and food expenses. However, financial assistance for food and medical care was only provided if the riders were in need. One respondent reported here under.

> *Of course, whenever they see us being attacked, they always come to rescue us. You see some of them will even scream for help if they happen to see something dangerous or suspicious. However, sometimes they will only support you if they know you but if they don't know you they are scared because they think you are colluding with criminals.* ***(Male Cyclist (84), Kigamboni, Dar es Salaam)***

In the side of non-Governmental Organisations NGOs; finding indicate that the NGOs had not offered much assistance to the community, it was stated.

> *There is nothing like NGOs that has attempted to support us in any way. Off course we very much expect non-profit organisations to support us in various ways but we rarely see them doing that. We have only been hearing about NGOs, but they have not supported us practically.* ***(Male Cyclist (29), Kinondoni, Dar es Salaam)***

In the context of Motorcycle riders themselves, finding revealed that motorcycle riders tended to support one another in dealing with crime to a large extent. Riders helped victims of crime not just by coming to their aid, but they also offered financial aid to any of their members who had either directly or indirectly suffered a loss as a result of the crime. They took involved in the motorcycles' recovery as well.

> *Yes, we do support each other even if we break the rules to protect our fellow we have to do so. We support one another, sometime we will go to look for the stolen motorcycle and contribute money to support one who is in trouble. Because we have our small groups, we even contribute some cash to support the riders who get injured or fall a victim of such crime situations* ***(Male Cyclist (73), Temeke, Dar es Salaam)***

The results show that a variety of actors, including the police, politicians, residents, and motorcycle riders, actively participated. All of them employ various strategies to help the victims of crime or ensure that the problem is reduced. Findings show that there was no active participation of NGOs in this area, contrary to expectations that they would have been in the forefront because of their active involvement in a variety of civil-related concerns.

Additionally, despite the participation of several actors being a double-edged sword, their participation has been noted. For instance, police forces have been reluctant and have not given the issue their full attention of curbing with crimes. It has also been stated that there is a perception that certain dishonest police personnel may be working together with criminals or may be falling into their traps thanks to financial inducements they receive from criminals. These results are consistent with Opondo and Kiprop (2018), who underline that corrupt police personnel who treat cases lightly and in favor of those who scratch their backs have contributed to the problem's gravity. Further findings suggest that people and riders take responsibility for solving their own problems. These results are in line with those made by Walwa (2017) in Dar es Salaam, who stated that the community supports crime reduction measures in a variety of ways, including through the collaboration of formal, informal, and non-profit groups.

3.2 Effectiveness of Various Initiatives Undertaken by the Government and Other Actors

It is common knowledge that criminality has threatened not only boda-bodas and their passengers but also the entire neighbourhood. Findings show that a number of actors launched numerous initiatives to reduce crime threats. These were the introduction of uniforms to Boda-boda, registration of cyclists and motorcycle stations, education, community policing, and donning uniforms with identifiable information. However, the researchers were curious to hear from respondents about how effective these initiatives have been in resolving the crimes as shown in Figs. 1 and 2 respectively.

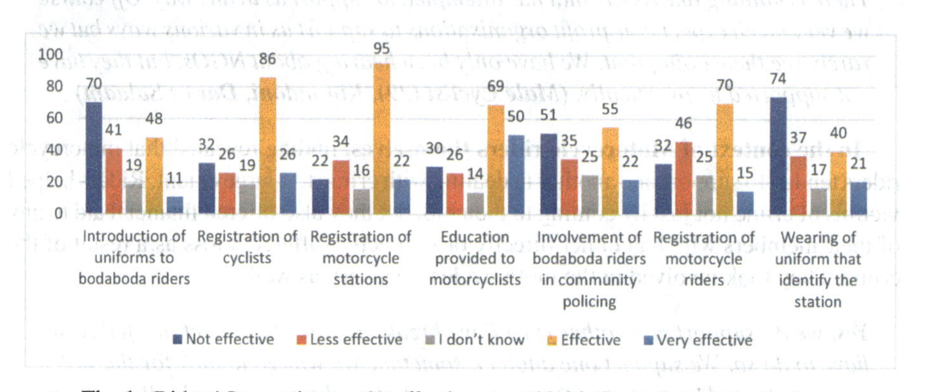

Fig. 1. Riders' Perception on the effectiveness of Initiatives taken to Curb crimes

Findings from Fig. 1 simply show that, for riders, registering motorbikes and cyclists has been more successful than other measures. On the other side, we offer Fig. 1, which displays passengers' opinions about the success of such projects. The two groups' explanations are then given jointly.

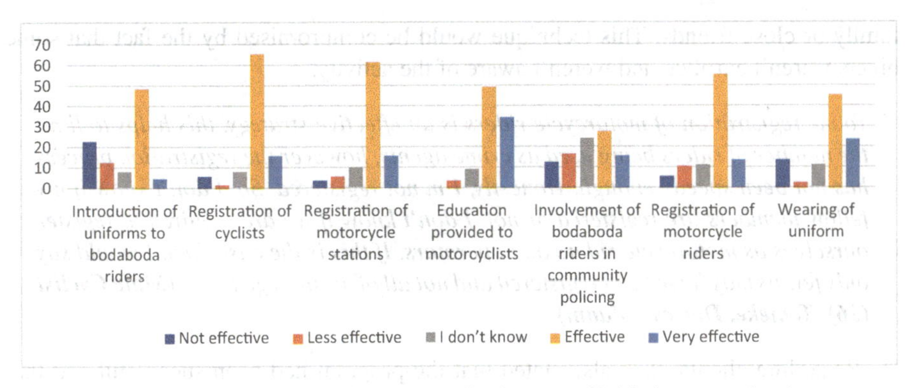

Fig. 2. Passengers' perception on the effectiveness of initiatives to curb crimes

3.3 Introduction of Uniforms to Boda-Boda Riders

This feature had two faces. For passengers and other potential customers who might become victims of such crimes. Findings show that the introduction of uniforms to boda-boda riders made it easier for them to be recognized and even when assaulted, they were able to readily obtain support from local residents but for them it was viewed as less effective. They further claimed that the introduction of uniforms made it easier for them to be recognized and even when assaulted, they were able to readily obtain support from local residents yet this was less effective. Some respondents added that this approach was up for debate because they did not see how wearing uniforms helped to reduce crime. Additionally, it was highlighted that the idea of introducing uniforms was still a recent one and had not been well received by many municipal administrations.

> *Basically, we don't have special uniforms here. But we need special uniforms specifically to identify us with our parking station. I hope our leaders are making follow up to make sure that every member at the station has one. All in all, we need them so that we can be easily be identified.* ***(Male Cyclist (32), Temeke, Dar es Salaam)***

Another cyclist simply stated that uniforms benefit customers, not riders, in presenting an idea that was also put forth by other customers. Customers can share the riders' information with anyone for security reasons because they can tell the rider's details just by looking at their clothing. "Uniforms have very poor effectiveness to riders' security," he continued. Such an instance demonstrates the truth that, while clients see uniform efficacy, riders perceive uniform initiative as less effective.

3.4 Registration of Cyclists

One of the most successful methods to reduce crime was the registration of bikers. In other words, there was a report that bikers were currently undergoing special registration. In this exercise, riders were required to provide information such as passport-size pictures, their country of origin, the names of their immediate family members, their mobile phone numbers, their place of permanent residence, and the phone numbers of their immediate family or close friends. This technique would be compromised by the fact that some bikers weren't enrolled and weren't aware of the activity.

> *To me registration of motorcycle riders is an effective strategy, this helps to limit the number of riders being used as crime agents, however the registration process has not been speedy enough. Honestly, I'm not registered. So, I don't know if my fellow members are registered or not. I don't know if we are required to register ourselves as motorcycle riders and operators. If this is the case then, I would say only few us may have been registered and not all of us are registered.* ***(Male Cyclist (26), Temeke, Dar es Salaam)***

Regarding clients, they also stated that the program had been successful and that motorcycle riders would be protected if they registered. Passengers will only employ registered riders in this case because their registration credentials may be displayed in a prominent location.

3.5 Registration of Motorcycle Stations

The registration of motorbike stations is a different crime-reduction technique that was looked at in this study. Because it has prevented the proliferation of unauthorized motorcycle riders and operators who participate in crime, both riders and consumers claimed that the registration of motorcycle stations was among the most successful methods for reducing crime. Findings show that the riders felt safer working in authorized stations because they were essentially protected from any type of work-related crime. Customers continued along the same lines by saying that having registered stations makes it simple for them to obtain safe transportation services since they only use motorcycles from their local, officially registered stations. One of the customers said, *"I just feel safe, because I know everyone is registered, and their station is registered as well. This has minimised the threat of crimes"*. This demonstrates that station registration has been successful in reducing crime. On the other hand, riders also commended the initiative's effectiveness and how it had helped them drive away people pretending to be riders but actually targeting them or their clients. One of them is reported to have said;

> *I thank God that our station here at Kivukoni is registered and we have permission from the police officers to operate our business in the area. Since we are now registered then we have the power to chase away those who are not registered or don't belong to this group because they are likely to propagate crimes here.* ***(Male Cyclist (22), Kivukoni, Dar es Salaam)***

3.6 Education Provided to Motorcyclists

On this, two sides came together. Customers claimed that the reduction in crime rates was the result of riders and customers receiving education, but riders claimed that the project was not very successful. While some riders in some stations stated that the police and local government officials had informed and sensitized them about crime as well as provided them with coping mechanisms, some riders in other areas, including Bagamoyo, Kibaha, and Temeke, stated that no one had ever given them such helpful information. They stated that no public or private group, including the police, had ever dared to educate them. Otherwise, they highlighted that they mostly distributed pertinent information about crime among themselves as a method of combating crime and to them this was effective.

Yes, sometimes but not often some local government officials and police officers may call us and provide us with some basic knowledge on safety issues. However, it is not effective enough. They focus on the importance of knowing and obeying traffic lights because this is a big problem that we rider have.

Cementing on this, another rider added;

Not real, here we only share knowledge among ourselves on what we know about crime and how to protect ourselves. However, we need experts like police officers to educate us on the safety issues detecting criminals and riding rules among others. ***(Male Cyclist (27), DUCE, Dar es Salaam).***

3.7 Involvement of Boda-Boda Riders in Community Policing

Even though research on riders and customers has shown that the score is practically identical to those opposed to the program, riders' involvement in community policing has been shown to be successful. This could lead to the realization that, despite the fact that police have deployed riders as community police officers, the rate of crime committed against riders and consumers has not decreased. Researchers had anticipated that this process would be more efficient than others, but they were completely wrong. According to reports, the effort was ineffective because the community policing members colluded with the police to arrest motorcycle riders for unrelated offenses despite lacking the necessary skills and participating in peace and security meetings. They are viewed as adversaries in this sense and receive no support from their other riders, who refer to them as "snitches." According to one of the cyclists;

The involvement of some of our fellow riders has not been effective, instead of hunting criminals who kill riders, take their motorcycles, or any other disguising motorcycles riders, they collude with the police to arrest us just for not wearing closed shoes! We thus end up isolating them. ***(Male Cyclist (84), Ubungo, Dar es Salaam)***

3.8 Wearing of Uniform that Identify the Station

It was claimed that encouraging customers to wear station-specific uniforms was ineffective. Customers claimed that the prevalence of riders wearing uniforms, registration numbers, and stations from which they park has reduced the amount of crimes. As a result, customers have been eager to work with riders who have complete registration information. Should something happen to the consumer in such a scenario, the rider will be found and held accountable. The motorcycle taxi industry as a means of combating crime since it would naturally distinguish between legitimate and unauthorized drivers. The vast majority of riders, however, felt that the program had no effect at all. Even one of them mockingly questioned., *What will wearing such a uniform do for me if I get attacked? Maybe if someone passes away, it helps to know where they came from, but that doesn't lessen the number of crimes.* Such a quote demonstrates the extent of unhappiness riders have with uniforms displaying their information.

By interpreting these results broadly, it can be stated that there have been certain trends in terms of the degree of success of actions made to reduce *'bodaboda'*-related crimes in the coastal region and Dar es Salaam's metropolitan environments. The results show that the most widely acknowledged successful initiatives have been the registration of motorcycle riders and the registration of stations. This suggests that registering riders and stations might simply lower the rate of crime because everyone working in the station would be aware of each other.

The results are consistent with those of Olvera (2020), who mentions that motorcycle registration, educating riders and incorporating them in neighorhood policing were useful strategy of reducing the incidence of crimes. In the similar vein, AFCAP (2019) reported that wearing a uniform is more important than ever in reducing the rate of crime among motorcycle riders. A rider's identity is revealed to the public via their uniform. On the other hand, the findings has opposed by by Olveira et al. (2020) and Opondo et al. (2018) who reported that low levels of education, police corruption, and ineffective community policing were factors in the commission of crime rather than its reduction.

3.9 Conclusion

This study sought to expose steps done by the government and other interested parties to reduce *'bodaboda'*-related crimes in Dar es Salaam, Tanzania. The study's findings show that, despite the involvement of several stakeholders and the deployment of various initiatives, the level of effectiveness of such initiatives is dual. While some people find them to be beneficial and in need of reform and enhancement, others feel them to be ineffective.

3.10 Recommendations

The following recommendations should be taken into account in order to assist the riders in more successfully coping with the crime and the nation eradicating crimes linked to motorcycle taxi riding operations in general.

- Authorities like the police and other security organizations need to comprehend that criminality among boda boda riders is a significant and complex issue, and that no one remedy will work in every circumstance. This indicates that, by and large, every region should assess the issue and take appropriate action to control it.
- While tracking devices are highly helpful, it is advised that they be made compulsory for everyone who registers to operate as a cab driver. However, the government should undertake the required measures through the ministries of trade, internal affairs, and other agencies to guarantee that the costs of these devices are kept to a minimum.
- Police officers must act swiftly to respond to crimes, conduct impartial investigations, and apprehend perpetrators. In a similar spirit, police personnel should refrain from pulling over boda boda users for insignificant offenses. As a result, there will be

more confidence between riders and police, and riders will feel free to report any questionable circumstances. In this way, it will be simple to stop crimes before they start.

- Authorities on community policing specifically the police and local government officials must restructure the community police's activities, assign them specific tasks, and instruct them on how to cope with lawlessness.
- It is advised that the police, LATRA, and any other relevant board maintain a nationwide register of motorbikes and the riders who use them in order to spot boda-boda riders who are hiding their identities. If the rider decides to cease using the bike, he or she should report so that the old rider can be erased and the new one added. In the same line, every motorcycle used as a taxi should have a sticker that is clearly visible and big enough to be recognized by anyone. The sticker's legitimacy may also be quickly verified in the system. This will eliminate cloaked riders who become thieves.
- The study suggests that the TRA, the police, and the government regulate used parts. For instance, every piece of spare equipment imported must have a certain number. This number will reveal where the component was sourced. A spare part should be doubted if it lacks such a number. Dealers in spare parts will avoid purchasing parts if they are aware of their origins in this way. This will reduce motorcycle theft, which ultimately only results in the sale of used spare parts.

Acknowledgement. We would like to recognize and appreciate invaluable financial support provided to us by Research on Poverty Alleviation (REPOA). The financial support obtained from REPOA facilitated the entire research cycle.

References

Adiambo, E.: The contribution of motorcycle business to the socio-economic wellbeing of operators in Kisumu County, Kenya. Int. J. Soc. Develop. Concerns **13**(November), 55–69 (2020)

AFCAP: Facilitation Services for Consultation on Motorcycles Operations in Ghana Final Report Quality assurance and review table ReCAP Database Details: Facilitation Services for Consultation on Motorcycles Operations in Ghana ReCAP Website Procurement Method Co. *ReCAP* | Facilitation Services for Consultation on Motorcycles Operation in Ghana, 1(July) (2019)

Astalin, P.K.: Qualitative research designs: a conceptual framework. Int. J. Soc. Interdiscipl. Res. **2**(1), 118–124 (2013)

Van Brakel, P.A., Town, C.: 12th annual conference on world wide web applications. In: Proceedings of the 14th Annual L Conference on World Wide Web Applications, 0–15 September (2010)

Caroline, B., Neil, R.: Tanzania motorcycle taxi rider training: assessment and development of appropriate training curriculum Final report Transaid May 2015. In: International Conference on Transport and Road Research (Issue May) (2015a)

Caroline, B., Neil, R.: Tanzania motorcycle taxi rider training : assessment and development of appropriate training curriculum Final report Transaid May 2015. In: International Conference on Transport and Road Research, 1–27 May (2015b)

Divall, A., et al.: Enhancing understanding on safe motorcycle and three-wheeler use for rural transport (Issue June) (2019)

Gibigaye, M.: Organising informal transport workers: The informal economy and the unionisation of motorcycle taxi drivers in Benin (2006)
Isaac, K.W.O., Benjamin, O., Opondo, F.: The effect of increased investment in *'bodaboda'* business on economic empowerment of people in Kisumu west district. Eur. J. Bus. Manage. **6**(39), 177–185 (2014)
Kariuki Nyaga, J.J., Gichuru Kariuki, J.: The influence of motorcycles/boda boda on community development in rural kenya: a study of the challenges facing motor cycle operators in meru south sub-county. J. Educ. Hum. Develop. **8**(1), 2334–2978 (2019). https://doi.org/10.15640/jehd.v8n1a10
Munishi, E.J.: Rural-urban migration and resilience implications on the Maasai Households ' in North - Eastern Tanzania. African J. Appl. Res. **5**(2), 24–44 (2019)
Ngemera, H.: Motocycle Taxi "Boda Boda " a New Paradigm for Livehood Diversification: A Case of Chamwino District. University of Dodoma, Tanzania (2017)
Nguyen-phuoc, D.Q., Gruyter, C.D., Anh, H., Nguyen, T., Ngoc, D.: Risky behaviours associated with traffic crashes among app-based motorcycle taxi drivers in Vietnam. Transp. Res. Part F: Psychol. Behav. **70**, 249–259 (2020). https://doi.org/10.1016/j.trf.2020.03.010
Nguyen, T.D., et al.: Injury prevalence and safety habits of boda boda drivers in Moshi, Tanzania: a mixed methods study. PLoS ONE **13**(11), 1–16 (2018). https://doi.org/10.1371/journal.pone.0207570
Nyachieo, G.M.M.: Socio-cultural and economic determinants of Boda-boda motorcycle transport safety in Kisumu county, Kenya. In: Hilos Tensados, vol. 1, Issue June. Kenyatta University (2015)
Nyaga, J.K.J.: The impact of motorcycle taxi transport (Boda Boda) in accessing the Kenyan rural areas: a case study of Meru South sub-county. University of Nairobu (2017)
Opondo, V.O., Kiprop, G.: Boda Boda Motorcycle Transport (2018)
Reisman, L., Mkutu, K., Lyimo, S.: Tackling the Dangerous Drift Assessment of Crime and Violence in Tanzania (Issue June) (2013)
REPOA: Afrobarometer: Briefing paper citizens' views on crime in Tanzania. REPOA, vol. 8 (2009)
Risdiyanto, Munawar, A., Irawan, M.Z., Biddinika, M.K., Alfed, J.: Importance performance analysis of online motorcycle taxi services: indonesian passenger perspective. In: International Conference on Applied Science, Engineering and Social Science, Icasess 2019, pp. 79–85 (2020). https://doi.org/10.5220/0009878600790085
Rollason, W.: Crisis as resource: entrepreneurship and motorcycle taxi drivers in Kigali drivers in Kigali. African Ident. **18**(3), 263–278 (2020). https://doi.org/10.1080/14725843.2020.1788919
Rose, S., Spinks, N., Canhoto, A.I.: Case study researh design. Manage. Res. Apply. Principles **1–11** (2015). https://doi.org/10.1097/FCH.0b013e31822dda9e
Starkey, P.: The benefits and challenges of increasing motorcycle use for rural access Animal production and health View project the benefits and challenges of increasing motorcycle use for rural access. In: International Conference on Transportation and Road Research, Mombasa, 15–17 March 2016 (2016). https://www.researchgate.net/publication/332413878
Transaid: An introductory webinar on the topic of motorcycle taxis in the rural context of Sub-Saharan Africa and South Asia (Issue April) (2017)
Urioh, N.: Contribution of 'bodaboda' Business on Improving the Standard of Living Among Youth in Tanzania : a Case of Ubungo District in Dar es Salaam. Masters Dissertation Mzumbe University (2020)

E-learning of Mathematics and Students' Perceptions in Public Secondary School, Oyo State, Nigeria

A. E. Kayode[1,2](✉) and E. O. Anwana[1]

[1] Faculty of Management Science, Durban University of Technology, Durban, South Africa
aderinsolaK@dut.ac.za

[2] Department of Educational Management, Faculty of Education, University of Ibadan, Ibadan, Nigeria

Abstract. Purpose: Integration of e-learning in classroom-based with mathematics courses/modules is recognized in all secondary schools, because mathematics subject have ability helps students to perform well in other subjects.

Design/Methodology/Approach: Descriptive survey research design with quantitative data approach through structured questionnaire. The reliability confirmed the Cronbach's Alpha and the coefficient was 0.78. 120 students were purposively selected from 36 secondary schools in Ibadan North, a local government area of Ibadan, capital of Oyo State in Nigeria. At a significance threshold of 0.05, four research questions and two hypotheses were developed.

Findings: The findings showed that e-learning platforms are not maximally utilized in mathematics among students in secondary school in Oyo State due to restrictions and difficulty of use as well as the students' preference for traditional learning.

Implications/Research Limitations: The study showed that there was a significant relationship between e-learning and students' perceptions of mathematics learning, though with weak correlation. The study was limited to only secondary school in Oyo State and therefore, could not be generalized to all other secondary schools in Nigeria.

Practical Implication: e-learning method serves as alternative to support teaching and learning using blended learning tools to support the classroom-based education, especially during this Post-COVID-19 pandemic.

Social Implication: This research study will encourage both teacher and students to embrace technological tool in teaching and learning for the improvement on students' academic performance.

Originality/Value: This study encouraged that mathematics teachers should make an effort to maximize the benefit of e-learning in mathematics to create a greater impact in teaching and learning. And also, students should use the opportunity to improve on their self-pace studies using asynchronous method of online.

Keyword: E-learning · Integration · Mathematics · Public secondary school · Students

C. Aigbavboa et al. (Eds.): ARCA 2022, *Sustainable Education and Development – Sustainable Industrialization and Innovation*, pp. 1077–1087, 2023.
https://doi.org/10.1007/978-3-031-25998-2_83

1 Introduction

Human beings have always devised ways to survive, from the Stone Age to present today, which is why human beings have always tried to make things easier for themselves by using technological tools in education (Edara 2021; Govender & Kayode 2020). This has led to the birth of ideas, a materialization of thoughts and the birth of technological ages; and has led us to where we now, the internet age, where everything and every resource needed for life is at an individual's fingertips (Gui & Büchi 2021).

Students have perceived mathematics as very difficult and hard to understand. Using online learning resources, the world has become connected into one global city where all kinds of political, economic, social and even educational sectors are interconnected. Students' perception towards school in general, and mathematics in particular, and ideas closely related to students' attitudes towards mathematics in this technological age have been studied globally for over four decades (Moreno-Guerrero et al. 2020).

The integration of E-learning integration is a dignified to traditional education process using electronic resources. Computers and the internet are the strengths of e-learning, even though instruction can take place in or outside of classrooms (Babu & Sridevi 2018). E-learning can also be a network-enabled method of education and skill transfer which can be provided to a large number of students either simultaneously or at different locations. Initially, E-learning was not frequently used with traditional education sector. Nevertheless, as educational technology advanced and learning strategies improved, teaching and learning positively (Govender & Kayode 2020).

Blending technology tool with traditional education formed the foundation for this digital revolution and made teaching and learning dependent on digital tools, which found an essential place in a learning environment like the classroom; likened to classroom-based learning, in which online learning is alleged to dearth of interactivity (Alfadil et al. 2020; Goverder & Kayode 2020). This is mostly caused by a lack of social engagement, social presence, and student pleasure. However, online learning has been marketed as being more affordable and practical than traditional educational settings and as giving more students the chance to further their education. (Bali & Liu 2018; Nafukho & Chakraborty 2014).

Previous studies have been examined on how students feel about online and tradition education learning and how satisfied they are with it. (Rathor 2022; Lazarevic & Bentz 2020; Berga et al. 2021; Nambiar 2020). In previous researchers, Fortune et al. (2011) looked into the learning preferences of the 156 students who participated in the two alternative learning modalities for a recreation and tourism course at a multicultural institution in Northern California, United States of America, did not differ statistically significantly from one another.

According to Bali and Liu's (2018), their study showed that online and traditional learning is unaffected by a university's courses level. There was no statistically significant difference in learning views between participants engaged in the e-learning and traditional learning courses. Although many people find face-to-face learning more gratifying, they prefer online learning because it is more convenient, saves time, and allows them to work when they want to rather than when they have to.

Some studies revealed that some students struggle with e-learning due to a lack of access to online facilities, thereby preventing them from learning through online modes during the Covid-19 lockdown. Therefore, some researchers have recommended using a blended method in teaching and learning (Aboagye et al. 2020; Aboagye 2021; Srinivasan et al. 2021).

Mathematics is one of the most dreaded subjects at almost all levels of schooling. Most students are necrophobic, resulting in many secondary school students finding it impossible to understand mathematics and in a low uptake in mathematics classes and high levels of failure (Mazana et al. 2020). As the world is changing, so is the dynamism in technology that cuts across every sphere, including education. For example, the Covid-19 lockdown has resulted in a rise in e-learning, especially in Africa, where teaching and learning moved to virtual methods and allowed students to learn from home.

2 Research Questions

The research questions addressed the problem of the study:

i. What are the e-learning tools available in public secondary schools? in Oyo State, Nigeria?
ii. To what extent are these e-learning platforms used among secondary students in Oyo State, Nigeria?
iii. What are the challenges of learning Mathematics through e- learning among public secondary school students?

3 Research Hypothesis

At a 5% level of significance, the following hypothesis was developed in an effort to fully meet the study's goal:

Ho1 - There is no significant relationship between e-learning and students' perceptions of mathematics learning in public secondary school.

4 Literature Review

Learning and technology tools are the two significant supports of e-learning. (Saeed Al-Maroof et al. 2020). Technology is an enabler of the learning process, which means that it is utilized in education just like any other tool, like a pencil or a notebook. Learning is a cognitive process for acquiring knowledge (Aparicio et al. 2016).

Francis Bacon, a British philosopher who lived in the 1600s, said, "knowledge is power" (Wojciuk & Górny 2018; Serjeantson 2014). Thus, the one who knows possesses power, while the uneducated are helpless. But knowledge is also power, irrespective of whether e-learning or traditional education. Education has a prominent role to play in people's lives (Algahtani & Rajkhan 2020). Previous research outlined the significance of education and information in influencing people's perceptions of economic and social growth. (Benjamin et al. 2021). 'Highly educated' groups of people are more fiercely competitive. And are more likely to get competent employment and, as a result, better opportunities to better their life. Education has the most profound implication for people's lives. Anderson et al. (2012) and Al Hadid (2022) stressed that the usefulness of the internet and the effect of e-learning in teaching and learning to impact more skills in technology to collaborate among students. These new technologies have and continue to offer non-traditional students educational options and higher education institutions the allure of economic riches.

Researchers emphasized that teachers and students should integrate the use of information and communication technologies (ICTs) with classrooms-based learning in today's educational system (Rahman et al. 2021). While e-learning is considered the best alternative for enabling students to use different type of ICT tools in their learning (Petretto et al. 2021; Valverde-Berrocoso et al. 2020; Wellington & Clarence (2021). E-learning method are vibrant for both teachers and students, according to numerous research studies (Bhuasiri et al. 2012; Arkorful & Abaidoo 2015). Accessibility to technological tools in education allows more flexible solutions for students to have an opportunity to study online to blend their learning with classroom-based. (Kayode 2019). According to recent study, improved flexibility is another significant improvement that ICT and e-learning bring, enabling universities all over the world to accept more students (Rakic et al. 2019, Govender & Kayode 2020). Integration of e-learning into students' learning modes has shifted traditional teaching and learning towards e-learning, though teaching during this period of Covid-19 has also proven relatively expensive (Kayode & Ekpenyong 2022; Srinivasan et al. (2021). However, there appears to be a consensus among academics that the usage of e-learning in higher education has several benefits and, given its several advantages and benefits, e-learning is considered among the best methods of education. Several studies and authors have provided benefits and advantages derived from adopting e-learning technologies in schools (Algahtani 2011; Reed et al. 2010; Even & Ball 2019; Collins & Halverson 2018; McGee et al. (2017).

5 Methodology

This study adopted a descriptive survey method to explore e-learning and students' perception of mathematics in public secondary schools in Oyo State, Nigeria. The reason for dopting descriptive survey research is to generalize the population so that inferences can be made, and to allows the researcher to collect data without manipulating any variables of interest to the study.

6 Study Population

The target population of this study is students in public senior secondary schools (SS1-SS3) students in Ibadan North Local Government Area, with 36 public senior secondary schools that range from Senior Secondary School One (SSS1) to Senior Secondary School Two (SSS2).

7 Sample and Sampling Procedure

Six senior secondary schools were chosen at random from the 36 senior secondary schools in the Ibadan North Local Government Area. (Polytechnic High School, Polytechnic Campus, Immanuel Grammar School, Orita U.I, Community Secondary School,

Sango, Community High School Agbowo, Bodija, Ijokodo High School, Ijokodo, and Abadina College Senior School). From each of the six schools selected, 20 students were randomly selected to arrive at 120 students who serve as participants in this research.

8 Research Instruments

The research instrument used for this study was a structured questionnaire which was used to collect information from the participants. The questionnaire consisted of statements drawn by the researcher in line with the entire concept of the study to bring into focus the problem under study. The questionnaire was divided into sections. Section A of the questionnaire was for the respondent's demographic information. The questionnaires were administered to school students in the above-mentioned local government area.

9 Quantitative Analysis

Frequency tables were used to extract the replies from the copies of the questionnaire given to the respondents and then code and present them using the statistical tools for social sciences (SPSS) and simple percentage. Furthermore, Chi-square distribution test was used to analyse the data and the results were reported with the aid of descriptive statistics.

10 Results and Discussion

The results are presented according to each research question and triangulated in the discussion (Table 1).

Table 1. Opinions of the respondents on the availability of e-learning platforms

Items	SA	A	U	D	SD	Total	X	SD
My School have provided us social media platforms for learning	56.3	37.0	5.0	1.7	–	100	1.52	.675
I have access to internet facilities in my school	27.7	46.2	20.2	5.0	0.8	100	2.05	.872
I have easy access to internet facilities and social media platforms at home which aids my learning in school	26.1	48.7	19.3	5.0	0.8	100	2.06	.857

X = mean; SD = Standard Deviation; *SA = Strongly Agree (1); A = Agree (2); U = Undecided (3); D = Decided (4); SD = Strongly Decided (5).

The results express the availability of e-learning platforms used amongst public secondary schools in Oyo State, Nigeria. The results show that for the majority of the respondents (93.3%, 73.9% and 74.8%, with mean and standard deviation scores of

1.52/0.675, 2.05/0.872 and 2.06/0.857 respectively) their schools provided them with social media platforms for learning, they had access to internet facilities in their schools and had easy access to internet facilities and social media platforms at home, all of which aided their learning in school. This shows that e-learning platforms are available for use in public secondary schools in Oyo State, Nigeria (Table 2).

Table 2. Opinions of the respondents on the extent of use of available resources

Items	SA	A	U	D	Total	X	SD
My teacher always gives us assignment that requires the use of internet	61.3	31.1	5.0	2.5	100	1.49	.711
My teacher always uses social media to give us learning instruction	31.9	54.6	8.4	5.0	100	1.87	.769
I always make use of social media platforms for learning in my school	22.7	48.7	23.5	5.0	100	2.11	.811
I have access to internet facilities in my school but I am forbidden to use them	21.8	35.3	30.3	12.6	100	2.34	.959
I have access to internet facilities but often find it difficult to use them	19.3	33.6	23.5	23.5	100	2.51	1.057
I have access to e-learning facilities but I prefer traditional learning to e-learning	25.2	33.6	25.2	16.0	100	2.32	1.025

X = mean; SD = Standard Deviation.

The results reveal the extent to which the available e-learning tools were used for mathematics learning among secondary school students in Oyo State, Nigeria. The results show that the majority of the respondents (92.4%, 86.5%, 71.4%, 57.1%, 52.9% and 58.8%, respectively) with mean and standard deviation scores of 1.49/0.711, 1.87/0.769, 2.11/0.811, 2.34/0.959, 2.51/1.057 and 2.32/1.025 agreed that their teachers always gave assignments that required the use of the internet; always used social media to give instructions; and always made use of social media platforms for learning in their school. They also agreed that their school provided access to internet facilities, but that they were often forbidden to use them or found it difficult to use them; and that they preferred traditional learning to e-learning. This shows that e-learning tools are not used to their full potential for mathematics learning among secondary students in Oyo State, Nigeria due to restrictions and difficulty of use, as well as a preference for traditional learning by the students (Table 3).

Table 3. Opinions of the respondents on the challenges in the use of e-learning

Items	SA	A	U	D	SD	Total	X	SD
Lack of feedback from peers	35.3	21.0	14.3	16.0	13.4	100	2.51	1.449
Lack of feedback from instructor in time	15.1	40.3	13.4	19.3	11.8	100	2.72	1.268
Workload not shared equally	12.6	36.1	17.6	22.7	10.9	100	2.83	1.230
Low or no collaboration of other classmates	11.8	30.3	26.1	22.7	9.2	100	2.87	1.168
Some set of students dominating the group discussion	9.2	30.3	21.8	28.6	10.1	100	3.00	1.172
Lack of time to participate in the group page	10.9	24.4	22.7	25.2	16.8	100	3.13	1.266

The results reveal the challenges of mathematics e-learning among public secondary school students. The result showed that the many of the students (56.3%, 55.4%, 48.7%, 42.1% and 39.5%, respectively), 2.51/1.449, 2.72/1.268, 2.83/1.230, 2.87/1.168, 3.00/1.172 and 3.13/1.266 agreed that lack of feedback from peers, lack of feedback on time from instructors, workloads not shared equally, low or no collaboration between classmates, with some groups of students dominating the group discussion, constitute challenges for mathematics e-learning among public secondary school students. A lower percent of the students (29.4%, 31.1%, 33.6%, 31.9% and 38.7%, respectively) felt that a lack of time to participate in the group page did not constitute a challenge to e-learning in mathematics as opposed to a lower percent (35.3%) that said it did. This shows that lack of feedback from peers and instructors on time, inability to share workloads equally, low or no collaboration with other classmates and some sets of students dominating the group discussion constituted the main challenges for e-learning in mathematics usage among public secondary school students.

11 Research Hypothesis

The following hypotheses was tested at 0.05 level of significance:

Ho_1 There is no significant relationship between e-learning and students' perception on mathematics learning in public secondary school.

The Spearman's rank-order correlation between e-learning and students' perceptions of mathematics learning in public secondary school is shown in Table 4. These results reveal weak positive correlation values between +0.116 to +0.285 which are insignificant at the p-value of 0.210 ($p < 0.05$). This shows that there is significant relationship between e-learning and students' perception of mathematics learning in public secondary school. Therefore, the null hypothesis was rejected.

Table 4. Results of Spearman's rank-order correlation analysis between e-learning and students' perception on mathematics learning in public secondary school.

Item	Choose only one e-learning platforms used mostly in your school	
	Correlation coefficient	Sig.
My school has provided us social media platforms for learning	0.242**	0.008
I have access to internet facilities in my school	0.244**	0.008
I have easy access to internet facilities and social media platforms at home which aids my learning in school	0.116	0.210
E-learning offer interesting instructional materials	0.285**	0.002

**$P < 0.01$.

12 Conclusion

The results show that e-learning platforms are available for use in public secondary schools in Oyo State, and there is an impact on students' perception.

However, the benefit of using e-learning to teach mathematics can be maximized if mathematics teachers effectively use e-learning for the benefit of students, provide feedback on time and improve collaboration. This was in line with the study of Babu and Srideve (2018) that teaching and learning based out of traditional education with the use of computer and internet in e-learning.

Respondents attested that their schools provided them social media platforms for learning, they have access to internet facilities in their schools and have easy access to internet facilities and social media platforms at home which aids their learning in school.

However, e-learning platforms are not used to their full potential for mathematics and technological tool in students' learning among secondary students in Oyo State, Nigeria due to restrictions and difficulty of use, as well as a preference for traditional learning by the students.

The majority of the students agreed that lack of feedback from peers, lack of timeous feedback from instructors, workloads not shared equally, low or no collaboration between classmates, with some students dominating the group discussion, constituted challenges for the e-learning of mathematics in public secondary schools.

Thus, it can be concluded that there is a significant relationship between e-learning and students' perceptions about mathematics learning in public secondary school (Bhuasiri et al. 2012; Arkorful & Abaidoo (2015). Kayode (2019) in the study supported that the accessibility to technological tools in education allows more flexible solutions for students to have an opportunity to study online – despite the student's geographical location. Therefore, the role of e-learning of mathematics in public secondary schools requires attention and further research.

Additionally, there is no significant difference between male and female students' perceptions of mathematics learning in public secondary school. The two genders do not have different perceptions pertaining to how e-learning impacts mathematics learning in public secondary schools.

Based on the findings and conclusions drawn, the following recommendations are hereby offered:

Mathematics teachers should be given opportunities to update their knowledge periodically through in-service training and retraining courses to enrich their knowledge on how to adequately maximize the opportunities e-learning has to offer in mathematics.

Mathematics teachers should make sure equal opportunities are given to all the students and also improve good communication and feedbacks.

Finally, students should maximize e-learning opportunities both in school and also in their respective homes.

13 Recommendation

The following recommendations are made:

- Improve feedback between peers, and encourage workload equity among students in public secondary schools.
- Mathematics teachers should improve their knowledge in e-learning platforms activities, so to impact positively to students learning.
- Students should use the e-learning platforms in school and at home for better understanding and for skills improvement; since findings shows that e-learning platforms are available for use in public secondary schools in Oyo State, Nigeria.

Acknowledgement. Sincere appreciation to the Management of Durban University of Technology for the 'Conference funding grant' for physical presentation and also, to the students who took their time to participate in this study.

References

Aboagye, E.: Transitioning from face-to-face to online instruction in the COVID-19 era: challenges of tutors at colleges of education in Ghana. Soc. Educ. Res. **2**(1), 8–17 (2021)

Aboagye, E., Yawson, J.A., Appiah, K.N.: Covid-19 and e-learning: the challenges of students in tertiary institutions. Soc. Educ. Res. **1**(1), 109–115 (2020)

Alfadil, M., Anderson, D., Green, A.: Connecting to the digital age: using emergent technology to enhance student learning. Educ. Inf. Technol. **25**(3), 1625–1638 (2020)

Algahtani, A.F.: Evaluating the effectiveness of the e-learning experience in some universities in Saudi Arabia from male students' perceptions. Doctoral Thesis, Durham University Durham e-Theses (2011)

Algahtani, A.Y., Rajkhan, A.A.: E-learning critical success factors during the covid-19 pandemic: a comprehensive analysis of e-learning managerial perspectives. Educ. Sci. **10**(9), 1–16 (2020)

Anderson, J.Q., Boyles, J.L., Rainie, L.: The Future Impact of the Internet on Higher Education: Experts Expect More Efficient Collaborative Environments and New Grading Schemes; They Worry about Massive Online Courses, the Shift Away from On-Campus Life. Pew Internet & American Life Project (2012)

Aparicio, M., Bação, F., Oliveira, T.: An e-learning theoretical framework. J. Educ. Technol. Syst. **19**(1), 292–307 (2016)
Arkorful, V., Abaidoo, N.: The role of e-learning, advantages and disadvantages of its adoption in higher education. Int. J.Instruct. Technol. Distance Learn. **12**(1), 29–42 (2015)
Babu, G.S., Sridevi, K.: Importance of e-learning in higher education: a study. Int. J. Res. Culture Soc. **2**(5), 84–88 (2018)
Bali, S.T., Liu, M.C.: Students' perceptions toward online learning and face-to-face learning courses. J. Phys. Conf. Ser. **1108**(1), 1–7 (2018)
Benjamin, S., Salonen, R.V., Gearon, L., Koirikivi, P., Kuusisto, A.: Safe space, dangerous territory: young people's views on preventing radicalization through education—perspectives for pre-service teacher education. Educ. Sci. **11**(5), 1–18 (2021)
Berga, K.-A., et al.: Blended learning versus face-to-face learning in an undergraduate nursing health assessment course: a quasi-experimental study. Nurse Educ. Today **96**, 1–6 (2021)
Collins, A., Halverson, R.: Rethinking Education in the Age of Technology: The Digital Revolution and Schooling in America. Teachers College Press (2018)
Edara, I.R.: Importance of holistic life education amid technology paradox. Int. J. Res. Stud. Educ. **10**(1), 1–12 (2021)
Fortune, M.F., Spielman, M., Pangelinan, D.T.: Students' perceptions of online or face-to-face learning and social media in hospitality, recreation and tourism MERLOT J. Online Learn. Teach. **7**(1), 1–16 (2011)
Govender, D.W., Kayode, A.E.: Examining availability and frequency use of computer-based technology resources among students in Nigerian Universities. PONTE. Academic J. **76**(4) (2020).https://doi.org/10.21506/j.ponte.2020.4.26.
Gui, M., Büchi, M.: From use to overuse: digital inequality in the age of communication abundance. Soc. Sci. Comput. Rev. **39**(1), 3–19 (2021)
Kayode, A.E.: Examining computer-based technology skill and academic performance of students in Nigerian universities. Doctoral Thesis, University of KwaZulu-Natal. University of KwaZulu-Natal ResearchSpace (2019)
Lazarevic, B., Bentz, D.: Student perception of stress in online and face-to-face learning: the exploration of stress determinants. Am. J. Distance Educ. **35**(2), 1–14 (2020)
Mazana, M.Y., Montero, C.S., Casmir, R.O.: Assessing students' performance in mathematics in Tanzania: the teacher's perspective. Int. Electron. J. Math. Educ. **15**(3), 1–28 (2020)
McGee, P., Windes, D., Torres, M.: Experienced online instructors: beliefs and preferred supports regarding online teaching. J. Comput. High. Educ. **29**(2), 331–352 (2017). https://doi.org/10.1007/s12528-017-9140-6
Moreno-Guerrero, A.-J., Aznar-Díaz, I., Cáceres-Reche, P., Alonso-García, S.: E-learning in the teaching of mathematics: an educational experience in adult high school. Mathematics **8**(5), 1–16 (2020)
Nafukho, F.M., Chakraborty, M.: Strengthening student engagement: what do students want in online courses? Eur. J. Train. Develop. **38**(9), 782–802 (2014)
Nambiar, D.: The impact of online learning during Covid-19: Students' and teachers' perspective. Int. J. Indian Psychol. **8**(2), 783–793 (2020)
Petretto, D.R., et al.: The use of distance learning and e-learning in students with learning disabilities: a review on the effects and some hint of analysis on the use during Covid-19 outbreak. Clin. Pract. Epidemiol. Ment. Health **17**, 92–102 (2021)
Rahman, M.A., Farooqui, Q.A., Sridevi, K.: IoT based comprehensive approach towards shaping smart classrooms. In: 2021 Fifth International Conference on I-SMAC (IoT in Social, Mobile, Analytics and Cloud) (I-SMAC), pp. 103–109. IEEE (2021)
Rakic, S., Pavlovic, M., Softic, S., Lalic, B., Marjanovic, U.: An evaluation of student performance at e-learning platform. In: 2019 17th International Conference on Emerging Elearning Technologies and Applications (ICETA), pp. 681–686. IEEE (2019)

Rathor, D.: Students' and teachers' perspectives on the impact of online education during Covid-19. Soc. Sci. J. Adv. Res. **2**(1), 1–5 (2022)

Reed, H.C., Drijvers, P., Kirschner, P.A.: Effects of attitudes and behaviours on learning mathematics with computer tools. Comput. Educ. **55**(1), 1–15 (2010)

Saeed Al-Maroof, R., Alhumaid, K., Salloum, S.: The continuous intention to use e-learning, from two different perspectives. Educ. Sci. **11**(1), 1–20 (2020)

Serjeantson, R.: Francis Bacon and the "interpretation of nature" in the late Renaissance. Isis **105**(4), 681–705 (2014)

Srinivasan, M., Dineshan, J., Ramappa, S.: Covid-19 and online education: digital inequality and other dilemmas of rural students in accessing online education during the pandemic. World Media J. Russian Media J. Stud. **4**, 34–54 (2021)

Valverde-Berrocoso, J., Garrido-Arroyo, M.del C., Burgos-Videla, C., Morales-Cevallos, M.B.: Trends in educational research about e-learning: a systematic literature review (2009–2018). Sustainability **12**(12), 1–23 (2020)

Wellington, R.J.O., Clarence, A.U.: Benefits of e-learning method as a pedagogical technique for secondary school education in nigeria in the face of covid-19 pandemic. J. Educ. Plan. Admin. **6**, 93 (2021)

Wojciuk, A., Górny, A.: Empires of Knowledge in International Relations: Education and Science as Sources of Power for the State. Routledge (2018)

Dynamics of Silica Nanofluid Under Mixed Electric Field Effect

R. N. A. Akoto[1,2(✉)], H. Osei[2], E. N. Wiah[3], and S. Ntim[4]

[1] School of Graduate Studies, University of Professional Studies, Accra, Ghana
nii.ayitey-akoto@upsamail.edu.gh
[2] Department of Petroleum and Natural Gas Engineering, University of Mines and Technology, Tarkwa, Ghana
[3] Department of Mathematical Sciences, University of Mines and Technology, Tarkwa, Ghana
[4] Institut für Physik, Johannes Gutenberg –Universität Mainz, Mainz, Germany

Abstract. Purpose: We study the motion of Silica nanoparticles in water subjected to superimposed a.c. driven field and a bias d.c. field.

Design/Methodology/Approach: We analytically solve the equation of translational motion of the silica nanoparticles in water subjected to bias dc and ac driven fields. The solutions of the equation of motion are confirmed with Molecular Dynamics simulations.

Findings: The results from the study shows that the motion of silica nanoparticles can be controlled by the use of mixed electric fields. We also show that, beyond the Brownian motion, the size of the nanoparticle, external electric field amplitude and frequency and surface charge are the main parameters that control the particle's motion within the fluid.

Research Limitation/Implications: This work considers the effect of the dielectrophoretic force on the motion of the nanoparticle to be negligible.

Practical Implication: The knowledge advanced in this work will afford sustainable nanomanufacturing application such as electrodepositing and the development of smart coolants with controllable thermal properties. An innovative application in enhanced recovery processes in the petroleum industry is also a possibility.

Social Implication: This work comes as an advantage to the nanomanufacturing and petroleum industry, as the knowledge advanced in this work lays basis for the sustainable design and manufacturing of various materials such as special paints, coolants, adhesives, petroleum reservoir drilling and injection fluid etc.

Originality/Value: The novelty of this work lies in the fact that, the mechanism of ceramic nanoparticle motion beyond Brownian limits using d.c./a.c fields is critically lacking for sustainable industrial applications.

Keywords: Brownian motion · Dynamics · Electric field · Nanofluid · Silica

C. Aigbavboa et al. (Eds.): ARCA 2022, *Sustainable Education and Development – Sustainable Industrialization and Innovation*, pp. 1088–1098, 2023.
https://doi.org/10.1007/978-3-031-25998-2_84

1 Introduction

Nanofluids are engineered suspensions of nanometer sized colloids in a fluid. These colloidal particles could be prepared from carbon nanotubes, carbides, oxides or metals with distilled water, oils or ethylene glycol as base fluids. These fluids have unique electrical, magnetic, transport and thermophysical properties for which reasons tremendous research interest have been seen recently. Amongst all the speculations aiming to explain such remarkable properties, that for the thermal conductivity has been widely accepted to be as a result of the surface charge states of the particles (Antoun et al. 2021; Lee et al. 2006; Lenin et al. 2021). The understanding of the interactions that occur amongst nanoparticles and that which exist between nanoparticles and their base fluids are key to explaining the other fascinating properties of nanofluids.

Nanofluids like all colloidal suspensions in the steady state may have some interesting and hidden characteristics which are unmasked upon agitation. It is therefore not surprising to see many experimental and numerical works in the quest to either disturb the fluid environment with external forces to investigate new features of the fluids (Liu et al. 2008; Min et al. 2005; Nawaz et al. 2018; Patriarca and Sodano 2017; Yakovlev et al. 2018; Wakif et al. 2018; Wei and Wang 2018). The transport properties of nanofluids have been an essential component for many proposed nanotechnologies including nanomanufacturing. Although the Brownian motion is the single most advanced classical theory to explain the transport of nanoparticles in a nanofluid, an understanding of effective mechanisms for the generation and control of motion to required speeds and directionality for practicality is lacking. The application of external electrical and magnetic forces has been used to generate nanoparticle motion beyond the Brownian limits and to also achieve stable particle dispersion (Dehkordi et al. 2020; Koutras et al. 2020; Ismail et al. 2020; Wei and Wang 2018; Espinoza Ortiz et al. 2016). These proposed methods largely affords limited control over the motion of particles.

The mechanism of the transport phenomena of nanoparticle motion generation and control by external forces, particularly electrical forces, has not been fully explained despite many attempts. The therefore the need to understand the actual molecular phenomena that gives rise to the physical manifestations of the transport phenomena of nanoparticles in nanofluids under external forces. In this paper we demonstrate analytically and compare results with a pseudo-experimental result via Molecular Dynamics (MD) simulation. Two phenomena of the motion SiO_2 nanoparticles in water that are exposed to the following cases of electric fields;

i. exposure to an external electric field i.e. $E(t) = E_0\cos(\omega t)$
ii. exposure to an a.c. and d.c. electric field i.e. $E(t) = E_{dc} + E_0\cos(\omega t)$.

We further study the motion dependence on nanoparticle dimension and temperature of the nanofluid. The study makes use of SiO_2-water nanofluid because, it belongs to a class ceramic nanofluids that are widely known to less costly, chemically stable in solutions and tolerate high temperatures (Bhardwaj et al. 2021; Nemade and Waghuley 2016).

2 Methodology

2.1 Analytical Model

Consider suspensions of SiO_2 nanoparticles in an isothermal incompressible base fluid such as water under the influence of external electric fields. A conceptual scheme is shown in Fig. 1. The motion of the nanoparticles can be described using the General Langevin Equation (GLE) (Medved et al. 2020) and given as.

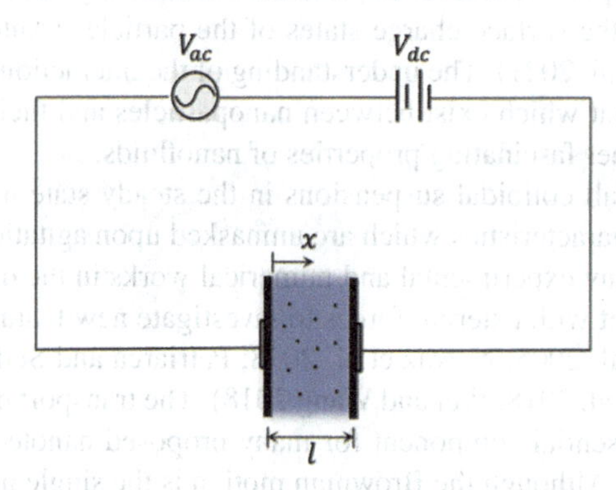

Fig. 1. SiO_2 nanoparticle suspended in water and under electric field influence.

$$m\frac{d\boldsymbol{v}(x)}{dt} = \boldsymbol{F}_{drag} + \sum \boldsymbol{F} \tag{1}$$

where, m is the mass of a single nanoparticle, $\boldsymbol{v}$ is the velocity of the nanoparticle, t is time elapsed and $\boldsymbol{F}_{drag}$ is the hydrostatic resistive force also known as the drag force. The second term of the equation represent random forces which encapsulates forces arising from thermal fluctuations in the nanofluid or Brownian motion of the nanoparticles, denoted by $\boldsymbol{\beta}$, and other external forces. For this study, the presence of an external electric force, $\boldsymbol{F}_e$, will imply a presence of the dielectrophoretic force. The dielectrophoretic force is only dominant in the presence of nonuniform electric fields and responsible for the movement of nanoparticles towards high or low electric field strength regions (Green and Nili 2012). This force is much prominent when the particle dimension is large enough to cause polarization near the electrodes due to electric field distortion that occurs between the ends of the channels (Cao et al. 2019; Imasato and Yamakawa 2008). However, the size of nanoparticles is in the order of 10^9 and the effect of the dielectric force will not be felt. As such, we ignore the effects of the dielectrophoretic force. Equation 1 is then expressed as

$$m\frac{d\boldsymbol{v}}{dt} = \boldsymbol{F}_{drag} + \boldsymbol{\beta} + \boldsymbol{F}_{Dep} + \boldsymbol{F}_e \tag{2}$$

The hydrostatic drag force is a deterministic resistance to motion of nanoparticles and expressed as

$$\boldsymbol{F}_{drag} = -\eta \boldsymbol{v} \tag{3}$$

where η is friction coefficient and is related to the viscosity of the nanofluid, μ and the Brownian relaxation time τ_B by $\eta = 6\pi\mu r = m/\tau_B$. r denotes the radius of a single nanoparticle. The Brownian force induces random forces on the nanoparticle due to thermal fluctuations owing to the size of the nanoparticle and given as (Azimi and Kalbasi 2017)

$$\boldsymbol{\beta} = \left(\frac{12\pi\mu r k_B T}{\Delta\tau}\right)^{\frac{1}{2}} \boldsymbol{\xi} \tag{4}$$

Here k_B represents the Boltzmann constant and the temperature of the nanofluid is denoted by T. ξ is the Gaussian vector and $\Delta\tau \gg \tau_B$. The action of the external electric forces on the nanoparticle suspension produces an interaction between the nanoparticles such that the external electric force acts as a body force on the nanoparticle's suspension per unit volume as (Sjöblom et al. 2021) given by

$$\boldsymbol{F}_e = e\boldsymbol{E}(t) \tag{5}$$

where $\boldsymbol{E}(t)$ is the time varying electric field as a function of the separation of the electrode plates, l and the surface charge, e, is given as

$$e = \pi\varepsilon\zeta r^2\left(\frac{1}{r} + \frac{1}{\lambda_D}\right). \tag{6}$$

From Eq. 6, ε denotes the dielectric constant of the nanofluid, ζ, the Zeta potential, and the electric double layer around the nanoparticle is λ_D. In this work, an a.c. signal is superimposed on a steady d.c. signal to produce a signal that vary in amplitude while its polarities either remain unchanged or changes asymmetrically. This signal mix can be achieved experimentally by a method called "coupling" through the connection of mutual inductors and capacitors. The coupling of a.c. and d.c. electric fields can simply be represented mathematically as

$$E(t) = E_{dc} + E_0 \cos(\omega t) \tag{7}$$

where E_{dc} is the constant field and E_o is the amplitude of the varying field with frequency, ω. It is expected that the electric field is induced between the electrodes or within the nanofluid set up which then influences the motion of the nanoparticles in the x–direction. The combined voltage which gives rise to the field is inversely proportional to the separation of the electrodes. Combining Eqs. 2–7, the 1-dimensional equations of motion of the nanoparticle in a nanofluid exposed to an external electric field, is expressed as

$$\frac{dv_x}{dt} + \frac{\eta}{m}v_x = \frac{\beta + eE(t)}{m}, \quad v_x(t=0) = 0 \tag{8}$$

with a standard solution (Di Terlizzi et al. 2020)

$$v_x(t) = v_x(0)e^{-\eta t/m} + \frac{1}{m}\int_0^t \left[\beta + eE(t')\right]e^{-\eta(t-t')/m}dt'. \tag{9}$$

Solving Eq. 8 via Eq. 9 gives

$$v_x(t) = \frac{\beta + eE_{dc}}{\eta} + \frac{\eta e E_0}{m^2\omega^2 + \eta^2}\left[\cos(\omega t) + \frac{m\omega}{\eta}\sin(\omega t)\right] - \left(\frac{\beta + eE_{dc}}{\eta} + \frac{\eta e E_0}{m^2\omega^2 + \eta^2}\right)e^{-\eta t/m} \tag{10}$$

Equation 10 reduces to a Brownian motion when $E_{dc} = E_0 = 0$, and the influence of a pure d.c. limit is established at $E_0 = 0$. For one to visualize the effect a monoharmonic wave, the d.c. wave would have to be removed via $E_{dc} = 0$.

2.2 Molecular Dynamics Simulation

We performed a pseudo-experiment, specifically MD simulations, to confirm the results of our study. Our system was made of a 19.887 × 19.887 × 20.767 Å^3 SiO_2-Cristobalite crystal solvated with 8, 605 molecules of water. Energy minimisation was performed to remove unphysical contacts after which a 1.6 ns NpT equilibration was run at 1 atm using a Nosé-Hoover barostat, allowing box fluctuations only in the x and y which were coupled. Figure 2 shows the simulation system.

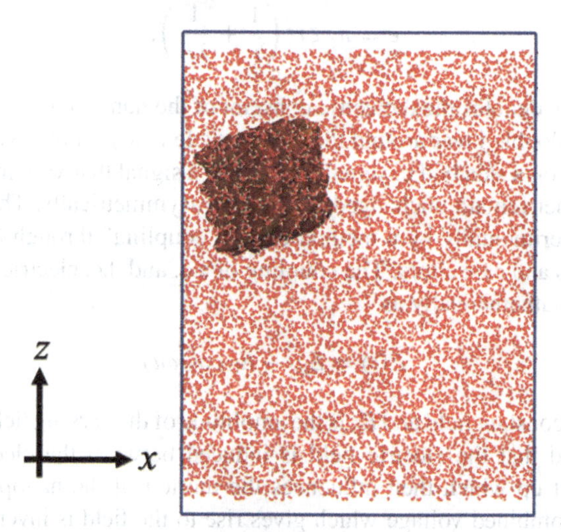

Fig. 2. Simulation box of SiO_2 nanoparticle in TIP3P water (red and white dots). The gaps between the water and the simulation boundaries in the z-direction are due to the Lennard-Jones walls.

2-D periodic boundary conditions were used, where flat walls that interact with the rest of the system via Lennard-Jones 12/6 potential were placed at the lower and upper bounds of the z-component of the simulation box. The Lennard-Jones potential parameters for gold (Heinz et al. 2008) were used for the walls with a cut-off of 2.5 Å. The TIP3P water potential was used for the water (Jorgensen et al. 1983) while the bonded and non-bonded parameters for silica were taken from Cruz-Chu et al. (2006). Lennard-Jones interactions were cut off at 15 Å. For Lennard-Jones interactions whose parameters were not explicitly defined, the Lorentz-Berthelot combination rules were employed. Long range Coulomb interactions were solved using the Particle-Particle Particle-Mesh solver with 1×10^{-4} relative error in the forces. Due to the slab geometry of our system (2-D periodicity), corrections were made to the 3-D PPPM solver (Yeh and Berkowitz 1999) using vacuum 3 times the box size in the z-direction. The simulation was run at a timestep of 1 fs, collecting system information every 1000 steps. Temperature was controlled via a 3-chain Nosé-Hoover chain thermostat set to 300 K. After equilibration, a 10 ns NVT production run was then performed, with the electric field turned on according to Eq. 7, along the z-axis where $E_{dc} = 1 \times 10^{-5}$ V/Å, $E_0 = 1$ V/Å and $\omega = 2 \times 10^3 \pi$ rad/ps. All simulations were performed using the LAMMPS simulation package.

2.3 Model Validation

The velocity model in Eq. 10 was validated against the MD simulation with input data shown in Table 1. The model verification was performed for four different cases – in the absence of electric fields, the effect of d.c. field only, the effect of an a.c. field only and the combined effect of both a.c. and d.c. fields. Figures 3, 4, 5 and 6 show centre of mass (COM) velocity comparisons of the SiO_2 nanoparticle, due to the absence and presence an external field. The magnitude of the Gaussian vector used in Eq. 4 is chosen to be in the order of 10^{-3} for fluid temperatures ranging from 300–310 K (Dhlamini et al. 2022; Li and Nie 2018; Li et al. 2016).

Table 1. Input parameters for model verification

Parameter	Value	Parameter	Value
μ(kg/Ås)	1.00×10^{-13}	λ_D(Å)	50
r(Å)	20	T(K)	303
m(kg)	1.96×10^{-23}	l(Å)	60.056
ε(C/V Å)	7.00×10^{-20}	E_{dc}(V/Å)	1.00
ζ(mV)	60	E_0(V/Å)	1.00×10^{-5}
ω(rad/ps)	$2 \times 10^3 \pi$		

Comparison between the theorical model of Eq. 10 and the pseudo-experimental results show evidence of a consistent phenomenon and close agreement is also observed (Figs. 3, 4, 5 and 6).

Therefore, the model derived can be used to predict the COM velocity profile at any given conditions.

3 Results and Discussion

The proposed model is utilized to investigate the influence of the radius of nanoparticles, Zeta potential, temperature and amplitude and frequency of electric field on the COM

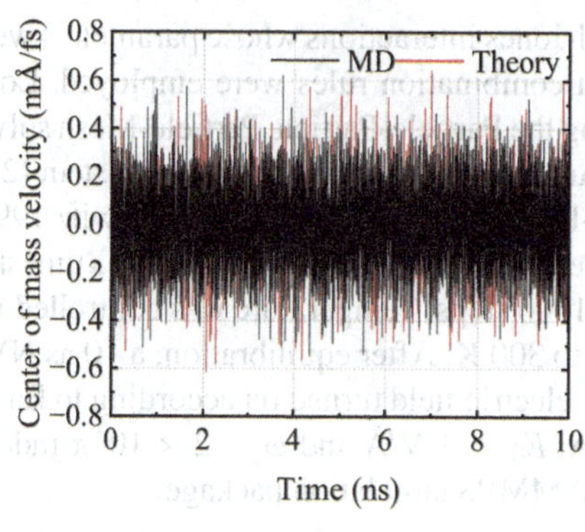

Fig. 3. COM velocity in Brownian theory.

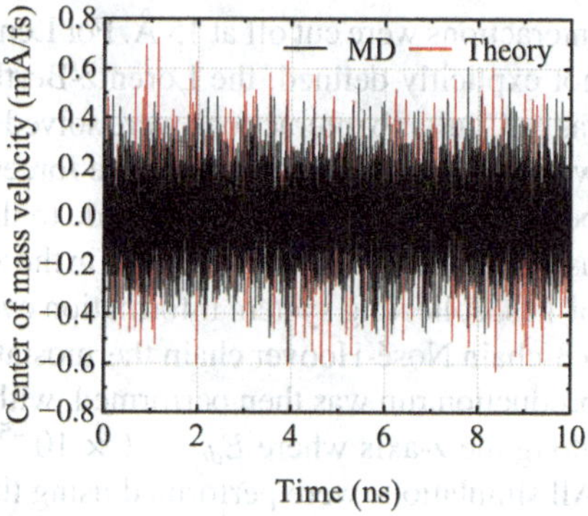

Fig. 4. COM velocity under pure d.c. limit.

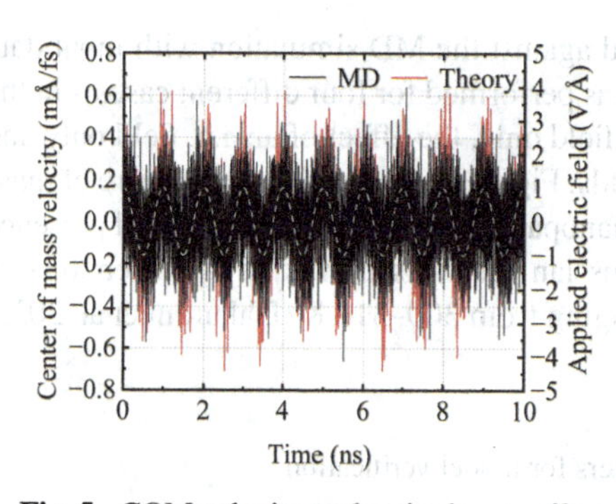

Fig. 5. COM velocity under single a.c. effect

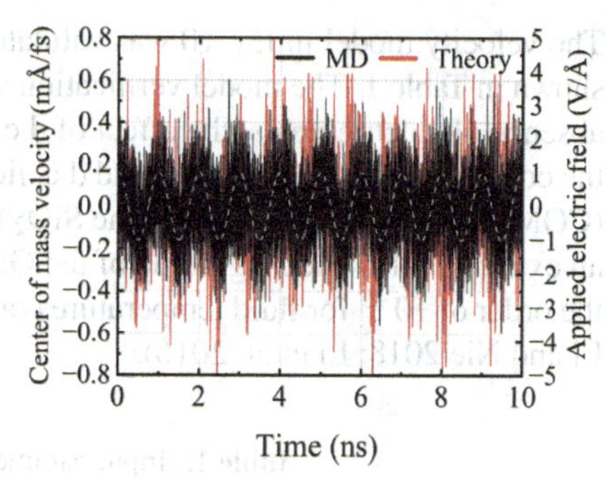

Fig. 6. COM velocity under a.c.-d.c. mix fields effect

velocity of SiO_2 nanoparticles. It is important to clarify that the Savitzky – Golay filter is used in order to smooth the COM velocity signal. In this work, the filter is not discussed in detail, but Persson and Strang (2003) and Ostertagová and Ostertag (2016) give an extensive overview. Prior to predicting, it should be noted that the presence of the d.c. field does not change the shape of the velocity profile. In fact, it appears from Figs. 3, 4, 5 and 6 that the presence of the d.c. field provides additional energy to the nanoparticles besides Brownian motion by shifting the cosnusoidal profile upward.

In Fig. 7, COM velocity is plotted against time for different amplitude ratios of applied a.c. and d.c. fields. A cosnusoidal waveform is evident in the velocity profile due to the a.c. signal. Higher displacements are observed as the ratio E_{dc}/E_0 increases with no peculiar change in the amplitudes of the velocity profile in the case where $E_{dc} = E_0$. To increase peaks and deepen troughs in a COM velocity profile, one has to increase the amplitude ratio.

The curves in Fig. 8 demonstrate the dependence of COM velocity profile on the frequency of the applied a.c. field. An increase in the frequency of the a.c. field promotes the generation of more velocity cycles beyond the Brownian limits. This is in agreement with results obtained by Min et al. (2005). However at low frequencies the motion of the nanoparticle is predomninatly Brownian.

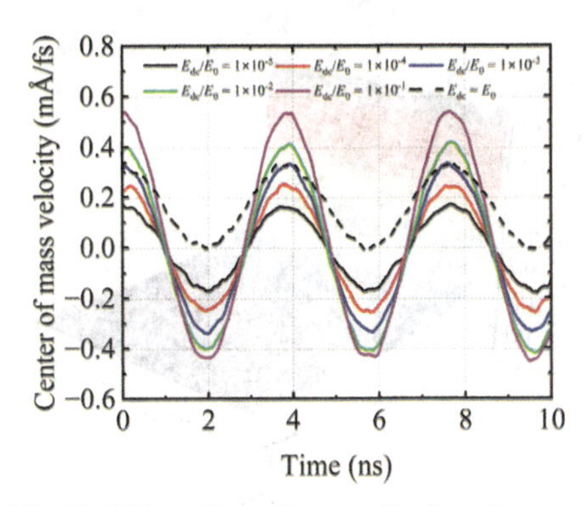

Fig. 7. Effect of a.c.-d.c. amplitude ratio ($\omega = 3 \times 10^{-2}\pi$ rad/ps, $\zeta = 60$ mV).

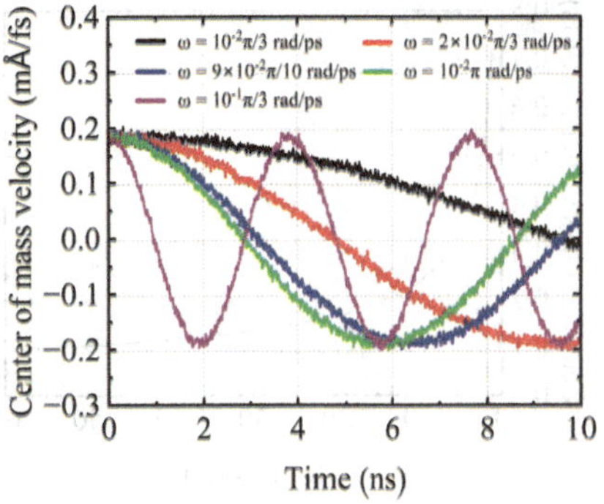

Fig. 8. Effect of frequency on COM velocity. ($E_{dc}/E_0 = 1 \times 10^{-5}$, $\zeta = 60$ mV).

The effect of the surface charge on the COM velocity is depicted in Fig. 9. It is observed that the motion of the nanoparticle increases linearly with the Zeta potential and that cosnusoidal motion become more pronounced at higher magnitudes of the Zeta potential.

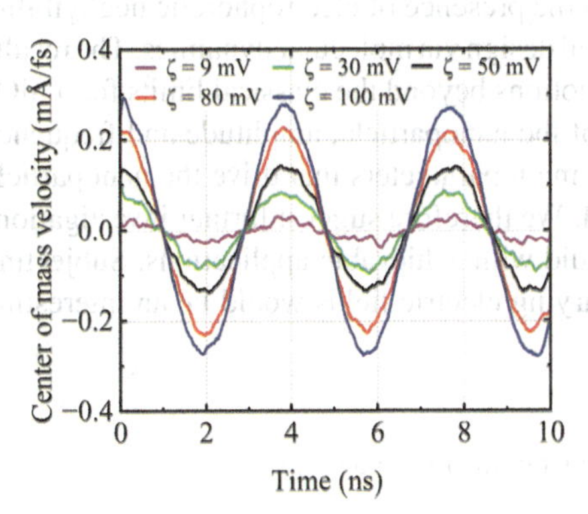

Fig. 9. Effect of Zeta potential on COM velocity ($\omega = 3 \times 10^{-2}\pi$ rad/ps, $E_{dc}/E_0 = 1 \times 10^{-5}$)

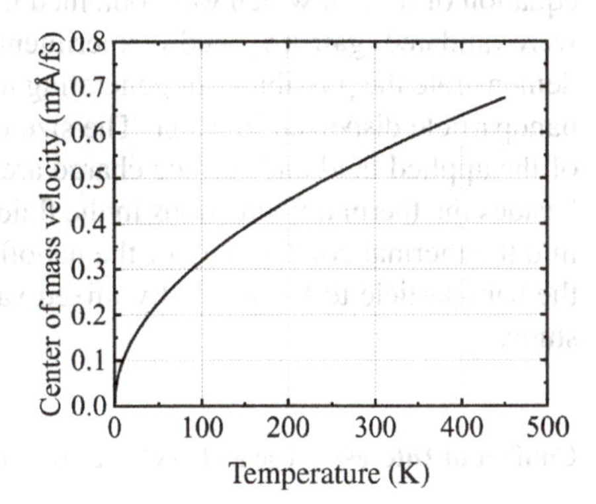

Fig. 10. Effect of fluid temperature COM velocity.

We also show from Fig. 10 with reference to Eq. 10, that the COM velocity dependence on temperature solely emanates from Brownian theory. As such an increase in temperature increases the motion due to thermal fluctuations in the nanofluid, which is in accordance with classical theory.

The size of the nanoparticle is a function of the motion of the nanoparticle. Figures 11a and b show 2D and 3D plots of the variation of the nanoparticle size with respect to its motion.

The curve depicts a nonlinear relationship such that the motion of the particle varies by the reciprocal of the root of its size. This observation has also been reported by Min et al. (2005). This understanding is essential in the preparaton of nanofluids to afford dispersion stability.

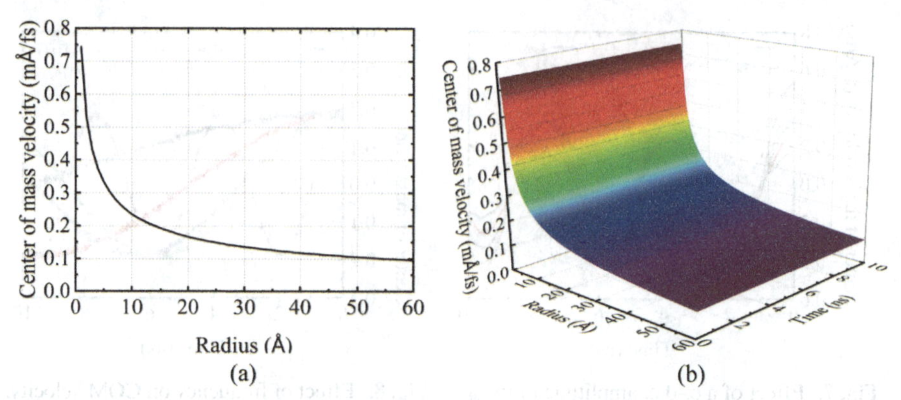

Fig. 11. Effect of particle radius on COM velocity ($\omega = 3 \times 10^{-2}\pi$ rad/ps, $E_{dc}/E_0 = 1 \times 10^{-5}$, $\zeta = 60$ mV): (a) 2D plot; (b) 3D plot.

4 Conclusion

The translational motion of a silica nanoparticle suspended in water and constrained by a mixed a.c. and d.c. fields has been analytically derived and solved. Solutions of the equation of motion which were obtained in the presence of electrophoretic negligibility were validated against a pseudo-experimental design via molecular dynamics. The results demonstrate the possibility of generating motions beyond the classical limits for a SiO_2 nanoparticle dispersed in water. The size of the nanoparticle, amplitude and frequency of the applied field and surface charge are main parameters that drive the nanoparticle besides the thermal fluctuations in the fluid. We therefore suggest further investigations into the thermal conductivity of the nanofluid with achievable applications. Subjecting the nanoparticle to harmonically mixed varying electric fields would be an interesting study.

Conflict of Interest. The authors have no conflict of interest to declare.

References

Antoun, S., Srinivasan, S., Saghir, M.Z.: A refined molecular dynamics approach to predict the thermophysical properties of positively charged alumina nanoparticles suspended in water. Int. J. Thermofluids **12**, 100114 (2021). https://doi.org/10.1016/j.ijft.2021.100114

Azimi, S.S., Kalbasi, M.: A molecular dynamics simulation of Brownian motion of a nanoparticle in a nanofluid. Nanoscale Microscale Thermophys. Eng. **21**(4), 263–277 (2017). https://doi.org/10.1080/15567265.2017.1286420

Bhardwaj, P., Singh, B., Behera, S.P.: Green approaches for nanoparticle synthesis: emerging trends. In: Kumar, R.P., Bharathiraja, B. (eds.) Nanomaterials: Application in Biofuels and Bioenergy Production Systems, pp. 167–193. Academic Press, Cambridge (2021). https://doi.org/10.1016/B978-0-12-822401-4.00015-5

Cao, W., Chern, M., Dennis, A.M., Brown, K.A.: Measuring nanoparticle polarizability using fluorescence microscopy. Nano Lett. **19**, 5762–5768 (2019). https://doi.org/10.1021/acs.nanolett.9b02402

Cruz-Chu, E.R., Aksimentiev, A., Schulten, K.: Water−silica force field for simulating nanodevices. J. Phys. Chem. B **110**(43), 21497–21508 (2006). https://doi.org/10.1021/jp063896o

Dehkordi, R.B., Toghraie, D., Hashemian, M., Aghadavoudi, F., Akbari, M.: Molecular dynamics simulation of ferro-nanofluid flow in a microchannel in the presence of external electric field: effects of Fe_3O_4 nanoparticles. Int. Commun. Heat Mass Transf. **116**, 104653 (2020). https://doi.org/10.1016/j.icheatmasstransfer.2020.104653

Dhlamini, M., Mondal, H., Sibanda, P., Motsa, S.: Activation energy and entropy generation in viscous nanofluid with higher order chemically reacting species. Int. J. Ambient Energy **43**(1), 1495–1507 (2022). https://doi.org/10.1080/01430750.2019.1710564

Di Terlizzi, I., Ritort, F., Baiesi, M.: Explicit solution of the generalised langevin equation. J. Stat. Phys. **181**(5), 1609–1635 (2020). https://doi.org/10.1007/s10955-020-02639-4

Espinoza Ortiz, J.S., Bauke, F.C., Lagos, R.E.: A Langevin Approach to a classical Brownian oscillator in an electromagnetic field. J. Phys.: Conf. Ser. **738**(1), 1–5 (2016). https://doi.org/10.1088/1742-6596/738/1/012032

Godin, B., et al.: Dielectrophoresis. In: Bhushan, B. (ed.) Encyclopedia of Nanotechnology, pp. 534–543. Springer Netherlands, Dordrecht (2012). https://doi.org/10.1007/978-90-481-9751-4_131

Heinz, H., Vaia, R.A., Farmer, B.L., Naik, R.R.: Accurate simulation of surfaces and interfaces of face-centered cubic metals using 12–6 and 9–6 lennard-jones potentials. J. Phys. Chem. C **112**(44), 17281–17290 (2008). https://doi.org/10.1021/jp801931d

Imasato, H., Yamakawa, T.: Measurement of Dielectrophoretic force by employing controllable gravitational force. J. Electrophor. **52**(1), 1–8 (2008). https://doi.org/10.2198/jelectroph.52.1

Ismail, H., Sulaiman, M.Z., Aizzat, M.A.: Qualitative investigations on the stability of Al2O3-SiO2 hybrid water-based nanofluids. IOP Conf. Ser.: Mater. Sci. Eng. **788**(1), 1–7 (2020). https://doi.org/10.1088/1757-899X/788/1/012091

Jorgensen, W.L., Chandrasekhar, J., Madura, J.D., Impey, R.W., Klein, M.L.: Comparison of simple potential functions for simulating liquid water. J. Chem. Phys. **79**(2), 926–935 (1983). https://doi.org/10.1063/1.445869

Koutras, K.N., et al.: The influence of nanoparticles' conductivity and charging on dielectric properties of ester oil based nanofluid. Energies **13**(24), 6540 (2020). https://doi.org/10.3390/en13246540

Lee, D., Kim, J., Kim, B.G.: A new parameter to control heat transport in nanofluids: surface charge state of the particle in suspension. J. Phys. Chem. B **110**(9), 4323–4328 (2006). https://doi.org/10.1021/jp057225m

Lenin, R., Joy, P.A., Bera, C.: A review of the recent progress on thermal conductivity of nanofluid. J. Mol. Liq. **338**, 116929 (2021). https://doi.org/10.1016/j.molliq.2021.116929

Li, C.-C., Hau, N.Y., Wang, Y., Soh, A.K., Feng, S.-P.: Temperature-dependent effect of percolation and Brownian motion on the thermal conductivity of TiO2–ethanol nanofluids. Phys. Chem. Chem. Phys. **18**, 15363–15368 (2016). https://doi.org/10.1039/C6CP00500D

Liu, Y., et al.: Manipulation of nanoparticles and biomolecules by electric field and surface tension. Comput. Methods Appl. Mech. Eng. **197**(25–28), 2156–2172 (2008). https://doi.org/10.1016/j.cma.2007.08.012

Li, Q., Nie, D.: Preferential Brownian motion induced by a fluid with inhomogeneous temperature. IOP Conf. Ser.: Mater. Sci. Eng. **417**(1), 1–5. (2018). https://doi.org/10.1088/1757-899X/417/1/012030

Medved, A., Davis, R., Vasquez, P.A.: Understanding fluid dynamics from Langevin and Fokker-Planck equations. Fluids **5**(40), 1–23 (2020). https://doi.org/10.3390/fluids5010040

Min, J.Y., Jang, S.P., Choi, S.U.: Motion of nanoparticles in nanofluids under an electric field. In: ASME 2005 International Mechanical Engineering Congress and Exposition. Heat Transfer, Part B, Orlando, pp. 497–501 (2005). https://doi.org/10.1115/IMECE2005-80139

Nawaz, M., Rana, S., Qureshi, I.H.: Computational fluid dynamic simulations for dispersion of nanoparticles in a magnetohydrodynamic liquid: a Galerkin finite element method. RSC Adv. **8**, 38324–38335 (2018). https://doi.org/10.1039/C8RA03825B

Nemade, K., Waghuley, S.: A novel approach for enhancement of thermal conductivity of CuO/H2O based nanofluids. Appl. Therm. Eng. **95**, 271–274 (2016). https://doi.org/10.1016/j.applthermaleng.2015.11.053

Ostertagová, E., Ostertag, O.: Methodology and application of Savitzky-Golay moving average polynomial smoother. Glob. J. Pure Appl. Math. **12**(4), 3201–3210 (2016)

Patriarca, M., Sodano, P.: Classical and quantum Brownian motion in an electromagnetic field. Fortschritte der Phys. **65**, 6–8 (2017). https://doi.org/10.1002/prop.201600058

Persson, P.-O., Strang, G.: Smoothing by Savitzky-Golay and Legendre Filters. In: Rosenthal, J., Gilliam, D.S. (eds.) Mathematical Systems Theory in Biology, Communications, Computation, and Finance, pp. 301–315. Springer, New York (2003). https://doi.org/10.1007/978-0-387-21696-6_11

Sjöblom, J., Mhatre, S., Simon, S., Skartlien, R., Sørland, G.: Emulsions in external electric fields. Adv. Colloid Interface Sci. **294**, 1–23 (2021). https://doi.org/10.1016/j.cis.2021.102455

Wakif, A., Boulahia, Z., Sehaqui, R.: A semi-analytical analysis of electro-thermo-hydrodynamic stability in dielectric nanofluids using Buongiorno's mathematical model together with more realistic boundary conditions. Results Phys. **9**, 1438–1454 (2018). https://doi.org/10.1016/j.rinp.2018.01.066

Wei, W., Wang, Z.: Investigation of magnetic nanoparticle motion under a gradient magnetic field by an electromagnet. J. Nanomater. **2018**, 1–5 (2018). https://doi.org/10.1155/2018/6246917

Yakovlev, E.V., et al.: Colloidal suspensions in external rotating electric field: experimental studies and prospective applications in physics, material science, and biomedicine. In: Tuchin, V.V., Genina, E.A., Derbov, V.L., Postnov, D.E. (eds.) Saratov Fall Meeting 2017: Optical Technologies in Biophysics and Medicine XIX, SPIE, Saratov, pp. 1–7 (2018). https://doi.org/10.1117/12.2315134

Yeh, I.-C., Berkowitz, M.L.: Ewald summation for systems with slab geometry. The J. Chem. Phys. **111**(7), 3155–3162 (1999). https://doi.org/10.1063/1.479595

Production and Marketing Strategies by Youth Vegetable Farmers in Urban Settlements, Tanzania

A. E. Maselle[1(✉)], D. L. Mwaseba[2], and C. Msuya-Bengesi[2]

[1] Institute of Development Studies, The University of Dodoma, Dodoma, Tanzania
atshalla_2000@yahoo.com

[2] Department of Agricultural Extension and Community Development, Sokoine University of Agriculture, Morogoro, Tanzania

Abstract. Purpose: This paper evaluates the strategies used by youth in vegetable farming for sustainable livelihood and growth in urban settings.

Design/Methodology/Approach: A multiple case studies design was adopted to explore the differences within and between cases.The approach allows the researcher to augur either contradicting findings for expected reasons or similar findings across cases. In selecting the participants for the study, a two-stage sampling technique was adopted. The four youth vegetable farmers who served as cases in this paper were selected through maximum variation purposive sampling procedure. Moreover, 9 FGDs and 14 KIs interviews were conducted. The transcripts from FGDs, KI and indepth were transcribed and coded into emerging themes. The content analysis techniques which were informed by the Interpretative phenomenological Approach (IPA) was employed for data analysis.

Findings: Research findings indicate that the strategies used in production and marketing of vegetable are interlinked. As such, the strategies used in production are conditioned by marketing strategies.However, proper timing of production season and producing crops that are beyond the farmers' comfort zone for attracting market were among the strategies which are linked to successful performance in the business.

Research Limitation/ Implications: The current study focused on production and maketing strategies by youth vegetable farmers in Dodoma city. This paper therefore concentrated mainly on crops with a short incubation period in urban settings rather than perennial or annual crops.

Practical Implication: The knowledge advanced in this study will inform the local governments and the youth on the need to borrow from the best practices examples of strategies used in vegetable marketing and production for formulating policies, interventions and support programmes for the youth.

Social Implication: The paper informs urban planners on the need to consider land for agricultural activities for the youth in particular as the informal sector in urban settlements is expected to continue and grow in Africa.

C. Aigbavboa et al. (Eds.): ARCA 2022, *Sustainable Education and Development – Sustainable Industrialization and Innovation*, pp. 1099–1108, 2023.
https://doi.org/10.1007/978-3-031-25998-2_85

Originality/Value: Novelty of this paper lies in youth livelihood opportunities in agriculture by focusing on fast-maturing crops since youth are interested in making quick money to meet their life desires.

Keywords: Vegetable farming strategies · Urban settings · Youth livelihoods in agriculture

1 Introduction

The engagement of youth in different agricultural activities such as vegetable farming, fruits and green maize production in urban areas has widely been documented (Ibidapo et al 2017; Agboola et al. 2015). It is largely contended that the agricultural sector has the possibility of creating employment opportunities for the youth and promote a sustainable livelihood provided that, most countries in Africa have ill functioning industries (Alliance for a Green Revolution in Africa (AGRA) 2015; URT 2016). However, evidence shows that the youth leave the agricultural sector prematurely in contrast to formal and other informal sectors mainly because of low financial revenues associated with the former (Leavy and Hossain 2014). Impliedly, for agricultural sector to achieve its potential in creating job opportunities for young people, the sector should first be lucrative, viable and intellectually motivating (Brooks et al. 2013, Proctor 2012). However, most of the conducted studies on youth and agriculture-focused on policies for engaging the youth (Ayele et al. 2017; Losch 2014); while other studies (Rutta 2012; Proctor *et al.* 2012) focused on the type of crops the youth are engaged within urban areas to generate quick income. Moreover, a lot of studies have been conducted on factors affecting youths' participation in agriculture (Mutua et al. 2017; Nguyen and Philippe 2016). However, the strategies employed in conducting agricultural activities have received low research interest. The study therefore has gone extra mile by analysing the strategies used by youth vegetable farmers in urban settings instead of the common focus as to why youth choose to exist agriculture prematurely.

Strategies refer to actions deliberately applied by the youth in vegetable farming inorder to attain a given performance with a projected effect from the business (Grando et al. 2016). The actions may include producing traditional farm crops efficiently, on-farm value addition and selling products to customers directly, to provide few examples. It is contented that, these actions are selected by youth vegetable farmers in particular contexts to expand farm undertakings and improve livelihoods or limiting the outcomes of possibly destructive changes (Grando et al. 2016). Nevertheless, similar conditions across geographical locations may principally not result into the selection of similar actions among the youth who are involved in vegetable farming (Brooks et al. 2013). With this view, evaluating different strategies that livelihoods are built upon is an imperative endeavour in establishing the mechanisms that enable few to improve their livelihood and the institutional aspects that may instead prevent others from attaing their goal. Accordingly, it provides a room for formulating youth policies and livelihood programmes and interventions in agriculture by drawing from cases of the best production and marketing strategies.

2 Theoratical Framework

The current study borrows from the Sustainable Livelihoods Framework (SLF). Sustainable livelihood framework is appropriate for the paper due to its significance in establishing the relationship between livelihood strategies and livelihood outcomes. The SLF identifies human, natural, financial, social and physical capitals as five of the assets in which production capabilities can be defined and livelihood formed (Ashley and Carney 1999). In particular, when individual qualities such as skills in farming, access to information and experience increase, efficiency in production increases too due to adoption of improved farming strategies and thus improving livelihoods. Furthermore, the well organized and networked the youth vegetable farmers are in viable farming schemes, the possible it becomes realizing their ambition and opportunities for engagement. However, the youth vegetable farmers are differently exposed to institutions and policies and they don't access livelihood assets equally. Consequently, these results into differences in livelihood outcomes as the choice of production and marketing strategies is determined by the resources endowed.

3 Methodology

This study was done in Ihumwa and Mtumba wards of Dodoma City[1]. Dodoma is among the fast growing cities in the country and due to that, the informal urban sectors is expected to grow and continue. Also, the opportunities for business has increased due to increased population following the government's decision to relocate it central administration from Dar es salaam to Dodoma. A combination of these factors has increased the demand of vegetable for household consumption and market.The situation attracts the youth as the city becomes strategically positioned for business. Purposive selection of Ihumwa and Mumba wards was done because these wards were potential areas in vegetable farming in the Dodoma city.

This paper used a case study design inorder to get holistic understanging of the strategies used by youth vegetable farmers to ensure sustainable livelihood and growth in the business. Qualitative case studies design provides the basis for applying solution to state of conditions by exploring the disparities between and within cases. Furthermore, case study allows the researcher to augur either similar findings across cases or contradicting findings for expected reasons (Yin 2014). The study adopted a two-stage sampling procedure to select the participants. In the first stage the production sites were purposively identified. Purposive sampling technique was also adopted in the second stage to select the participants for the study. Data were collected through in-depth interviews, focus group discussions and key informants interview. From Ihumwa and Mtumba wards nine production sites were identified and all included in the study. This was done to ensure that the discrepancies in strategies and livelihood outcomes that are linked to production areas are well captured.

A group of 9–12 youth vegetable farmers was selected in each production area for conducting focus group discussion (FGD). This is a recommendable number for obtaining meaningful information through (Barbour and Kitzinger 2011). This means nine

[1] Formerly a Municipality and only attained the City status in 2018.

FGDs were conducted for the study. To identify the most or less successful youth vegetable in area a wealth ranking exercise was done in each production site. The inclusion and exclusion criteria were; the variety of crops produced, purchasing inputs on time, to engage in production in all seasons and the use of modern farming equipments. The inclusion and exclusion criteria were established in the field.

Using the established criteria, five (5) youth farmers were recognized as successful cases in the study area and four youth farmers identifies as less successful cases. To select the four farmers who served as case studies in this paper a maximum variation purposive sampling technique was adopted. The approach was applied for ensuring that the diversity of the startegies employed by individual farmers is captured (Cresswell 2011). According to the inclusion and exclusion criteria established two cases among the four were identified as successful ones and the remaining two as less successiful. FGD and interviews checlist were prepared to guide the discussions (Rubin and Babbie 2011). Key informant interviews involved the city Agricultural Extension Officers, Horticulturalists, Agro-input dealers, Community development Officers and Ward Executive Officers.Indepth interviews with the selected cases were conducted at participants households while FGDs and KIs were conducted at participants working places.The FGDs and Interviews lasted for 45 to 90 min. For recording purposes of the the discussions the participant were consulted for consent.

The transcripts from FGDs, key informants and indepth interviews were transcribed and coded into emerging themes and sub-themes.Data analysis was done through content anlaysis techniques that was informed by interpretative phenomenological approach (IPA) in order to understand the investigated phenomenon from the farmers point of view (Mayring 2014).Pseudonyms have been used throughout the paper to make sure that the confidentiality of participants account is maintained. John and Herman were being referred to as successful cases while Daniel and Pendo less successful.

4 Results

Drawing on both cases, FGDs and KIIs, vegetable farming was mainly based on production and marketing strategies as presented in Table 1.

4.1 Discussion

As is the case in any other business three sub-themes emerged as critical issues influencing sustainable livelihood and growth under production strategies. These include: strategies used for funds accessibility, strategies for accessing inputs and the types of produced crops.

4.1.1 Strategies for Funds Accessibility

In this sub-theme two strategies were identified. The first strategy was to join farmers' organizations and the second one was to sale labour strategically. It was revealed that the youth vegetable farmers who invested in social capital had a better chance of accessing funds as a working capital. In particular, John and Herman were identified as successful

Table 1. Strategies employed by the youth in vegetable farming

Production strategies		
Funds for working capital	Access to inputs	Type of crops produced
1. Membership in FO 2. Selling labour 3. Business diversification	1. Agro dealer companies 2. Retail shops	1. Production of vegetables that needed high/low investment
Marketing strategies		
Mode of sale	Choice of vegetable	Production season
1. Online 2. Market vending 3. Farmgate	1. Rarely produced 2. Easily stored	1. Dry season 2. Rainy season

in vegetable enterprise unlike Daniel and Pendo who identified as less successful. This was due to the reason that the first two had been able to invest into non agricultural activites and gained assets that are substantial. Nevertheless, it has taken John seven years and seventeen years for Herman to excel in the business. John being a member of farmers organization (FO[2]) it was possible for him to succeed in the busisess as through the organization he acquired some funds and training on better farming practices. On the other side, no any agricultural related opportunities that was accessed by Herman but invested in other business. However, diversification has implications on finance and time that limited him from undertaking better farming practices for increased productivity.

The discrepancy observed between John and Herman livelihood outcomes could be attributed to the strategies employed for funds accessibility by each of them. Literature indicates that to financial assets is crucial for the youth because they are interested in modernized kind of farming (Akpan et al. 2015). Moreover, all participants used in this paper had land but could not use it as collateral for funds accessibility from financial institutions.This was due to the difficulties encountered in obtaining the legal rights over the owned land. Implicitly, FO serves as a means of communication to funding agencies and other development partners. Nevertheless, evidence suggests that for sustainable FOs the formulation of farming schemes' must take into account youth commitment by considering those who have established themselves as farmers. In the same vein, if farmers' livelihood is to be improved, FOs' capability to served members should be critically evaluated. As it could be learnt from Pendo who was a member of self-help group (SHG) could only receive a token amount of money as loan for working capital. This means that the income obtained through SHG could not adequately cater for participants production needs. As a result, she was using a bucket still for watering her garden even after seventeen years of farming.

[2] FO has been defined as a formal or informal membership-based collective action institution serving its members that get their entire livelihood from agriculture (FAO, 2014).

4.1.1.1 Strategic Sale of Labour

Strategic sale of labour was idendified as an important strategy for funds accessibility among the youth. It was strategically done just before production starts to make sure the income obtained is efficiently used to purchase all essential inputs for the specific season. The youth worked jointly in groups of 5 to10 members. These groups were formed according to the type of crops produced to ensure that members produced similar crops and within the same period for the inputs to be equally shared. The strategy created a good space for the participants to work in their own vegetable gardens during the production season. However, evidene show that with this strategy it might take long for the youth to realise ambition in vegetable farming. This is due to the reason that agricultural labourers generally face meagre and insecure wages. Implicitly, the vegetable farming needs could not be sufficiently met by the income genetated through this approach. This also explains why Daniel was considered one of the less successiful cases in comparison to John and Herman. In many African countries agricultural workers are normally confronted with high incidences of poverty as descent levels of income could not be guaranteed (Shoaib *et al.* 2016).

4.1.2 Production of Crops that Are Beyond "Comfort Zone"

Research results indicate that the youth vegetable farmers preferred production of crops that required minimal investment such as *Amaranthus* and Chinese cabbage. These crops are grown within a short period of time as in one to two months Chinese cabbage can mature and be picked. On the other hand, crops such as onions, zucchini and beetroots are less produced since they require three to four months of resource expenditures in terms of time, labour and income. However, evidence show that onions and other crops of the sort were high income generating vegetable comparing to the vegetable that are cited as low investment crops. This means that, the tendency of youth vegetable farmers to grow crops that need less capital investment without considering the economic capability of the crops makes farmers end up as subsisitence producers. Evidently, it takes more than producing vegetable that are within the farmers comfort zone in order to transform subsistence farming into business farming. It is also argued by other scholars that for subsistence producers the costs of investment and taste of the crops are important while for commercial producers the cost effective crops matters (Greig 2009).

4.1.3 Strategies for Accessing Inputs

Research findings identified two strategies that were used by the youth in accessing inputs.The first approach involved the agro companies while the second involved agro dealer shops. John employed the first strategy and happened to be among the successiful youth farmers in the study area. It was noted that these companies provide trainings on the appropriate use of inputs and the performance of their product is monitored in the field. As a result, the costs of production that might increase due to purchase of spurious inputs and inappropriate use of inputs is reduced. It was also revealed in a male FGDs conducted at Mtumba that in 2016 some youth farmers planted tomatoes but the plants could not bear fruits at all, farmers had to replant using different types seeds. However, the responsible government offices were consulted about the issue and it was argues;

"It is has become difficult to control for the quality and price since we are in a free market era. Instead, we have been advocating for farming schemes for farmers to access inputs from agro-companies direct, it becomes possible to hold down the responsible companies for the substandard performance of the products supplied".

Despite the reservations, evidence shows that the youth depend on agro-shop dealers for advisory services. For instance, it was revealed in a male FGD done at Ihumwa on the 11th of July 2017 that farmers were concerned with Chinese cabbage hybrid seeds. This is due to the reason that these seeds make them end up picking Chinese leaves once or twice and the plant matures unlike traditional seeds, which farmers could pick the leaves up to four times (cycles). The consulted input dealers were not able to provide a better clarification for the scenario. Obviously, farmers did not understand that crops from traditional seeds matures slowly and at different times in contrary to plants from hybrid seeds which mature at the same time. The situation indicates that, for the farmers to make an informed decision the input dealers should be knowledgeable about the inputs they sell. Impicitly, proper training to agricultural input dealers on a normative basis is instrumental in transforming smallholder farmers from subsistence production to commercial farming (Waghmode *et al.* 2014).

4.1.4 Marketing and Trading Strategies

Results revealed two types of maketing strategies. These include, farm gate sales and direct market channel. In direct market channel two approaches of selling produce to consumers were also discovered. Establishement of buyers database and selling the products online was the first strategy. The second one entailed bundling the products into small bunches of 200–500 g then selling them physically to consuners. The advantages of marketing the products to consumers direct has been acknowkedged by different scholars (see for example, Karaxha et al. 2016). However, it has been revealed in the current study that there are important matters including location that need to be critically analysed when opting for this channel. It can be learnt from the case studies that both Pendo and John used direct channel strategy but John was successful compared to Pendo according to the criteria established. The reasons for the welfare discrepancy between the two cases who used same strategy could be numerous but from Pendo's own story one is the fact that she had no permanent location for the products. Due to that she was required to sell her products at low prices to make sure that everything was sold on the same day as she had no proper storage facilities for storing unsold products at the end of the day. This resulted into freely giving away some products in the following day or selling them at a throw away price.The situation makes her waste a substantive income as her potential customers prefer fresh produce.

On the other hand, Herman employed the farmgate sales strategy and considered more successful than Pendo but less successful to John. John used the direct channel through buyers data base establishment. With the approach John had constant communications with the buyers and hence even before the actual production he knew what to produce and to whom to sell the produce. He could easily get information on customers changing preferences and strategize accordingly. The findings are corroborated by Mittal and Tripathi (2009) that farmers can raise income and increase productivity when have access to appropriate and reliable information about proction and marketing. Also, disparities between the final market price and farm gate price is a critical issue in marketing of agricultural products as only informed farmers are in a better position of asking for a better price at farm gate sale.

4.1.4.1 Beyond Traditional Crops for Market Targets

The paper findings revealed that types of crop produced by vegetable farmers had effects on marketing. It was opined in female FGD conducted at Mtumba that African spinach (*Amaranthus* spp.) was only produced when the season was low because the crop was not doing well in the market interms of income generation. Thus, the youth who invested in production of this type of crop were associated low levels of livelihood outcomes. Herman for instance was a prosperous case among the youth because he managed to produce vegetable that could fetch a better price in the markets, hotels and super markets. These crops included red/yellow sweet pepper and beetroots. Also, crops like onions that could be stored after harvesting until the price is conducive for the farmers. This means the price of these products is guaranteed if the farmer is well equipped with skills on food processing and marketing.

4.1.4.2 Proper Timing in Vegetable Production

In the study area vegetable production was conducted in two approaches: Firsly, producing vegetable seasonally and secondly producing vegetable throughout the year. Vegetable producers who purposively conducted faming activities in the rainy season or dry season mainly were in the seasonal production category. The presence of limited crops in the market during the dry season causes the price of products to rise due to high demand. With this view Herman undertook his vegetable farming activities in the dry season mainly to attract good price as many vegetable farmers do not produce in this season because they depend on raifed agriculture. John was producing throughout the year produced and was considered the most successiful compared to the rest. The differences noted in the welfare outcomes between the two cases is attributed to the fact that, with Herman's approach the market opportunities that are significant and required bulk buying and continuous supply could not be accessed despite that the strategy worked better for him. Evidently, aggregation of crops for smallholder farmers is inevitable especially when producers are to meet the economically feasible capacity that can enhance access to international markets and increase profit.

The second production season involved vegetable farmers who conducted their farming activities during the rainy season due to lack of water facilities for production. However this study indicates that with this approach it is difficult to excel in this business. Learning from the experience of Pendo and Daniel it is obvious that water is a strategic resource in vegetable business given that Dodoma is a dry area and rainfall is therefore not reliable.With this view irrigation becomes crucial if the youth vegetable farmers in the area are to realise their ambition in this enterprise. As reported by other scholars that the expected returns in irrigation farming are higher by around 30 percent when comparing with farming without irrigation, and in areas with longer dry seasons can rise to 100 percent (Mihailovic *et al.* 2014).

5 Conclusion and Recommendations

Evidently, production strategies are conditioned by marketing opportunities. This was evident to youth vegetable farmers who invested in production of crops that required low investment, as they do so mainly to minimize the risks associated with marketing. However, the challenge that farmers among the youth face are how to increase their livelihood assets as this could be instrumental in determining the production and marketing strategy to employ. The current study, therefore, recommends the youth vegetable farmers to operate in viable farming schemes. The approach can enable the youth gain access to bulk buyers by aggregating their crops and enhance negogiating power. However, the formation of groups need pertinence and commitment before establishing themselves as vegetable producers for commercial purposes. To realise the youth ambition in farming takes time given that the resousec are scarce, this is where the non-committed youth leave farming prematurely and thus weaken the groups. The government is urged to ensure that the land usage and rights facets are recognized by policies and legislations. This is to enable the youth vegetable farmers access improved farming technologies for an enhanced productivity and livelihoods.

References

Agboola, A.F., Adenkule, I.A., Ogunjimi, S.I.: Assessment of youth participation in indigenous farm practices of vegetable production in Oyo State, Nigeria. J. Agric. Extens. Rural Dev. **7**(3), 63–72 (2015)

Akpan, S.B., Inimfon, V.P., Samuel, U.J., Agom, D.I.: Determinants of decision and participation of rural youth in agricultural production: a case study of youth in southern region of Nigeria. Russ. J. Agric. Socio-Econ. Sci. **43**(7), 1–14 (2015)

Alliance for a Green Revolution in Africa (AGRA). Youth in Agriculture in Sub-Saharan Africa. Africa Agriculture Status Report: Nairobi, Kenya. Issue No. 3(2015)

Ashley, C., Carney, D.: Sustainable Livelihoods: Lessons from Early Experience. DFID, London (1999). http://www.eldis.org/document/A28067. Accessed 12 June 2018

Ayele, S., Khan, S., Sumberg, J.: Africa's youth employment challenge: new perspectives. IDS Bull. **3**(48), 1–14 (2017)

Brooks, K., Zorya, S., Gautam, A., Goyal, A.: Agriculture as a Sector of Opportunity for Young People in Africa. Policy Research Working Paper No. 6473. World Bank, Washington, DC, p. 41 (2013)

Bullen, D., Sokheang, H.: Identification and effectiveness of self-help groups in Cambodia. Penang, Malaysia: CGIAR Research Program on Aquatic Agricultural Systems. Program Report: AAS-2015-11 (2015)

Creswell, J.W., Plano, C.V.L.: Designing and Conducting Mixed Methods Research, 2nd edn., p. 457. Sage, Thousand Oaks (2011)

FAO. *Youth and Agriculture: Key Challenges and Concrete Solutions.* Food and Agriculture Organization, Accra, Ghana, p. 128 (2014)

Gabagambi, D.M.: Empowering small holder farmers in Eastern Africa to access agro-markets and secure agricultural land: Agricultural market policy study on barriers to trade for Smallholder farmers in Tanzania. Final Report submitted to PELUM Tanzania, p. 37 (2011)

Grando, S., et al.: Strategies for sustainable farming: an overview of theories and practices. J. Rural. Stud. **7**(3), 207–218 (2016)

Greig, L.: An analysis of the key factors influencing farmer's choice of crop, Kibamba Ward, Tanzznia. J. Agric. Econ. **60**(3), 699–715 (2009)

Ibidapo, I., Faleye, O.M., Akintade, T.F., Oso, O.P., Owasoyo, E.O.: Determinants of youth participation in dry season vegetables cultivation in urban areas of Ondo State, Nigeria. J. Agric. Sci. Pract. **2**, 109–114 (2017)

Karaxha, H., Tolaj, S., Kristo, I.: Promotion through marketing channels: the case of Kosovo. ILIRIA Int. Rev. **6**(2), 1–15 (2016)

Leavy, J., Smith, S.: Future farmers: youth aspirations, expectations and life choices. J. Res. Rural. Educ. **15**(3), 141–156 (2010)

Losch, B.: Background paper for the FAO Regional Conference for Africa 28th session – Tunis, Tunisia, 24–28 March 2014, p. 9 (2014)

Mayring, P.: Qualitative Content Analysis. Theoretical Foundation, Basic Procedures and Software Solution, Klagenfurt, Austria (2014). http://nbn-resolving.de/urn:nbn:de:0168-ssoar-395173. Accessed 12 July 2018

Mihailović, B., Cvijanović, D., Milojević, I., Filipović, M.: The role of irrigation in development of agriculture in Srem District. Econ. Agric. **61**(4), 829–1088 (2014)

Mutua, E., Bukachi, S., Bett, B., Estambale, B., Nyamongo, I.: Youth participation in smallholder livestock production and marketing. IDS Bull. **3**(48), 95–108 (2017)

Nguyen, T.M.K., Nguyen, T.D., Philippe, L.: Smallholder farming and youth's aspirations: case study in Bacninh province, Red River Delta, Vietnam. Int. J. Agric. Manag. Dev. **1**(1), 11–19 (2016)

Njenga, P., Mugo, F., Opiyo, R.: Youth and women empowerment through agriculture. Int. J. Dev. Econ. Sustainabil. **2**(3), 1–8 (2013)

Proctor, F.J., Lucchesi, V.: Small-Scale Farming and Youth in an Era of Rapid Rural Change. IIED/HIVOS, London, The Hague, p. 74 (2012)

Rubin, A., Babbie, E.: Research Methods for Social Work, p. 378. Brooks/Cole Cengage Learning, New York (2011)

Rutta, E.: Current and emerging youth policies and initiatives with a special focus and links to agriculture. J. Mod. Afr. Stud. **38**(4), 683–712 (2012)

Scoones, I.: Sustainable Rural Livelihoods: A Framework for Analysis. Institute for Development Studies, Working Paper No. 72, Brighton, UK, p. 48 (1998)

Sheheli, S.: Improving Livelihood of Rural Women through Income Generating Activities in Bangladesh. PhD Dissertation for Award Degree of Doctorate at Humboldt University, Berlin Germany, p. 247 (2012)

Shoaib, A.W., Luan, J., Xiao, S., Sanaullah, N., Qurat, U., Moula, B.: Significance of agricultural finance in agricultural and rural development of Pakistan: a case study of Qambar Shahdadkot District. Res. J. Finan. Account. **9**(7), 86–94 (2016)

URT, United Republic of Tanzania. Annual Agricultural Sample Survey 2014–2015: National Bureau of Statistics. Dar es Salaam (2016). http://www.tanzania.go.tz/agriculturef.html. Accessed 12 Aug 2018

World Bank. Financing Agribusiness in Sub-Saharan Africa: Opportunities, Challenges, and Investment Models. Washington, DC, p. 95 (2011)

Yin, R.K.: Case Study Research Design and Methods, 5th edn., p. 282. Sage, Thousand Oaks (2014)

Conceptualising Technology Exchange as a Critical Gap for Higher Education and Industry Collaborations in Ghana

M. Alhassan[1(✉)], W. D. Thwala[2], and C. O. Aigbavboa[2]

[1] Department of Welding and Fabrication Engineering, Tamale Technical University, Tamale, Ghana
munkhas@yahoo.com
[2] Department of Construction Management and Quantity Surveying, University of Johannesburg, Johannesburg, South Africa
{didibhukut,caigbavboa}@uj.ac.za

Abstract. Purpose: This study is designed to develop a model for higher education institutions and industrial firms for effective collaborations. It will identify and evaluate technology exchange determinants upon which a model will be developed. Effective collaborations between higher education and industry will improve product development between them maximize the technical capacity of the two partners, as well as improve interaction.

Research Design: The survey research questionnaire was used, of which data were collected in the form of field research by distributing self-administered questionnaires.The data were gathered from two technical universities; Kumasi and Accra technical universities, including 20 public and 20 private manufacturing firms within these two metropolises. The respondents were solely from the manufacturing engineering background. A sample of 400 was selected from both higher education and industry in equal proportions. The data collected were entered into SPSS and analysed.

Findings: The study reveals that the commercialisation of technology output to the industry by higher education institutions is much more important in strengthening and improving collaborations. Joint new product development between university and industry will surely enhance interaction and maximise the technical skills of the two partners.

Research Limitation: The study focused on forty (40) manufacturing firms of two metropolises and two technical universities out of the ten (10) technical universities in Ghana. These are the communities where there is a cluster of firms. Also, these are areas where much of the university-industry collaborations are entered into.

Practical Implication: The findings of this study will equip higher education institutions as well as manufacturing firms with the factors that significantly strengthen collaborations between the two partners.

Social Implication: The knowledge advanced by this study will help higher education institutions and manufacturing firms to review their policies regarding collaborations with each other. These collaborations are inclined towards building indigenous capacity and improving productivity in the manufacturing industry.

C. Aigbavboa et al. (Eds.): ARCA 2022, *Sustainable Education and Development – Sustainable Industrialization and Innovation*, pp. 1109–1121, 2023.
https://doi.org/10.1007/978-3-031-25998-2_86

Originality: This study proposed a conceptual description of technology exchange (TE) which is viewed as a bidirectional (two-way flow) approach to technology acquisition. The novelty of this study centres on the new model built for higher education and industry for effective collaborations. This manifests the fact that technology should be based on two-way (bidirectional) flow in every higher education and industry collaboration as this is necessary for the capacity maximisation of both partners. Technology transfer should no longer be a one-way flow (unidirectional) as it is now.

Keyword: Collaboration · Higher education · Industry · Technical capacity · Technology exchange

1 Introduction

The recent phenomenon of universities forging links with industry for effective innovation on technology transfer is quite essential. Collaborations between higher education institutions and industry are viewed as a leverage to capacity maximization with mutual benefits accruing to both collaborative partners (Muscio 2010). In most literature, Higher Education and Industry collaboration is invariably referred to as University-Industry collaboration. Many varied factors constitute higher education and industry collaborations. Indeed, higher education institutions and industrial firms could collaborate on matters of sponsored research, contract or joint research. It could also be in the form of professional courses, consultancy services. This includes student placements during attachment, mobility of staff as well as developing a curriculum for the university (Ssebuwufu et al. 2012). ADEA (2012) also noted that higher education and industry collaborations may be entred into in a form of sponsored research, or contract.

This study focuses on technology exchange which is viewed as a critical gap existing in the previously developed models. As far as higher education and industry technology upgrading is concerned; it is viewed that collaborating partners would be more successful if they partner on technology exchange. The concept of technology transfer is seen as nebulous; thus, difficult to define (Robinson 1988; Spivey *et al.* 1997). The definition largely depends on the context in which the user defines technology (Chen 1996; Bozeman 2000). Bozeman (2000) defines technology transfer as moving technology from one entity to another. Lundquist (2003) contend that technology could be the movement of a specific set of capabilities. OECD (2005) developed a policy for technology transfer that is geared towards infrastructure, especially in economically developed countries. With regard to technology acquisition, It may be a necessary to acquire technology, sepecially when firms have a strong familiarity with a particular market or product but are unfamiliar with the technologies embedded into the product (Roberts and Berry 1985). Further, if firms do not have all the resources and the required technology to develop a product or process or when no commercial solution addresses a particular business problem acquisition of technology might be a necessity. (Steensma and Corley 2000).

The most common form of collaboration that exists between Ghanaian higher education institutions and firms is student attachments to industry and internships. These usually range from few weeks to about one year (Mamudu and Hymore 2016). It is usually hoped that this attachments after graduation may earn them employment. There exists also a government instituted training programme referred to as national service training scheme. The training period is usually one year with the aim of discharging their civil duty to the nation and not necessarily obtaining on-the-job experience. Mamudu and Hymore (2016) clearly indicated that collaborations between higher education and industry in Ghana tends to be informal, and manifest curricula development. Literature informed that universities in Ghana usually partner with firms in the form of conferences, seminars, workshops, projects and consultancies. In addition students training programmes during attachment are not directed towards research innovations and spin-offs. Rather the informal sector of the economy normally collaborate on matters of innovation for competitive reasons. There is scarce literature on technology exchange collaborations between higher education and industry in developing countries. Technology exchange is viewed as a catalyst to maximising technical capacity and innovation especially in developing countries. The focus of this study was to examine the effectiveness of collaboration between higher education institutions and industry in relation to technology exchange. The specific objectives were to identify and evaluate technology exchange determinants and to develop a model for higher education institutions and industrial firms for effective collaborations.

2 Theoretical Considerations

The idea of higher education and industry collaborating on technology exchange, on a two-way flow, is quite essential for the growth of higher education institutions and industry (Omar et al. (2010). Technology exchange is viewed as a bidirectional (two-way) flow which includes exchange processes of different components of technology. Technology exchange involves knowledge transfer, sharing of equipment and skills (Omar et al. 2010). Technology acquisition for firms is a necessity for technological growth, gaining efficiency and improvement of technical capacity (Ford and Probert 2010). Technology exchange is supposedly a flip side of both knowledge transfer and technology transfer (Omar et al. 2010), especially for developing countries. It is apparent from literature that knowledge transfer was used to describe a unidirectional (one-way) flow of knowledge; usually transmitted from researchers to receipients. Johnson (2005) contend that knowledge transfer was considered the sole duty of researchers. According to Omar et al. (2010), the idea of technology transfer and technology exchange is a bidirectional flow, especially in construction projects. In a small way, technology exchange includes transfer and sharing of equipment (Omar et al. 2010).

Technology exchange is conceptualised in two major areas such as entrepreneurship and technology innovation (UNCTAD 2013). Implying entrepreneurship and technology innovation are interdependent and mutually supportive. Clearly, the exchange manifestation is the ability of technology to provide entrepreneurs with new tools to improve efficiency while entrepreneurs assists technology by improving new products and ensuring commercialisation (UNCTAD 2013). Technology exchange was also used in the US prior to the patenting activities of 1980. Basically, universities of US collaborated with industrial firms on research, including a number of channels such as publishing, training of industrial researchers, faculty consulting, and other activities (Mowery and Sampat

2005). Interestingly, technology exchange accelerates technology acquisition through exchange of infrastructure and partnership in new product development (Durrani et al. 1999; Bines 2004, Daim and Kocaoglu 2008).

This research study, therefore, views technology exchange as a bidirectional activity between higher education and industry where it is expected that a complete synergy of technology innovation may be actualised. The phenomenon of technology exchange is a proposed concept by this study for developing countries such as Ghana, since there is the need to raise the capacity of higher education as well as industry alongside technology innovation. Essentially, HE&I collaborations on technology maximise skills acquisition, promote knowledge transfer as well as entrepreneurship in a form of start-ups and spin-offs (Dooley and Kirk 2007, Mgonja 2017). As has been mentioned, the concept of technology exchange aims at maximising the innovative capacity of higher education institutions and industry simultaneously and to facilitate technology acquisition. Technology exchange bears similar characteristics as knowledge transfer and technology transfer but with emphasis on "bidirectional flow" or "two-way flow" (Omar et al. 2010) in terms of technical know-how, sharing of technological infrastructure and new product manufacture between the two collaborating partners. Technology exchange is supposedly more interactive and more sharing between university and industry than technology transfer which is unidirectional from researchers to potential users including policy makers, technicians and clients. Knowledge transfer is also considered the sole responsibility of researchers (Johnson 2005). Obviously, technology exchange is geared more towards applied research and practical solutions than just generating research output for commercialisation to industry. The assumption is that technology exchange mechanisms can improve technical capacity, management of technological infrastructure of partners and

Table 1. Postulated technology exchange model

Latent construct	Indicator variables	Label
Technology Exchange (TEX)	Industry assists the university in maintaining broken down specialized machines/equipment	TEX 1
	University and industry share technology experiences	TEX 2
	Industry helps university staff develop technical capacity	TEX3
	University and industry share some of their specialised machines with each other	TEX 4
	There is joint new product development between university and industry	TEX 5
	University commercialises its technology output for use by industry	TEX 6
	University generates enough technology with industrial relevance	TEX 7
	University conducts prototype testing with industry	TEX 8

facilitate the production of indigenous products. Table 1 presents a postulated model of technology exchange with eight (8) determinants.

3 Methodology

3.1 Research Design

In considering the research design, the survey research questionnaire was used, of which data were collected in the form of field research by distributing self-administered questionnaires. The data were gathered from two technical universities; Kumasi and Accra technical universities, including 20 public and 20 private manufacturing firms within these two metropolises. The respondents were solely from the manufacturing engineering background. A sample of 400 was selected from both higher education and industry in equal proportions. The sample was deemed a representation of the population that had adequate knowledge and understanding of higher education and industry collaborations. This study employed a two-stage research approach to measure the determinants of technology exchange. The first stage involved the review of the literature to identify those determinants. With extensive review of literature, the researcher had been informed of various avriables, and a careful selection and adaptation of those variables had been done. The second part involved the use of survey to evaluate the measurement model and establish the determinants of technology exchange. A five-point likert scale was rated by respondents on an eight (8) identified determining factors where 1 represents "not at all important", 2 "little important", 3 "moderately important", 4 "very important" and 5 "critically important". The data collected were entered into SPSS and transported to structural equation modelling (SEM) for analysis. The SEM software of EQS version 6.2 was used for the analysis. SEM is noted for being the most inclusive statistical procedure in social and scientific research; it also requires a large sample size. SEM consists all the indices of general linear modelling (GLM) of statistical operations.

4 Results

4.1 Data Presentation

Table 2. Characteristics of respondents

Respondent	Frequency	Percentage
Gender (industry)		
Male	176	88.0%
Female	24	12.0%
Gender (Academia)		
Male	169	84.5%

(continued)

Table 2. (*continued*)

Respondent	Frequency	Percentage
Female	31	15.5%
Qualification(Academia)		
Doctoral (PhD)	39	19.5%
Masters	161	80.5%
Industry		
Doctoral (PhD)	0	0.0%
Masters	19	9.5%
Bachelors	110	55.0%
Higher national diploma	71	35.5%
Organization (industry)		
Private companies	140	70.0%
Government institutions	60	30.0%
Institution (University)		
Lecturers	160	80.0%
Senior management	40	20.0%

The survey considered a total of 400 responses, and out of that 345 were males from both university and industry, representing 86.25%. Female respondents were 55 from both university and industry, representing 13.75%. From Table 2, gender distribution on the part of male participants indicates that 176 were from industry, representing 88.0%. While 169 were from higher education, representing 84.5%. With industry, 24 female respondents were from industry, representing 12.0%, while 31 female respondents, representing 15.5% were from higher education. With the industry, Bachelor's degree holders were 110, representing 55.0%; this category of personnel was in the majority. Those with higher national diplomas were 71, representing 35.5%, however, master's degree holders in the industry were the least (N = 19), representing 9.5%. Industry is interested in skilled labour and professionalism and does not lay much emphasis on the highest academic qualifications. The academic qualification for higher education is both a PhD and a Master's Degree. PhD holders were 39, representing 19.5% while those holding master's degrees were 161, representing 80.5%, this suggests that those with master's degrees are in the majority within the Technical Universities in Ghana. In considering the industrial sector, a total of 140 respondents were from private companies, representing 70.0% while 60 respondents were from government institutions, representing 30.0%. With higher education, both academia and senior management were considered. Whereas academia represented 80.0% (N = 160), senior management represented 20.0% (N = 40), indicating that lecturers were in the majority. The technology exchange (TEX) construct constitutes eight (8) determinants as presented in Table 2. These determinants were identified from an extensive review of the literature. Respondents were allowed to indicate the level of importance of the determinants on a five-point rating scale regarding the influence of technology exchange. The rating scales took the form of Not at all Important (NI = 1), (Little Important (LI = 2), Moderately Important (MI = 3), (Very Important (VI = 4) and Critically Important (CI = 5). With 1 being the lowest and 5 being the highest. Clearly, from preliminary confirmatory factor (CFA) analysis, all 8 variables were measured. No indicator variable was found to be unacceptably high unstandardised and standardised residual covariance matrix. Table 3 presents the rating of respondents on technology exchange determinants.

Table 3. Participants' response on technology exchange determinants

Technology Exchange (TEX) determinants	1	2	3	4	5	Total
TEX1. Maintaining broken down machines	7 1.8%	37 9.3%	118 29.5%	161 40.3%	77 19.3%	400 100%
TEX2. Share technology experiences	6 1.5%	36 9.0%	111 27.8%	177 44.3%	70 17.5%	400 100%
TEX3.University staff develop technical capacity	7 1.8%	33 8.3%	97 24.4%	174 43.8%	86 21.7%	400 100%
TEX4.Share some of their specialised machines	15 3.8%	24 6.0%	100 25.0%	176 44.0%	85 21.3%	400 100%
TEX5. Joint new product development	3 0.8%	34 8.5%	92 23.0%	180 45.0%	91 22.8%	400 100%
TEX6. University commercialises its technology output	9 2.3%	22 5.5%	111 27.8%	162 40.6%	95 23.8%	399 100%
TEX7. Generates enough technology for industry	8 2.0%	30 7.5%	98 24.5%	168 42.0%	96 24.0%	400 100%
EX8. Joint prototype testing	14	29	81	171	105	400

4.2 Analysis and Discussion

4.2.1 Statistical Significance of Parameter Estimate for the Various Determinants

The results from Table 4 showed that all correlation (standard coefficient) values were less than 1.0 while, unstandardised coefficient values were greater than 1.00. The parameter with the highest standardised coefficient was the determinant, TEX 6 (University commercialises its technology output for use by industry). The next parameter with the highest standardised coefficient was found to be 0.799 thus, TEX 5 (Joint new product development between university and industry), as presented in Table 4.

The determinant TEX6, asked the participants if "university commercialises its technology output for use to industry" whether this could strengthen university-industry collaboration. With the R^2 values, only TEX 1, TEX 2 and TEX 3 were below 0.50.

Table 4. Factor loadings and Z-statistics of TEX model

Indicators	Unstandardised coefficient	Standardised coefficient	Z-Statistics	R-Square	Sig (5%)
TEX1	1.000	0.685	–	0.422	Significant
TEX2	0.941	0.665	13.129	0.397	Significant
TEX3	1.071	0.718	13.248	0.483	Significant
TEX4	1.146	0.715	12.198	0.507	Significant
TEX5	1.233	0.799	13.537	0.697	Significant
TEX6	1.256	0.811	13.534	0.660	Significant
TEX7	1.145	0.772	12.251	0.536	Significant
TEX8	1.203	0.717	12.039	0.528	Significant

The other determinants indicate that the factor explained the variance in the indicators. Therefore suggesting that the determinants significantly predict technology exchange.

In ranking the determinants, the presentation in Table 5, indicates that technology commercialisation was ranked first, having the highest standardised coefficient (correlation value) of 0.811 and an unstandardised coefficient of 1.256. Joint new product development was ranked second with a correlation value of 0.799 and an unstandardised coefficient of 1.233. Technology generation with industrial relevance was ranked third with a correlation value of 0.772 and an unstandardised coefficient of 1.145. Development of technical capacity was ranked fourth with a standardised coefficient of 0.718 and an unstandardised coefficient of 1.071. While Joint prototype testing has a standardised coefficient of 0.717 and an unstandardised coefficient of 1.203 and ranked fifth. Sharing of specialised equipment yielded a standardised coefficient of 0.715 and an unstandardised coefficient of 1.146 and ranked sixth. Maintenance of broken down specialised machines yielded a standardised coefficient of 0.685 and unstandardised coefficient of 1.000 and ranked seventh while sharing of technology experiences yielded a standardised coefficient of 0.717 and unstandardised coefficient of 0.941 and ranked eighth.

4.3 Technology Commercialisation

In recent years, university-industry collaborations have been particularly focused upon technology innovation systems as well as entrepreneurship and commercialisation of research output. The collaborative nature of universities with firms triggered public policies to focus on how to commercialise research output in higher education. Commercialisation of university research basically embraces patents and spin-offs, though there are other variables (Todorovic et al. 2011).

4.4 New Product Development

There is the need to contextualise what constitutes a new product; many researchers have conceived this phenomenon to constitute different elements (Stendahl 2009; Garcia and

Table 5. Ranking of determinants

Determinant	Unstandardised coefficient	Standardised coefficient	Z-Statistics	R-Square	Rank
University commercializes its output	1.256	0.811	13.534	0.660	1st
Joint new product development	1.233	0.799	13.537	0.697	2nd
University generates enough technology	1.145	0.772	12.251	0.536	3rd
Industry helps university to develop technical capacity	1.071	0.718	13.248	0.483	4th
Joint prototype testing between partners	1.203	0.717	12.039	0.528	5th
Jointly share specialized machines	1.146	0.715	12.198	0.507	6th
Industry assists university in maintaining broken down specialized machines	1.000	0.685	-	0.422	7th
University and industry share technology experiences	0.941	0.665	13.129	0.397	8th

Calantone 2002). What constitute a new product or innovation is when there is some uniqueness, improvement or modification concerning its characteristics or intended uses (Stendahl 2009; Garcia and Calantone 2002). Amoah and Fordjour (2012) argue that new product development is a matter of experience.

4.5 Creation of Technology with Industrial Relevance

Every research leading to technological innovation should manifest new ideas. Therefore both basic research and technological projects in the higher education institutions laboratories should reflect incubation of technology-based activities (Valavanidis and Vlachogianni 2016). In addition, higher education must be proactive in ensuring the application of knowledge generated as well as creating and building a pool for academic excellence (Valavanidis and Vlachogianni 2016). In order to facilitate application of research output, higher education has to discourage the linear model of technology innovation and rather embrace collaborative innovation with industrial firms which encourages sharing of ideas, technical infrastructure and technology. In the normal course of events, industry is expected to operate as a locus of production (machines, consumer products, electronic instruments, etc.) while the government operates as a source of contractual relations that guarantee stable interactions. The universities operate as a source of new knowledge and technological discoveries (Etzkowitz 2003; Etzkowitz 2008; and Viale and Etzkowitz 2010).

4.6 Industry Develops Technical Capacity for University Staff

Firms must assist university lecturers to maximise their technical capacity, especially in technical universities through mobility (Colucci et al. 2012). This will go a long way to improve the relationship between the two partners. Currently, what is observed in higher education institutions is the absence of staff movement to local firms. Currently, there is the mobility of students only when they are placed for an internship or industrial attachment. It is pertinent to mention several factors that drive academic mobility to industry; the student who chooses to be mobile, the government and the higher education institution that structure and support such projects (Colucci et al. 2012).

4.7 Joint Prototype Testing

Prototypes are quite essential in product development. They can help to create, explore, describe, test and analyse the item being designed (Jensen et al. 2016). Prototypes vary from industry to industry and from university to university. Industrial designers produce prototypes of conceptual ideas to explore form and geometry while engineers prototype is designed to validate a functional principle or to benchmark performance (Jensen et al. 2016). Joint prototype testing between university and industry is a way of sharing design ideas regarding technical skills and know-how. It involves sharing some of their specialised equipment/machines, this noble idea will surely maximise the technological capability of partners and allow both partners to incorporate certain essential ideas into the design during prototype testing.

4.8 Sharing of Specialised Machines

It is common knowledge that higher education institutions have special equipment in their laboratories that are not found in the local firms likewise the industry; they possess certain machines that are not found in the universities. Therefore exposure to special equipment or machines through mobility will certainly improve the practical knowledge of staff and students as well.

5 Conclusion

In conclusion, the concept of technology exchange as proposed in this study is aimed at building an innovative and technical capacity of higher education institutions and industry through mutual interaction. Technology exchange lays emphasis on "bidirectional flow" or "two-way flow". The study clearly outlined the key determinants of technology exchange suitable for building a model. Eight (8) variables determined the effective implementation of technology exchange. University commercialises its technology output for use by industry as determinant from the finding. Implying this determinant could strengthen the collaboration between university and industry. Followed by the determinant; "collaboration in new product development". Indeed, all determinants of the model showed high standardised coefficient values. The rest are; "university generates enough technology with industrial relevance", "Industry helps university staff to develop technical capacity", "University conducts prototype testing with industry", "university and industry share some of their specialised equipment/machines" and so on.

The study reveals that commercialisation of technology output to the industry is much more important in strengthening and improving collaborations. Joint new product development between university and industry will surely enhance interaction and maximise the technical skills of the two partners. Also if higher education institutions generate enough technology with industrial relevance, firms shall be willing to go into partnership with them for mutual benefits. Another important element is the role of industry in assisting the staff of higher education institutions to develop technical capacity through staff mobility. Joint prototype testing between higher education and industry is a way of sharing design ideas regarding technical skills and know-how. Clearly, an effective element that is missing in the higher education and industry collaboration model is technology exchange. Essentially, higher education and industry collaborating on technology exchange is surely a manifestation of a two-way flow of technology which is necessary for the growth of both partners. In addition, technology exchange is geared towards enhancing the technical capacity and knowhow of higher education and industry. Technology exchange is seen in this context to be more interactive between university and industry than technology transfer which reflects give and take attitude from university to industry and to other stake holders. Obviously, technology exchange is directed towards applied research and practical solutions of problems. As a result, the study recommends that higher education and industry collaborate on technology exchange activities; regarding commercialization, joint product development and manufacture, sharing of specialised equipment as well as joint prototype testing.

References

Adam, B.J., Lerner, J.: Reinventing Public R&D: patent policy and the commercialisation of national laboratory technologies. RAND J. Econ. **32**(1), 167–197 (2001). https://doi.org/10.2307/2696403

ADEA. Triennale on Education and Training in Africa (Ouagadougou, Burkina Faso, February 12–17 2012) Association for the Development of Education in Africa (ADEA) African Development Bank (AfDB)Temporary Relocation Agency (ATR)13 avenue du Ghana BP 3231002 Tunis BelvédèreTunisia (2012)

Amoah, M., Francis, F.: New product development activities among small and medium-scale furniture enterprises in Ghana: a discriminant analysis. Am. Int. J. Contemp. Res. **2**(12), 41–53 (2012)

Arranz, N., Fernandez, D.E., Arroyabe, J.C.: The choice of partners in R&D cooperation: an empirical analysis of Spanish firms. Technovation **28**(1–2), 88–100 (2008)

Bines, T.: An integrated process for forming manufacturing technology acquisition decisions. Int. J. Oper. Prod. Manag. **24**(5), 447–467 (2004)

Bozeman, B.: Technology transfer and public policy: a review of research and theory. Res. Policy **29**, 627–655 (2000)

Cetindamar, D., Phaal, R., Probert, D.: Technology Management: Activities and Tools. Palgrave Mcmillan, Hampshire (2010)

Chen, M.: Managing International Technology Transfer. International Thomson Business Press, London (1996)

Colucci, E., Davies, H., Korhonen, J., Gaebel, M.: Mobility: Closing The Gap Between Policy And Practice EUA Publications (2012. ISBN: 9789078997351

Daim, T.U., Kocaoglu, D.F.: How do engineering managers evaluate technologies for acquisition? a review of electronics industry. Eng. Manag. J. **20**(3), 44–52 (2008)

Dooley, L., Kirk, D.: University-industry collaboration-grafting the entrepreneurial paradigm onto academic structures. Eur. J. Innov. Manag. **10**(3), 316–332 (2007)

Durrani, T.S., Fobes, S.M., Broadfood, C.: An integrated approach to technology acquisition management. Int. J. Technol. Manag. **17**(6), 597–618 (1999)

Etzkowitz, H.: Innovation in innovation: the triple helix of university-industry- government relations. Social Sci. Informat. Sci. Sociales **42**(3), 293–337 (2003)

Ford, S., Probert, D.: Why do firms acquire external technologies? understanding the motivations for technology acquisitions. In: The 2010 Portland International Conference on Management of Engineering & Technology (2010)

Garcia, R., Calantone, R.: A critical look at technological innovation typology and innovativeness terminology: a literature review. J. Prod. Innov. Manag. **19**, 110–132 (2002)

Gregory, M.: Technology management: a process approach. Proc. Inst. Mech. Engineers Part B: J. Eng. Manuf. **209**, 347–356 (1995)

Henderson, R., Jaffe, A.B., Trajtenberg, M.: Universities as a source of commercial technology: a detailed analysis of university patenting, 1965–1988. Rev. Econ. Stat. **80**(1), 119–127 (1998)

Jensen, L.S., Özkil, A.G., Mortensen, N.H.: Prototypes in engineering design: definitions and strategies. In: International Design Conference - Design 2016, Dubrovnik, Croatia, 16–19 May 2016 (2016)

Johnson, L.S.: From knowledge transfer to knowledge translation: applying research to practice, developing expert practice, pp. 11–14 (2005)

Kline, R.B.: Principles and Practice of Structural Equation Modeling, 1st edn. Guilford Press, New York (2005)

Klofsten, M., Jones-Evans, D.: Comparing academic entrepreneurship in europe – the case of Sweden and Ireland. Small Bus. Econ. **14**(4), 299–309 (2000). https://doi.org/10.1023/A:1008184601282

Lundquist, G.: A rich vision of technology transfer technology value management. J. Technol. Transfer **28**(3–4), 284 (2003)

Mamudu, A., Hymore, K.: Enhancing university-industry (Ui) collaboration in ghana for improved skilled labour. In: Proceedings Of Incedi 2016 Conference, Accra, Ghana, 29th–31st August 2016 (2016)

Mgonja, C.T.: Enhancing the university –industry collaboration in developing countries through best practices. Int. J. Eng. Trends Technol. (IJET) **50**(4), 216–225 (2017)

Mowery, D.C., Sampat, B.N.: Universities in National Innovation Systems. In: Fagerberg, J., Mowery, D., Nelson, R. (eds.) The Oxford Handbook of Innovation, pp. 209–239. Oxford University Press, Oxford (2005)

Muscio, A.: What drives the university access to technology transfer offices? evidence from Italy. J. Technol. Transfer **35**, 181–202 (2010)

OECD. Achieving the Successful Transfer of Environmentally Sound Technologies: Trade-Related Aspects Trade and Environment Working Paper No. 2005–02 (2005)

OECD: New Nature of Innovation. OECD, Copenhagen (2009)

Omar, R.M., Takim, R., Nawawi, A.H., Hassan, F.: Technology Transfer (TT) and Technology Exchange (TE) in Malaysia. In: International Conference on Education and Management Technology (ICEMT) (2010)

Osaman-gani, A., Ahad, M.: International technology transfer for competitive advantage: a conceptual analysis of the role of HRD. Compet. Rev. **9**(1), 9–18 (1999)

Phillips, R.G.: Technology business incubators: how effective as technology transfer mechanisms. Technol. Soc. **24**, 299–316 (2002)

Pries, F., Guild, P.: Commercializing inventions resulting from university research: analyzing the impact of technology characteristics on subsequent business models. Technovation **31**(4), 151–160 (2011). https://doi.org/10.1016/j.technovation.2010.05.002

Ramanathan, K.: The polytrophic components of manufacturing technology. Technol. Forecast. Soc. Chang. **46**, 221–258 (1994)

Rasmussen, E., Moen, Ø., Gulbrandsen, M.: Initiatives to promote commercialisation of university knowledge. Technovation **26**, 518–533 (2006)

Roberts, E.B., Berry, C.A.: Entering new businesses: selecting strategies for success. Sloan Manag. Rev. **26**(3), 3–17 (1985)

Robinson, R.D.: The International Transfer of Technology, Theory, Issues, and Practice. Ballinger Publishing Company, Cambridge (1988)

Refn, A.: Research Collaboration, Knowledge and Innovation. Technical University of Denmark (2018)

Spivery, W.A., Munson, J.M., Nelson, M.A., Dietrich, G.B.: Coordinating the technology transfer and transition of information technology: a phenomenological perspective. IEEE Trans. Eng. Manag. **44**(4), 359–366 (1997)

Ssebuwufu, J., Ludwick, T., Béland, M.: Strengthening university-industry linkages in Africa: a study on institutional capacities and gaps. Association of African Universities (AAU). 11 Aviation Road Extension. P.O. Box 5744, Accra-North Ghana (2012)

Steensma, H.K., Corley, K.G.: On the performance of technology-sourcing partnerships: the interaction between partner interdependence and technology attributes. Acad. Manag. J. **43**(5), 1045–1067 (2000)

Stendahl, M.: Product Development in the Wood Industry: Breaking Gresham's Law. Doctoral Thesis. Swedish University of Agricultural Sciences, Uppsala (2009)

Todorovic, Z.W., Rod, B., McNaughton, B., Guild, P.: ENTRE-U: an entrepreneurial orientation scale for universities. Technovation **31**(2011), 128–137 (2011)

UNCTAD. Transfer of Technology and Knowledge-Sharing for Development: Science, Technology and Innovation Issues for Developing Countries. United Nations Conference on Trade and Development. UNCTAD Current Studies on Science, Technology and Innovation, No 8 (2013)

Valavanidis, A., Vlachogianni, T.: Research and Development. The Role of Universities for the Knowledge-Based Society and Technological Innovations. Expenditure in Scientic Research and Applications as Crucial Factors for Economic Growth and the New Technological Frontiers (2016)

Van Haverbeke, W., Duysters, G., Noorderhaven, N.: External technology sourcing through alliances or acquisitions: an analysis of the application specific integrated circuits industry. Organ. Sci. **13**(6), 714–733 (2002)

Viale, R., Etzkowitz, H. (eds.): The Capitalisation of Knowledge: A Triple Helix of University-Industry-Government. EE Edward Elgar, Cheltenham (2010)

Zucker, L.G., Darby, M.R., Armstrong, J.: Commercialising knowledge: university science, knowledge capture, and firm performance in biotechnology. Manag. Sci. **48**(1), 138–153 (2002)

Workplace Health and Safety Procedures and Compliance in the Technical and Vocational Institutions Workshop in Ghana

T. Adu Gyamfi[1](✉), S. K. Akorli[1], E. Y. Frempong-Jnr[1], and M. Pim-Wusu[2]

[1] Department of Building Technology, Faculty of Built and Natural Environment, Koforidua Technical University, Koforidua, Ghana
agttimo78@gmail.com

[2] Department of Building Technology, Faculty of Built Environment, Accra Technical University, Accra, Ghana

Abstract. Purpose: Adherence to safety procedures is necessary for the effective management of health and safety in occupations and the industrial world. The purpose of this study is to evaluate the workshop safety procedure and compliance at Technical and Vocational institutions in Ghana.

Design/Methodology/Approach: A descriptive research design and quantitative methodology were employed in the study. 200 people took part in the study at the TVET Institute Training Center. The respondents for the study were chosen through the use of purposeful sampling. Descriptive statistics were used to analyze the data.

Findings: The study's main finding was that the TVET workshops have established protocols. The study found that there are some elements that work against trainee adherence to workshop safety procedures. The study also identified ways to lessen occupational risks in TVET institutions' workshops.

Implications/Research Limitations: The management of TVET institutions is expected to provide strict safety measures at various workshops to ensure the safety and well-being of the trainees. The present study was limited to pre-tertiary TVET institutions in the Greater Accra and Ashanti Region.

Practical Implications: The government agencies are required to provide time-to-time inspection as to the implementation of safety rules operationalisation in the TVET Workshops in Ghana.

Originality/Value: At a TVET pre-tertiary institution workshop in Ghana, there is no information in the extant literature on workplace safety policies and adherence. The findings of this study, which are based on prior empirical and theoretical research, help people understand the importance of TVET institutions strictly adhering to safety rules and regulations.

Keywords: Compliance · Health and safety procedures · Workshop · TVET

C. Aigbavboa et al. (Eds.): ARCA 2022, *Sustainable Education and Development – Sustainable Industrialization and Innovation*, pp. 1122–1134, 2023.
https://doi.org/10.1007/978-3-031-25998-2_87

1 Introduction

The rate of globalisation has helped to boost employment, which is typically flexible and involves risky, insecure work that puts workers' well-being at serious risk (Amfo-Out and Agyemang 2016). One of the characteristics of technical vocational education and training (TVET), according to Okwelle and Okeke (2016), is its focus on the development of employable skills and its orientation towards the workplace. Saskatchewan Polytechnic (2016) asserts that environmental awareness and safety training should also be a part of TVET. This demonstrates the importance of having a clear health, safety, and environmental plan that considers how important it is for both teachers and students to adhere to safety rules and guidelines when using these workshops (Amenger 2013). A key factor in the establishment of an occupational health and safety (OHS) culture at TVETs can be found in the gradual enhancement of social and professional skills and in the gradual rise in staff and student awareness of OHS (Zakir et al. 2021). There are risks involved with teaching in TVET institutions like vocational colleges. To ensure workers' health and safety, the workplace must adhere to all safety regulations. To improve the conditions for instructor and student safety, particularly in workshops and laboratories, appropriate OHS procedures and guidelines must be developed and put into operation (Zakir et al. 2021).

In most industrialised nations over the past 20 to 30 years, occupational health and safety has improved (International Labour Organization 2013). But it's not entirely apparent how things stand in developing nations. Van-Bommel (2006) highlighted that developing countries have a disproportionately greater number of fatal accidents compared to developed countries. There are many different forms of literature that demonstrate studies in TVET workshops, with Itohan (2018) being just one example. In his study, Itohan (2018) examined strategies for reducing accidents and upholding industrial safety in technical education workshops in Nigeria. In order to better satisfy the safety practice skill requirements of sawmill workers, Anaele et al. (2014) performed a study on re-engineering technical vocational education and training (TVET) in Nigeria. Also, in the study of Hettiarachchi and Coomasaru (2021) they discovered common causes of accidents and safety procedures in their paper titled difficulties, trends, and opportunities in training technical and vocational schools. For the purpose of managing the technical college workshop effectively, Omeje and Osita (2014) looked into the safety skills needed of workshop technicians and assistants. In Nigeria's rivers state technical colleges' workshops, Okwelle and Normakoh (2019) evaluated the health, safety, and environmental practises. Although there is a knowledge gap regarding health and safety issues associated with OHS process and adherence at Technical and vocational institutes workshops in Ghana, the study was carried out to fill this gap. The study's goal is to evaluate pre-tertiary TVET institution workshops' adherence to health and safety policies as well as methods to lower occupational risks.

2 Health and Safety Theory

The 1931 invention of Heinrich's domino theory, according to Micah and Aikins (2002), states that an occurrence is one element in a chain that may lead to an injury. According to the hypothesis, 1) a prospective injury develops as a result of an incident. 2) Only a personal dangerous behaviour, a mechanical fault, or a physical danger can cause an event. 3) Human error is the only reason why personal or mechanical risks exist. 4) Flaws can be inherent or developed as a consequence of the social environment in which a person was born, raised, or schooled. Contrary to popular belief, all four criteria must be present for there to be an injury or property damage. It follows that the loss can be averted if one of the components in the series of conditions that contribute to the accident may be changed. The issue that caused the accident should receive the most attention.

A successful safety system should be founded on policies, according to Dorji and Hadikusumo's (2006) argument. This suggests that the foundation for creating and putting into practise a safety management system is having a clearly articulated safety policy.

For employees in TVET seminars, Rowe (2001) created safe working practises. These safe working practises comprised the proper instruction and risk management techniques. The risk management plan included safety inspection, onboarding, training, and continual tool and equipment improvement supported by staff engagement. Business owners, such as organisations, industries, and enterprises, must have safe operating policies or codes of practise for all organisation ranks that will ensure that employees are given enough guidelines and instructions on how to work safely. Amongst some of the safety protocols that need to be pursued at workshops is the orientation (WorkSafe Saskatchewan 2014). The process of familiarising new, inexperienced, and transferred individuals with other workers, including supervisors and coworkers, as well as their workplace, is known as orientation,

(WorkSafe Saskatchewan 2021). Health and safety considerations are important for TVET trainees' performance in their individual workshops. Developing TVET trainees' knowledge, skills, attitudes, and aptitudes at an early stage of work are crucial. Orientation gives new hires the chance to learn about their employer and coworkers while also providing them with essential safety information regarding their position and tasks (WorkSafe Saskatchewan 2014). As opined by WorkSafe Saskatchewan (2021) unsuitable information regarding workplace dangers or insufficient safety training, regrettably, results in employee injuries or fatalities every year. In order to keep individuals safe at work, the Saskatchewan Occupational Health and Safety Regulations (2020), require that all novice, incompetent, and transfer employees from one workstation or workplace to another get a fundamental orientation. In especially when a person begins a new job or returns after a long absence, planning and arranging an efficient orientation is essential to preventing injuries and fatalities.

The study by Aluko et al. (2016) found that occupational hazards should be addressed and taken into account right away as part of the requirements for safe working conditions and that management and employees should share responsibility for preventing and controlling hazards.

Hazards can be reduced at their sources, on their routes, and at the worker level. Elimination, substitution, and isolation are methods used to control hazards at their sources. Ventilation, barriers, and housekeeping are methods used to control hazards along the path. Personal protective equipment is used to control hazards at the worker, and policies, procedures, and rules are established (Carleton University 2013). Aluko et al. (2016) observed the following hazard control measures at TVET workshops, such as staff training and protective equipment provision as required to minimise the risk of exposure to the occupational hazard, report and document all occupational hazard exposure levels to appropriate agencies, and provision of stringent severe sanctions to persons that contravene safety protocols to discourage everyone else. Workshops for technical education that lack proper safety precautions expose students to risks that ultimately hinder their ability to learn practical skills. According to Anaele et al. (2014), implementing safety measures in TVET workshops is a strategic policy and plan to prevent accidents involving employees and bystanders as well as the breakdown of supplies, tools, and equipment. All measures and actions performed by workshop staff to protect workplace safety from harm to life or health are included in safety policies. Employers must still supply personal protective equipment and clothing (PPE) when risks in the workplace or workshop cannot be prevented or mitigated by other risk measures (State Government of Victoria 2019). Laws governing the workplace mandate PPE, which is made to shield the wearer from potential danger. It is a control mechanism for a known hazard; it neither removes the hazard from the workplace nor provides protection for other workers; all it does is keep the user safe (2013). (WorkSafe Saskatchewan (2014) According to the Occupational Health and Safety Act, businesses are required to give "new employees" training on the PPE necessities for their job duties. Additionally, as mandated by the Act, all companies must give employees with appropriate PPE and oversee how they use it. Employees must use PPE as prescribed by law, wear it, and notify their employers of any potential hazards (State Government of Victoria 2019). According to Taylor (2011), it is crucial for TVET workshops to choose the proper PPE for the right jobs. Pruss-Ustun et al. (2003) claim that wearing protective barriers in TVET workshops is beneficial while working with potentially infectious materials and heated environments. Before usage and on a regular basis, PPE must be examined to make sure it is in good functioning condition.

It is crucial to prioritise employee safety at the TVET workshop by educating trainees about safety in their everyday tasks, which includes giving daily safety talks. A safety lecture is a practical approach to address specific issues in the workshops and remind employees that health and safety are crucial at work (WorkSafe NB 2020). Regular safety meetings and toolbox talks should always be given to educate personnel on workplace safety and enforce compliance with safety and training regulations (Kendall 2020). The TVET workshop's inclusion of Safety meetings and toolbox lectures fosters a strong safety culture and strengthens trainees' resolve to take care of themselves. Regular toolbox talks and safety meetings can help workers avoid becoming satisfied and not play with safety issues. To assist the business in adhering to Occupational Safety and Health Administration (OSHA) standards, they can also adopt innovative safety policies and offer training on new safety laws and regulations (Kendall 2020). WorkSafe NB (2020) asserts that when trainees start using a particular tool, piece of equipment, kind of material, or work approach on the project, the knowledge provided in a safety session may be the only information they remember. However, it is essential that the safety discussion leave TVET trainees with a lasting impression that will guide them in their daily activities at the workshop. Employee engagement in health and safety education, according to Kao et al. (2019), will significantly increase their level of awareness and provide them with a better understanding of the possible hazards they face at work.

2.1 Technical and Vocational Trainees' Compliance with Occupational Health and Safety Protocols

A significant accident that occurred in the TVET workshop was caused by a failure to follow safety procedures, raising concerns about the environment's overall safety (Omeje and Osita 2014). Gloves, masks, goggles, and safety clothing are all required for everyone attending the TVET workshop. Wearing gloves does not make handwashing unnecessary because they may have tiny, unnoticeable defects or may rip while being worn. Additionally, hands could become soiled when the gloves are removed (Smelzer and Bare 2003). The degree of non-adherence among workers can occasionally be significant because they are unsure of how to deploy protective barriers in the right way (Janjua et al. 2007). Limited comprehension, a large workload, and forgetfulness are additional factors that affect employee non-compliance with safety measures (Janjua et al. 2007). Amenger (2013) claimed that repeated workshop mishaps cause students and employees to lose confidence, which inhibits learning and production processes. A study was carried out by Rizwan and colleagues (2010) to examine the issue of PPE compliance (PPE). According to the study, although employers offer some form of PPE, workers are not devoted to using it due to factors such as uncomfortable or inadequate fitting clothing, discomfort from the heat, decreased productivity, absence of PPE, absence of employer enforcement, and absence of training in proper usage. For employees to adhere to safety procedures better and feel more satisfied with their work, they must have high levels of understanding (Khan et al. 2012). The lack of this knowledge during initial training courses and orientation programmes may be responsible for the low level of awareness and comprehension of general precautions among employees. Despite the fact that instruction and training have been given, it is important to keep in mind that understanding and adherence to recommended practises may still be inadequate due to poor information retention (Stein et al. 2003). The development of worker awareness and increasing adherence to best practises have been proven to be greatly aided by training and education (Wang et al. 2003). On the other hand, Tesfay and Habtewold (2014) pointed out in their research that an absence of or insufficiency of basic safety equipment, consisting of masks, gloves, and goggles, has been highlighted as a factor impeding compliance in multiple studies.

3 Research Methods

3.1 Research Design

The study was carried out with utilisation of quantitative research methods. The study's chosen design is a descriptive one, which, in accordance with Grimes and Schulz (2002), is simply concerned with and intended to characterise the distribution of variables as they currently exist, without consideration for causal or other hypotheses. Because the study's main objective was to describe in great detail the occupational health and safety practises used in TVET Institute training workshops, it was descriptive in nature.

3.2 Population and Sample

All students who attended the TVET Institution Workshop were the target population for the research. The study included 200 trainees at the four TVET institutions in Greater Accra and Ashanti region training facility as part of a predetermined sample size. The participants of the research were chosen with purposive sampling. The critical case sampling method is the kind of purposive sampling approach employed in this study, in which a typical case is selected to explore a certain phenomenon in depth. While logical inferences can be drawn with caution from this type of sampling, statistical generalisations cannot be made. As a result, the findings of this study are largely transferable to other TVET institution workshops throughout Ghana. It is possible to draw conclusions and make projections for different types of TVET institution workshops, but only after carefully weighing the system variations of such workshops.

3.3 Data Collection Instruments and Procedures

A questionnaire was used in the research. The survey had both closed- and open-ended questions. Four sections, A through D, made up the questionnaire. Section A compiles the respondents' demographic information. Information on the occupational health and safety policies at TVET institutions workshops is gathered in Section B, while trainees' compliance with health and safety policies is elicited in Section C, and information on safety measures taken at TVET institution workshops is assembled in Section D. Instructors from the TVET Institution Workshop participated in the instrument's pilot test. Some of the questions had to be restructured as a result. With the assistance of several workshop assistants, the questionnaires were distributed to the trainees. At several workshops, the researchers immediately gathered all 200 questionnaires. The study's reliability test, however, produced a Cronbach alpha result of 0.85. The analysis was done using SPSS Version 21 (Statistical Product and Service Solutions). Utilising descriptive statistics like frequencies and percentages, the data were statistically analysed.

4 Result Analysis and Discussion

4.1 Demographic Characteristics of Respondents

The respondents' demographic details, as stated in Table 1, are presented in this section. The study revealed that male respondents made up 90.5% of the total sample. About 28% of them were under the age of 19, 26.5% were in the 20 to 26-year range, and 19.5% were above the age of 40. In addition, the findings revealed that 33.5% of the respondents had been trained at TVET colleges for three years.

Table 1. Respondents demographic characteristics (N = 200)

Variables	Frequency	Percentage (%)
Gender		
Male	181	90.5
Female	19	9.5
Age		
Below 19 years	56	28.0
20–26 years	53	26.5
27–31 years	18	9.0
32–40 years	34	17.0
Above 40 years	39	19.5
Number of years of training		
1 year	25	12.5
2 years	46	23.0
3 years	67	33.5
4 years and above	62	31.0

4.2 Procedures for Occupational Health and Safety in TVET Institutions

The outcomes of the occupational health and safety procedures in the TVET workshop are displayed in Table 2 below. The majority of participants 65%, believed that the workshop had policies in place regarding occupational health and safety standards. Additionally, roughly 62% of those surveyed said the institution gives new hires a safety orientation. Additionally, it was discovered that more than half of participants 57% said the TVET management did not directly talk to trainees about health and safety. The majority of participants 66.5%, responded negatively when asked if personal protection equipment is provided to trainees in an acceptable and timely manner.

Table 2 further revealed that 60% of respondents said that the TVET Institution session provided them with information on how to prevent workplace dangers. 70% of those surveyed said that monthly training sessions on workplace dangers are organised for trainees. The most common piece of personal protective equipment given to trainees is an overall or overcoat (28%), safety gloves (17.5%), boots (14.3%), a helmet (10.9%), glasses (10.9%), ear protection (9.2%), and a nose or dust mask (9.2%).

Table 2. TVET institution occupational health and safety procedures (N = 200)

Variables	Frequency	Percentage (%)
Establishing policies that address OHS procedures		
Yes	130	65.0
No	40	29.0
Don't know	30	15.0
Giving trainees a safety orientation		
Yes	123	61.5
No	64	32.0
Don't know	13	6.5
Management gives learners direct health and safety talks		
Yes	86	43.0
No	114	57.0
PPEs are provided to trainees		
Yes	67	33.5
No	133	66.5
Information sharing about how to avoid workplace dangers		
Yes	120	60.0
No	68	34.0
Don't know	12	6.0
How often do trainees have training sessions? *		
Every morning	5	2.5
Every week	10	5.0
Every two weeks	45	22.5
Every month	140	70.0
PPE options available to students *		
Protective boots	65	14.3
Protective gloves	80	17.5
Overall/overcoat	128	28.0
Ear plug	42	9.2
Nose mask or dust mask	42	9.2
Hard hat	50	10.9
Safety googles	50	10.9

** Multiple response, n = 457

Study results revealed that TVET Institution's occupational health and safety practises were subpar. In a survey of respondents, it was shown that more than half (55.3%)

said management did not directly talk to trainees about health and safety. The majority of them (68%) also claimed that they did not receive personal protection equipment in a sufficient or timely manner. This TVET Institution workshop's subpar occupational health and safety procedures are consistent with those at other institutions, according to a study by Adebola (2014).

Trainees' compliance with health and safety guidelines. Determine the degree of trainee compliance with occupational health and safety procedures was the study's third goal. Table 3 displays the responses they provided. It was discovered that most respondents (77%) wear their personal protective equipment when working in the workshop. Approximately 45% of the respondents said they use it constantly. 78.5% of those polled said that the institution should forbid trainees without personal safety equipment from entering the workshop.

Nearly 65.5% of the respondents claimed that there are a few things that influence how much they adhere to the workshop's OHS policies. They identified the main obstacles to their level of compliance with occupational health and safety procedures at the workshop as the absence of personal protective equipment (PPE) (36.1%), a lack of knowledge of such procedures (23.1%), ignorance of such procedures (23.1%), and insufficient workshops (17.7%).

Table 3. Trainees compliance to health and safety procedures (N = 200)

Variables	Frequency	Percentage (%)
Respondents using protective equipment		
Yes	154	77.0
No	46	23.0
How frequently do respondents wear protective gear?		
Always	90	45.0
Seldom	77	38.5
Never	33	16.5
Participants concur that the institution should prohibit trainees from entering without PPE		
Agree	157	78.5
Disagree	43	21.5
OHS compliance obstacles are present		
Yes	131	65.5
No	69	34.5

(continued)

Table 3. (*continued*)

Variables	Frequency	Percentage (%)
Barriers to compliance with OHS*		
PPE is not available	86	36.1
Information regarding OHS is insufficient	55	23.1
Unsuitable workshops	42	17.7
inadequate OHS knowledge	55	23.1

* Multiple response, n = 289

The purpose of the survey was to get respondents to recommend ways that the TVET Institution may lessen workplace dangers for trainees in the workshop. Table 4's results provide a variety of suggestions. The most common tactics include posting safety warnings, rules, and regulations in the workshop (16.8%), arranging for regular health and safety training for trainees (14.9%), holding health and safety meetings (13.7%), encouraging trainees to practise OHS (13.4%), and giving trainees personal protective equipment by TVET management.

Table 4. Addressing occupational hazards in TVET institution workshop

Variables	Frequency	Percentage (%)
Management provides learners with regular PPE	52	12.7
Trainees receive regular health and safety training	61	14.9
OHS measures are offered and improved	3	0.7
Compliance with safety laws and regulations	23	5.6
The institution should hire OHS officials	11	2.7
Concerns about safety should interest the institution's management	4	1
Regular meetings for health and safety should be held	56	13.7
PPEs and safety orientations ought to be required	19	4.6
The workshop should conspicuously display safety notices, rules, and regulations	69	16.8
In order to certify trainers, management should offer OHS training	1	0.2
Educating learners about dangers in the workshop	55	13.4
Arranging routine medical exams to determine the trainees' health	2	0.5

(*continued*)

Table 4. (*continued*)

Variables	Frequency	Percentage (%)
The workshops require that trainees adhere to health and safety regulations	54	13.2
Total	**410***	**100.0**

* Multiple response, N = 200

According to the results of Adebola's (2014) study, the trainees in this study showed a high level of commitment to occupational health and safety rules.

Numerous studies also yield contradictory results. According to Chadir et al. in 2007, a sizable amount of non-compliance with occupational health and safety rules may occur occasionally.

5 Conclusion

It was necessary for trainees to follow health and safety procedures in order to reduce the likelihood of fatalities at the TVET workshops because the tasks they performed in the workshops at the TVET institutions exposed them to risks, including handling materials, using tools and equipment, operating machines, and dealing with dangerous situations. The study's findings offer important discoveries about how to lower the accident rate at the TVET workshops, the study's authors can infer. Despite the fact that there are established safety procedures at the TVET workshop, managers of TVET institutions frequently fail to effectively implement them. This covers protocols for conducting safety orientations, giving direct health and safety lectures, giving trainees PPE, disseminating knowledge to help prevent work dangers, and planning educational training. Following the disclosures, the vast majority of trainees proposed that the institutions limit the number of trainees who attend TVET workshops without wearing protective equipment (PPE). As a result of the study, it is also possible to draw the conclusion that there are several variables that influence how well trainees adhere to workplace safety and health regulations in the workshop.

The study also discovered that posting safety signs, rules, and regulations, offering regular health and safety training, holding regular health and safety meetings, raising awareness of potential hazards, adhering to health and safety procedures, and having management provide sufficient and appropriate PPE to trainees when necessary are all essential for reducing occupational hazards among trainees in the TVET workshop. The study made the following recommendations: TVET institution managers should guarantee that operations at TVET workshops are effectively supervised, and government organizations should be urged to conduct routine inspections of TVET workshops to ensure that safety regulations are being followed.

References

Adebola, J.O.: Knowledge, attitude and compliance with occupational health and safety practices among pipeline products and marketing company (PPMC) staff in Lagos. Merit Res. J. Med. Med. Sci. **2**(8), 158–173 (2014). http://www.meritresearchjournals.org/mms/index.htm

Aluko, O.O., Adebayo, A.E., Adebisi, T.F., Ewegbemi, M.K.: Knowledge, attitudes, and perceptions of occupational hazards and safety practices in Nigerian healthcare workers knowledge, attitudes, and perceptions of occupational hazards and safety practices in Nigerian healthcare workers. BMC. Res. Notes (2016). https://doi.org/10.1186/s13104-016-1880-2

Amenger, M.: Workshop Management Techniques Needed for Improving the Teaching of Electrical Technology in Technical Colleges in Benue State. Unpublished M.Ed dissertation, University of Nigeria, Nsukka (2013)

Amfo-Otu, R., Agyemang, K.J.: Occupational health hazards and safety practices among the informal sector auto mechanics. Appl. Res. J. **1**(4), 59–69 (2016)

Anaele, E.O., Adelakun, O.A., Olumoko, B.O.: Re-engineering technical vocational education and training (TVET) towards safety practice skill needs of sawmill workers against workplace hazards in Nigeria. J. Educ. Pract. **5**(7), 150–157 (2014). http://www.iiste.org/Journals/index.php/JEP/article/view/11606. Accessed 22 Mar 2022

Carleton University: Employee Health and Safety Orientation Manuals: A Guide to Health and Safety in The Workplace (2013)

Dorji, K., Hadikusumo, B.H.W.: Safety management practices in the Bhutanese construction industry. J. Constr. Dev. Ctries. **11**(2), 53–75 (2006)

Grimes, A., Schulz, M.D.: Descriptive studies: what they can and cannot do. Lancet **359**(9301), 145–149 (2002)

Hettiarachchi, G.H.T.H., Coomasaru, P.: Challenges, Trends, and Opportunities of Technical Vocational Education and Training Common Causes of Accident and Safety Precaution - A Review. International Research Symposium – University of Vocational Technology (2021)

International Labour Organisation: Guidelines on occupational safety and health management systems. International Labour Organisation, Geneva (2013)

Itohan, O.J.: Strategies for preventing accidents and maintaining industrial safety in technical education workshops. J. Sci. Technol. Educ. (JOSTE) **6**(4), 217–226 (2018)

Janjua, N.Z., Razaq, M., Chandir, S., Rozi, S., Mahmood, B.: Poor knowledge predictor of non-adherence to universal precautions for blood borne pathogens at first-level care facilities in Pakistan. BMC Infect. Dis. **7**(1), 1–11 (2007)

Kao, K.-Y., Spitzmueller, C., Cigularov, K., Thomas, C.L.: Linking safety knowledge to safety behaviours: a moderated mediation of supervisor and worker safety attitudes. Eur. J. Work Organ. Psychol. **28**(2), 206–220 (2019)

Kendall, J.: Toolbox Talks and the Importance of Safety Meetings in Construction(2020). https://www.constructconnect.com/blog/importance-safety-meetings-toolbox-talks-construction. Accessed 24 Mar 2022

Khan, N., Khowaja, K.Z.A., Ali, T.S.: Assessment of knowledge, skill, and attitude of nurses in chemotherapy administration in tertiary hospital Pakistan. Open J. Nurs. **2**, 97–103 (2012)

Occupational and Environmental Safety Manual (2013). www.safety.duke.edu/pdf. Assessed 23 Mar 2022

Micah, J.A., Aikins, K.S.: Safety training in Ghanaian Industries Cape Coast: Institute for Development Studies, University of Cape Coast (2002)

Okwelle, P.C., Okeke, B.C.: An overview of the role of technical vocational education and training (TVET) in the national development of Nigeria. Afr. J. Hist. Sci. Educ. **12**(1), 167–182 (2016)

Okwelle, P.C., Normakoh, J.: Assessment of health, safety, and environment procedures in technical colleges' workshops in rivers state. Int. J. Innov. Sci. Eng. Technol. Res. **7**(1), 1–6 (2019)

Omeje, P.U., Osita, H.O.: Safety Skills Desired of Workshop Technicians/Assistants for Effective Management of Technical College Workshop (2014)
Rizwan, U., Ahmed, S.M., Azhar, S.: Safety management practices in Florida construction industry. In: Proceedings of the Associated Schools of Construction (ASC) 43rd International Conference, Flagstaff, Arizona, USA, 11–14 April 2007 (2010)
Rowe, H.: Best practice in health and safety through staff involvement. Conference Papers Safety in Action 2001. Safety Institute of Australia, Melbourne (2001)
Saskatchewan Polytechnic: Needs Assessment of the TVET System in Ghana as it relates to the Skill Gaps that Exist in the Extractive Sector (2016)
Saskatchewan Occupational Health and Safety Regulations (2020)
WorkSafe Saskatchewan: Health and Safety Orientation Guide for employers (2014)
Smelzer, S.C., Bare, B.: Brunner and Suddarth's Textbook of medical surgical nursing, 10th edn. Lippincott Williams and Wiking, Philadelphia (2003)
Stein, A.D., Makarawo, T.P., Ahmad, M.F.: A survey of doctors' and nurses' knowledge, attitudes, and compliance with infection control guidelines in Birmingham teaching hospitals. J. Hosp. Infect. **54**(1), 68–73 (2003)
Tesfay, F.A., Habtewold, T.D.: Assessment of prevalence and determinants of occupational exposure to HIV infection among healthcare workers in selected health institutions in Debre Berhan Town, North Shoa Zone, Amhara Region, Ethiopia. AIDS Res Treat. 1–11 (2014)
The Occupational Health and Safety (OHS) Act State Government of Victoria: Personal protective equipment (PPE) Australia (2019). https://www.education.vic.gov.au/school/students/beyond/. Accessed 23 Mar 2022
Van-Bommel, W.J.M.: Non-visual biological effect of lighting and the practical meaning of lighting for work. Appl. Ergon. **16**, 258–265 (2006)
Wang, H., Fennie, K., He, G., Burgess, J., Williams, A.B.: A training programme for presentation of occupational exposure to blood borne pathogens impact on knowledge, behaviour, and incidence of needle stick injuries among student nurses in Changsha People's Republic of China. J. Adv. Nurs. **41**(12), 187–194 (2003)
WorkSafe NB 2020 safety talk? Guidelines for use. https://www.worksafenb.ca/safety-talk/en/guidelinesforuse.html. Accessed 24 Mar 2022
WorkSafe Saskatchewan: Health and safety of worker's orientation & training: Guide for employers (2021)
Zakir, H., Burhanuddin, M.A., Khanapi, A.G.: Measure of awareness on occupational health and safety vulnerability in technical and vocational education and training institutions. Turk. J. Comput. Math. Educ. **12**(9), 1093–1103 (2021)

The Effect of Magnetic Field on the Motion of Magnetic Nanoparticles in Nanofluid

R. N. A. Akoto[1,2(✉)] and L. Atepor[3]

[1] School of Graduate Studies, University of Professional Studies, Accra, Ghana
nii.ayitey-akoto@upsamail.edu.gh

[2] Department of Petroleum and Natural Gas Engineering, University of Mines and Technology, Tarkwa, Ghana

[3] Department of Mechanical Engineering, Cape Coast Technical University, Cape Coast, Ghana

Abstract. Purpose: An external magnetic field is presented in this paper as a mechanism for regulating the motion of magnetic nanoparticles in nanofluids. An analytical solution is also presented for the motion of these particles in fluids.

Design/Methodology/Approach: We develop and solve the equation of translational motion for a magnetic nanoparticle in water subjected to an external magnetic field via a simplified description of the field gradient. The effect of some parameters on the motion are also simulated with MATLAB and displayed on graphs.

Findings: The results shows that beyond the thermal fluctuations of the fluid which gives rise to Brownian motion, the size of the nanoparticle, the characteristics of the magnetic, separation of the magnetic poles as well as the polarization of the nanofluid are the main parameters that control the particle's motion.

Research Limitation/Implications: The research approach to the development of the velocity model was entirely analytical and devoid of experimental or pseudo-experimental data.

Practical Implication: The control of magnetic nanoparticle suspensions in fluids can be achieved for magnetoplating and the manufacturing of industrial fluids with alterable thermal properties.

Social Implication: This work comes as an advantage to the nanomanufacturing industry, as the knowledge advanced in this work lays basis for the sustainable design and manufacturing of various materials such as special/smart paints, coolants, adhesives etc.

Originality/Value: The novelty of this work lies in the fact that, the knowledge of magnetic nanoparticle motion control using magnetic fields is critically lacking for sustainable industrial applications.

Keywords: Magnetic field · Model · Motion · Nanofluid · Nanoparticles

C. Aigbavboa et al. (Eds.): ARCA 2022, *Sustainable Education and Development – Sustainable Industrialization and Innovation*, pp. 1135–1142, 2023.
https://doi.org/10.1007/978-3-031-25998-2_88

1 Introduction

Advances in nanotechnology have made working at the nanoscale simpler, and recent years have seen a rise in the importance of nanoscience research. This has actioned the boundaries of science and engineering study to be pushed by technology. Nanofluids are fluids that may have their properties, such as viscosity, thermophysical, and stability, changed at will to suit specific applications (Choi et al. 2001; Ouabouch et al. 2021). They are created by distributing nanometer-sized particles in a base fluid in the form of a colloidal solution.

The Magnetic Nanofluids (MNfs) or simply ferrofluid is a type of nanofluid with magnetic nanoparticles (MNP) scattered in a base fluid and has acquired widespread use in the biological, biomedical, petroleum and other engineering communities. A detailed write up on their fascinating properties and applications can be found in Hassan, et al. (2021), Kumar and Subudhi (2018), Pîslaru-Dănescu, et al. (2017), Pîslaru-Dănescu, et al. (2013) and Vékás (2004). Although the magnetophoretic mobility and magnetic polarization serve as the foundation for MNP transport in the presence of external magnetic field gradient (Han et al. 2015; Nacev et al. 2011), the efficiency to effectively manipulate the MNP in fluids in a fabricated system is a challenge (Wei and Wang 2018). Given that Brownian motion of nanoparticles induced by thermal energy plays a crucial role in determining the thermal behavior of nanofluids at the molecular and nanoscale level, the motion of magnetic nanoparticles could be modified by other driving forces, such as electrical or magnetic fields, enabling nanofluids to be controlled and enhanced further. According to Min, et al. (2005) and Jang and Choi (2004), Maxwel's theory and other macroscale theoretical approaches are insufficient to explain the many characteristics and difficulties of the nanofluid system, particularly when external forces are used to generate transport that exceeds Brownian limits. The underlying mechanism of MNP in ferrofluids beyond Brownian bounds has not been detailed in any published works, according to a literature search. It is therefore the aim of this work, to present a basic comprehension of the properties of MNP mobility in nanofluids in the presence of an external magnetic field gradient. We offer an analytical solution for the MNP motions in ferrofluids based on the force equilibrium affecting the MNPs under the magnetic field.

The remaining work is divided up as follows. A description of the magnetic force field around the ferrofluid is given in Sect. 2. The 1-D equation of motion of the MNP in the ferrofluid is stated using the Langevin equation in the Ornstein-Uhlenbeck theory. An analytical solution of the equation of motion is achieved via a simplification of the description of the magnetic field gradient in Sect. 2. From the simplification, it is possible to read off other directions to allow for a one-dimensional statement of the magnetic force. A discussion of the results is presented in Sect. 3, which presents the motion of the MNP under various conditions. Conclusion is found in Sect. 4.

2 Mathematical Modeling

2.1 Magnetic Force

The single-domain MNPs suspended in the base fluid become magnetized when they are subjected to an external magnetic field. In ferrofluid systems, magnetization, also known as polarization, produces attractive forces on each MNP to cause a body force on the base fluid. This body force, also known as the magnetic force operating on the ferrofluid in the present context is given by (Oldenburg et al. 2000)

$$F_m = \mu_0 M \nabla_{\hat{r}} H, \tag{1}$$

where M is the magnetization, μ_0 is the magnetic permeability and H is the total magnetic field streamlined in rectangular coordinates. The non-linear relationship,

$$M = \alpha \tan^{-1}(\omega H), \tag{2}$$

shows the rise in a ferrofluid's magnetization until it reaches saturation, beyond which time further increases in H have no effect on the change in M. Also, α and ω characterizes the ferrofluid so that their magnitudes regulate the initial susceptibility and saturation magnetization levels, respectively with $10^4 \leq \alpha \leq 10^5$ A/m and $10^{-6} \leq \omega \leq 10^{-5}$ m/A.

In order to define H as functions of the characteristics and geometry of a permanent magnet, we use McCaig and Clegg (1987) approach. Although the derivation is rather lengthy, the magnetic field strength with magnetic polarization, B_r at any point in space is the difference between the fields due to each pole such that,

$$H_n = G_n(x, y, z) - G_n(x + l, y, z), \tag{3}$$

where the subscript n denote the choice to render definitions either in the x, y or z directions and the functions

$$\begin{aligned} G_x(x, y, z) = \frac{B_r}{4\pi\mu_0}\Bigg\{ & \tan^{-1}\left[\frac{(y+a)(z+b)}{x\left[(y+a)^2+(z+b)^2+x^2\right]^{1/2}}\right] \\ & + \tan^{-1}\left[\frac{(y-a)(z-b)}{x\left[(y-a)^2+(z-b)^2+x^2\right]^{1/2}}\right] \\ & - \tan^{-1}\left[\frac{(y+a)(z-b)}{x\left[(y+a)^2+(z-b)^2+x^2\right]^{1/2}}\right] \\ & - \tan^{-1}\left[\frac{(y-a)(z+b)}{x\left[(y-a)^2+(z+b)^2+x^2\right]^{1/2}}\right]\Bigg\}, \end{aligned} \tag{4}$$

$$\begin{aligned} G_y(x, y, z) = \frac{B_r}{4\pi\mu_0} \ln\Bigg\{ & \frac{(z+b)+\left[(z+b)^2+(y-a)^2+x^2\right]^{1/2}}{(z-b)+\left[(z-b)^2+(y-a)^2+x^2\right]^{1/2}} \\ & \times \frac{(z-b)+\left[(z-b)^2+(y+a)^2+x^2\right]^{1/2}}{(z+b)+\left[(z+b)^2+(y+a)^2+x^2\right]^{1/2}}\Bigg\}, \end{aligned} \tag{5}$$

$$G_z(x, y, z) = \frac{B_r}{4\pi\mu_0} \ln\left\{\frac{(y+a)+\left[(z-b)^2+(y+a)^2+x^2\right]^{1/2}}{(y-a)+\left[(z-b)^2+(y-a)^2+x^2\right]^{1/2}}\right.$$

$$\times \frac{(y-a)+\left[(z+b)^2+(y-a)^2+x^2\right]^{1/2}}{(y+a)+\left[(z+b)^2+(y+a)^2+x^2\right]^{1/2}}\Bigg\}. \tag{6}$$

The dimensions of the permanent magnet are l, $2a$ and $2b$ in the x, y and z directions respectively.

Substituting Eq. (3) into Eq. (1) via Eqs. (4–6) and applying them to ferrofluid flow problems result in complications that can only be resolved by numerical methods. However, assuming a ferrofluid with a uniform magnetic permeability and a magnetic field source emanating from a permanent magnet, a simpler expression for H can be obtained to allow for analytical solution methods (Ida 1995; Oldenburg et al. 2000).

2.2 Motion of Magnetic Nanoparticles in Water

The one-dimensional equation of motion for the MNP in water subjected to an external magnetic force can be straightforwardly derived following the Langevin equation in the Ornstein-Uhlenbeck theory (Medved et al. 2020; Nelson 2001) as

$$m\frac{d\boldsymbol{v}}{dt} = \boldsymbol{F}_d + \boldsymbol{F}_B + \boldsymbol{F}_N + \boldsymbol{F}_r + \boldsymbol{F}_m, \tag{7}$$

where, m is the mass of a single MNP, $\boldsymbol{v}$ is the velocity of the nanoparticle, $\boldsymbol{F}_d$ is the hydrostatic resistive force also known as the drag force, $\boldsymbol{F}_B$ is the force arising from thermal fluctuations in the nanofluid or Brownian motion of the MNP, $\boldsymbol{F}_N$ is the force that manifests as a result of the integration of the magnetic stress tensor at the boundary between a magnetic fluid and a non-magnetic fluid, and $\boldsymbol{F}_r$ is the magnetorestrictive force and $\boldsymbol{F}_m$ is the magnetic force. The hydrostatic resistive force is defined as

$$\boldsymbol{F}_d = \eta \boldsymbol{v}, \tag{8}$$

where friction coefficient, η, is related to the viscosity of the nanofluid, μ and the Brownian relaxation time τ_B by $\eta = 6\pi\mu r = m\tau_B$ The radius of a single MNP is denoted by r. The Brownian force is

$$\boldsymbol{F}_B = \left(\frac{12\pi\mu r k_B T}{\Delta\tau}\right)^{\frac{1}{2}} \boldsymbol{\xi}. \tag{9}$$

Here k_B represent the Boltzmann constant, T is the temperature of the ferrofluid, $\boldsymbol{\xi}$ is the Gaussian vector and $\Delta\tau \gg \tau_B$. In this work, it is assumed that the ferrofluid is single-phase and incompressible, and that the magnetization is independent of the density of the ferrofluid. Therefore, the effects of $\boldsymbol{F}_N$ and $\boldsymbol{F}_r$ can reasonably be ignored. Substituting the 1-dimensional scalar forms of Eqs. (1) and (8) into that of Eq. (7) gives the equations of motion of a MNP in a base fluid exposed to an external magnetic field as

$$\frac{dv_x}{dt} + \frac{\eta}{m}v_x = \frac{F_B}{m} + \frac{\mu_0 M}{m}\frac{\partial H}{\partial x}, \qquad v_x(t=0) = 0. \tag{10}$$

2.3 Solution of Equation of Motion

The formal solution of Eq. (10) is given by

$$v_x(t) = v_x(0)e^{-\eta t/m} + \frac{1}{m}\int_0^t e^{-\eta(t-t')/m}\left[F_B + \mu_0 M\frac{\partial H}{\partial x}\right]dt'. \tag{11}$$

Except for the specification of the magnetic field gradient, $\partial H/\partial x$, whose description complexity as already noted in the last paragraph of Sect. 2.1, this solution appears to be simple. We therefore simplify the problem further by considering the x – axis of the magnet to be aligned with the direction of MNP motion, so that $y = z = 0$ and Eq. (3) is rewritten as

$$H_x = G_x(x, 0, 0) - G_x(x + l, 0, 0), \tag{12}$$

and $G_x(x, 0, 0)$ is obtained from Eq. (4) as

$$G_x(x, 0, 0) = \frac{B_r}{\pi\mu_0}\left\{\tan^{-1}\left[\frac{ab}{x\left[a^2 + b^2 + x^2\right]^{1/2}}\right]\right\}. \tag{13}$$

The magnetic field gradient is then deduced by differentiating Eqs. (12) to have

$$\frac{\partial H}{\partial x} = \frac{B_r}{\pi\mu_0}\left\{\frac{ab\left[2(x+l)^2 + c\right]}{\sqrt{(x+l)^2 + c}\left\{(x+l)^2\left[(x+l)^2 + c\right] + a^2b^2\right\}} - \frac{ab\left(2x^2 + c\right)}{\sqrt{x^2 + c}\left(x^4 + cx^2 + a^2b^2\right)}\right\}. \tag{14}$$

where $c = a^2 + b^2$. Substituting Eq. (14) into Eq. (11) and performing the integration, we arrive at the velocity of the MNP in the ferrofluid under external magnetic field influence as

$$v_x(t) = \frac{1}{\eta}\left\{F_B + \frac{MB_r}{\pi}\left[\frac{ab\left[2(x+l)^2 + c\right]}{\sqrt{(x+l)^2 + c}\left\{(x+l)^2\left[(x+l)^2 + c\right] + a^2b^2\right\}} - \frac{ab\left(2x^2 + c\right)}{\sqrt{x^2 + c}\left(x^4 + cx^2 + a^2b^2\right)}\right]\right\}\left(1 - e^{-\eta t/m}\right) \tag{15}$$

3 Discussion of Results

The following parameter values are used in the computation of velocities via Eq. (15); namely Boltzmann constant, $k_B = 1.38064852 \times 10^{-23}$ [kgm^2/s^{2}K], half height of the magnet, $b = 0.1$ [mm], half width of magnet, $a = 0.1$ [mm], magnetic permeability, $\mu_0 = 4\pi 10^{-7}$ [Tm/A], mass of the MNP, $m = 1.96\times 10^{-20}$ [kg], residual magnetization,

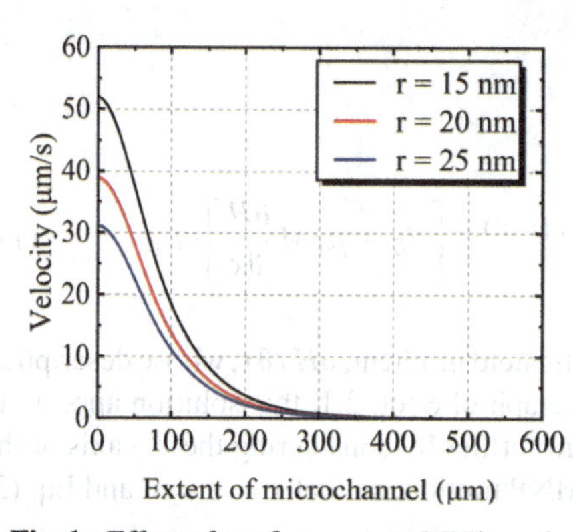

Fig. 1. Effect of confinement on MNP motion ($\alpha = 1.5 \times 10^7$ A/mm, $l = 0.5$ mm, $\omega = 2.4 \times 10^{-2}$).

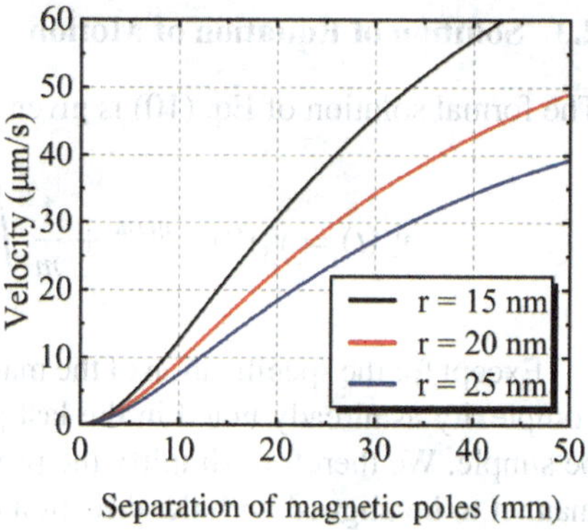

Fig. 2. Effect of poles' separation on MNP motion ($\alpha = 1.5 \times 10^7$ A/mm, $x = 100\,\mu$m, $\omega = 2.4 \times 10^{-2}$)

$B_r = 11.9$ [mT], temperature of ferrofluid, $T = 303$ [K] and the viscosity of base fluid, $\mu = 8.9 \times 10^{-4}$ [kg/m · s].

The motion of MNP is plotted against its spatial confinement in Fig. 1 for three different sizes of the MNPs. Even though the 15 nm MNP manifests higher velocities, followed by the 20 nm MNP and then the 25 nm MNP, a decline in the motions is observed for all the three sizes as the spatial confinement increases. This indicates that the field strength has less effect as the MNP moves away from the magnetic source, thereby indicating weaker polarization around the vicinity of the MNP. This phenomenon also confirms that the farther the MNP is from the magnetization source, the lesser the effect of the magnet field. This is further explained as residual magnetization immediately being lost as one moves farther from magnetic source so far as the ferrofluid is concerned. Unlike other materials that retain polarization for longer times even after the withdrawal of the magnet field, ferrofluids tend to lose theirs within a shorter time (Li et al. 2014; Muhammad et al. 2018). The motion of the MNP in essence is then controlled by Brownian motion as magnet is moved farther away from the test particle.

At a constant temperature and confinement, the effect of magnetic poles separation on velocity is studied for varying sizes of the MNP and shown in Fig. 2. It is observed that the motion of the MNP, as controlled by the magnetic force, increases nonlinearly as the separation of the magnet's poles increase. This may be due to the proportionally higher number of aligned magnetic domains and field lines as the poles' separation increase (indicative of increasing size of magnet) thereby producing stronger fields - a phenomenon also reported in the case of Žežulka and Straka (2016). As already mentioned, the less weighty MNP experiences greater perturbation in the face of the external field and thus the 15 nm particles experiences elevated velocities comparable to that of the 20 and 25 nm particles.

In Fig. 3, we show the effect of the parameters that control magnetization on the motion of the MNP in a three-dimensional surface plot. The MNP velocity is shown on the vertical axis while the parameters that regulate the initial susceptibility and saturation

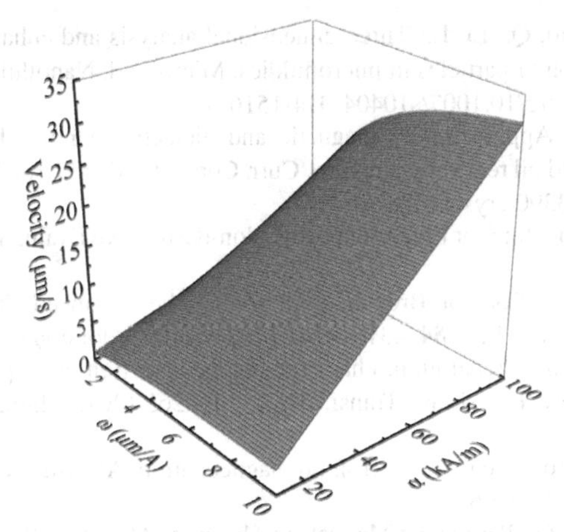

Fig. 3. Effect of magnetization parameters on MNP motion ($l = 0.5$ mm, $r = 25$ nm, $x = 100$ μm)

magnetization levels of the magnetic are depicted on the horizontal axes. The increasing motion can be explained via Eq. (2). Magnetization increases linearly with α. Also, since $\tan^{-1}(\omega)$ is an ever-increasing function, the magnetization will always increase. Therefore, an increase in α and ω simply shows an increasing supply of magnetic force beyond the Brownian limits which consequently manifests in increased motion.

4 Conclusions

A number of fields in the engineering, the medical, the pure and applied sciences are affected by the issue of MNP motion in a ferrofluid in an external electromagnetic field. In this paper, the fundamental equations that govern the mobility of a MNP in a ferrofluid in a non-uniform external magnetic field are proposed. According to the study's findings, magnetic fields can be used to regulate how MNP moves in a ferrofluid. MNPs in ferrofluids are controlled by several variables, including their size, magnetic property, separation of magnetic poles, and polarization (magnetization), in addition to thermal fluctuations, which cause Brownian motion. It is recommended that the theoretical results from this study be confirmed by experimental or pseudo-experimental means such as Molecular Dynamics simulation and/or purely numerical methods. An evaluation of other electromagnetic formalisms as external forces for MNP motions in ferrofluids would be an interesting study. It would also be interesting to explore other electromagnetic formalisms as possible external forces for MNP motions in ferrofluids.

Conflict of Interest. The authors declare that they have no conflict of interest.

References

Choi, S.U., Zhang, Z.G., Yu, W., Lockwood, F.E., Grulke, E.A.: Anomalous thermal conductivity enhancement in nanotube suspensions. Appl. Phys. Lett. **79**(14), 2252–2254 (2001). https://doi.org/10.1063/1.1408272

Han, X., Feng, Y., Cao, Q., Li, L.: Three-dimensional analysis and enhancement of continuous magnetic separation of particles in microfluidics. Microfluid. Nanofluid. **18**(5–6), 1209–1220 (2015). https://doi.org/10.1007/s10404-014-1516-6

Hassan, Y.M., et al.: Application of magnetic and dielectric nanofluids for electromagnetic-assistance enhanced oil recovery: a review. Curr. Comput.-Aided Drug Des. **11**(2), 106 (2021). https://doi.org/10.3390/cryst11020106

Ida, N.: Numerical Modeling for Electromagnetic Non-destructive Evaluation. Chapman and Hall, New York (1995)

Jang, S.P., Choi, S.U.: Role of Brownian motion in the enhanced thermal conductivity of nanofluids. Appl. Phys. Lett. **84**, 4316–4318 (2004). https://doi.org/10.1063/1.1756684

Kumar, A., Subudhi, S.: Preparation, characteristics, convection and applications of magnetic nanofluids: a review. Heat Mass Transf. **54**(2), 241–265 (2018). https://doi.org/10.1007/s00231-017-2114-4

Li, J., et al.: Investigation into loss in ferrofluid magnetization. AIP Adv. **4**(7), 1–7 (2014). https://doi.org/10.1063/1.4890866

McCaig, M., Clegg, A.G.: Permanent Magnets in Theory and Practice, 2nd edn. Wiley, New York (1987)

Medved, A., Davis, R., Vasquez, P.A.: Understanding fluid dynamics from Langevin and Fokker-Planck equations. Fluids **5**(1), 40 (2020). https://doi.org/10.3390/fluids5010040

Min, J.Y., Jang, S.P., Choi, S.U.: Motion of nanoparticles in nanofluids under electric field. In: ASME 2005 International Mechanical Engineering Congress and Exposition. Heat Transfer, Part B. Orlando, pp. 497–501 (2005). https://doi.org/10.1115/IMECE2005-80139

Muhammad, N., Nadeem, S., Mustafa, M.T.: Analysis of ferrite nanoparticles in the flow of ferromagnetic nanofluid. PLoS ONE **13**(1), 1–23 (2018). https://doi.org/10.1371/journal.pone.0188460

Nacev, A., Beni, C., Bruno, O., Shapiro, B.: The behaviors of ferromagnetic nano-particles in and around blood vessels under applied magnetic fields. J. Magn. Magn. Mater. **323**(6), 651–668 (2011). https://doi.org/10.1016/j.jmmm.2010.09.008

Nelson, E.: Dynamical Theories of Brownian Motion, 2nd edn. Princeton University Press, New Jersey (2001)

Oldenburg, C.M., Borglin, S.E., Moridis, G.J.: Numerical simulation of ferrofluid flow for subsurface environmental engineering applications. Transp. Porous Media **38**, 319–344 (2000). https://doi.org/10.1023/A:1006611702281

Ouabouch, O., Kriraa, M., Lamsaadi, M.: Stability, thermophsical properties of nanofluids, and applications in solar collectors: a review. AIMS Mater. Sci. **8**(4), 659–684 (2021). https://doi.org/10.3934/matersci.2021040

Pîslaru-Dănescu, L., Morega, A.M., Telipan, G., Morega, M., Dumitru, J.B., Marinescu, V.: Magnetic nanofluid applications in electrical engineering. IEEE Trans. Magn. **49**(11), 5489–5497 (2013). https://doi.org/10.1109/TMAG.2013.2271607

Pîslaru-Dănescu, L., Telipan, G., Stoian, F.D., Sorin Holotescu, S., Marinică, O.M.: Nanofluid with colloidal magnetic Fe_3O_4 nanoparticles and its applications in electrical engineering. In: Kandelousi, M.S. (ed.) Nanofluid heat and Mass Transfer in Engineering Problems. IntechOpen, London (2017). https://doi.org/10.5772/65556

Vékás, L.: Magnetic nanofluids properties and some applications. Rom. J. Phys. **49**(9–10), 707–721 (2004). https://doi.org/10.5772/65556

Wei, W., Wang, Z.: Investigation of magnetic nanoparticle motion under a gradient magnetic field by an electromagnet. J. Nanomater. 6246917, 1–5 (2018). https://doi.org/10.1155/2018/6246917

Žežulka, V., Straka, P.: The creation of a strong magnetic field by means of large magnetic blocks from NdFeB magnets in opposing linear Halbach array. J. Magn. **21**(3), 364–373 (2016). https://doi.org/10.4283/JMAG.2016.21.3.364

Ultrasound-Assisted Alkaline Treatment Effect on Antioxidant and ACE-Inhibitory Potential of Walnut for Sustainable Industrialization

M. K. Golly[1(✉)], H. Ma[1,2,3(✉)], D. Liu[2], D. Yating[2], A. S. Amponsah[1,3], and K. A. Duodu[1]

[1] Faculty of Applied Science and Technology, Sunyani Technical University, Box 206, Sunyani, Ghana
moses.golly@stu.edu.gh

[2] School of Food and Biological Engineering, Jiangsu University, 301 Xuefu Road, Zhenjiang 212013, Jiangsu, People's Republic of China
mhl@ujs.edu.cn

[3] Key Laboratory for Physical Processing of Agricultural Products, Jiangsu University, Zhenjiang, China

Abstract. Purpose: Bioactive peptides from food industry by-products can contribute to good health and well-being (SDG 3) and ensure sustainable industrialization. We investigated the influence of ultrasound-assisted alkali action on the bioactive potencies of the walnut by-product peptide fractions (WMPFs).

Design/Methodology/Approach: The walnut meal protein was extracted via ultrasonic-assisted alkaline treatment and then hydrolyzed with trypsin. Peptide fractions were obtained via an ultrafiltration process using specific molecular weight cut-off membranes. The amino acid, antioxidant and ACE-inhibitory activities were assayed using standardized protocols.

Findings: A fairly balanced amino acid profile was observed. Glu, Arg, and Asp were the dominant amino acids measuring 23.5–25.7, 14.2–14.9 and 9.9–11.5 g/100g protein respectively. Generally, peptides with molecular weight <5 (<1, 1–3, and 3–5) kDa displayed higher antioxidant and ACE repressive capabilities. DPPH scavenging activity also ranged between 61.59–84.19% for the ultrasound samples compared with 56.98–75.56% for the control. The ACE activity was concentration-dependent with >1 (1–3) kDa peptide fraction inhibiting 50% ACE (IC50) at 0.2 mg/mL. The antioxidant and ACE repressive actions were generally higher in the ultrasonicated samples compared to the control samples.

Research Limitation: The study involved the use of a specific type of counter-current ultrasound pretreatment machine. The antioxidant and antihypertensive potencies of the peptide fractions were not compared with existing artificial antihypertensive drugs.

Practical Implications: This study demonstrates that walnut meal protein could be fashioned as a commercial value-added functional food ingredient as evidenced by the antioxidant and antihypertensive activities of the peptide fractions. These bioactive peptides could be useful in the pharmaceutical as well as food processing

C. Aigbavboa et al. (Eds.): ARCA 2022, *Sustainable Education and Development – Sustainable Industrialization and Innovation*, pp. 1143–1163, 2023.
https://doi.org/10.1007/978-3-031-25998-2_89

industries. The ultrasonic pretreatment could help extract vital components from food industry by-products like groundnut and soybean meals.

Social Implication: Besides generating additional literature on the subject area, the study highlights information that will help policy formulation to promote the use of natural food resources to enhance good health and well-being as set out in sustainable development goal three.

Originality/Value: Previous works provided insufficient information on the ultrasound-assisted alkaline pretreatment and subsequent trypsin enzymolysis to improve walnut meal peptide antioxidant and antihypertensive activities. This study fills that gap and provides the basis for further study into the area.

Keywords: ACE inhibition · Antihypertensive activity · Antioxidant · Peptides · DPPH

1 Introduction

Sustainable development goal three (SDG 3) emphasizes good health and well-being. However, free radicals, also known as reactive oxygen species (ROS) are noxious derivatives of oxygen breakdown that can cause significant damage to living cells and tissues in the human body thereby hampering good health. In what could be described as irony, Lobo et al. (2010) detailed the detrimental action of ROS in essentially required biomolecules in the human body, consequential to cell and tissue malfunction. This destructive tendency of free radicals in the human body amounts to what is known as 'oxidative stress.' One effective way to counteract the action of ROS in reducing oxidative stress is the use of antioxidants. Both artificial and natural antioxidants from various materials have been explored for good health (Lobo et al. 2010) but fruits and vegetables are widely appraised as being the primary sources (Daliri et al. 2017; Udenigwe et al. 2009). Numerous artificial antioxidants have been used to control the detrimental action of ROS; nevertheless, the use of synthetic antioxidants presents some potential health hazards (Park et al. 2001). Thus, the increased demand for natural products in recent times has witnessed the production of antioxidants from proteins of plants and animal origins (Yang et al. 2014). Another cluster of health conditions known as cardiovascular diseases (CVDs) is gradually gaining recognition as a leading cause of death worldwide being responsible for 30% of global death (Wang et al. 2016; Pagidipati and Gaziano 2013). A critical factor enhancing these CVDs is when angiotensin I is converted into the vasoconstrictor angiotensin II (hypertensive peptide) which also promotes the inactivation of the vasodilator bradykinin (hypotensive peptide) in the Renin-Angiotensin and kallikrein-kinin systems (Cushman et al. 1982; Ganten et al. 1984; Jia et al. 2010; Wang et al. 2016). One significant and relevant enzyme that facilitates the transformation of the angiotensin I is the Angiotensin-I-converting enzyme (ACE). Therefore, inhibition of the ACE may exercise an antihypertensive outcome. Literature opined that the ACE inhibitors currently being used such as alacepril, captopril, and lisinopril for the treatment of hypertensive patients incite objectionable outcomes like coughing, loss of taste, and renal damage (Jia et al. 2010).

To reduce if not eliminate the adverse effects of artificial antioxidants and antihypertensive drugs, the use of proteins from numerous animal, plant and other natural sources have been found to have antioxidant and antihypertensive properties (Dadzie et al. 2013; Udenigwe et al. 2009; Wongekalak et al. 2011). Plant products such as soy proteins (Dabbour et al. 2019), peanut proteins (Ji et al. 2014), flaxseed proteins (Udenigwe et al. 2009), arrowhead proteins (Wen et al. 2018a, b), mungbean protein (Wongekalak et al. 2011), walnut protein (Gu et al. 2015; Lai et al. 2016; Wu et al. 2014), rapeseed (Zhang et al. 2009), canola meal protein (Aider and Barbana 2011; Alashi et al. 2014), and wheat gluten (Dadzie et al. 2013), have received much attention as sources of bioactive peptides. Food proteins and their isolates are essential biomolecules for tissue and cell growth by providing nutrition via crucial amino acids as well as playing various useful functions in food quality (Tapal and Tiku 2019). Meanwhile, in their native forms, food proteins are less active biologically due to their high molecular weights and structural conformities, but through a process of hydrolysis, the molecular weights of the proteins are reduced and are made biologically more active, possessing antioxidant abilities (Dabbour et al. 2019; Korhonen and Pihlanto 2006; Wongekalak et al. 2011). Peptide/protein fragments with antioxidant activities and other bioactivities (antihypertensive activity) which ultimately impact health positively are called bioactive peptides (Tapal and Tiku 2019).

Bioactive peptides can be produced through various hydrolysis techniques with digestive enzymes derived from plant or animal sources, by microbial fermentation and acid hydrolysis. The easiest, most cost-effective, safest and most preferred way is hydrolysis with enzymes (Alashi et al. 2014; Aluko 2015; Daliri et al. 2017; Korhonen and Pihlanto 2006; Tapal and Tiku 2019). The enzyme hydrolysis process entails the proteolysis of the protein material with an enzyme at a specified pH and temperature (Tapal and Tiku 2019). Its preference over other methods is underpinned by the shortened period of reaction, ease of prediction and scalability. Additionally, enzymatically treated proteins present superior functionality such as solubility, flavour, and digestibility while exhibiting physiologically beneficial activities (Tapal and Tiku 2019) and provide relatively similar amino acid profiles to the parent chain (Alashi et al. 2014). Since enzymes are specific and split specific peptide bonds, their access to the bonds in the protein parent chain is crucial to effective and enhanced proteolytic processes (Aluko 2018). Varied pretreatments given to protein material before enzyme hydrolysis to improve the process include thermal (Adam Yahya Abdulrahman et al. 2017; Chilakala et al. 2018; Yu et al. 2007) and non-thermal, a typical example being ultrasound (Chandrapala et al. 2011; Dabbour et al. 2019, 2018; Golly et al. 2019; Huang et al. 2017; Ma et al. 2015).

Numerous researchers have demonstrated the beneficial effect of ultrasonic pretreatment on the enzymatic hydrolysis efficiencies of food materials such as garlic powder, sunflower and walnut meals and zein protein with improved bioactivity (Dabbour et al. 2018; Golly et al. 2019; Ma et al. 2015; Ren et al. 2013; Wen et al. 2019; Wen et al. 2018a, b). Also, ultrasound technology has been widely recognized for inducing structural changes in food material before enzymolysis, which are caused by a swift impact between enzyme and substrate, consequential in massive bond cleavage (Dabbour et al. 2018; Golly et al. 2019; Ma et al. 2015). This phenomenon has been demonstrated in the previous chapter of this document, where the effect of different ultrasonic frequency modes was shown to have impacted positively on the enzymolysis of the walnut meal. In the current chapter, the researchers desire to evaluate the influence of dual-frequency counter-current ultrasound treatment on the bioactivity of the peptide fractions from the walnut meal.

The walnut meal contains nearly 80% protein which could be harnessed for use in food and pharmaceuticals (Lai et al. 2016). Hu et al. (2015) stated that, despite its numerous applications, walnut meal protein is underutilized, which is due in part to its low functionality and solubility. Meanwhile, some research works on walnut meal for bioactive peptides production through pancreatin hydrolysis (Gu et al. 2015; Lai et al. 2016; Wang et al. 2016; Wu et al. 2014) or trypsin (Lai et al. 2016) separation/purification of the ACE (Liu et al. 2013; Wang et al. 2016), and production of antioxidant peptides using solid-state fermentation (Wu et al. 2014) exist. Likewise, Zhu et al. (2018) studied the influence of ultrasonication on the useful physical and chemical characteristics of walnut protein extracts, rather than the meal. To the best of our knowledge, there is inadequate data on the countercurrent ultrasound-assisted alkaline pretreatment and subsequent enzymolysis to improve walnut meal peptide antioxidant and antihypertensive activities. Because of this, this study aimed at the fractionation of trypsin hydrolysates with ultrasound pretreatment of the walnut meal using ultrafiltration and assessing the antioxidant along with the ACE inhibitory activities of the walnut meal peptide fractions.

2 Materials and Methods

2.1 Materials

Walnut meal (WM) with an initial protein content of 72.71% was acquired from a factory in Liaoning Province, China. Trypsin (50,000 U/g) was from Sinopharm Chemical Reagent Co. Ltd. (Shanghai Shi, China). Unless otherwise stated, the entire chemicals used were presumed to be of analytical grade.

2.2 Ultrasound Pretreatment

The ultrasonic treatment followed the steps outlined in our earlier research (Golly et al. 2020) using the dual-frequency 20/40 kHz/kHz. The pretreated sample was used for the trypsin hydrolysis and was centrifuged (5000 x g, 15 min), precipitated at pH 4.5, freeze-dried, and preserved for further use.

2.3 Trypsin Hydrolysis

The ideal conditions for hydrolyzing the walnut meal using trypsin were established previously (Golly et al. 2019) as; enzyme concentration of 2000 U/g, substrate concentration of 20 g/L, hydrolysis temperature and time of 50 °C and 150 min respectively.

2.4 Ultrafiltration of the Walnut Meal Hydrolysate

The Walnut meal peptide hydrolysates (WMPHs) were separated using an ultrafiltration unit (UFU) (Pellicon, Millipore Corporate, Billerica, MA, USA). The molecular weights cut-off (MWCO) of the membranes used sequentially were 100, 30, 10, 5, 3 and 1 kDa respectively with a working area of 0.5 m^2. Thus, the retentate of the 100 kDa (>100 kDa) assumed to contain a mixture of higher molecular weight biomolecules besides the enzymes was discarded. The permeate from the 100 kDa membrane (<100 kDa) concentration step was collected and was taken through the next membrane at 30 kDa; the retentate (30–100 kDa) was collected, and the permeate infiltrated the 10 kDa membrane to obtain once more a retentate (10–30 kDa). Additionally, the permeate of the 10 kDa was infiltrated by the 5 kDa membrane to accumulate the retentate (5–10 kDa) while the permeate was screened once more by the 3 kDa membrane to accumulate the retentate (3–5 kDa). Finally, the 3 kDa membrane permeate was screened by the 1 kDa membrane to obtain both the retentate (1–3 kDa) and the permeate (<1 kDa). For further investigation, all peptide fractions were lyophilized and stored at −20 °C. To ensure efficiency, the membranes were washed (cleaned) appropriately before first use and after every filtration run, following commendations of the membrane supplier, sluicing meticulously with distilled water before and after every cleaning process. Cleaning agents used included 1M NaOH and 2% H_3PO_4, and the washing procedure was a 10 min rinsing followed by a 30 min recirculation at 40 °C respectively. A persistent 0.5 MPa and 1.0 MPa for the feed and retentate pressures were ensured respectively. Figure 1 represents the flow chart for the ultrafiltration sequence adopted in this study.

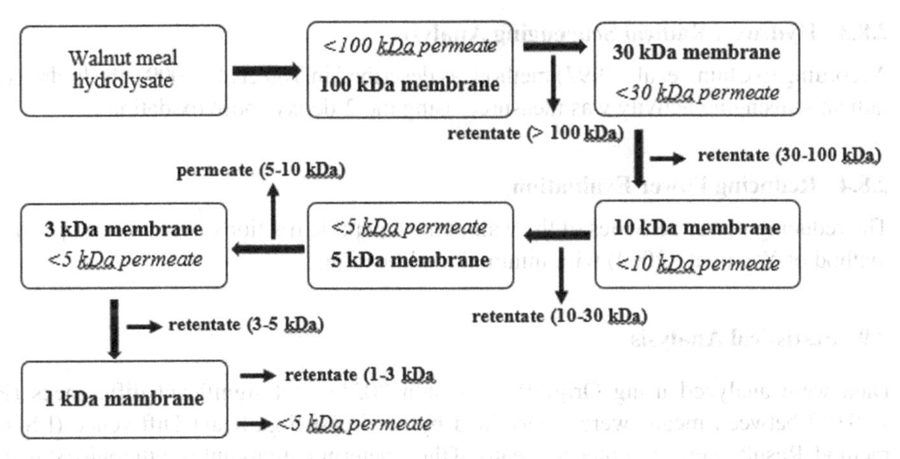

Fig. 1. Graphical representation of ultrafiltration sequence for the retentate collection method

2.5 Color Measurement

A laboratory colourimeter (Konica Minolta, Inc., Japan) was used to acquire the colour profile of the protein isolates according to Sarpong et al. (2018).

2.6 Amino Acid (AA) Evaluation

The freeze-dried peptide fractions AA contents were evaluated via Ketnawa et al. (2016) technique.

2.7 Antihypertensive Activity Determination

A modified spectrophotometric method was used to assess ACE inhibitory activity (Li et al. 2018).

2.8 Antioxidant Activity Determination

2.8.1 ABTS Radical Scavenging Activity

A minimally modified method was used to conduct the ABTS radical foraging activity assay Alashi et al. (2014) and Ilyasov et al. (2018). This test is based on the percentage inhibition of peroxidation of the $ABTS^{\cdot+}$ radicals. The reduction of the $ABTS^{\cdot+}$ back to its normal state ABTS is detected as discolouration of blue-green colour at a spectrophotometric wavelength of 734 nm.

2.8.2 DPPH Radical Scavenging Activity

A modified technique by You et al. (2009) was adopted for the determination of the scavenging activity of the walnut meal peptide fractions against the DPPH radical.

2.8.3 Hydroxyl Radical Scavenging Analysis

According to Chung et al. (1997) method, as described in You et al. (2009), the hydroxyl radical scavenging activity was measured using the 2-deoxyribose oxidation.

2.8.4 Reducing Power Evaluation

The reducing power activities of the walnut meal peptide fractions were assayed per the method of You et al. (2009) with minimal modification.

2.9 Statistical Analysis

Data were analyzed using OriginPro (version 2021), and significant differences ($P < 0.05$) between means were determined by the Least Significant Difference (LSD) method. Results were presented as means of three determinations unless otherwise stated.

3 Results and Discussion

3.1 Colour Properties of the Peptide Fractions

The results of the colour determination of the walnut meal peptide fractions (WMPFs) are displayed in Table 1. When the molecular weight decreased from >30 kDa to <1 kDa,

L^* values of WMPF meaningfully ($p < 0.05$) increased from 19.3% to 30.1% and 20.7–35.2% for the control and ultrasound samples respectively. Similarly, the b^* index increased when the molecular weight of peptide fractions reduced from >30 kDa to <1 kDa. The b^* values significantly increased from 2.5–13.4% and 6.1–17.9% for control and ultrasound samples, respectively.

Table 1. Effect of sonication and molecular weight cut-off (MWCO) on the colour of the peptides

Colour parameter	Sample	MWCO (kDa)					
		>30	10–30	5–10	3–5	1–3	<1
L*	Control	$^{b}19.3 \pm 0.5^{A}$	$^{b}21.2 \pm 0.5^{A}$	$^{b}23.7 \pm 0.5^{A}$	$^{b}26.2 \pm 0.3^{A}$	$^{b}28.5 \pm 0.4^{A}$	$^{b}30.1 \pm 0.5^{B}$
	MFSU	$^{a}20.7 \pm 0.5^{B}$	$^{a}23.9 \pm 0.5^{B}$	$^{a}25.9 \pm 0.5^{B}$	$^{a}27.6 \pm 0.9^{B}$	$^{a}30.1 \pm 0.7^{B}$	$^{a}35.2 \pm 0.7^{A}$
a*	Control	$^{a}8.2 \pm 0.3^{A}$	$^{a}6.6 \pm 0.3^{A}$	$^{a}4.6 \pm 0.3^{A}$	$^{a}4.5 \pm 0.7^{A}$	$^{a}2.9 \pm 0.3^{A}$	$^{a}1.5 \pm 0.1^{A}$
	MFSU	$^{b}5.9 \pm 0.2^{B}$	$^{b}4.7 \pm 0.2^{B}$	$^{b}3.8 \pm 0.2^{B}$	$^{b}2.7 \pm 0.2^{B}$	$^{b}1.5 \pm 0.1^{B}$	$^{b}1.4 \pm 0.1^{A}$
b*	Control	$^{b}2.5 \pm 0.2^{B}$	$^{b}4.9 \pm 0.2^{B}$	$^{b}6.9 \pm 0.2^{B}$	$^{b}10.7 \pm 0.5^{A}$	$^{b}11.9 \pm 0.5^{A}$	$^{b}13.5 \pm 0.9^{A}$
	MFSU	$^{a}6.1 \pm 0.3^{A}$	$^{a}7.4 \pm 0.3^{A}$	$^{a}9.4 \pm 0.3^{A}$	$^{a}12.6 \pm 1.7^{A}$	$^{a}14.2 \pm 0.9^{B}$	$^{a}17.9 \pm 0.6^{B}$

For a given parameter, different lower-case letters in the same row are considerably different ($p < 0.05$). The values of L^*, a^* or b^* with different upper-case letters in the same column are noticeably ($p < 0.05$) different. Positive L^*, a^*, and b^* factors designate lightness, redness, and yellowness, respectively. ($n = 3$). MWCO – molecular weight cut off.

On the other hand, a^* values of the WMPFs reduced along with the peptide fractions' decreasing molecular weight. The values of a^* decreased from 8.2 to 1.5% and 5.9–1.4% for the control and ultrasound samples correspondingly (Table 1). These findings show how peptide molecular weight affects the chromatic characteristics of protein hydrolysates (peptide fractions). Generally, the L^* and b^* values for the samples that had ultrasonic treatment were substantially higher ($p < 0.05$) while the a^* values were substantially lower ($p < 0.05$) than the control signifying a lighter colour in the ultrasound samples compared to the control.

3.2 Peptide Fractions' Amino Acid Composition

Bioactive potentials of protein hydrolysates have been linked to amino acid constituents or profiles of the hydrolysates. The amino acid profile of the Walnut meal peptide fractions (WMPFs) was determined to establish their relationship with the bioactive (antioxidative and antihypertensive) properties, and the results are displayed in Table 2. It is visible from Table 2 that Glu, Arg, and Asp are the predominant amino acids in WMPFs with corresponding values ranging between 23.5–25.7, 14.2–14.9 and 9.9–11.5 g/100g protein. These results are comparable to previously reported works by Amza et al. (2013) for gingerbread plum seed protein isolates, Zhu et al. (2006) for wheat germ protein isolate and associated hydrolysates, Gu et al. (2015) for reduced-fat walnut meal proteins besides Nourmohammadi et al. (2017) for pumpkin oil cake protein hydrolysate. Furthermore, Thr, Met, and Tyr were present in limited amounts with Lys being the most limiting amino acid in the WMPFs in both the ultrasound and control samples. It is, however, worth noting that the amino-acids profile is relatively similar comparing the protein isolate and the trypsin peptide fractions (Table 2). Also, the ultrasound and the control samples do not differ so significantly. That means both ultrasonication and enzyme hydrolysis processes have no adverse impact on the amino-acids composition of the sample and could be safely applied in food product formulation.

Specific amino acids like Asp, Glu Pro, Arg, His, Met, Leu, Ile, Ala, Tyr, and Val have been recognized to demonstrate antioxidative properties (Zhu et al. 2008). Comparing the amino acid profiles of the ultrasound and the control samples for WMPFs, it is possible to correlate their antioxidative activities to the amino acid distribution owing to an increase in hydrophobicity. Hydrophobic amino acid (HAA) content was generally higher in the ultrasound samples compared to the control and was higher in lower molecular weight fractions than higher molecular weight fractions. The HAA contents were 24.6, 25.7, 25.2, 25.6, 25.1, and 24.7 g/100g protein for the control and 26.3, 26.2, 25.6, 25.7, 24.8, and 24.8 g/100g protein for ultrasound sample corresponding to <1, 1–3, 3–5, 5–10, 10–30, and >30 kDa molecular weight cut off (MWCO). It has been reported that increased hydrophobicity will enhance the solubility of the amino acid in lipids and subsequently increase their antioxidative activity (Li et al. 2008; Nourmohammadi et al. 2017). The measure of the proportion of branched-chain amino acids (BCAA: leucine, valine, and isoleucine) to aromatic amino acids (phenylalanine and tyrosine) called the Fischer's ratio is occasionally used as a yardstick to determine the biological activity of proteins (Aryee and Boye 2017). In Table 2 below, Fischer's ratio in both the control and ultrasound samples across the different MWCOs was higher than 1.8. Fischer's Ratio is closely related to the BCAA to tyrosine ratio (BTR) is another useful indicator. The BTR in the current study ranges between 4.23–4.76 and 4.40–4.94 for control and ultrasound respectively which is regarded as normal (Ishikawa 2012a, b) though relatively lower than the 5.50 previously reported (Aryee and Boye 2017). Aryee and Boye (2017) stated that Fischer's ratio and BTR decline as the severity of hepatic injury grows, and it has been shown that giving BCAA supplements can help (Ishikawa 2012a, b). It has been demonstrated that the proportion of hydrophobic and hydrophilic amino acid residues influences and modulates taste perception, and this indicator has been adopted to predict sensory properties (Humiski and Aluko 2007).

Table 2. Amino acid profile (g/100 g) of WMPF as affected by sonication with dual-frequency, (n = 1)

Amino acid	The molecular weight cut-off (MWCO)											
	<1		1–3		3–5		5–10		10–30		>30	
	Control	DFSU	Control	DFSU	Control	DFSU	Control	DFSU	Control	DFSU	Control	DFSU
Asp	**10.6**	**10.6**	**10.8**	**11.2**	**10.9**	**10.0**	**10.8**	**10.6**	**10.6**	**9.9**	**10.7**	**10.8**
Glu	**22.4**	**23.4**	**23.8**	**25.0**	**23.7**	**23.6**	**24.5**	**25.2**	**23.6**	**24.9**	**23.5**	**24.5**
Ser	4.9	6.2	5.2	6.5	5.2	5.3	6.5	5.8	5.8	5.4	5.6	5.6
His* (1.9)	3.2	3.1	3.2	3.3	3.0	3.1	2.9	2.7	2.6	2.6	2.7	2.7

(continued)

Table 2. (*continued*)

Amino acid	The molecular weight cut-off (MWCO)											
	<1		1–3		3–5		5–10		10–30		>30	
	Control	DFSU	Control	DFSU	Control	DFSU	Control	DFSU	Control	DFSU	Control	DFSU
Gly	4.6	5.1	5.0	5.0	4.7	4.8	5.8	5.5	5.5	4.9	5.0	5.0
Thr* (1.4)	3.0	3.4	3.2	3.4	3.1	3.1	3.3	3.2	3.0	2.7	2.8	2.8
Arg	**14.7**	**14.9**	**14.9**	**14.8**	**14.4**	**14.5**	**14.4**	**14.6**	**14.4**	**14.5**	**14.3**	**14.6**
Ala	4.1	4.8	4.4	4.7	4.3	4.5	4.9	4.5	4.3	4.6	4.2	4.5
Tyr	3.5	3.5	3.4	3.4	3.3	3.5	3.5	3.3	3.3	3.2	3.3	3.2
Cys	1.2	1.2	1.2	1.3	1.4	1.5	1.6	1.4	1.5	1.7	1.9	1.9
Val* (3.5)	4.3	4.6	4.6	4.7	4.6	4.5	4.6	4.7	4.7	4.6	4.7	4.7
Met* (2.5)	1.9	1.9	2.1	2.2	1.9	2.1	2.1	2.1	2.0	1.8	1.8	1.8
Phe* (6.3)	4.4	4.6	4.6	4.5	4.6	4.6	4.2	4.4	4.1	4.0	4.0	4.0
Ile* (2.8)	3.5	4.0	3.8	3.9	3.8	3.8	3.8	3.8	3.7	3.8	3.7	3.8
Leu* (6.6)	7.0	7.7	7.2	7.5	7.0	7.1	7.4	7.4	7.3	7.4	7.2	7.3
Lys* (5.8)	2.6	2.6	2.5	2.9	2.5	2.5	2.4	2.5	2.4	2.4	2.3	2.3
Pro	3.7	3.7	3.8	3.8	3.7	3.7	3.6	3.8	3.7	3.8	3.8	3.7
ΣAA	99.6	105.3	103.7	108.1	102.1	102.2	106	105.4	102.5	102.2	101.5	103.3
HAA	24.6	26.3	25.7	26.2	25.2	25.6	25.6	25.7	25.1	24.8	24.7	24.8
AOAA	11.2	11.1	11.2	11.8	10.7	11.2	10.8	10.5	10.3	10.0	10.1	10.0
AAA	7.9	8.1	8.0	7.9	7.9	8.1	7.7	7.7	7.4	7.2	7.3	7.2
SCAA	3.1	3.1	3.3	3.5	3.3	3.6	3.7	3.5	3.5	3.5	3.7	3.7
BCAA	14.8	16.3	15.6	16.1	15.4	15.4	15.8	15.9	15.7	15.8	15.6	15.8
FR	1.87	2.01	1.95	2.04	1.95	1.90	2.05	2.06	2.12	2.19	2.14	2.19
BCAA/Tyr	4.23	4.66	4.59	4.74	4.67	4.40	4.51	4.82	4.76	4.94	4.73	4.94
ACE[a]	9.6	10.6	10.4	11.0	9.9	10.0	10.8	10.6	10.6	9.9	10.0	10.0

ΣAA, Total amino acids; HAA, Hydrophobic amino acids; AAA, Aromatic amino acids (Tyr + Phe); AOAA, Antioxidant amino acids (Tyr + Met + Lys + His); BCAA, Branched-chain amino acids (Ile + Leu + Val); SCAA, Sulphur-containing amino acids (Cys + Met); FR, Fischer's ratio (BCAA/AAA); BCAA to Tyrosine ratio (BCAA/Tyr); WMPF, Walnut meal peptide fractions. [a]ACE inhibition-related amino acids (Pro + Lys + Arg) (Ferri et al. 2017). * Essential amino acids and values in parenthesis are values recommended by FAO for preschool children.

3.3 Antihypertensive Activity Assay

The results of the ACE inhibitory activity test in this study indicate that all peptide fractions have the potential to be anti-hypertensive (Fig. 2). The results obtained herein appear to affirm the fact that peptide fractions containing lower molecular weight cut-offs (<1, 1–3, and 3–5 kDa) posse and exhibit more or higher ACE inhibitory activity compared to the peptides of high MWCO (>5 kDa). In general, the most potent ACE inhibitory action was reported for 1–3 kDa MWCO peptide fractions (Fig. 2) with values ranging between 22.86 and 66.03%. Similar results were reported in previous works (Amado et al. 2014; Ferri et al. 2017; Li et al. 2007; Segura-Campos et al. 2013). The ACE inhibitory action herein reported exhibited concentration-dependency such that inhibitory activity increased with increasing sample concentration for all peptide fractions. Significant variations ($p < 0.05$) were discovered between the concentrations used in the ultrasonography and control samples.

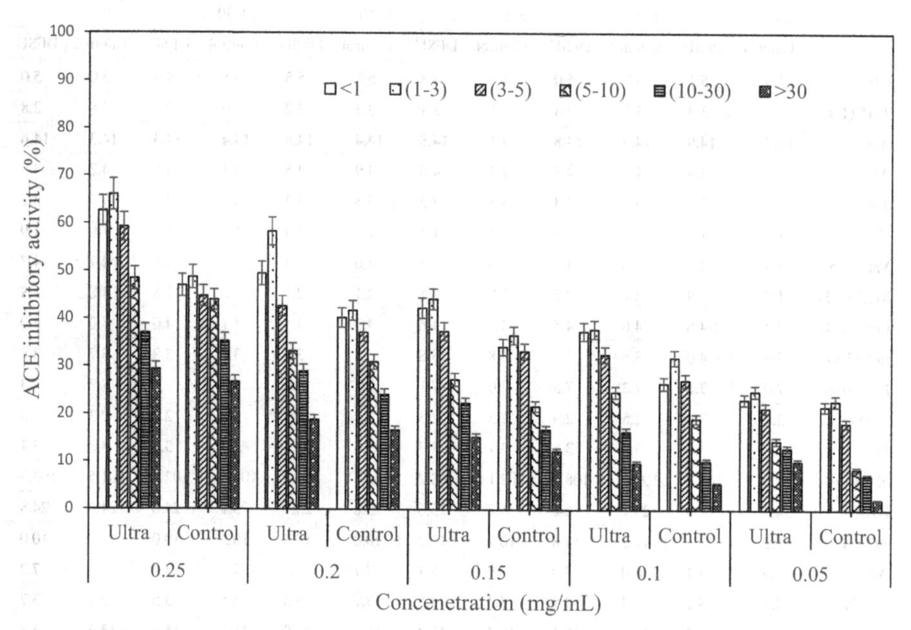

Fig. 2. Effect of ultrasound and peptide molecular weight on angiotensin I conversion enzyme (ACE) inhibitory activity of walnut meal peptide fractions.

The 1–3 kDa peptides portion of the ultrasound treatment almost achieved 50% inhibition (44.14% inhibition) at a sample concentration of 0.15 mg/mL. Meanwhile, peptide fractions with MWCO less than 5 kDa produced more than 50% inhibition when the sample concentration reached 0.2 mg/mL whereas the high MWCO peptide fraction produced less than 50% inhibition (Fig. 2). When the concentration reached 0.25 mg/mL, all peptide fractions below 5 kDa produced more than 50% inhibition while the others did not. Largely, the ultrasound-treated samples exhibited a higher capability of inhibiting ACE than the control samples. Ultrasound pretreatment has been reported to exert shear force through cavitation causing the breakup of the peptide bonds in the protein molecules resulting in improved enzyme hydrolysis leading to the release of bioactive peptides which hitherto were embedded in the parent protein chain (Bosiljkov et al. 2011; Ferri et al. 2017; Guerra et al. 2019; Liu et al. 2018; Ren et al. 2013). The IC_{50} observed here in this study is low compared with previous reports indicating a high ACE inhibitory activity (Chirinos et al. 2018; Guerra et al. 2019; Hanafi et al. 2018; Torruco-Uco et al. 2009). Ferri et al. (2017) and Amado et al. (2014) stated that the ACE inhibitory effect is linked to certain structural amino acids at the C-terminus, such as proline, lysine, or arginine, and that short peptides (2–5 amino acids) are the most active, and that peptide potency diminishes with increasing chain length. Evidentially, these positively charged amino acids were available in the peptide fractions as presented in Table 2 above. Consequently, it was determined that the ultrasonication significantly amplified the ACE-repressive action of peptides fractions from walnut meal.

3.4 WMPF Antioxidant Properties

3.4.1 Free Radical-Scavenging Capability of WMPFs

The results of the ABTS*+ scavenging ability of the walnut meal peptide fractions (WMPFs) are presented in Fig. 3. Figure 3 clearly shows that the ability of fractionated peptides of the walnut meal with low molecular weight to scavenge ABTS free radicals was considerably higher ($p < 0.05$) than higher molecular weight peptide fractions. The highest ABTS*+ scavenging activity was highest in peptides with MWCO of 1–3 kDa [2.59 and 2.78 TEAC value (mM) for the control and ultrasound samples respectively]. The least scavenging activity of ABTS*+ cations was observed with peptides having MWCO of >30 kDa as can be seen in Fig. 3. The reduced ABTS free radical scavenging action of higher MW peptides may be related to steric restraint (Hwang et al. 2016). These results substantiate the findings that short peptides mostly display more robust antioxidant activity than large polypeptides (Ajibola et al. 2011; Ferri et al. 2017; You et al. 2010). Wattanasiritham et al. (2016) and Hwang et al. (2016) in their respective works established the fact that peptide fractions with molecular weights up to 2.1 kDa

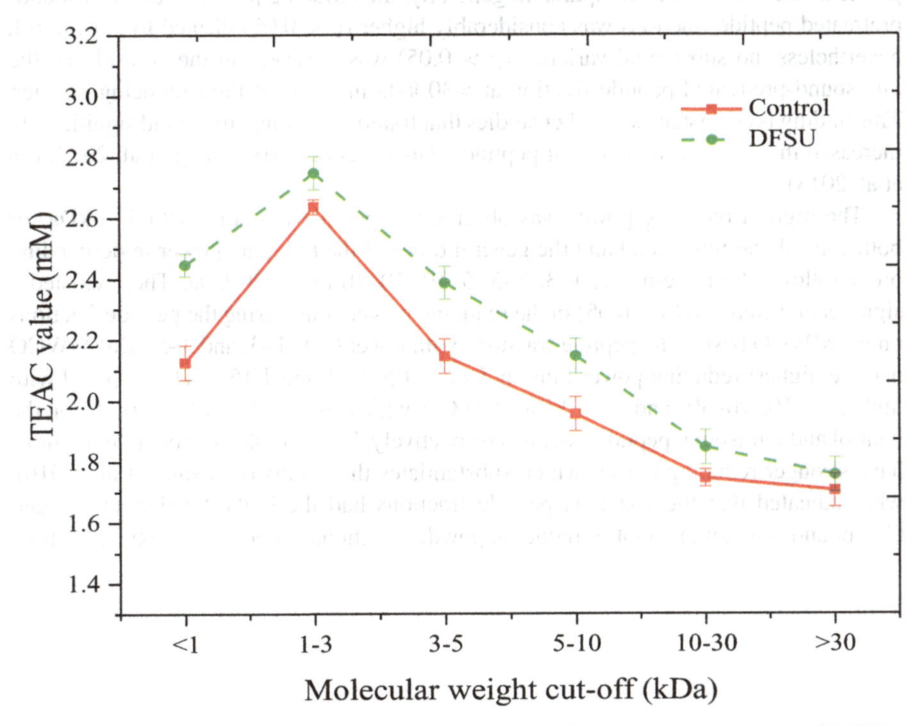

Fig. 3. Effect of the molecular weight cut-off and ultrasound treatment on scavenging of ABTS*+ cation ion free radicals by walnut meal peptide fractions.

demonstrated superior antioxidant capacities and a similar result was also reported (Ferri et al. 2017).

The elevated ABTS*+ scavenging activity of peptides <1, 1–3 and 3–5 kDa peptides compared to the high MWCO peptides in this study may be related to the total hydrophobic and aromatic amino acid levels observed in Table 2 and the cationic properties (Ajibola et al. 2011; Hwang et al. 2016). Furthermore, it could be found in Fig. 3 that the ultrasound treatment caused increased ABTS*+ scavenging activity compared to the control. There was considerably higher ($p < 0.05$) scavenging of ABTS*+ by the ultrasound peptide fractions than in the control samples except at >30 kDa peptide fractions with no significant difference ($p > 0.05$). As documented in literature, these variations may be due to shear forces created by cavitation during sonication, which might disrupt the bonds between protein molecules, enhancing protein hydrolysis and peptide production (Liu et al. 2018; Ren et al. 2013). Furthermore, the released peptides could be electron donors that convert unstable radicals to more stable compounds, hence, truncating the radical cyclical reaction (Guerra et al. 2019; Intiquilla et al. 2019; Schaich et al. 2015).

3.4.2 Lipid Peroxidation Inhibition and Reducing Powder of WMPFs

Figure 4 depicts the results of the effect of ultrasound and MWCO on the reducing power of the WMPFs. In comparison, generally, the reducing power of the ultrasound-pretreated peptide fractions was considerably higher ($p < 0.05$) likened to the control, nevertheless, no substantial variance ($p > 0.05$) was observed in the control and the ultrasound-pretreated peptide fraction at >30 kDa in terms of their reducing powder. This finding is consistent with other studies that found that using ultrasound significantly increased the antioxidant action of peptides (Guerra et al. 2019; Liang et al. 2017; Ma et al. 2018).

The highest reducing power was observed in peptide fractions with 1–3 kDa for both control and ultrasound and the general order of the reducing power in descending order follows the pattern <1, 1–3, 3–5, 5–10, 10–30 and >30 kDa. There existed a significant difference ($p < 0.05$) in the reducing power considering the peptide fractions on the MWCO basis. The peptide fractions with lower (<1, 1–3, and 3–5 kDa) MWCO had the highest reducing power ranging from 1.05–1.31 and 1.15–1.41 compared with higher (5–10, 10–30, and >30 kDa) MWCO with 0.84–0.49 and 0.92–0.65 for the control and ultrasound peptide fractions respectively. Lower molecular peptide fractions have stronger reducing power, which substantiates the results of Hwang et al. (2016), who indicated that the 1–3 kDa peptide fractions had the highest reducing powder. Compounds containing a lot of reducing powder might have a lot of antioxidant action.

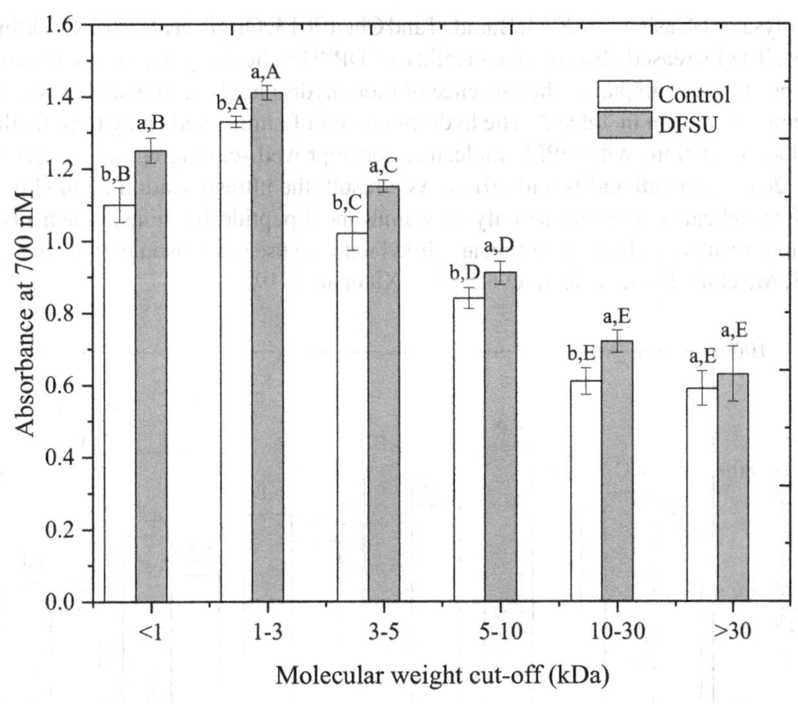

Fig. 4. Reducing power activity of WMPF prepared by trypsin hydrolysis (2000 U/g enzyme concentration, pH 7.5, 50 °C, 1:20 g/mL ratio of materials to liquids, 150 min hydrolysis time). A, B and C are significant differences ($p < 0.05$) with the same treatment while a, b and c are considerably variant ($p < 0.05$) at the same molecular weight. DFSU, dual-frequency (20/40 kHz/kHz) countercurrent S-type ultrasound (80 W/L power density, 45 min sonication time, 1:20 g/mL ratio of materials to liquids, pH of 9.5). Bars are the standard deviations ($n = 3$).

3.4.3 DPPH Radical Scavenging Capability of WMPFs

Free radicals and antioxidants can interact to produce stable species, which puts a halt to the oxidation process and reduces oxidative stress (You et al. 2009). The DPPH* radical scavenging evaluation has been routinely used to evaluate materials' capacity to scavenge free radicals (Aderinola et al. 2018; Bae and Suh 2007). The DPPH* is a persistent free radical that has a high absorbance at 517 nm in methanol/ethanol and a proclivity to receive protons; as a result, when an antioxidant or other chemical has an extra proton to give, the DPPH free radical is scavenged, causing the measured absorbance to decrease (Aderinola et al. 2018). Figure 5 displays the influence of the ultrasonication and molecular weight cut-off on the radical-trapping activity (DPPH*) of walnut meal hydrolysates. The antioxidant activities of both the control and the ultrasound samples tended to decrease with an increased molecular weight of peptide fractions. Significantly higher values ($p < 0.05$) of the trapping activity of DPPH* radicals were found for the ultrasound samples ranging from 61.59–84.19% in comparison with 56.98–75.56% for the control (Fig. 5). Similar activities and trends were identified in protein fragments of canola meal, barley hordein, hemp protein hydrolysate, and mungbean meal protein hydrolysates (Alashi et al. 2014; Bamdad and Chen 2013; Girgih et al. 2012; Sonklin et al. 2018). The increased absorption capability of DPPH radicals by the ultrasonic samples may be attributed in part to the presence of more hydrophobic amino acid residues than the control, as seen in Table 2. The hydrophobicity of amino acids may have facilitated peptide interactions with DPPH molecules for improved scavenging activity (Oluwole et al. 2018; Sarmadi and Ismail 2010). As a result, the ultrasonication method boosted the free radical scavenging activity of walnut meal peptide fractions, which fits with previous results in which the action amplified with the use of ultrasonication (Jiang et al. 2018; Ma et al. 2018; Nadeem et al. 2018; Xu et al. 2019).

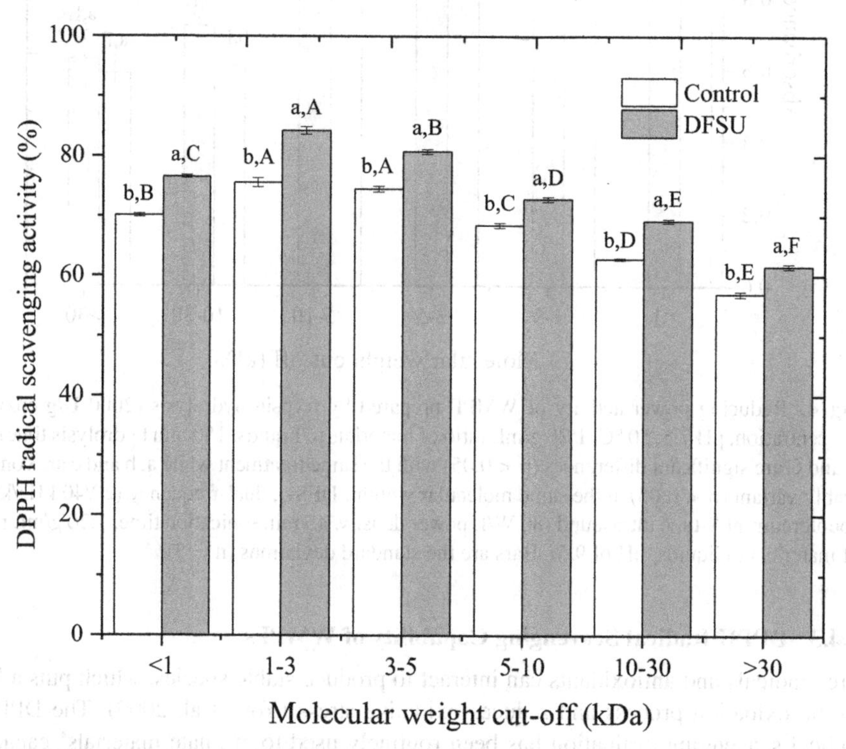

Fig. 5. DPPH radical scavenging activity of WMPF prepared by trypsin hydrolysis (2000 U/g enzyme concentration, pH 7.5, 50 °C, 1:20 g/mL ratio of material to liquid, 150 min hydrolysis time). A, B and C are considerably variant ($p < 0.05$) with the same treatment while a, b and c are significantly different ($p < 0.05$) at the same molecular weight. DFSU, dual-frequency (20/40 kHz/kHz) countercurrent S-type ultrasound (80 W/L power density, 45 min sonication time, 1:20 g/mL ratio of material to liquid, pH of 9.5). Bars are the standard deviations ($n = 3$).

On a molecular weight basis, the significant difference ($p < 0.05$) in the radical-trapping activity (DPPH*) of the fractions for the ultrasound treatment. Though a similar trend existed in the control sample, however, there was no discernible variation ($p > 0.05$) amid the radical-trapping activity (DPPH*) of 1–3 and 3–5 kDa molecular weight fractions. Other researchers also reported similar trends in their works (Aderinola et al. 2018; Yang et al. 2011, 2008, 2014). By extrapolation, the WMH peptide fractions with molecular weights less than 3 kDa were efficient antioxidant peptides capable of scavenging DPPH radicals and preventing the free radicals associated with oxidative stress (Sonklin et al. 2018). Table 2 shows that the peptide fractions with the strongest DPPH scavenging activity had significant levels of total HAA and AAA, which is consistent with previous findings (Kim et al. 2007; Sonklin et al. 2018). The ultrasonic pretreatment may have increased enzyme hydrolysis, causing high-molecular-weight proteins to be reduced into fractions with comparatively low molecular weights. These low molecular weight peptide fractions might then give electrons to free radicals (DPPH*) and convert them into more stable molecules, effectively breaking the chain of their processes (Liu et al. 2010).

3.4.4 WMPF Hydroxyl Radical Scavenging Action

Zhu et al. (2006) state that the hydroxyl radical is among one of the furthermost unstable as well as reactive oxygen radicals, as such brutally harms any biological molecule (proteins, DNA, PUFA, nucleic acid) and cell that it comes into contact with (Sonklin et al. 2018). This voracious reactive activity of hydroxyl radicals and other reactive oxygen radicals due to their unstable nature results in cell damage and consequential humanoid ailments (Arouma 1998; Halliwell 1989, 2006). It is therefore eminent to eliminate reactive oxygen species, notably hydroxyl radicals from the human body which could serve as very active protection against varied illnesses. Figure 6 displays the hydroxyl radical-scavenging effects of WMPF as affected by ultrasound and molecular MWCO. The peptide fragments of MWCO 1–3 kDa displayed a higher hydroxyl radical scavenging activity compared to the other peptide fractions for both the control and ultrasound-treated samples. The hydroxyl radical scavenging activity of the ultrasound and control samples differed significantly ($p < 0.05$), with the ultrasound sample exhibiting more scavenging activity than the control. The ultrasound sample exhibited the highest (56.28%) and lowest (39.49%) hydroxyl radical scavenging activities at MWCOs of 1–3 and >30 kDa respectively (Fig. 6). A similar trend was observed with the control sample with the highest and lowest hydroxyl radical scavenging activities of 53.51 and 35.71% respectively. Previous researchers have reported similar results where peptide fractions with lower molecular weight exhibited higher scavenging activity (Ajibola et al. 2011; Cheung et al. 2012; Sonklin et al. 2018).

In descending order, the sequence of hydroxyl radical scavenging activities of both the ultrasound and the control samples was >1, 1–3, 3–5, 5–10, 10–30, and >30 kDa peptide fractions. The ultrasonic pretreatment may have boosted the enzyme hydrolysis causing the proteins with higher molecular weight to be cleaved into portions of moderately small molecular weights with sequences that induced higher antioxidant activity. Table 2 shows that the overall HAA and AAA content possibly played a significant role in the hydroxyl radical scavenging performance of the WMPFs. Hernández-Ledesma et al. (2005) discovered that hydroxyl radical scavenging activity depends on the type of amino acid, such as His and Met, and particular peptide architectures and amino acid sequences.

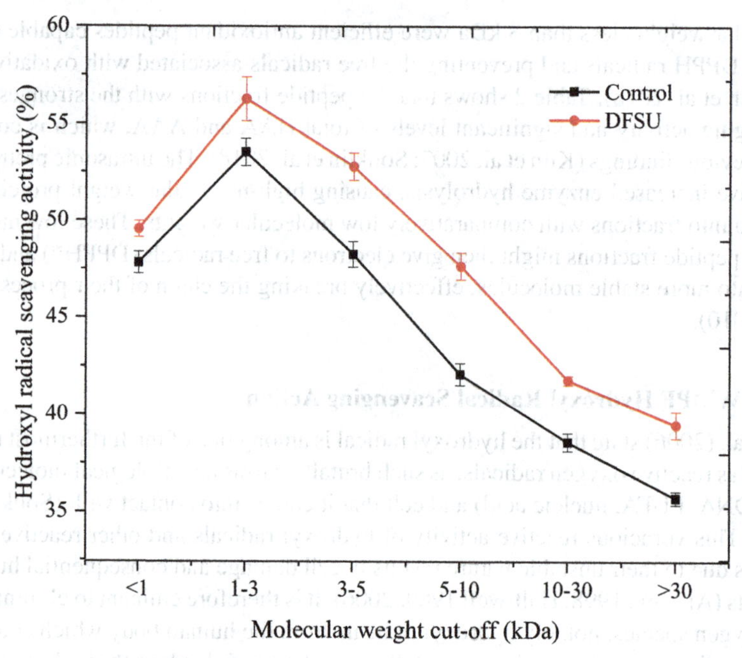

Fig. 6. Hydroxyl radical scavenging activity of WMPF prepared by trypsin hydrolysis (2000 U/g enzyme concentration, pH 7.5, 50 °C, 1:20 g/mL ratio of material to liquids, 150 min hydrolysis time). A, B and C are significant differences ($p < 0.05$) with similar treatment while a, b and c are considerably diverse ($p < 0.05$) at the same molecular weight. DFSU, dual-frequency (20/40 kHz/kHz) countercurrent S-type ultrasound (80 W/L power density, 45 min sonication time, 1:20 g/mL ratio of material to liquids, pH of 9.5). Bars are the standard deviations ($n = 3$).

4 Conclusion

In pursuance of good health and well-being (SDG3), consumer demand for naturally derived functional foods has been increasing, and therefore walnut meal as a low-cost and readily available raw material may be used to develop functional food ingredients in the pharmaceutical and food industries. The antioxidant and ACE inhibitory properties of the peptide portions were evaluated after fractionating trypsin hydrolysates with ultrasound pretreatment of the walnut meal. Through ultrafiltration, peptide fractions with MW less than (1–3) kDa generally displayed the highest ABTS, DPPH, and hydroxyl radical scavenging activities as well as reducing power ability compared to peptide fractions of the molecular weight cut-off >5 kDa. Furthermore, angiotensin 1 inhibitory enzyme (ACE) inhibitory activity also followed a similar pattern of higher inhibitory activity in low molecular weight peptides fractions (<1, 1–3 and 3–5). Generally, the ultrasound samples exhibited higher antioxidant and antihypertensive capabilities than the control without ultrasound treatment. Given that the study revealed that the peptides were abundant in hydrophobic, antioxidant, and aromatic amino acids, it may be confirmed that the amino acid compositions of the peptide fractions play a crucial role in the antioxidant and antihypertensive effects of walnut meal peptide fractions. Ultrasound pretreatment generally increased the bioactivity (antioxidant and antihypertensive activities) of the walnut meal trypsin hydrolysates compared to the control sample without sonication. It is possible to create walnut meal protein as a commercial functional food ingredient with added value, which would boost the economic value of this crop while also broadening its applications in agriculture and industry. Furthermore, peptide production from the walnut meal will help in sustainable environmental practices.

References

Adam Yahya Abdulrahman, M., Zhou, C., Zhang, Y., et al.: Effects of ultrasound pretreatment on enzymolysis of sodium caseinate protein: kinetic study, angiotensin-converting enzyme inhibitory activity, and the structural characteristics of the hydrolysates. J. Food Process. Preserv. **41**, e13276 (2017)

Aderinola, T.A., Fagbemi, T.N., Enujiugha, V.N., et al.: Amino acid composition and antioxidant properties of Moringa oleifera seed protein isolate and enzymatic hydrolysates. Heliyon **4**(10), e00877 (2018)

Aider, M., Barbana, C.: Canola proteins: composition, extraction, functional properties, bioactivity, applications as a food ingredient and allergenicity – a practical and critical review. Trends Food Sci. Technol. **22**(1), 21–39 (2011)

Ajibola, C.F., Fashakin, J.B., Fagbemi, T.N., et al.: Effect of peptide size on antioxidant properties of African yam bean seed (Sphenostylis stenocarpa) protein hydrolysate fractions. Int. J. Mol. Sci. **12**(10), 6685–6702 (2011)

Alashi, A.M., Blanchard, C.L., Mailer, R.J., et al.: Antioxidant properties of Australian canola meal protein hydrolysates. Food Chem. **146**, 500–506 (2014)

Aluko, R.E.: antihypertensive peptides from food proteins. Annu. Rev. Food Sci. Technol. **6**(1), 235–262 (2015)

Aluko, R.E.: Food protein-derived peptides: production, isolation, and purification. In: Yada, R.Y. (ed.) Proteins in Food Processing, 2nd edn., pp. 389–412. Woodhead Publishing (2018)

Amado, I.R., Vázquez, J.A., González, P., et al.: Identification of the major ACE-inhibitory peptides produced by enzymatic hydrolysis of a protein concentrate from cuttlefish wastewater. Mar. Drugs **12**(3), 1390–1405 (2014)

Amza, T., Balla, A., Tounkara, F., et al.: Effect of hydrolysis time on nutritional, functional and antioxidant properties of protein hydrolysates prepared from gingerbread plum (Neocarya macrophylla) seeds. Int. Food Res. J. **20**, 2081–2090 (2013)

Aruoma, O.I.: Free radicals, oxidative stress, and antioxidants in human health and disease. J. Am. Oil. Chem. Soc. **75**(2), 199–212 (1998). https://doi.org/10.1007/s11746-998-0032-9

Aryee, A.N.A., Boye, J.I.: Comparative study of the effects of processing on the nutritional, physicochemical and functional properties of lentil. J. Food Process. Preserv. **41**(1), e12824 (2017)

Bae, S.-H., Suh, H.-J.: Antioxidant activities of five different mulberry cultivars in Korea. LWT Food Sci. Technol. **40**(6), 955–962 (2007). https://doi.org/10.1016/j.lwt.2006.06.007

Bamdad, F., Chen, L.: Antioxidant capacities of fractionated barley hordein hydrolysates in relation to peptide structures. Mol. Nutr. Food Res. **57**(3), 493–503 (2013)

Bosiljkov, T., Tripalo, B., Brnčić, M., et al.: Influence of high-intensity ultrasound with different probe diameter on the degree of homogenization (variance) and physical properties of cow milk. Afr. J. Biotech. **10**(1), 34–41 (2011)

Chandrapala, J., Zisu, B., Palmer, M., et al.: Effects of ultrasound on the thermal and structural characteristics of proteins in reconstituted whey protein concentrate. Ultrason. Sonochem. **18**(5), 951–957 (2011)

Cheung, I.W.Y., Cheung, L.K.Y., Tan, N.Y., et al.: The role of molecular size in antioxidant activity of peptide fractions from Pacific hake (Merluccius productus) hydrolysates. Food Chem. **134**(3), 1297–1306 (2012)

Chilakala, S., Li, L., Feng, Y., et al.: Microwave-assisted enzymatic hydrolysis of DNA for mass spectrometric analysis: a new strategy for accelerated hydrolysis. Anal. Biochem. **546**, 28–34 (2018)

Chirinos, R., Ochoa, K., Aguilar-Galvez, A., et al.: Obtaining of peptides with in vitro antioxidant and angiotensin I converting enzyme inhibitory activities from cañihua protein (Chenopodium pallidicaule Aellen). J. Cereal Sci. **83**, 139–146 (2018)

Chung, S.-K., Osawa, T., Kawakishi, S.: Hydroxyl radical-scavenging effects of spices and scavengers from brown mustard (Brassica nigra). Biosci. Biotechnol. Biochem. **61**(1), 118–123 (1997)

Cushman, D.W., Cheung, H.S., Sabo, E.F., et al.: Development and design of specific inhibitors of angiotensin-converting enzyme. Am. J. Cardiol. **49**(6), 1390–1394 (1982)

Dabbour, M., He, R., Mintah, B., et al.: Antioxidant activities of sunflower protein hydrolysates treated with dual-frequency ultrasonic: optimization study. J. Food Process Eng. **42**(5), e13084 (2019)

Dabbour, M., He, R., Mintah, B., et al.: Ultrasound-assisted enzymolysis of sunflower meal protein: kinetics and thermodynamics modelling. J. Food Process Eng. **41**(7), e12865 (2018)

Dadzie, R.G., Ma, H., Abano, E.E., et al.: Optimization of process conditions for the production of angiotensin I-converting enzyme (ACE) inhibitory peptides from vital wheat gluten using response surface methodology. Food Sci. Biotechnol. **22**(6), 1531–1537 (2013)

Daliri, E.B.-M., Oh, D.H., Lee, B.H.: Bioactive peptides. Foods (Basel Switz.) **6**(5), 32 (2017)

Ferri, M., Graen-Heedfeld, J., Bretz, K., et al.: Peptide fractions obtained from rice by-products by means of an environment-friendly process show in vitro health-related bioactivities. PLoS ONE **12**(1), e0170954 (2017)

Ganten, D., Unger, T., Lang, R.E.: Pharmacological interferences with the renin-angiotensin system. Arzneimittelforschung **34**(10B), 1391–1398 (1984)

Golly, M.K., et al.: Effect of multi-frequency countercurrent ultrasound treatment on extraction optimization, functional and structural properties of protein isolates from walnut (Juglans ragia L.) meal. J. Food Biochem. **44**(6), e13210 (2020)

Golly, M.K., Ma, H., Yuqing, D., et al.: Enzymolysis of walnut (Juglans regia L.) meal protein: ultrasonication-assisted alkaline pretreatment impact on kinetics and thermodynamics. J. Food Biochem. **43**(8), e12948 (2019)

Girgih, A., Udenigwe, C., Aluko, R.: In vitro antioxidant properties of hemp seed (Cannabis sativa L.) protein hydrolysate fractions. J. Oil Fat Ind. **88**, 381–389 (2012)

Gu, M., Chen, H.-P., Zhao, M.-M., et al.: Identification of antioxidant peptides released from defatted walnut (Juglans Sigillata Dode) meal proteins with pancreatin. LWT Food Sci. Technol. **60**(1), 213–220 (2015)

Guerra, C., Torruco-Uco, J., Arango, W., et al.: Effect of ultrasound pretreatment on the antioxidant capacity and antihypertensive activity of bioactive peptides obtained from the protein hydrolysates of Erythrina edulis. Emir. J. Food Agric. **31**, 288–296 (2019)

Halliwell, B.: Free radicals, reactive oxygen species and human disease: a critical evaluation with special reference to atherosclerosis. Br. J. Exp. Pathol. **70**(6), 737–757 (1989)

Halliwell, B.: Reactive species and antioxidants. Redox biology is a fundamental theme of aerobic life. Plant Physiol. **141**(2), 312–322 (2006)

Hanafi, M.A., Hashim, S.N., Chay, S.Y., et al.: High angiotensin-I converting enzyme (ACE) inhibitory activity of Alcalase-digested green soybean (Glycine max) hydrolysates. Food Res. Int. **106**, 589–597 (2018)

Hernández-Ledesma, B., Miralles, B., Amigo, L., et al.: Identification of antioxidant and ACE-inhibitory peptides in fermented milk. J. Sci. Food Agric. **85**(6), 1041–1048 (2005)

Hu, H., Sun, Y., Zhao, X., et al.: Functional and conformational characterisation of walnut protein obtained through AOT reverse micelles. Int. J. Food Sci. Technol. **50**(11), 2351–2359 (2015)

Huang, L., Ding, X., Dai, C., et al.: Changes in the structure and dissociation of soybean protein isolate induced by ultrasound-assisted acid pretreatment. Food Chem. **232**, 727–732 (2017)

Humiski, L.M., Aluko, R.E.: Physicochemical and bitterness properties of enzymatic pea protein hydrolysates. J. Food Sci. **72**(8), S605–S611 (2007)

Hwang, C.-F., Chen, Y.-A., Luo, C., et al.: Antioxidant and antibacterial activities of peptide fractions from flaxseed protein hydrolysed by protease from Bacillus altitudinis HK02. Int. J. Food Sci. Technol. **51**(3), 681–689 (2016)

Ilyasov, I.R., Beloborodov, V.L., Selivanova, I.A.: Three ABTS•+ radical cation-based approaches for the evaluation of antioxidant activity: fast- and slow-reacting antioxidant behavior. Chem. Pap. **72**(8), 1917–1925 (2018)

Intiquilla, A., Jimenez-Aliaga, K., Guzman, F., et al.: Novel antioxidant peptides obtained by alcalase hydrolysis of Erythrina edulis (pajuro) protein. J. Sci. Food Agric. **99**(5), 2420–2427 (2019)

Ishikawa, T.: Branched-chain amino acids to tyrosine ratio value as a potential prognostic factor for hepatocellular carcinoma. World J. Gastroenterol. **18**(17), 2005–2008 (2012a)

Ishikawa, T.: Early administration of branched-chain amino acid granules. World J. Gastroenterol. **18**(33), 4486–4490 (2012b)

Ji, N., Sun, C., Zhao, Y., et al.: Purification and identification of antioxidant peptides from peanut protein isolate hydrolysates using UHR-Q-TOF mass spectrometer. Food Chem. **161**, 148–154 (2014)

Jia, J., Ma, H., Zhao, W., et al.: The use of ultrasound for enzymatic preparation of ACE-inhibitory peptides from wheat germ protein. Food Chem. **119**(1), 336–342 (2010)

Jiang, Y., Li, D., Ma, X., et al.: Ionic liquid–ultrasound-based extraction of biflavonoids from selaginella helvetica and investigation of their antioxidant activity. Mol. (Basel Switz.) **23**(12), 3284 (2018)

Kim, S.-Y., Je, J.-Y., Kim, S.-K.: Purification and characterization of an antioxidant peptide from hoki (Johnius belengerii) frame protein by gastrointestinal digestion. J. Nutr. Biochem. **18**(1), 31–38 (2007)

Korhonen, H., Pihlanto, A.: Bioactive peptides: production and functionality. Int. Dairy J. **16**(9), 945–960 (2006)

Lai, T., Lin, Z., Zhang, R., et al.: Processing stability of antioxidant protein hydrolysates extracted from degreased walnut meal. Int. J. Food Eng. **2**(2), 155–161 (2016)

Li, G.H., Qu, M.R., Wan, J.Z., et al.: Antihypertensive effect of rice protein hydrolysate with in vitro angiotensin I-converting enzyme inhibitory activity in spontaneously hypertensive rats. Asia Pac. J. Clin. Nutr. **16**(1), 275–280 (2007)

Li, M., Xia, S., Zhang, Y., et al.: Optimization of ACE inhibitory peptides from black soybean by microwave-assisted enzymatic method and study on its stability. LWT **98**, 358–365 (2018)

Li, Y., Jiang, B., Zhang, T., et al.: Antioxidant and free radical-scavenging activities of chickpea protein hydrolysate (CPH). Food Chem. **106**(2), 444–450 (2008)

Liang, Q., Ren, X., Ma, H., et al.: Effect of low-frequency ultrasonic-assisted enzymolysis on the physicochemical and antioxidant properties of corn protein hydrolysates. J. Food Qual. **10**, 71–80 (2017)

Liu, C., Fang, L., Min, W., et al.: Exploration of the molecular interactions between angiotensin-I-converting enzyme (ACE) and the inhibitory peptides derived from hazelnut (Corylus heterophylla Fisch.). Food Chem. **245**, 471–480 (2018)

Liu, M., Du, M., Zhang, Y., et al.: Purification and identification of an ACE inhibitory peptide from walnut protein. J. Agric. Food Chem. **61**(17), 4097–4100 (2013)

Liu, Q., Kong, B., Xiong, Y.L., et al.: Antioxidant activity and functional properties of porcine plasma protein hydrolysate as influenced by the degree of hydrolysis. Food Chem. **118**(2), 403–410 (2010)

Lobo, V., Patil, A., Phatak, A., et al.: Free radicals, antioxidants and functional foods: impact on human health. Pharmacogn. Rev. **4**(8), 118–126 (2010)

Ma, H., Huang, L., Peng, L., et al.: Pretreatment of garlic powder using sweep frequency ultrasound and single frequency countercurrent ultrasound: optimization and comparison for ACE inhibitory activities. Ultrason. Sonochem. **23**, 109–115 (2015)

Ma, S., Yang, X., Wang, C., et al.: Effect of ultrasound treatment on antioxidant activity and structure of β-Lactoglobulin using the Box-Behnken design. CyTA J. Food **16**(1), 596–606 (2018)

Nadeem, M., Ubaid, N., Qureshi, T.M., et al.: Effect of ultrasound and chemical treatment on total phenol, flavonoids and antioxidant properties on carrot-grape juice blend during storage. Ultrason. Sonochem. **45**, 1–6 (2018)

Nourmohammadi, E., SadeghiMahoonak, A., Alami, M., et al.: Amino acid composition and antioxidative properties of hydrolysed pumpkin (Cucurbita pepo L.) oil cake protein. Int. J. Food Prop. **20**(12), 3244–3255 (2017)

Oluwole, S.I., Sunday, A.M., Adeola, M.A., et al.: Antioxidant and antihypertensive activities of wonderful cola (Buchholzia coriacea) seed protein and enzymatic protein hydrolysates. J. Food Bioact. **3**, 133–143 (2018)

Pagidipati, N.J., Gaziano, T.A.: Estimating deaths from cardiovascular disease: a review of global methodologies of mortality measurement. Circulation **127**(6), 749–756 (2013)

Park, P.J., Jung, W.K., Nam, K.S., et al.: Purification and characterization of antioxidative peptides from protein hydrolysate of lecithin-free egg yolk. J. Am. Oil Chem. Soc. **78**(6), 651–656 (2001)

Ren, X., Ma, H., Mao, S., Zhou, H.: Effects of sweeping frequency ultrasound treatment on enzymatic preparations of ACE-inhibitory peptides from zein. Eur. Food Res. Technol. **238**(3), 435–442 (2013). https://doi.org/10.1007/s00217-013-2118-3

Sarmadi, B.H., Ismail, A.: Antioxidative peptides from food proteins: a review. Peptides **31**(10), 1949–1956 (2010)

Schaich, K.M., Tian, X., Xie, J.: Hurdles and pitfalls in measuring antioxidant efficacy: a critical evaluation of ABTS, DPPH, and ORAC assays. J. Funct. Foods **14**, 111–125 (2015)

Segura-Campos, M.R., Peralta González, F., et al.: Angiotensin I-converting enzyme inhibitory peptides of chia (Salvia hispanica) produced by enzymatic hydrolysis. Int. J. Food Sci. **8**, 597–604 (2013)

Sonklin, C., Laohakunjit, N., Kerdchoechuen, O.: Assessment of antioxidant properties of membrane ultrafiltration peptides from mungbean meal protein hydrolysates. Peer J **6**, e5337–e5337 (2018)

Tapal, A., Tiku, P.K.: Nutritional and nutraceutical improvement by enzymatic modification of food proteins. In: Enzymes in Food Biotechnology, pp. 471–481. Elsevier (2019)

Torruco-Uco, J., Chel-Guerrero, L., Martínez-Ayala, A., et al.: Angiotensin-I converting enzyme inhibitory and antioxidant activities of protein hydrolysates from Phaseolus lunatus and Phaseolus vulgaris seeds. LWT Food Sci. Technol. **42**(10), 1597–1604 (2009)

Udenigwe, C.C., Lu, Y.-L., Han, C.-H., et al.: Flaxseed protein-derived peptide fractions: antioxidant properties and inhibition of lipopolysaccharide-induced nitric oxide production in murine macrophages. Food Chem. **116**(1), 277–284 (2009)

Wang, F.-J., Yin, X.-Y., Regenstein, J.M., et al.: Separation and purification of angiotensin-I-converting enzyme (ACE) inhibitory peptides from walnuts (Juglans regia L.) meal. Eur. Food Res. Technol. **242**(6), 911–918 (2016)

Wattanasiritham, L., Theerakulkait, C., Wickramasekara, S., et al.: Isolation and identification of antioxidant peptides from enzymatically hydrolyzed rice bran protein. Food Chem. **192**, 156–162 (2016)

Wen, C., Zhang, J., Zhang, H., et al.: Advances in ultrasound-assisted extraction of bioactive compounds from cash crops - a review. Ultrason. Sonochem. **48**, 538–549 (2018a)

Wen, C., Zhang, J., Zhou, J., et al.: Antioxidant activity of arrowhead protein hydrolysates produced by a novel multi-frequency S-type ultrasound-assisted enzymolysis. Nat. Prod. Res., 1–4 (2019)

Wen, C., Zhang, J., Zhou, J., et al.: Effects of slit divergent ultrasound and enzymatic treatment on the structure and antioxidant activity of arrowhead protein. Ultrason. Sonochem. **49**, 294–302 (2018b)

Wongekalak, L.-O., Sakulsom, P., Jirasripongpun, K., et al.: Potential use of antioxidative mungbean protein hydrolysate as an anticancer asiatic acid carrier. Food Res. Int. **44**(3), 812–817 (2011)

Wu, W., Zhao, S., Chen, C., Ge, F., Liu, D., He, X.: Optimization of production conditions for antioxidant peptides from walnut protein meal using solid-state fermentation. Food Sci. Biotechnol. **23**(6), 1941–1949 (2014). https://doi.org/10.1007/s10068-014-0265-3

Xu, X.-Y., Meng, J.-M., Mao, Q.-Q., et al.: Effects of tannase and ultrasound treatment on the bioactive compounds and antioxidant activity of green tea extract. Antioxidants **8**(9), 362 (2019)

Yang, B., Yang, H., Li, J., et al.: Amino acid composition, molecular weight distribution and antioxidant activity of protein hydrolysates of soy sauce lees. Food Chem. **124**(2), 551–555 (2011)

Yang, B., Zhao, M., Shi, J., et al.: Effect of ultrasonic treatment on the recovery and DPPH radical scavenging activity of polysaccharides from longan fruit pericarp. Food Chem. **106**(2), 685–690 (2008)

Yang, R.W., Wang, J., Lin, S.Y.: Isolation and purification of soybean antioxidant peptides. Adv. Mater. Res. **881–883**, 811–814 (2014)

You, L., Zhao, M., Cui, C., et al.: Effect of degree of hydrolysis on the antioxidant activity of laoch (Misgunus anguillicaudatus) protein hydrolysates. Innov. Food Sci. Emerg. Technol. **10**, 235–240 (2009)

Yu, J., Ahmedna, M., Goktepe, I.: Peanut protein concentrate: production and functional properties as affected by processing. Food Chem. **103**(1), 121–129 (2007)

You, S.-J., Udenigwe, C., Aluko, R., et al.: Multifunctional peptides from egg white lysozyme. Food Res. Int. (ott. Ont.) **43**(3), 848–855 (2010)

Zhang, S.B., Wang, Z., Xu, S.Y., et al.: Purification and characterization of a radical scavenging peptide from rapeseed protein hydrolysates. J. Am. Oil Chem. Soc. **86**(10), 959–966 (2009)

Zhu, K., Zhou, H., Qian, H.: Antioxidant and free radical-scavenging activities of wheat germ protein Hydrolysates (WGPH) prepared with Alcalase. Process Biochem. **41**, 1296–1302 (2006)

Zhu, L., Chen, J., Tang, X., et al.: Reducing, radical scavenging, and chelation properties of in vitro digests of alcalase-treated zein hydrolysate. J. Agric. Food Chem. **56**(8), 2714–2721 (2008)

Zhu, Z., Zhu, W., Yi, J., et al.: Effects of sonication on the physicochemical and functional properties of walnut protein isolate. Food Res. Int. **106**, 853–861 (2018)

Achieving Sustainable Housing in Nigeria: A Rethink of the Strategies and Constraints

I. R. Aliu(✉)

Department of Geography and Planning, Faculty of Social Sciences, Lagos State University, Ojo, Lagos, Nigeria
ibrahim.aliu@lasu.edu.ng, ibrolordtimi@yahoo.co.uk

Abstract. Purpose: Housing is central to sustainable human development which in recent times has been the frontier of debate at global and local levels. In Nigeria, housing has been routinely provided by the governments (public-social housing), the developers (private housing) and the individual citizens (informal housing) with little concern for sustainability. The housing deficit in the country remains at an all-time high of about 20 million, leading to housing deprivation, homelessness and non-affordability. The paper focused on sustainable housing strategies and constraints in Nigeria.

Design/Methodology/Approach: This article employs a perspective discourse approach using some existing works to make arguments on inherent issues in achieving housing sustainability in Nigeria. The discourse on sustainable housing strategies and constraints in Nigeria are made along three lines: the discussion of the theories of housing and housing sustainability, the strategies for sustainable housing provision and the constraints militating against sustainable housing provision in Nigeria.

Findings: Sustainable housing strategies normally involve the provision of affordable dwellings that ensure energy efficiency, resilience against natural and human hazards, healthy physical environment, overall access of citizens to adequate, good-quality, secure housing and utility services with particular attention to vulnerable groups and private investment in the housing sector. The constraints to sustainable housing production are funding, inestimable population growth, poor economy, urban poverty, rapid urbanization, inflation, non-use of indigenous knowledge, lack of local materials, high construction cost and poor housing policy.

Research Limitation/Implications: This research focused on sustainable housing strategies and constraints in Nigeria.

Practical implication: This paper has potential implications for understanding sustainable housing dynamics in Nigeria and other developing countries.

Social Implication: This study will assist development policy-makers in addressing sustainable housing through social, economic and environmental stability for the present and future generations.

Originality/Value: This study is based on the design and implementation of smart growth cities that make proper use of spaces.

Keyword: Housing policy · Housing strategies · Sustainable development · Sustainable housing · Nigeria

C. Aigbavboa et al. (Eds.): ARCA 2022, *Sustainable Education and Development – Sustainable Industrialization and Innovation*, pp. 1164–1179, 2023.
https://doi.org/10.1007/978-3-031-25998-2_90

1 Introduction

Housing and sustainability are two development-related issues that have received tremendous global attention in the last one-half decades of the 21st century. This is partly because of their relevance to human development and partly because of their inexorable significance to environmental stability. The Nigerian housing deficit is the worst in the African continent and the country faces the most horrible housing crisis. With a population of about 200 million, an urban growth rate of 4.9%, and annual housing production of less than 100,000 units, Nigeria faces the most daunting housing needs and demand on the continent (Ibem 2011; Makinde 2014; NBS 2020; UN-HABITAT 2003, 2008; Aliu et al. 2018). In addition, with a housing deficit of nearly 20 million units that require about 700,000 housing units per annum (Bah et al. 2018; CAHF 2018), Nigeria is engulfed in an imminent housing crisis. Over 50% of the Nigerian urban population lives in substandard housing and slum neighbourhoods with poor facilities and structural qualities (CAHF 2018). Due to critical housing deficits and pressure in providing accommodation for the less privileged urban poor, the use of substandard materials and illegally constructed structures has led to incessant housing collapses and heightened public health risks to the residents in the Nigerian cities. According to Fadamiro (2002), the incessant collapses that characterize the Nigerian building industry are due to the pressure of providing shelter for the teeming populations, relegation of building regulations and use of substandard materials and designs. The prevalence of inadequate housing, informal settlements, poor organized neighbourhoods, illegal structures and collapsed buildings has negative effects on environmental sustainability in Nigeria.

The centrality of housing as a basic social need to man and his continued existence warrants no polemics. Housing does not only provide for man shelter against natural elements and climatic vagaries, built also provides for him varying biological, health and psychological benefits (Ibem 2011; Gbadeyan 2011; Mabogunje et al. 1978; Aliu 2012; Ajala et al. 2010; Aliu and Adebayo 2013). In addition, housing performs cultural and economic roles within the society; hence housing is more than mere shelter but an economic property and a measure of social status in the society (Rapoport 2001). Because of the vital position of housing in society, it has become in modern times a pervasive target of the collective and individual pursuit. Although there is a general sense of residential inadequacies in all societies of the world, the situation in the developing economies especially Nigeria is more precarious (Arku 2009; Aribigbola 2008; Headey 1978). Generally speaking, housing represents a major focus of household expenditure and a source of social deprivation. This is perhaps due to the expensive housing budget, the socioeconomic peculiarities of people and the policy orientation operating within different jurisdictions (Aliu et al. 2018; Huang 2012). There are two indicators of housing deprivations – the proportion of homeownership and housing deficit or homelessness. In developed economies homeownership is quite easier and more feasible to achieve than in the developing economies where people struggle to own or rent an apartment and aspire throughout their life span to build at least a housing unit. Homelessness is a common denominator of developing African countries and housing deprivation is more like a rule than an exception (UN-HABITAT 2008).

Incidentally, housing has been elevated to the status of a fundamental human right and has been duly recognized by United Nations Organization (UNO) article 25 as "an inalienable right of individual to decent, affordable and safe housing" (UN 1965; Ajala et al. 2010). Despite this, housing deprivation has continued to stare the individuals in the developing economies in the face. In contemporary Nigeria, housing provision is more or less a debacle (Agbola 2005). This is partly due to the inability of the stakeholders to harness properly all possible resources for providing adequate housing and partly because of the levels of population growth (demand), socio-economic variations (affordability) and deficit financing (Buckely et al. 1993; Onibokun 1985; Sule 1981; Jiboye 2011; Towry-Coker 2012). In a recent article, Aliu et al. (2018) estimated the factors that lead to housing failure in Nigeria emphasizing the common issues of land, affordability, policy and mortgage finance. A similar argument has been made of Chinese housing policy frailties by Huang (2012, p. 941) when she observed that housing policies in China for

Table 1. Regional dimensions of housing deficits in Africa

Sn	A. West African region		Sn	B. South African region	
	Country	Housing deficit		Country	Housing deficit
1	Nigeria	17,000,000	1	South Africa	2,300,000
2	Ghana	1,700,000	2	Mozambique	2,000,000
3	Cotedivoire	600,000	3	Madagascar	2,000,000
4	Mali	400,000	4	Zambia	1,500,000
5	Liberia	200,000	5	Zimbabwe	1,250,000
6	Guinea	140,000	6	Namibia	80,000
	Total	20,040,000 (40%)		Total	9,230,000 (18%)

Sn	C. Central African region		Sn	D. East African region	
	Country	Housing deficit		Country	Housing deficit
1	Angola	1,900,000	1	Kenya	2,000,000
2	Dr Congo	1,200,000	2	Uganda	1,600,000
3	Cameroon	1,200,000	3	Ethiopia	1,000,000
4	Car	1,000,000	4	Rwanda	109,000
5	Capeverde	82,000	5	Total	4,709,000 (9%)
	Total	5,582,000 (11%)			

Sn	E. North African region				
	Country	Housing deficit	Sn	Country	Housing deficit
1	Egypt	3,500,000	4	Libya	350,000
2	Algeria	1,200,000	5	Mauritania	50,000
3	Morocco	600,000		Total	5,700,000 (11%)

Source: Bah et al. (2018)

the low-income groups have failed for three vital reasons – lack of clear mission, lack of local government commitment and an exclusionary policy towards migrants.

It is important to note that adequate housing provision is an integral part of the sustainable development goals. Sustainable development is a concept that universally focuses on striking a balance between economic growth and ecological stability. The proponents of sustainable development have argued that all stakeholders in economic production including housing should exercise great caution in using environmental resources for man's consumption, urging a culture of frugality. Housing sustainability is in the larger spirit of sustainable development all about providing adequate housing for the present and future generations. Sustainable housing is therefore a planning idea that seeks adequate shelter for humanity at all-time scales. It is an attempt at banishing homelessness in every state of human existence. Lack of adequate and sustainable home provision has critical implications for residents health, the general quality of life, satisfaction, preferences and choices (UN-HABITAT 2012; Aliu and Adebayo 2013; Aliu and Ajala 2014). As Nigeria's population continues to grow and the quality of life is increasing daily, the demand for housing inevitably grows. To meet everyone's housing needs in the nearest future millions of housing units need to be produced. As indicated in Table 1 and Fig. 1 above, the housing deficit in Nigeria and many other African countries have been appalling. To worsen the situation, housing deprivation has been escalated by a high number of obsolescent and blighted properties in widely unregulated housing markets. Unfortunately, the Nigerian housing and construction sectors have been mired by professional incompetence and critical underfunding (Aliu et al. 2018).

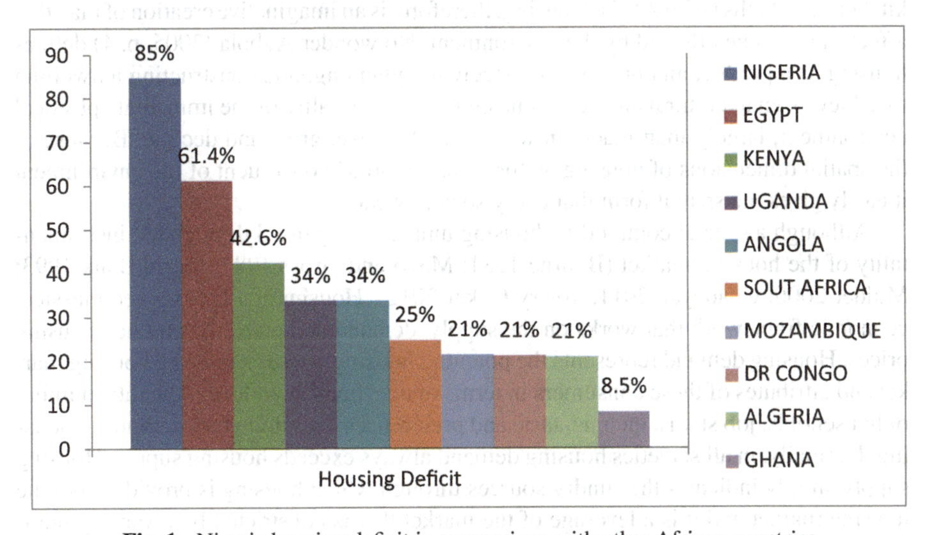

Fig. 1. Nigeria housing deficit in comparison with other African countries

While recent studies have widely explored the social and economic dimensions of sustainability, the intricate relationships between housing and sustainable development have been severely glossed over. Little has also been done to explicate the intrinsic components of sustainable housing development generally. This paper is organized into four sections. The first is the introduction section which gives the background to housing problems in Nigeria. The second section reviews the conceptual aspects of housing, sustainable development and sustainable housing. The third section considers the strategies through which sustainable housing is provided in Nigeria. The fourth section takes a look at the many constraints militating against an adequate provision of housing in the country. The paper concludes with some policy suggestions on how sustainable housing could be achieved in Nigeria and other developing economies generally.

2 Housing and Sustainable Housing Theories

Housing has a multiplicity of connotations and it is only for purposes of convenience that housing is defined simply as mere shelter. Indeed housing is a special product that represents the shell or structure of the dwelling, designed and basic built-in equipment for space allocation to heating, sanitary, sleeping, resting and a process by which shelter is provided in a safe and comfortable location relative to transit and transportation, places of work, schools and hospitals, religious and recreational centres (Aliu 2012). According to Beyer (1965), housing is a complex economic product that is bulky, immovable and serves as a store of economic value in that it could be traded in the stock markets. From Rapoport's (2001) point of view, housing is more like a cultural artefact indicating the level of socio-cultural development of the society at a point in time. Housing is a bundle of services consisting of neighbourhood services – schools, parks, location services - accessibility to employment, amenities; and structural services-living rooms, kitchens and baths (Aliu 2012). Housing, therefore, is an imaginative creation of man that affects and can be affected by the environment. No wonder Agbola (2005, p. 4) defines housing as a product and process of conceiving, planning, and constructing a dwelling to achieve social, cultural and economic ends and the totality of the immediate physical environment, largely man-made, in which families live, grow and decline. Because of the spatial dimensions of housing as 'brick and mortar' constituent of the environment it easily gives the spatial form that every society wears.

Although a special commodity, housing units are dispensed through the instrumentality of the housing market (Bourne 1981; Mayo and Gross 1987; World Bank 1993; Mulder 2006; Gbadeyan 2011; Towry-Coker 2012). Housing market is a neo-classical economic framework that works on the supply, demand and policy to produce housing prices. Housing demand represents the potential housing consumers in the housing market and attributes of these consumers in terms of age, household formation rate, number of households, job status, income, taste and preferences determine the demand for housing. Normally, in all societies housing demand always exceeds housing supply. Housing supply simply indicates the sundry sources through which housing is provided into the housing market and it is a leverage of the market that is constricted by a varying number of issues- land, fund, and access to the mortgage, interest rate and cost of building materials. This, therefore, precludes a balance of strength between supply – pervasively in shortage and demand which is pervasively in surplus. Housing is inherently spatial and the discrepancies among housing qualities and market outcomes within a large built environment have been a major policy concern for decades (Buist et al. 1993; Aliu and Ajala 2014, 2015). Housing markets generally display segmentation or variation and therefore require elemental analysis for qualitative and quantitative differentials and similarities. As observed by Aliu and Ajala (2014, p.11), 'the housing market is not a monolithic phenomenon, but a framework that displays wide variations in quantity, quality and prices. The housing market area (HMA) analysis is simply the explanation and description of the various housing segments within a given spatially bounded region or place'. This exposes the quality and quantity of polarization within a big city (Aliu and Ajala 2013, 2014).

Sustainable development, on the other hand, is an environmental, social, economic and ethical concept that seeks to achieve economic growth and ecological stability simultaneously. It derives from the long tradition of development scholars' idea that the human perpetual occupation of the earth planet will depend more on the preservation and conservation of the earth resources such that generational requirements for survival are at the least risk of compromise. In order words, unhindered human development depends on the sustainability of resources. Sustainability requires that development will not be detrimental to the stock of resources of the environment and will infinitely occur through generations. Development is a multi-dimensional concept not confined to the realization of sustained economic growth alone but a drastic improvement in the general well-being of the people, socio-political transformation and reduction in inequality (Todaro and Smith 2003). Hence, development is not just one indicator concept but a combination of varying socio-economic and political indicators. Part of these socio-economic indicators of development is housing and its quality. According to WCED (1987) Brundtland's report, sustainable development shares three fundamental realities of human development namely environmental protection, economic growth and social equity. Hence sustainable development requires integrating policies related to social justice, environmental protection and economic development taking into consideration interests of future generations, transparency and public participation at all levels of decision-making from local to global (Motsoene 2014). Sustainable development presupposes the sustainability of natural and anthropogenic resources including housing.

Therefore, in perfect agreement with WCED Brundtland's rationalizations housing sustainability is simply the process of providing shelter in such a way that both the present and future generations' housing needs are taken into consideration. Sustainable housing is a gradual, continual and replicable process of meeting the housing needs of the populace majority of who are poor and incapable of providing adequately for themselves (Olotuah and Bobadoye 2009). It is apparent that to attain sustainable development or sustainable housing, the housing sector must be efficient at resource utilization for optimal housing provision and the prediction of housing needs must envisage the changes in natural resources, socio-economic and demographic attributes of people. Sustainable houses are 'those that are designed, built and managed as healthy, durable, safe and secure; affordable for the whole spectrum of incomes; using ecological low-energy technology; resilient to potential natural disasters; connected to decent, safe and affordable energy water, sanitation and recycling facilities; not polluting the environment, well connected to jobs, shops, health and child care, schools; properly integrated into the social cultural and economic fabric of the local neighborhood. (UN-HABITAT 2012 p. 9). However, adequate housing, it must be noted, is more than an exponential increase in housing provision but an increase in more qualitative dwellings for residents as well. Housing has great implications for the physical environment in positive and negative ways. Hence, housing sustainability is viewed from two perspectives – how to produce adequate housing sustainably and how to produce quality housing sustainably with minimal negative influence on the physical environment.

3 Sustainable Housing Provision Strategies in Nigeria

3.1 Sustainable Affordable Housing Provision

Figure 2 graphically displays the normative components and requirements of sustainable housing provision. According to UN-HABITAT (2012, p. 8) sustainable housing provision policies and strategies are made to achieve three developmental ends namely environmental protection through quality neighborhood and structural design, social-cultural equity through dwellings that protect people's socio-cultural heritage and economic growth through affordable housing. Affordable housing provision is one of the fundamental strategies for achieving sustainable housing in any society. However, in Nigeria and all societies of the world, housing is always provided by the tripartite framework comprising the individuals, the private organized sector often regarded as developers and the government (Mabogunje et al. 1978; Sule 1981; Aliu et al. 2018). The first source of housing provision is through the government or public agencies. It is widely believed that housing is a fundamental human need that can dictate the course of human progress, social stability, socio-economic and health wellbeing. Hence, the government of every jurisdiction in Nigeria supports the human community to grow by rendering welfare services including housing (see Table 2). In most cases governments play a vital role in giving the direction of housing provision using its policy frameworks (Aliu et al. 2018; Ibem 2011; Ibem and Amole 2010; Towry-Coker 2012). The second source of housing is the organized private sector often regarded as property developers. This housing provider group serves as the intermediate between the public and the individual residential provision. The intervention of the private sector in housing provision is more attributed to the role of housing in the integrated economy of every society than the social responsibility of public housing.

The organized private sector consists of the developers, primary mortgage institutions (PMIs), cooperative organizations. All these privately driven organizations provide housing for residents within the society and assist the public efforts by adding a sizeable number of housing units to the housing stock. However, unlike the public housing system, private sector housing is often costly and insensitive to the variegated spectrum of the society whose majority is poor. The third source of housing and the most important source of housing provision in any society is individual housing. Regardless of the operating ideological or cultural persuasions, housing is generally viewed as a private responsibility for which individuals must be first responsible. One of the basic reasons for working and engaging in any economic activity is to achieve the life goal of shelter. Apart from public and organized private sectors, housing is more produced through individual efforts. Individual owner-occupier housing is no doubt the largest housing in any society. People normally from the proceeds of their economic functions or jobs save and construct housing mostly for the family members and sometimes for letting. Individual housing is flexible and takes a longer time to complete. This is otherwise known as

Table 2. Public housing provision in Nigeria (1975–1980) & (1980–1989)

S/n	State	1975–1980* Housing units	1980–1989** Housing units	Difference (1970s–1980s)	% Change rate
1	Bauchi	1819	2816	997	54.81
2	Bendel	250	1422	1172	468.80
3	Benue	38	1980	1942	5110.53
4	Borno	2480	2808	328	13.23
5	Anambra	400	2400	2000	500.00
6	Cross river	525	2258	1733	330.10
7	Gongola	382	3038	2656	695.29
8	Imo	488	2758	2270	465.16
9	Kaduna	1620	2716	1096	67.65
10	Kano	976	1590	614	62.91
11	Kwara	941	2462	1521	161.64
12	Niger	520	2692	2172	417.69
13	Ogun	512	2160	1648	321.88
14	Ondo	1378	2930	1552	112.63
15	Oyo	323	2128	1805	558.82
16	Plateau	1000	2546	1546	154.60
17	Rivers	281	1580	1299	462.28
18	Sokoto	1000	2314	1314	131.40
19	Lagos	8616	1908	−6708	−77.86
	Total	**23549**	**47500**	**23951**	**101.71**

Source: Federal Housing Authority; Towry-Coker (2012)* and Jinadu (2007)**

incremental housing. However, in developing economies the quality of individual housing is quite suspect and varied. Individual housing evolves the process of slum housing which is an informal set of spatial practices and tactics, a makeshift approach to housing and shelter and a precarious form of inhabiting the city (Vasudevan 2015). While public housing is focused on producing affordable housing the private organized housing sector is not too keen on this social inclusiveness in affordable housing rather it concentrates on providing housing for both high and medium-income groups. Unfortunately, all these strategies of housing provision have not led and will not lead to sustainable affordable housing in Nigeria.

3.2 Sustainable Residential Neighbourhoods

As indicated in Fig. 2 an important aspect of sustainable housing production is the provision of virile built environment with space efficiency, sustainable neighborhood design, paved roads, parks, street lights, sanitary facilities, community facilities and moderate residential density. The built environments at local, regional and global scales make impacts on their surroundings (UN-HABITAT 2012). In Nigeria, most of the estates and residential new neighborhoods are designed to have convivial physical environment but in the end, are not strictly provided. Universally designed neighborhoods often reflect multidimensional needs and abilities and environments that meet the needs of all residents irrespective of their natural incapabilities and social status. Sustainable neighborhoods usually lead to the realization of efficient communities with mixed-land uses, manageable design that allows walkability in all directions, inclusive social integration, good physical layout and accessibility to public facilities (UN-HABITAT 2012). In Nigeria most of the public housing development projects seek to achieve this environmental objective but the private housing does not reflect universal design principles especially in terms of lot size and public spaces. Secondly, sustainable neighbourhoods require the presence of a range of community facilities such as open spaces, community halls, recreational facilities, sporting arenas, schools, street lights, health care centres, religious praying grounds or halls, and parks. Unfortunately due to the level of development in Nigeria only few of these community facilities are usually provided by the real estate development firms. Most of these community facilities are being produced by the residents and in very inadequate manner leading to a disorganized physical environment that eventually creates slum growth. In order to achieve sustainability the housing development agencies must be encouraged to provide these community facilities at the right places and in standard condition.

Thirdly, a sustainable neighbourhood envisions sustainable energy management. The built environment accounts normally for about 25% of the carbon dioxide (CO_2) emissions in any settlement (UN-HABITAT 2012). This has strong implications for global warming and climate change. The green house gases contributed by the built environment comes from varying sources such as domestic household cooking, industrial activities, institutional and commercial activities as well as from waste generation. These different green house gas emission sources could be managed to reduce through the use of clean energy systems, recycling of waste and reuse of solid waste. Also the dwelling construction may be used to manage energy inefficiencies by having wide windows, wide doors and big attic which will allow more air circulations and less dependent on electricity or gas. Nigeria is located in the tropics and the housing residents therefore need cooler weather and airflow within the dwellings. Using the air conditioner is not a good alternative to this problem but creating suitable spatial design of rooms and living apartments as well as green energy, biomass wind energy plants and renewable energy technologies. Fourthly, a sustainable neighbourhood should make provision for life quality enhancing amenities such as open green fields, walkways, public transport corridors. These community amenities have a strong impact on the social and health well-being of people. Lastly, the neighbourhood should consist of not only the residences but the natural environments, historical buildings, flood plains and sustainable drainage systems. Incidentally, all these neighbourhood design features have been shown to influence housing satisfaction, choice decision making processes and residential values (Jiboye 2011; Aliu and Ajala 2014).

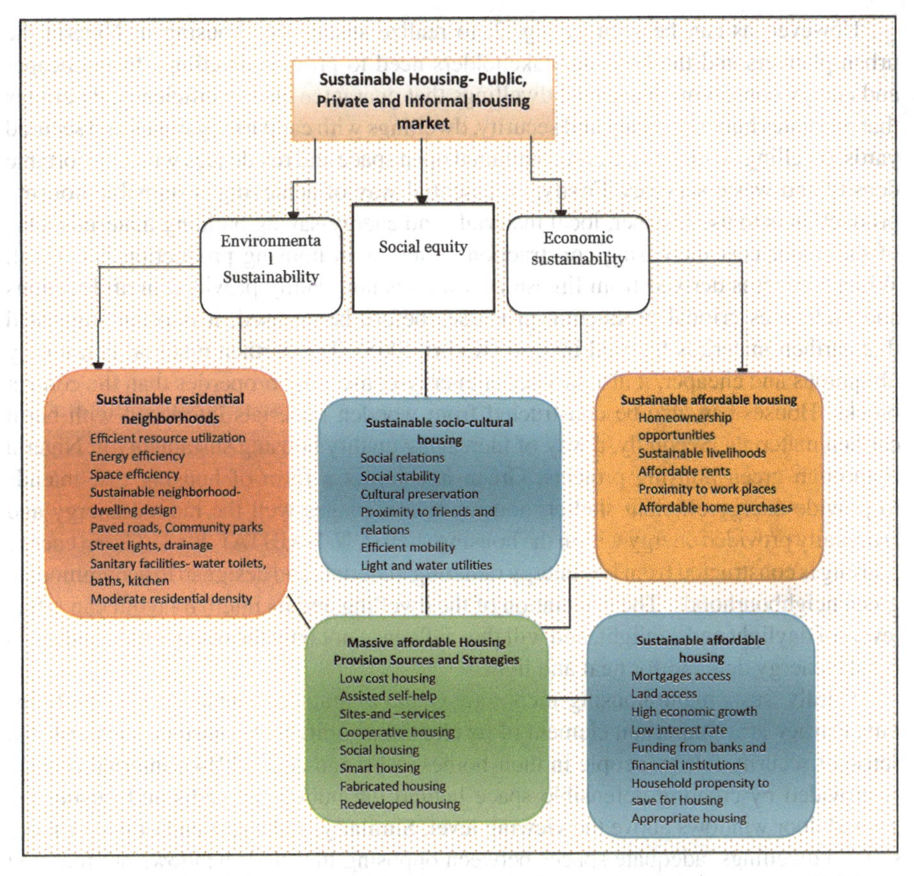

Fig. 2. Sustainable housing provision framework (Source: Author's impression, 2021)

3.3 Sustainable Structural Quality Dwellings

Another crucial aspect of sustainable housing provision is the production of housing units with appropriate structural designs that can enhance cultural and social backgrounds of the people. In Nigeria, public and private housing display different structural qualities in terms of housing types, number of rooms, room sizes, number of toilets and baths, kitchen types and quality, water provision and energy sources (Aliu and Ajala 2014; Ibem 2011; Olotuah and Bobadoye 2009). While public housing has been constructed to reflect pro-type, homogenous structures often in two to three-storey buildings, the private organized housing has been usually designed to be majorly pro-type bungalow housing. The private housing units have largely been two-three bedrooms which in most cases are beyond the reach of the low-income groups. Very few of the housing being provided are one-room self-contained types that suit the poor urban residents in Nigerian cities. While formal housing has greatly assisted in improving the structural quality of housing in Nigeria, personal individual housing has often led to slum and poor structural quality housing as many of the houses are built on an incremental basis taking between 5 years and 15 years to complete.

However, as can be seen in Fig. 2, to realize sustainable housing in Nigeria the urban planners and the housing stakeholders need to provide dwellings from cheaper and energy conserving materials, dwellings that guarantee energy efficiency, dwellings that promotes human privacy and security, dwellings with car parks, garages and secured yards, dwellings with private and communal open space and residences with comfortable density and waste services. Firstly, an essential way of achieving sustainable housing production is to use cheaper, local materials and energy-saving building materials. One of the major cost increasing construction materials in housing production is cement. While cement is derived from limestone which is not readily provided in all locations and undergoes critical production processes before being used, it is often suggested that earthen materials be used instead (Kadiri 2005). The earthen material is not only ubiquitous and cheaper, it also has better energy-conserving properties than the cement blocks. Houses can also be constructed from wooden materials interlaced with burnt earthen materials. Secondly, a way of increasing quality housing sustainably in Nigeria is through green housing projects. Green housing is a form of housing that intends to provide energy-efficient shelter, housing that manages well the radiant energy and artificially provided energy within the housing shell (UN-HABITAT 2012). Green energy housing is constructed from low energy radiating materials and designed to accommodate green neighbourhoods. The orientation of the dwelling and its internal layout can affect levels of daylight and sunlight, and will thus influence not only the occupants' conditions but the energy demand for heat and light.

Thirdly, sustainable housing incorporates and guarantees human privacy and dignity. Privacy is an important element of residential amenity that contributes towards the sense of security felt by people in their homes (Aliu and Ajala 2014). Internal privacy is provided by creating defensible space behind the public footpaths and raising the ground floor windows above the viewing level. Similarly, at the rear end of two or more storyed dwellings, adequate spaces between opposing first-floor windows are required to guarantee sufficient social distance and privacy of residents.

The dwelling features of sustainable housing like the neighbourhood features have effects on the housing choices and preferences of residents (Aliu and Ajala 2014).

4 Sustainable Housing Provision Constraints

A patient look at the requirements for the attainment of sustainable housing generally, makes one pause awhile and ask a cogent poser whether sustainable housing would ever be achieved in Nigeria or other developing economies in Africa. This poser has become almost inevitable because of the numerous challenges that have stood in the paths of mass and qualitative housing development in these countries. Out of the many problems facing sustainable housing provision in Nigeria five of them are quite outstanding and these are ponderous and inestimable population, costly building materials and high professional fees, profound social polarization and marginality, unorganized and dishevelled housing policy and undependable and non-inclusive financial system. The most critical challenge to sustainable housing production in Nigeria is the proportion and rate of population growth (see Table 4 on Nigeria's and global city population growth respectively). Nigeria is also characterized by rapid and population induced urbanization. In most Nigerian cities the growing rate of urban population is worrisome leading to high household formations that require adequate housing units (Ajala et al. 2010; UN-HABITAT 2008). Unfortunately, this rapid urban growth is not commensurate with per capita infrastructure and income. The size of a population and particularly the number of households, determine the demand for housing (Mulder 2006).

Housing is a massive user of diverse environmental resources ranging from liquid to solid and renewable to non-renewable resources (Kadiri 2005; Towry-Coker 2012). The lack of affordable and sustainable housing in Nigeriais is much related to the high and multiple costs of building materials and professional charges. To be candid sustainable housing construction requires threc components namely the construction material deployment, the professional material deployment and the maintenance material deployment. The construction of dwellings makes use of many resources such as granite, sand, iron or steel, limestone cement, zinc roof, woods, ceramic, polyvinyl products, and synthetic materials. While the majority of these products are obtainable within the immediate local environment, many of them are also imported from other countries. These materials are not ubiquitous and therefore are competitive and expensive. The international monetary system also affects the imported building materials very unfairly. The combined effect of these escalates the total budget for housing construction. Much more worrisome for the housing development is the experts' quality and increasing trend of fees charged by the planners, builders, artisans and other players in housing construction. The total money required to overcome construction materials, professional charges and maintenance costs are often huge making housing produced very unaffordable.

Social polarization or social inequality in Nigeria is widespread (Alu and Ajala 2015). Social polarization is engendered by inequality in social, economic and political opportunities. Access to resources in society is skewed towards a group of people and against some people. The lack of equity in resource distribution leads to acute polarization of society. Socio-economic polarization affects sustainable housing development directly by making access to adequate housing by the less privileged more herculean. Extreme poverty which reflects in multiple deprivations in terms of access to basic needs of a man like food, clothing and shelter can lead to advertent marginalization of the poor in society as recently observed in Maseru Lesotho (Motsoene 2014). Poverty and marginalization are social traits that can predispose individuals or groups to some feelings of alienation or non-inclusivity. Another factor undermining sustainable housing development in Nigeria is the uncoordinated housing policy that predominates in the region. Housing policy in Nigeria, like all public policies, is often bedevilled by a lack of organization, clarity and reality. The policy tools through which housing development can be achieved in third world countries are not only limited but ineffective. Ordinarily, policy tools such as mortgage finance, land subsidies, soft bank loans, stock exchange market and multilateral organizations should be meticulously fashioned to capture in holistic ways the target problems and populations. But alas, in developing countries policies are made without recourse to proper base data and without due consideration for other factors that could aid or compromise the realization of policy objectives (Towry-Coker 2012; Huang 2012). What thereafter results, is the situation where every part of the policy works at cross purposes to one another. A strong and effective housing policy is central to the realization of sustainable housing development.

Table 4. Nigerian population 1952–2015

Sn	Country	Year	Population	Rate of change
1	Nigeria	1952	30,403,305	–
2	Nigeria	1963	54,959,426	5.53
3	Nigeria	1991	88,992,220	1.74
4	Nigeria	2006	140,431,790	3.09
5	Nigeria	2015	174,507,539	3.15
Z	City/country	Population (million) 1995	City/country	Population (million) 2015
1	Tokyo, Japan	26.9	Tokyo, Japan	29.0
2	Mexico City, Mexico	16.6	Mexico City, Mexico	25.1
3	Sao Paulo, Brazil	16.5	Sao Paulo, Brazil	20.3
4	New York, USA	16.3	New York, USA	17.6
5	Mumbai, India	15.6	Mumbai, India	26.2
6	Shanghai, China	13.6	Shanghai, China	18.0
7	Los Angeles, USA	12.4	Los Angeles, USA	14.2
8	Calcutta, India	11.9	Calcutta, India	17.3
9	Seoul, South Korea	11.6	Buenos Aires, Argentina	13.9
10	Beijing, China	11.3	Beijing, China	16.0
11	Osaka, Japan	10.6	Osaka, Japan	10.6
12	Lagos, Nigeria	10.3	Lagos, Nigeria	24.6
13	Rio de Janeiro, Brazil	10.2	Rio de Janeiro, Brazil	11.9

Source: National Population Commission (NPC 2015) Lagos; United Nations Population Division (2015)

The Nigerian housing debacle is due mostly to poor access by the individual, private and public sectors to fund (Agbola 2005; Aliu et al. 2018). In well-organized economies ordinarily, the housing finance system is often organized around a mortgage system- primary and secondary (Megbolugbe and Cho 1993; Buist et al. 1994). Beyond the mortgage system housing finance is also augmented by the banks, insurance organizations, government financial subsidies in terms of budgetary allocation to housing construction, CBOs and NGOs. The truth about the housing fiancé system in developing countries is that sources through which housing is financed are very limited, weak and grossly inadequate in terms of the amount of money committed. Whether the individual or private or public housing, access to copious and sustainable finance is limited. In most developing countries the mortgage system is so organized as to take care of the workers alone leading to the exclusion of the majority of people in the informal sector.

5 Conclusions

From the detailed discussion rendered above, it is apparent that housing and sustainability have a strong relationship and achieving sustainable development is rather improbable without realizing sustainable housing. The difficulties in attaining sustainable housing in Nigeria can be ascribed to several reasons which include rapid urbanization, high cost of building materials, social polarization, a poor financial system for housing, a poor mortgage system and an uncoordinated housing policy. The implications of the revelations about the underlining relationship between housing and sustainable development call for several policy options that will support and facilitate the achievement of sustainable housing provision in Nigeria within the context of available natural environmental resources and socioeconomic realities. These alternative sustainable housing options include:

- The design and implementation of smart growth cities that make proper use of spaces
- The adoption and operationalization of green energy housing
- The strengthening of public and private sector participation (PPP) frameworks in housing provision
- The restructuring of housing policy and mortgage finance system to include the workers in the private sectors
- The development of green building materials that are derivable from the immediate local environment and less reliance on imported materials
- The control of rates of urbanization, population growth and social polarization within the urban centres

It is quite possible, that if these highlighted policy suggestions are adopted and implemented, the course of sustainable housing delivery will be forced to change positively in all developing economies.

References

Agbola, S.B.: The housing debacle. An Inaugural Lecture, University of Ibadan (2005)

Ajala, O.A., Aigbe, G.O., Aliu, I.R.: Affordable housing and urban development in Nigeria: contemporary issues, challenges and opportunities. Ilorin J. Bus. Soc. Sci. **14**(1), 1–13 (2010)

Aliu, I.R., Towry-Coker, L., Odumosu, T.: Housing policy debacle in sub-Saharan Africa: an appraisal of three housing programs in Lagos Nigeria. Afr. Geogr. Rev. **37**(3), 241–256 (2018)

Aliu, I.R., Adebayo, A.: Establishing a nexus between residential quality and health risks: an exploratory analytical approach. Indoor Built Environ. **22**(6), 852–863 (2013)

Aliu, I.R., Ajala, O.A.: Understanding residential polarization in a globalizing city: a study of Lagos. Sage Open J. Soc. Sci. Human. **3**(4), 1–15 (2013)

Aliu, I.R., Ajala, O.A.: Intra-city polarization, residential type and attribute importance: a discrete choice study of Lagos. Habitat Int. **42**(2), 11–20 (2014)

Aliu, I.R.: Spatial patterns of residential quality and housing preferences in selected Local Government Areas of Lagos Nigeria. Unpublished Ph.D. thesis, Obafemi Awolowo University Ile-Ife (2012)

Aliu, I.R., Ajala, O.A.: Residential polarization in a sub-Sahara African megacity: an exploratory study of Lagos. S. Afr. Geogr. J. **97**(3), 264–286 (2015)

Aribigbola, A.: Housing policy formulation in developing countries: evidence of program implementation from Akure Ondo State, Nigeria. J. Hum. Ecol. **23**(2), 125–134 (2008)

Arku, G.: Housing policy changes in Ghana in the 1990s: policy review. Hous. Stud. **24**(2), 261–272 (2009)

Bah, E.M., Faye, I., Geh, Z.F.: Housing market dynamics in Africa, p. 285. Palgrave-Macmillan, London (2018). https://doi.org/10.1057/978-1-137-59792-2

Beyer, G.H.: Housing and Society. Macmillan, London (1965)

Bourne, L.S.: The Geography of Housing. Edward Arnold, London (1981)

Buckley, R.M., Faulk, D., Olajide, L.: Private sector participation, structural adjustment, and Nigeria's new housing policy: Lesson from foreign experience. World Bank, Washington DC (1993)

Buist, H., Megbolugbe, I.F., Trent, T.R.: Racial home ownership patterns, the mortgage market and public policy. J. Hous. Res. **5**(1), 91–116 (1994)

Centre for Affordable Housing Finance CAHF.: Housing finance in Africa: A review of African housing finance markets. CAHF, Johannesburg, South Africa (2018)

Fadamiro, J.A.: An assessment of building regulations and standards and the implication for building collapse in Nigeria. In: Ogunsemi, D.R. (ed.) Building Collapse: Causes, Prevention and Remedies, pp. 28–39. The Nigerian Institute of Building, Ondo State, Nigeria (2002)

Gbadeyan, R.A.: Private sector's contributions to the development of the Nigerian housing market. Curr. Res. J. Soc. Sci. **3**(2), 104–113 (2011)

Headey, B.: Housing Policy in the Developed Economy. Croom Helm, London (1978)

Ibem, E.O.: Public-private partnership in housing provision in Lagos megacity region Nigeria. Int. J. Hous. Policy **11**(2), 133–154 (2011)

Ibem, E.O., Amole, O.O.: Evaluating of public housing programs in Nigeria: a theoretical and conceptual approach. Built Hum. Environ. Rev. **3**, 88–117 (2010)

Jiboye, A.D.: Urbanization challenges and housing delivery in Nigeria: the need for an effective policy framework for sustainable development. Int. Rev. Soc. Sci. Humanit. **2**(1), 176–185 (2011)

Jinadu, A.M.: Understanding the Basics of Housing: A Book of Study for Tertiary Institutions. University of Jos Press, Jos Nigeria (2007)

Kadiri, K.: Mass housing through earth construction technology in Nigeria. Pak. J. Soc Sci. **3**(5), 755–760 (2005)

Mabogunje, A.L., Hardoy, J.E., Misra, R.P.: Shelter provision in developing countries. Scientific Committee on Problems of the Environment SCOPE. Wiley, New York (1978)

Makinde, O.O.: Housing delivery systems need and demand. Dev. Environ. Sustain. **16**, 49–69 (2014)

Mayo, S., Gross, D.: The demand for housing in developing countries. Econ. Dev. Cult. Chang. **35**(4), 687–721 (1987)

Megbolugbe, I.F., Cho, M.: An empirical analysis of metropolitan housing and mortgage markets. J. Hous. Res. **4**(2), 191–224 (1993)

Motsoene, K.A.: Housing implications on sustainable development in Maseru-Lesotho. J. Sustain. Dev. Afr. **16**(4), 84–101 (2014)

Mulder, C.H.: Population and housing: a two-sided relationship. Demogr. Res. **15**, 401–412 (2006). https://doi.org/10.4054/DemRes.2006.15.13

National Bureau of Statistics NBS.: National Bureau of Statistics: Abuja Nigeria (2020)
National Population Commission NPC.: National Population Estimates. NPC, Abuja, Nigeria (2015)
Olotuah, A.O., Bobadoye, S.A.: Sustainable housing provision for the urban poor: a review of public sector intervention in Nigeria. Built Hum. Environ. Rev. **2**, 51–62 (2009)
Onibokun, A.G.: Housing in Nigeria. Nigerian Institute of Social and Economic Research, Ibadan (1985)
Rapoport, A.: Theory, culture and housing. Hous. Theory Soc. **17**, 145–165 (2001)
Sule, R.A.: The future of Nigerian housing subsidy- the unanswered question. Niger. J. Econ. Soc. Stud. **23**(1), 109–128 (1981)
Todaro, M., Smith, S.C.: Economic Development, 8th edn. Pearson, Addison Wesley, London (2003)
Towry-Coker, L.: Housing Policy and Housing Delivery in Nigeria: A Case of Lagos. Makeway Publication, Ibadan (2012)
UN-HABITAT: Sustainable housing for sustaianable cities: a policy framework for developing countries. HABITAT, Nairobi (2012)
UN-HABITAT: The state of African cities. HABITAT, Nairobi (2008)
UN-HABITAT: National trends in housing production practices, Nigeria, vol. 4. HABITAT, Nairobi (2003). http://hq.habitat.org. Accessed 5 Jan 2021
United Nations Population Division.: United Nations World Population Prospects Estimates 1950–2021. United Nations Department of Economic and Social Affairs of the United Nations Secretariat Washington DC, USA (2015)
Vasudevan, A.: The makeshift city: towards a global theory of squatting. Prog. Hum. Geogr. **39**(3), 338–359 (2015)
WCED: Our Common Future. World Commission on Environment and Development, Washington, DC (1987)
Huang, Y.: Low-income housing in Chinese cities: policies and practices. China Q. **212**, 941–964 (2012). https://doi.org/10.1017/S0305741012001270

Minimization of Transportation Cost for Decision Making on Covid-19 Vaccines Distribution Across Cities

H. T. Williams[1](✉), J. N. Mojekwu[2], and T. D. Ayodele[1]

[1] Department of Finance, Redeemer's University, Ede, Nigeria
{williamsh,ayodeleduro}@run.edu.ng

[2] Department of Actuarial Science and Insurance, University of Lagos, Lagos, Nigeria

Abstract. Purpose: The study demonstrates that population statistics and distance between points in kilometers are essential variables in modelling a transportation problem for vaccines distribution.

Design/Methodology/Approach: The study adopts Stratified sampling, quantitative research design and a transportation model of linear programming as methodology. 193,500,543 population data and a sample size of 4,705,800 were used to allocates COVAX 3.94 million doses of AstraZeneca/Oxford vaccines across different strata in Nigeria during the early period of COVID-19. Uber standard cost per kilometer and total quantities of vaccines available were used to compute the cost per unit of distributing a vaccine.

Findings: The results reconfirmed transportation model as an appropriate model suitable to solve transportation problem and minimizes the cost of distributing vaccines across cities. The results also shows the appropriate allocations of vaccines to demand and supply location at a given cost across each stratum of the population.

Research Limitations/Implications: Numerous constraints variables and population stratum.

Practical Implication: The quantitative research design used in the transportation model would informed policy holders that accurate statistics and cities topology are essential to minimize budgeted cost for vaccines distributions.

Social Implication: The variables used in the study practically tested transportation model on transportation problem.

Originality/Value: The originality of the study lies on the population data used, distance in kilometers and the mathematical techniques and assumptions adopted from the field of operations research.

Keyword: Cities · Decision-making distribution · Minimization · Vaccines

C. Aigbavboa et al. (Eds.): ARCA 2022, *Sustainable Education and Development – Sustainable Industrialization and Innovation*, pp. 1180–1189, 2023.
https://doi.org/10.1007/978-3-031-25998-2_91

1 Introduction

This study emanated as a result of the federal government of Nigeria's interest in planning and budgeting cost for the transportation of Corona Virus Pandemic (COVID-19) vaccines (COVAX 3.94 million doses of the AstraZeneca/Oxford) to the 36 States and the Federal Capital Territory in Nigeria. Financial accounting states that appropriate budgeting is essential while forensic accounting opined that the variance between budgeted amount and the actual amount is an essential variable for applied auditing hence it is essential for researchers and scholars in the field of Operation Research in Nigeria to look at transportation cost and the transportation modeling techniques.

Transportation model is a mathematical model used to minimize cost. Therefore a transportation problem should be solved with a transportation model. As the Federal Government of Nigeria prepares to receive the vaccine for Coronal Virus Pandemic (COVID-19), the need for adequate distribution of the vaccine from source to destination becomes necessary for a study of this nature. Due to the complexity in optimizing transportation problem in Nigeria, there is need for adequate preparedness in the Nigeria transportation industry because the usual analogue pattern of transportation modeling may not be suitable to solve and model transportation problems for COVID-19 vaccine distribution across all States in Nigeria. The COVID-19 pandemic has affected all forms of economic activities in the world (The World Bank - Global Economic Prospect 2020), and Nigeria which is a major player in the Africa continent was not exempted. Many were affected medically and some were affected economically by the COVID-19. The process of preparing for decision-making by government on COVID-19 vaccine supply and allocation is a challenge that required strategic planning. Most developing nations are faced with the issue of uncertainties and thus affecting decision process. A transportation model to optimize cost for COVID-19 vaccine required management to apply alertness and effective feedback segment. In order to control the effect of COVID-19 in Nigeria in the year 2020, the Federal Government of Nigeria declares a total lock down of her economy because the World Health Organization (WHO) categorized COVID-19 as a pandemic and a public health crisis. Essentially, stakeholders have already engaged in one form of movement or the other but the impact of COVID-19 make a clarion call for all stakeholders to leave up to expectations and follow the COVID-19 protocols. The call for social distancing is a mathematical modeling to minimize the spread of the virus. As the world have developed vaccine to combat the deadly COVID-19, the need for the vaccine to be appropriately distributed is essential to the world hence a level of preparedness is required on every elements of the world population. Managing COVID-19 vaccine shipment from source to destination involves planning and coordination otherwise additional complexity might occur in the process. This study focused on the optimal distribution of the AstraZeneca/Oxford vaccine across Osun State using transportation model with R-programming.

2 Theories Underpinning the Study

The theory of Mechanism and that of the theory of game was adopted as the theory for the study. The mechanism theory opined that a form of compensation can be used to structure an act of behavior. The grouping of local governments area in Osun State Nigeria to form a demand point and supply helps in avoiding maginalisation of local government areas. Mechanism theory was used to determine whether the government of Osun State can achieve the desired result. by implementing specific strategies and incentives to obtain the desired outcome. Batkovskiy et al. (2016) concluded that the use of mechanism theory in mathematical based analysis could help in effective structuring of health services within a location where healthcare is essential. Craven and Islam (2005) states that the use of game theory in health research can lead to a zero sum game where the health professional render all essential health services to keep a person alive.

Muhammed (2020) looked at the impact of appropriate modeling of logistic problem using discrete data for South Africa companies listed in the capital market and found out that transportation modeling is essential for finding optimal cost in logistic problem. James and George (2021) states that the use of transportation models in management sciences to solve logistics problem has not been fully utilized hence an urgent request for the use of transportation model in logistic problems. Bahmani et al. (2013) studied some parts of stochastic transportation problem and the transportation model using machine learning and conclude that the transportation model is not only suitable to solve the problem of logistics but also the problem of cost minimization in any field of research. Williams et al. (2021) transform a linear programming model to a transportation model to solve the problem of healthcare cost minimization in tertiary institutions in Nigeria. Semad and Irfan (2017) construct transportation model using stochastic analysis to determine the cost per unit of transporting an object from a particular source to a particular destination. The MS excel spreadsheet, R and python are robust tools in which transportation model can be carried out. Dedu and Sarban (2015) states that the application of transportation model in real life case is the same in academic case and vital variables such as the distance between sources and destinations most be known as well as the total units of supply and demand.

3 Methodology

The study adopts Stratified sampling, quantitative research design and a transportation model of linear programming as methodology. 193,500,543 population data and a sample size of 4,705,800 were used to allocates COVAX 3.94 million doses of AstraZeneca/Oxford vaccines across different strata in Nigeria during the early period of COVID-19. Uber standard cost per kilometer and total quantities of vaccines available were used to compute the cost per unit of distributing a vaccine.

Table 1. 2016 Nigeria projected population data

States	Total	Joint prob.	COVAX 3,940,000. million doses of the AstraZeneca/Oxford vaccine
Kano State	13,076,892	0.067581	
Borno State	5,860,183	0.030285	
Delta State	5,663,362	0.029268	
Niger State	5,556,247	0.028714	
.	.	.	
.	.	.	
.	.	.	
.	.	.	
Ondo State	4,671,695	0.024143	
Osun State	**4,705,589**	**0.024318**	**95813.79 = 95814**
Kogi State	4,473,490	0.023119	
.			
.			
.			
Ebonyi State	2,880,383	0.014886	
Nasarawa State	2,523,395	0.013041	
Bayelsa State	2,277,961	0.011772	
Federal Capital Territory	3,564,126	0.018419	
Total	193,500,543	1	

Source: Nigeria Bureau of Statistics (projected population for the year 2016).

Summary: from the above table, 95,814 doses of COVAX of the AstraZeneca/Oxford vaccine should be allocated to Osun State (Fig. 1, Tables 2, 3, 4 and 5).

Table 2. Osun state 2016 projected population data across local government areas

S/n	LGA	Headquarters	Population
1	Aiyedaade	Gbongan	206,000
2	Aiyedire	Ile Ogbo	105,100
3	Atakunmosa East	Iperindo	104,800
4	Atakunmosa West	Osu	94,100
5	Boluwaduro	Otan Ayegbaju	97,700

(*continued*)

Table 2. (*continued*)

S/n	LGA	Headquarters	Population
6	Boripe	Iragbiji	191,100
7	Ede North	Oja Timi	115,400
8	Ede South	Ede	104,000
9	Egbedore	Awo	101,900
10	Ejigbo	Ejigbo	182,500
11	Ife Central	Ile-Ife	230,300
12	Ife East	Oke-Ogbo	51,700
13	Ife North	Ipetumodu	259,700
14	Ife South	Ifetedo	132,800
15	Ifedayo	Oke-IlaOrangun	211,100
16	Ifelodun	Ikirun	185,200
17	Ila	IlaOrangun	85,500
18	Ilesa East	Ilesa	145,200
19	Ilesa West	Ereja Square	147,100
20	Irepodu	Ilobu	164,700
21	Irewole	Ikire	196,700
22	Isokan	Apomu	140,500
23	Iwo	Iwo	263,500
24	Obokun	Ibokun	160,900
25	Odo Otin	Okuku	181,900
26	Ola Oluwa	Bode Osi	105,000
27	Olorunda	Igbonna Osogbo	181,300
28	Oriade	Ijebu-Jesa	204,300
29	Orolu	Ifon-	141,600
30	Osogbo	Osogbo	214,200
	Total state population (TSP)		4705800

Source: Osun State government secretariat, Abere (2021)

The parameters are as follows

$W_{1,4}$ is the set of supply location across Osun State- Head Quarter of each senatorial district.

D is the demand for the COVID-19 vaccine at each location.

S is the supply of COVID-19 vaccine in each senatorial district across the state.

CPU = Using Uber standard cost of N550 per kilometer. (Cost per unit = distance covered * standard Uber cost per KM/Exp.SS)

Table 3. Distributing COVAX 95,814 doses of the AstraZeneca/Oxford vaccine using the three senatorial districts and the state capital as supply locations.

District	HQ	Supply location	Population	Joint prob.	Storage (house)
Central district (exclude Osogbo)	Ilobu	W_1 - Ilobu	1647500	0.350	33,535
West district	Iwo	W_2 - Iwo	1630900	0.347	33,247
East district	Ile ife/ijesha	W_3 – Ile ife	1213200	0.258	24,720
State capital	Osogbo	W_4 – Osogbo	214,200	0.045	4,312
		Total	**4705800**	**1**	**95,814 doses**

Source: Computed

Table 4. Determining expected demand of AstraZeneca/Oxford vaccine, across Osun State using the state joint probability obtained from Table 1 and the clustered.

S/n	LGA	Headquarters	Population	Demand	Expected Demand
1	Aiyedaade	Gbongan	206,000	$\sum_{1}^{3} population$ = 415,900 * 0.024318	D_{1HG} = 10,114
2	Aiyedire	Ile Ogbo	105,100		
3	Atakunmosa East	Iperindo	104,800		
4	Atakunmosa West	Osu	94,100	$\sum_{4}^{6} population$ = 382,900 * 0.024318	D_{2HI} = 9,311
5	Boluwaduro	OtanAyegbaju	97,700		
6	Boripe	Iragbiji	191,100		
7	Ede North	OjaTimi	115,400	$\sum_{7}^{9} population$ = 321,300 * 0.024318	D_{3HO} = 7,813
8	Ede South	Ede	104,000		
9	Egbedore	Awo	101,900		
10	Ejigbo	Ejigbo	182,500	$\sum_{1o}^{12} population$ = 464,500 * 0.024318	D_{4HI} = 11,296
11	Ife Central	Ile-Ife	230,300		
12	Ife East	Oke-Ogbo	51,700		
13	Ife North	Ipetumodu	259,700	$\sum_{13}^{15} population$ = 603,600 * 0.024318	D_{5HI} = 14,678
14	Ife South	Ifetedo	132,800		
15	Ifedayo	Oke-IlaOrangun	211,100		

(continued)

Table 4. (*continued*)

S/n	LGA	Headquarters	Population	Demand	Expected Demand
16	Ifelodun	Ikirun	185,200	$\sum_{16}^{18} population$ = 415,900 * 0.024318	D_{6HI} = 10,114
17	Ila	IlaOrangun	85,500		
18	Ilesa East	Ilesa	145,200		
19	Ilesa West	Ereja Square	147,100	$\sum_{19}^{21} population$ = 508,500 * 0.024318	D_{7HI} = 12,366
20	Irepodun	Ilobu	164,700		
21	Irewole	Ikire	196,700		
22	Isokan	Apomu	140,500	$\sum_{22}^{24} population$ = 564,900 * 0.024318	D_{8HI} = 13,737
23	Iwo	Iwo	263,500		
24	Obokun	Ibokun	160,900		
25	OdoOtin	Okuku	181,900	$\sum_{25}^{27} population$ = 468,200 * 0.024318	D_{9HO} = 11,386
26	Ola Oluwa	Bode Osi	105,000		
27	Olorunda	Igbonna, Osogbo	181,300		
28	Oriade	Ijebu-Jesa	204,300	$\sum_{28}^{30} population$ = 560,100 * 0.024318	D_{10HO} = 13,621
29	Orolu	Ifon-Lagos	141,600		
30	Osogbo	Osogbo	214,200		
				Total	114,436

Source: Computed

$$C11 = \frac{Distance\ covered * Standard\ Uber\ Cost\ Per\ Kilometer}{Expected\ Supply} = \frac{55.3\,KM * 550}{33{,}535} = 0.907$$

Transforming Transportation Model to Linear Programming Model.

The transportation model is minimizations model hence a Linear Programming Model for minimization was adopted.

Minimize Cost (C) = $C_{11}+C_{12}+C_{13}+C_{14}+C_{15}+C_{16}+C_{17}+C_{18}+C_{19}+C_{10}+C_{21}+C_{22}+C_{23}+C_{24}+C_{25}+C_{26}+C_{27}+C_{28}+C_{29}+C_{210}+C_{31}+C_{32}+C_{33}+C_{34}+C_{35}+C_{36}+C_{37}+C_{38}+C_{39}+C_{310}+C_{40}+C_{41}+C_{42}+C_{43}+C_{44}+C_{45}+C_{46}+C_{47}+C_{48}+C_{49}+C_{410}$

Subject to:

$C_{11} + C_{12} + C_{13} + C_{14} + C_{15} + C_{16} + C_{17} + C_{18} + C_{19} + C_{110} \geq 33535$ (Ilobu SS)

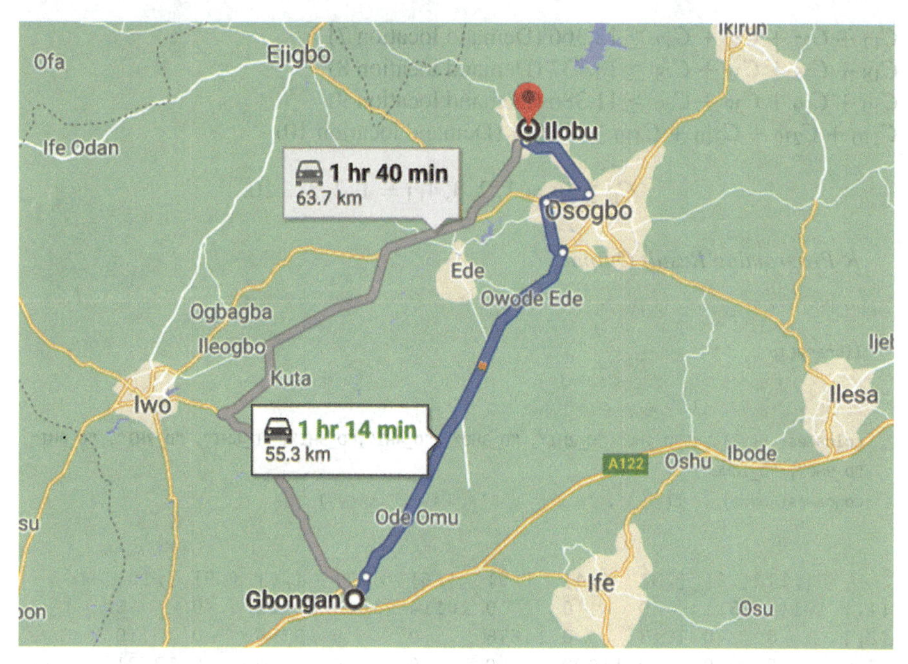

Fig. 1. Distance between Two Locations (W) and (D) across Osun State with Google Map.

Table 5. Determining the distance, Cost Per Unit (CPU) of distributing the AstraZeneca/Oxford vaccine from location (W) to demand location (D) across Osun State.

To From	D_{1HG}	D_{2HI}	D_{3HO}	D_{4HI}	D_{5HI}	D_{6HI}	D_{7HI}	D_{8HI}	D_{9HO}	D_{10HO}	Exp. SS
W_1	55.3km $C_{11} = 0.907$	C_{12}	C_{13}	C_{14}	C_{15}	C_{16}	C_{17}	C_{18}	C_{19}	C_{110}	33,535
W_2	C_{21}	C_{22}	C_{23}	C_{24}	C_{25}	C_{26}	C_{27}	C_{28}	C_{29}	C_{210}	33,247
W_3	C_{31}	C_{32}	C_{33}	C_{34}	C_{35}	C_{36}	C_{37}	C_{38}	C_{39}	C_{310}	24,720
W_4	C_{41}	C_{42}	C_{43}	C_{44}	C_{45}	C_{46}	C_{47}	C_{48}	C_{49}	C_{410}	4,312
Dum	0	0	0	0	0	0	0	0	0	0	18,622
Exp.D	10,114	9,311	7,813	11,296	14,678	10,114	12,366	13,737	11,386	13,621	114,436

Source: researcher computation 2021

$C_{21} + C_{22} + C_{23} + C_{24} + C_{25} + C_{26} + C_{27} + C_{28} + C_{29} + C_{210} \geq 33{,}247$ (Iwo SS)
$C_{31} + C_{32} + C_{33} + C_{34} + C_{35} + C_{36} + C_{37} + C_{38} + C_{39} + C_{310} \geq 24{,}720$ (Ile-ife SS)
$C_{41} + C_{42} + C_{43} + C_{44} + C_{45} + C_{46} + C_{47} + C_{48} + C_{49} + C_{410} \geq 4{,}312$ (Osogbo SS)
$C_{11} + C_{21} + C_{31} + C_{41} \geq 10{,}114$ (Demand location 1)
$C_{12} + C_{22} + C_{32} + C_{42} \geq 9{,}311$ (Demand location 2)
$C_{13} + C_{23} + C_{33} + C_{43} \geq 7{,}813$ (Demand location 3).
$C_{14} + C_{24} + C_{34} + C_{44} \geq 11, 296$ (Demand location 4)
$C_{15} + C_{25} + C_{35} + C_{45} \geq 14{,}678$ (Demand location 5)
$C_{16} + C_{26} + C_{36} + C_{46} \geq 10{,}114$ (Demand location 6)

$C_{17} + C_{27} + C_{37} + C_{47} \geq$ 12,366 (Demand location 7)
$C_{18} + C_{28} + C_{38} + C_{48} \geq$ 13,737 (Demand location 8)
$C_{19} + C_{29} + C_{39} + C_{49} \geq$ 11,386 (Demand location 9)
$C_{110} + C_{210} + C_{310} + C_{410} \geq$ 13,621 (Demand location 10)

$$C_{ij} \geq 0 \mathrm{i} = 1, 2, 3, 4, \mathrm{j} = 1, 2, 3.....10.$$

R-Programing Result Output.

```
library(lpSolve)
# Set transportation costs matrix
# Set customers and suppliers' names
colnames(costs)<-c("D_1HG","D_2HI", "D_3HO", "D_4HI","D_5HI", "D_6HI", "D_7HI", "D_8HI",
"D_9HO", "D_10HO")
rownames(costs) <- c("W 1", "W 2", "W 3", "W4", "Dummmy")
```

```
      [,1] [,2] [,3]  [,4]  [,5]  [,6]  [,7]  [,8] [,9] [,10]
[1,] 10114 9311    0     0     0 10114  1015  2981    0     0
[2,]     0    0 7813     0 14678     0     0 10756    0     0
[3,]     0    0    0 11099     0     0     0     0    0 13621
[4,]     0    0    0     0     0     0     0     0 4312     0
[5,]     0    0    0   197     0     0 11351     0 7074     0
```

4 Findings and Discussion

The result shows that was divided in strata and 95814 doses of the AstraZeneca/Oxford vaccines should be allocated to a particular stratum of the population based on population statistics and appropriate proportions for optima results. The results also shows four storage locations, one dummy location and ten demand locations for the AstraZeneca/Oxford vaccines distribution across cities. The cost per unit of distributing AstraZeneca/Oxford vaccine was obtained using Uber standard average cost of ₦550 per kilometer, projected population statistics and quantity of vaccines available for distributions. This results of this study support the work of Semad and Irfan (2017), who show the distribution of medical equipment using stochastic programming. The study also corroborates with the findings of Williams et al. (2021) and Bahmani et al (2013) that transportation model of linear programming minimized cost. It also support the work of Muhammed (2020), James and George (2021) in logistic and distribution. The study does not support the work Batkovskiy et al (2016) and Craven and Islam (2005) by testing mechanism and game theory on linear programming model. The study support the work of Dedu and Sarban (2015) that the application of transportation model is different from real world situation and in theory. Due to the complexity of this study, the results does not support the classroom use of Vogel and minimum cost methods to solve a problem of this nature hence the use of R and Python programming to run transportation model.

5 Conclusion

The study aimed to answer the research question and established empirical and statistical evidence for obtaining the optimal cost and allocations of the AstraZeneca/Oxford vaccines across a given population within a communities using population statistics and the cost per unit of transporting the vaccine from the storage (warehouse or supply) to demands locations across Osun State, Nigeria. The relevance of transportation modelling in solving logistics problem in time of pandemic have not been utilized in the Nigeria decision making process for vaccines distribution and other logistics problem that relates to health across states and communities in Nigeria. The study concluded by validating transportation model as a model to determine optimal cost. The results of this study is used to benchmark the cost of transporting of the 3.9 million doses of the AstraZeneca/Oxford vaccines across the population strata in Nigeria. The study failed to account for all the strata of the population but recommended that if the transportation model is replicated in other statum of the population, an optimal results would be achieved that minimized cost.

References

Bahmani, S., Raj, B., Boufounos, P.: Greedy sparsity-constrained optimization. J. Mach. Learn. Res. **14**, 807–841 (2013). https://doi.org/10.1109/ACSSC.2011.6190194

Batkovskiy, A.M., Semenova, E.G, Trofimets, V.Ya., Trofimets, E.N., Fomina, A.V.: Method for adjusting current appropriations under irregular funding conditions. J. Appl. Econ. Sci. **XI**(5(43)), 828–841 (2016). https://www.ceeol.com/search/article-detail?id=534283

Cornuejols, G., Tutuncu, R.: Optimization Methods in Finance. Carnegie Mellon University, Pittsburgh, p. 349 (2006)

Craven, B.D., Islam, S.M.N.: Optimization in Economics and Finance: Some Advances in Nonlinear, Dynamic and Stochastic Models, p. 163. Springer, New York (2005). https://doi.org/10.1007/b105033

Dedu, S., Şerban, F.: Multiobjective mean-risk models for optimization in finance and insurance. Procedia Econ. Financ. **32**, 973–980 (2015). https://doi.org/10.1016/S2212-5671(15)01556-7

James, E., George, C.: Cost optimization with transportation modeling. Int. J. Account. Econ. **1**(1), 21–26 (2021)

Mohammed, A.: Transportation modeling in logistics: the South Africa experience. J. Logist. Transp. **12**(3), 29–36 (2020)

Semad, B., Irfan, F.: A stochastic programming model for decision-making concerning medical supply location and allocation in disaster management. J. Disaster Med. Public Health Prep. **11**, 747–755 (2017)

Williams, H.T., Abiola, B., Ojikutu, R.K.: Minimizing healthcare cost in selected tertiary institutions in Nigeria. J. Innov. **64**(1), 544–554 (2021)

Promoting Sustainable Industrialization in Tanzanian Agro-Processing Sector: Key Drivers and Challenges

M. A. Tambwe(✉) and M. A. Mapunda

Department of Marketing, College of Business Education, P.O. Box 1968, Dar es Salaam, Tanzania
Mariam.tambwe@cbe.ac.tz

Abstract. Purpose: The main purpose of the paper is to address the key drivers and challenges to sustainable industrialization in Tanzania. The authors explore how attention to these challenges and drivers can enhance sustainable industrialization.

Design/Methodology/Approach: This paper uses a resource-based view theoretical lens to inform the author on the key drivers and challenges to sustainable industrialization. A mixed-method research design was adopted and data were drawn from 150 respondents in agro-processing industries from five regions in Tanzania selected by cluster and random sampling technique. Questionnaires, key informant interviews, and documentary reviews were used in data collection. Regression analysis and thematic data analysis strategy were adopted whereby quantitative data gathered were analysed by regression analysis using SPSS version 23 and qualitative data analyzed by content analysis using MAXQDA 2020 software.

Findings: The main findings based on the resource-based view indicated that the key drivers include; physical, human, organizational, financial and technological resources which have a positive influence on sustainable industrialization in Tanzania. The key challenges to industrialization in Tanzania include; a lack of long-term loans for investment in industrialization, difficult government regulations and taxation, poor agricultural technology, poor infrastructure, and lack of compliance with standards to mention just a few.

Research Limitation: The study focused on the key drivers and challenges to sustainable industrialization in Tanzania based on agro-processing industries.

Practical Implication: These findings contribute to the existing knowledge in the sector and insist on stakeholders' participation in the improvement of the physical, human, organizational, financial and technological resources to reach sustainable industrialization.

Social Implication: Solving the identified challenges in the sector will help in improving people's livelihood, the country's economy and employment in general.

Originality/Value: The main contribution is the improvement of our knowledge about the drivers and challenges to sustainable industrialization. Furthermore, the study contributes to the extension of RBV theory by adding two more variables obtained from the literature.

C. Aigbavboa et al. (Eds.): ARCA 2022, *Sustainable Education and Development – Sustainable Industrialization and Innovation*, pp. 1190–1208, 2023.
https://doi.org/10.1007/978-3-031-25998-2_92

Keywords: Challenges · Drivers · Economic growth · Industrialization · Tanzania

1 Introduction

Industrialization has an essential role to play in helping Tanzania increase its economic growth and development (URT 2016). The industry continues to be a proven and critically important source of employment, accounting for nearly 500 million jobs worldwide, representing about one-fifth of global employees (Guadagno et al. 2019). Industrial development and trade in industrial goods have proven successful in reducing poverty. Industrialization is taken as a significant substance in economic growth promotion. (UNIDO 2017; Moyo 2017). Of late, the Tanzania government has declared industrialization theme as a driver towards a middle-income economy (Kirumirah et al. 2020). It is noted that currently industry contributed approximately 29.28% of the GDP (O'Neill 2023).

Over the past decade, Tanzania has experienced an average annual growth of 3.1% (Global Development Indicators 2018). This is below the 8–10% target set in several government plans, which is essential to maintaining strong long-term growth in GDP per capita (Oqubay 2015). Unfortunately, since 1971 Tanzania has been recorded as a slightest industrialized country by having an estimate of $900 as a per capita income (Guadagno et al. 2019). This was due to low production as well as low production revolution leading to sluggish economic development (Global Development Indicators 2018). Although since the early 2000s the industry has developed significantly, contribution of manufacturing equated to 6% of the total value added remains to have minor performance. For that reason, Tanzania Development vision 2025 is geared to champion industrialization for the attainment of middle-income status by 2025. For this case the Tanzania industrial policy strategy discloses that the agro-processing industry deserves the attention of the government due to the importance of agriculture in the country's economy.

Realizing the importance of industrialization, various initiatives were made globally, regionally and nationally. The global efforts include the Sustainable Development Goals (SDG) number 9 restates sustainable industrialization importance in attaining innovativeness as well as reaching economic growth (Kirumirah et al. 2020; Osborn et al. 2015). Correspondingly, the agenda 2063 the Africa we want, strengthening industrialization through value addition was highlighted extremely. Hence, seriousness to its achievement is needed by developing countries (Moyo 2017; African Union Commission 2015). In 2015 Tanzania set an industrialization agenda forecasting the realization of its long-waited goal of becoming a middle-income country (Moyo 2017). The adoption of the integrated industrial development strategy 2025 aiming to execute the policy based on sustainable industrial development. In addition, the policy was geared to create viable industries to achieve a middle-income economy by 2025. The sustainable industrialization expected to create jobs leading to economic development. The key drivers to industrial development was envisioned to include; access to long-term credits, energy supply, infrastructure, skilled labour, supportive systems, and technology to mention the few. In addition, the capacity of the firms to manufacture competitive products for the domestic and foreign markets are to be considered. However, despite of the efforts made, still the economy of Tanzanian as of now are essentially agricultural dependent, employing 66% of its labour force. The agriculture also produces 30% of its value added.

Due to the contribution and importance of industrialization to the different countries' economic growth, various researches were done (Wangwe et al. 2014; Etuk et al. 2014; Moyo 2017; Pérez-Fuentes 2013). Despite, the importance of industrialization for the country's economic growth, few researches have been conducted on industrialization promotion in Tanzania and none has used resource-based theory. The paper's goal is to assess the key drivers and challenges to promote sustainable industrialization for improving Tanzania economy and peoples' livelihood. Likewise, the study identified a gap in theory to amplify the key drivers and challenges in promoting sustainable industrialization. Therefore, this study aimed at assessing key drivers and challenges to promote sustainable industrialization in Tanzania agro-processing sector, through the resource-based view theoretical lens. The study objective was to assess key drivers and challenges to promoting sustainable industrialization in Tanzanian agro-processing sector. The following was the study's specific objectives: to determine the effect of physical resources on sustainable industrialization in Tanzanian agro-processing sector; to determine the influence of human resources on sustainable industrialization in Tanzanian agro-processing sector; to establish the effects of organizational resources on sustainable industrialization in Tanzanian agro-processing sector and to explore the key challenges hindering sustainable industrialization in Tanzanian agro-processing sector.

The research results will enlighten stakeholders such as policy makers about key drivers and challenges towards sustainable industrialization. This knowledge is expected to assist them in formulating responsive achievable enabling policies that will aid the industrial sector and practitioners to acquire the needed backing in their struggles to reach economic growth.

The study's results will offer understandings to industrial sector, by developing industrial ready entrepreneurs either by being employed in the existing industries or being owners by establishing their own industries and employ others.

2 Theoretical Framework

Several studies (Seetoh and Ong 2008; Dinh n.d.; Guadagno et al. 2019; Haraguchi et al. 2019) have demonstrated that sustainable industrialization can be achieved through the availability of various factors and resources at the country level as well as the firm level. For instance, Dinh (n.d.) identified market, capital, technologies, natural resource, development of human resources and protection of environment to be the key drivers to sustainable industrialization. Similarly, Wangwe et al. (2014) reported product quality, investment in technology, strategic marketing, investment in human resources, infrastructure, and improved access to finance as key drivers to sustainable industrialization. In addition, Haraguchi et al. (2019) study reveal that sustainable industrialization is drven by various determinants for instance, factor bequests, preliminary economic situations of a country, and other features, like demographics and natural features. Other factors were investment promotion, entrepreneurship training, management of business and capital accessibility, institutional constancy, financial sector development and macroeconomic promotion.

Despite the availability of the above-mentioned studies, these studies fall short in scope, context and approach used. The studies differ with this study's context and approach used. While the study of Dinh (n.d.) was conducted in Asian economies using secondary data, and the study by Haraguchi et al. (2019) studied developing countries in general using secondary data. This research differ in approach where mixed methods have been applied using the context of agro-processing in Tanzania. Furthermore, the research explored the challenges hindering sustainable industrialization to increase understanding of the phenomenon studied.

Additionally, the revised literatures depicted many challenges hindering sustainable industrialization. For example, Guadagno et al. (2019) revealed the following challenges to sustainable industrialization; poor infrastructure, lack of access to credit for industrial investment, unfriendly government regulations and tax issues, lack of agro-processing technology, under-developed linkages with agro-processing suppliers and support organizations which reduce productivity just to metion the few. In tandem with that, Sampath (2016) showed the challenges to sustainable industrialization as poor technology, lack of finance, poor infrastucture, unsupportive government regulations and ploicies, lack of markets due to standard compliance issues etc. Wangwe et al. (2014) identified various technological, financial, policy, and administrative constraints to hinder sustainable industrialization. Mkwizu and Monametsi (2021) identified poor implementation of policies, inadequate supply of energy, lack of innovation, lack of informal sector involvement and labour market skills gap to be the major challenges to industrialization. The current research uses these foundations obtained from the previous literatures to shape thoughtful of the many challenges hindering sustainable industrialization from the empirical studies earlier conducted. It also increases our knowledge concerning diverse challenges impeding sustainable industrialization.

2.1 Theoretical Stance and Development of Hypotheses

The resource-based theory emphasizes the strategic used of the firms' resources that are "valuable", "heterogeneous", "immobile", and "inimitable resources" to produce a competitive advantage (Barney 1991; Kumar and Rodrigues 2020). There are two types of resources intangible and tangible. Machines, assets, etc. form tangible resources while staff skills, knowledge, intellectual capital, learning etc. are part of intangible resources (Barney 1991). Competitive advantage can be obtained by firms through utilization of these resources in an inimitable way (Barney and Clark 2007). Furthermore, the classification of resources are as follows: human, physical, and organizational capital (Grant 1991). Then later resources were extended by Sony and Aithal (2020) to financial, reputational and technological, capital. Capabilities to acquisition of these resources differs from one firm to the other. Also, the acquisition of these resources is difficult. Hence, to reach competitive advantage firms have to package these resources in a complex mixture (Ulrich et al. 1995).

However, resource-based theory deals with organizational value to meet competitive advantage. But in this paper resource-based theory was applied to explain the key drivers and challenges to sustainable industrialization and how Tanzania as a country should craft its strategies to reach sustainable industrialization using the available capabilities and resources. In addition, the original resource-based view considers the three resources namely, physical, human and organizational resources. But from the literatures (Dinh n.d.; Haraguchi et al. 2019; Guadagno et al. 2019; Wangwe et al. 2014), it was found that, there are other resources like financial and technological which are necessary especially in studying industrialization.

The study's theoretic framework is shown in Fig. 1. Following the resource-based perspective, we hypothesize that physical resources, the human resources as well as organizational resources influence the sustainability of the industries.

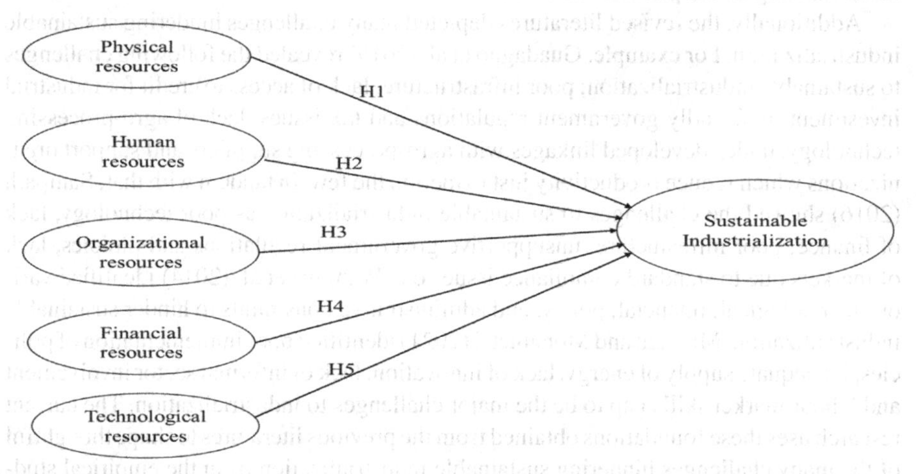

Fig. 1. Drivers to sustainable industrialization constructed from the extended resource-based theory

2.1.1 Physical Resources and Sustainable Industrialization

Physical resources are what the country owns and uses including; production facilities, equipment, geographic location, infrastructure and natural resources (Madhani 2010). The research demonstrated an association between physical resources and industrialization. Goodwin (2003) studied the five capital useful for sustainable industrialization using secondary data from American experience. The study found positive relationship between physical resources in terms of infrastructure, equipments, machines, natural resources and sustainable industrialization. Similarly, Sony and Aithal (2020) used resource based view to examine how the "physical, human, organizational, technological, financial and reputational capital" can be used in Engineering Industry in India to achieve competitive advantage. The study found that, physical resources positively influence sustainable industrialization. Also, Rahim et al. (2021) conducted a study to eleven countries including; Iran, South Korea, Banglades, Nigeria, Pakistan, Mexico, Vietnam, Indonesia, Philippines, Turkey, and Egypt on resources and industrialization issues using secondary data. Their study revealed that natural as well as physical resources have an impact to sustainable industrialization.

From the foregoing discussion, we make the following assumption:

H_1: Physical resources have a positive relationship to sustainable industrialization in Tanzania.

2.1.2 Human Resources and Sustainable Industrialization

Human capital capabilities include intangible resources like, intelligence, people experience, relationships, training, judgment, and understanding of the company's managers and employees (Madhani 2010). Various studies (Rahim et al. 2021; Babasanya et al. 2018; Goodwin 2003) have exerted a link between human resources and sustainable industrialization. For instance, Rahim et al. (2021) made a survey to eleven countries and discovered that education and skills positively impact sustainable industrialization. Also, (Goodwin 2003), found that knowledge, skills and experience have a positive relationship with sustainable industrialization. From these findings it is hypothesized that;

H_2: Human resources positively affect sustainable industrialization in Tanzania.

2.1.3 Organization Capital Resources and Sustainable Industrialization

Organizational resources are tangible resources which includes formal structure, coordinating systems, planning, controlling, informal relations within a firm and amongst its environment (Madhani 2010; Barney 1991). According to Barney, this construct influences sustainability and competitiveness. The construct has been extensively explored by previous researchers (Goh and Loosemore 2017; Babasanya et al. 2018; Sony and Aithal 2020) and found to have a positive effect to industrialization. For example, Babasanya et al. (2018) found that organizational plans have a positive influence on sustainable industrialization in the Ethiopian environment. In addition, Goh and Loosemore (2017) found that structures and systems are linked to the sustainable industrialization in construction industry on Australian context. Thus, the following assumption is proposed;

H_3: Organizational resources have a positive association to sustainable industrialization in Tanzania.

2.1.4 Financial Resources and Sustainable Industrialization

Financial resources are tangible resources which consist of ability to generate internal funds and raise external capital (Madhani 2010) to improve business performance. Access to financial resources is proposed to be among the drivers to sustainable industrialization (Rahim et al. 2021). The variable has been extensively used by the past empirical literature (Goh and Loosemore 2017; Dinh n.d.; Sony and Aithal 2020) and indicates a significant positive connection among accessibility of financial resouces and sustainable industrialization. For instance, a study by Goh and Loosemore (2017) revealed that availability of financial resources was one among the drivers that have a positive and significant influence to construction industries in Australian context. Thus, leading to the following proposition;

H_4: Financial resources have a positive relationship to sustainable industrialization in Tanzania.

2.1.5 Technological Resources and Sustainable Industrialization

Technology has an impact to industrialization. Most of the developing countries are still using traditional technologies which could not assist them to increase productivity (Dinh n.d.). To strengthen industrial competitiveness, there is a need for transformation from the current local technology to advanced modern technology. Due to technology importance, various studies (Dinh n.d.; (Goodwin 2003; Goh and Loosemore 2017) have exerted a link between technological resources and sustainable industrialization. For instance, Goodwin (2003) studied the five capital resources useful for sustainable industrialization using secondary data from American experience. The study found that technology has a positive influence to sustainable industrial development by increasing productivity and reducing cost of production. Similarly, Goh and Loosemore (2017) study in Australian construction sector found that technology positively impact the industrial development. From the above-mentioned evidence, we came to an assumption that;

H_5: Technological resources have a positive relationship to sustainable industrialization in Tanzania.

3 Methods

3.1 Survey Approach

The mixed approach triangulation method was used in this study (Wang and Cheng 2020). The method applied by the researchers to discover participants' views concerning key drivers and challenges of sustainable industrialization (Creswell 2014). Yet, quantitative and qualitative data collection on the key drivers and challenges to sustainable industrialization assisted the researchers to acquire deeper knowledge on the studied agenda. To solve the weakness of a single method, different data collections methods were used through concurrent triangulation strategy in order to substantiate the authenticity of the results (Hair et al. 2016). As noted by Creswell (2014), this model is most appropriate for studies to determine the pervasiveness of a problem through collecting a representative sample.

3.2 Area of the Research and the Population

A survey was carried out in Dar Es Salaam, Dodoma, Mwanza, Kilimanjaro, Mtwara and Mbeya. The choice of these regions includes all areas of mainland Tanzania. The study areas have also taken account of the growing number of industries. The target group was the owners of agro-industrial industries. These have been targeted because they have extensive knowledge of the main drivers and challenges of sustainable industrialization.

3.3 Sample Size/Sampling

The sampling techniques used by this study was simple random. Regions studied were purposively selected based on the number of agro-processing industries found from the Industrial Production (CIP) Census of 2013 conducted by the Tanzania National Bureau of Statistics (NBS). Respondents were randomly selected from the regional trade officers' databases. The selection was done based on their understanding and capability of information sharing concerning the topic under the study (Fowler 2014). 150 owners of agro-processing industries were taken as a sample and the study's respondents.

3.4 Collection of Data

The questionnaire was applied in data collection for objective 1–5 and interview for objective 6. Five-point Likert scale questionnaire was developed with 1 to 5 items meaning "1 = strongly disagree" to "5 = strongly agree" for the six constructs which were adopted from (Sony and Aithal 2020). The six variables included physical, human, organizational, financial, technological resources and sustainable industrialization. Sustainable industrialization was measured by the number of employments created, specific industry productivity improvement and environment sustainability. The questionnaire was back translated from an English version into Swahili - a language regularly spoken in Tanzania. To ensure conversion equivalency two freelance proficient translators were consulted.

Interviews with owners of agro-industrial industries in the selected regions were conducted for qualitative data collection. Interviews were performed in Swahili and audio recorded ready for analysis. The collected data were transliterated verbatim into Swahili, and excerpts of text supporting the conclusions were translated into English.

3.5 Validity and Reliability

This study tested the scale's reliability and validity. The Cronbach alpha was applied to assess the elements reliability. Instruments validity was determined by three professionals. Data were analysed through correlations, descriptive statistics and regressions.

3.6 Analysis of the Data

Data which is quantitative in nature were analysed through descriptive statistics and multiple regression by version 25 of the Social Sciences Statistical Package (SPSS). This method was used to test hypotheses 1 to 5 the key drivers of sustainable industrialization.

Data in qualitative nature were analyzed by means of content analysis, whereby identification of themes and illustration of excerpts from respondents' transcripts was done. This technique served to address objective 6 of the study about the main challenges of sustainable industrialization.

3.7 Ethics Considerations

The search authorization was requested from the authorities, at the district and region level. Additionally, informed consent was respected. Respondents' anonymity and privacy were guaranteed.

4 Results

4.1 Demographic Background

The results showed that all these enterprises (100%) belong to the domestic private sector. The studied agro-processing industries are largely small, employing between 3–200 workers. Half of the studied industries have less than 3 employees. Among the 150 agro-processing industries participated in the study only one firm has over 300 employees while three have amongst 100 to 200 workers. Few businesses have received government support in the early stages of their establishments. Although few studied industries have informed the lack of government support issue, most of them do not give the impression to show significant under-utilization of capacity. The results indicate that 55% of all the respondents state that there is unsatisfactory electricity source depicting that the issue of capacity underutilization is hectic.

Additionally, high cost of production was report as the main challenge for business operation in agro-processing sector. According to their demographic nature, these agro-processing industries normally have not engaged into exportation. For the few agro-processors who export, consider inadequate supply capacity as the main obstacle to export expansion. They work with manual and semi-automatic machines. Fifty-six% of these machines are imported. The findings also indicate that, 95% of these industries are not bothered with their products quality standards. Besides, the results indicate that over 95% of them have none of their products certified by regulatory authorities. Of these, 45 out of 150 industries conduct quality control checks the raw materials purchased. Also, few agro-processors have commercialization plans. Based on these facts, poor quality products due to lack of technology and low production capacity was cited as a barrier to export growth.

4.2 Quantitative Findings

In the current review, inferential statistics were conducted through regression and correlation analysis to assess the key drivers to sustainable industrialization.

4.2.1 Correlation Investigation

To measure the trajectory and strength of the association among the correlative Pearson R constructions was used. The correlation was deliberated as weak once $r = \pm 0.1$ to ± 0.29, whereas correlation was seen moderate once $r = \pm 03$ to ± 0.49, as well as once r is ± 0.5 or higher than 0.5, the correlation was seen as strong as presented in Table 1.

The study reviewed the relationships between the six variables reported in Table 1. The key drivers showed a statistically significant correlation between physical, human, organizational, financial and technological capital resources and the sustainable industrialization of Tanzanian agro-processing sector, revealing a high positive correlation between the four concepts. These results suggest that further analysis could be undertaken to substantiate these conclusions.

Findings from Table 1 illustrates the weak, moderate and strong correlations among the study concepts. The significance test (p) shows the correlation of ($p = 0.000$) which

Table 1. Correlation analysis findings.

Correlations		PC	HC	OC	FC	TC	SI
PC	Pearson correlation	1	.432**	.439**	.410**	.335**	.231**
	Sig. (2-tailed)		.000	.000	.000	.000	.004
	N	150	150	150	150	150	150
HC	Pearson correlation	.432**	1	.479**	.321**	.248**	.252**
	Sig. (2-tailed)	.000		.000	.000	.002	.002
	N	150	150	150	150	150	150
OC	Pearson correlation	.439**	.479**	1	.666**	.622**	.575**
	Sig. (2-tailed)	.000	.000		.000	.000	.000
	N	150	150	150	150	150	150
FC	Pearson correlation	.410**	.321**	.666**	1	.815**	.689**
	Sig. (2-tailed)	.000	.000	.000		.000	.000
	N	150	150	150	150	150	150
TC	Pearson correlation	.335**	.248**	.622**	.815**	1	.838**
	Sig. (2-tailed)	.000	.002	.000	.000		.000
	N	150	150	150	150	150	150
SI	Pearson correlation	.231**	.252**	.575**	.689**	.838**	1
	Sig. (2-tailed)	.004	.002	.000	.000	.000	
	N	150	150	150	150	150	150

** Correlation is significant at the 0.01 level (2-tailed).
Source: Fieldwork data, 2022.

is lower than the guideline of 0.5 representing statistical significance at a 99% confidence interval. The results show physical capital resource challenges are moderate positively correlated with sustainable industrialization (r = 0.231, p-value = 0.000); human capital resources (r = 0.252, p-value = 0.000); organizational capital resource are seen to be strongly and positively correlated (r = 0.575, p-value = 0.000); financial capital resources (r = 0.689, p-value = 0.000) and technological capital resources (r = 0.838, p-value = 0.000) are very strongly corrected with sustainable industrialization. These results suggest that all the studied resources are correlated with sustainable industrialization in Tanzanian agro-processing.

4.2.2 Multivariate Regression Analysis

To scrutinize if there exist indirect and straight connection among the study constructs and sustainable industrialization, the multiple regression models at 5% level of significance was implemented as explained in following subsections.

4.2.2.1 Model Summary

The changes that occur in dependent constructions are caused by the modifications in independent constructions. The model summary in the current study shows the support that sustainable industrialization has as a consequence of deviations in physical, human, organizational, financial and technological capital resources offered. Meanwhile, in order to obtain multivariate results, a regression analysis was used. Table 2 presents the results of the prediction model to confirm the importance of all the capital resources to the sustainable industrialization.

Table 2. Model summary

Model summary[b]				
Model	R	R Square	Adjusted R Square	Std. error of the estimate
1	.845[a]	.714	.704	1.07204

[a] Predictors: (Constant), TC, HC, PC, OC, FC
[b] Dependent Variable: SI
Source: Fieldwork data (2022)

The model in Table 2 shows that the total adjusted R-square (R^2) contribution to explain the change in sustainable industrialization for agro-processing industry in Tanzania is 70.4% (0.704). The interpretation is that 29.6% of the agro-processing sustainable industrialization is explained by other factors rather than resources including physical, human, organizational, financial and technological capital resources. Furthermore, the results show that the variables studied show a strong positive relation, as shown by the R value of 0.845 (correlation coefficient).

4.2.2.2 Analysis of Variance

In deciding how the developed model fits with the analysis of variance data is used because it tests the meaning of the model. The model for the current research was confirmed at a level of 5% significance (Table 3).

The results of the ANOVA analysis exhibited the proportion of variance had a p-value of 0.000, suggesting the model significance. The result suggests that it is possible to draw conclusions from the data tested because the p-value was below 0.05. This demonstrates that physical, human, organizational, financial and technological capital resources have a significant impact on the sustainable industrialization in Tanzanian agro-processing sector.

4.2.2.3 Coefficient Model

Table 4 shows the coefficients of the multi-linear regression analysis. Standardized

Table 3. Variance analysis findings.

ANOVA[a]						
Model		Sum of squares	df	Mean square	F	Sig.
1	Regression	412.980	5	82.596	71.868	.000[b]
	Residual	165.495	144	1.149		
	Total	578.475	149			

[a] Dependent Variable: SI
[a] Predictors: (Constant), TC, HC, PC, OC, FC
Source: Fieldwork data (2022)

coefficients were used to construct the regression model, since standardized data can be compared directly and easily (Hair, Money, Samoul, Page, and Celsi 2016). Conversely, non-standard coefficients are expressed in related building units, resulting in inappropriate judgments.

$$\mathbf{Y = \alpha + \beta_1 X_1 + \beta_2 X_2 + \beta_3 X_3 + \beta_4 X_4 + \beta_5 X_5 + \varepsilon} \tag{1}$$

Table 4. Coefficient model.

Coefficients[a]						
Model		Unstandardized coefficients		Standardized coefficients	T	Sig.
		B	Std. error	Beta		
1	(Constant)	1.769	.521		3.393	.001
	PC	.251	.269	.203	2.912	.058
	HC	.226	.228	.249	2.564	.045
	OC	.256	.369	.407	4.930	.033
	FC	.302	.359	.303	3.510	.041
	TC	.769	.076	.799	10.166	.000

[a] Dependent Variable: SI
Source: Fieldwork data (2022)

The regression model that followed was adjusted based on the results of the study;

$$\mathbf{Y = 1.769 + .203 + .249 + .407 + .303 + .799 + \varepsilon} \tag{2}$$

The preceded equation shows that when the constructs; Physical, Human, Organizational, Financial and Technological capital resources are detained to a zero constant, Tanzanian agro-processing sustainable industrialization would have a constant value of **1.769**. From these findings shown in Table 4 above, it is clear that physical, human, organizational, financial and technological capital resources resolute the variance in sustainable industrialization to 70.4% (Adjusted R Square = .704) approving that an alteration made in physical, human, organizational, financial and technological capital resources could lead to a 70.4% change in Tanzanian agro-processing sustainable industrialization. These findings confirm the importance and reliability of the regression model in which assumptions and references can be formulated because the significance level was below 0.5. The most important factor behind Tanzanian agro-processing sustainable industrialization has been technological capital resources (β = .799, t = 10.166, p = 0.000), followed by organizational capital resources (β = .407, t = 4.930, Sig. = .033). Financial capital resources were ranked third with (β = .303, t = 3.510, Sig. = .041). Fourth was human capital resources with (β = .249, t = 2.564, Sig. = .054). The last was physical capital resources (β = .203, t = 2.912, Sig. = .045). All the tested constructs were significant.

4.3 Qualitative Findings

In objective six we aimed to explore the key challenges hindering sustainable industrialization in Tanzania and the outcome is shown in Table 5. The interview responses show that the biggest challenge is inadequate capital, followed by poor technology and insufficient power supply. The following quotes justify the results. "... *to my side, inadequate capital hinders me to buy sophisticated technology and hire experts in agro-processing*" (female, Dodoma).

Table 5. Challenges to sustainable industrialization

Theme	Sub-theme (challenges)	Sample quotes
Physical capital resources **(17 cases)**	Poor infrastructure	'...*poorly developed roads... Increases the cost of transports...difficulty products to timely reach markets*'
	Lack of appropriate equipment	"*.... Inadequate equipment and facilities hinder efficiency of agro-processing*"
	Insufficient power supply	'...*we suffer mostly from insufficient power supply ...*'
	Difficult in accessing raw materials	'*Difficulty of timely access to raw materials is a great challenge to us*
Human capital resources **(12 cases)**	Lack of proper knowledge and skills on agro-processing management	"*... I suggest more training to equip staff with proper knowledge....*"

(*continued*)

Table 5. (*continued*)

Theme	Sub-theme (challenges)	Sample quotes
	Low experience on agro-processing management	'*......employing experienced staff (experts) on agro-processing is difficult...:*"
	Negative insights from managers and staff	'*... some managers and staff have negative insights ....*"
	Improper staff handling	*'The management has to handle staff properly ...helping reach high productivity.'*
	Lack of staff training on agro-processing	*".... Small industries do not have enough budgets to train staffs on agro-processing"*
Organizational capital resources **(10 cases)**	Poor time management	'... improvement of the processes ...'
	Stringent government regulations	'... sometimes the government create difficult regulations for small entrepreneurs...'
	Being unprepared for the demand	'*...timely information... to avoid excess inventory of stock ...*'
	Poor quality control	'*...we produce low quality products ... poor technology*'
Financial resources **(22 cases)**	Lack of access to credits	' '*...access to long-term loans for investment ... hampers our development ....*'
	Huge Tax	'*...tax exemptions offered ... NOT honoured by implementers*'
	Huge interest rates for investment loans	'*...... interest rates are huge to bare for small industries*'
Technological resources **(19 cases)**	Poor technology	'*... proper technology use will enhance sustainable industrialization ...).*'
	Low traceability and Connectivity	'*.... Modern technology application is mandatory in agro-processing...makes the process easier...*'
	Poor compliance to standards	*'Little support from TBS to control and comply to quality standards ......'*

The results presented in Table 5 reveal that the issue of financial capital resources is the major key driver to sustainable (22 cases) followed by technological resources

(19 cases), physical resources (17 cases), human resources (12 cases) and lastly organizational resources (10 cases).

5 Discussion of Findings

5.1 Physical Capital Resources Have an Impact on the Sustainable Industrialization of Tanzanian Agro-Processing Sector

The findings of the regression analysis show a positive relationship among physical capital resources and the sustainable industrialization of Tanzanian agro-processing sector. This study's findings were consistent with earlier documentation that was looked at. For example, researcher such as Goodwin (2003), found positive relationship between physical resources in terms of infrastructure, equipment, machines, natural resources and sustainable industrialization. Similarly, such findings were in consistence with Sony and Aithal (2020) who used resource-based view and their study revealed that, physical resources positively influence sustainable industrialization. Also, Rahim et al. (2021) study discovered that natural as well as physical resources have an impact to sustainable industrialization.

5.2 Human Capital Resources Affect the Sustainable Industrialization of Tanzanian Agro-Processing Sector

Similarly, the results from regression analysis shows that Human capital resources have positive influence on the sustainable industrialization of Tanzanian agro-processing sector meaning that the hypothesis stating that there is positive relationship and significant influence among human capital resources and the sustainable industrialization of Tanzanian agro-processing sector is accepted. The study findings were in line with Rahim et al. (2021) who found that education and skills positively impact sustainable industrialization. Also, (Goodwin 2003), found that knowledge, skills and experience have a positive relationship with sustainable industrialization. Hereafter, we affirm that human capital advancement contributes to lessen the resource profanity effects in the Tanzanian agro-processing sector. Additionally, human capital expansion ought to be arranged for changing the obscenity of the natural resources to dedication for the Tanzanian agro-processing sector. Therefore, these findings necessitate the pertinence of boosting investments in human capital development to enhance sustainable industrialization in the Tanzanian agro-processing sector.

5.3 Organizational Capital Resources Affect the Sustainable Industrialization of Tanzanian Agro-Processing Sector

The results from regression analysis shows that organizational capital resource challenges have positive and significant influence on the sustainable industrialization portraying that the proposition asserting a significant positive connection amongst organizational capital resources and sustainable industrialization in Tanzanian agro-processing sector is accepted. The above findings are consistent with those of Babasanya et al. (2018)

who found that organizational plans have a positive influence on sustainable industrialization in the Ethiopian environment. In addition, Goh and Loosemore (2017) found that structures and systems are linked to the sustainable industrialization in construction industry on Australian context.

5.4 Financial Resources Impact the Sustainable Industrialization of Tanzanian Agro-Processing Sector

The results show that financial resources were positively and significantly impact sustainable industrialization of Tanzanian agro-processing sector. However, the hypothesis stating that there is positive relationship and significant influence among financial capital resources and the sustainable industrialization of Tanzanian agro-processing sector is accepted. In some ways, our results support a study by Goh and Loosemore (2017) which discovered that the availability of financial resources was one among the drivers that have a positive and significant influence to construction industries in Australian context. In tandem with that, Sony and Aithal (2020) found that financial capital resources positively influence sustainable industrialization.

5.5 Technological Resources Affect the Sustainable Industrialization of Tanzanian Agro-Processing Sector

The results revealed that technological resources positively affect the sustainable industrialization of Tanzanian agro-processing sector. In some ways, our results support Goodwin (2003) study which found that technology has a positive influence to sustainable industrial development by increasing productivity and reducing cost of production. Similarly, Goh and Loosemore (2017) study in Australian construction sector found that technology positively impact the industrial development.

5.6 Challenges to Sustainable Industrialization

The findings presented in Table 5 disclose the issue of financial capital resources to be the major key driver to sustainable (22 cases) followed by technological resources (19 cases), physical resources (17 cases), human resources (12 cases) and lastly organizational resources (10 cases). These findings are in consistent with the RBV dimensions (Haraguchi et al. 2019). They imply that improvements in the technology, finance, physical, organizational (structures, processes and systems), and human development are needed in order to ensure sustainable industrialization.

6 Policy Implications

The study informs stakeholders - policy makers on the key drivers and challenges towards sustainable industrialization. Knowledge will drive them to put in place user-friendly empowerment policies that will help the industrial sector and practitioners get support in their economic development activities. The results of the study provide information to the industry sector through the creation of entrepreneurs more motivated to participate in industrialization as employees or owners.

7 Concluding Remarks and Contribution

The study's aim was to assess the key drivers and challenges of sustainable industrialization in the agro-processing sector in Tanzania. The study indicated the value and RBV applicability in understanding effects of physical, human, organizational, financial and technological resources on sustainable industrialization in the Tanzanian agro-industrial sector. Findings from this study specify how the agri-based industry develops novel models and conducts for sales increase in agro-processing business. To guarantee a competitive advantage agro-processing sector should ensure creation of new knowledge, human resource development skills, partnerships, relationships and approaches in the innovative value chains. In the elongated timeframe, agro-processors should think of adjusting their organizational and physical resource development approaches allowing them to shift to modern technological-based organizations. The future of agribusiness is requiring abrupt changes and adapting the available resources in response to those variations. Therefore, the aggressiveness of agro-processors in managing their resources will impact industrialization. However, the study's findings recommend that agro-processors should keep pace with the advances occurring in the industrialization by changing their old structures in anew developed production and high-tech world.

Based on the current study's results, it is recommended that for sustainable industrialization in Tanzania to occur industries should; 1) create enough decent jobs leading to sustainable livelihoods; 2) have capability of increasing returns on investment; 3) be efficient to use finite resources; 4) reduce damaging the environment; 5) technology, innovation, and skills development 6) be fair in global competitiveness; 7) ensure the long-term business sustainability.

The study contributes to the existing knowledge in the sector and insist on stakeholders' participation in the improvement of the physical, human, organizational, financial and technological resources to reach sustainable industrialization. The authors provide quantitative and qualitative data on the key drivers and challenges to sustainable industrialization in Tanzanian context and thus informs the debate on the use of resource-based theory in industrialization study. In addition, the study contributes to the extension of the BRV theory by adding two additional variables obtained from the literatures.

The current study was limited to small sample size, which reflects few agro-processors with enough experience to provide reliable information. This limitation restricts generalization of the study. Therefore, we recommend further research to expand the understanding of agro-processor strategies provided by this research. Research has also been restricted to a few areas in Tanzania. It is thus necessary to continue research in this topic to widen the knowledge in other geographic, sectoral and industrial relations contexts.

References

African Union Commission: Common African Position (CAP) on the post-2015 development agenda. Addis Ababa (2014)

Babasanya, A.O., Oseni, I.O., Awode, S.S.: Human capital development: a catalyst for achieving SDGs in Nigeria. Acta Universitatis Danubius. Oeconomica **14**(4), 25–41 (2018)

Barney, J.: Firm resources and sustained competitive advantage. J. Manag. **17**(1), 99–120 (1991)

Barney, J.B., Clark, D.N.: Resource-Based Theory: Creating and Sustaining Competitive Advantage. Oxford University Press on Demand, Oxford (2007)

Creswell, J.W.: A Concise Introduction to Mixed Methods Research, 4th edn. Sage Publications, Thousand Oaks (2014)

Dinh, D.: Sustainable Industialization: A New Trends in Asian Developing Economies. Institute of World Economy, Hanoi, Vietnam (n.d.)

Etuk, R.U., Etuk, G.R., Michael, B.: Small and medium scale enterprises (SMEs) and Nigeria's economic development. Mediterr. J. Soc. Sci. **5**(7), 656–656 (2014)

Fowler, F.: Survey Research Methods for Managers. Sage Publications, Los Angeles (2014)

Goh, E., Loosemore, M.: The impacts of industrialization on construction subcontractors: a resource based view. Constr. Manag. Econ. **35**(5), 288–304 (2017). https://doi.org/10.1080/01446193.2016.1253856

Goodwin, N.: Five Kinds of Capital Useful for Industrial Development. Global Development and Environmental Institute, Medford MA, USA (2003)

Grant, R.M.: The resource-based theory of competitive advantage: implications for strategy formulation. Calif. Manag. Rev. **33**(3), 114–135 (1991)

Guadagno, F., Wangwe, S., Delera, M., de Castro, A.: Horticulture, and wood and furniture industries in Tanzania: performance, challenges and potential policy approaches. International Growth Centre (2019)

Hair, J., Money, M., Samoul, P., Page, M., Celsi, M.: Essentials of Business Research Methods, 3rd edn. Routledge, New York (2016)

Haraguchi, N., Martorano, B., Sanfilippo, M.: What factors drive successful industrialization? Evidence and implications for developing countries. Struct. Chang. Econ. Dyn. **49**, 266–276 (2019)

Kirumirah, M.H., Tambwe, M.A., Mapunda, M., Mazana, M.Y.: An Assessment of women and youth participation in implementing the industrialization theme in Tanzania. Bus. Educ. J. **4**(1), 11 (2020)

Kumar, M., Rodrigues, V.S.: Synergetic effect of lean and green on innovation: a resource-based perspective. Int. J. Prod. Econ. **219**, 469–479 (2020)

Madhani, P.M.: Resource-based view (RBV) of competitive advantage: an overview. In: Resource-Based View: Concepts and Practices, pp. 3–22 (2010)

Mkwizu, K.H., Monametsi, G.L.: Impacts and challenges of Southern African Development Community's industrialization agenda on Botswana and Tanzania. Public Adm. Policy **24**, 212–223 (2021)

Moyo, T.: Promoting inclusive and sustainable industrialisation in Africa: a review of progress, challenges and prospects. In: The 2nd Annual International Conference on Public Administration and Development Alternatives, Tlotlo Hotel, Gaborone, Botswana, pp. 365–375 (2017). http://ulspace.ul.ac.za/bitstream/handle/10386/1881/moyo_promoting_2017.pdf?sequence=1&isAllowed=y

O'Neill , A.: Share of Economic Sectors in the GDP in Tanzania 2021 (2023). https://www.statista.com/statistics/447719/share-of-economic-sectors-in-the-gdp-in-tanzania/

Oqubay, A.: Made in Africa: Industrial Policy in Ethiopia, p. 374. Oxford University Press (2015)

Osborn, D., Cutter, A., Ullah, F.: Universal sustainable development goals. Understanding the Transformational Challenge for Developed Countries (2015)

Pérez-Fuentes, P.: Women's Economic Participation on the Eve of Industrialization: Bizkaia, Spain, 1825. Feminist Economics **19**(4), 160–180 (2013)

Rahim, S., et al.: Do natural resources abundance and human capital development promote economic growth? A study on the resource curse hypothesis in Next Eleven countries. Resour. Environ Sustain. **4**, 100018 (2021)

Sampath, P.G.: Sustainable Industrialization in Africa: Toward a New Development Agenda. Palgrave Macmillan, London (2016)

Seetoh, K.C., Ong, A.H.F.: Achieving sustainable industrial development through a system of strategic planning and implementation: the Singapore model. In: Wong, TC., Yuen, B., Goldblum, C. (eds.) Spatial Planning for a Sustainable Singapore, pp. 113–133. Springer, Dordrecht (2008). https://doi.org/10.1007/978-1-4020-6542-2_7

Sony, M., Aithal, P.: A resource-based view and institutional theory bases analysis of Industry 4.0 implementation in the Indian engineering industry. Int. J. Manag. Technol. Soc. Sci. **5**(2), 155–166 (2020)

Ulrich, D., Brockbank, W., Yeung, A.K., Lake, D.G.: Human resource competencies: an empirical assessment. Hum. Resour. Manage. **34**(4), 473–495 (1995)

UNIDO: UNIDO Annual Report 2016 (2017). https://sustainabledevelopment.un.org/content/documents/16781UNIDOAnnual_Report_2016_EN.pdf

United Republic of Tanzania (URT): National Five-Year Development Plan 2016/17 - 2020/21: "Nurturing Industrialization for Economic Transformation and Human Development" Ministry of Finance and Planning (2016)

Wang, X., Cheng, Z.: Cross-sectional studies: strengths, weaknesses, and recommendations. Chest **1**, S65–S71 (2020)

Wangwe, S., Mmari, D., Aikaeli, J., Rutatina, N., Mboghoina, T., Kinyondo, A.: The performance of the manufacturing Sector in Tanzania. UNU-WIDER, Helsinki (2014)

Female Social Entrepreneurship in Male-Dominated Industries in Ghana and Agenda 2030

S. Dzisi(✉)

Data Link Institute of Business and Technology, Tema, Ghana
smiledzisi@gmail.com

Abstract. Purpose: This study intends to provide an empirical understanding of female social entrepreneurship in industries with a male preponderance in a setting of emerging economies and their contribution to the achievement of Agenda 2030.

Design/Methodology/Approach: A qualitative research approach was employed using an exploratory research design to conduct interviews with thirty (30) women entrepreneurs in the IT/Software Development, Mining and Construction industries in Ghana. The participants were purposively selected.

Findings: The survey found that sociocultural issues, sexual harassment, and a lack of support resources are the main problems these women face. Due to societal preconceptions and other established ideas about the kinds of industries in which women might successfully launch their own businesses, the findings highlight gender disparities in access to land. Without a question, the initiatives of these female social entrepreneurs have helped to solve societal issues, combat gender inequality, and enhance the quality of life in their communities.

Research Limitations: The present study can be broadened further by considering other industries to explore more information from the females as well as challenges that they face.

Practical Implications: The current awareness in society about inequalities and need for women to be empowered is gradually opening doorways for society to support their initiatives. Support is now being provided locally, nationally and internationally but requires these women's personal initiatives.

Social Implications: More attention thus needs to be paid to social enterprise development as a means to achieving gender equality and women's empowerment by 2030.

Originality/Value: The findings of this study contribute to a better understanding of the structural challenges women social entrepreneurs encounter.

Keywords: Developing economies · Male-dominated industries · SDGs · Social entrepreneurs · Women

1 Introduction

The benefits associated with social entrepreneurship have been touted in many economies. Though still dominated by males, female participation is gradually increasing

C. Aigbavboa et al. (Eds.): ARCA 2022, *Sustainable Education and Development – Sustainable Industrialization and Innovation*, pp. 1209–1216, 2023.
https://doi.org/10.1007/978-3-031-25998-2_93

with many of these women social entrepreneurs operating in areas previously dominated by males. Despite the challenges in gender equality and women empowerment, something that prevails in both developing and developed economies, social enterprises are providing mechanisms to address these maladies in our societies (Yunis et al. 2018). In male dominated industries such as construction, information technology and mining, social enterprises are affording women the opportunity to lead, address social and economic challenges, earn income, empower other women and generally participate in decision making. These developments are welcoming as they offer a path for women's self-expression, gradually contending the disbelief in their potentials and capabilities in society. Though the known challenges are still inherent, the success stories of some of these women coupled with the impact their works are having on society are beginning to highlight the potentials inherent in social enterprises. The field is slowly but consistently changing the way entrepreneurship is perceived in developing economies and altering the perception of women about male dominated fields (Rashid and Ratten 2020). Social entrepreneurship offers many women the opportunity to do what they want and feel like doing.

Gradually, the number of women operating in male dominated industries is growing and breaking the age-old perceptions. These discussions will hopefully redirect efforts aimed at addressing lingering gender inequalities and disempowerment traces in Ghanaian societies and other parts of the world.

The paper discusses how women social enterprises are providing solutions to gender inequalities, some of the practical injustices against gender that social enterprises are helping to solve, how participation of women in social enterprise are helping address these challenges in Ghana.

2 Literature Review

It is believed that, the digital technology space in Ghana is dealing with '…an obvious gender divide' of about 80:20 men to women ratio. Although the world has seen increase in the adoption and use of digital technologies over the past 20 decades, most developing economies, including Ghana are yet to fully incorporate them into every area of their development efforts. In terms of usage and work, women are believed to be more disadvantaged due to unequal access and opportunities to digital technologies and training (Zhang et al. 2020). Perception in society and challenges with infrastructure has all contributed to this situation. In recent times however, due to efforts aimed at tackling existing inequalities between men and women, several organisations including government, international agencies, local telecommunication industries and social enterprises have rallied to the aid of women and girls. This has seen increased training for women in ICT both in life skills and for businesses. Several of the social enterprises engaged in these services are owned and operated by women with an agenda to increase the number of women in the digital space. Women are becoming more active in the digital space, an indication of their determination not to be left out.

One of the key contributors to the economy of Ghana is the mining industry. The mines (gold alone) provide about 5,300 direct employment and over 83,000 indirect employments. The industry contributed about 38% direct corporate tax and 27.6% government revenue and 6% of GDP in 2011 (Aryee 2012). The industry invests in Corporate

Social Responsibility (CSR) programmes especially in host communities as a means of contributing to sustainable development. Like the trend in many mining communities and countries, Kwami (2007) in his study identifies some obstacles towards the participation of the Ghanaian woman in mines. Among the reasons identified included cultural restrictions, prejudice and discrimination. His study concluded that although women formed the majority in the population of Ghana, they represented the minority in the sector that was singularly contributing so much in employment creation and GDP. This unfortunate reality and trend are a concern for stakeholders in the mining industry and the agenda 2030. Scanty literature exists on the studies on women in mining in the south Saharan Africa and the Ghana context, most of the available literature are foreign and the few research available on Ghana are mainly on small scale mining (Kilu et al. 2014). In Ghana, the industry of mining is perceived to require physical manly strength, and as such, a preserve for men. Sadly, this perception is not only in Ghana, but can be traced world-wide (Abrahamsson et al. 2014; Bryant and Jaworski 2011; Purevjah 2010).

Historic antecedence about male-dominance in the mining industry had both formal and informal barriers. The formal barriers are attributed to legal and legislative instruments, as well as educational traditions that STEM is a preserve for males. For example, in Britain, women were legally excluded from working underground by the 1842 Mines Act. Similarly, in early 1900s, Article 2 of the International Labor Organization (ILO) convention 45 of 1935 came into play, forbidding women in underground mining. The informal barriers, on the other hand, are results of long-standing cultures and traditions that discriminated and stereotyped women from certain employment and jobs as evidence in the study by Tabassum, and Nayak (2021). A combination of these factors (formal and informal) makes the mine industry market unfriendly for women. Though many of these formal barriers have been broken, a lot of the informal barriers still persist.

3 Methodology

This study explores women social entrepreneurship in the Information Technology (IT)/Software Development, Construction and Mining industry in Ghana. The study largely focuses on the backgrounds of these women, the difficulties they faced, their triumphs, and the lessons they learned.In order to better understand the world of women social entrepreneurs in Ghana, a qualitative study methodology was used. Thirty (30) female social entrepreneurs were purposefully chosen from the list of women working in Ghana's mining, construction, and IT/software development sectors (10 respondents from each industry).

To gather comprehensive and rich information, interviews with these thirty female social entrepreneurs were undertaken. Purposive sampling is recommended by researchers (Kumar 2018; Neuman 2006) when the researcher can identify participants who will be able to provide information to meet the research objectives. This method was also endorsed by Merriam and Tisdell (2015), who noted that "the key factors are not the number and representativeness of the sample, but rather the capacity of each individual to contribute to the growth of insight and knowledge."

In order to confirm and authenticate the 30 respondents' responses, three (3) supervisors from each of the industries of IT/Software Development, Mining, and Construction were also questioned.

For validation of what happened, the interviewers received the transcripts of their conversations.

Before the contents were evaluated, corrections were made afterward.

4 Results and Discussion

It is very important to know the various backgrounds of the women social entrepreneurs in Ghana. The variables investigated in the study include their age, educational background, the number of years they have been entrepreneurs, and whether they have a mentor.

Demographics
A majority of them (60%) aged between 41 to 50; some of them (20%) aged between 31 to 40 years. Also, a fair number of the respondents (13%) aged between 51 to 65 years and a few of them (7%) falling between 20 to 30 years. The findings also revealed that, majority of the women in IT/Software Development are youthful (aged between 20 – 30 years) with only one woman at age between 51 to 65 years. This is an indication that the younger female population of females is venturing into IT/Software Development industry. Again, the construction industry is represented by women above 41 years this is because the industry is capital intensive and most youth are unable to enter into it at their youthful age. Table 1, also revealed that the age distribution for mining is in the middle age. The frequency dropped to 1 from 5. The mining industry is an energetic industry and hence many people leave the industry as they age. The study also revealed that the years these women have been in the various industries ranges from 3 to 20, with majority (40%) between 5 to 10 years and a minimum (13%) between 16 to 20 years.

Table 1. Age distribution

Parameters (years)	Type of industry			Frequency	Percentage
	IT/software development	Construction	Mining		
20–30	4	–	–	4	13
31–40	2	1	4	7	23
41–50	2	5	5	13	43
51–65	1	4	1	6	20
Total	**10**	**10**	**10**	**30**	**100**

Parameters (years)	Frequency	Percentage
Below 5	6	20
5–10	11	37
11–15	8	27
16–20	5	17
Total	**30**	**100**

Source: Field Data, 2022

Exploring further, it was realized that, the women in the IT industry were mostly software developers and programmers. Exploration, production, supplys were areas of specialties for the women social entrepreneurs in the mining industries. This still shows how males still play a major role in developing and supporting women's growth and independence.

Challenges
A respondent in the mining industry (exploration) indicated that, there are many instances she was refused access to parcels of land to mine just because she was a woman and that the custom of the area does not allow women access to the land and that if a woman mines the land, the minerals will disappear. *I have to get a male face to usually do the negotiation of lands for mining. As a woman in small scale mining, I am always refused access to lands anytime I front it myself.*

Another respondent also lamented how often while she was learning to programme, her male counterpart often teased her with comments like '... *are you sure you can do this?'*. Others lamented about how difficult it was to stay long hours learning with their male counterparts. As one woman commented, *'it was very intimidating', 'some of the ladies we started with even dropped out after a while. Some even never managed to get to programme well'*. IT generally is believed by any to be technical and something for men only, but this belief is gradually changing although at a slow pace.

A participant indicated that, many of the women involved in the IT field generally felt society perceives them to be tough once they learn they are into computing. One participant commented that '... *they almost appear to be shying away from you when they learn you are into IT. They make comments that almost sound like they need to be cautious around you as it sounds like you are trying to be like a man', they literally call you a witch.*

A woman in the Mining Suppliers industry interviewed said that, most of the men in the mining industry see any woman in the industry as a sex object, this makes it very difficult staying in the industry for most women and the few that struggle to stay are tagged with negatively to sex. A participant in the IT industry had this to say "...*mentors are good but not all of them come with a good thought. When I began webpage development, I had a male I perceived as my mentor, he started making sexual advances towards me when I refused his advances, he sidelined me from work, withdrew all the professional support he was giving me and emotionally harassed me until I resigned from that company..."*.

Some participants in the IT industries indicated that, even though their main intentions were to raise funds to help women in need in their societies, through training and empowerment in ICT skills, they have had to start charging for their services to remain sustainable and even this they claimed was still insufficient. Some also claimed monies promised by donners sometimes delayed causing their projects to delay and sometimes get postponed. They contended that even though there was a heightened awareness of the need to train more women to be in male-dominated industries, support services to help to achieve this was simply not available.

The women interviewed believed that they were contributing significantly to the attainment of the SDGs through empowering women, including themselves with jobs, digital skills for living and entrepreneurship, providing leadership etc.

Many of these women social entrepreneurs have employed women and trained women who currently work either by themselves or with reputable organizations. Some of these women have been employed by international organizations and many have been exposed to the world of work with reputable international organizations. In general, they have become independent of their men and are contributing to the GDP through the payment of taxes.

A participant who is into software development indicated that she has developed a webpage for free for youth of her community to help them access for jobs online SDG 8.6).

In addition, one distinctive feature among these entrepreneurs was their provision of free training to women and girls in their communities. By soliciting for support, especially from international organizations, these social entrepreneurs have dedicated themselves to increasing the number of women and girls in the digital space. Among them, a total of close to 4000 girls and women have been equipped with digital knowledge and skills for living and entrepreneurship.

Social entrepreneurship clearly is enabling women empowerment and gender equality. Many of these women are going out of their space to set up businesses that both provides them with income that makes them independent of their men and parents and allow them to operate in male-dominated industries. They are also able to focus on training and empowering more women so as to reduce the gender divide. These women are thus able to provide leadership in their chosen sectors, gradually but surely changing the fabric of their societies in terms of the social and economic directions they pursue. Although this may appear to be in small leaps, it is evident that social entrepreneurship has the potential to contribute to women's empowerment and equality albeit not the only panacea. In other words, it is clear that social enterprise alone will not be able to solve all gender inequality challenges in Ghana due to the deep-seated cultural mindset and beliefs inherent in the society.

5 Conclusion

From the above findings and discussions, it is evident that social enterprise is helping provide solutions to gender inequality by enabling these women to set up their own businesses where they are their own bosses. Subsequently they take instructions from no one, manage their own processes and activities, employ their own staff and decide what they want to do. They also are able to decide on which area of the society's challenges they want to address, providing leadership in the process. Thus, in male-dominated industries, these women ae still able to identify sections where society is still not addressing and subsequently take advantage of the opportunity to address them in their own way. Clearly women are participating in decision making, earning their income and becoming independent, providing leadership, engaging and contributing to decision making in society and many more through their involvement in social enterprise projects. They are thus being empowered and reducing the inequality inherent in the Ghanaian society. Social enterprise is thus, providing machinery for empowering women in male-dominated industries as well as society in general. This is made possible through self-initiatives in their ability to identify problems in society and take proactive actions to address them. The current awareness in society about inequalities and need for women to be empowered is gradually opening doorways for society to support their initiatives. Support is now being provided locally, nationally and internationally but requires these women's personal initiatives. These women subsequently are able to support themselves and others through the incomes they are able to generate. These incomes help them sustain their businesses, support more women in their societies, also become independent and gradually bridge the gap in gender equality. More attention thus needs to be paid to social enterprise development as a means to achieving gender equality and women's empowerment by 2030. This however, is not to say is the only measure that will address the imbalance. Other strategies must also be relooked at and supported and this calls for more research into social enterprises to identify extensions to it and understand more about women's abilities to self-initiate.

References

Abrahamsson, L., et al.: Mining and sustainable development: gender, diversity and work conditions in mining (2014)

Agarwal, B.: Gender equality, food security and the sustainable development goals. Curr. Opini. Environ. Sustain. **34**, 26–32 (2018)

Aryee, B.N.A.: Contribution of the minerals and mining sector to national development: Ghana's experiment. Great Insights **1**(5), 14–15 (2012)

Begashaw, B.: Strategies to deliver on the Sustainable Development Goals in Africa. In: Caulibaly, B. (ed.) Foresight Africa. Africa Growth Initiative at Brookings (2020)

Bornstein, D., Davis, S.: Social Entrepreneurship: What Everyone Needs to Know®. Oxford University Press, Oxford (2010)

Bymolt, R., Laven, A., Tyzler, M.: Demystifying the cocoa sector in Ghana and Côte d'Ivoire. The Royal Tropical Institute (KIT), Amsterdam, The Netherlands (2018)

Kania, J., Kramer, M.: Collective impact, pp. 36–41. FSG (2011)

Karki, S.T., Xheneti, M.: Formalizing women entrepreneurs in Kathmandu, Nepal: pathway towards empowerment? Int. J. Sociol. Soc. Policy (2018)

Kobia, M., Sikalieh, D.: Towards a search for the meaning of entrepreneurship. J. Eur. Ind. Train. **34**, 110–127 (2010)

Kumar, R.: Research Methodology: A Step-by-Step Guide for Beginners. Sage, Thousand Oaks (2018)

Kumar, S., Gupta, K.: Social entrepreneurship: a conceptual framework. Int. J. Manag. Soc. Sci. Res. **2**(8), 2319 (2013)

Merriam, S.B., Tisdell, E.J.: Qualitative Research: A Guide to Design and Implementation. Wiley, New York (2015)

Neuman, W.L.: Workbook for Neumann Social Research Methods: Qualitative and Quantitative Approaches. Allyn & Bacon, Boston (2006)

Nicholls, A. (ed.): Social Entrepreneurship: New Models of Sustainable Social Change. OUP, Oxford (2008)

Rashid, S., Ratten, V.: A systematic literature review on women entrepreneurship in emerging economies while reflecting specifically on SAARC countries. In: Ratten, V. (ed.) Entrepreneurship and Organizational Change, pp. 37–88. Springer, Cham (2020). https://doi.org/10.1007/978-3-030-35415-2_4

Tabassum, N., Nayak, B.S.: Gender stereotypes and their impact on women's career progressions from a managerial perspective. IIM Kozhikode Soc. Manag. Rev. **10**(2), 192–208 (2021)

Titscher, S., Meyer, M., Wodak, R., Vetter, E.: Methods of text and discourse analysis (2015)
Yunis, M.S., Hashim, H., Anderson, A.R.: Enablers and constraints of female entrepreneurship in Khyber Pukhtunkhawa, Pakistan: institutional and feminist perspectives. Sustainability **11**(1), 27 (2018)
Zhang, M., Zhao, P., Qiao, S.: Smartness-induced transport inequality: privacy concern, lacking knowledge of smartphone use and unequal access to transport information. Transp. Policy **99**, 175–185 (2020)

Author Index

C. Aigbavboa et al. (Eds.): ARCA 2022, *Sustainable Education and Development – Sustainable Industrialization and Innovation*, pp. 1217–1219, 2023.
https://doi.org/10.1007/978-3-031-25998-2

O

P

Q

R

S

T

U

W

Y

9783031259975VOL02